Fundamentals of Momentum, Heat, and Mass Transfer

7th Edition

James R. Welty

Department of Mechanical Engineering
Oregon State University

Gregory L. Rorrer

Department of Chemical Engineering
Oregon State University

David G. Foster

Department of Chemical Engineering
University of Rochester

VP AND EDITORIAL DIRECTOR	Laurie Rosatone
EDITOR	Jennifer Brady
SENIOR DIRECTOR	Don Fowley
EDITORIAL MANAGER	Judy Howarth
CONTENT MANAGEMENT DIRECTOR	Lisa Wojcik
CONTENT MANAGER	Nichole Urban
SENIOR CONTENT SPECIALIST	Nicole Repasky
PRODUCTION EDITOR	Umamaheswari Gnanamani
COVER PHOTO CREDIT	© Wiley

This book was set in 10/12 TimesLTStd by SPi Global and printed and bound by Quad Graphics.

Founded in 1807, John Wiley & Sons, Inc. has been a valued source of knowledge and understanding for more than 200 years, helping people around the world meet their needs and fulfill their aspirations. Our company is built on a foundation of principles that include responsibility to the communities we serve and where we live and work. In 2008, we launched a Corporate Citizenship Initiative, a global effort to address the environmental, social, economic, and ethical challenges we face in our business. Among the issues we are addressing are carbon impact, paper specifications and procurement, ethical conduct within our business and among our vendors, and community and charitable support. For more information, please visit our website: www.wiley.com/go/citizenship.

ISBN: 978-1-119-72354-7 (PBK)
ISBN: 978-1-119-49556-7 (EVAL)

Library of Congress Cataloging-in-Publication Data

Names: Welty, James R., author. | Rorrer, Gregory L., author. | Foster, David G., author.
Title: Fundamentals of momentum, heat, and mass transfer / James R. Welty,
 Department of Mechanical Engineering, Oregon State University, Gregory L. Rorrer,
 Department of Chemical Engineering, Oregon State University, David G. Foster,
 Department of Chemical Engineering, University of Rochester.
Description: 7th edition. | Hoboken, NJ : John Wiley & Sons, [2019] |
 Includes index. |
Identifiers: LCCN 2018045781 (print) | LCCN 2018047118 (ebook) | ISBN
 9781119495581 (Adobe PDF) | ISBN 9781119495413 (ePub) | ISBN 9781119495529
 (pbk.)
Subjects: LCSH: Fluid mechanics. | Heat–Transfer. | Mass transfer.
Classification: LCC TA357 (ebook) | LCC TA357 .W45 2019 (print) | DDC
 532–dc23
LC record available at https://lccn.loc.gov/2018045781

The inside back cover will contain printing identification and country of origin if omitted from this page. In addition, if the ISBN on the back cover differs from the ISBN on this page, the one on the back cover is correct.

Preface to the Seventh Edition

The first edition of *Fundamentals of Momentum, Heat, and Mass Transfer*, published in 1969, was written to become a part of what was then known as the engineering science core of engineering curricula. Requirements for ABET accreditation have continued to stipulate that a significant part of all engineering curricula be devoted to fundamental subjects.

Applications of engineering science fundamentals have changed in many ways since the first edition was published. We now consider devices such as inkjet printers, macro- and micro-scale chemical reactors, and a myriad of biological and physical processes that were unheard of over 45 years ago. These and other applications are considered in the seventh edition; however, the fundamentals of momentum, heat, and mass transport used in analyzing these processes remain unchanged.

This text is intended for use by sophomore- or junior-level engineering students whose technical interests require an understanding of transport phenomena. The most likely users include students majoring in Chemical, Mechanical, Environmental, and Biochemical engineering. Other engineering majors will also find the ideas of commonality among fluid mechanics, heat and mass transfer to be useful and important. It is assumed that students using this book will have completed courses in calculus and in mass and energy balances. A rudimentary knowledge of differential equations is also recommended.

Computer competency among students using this text is assumed. Students may find it useful to solve some assigned homework problems using numerical computing packages; however, the bulk of our problems can be solved using fundamental mathematical methods.

Resources for adopting instructors, including the Instructor Solutions Manual and electronics images from the book suitable for use in lecture slides, are available on the Instructor Companion Site at www.wiley.com/go/welty/fundofmomentum7e. Please visit the Instructor Companion Site to register for a password to access these resources.

We wish to thank members of the John Wiley & Sons editorial and production staffs for their professionalism, continuing support, and congenial working relationships that have continued since the publication of the first edition.

The e-book organization includes problems at the beginning of each chapter and a select number of student practice problems with the solution as a drop down selection.

Corvallis, Oregon and Rochester, New York
December, 2018

Contents

CHAPTER 1

Introduction to Momentum Transfer

Momentum transfer in a fluid involves the study of the motion of fluids and the forces that produce these motions. From Newton's second law of motion it is known that force is directly related to the time rate of change of momentum of a system. Excluding action-at-a-distance forces, such as gravity, the forces acting on a fluid, such as those resulting from pressure and shear stress, may be shown to be the result of microscopic (molecular) transfer of momentum. Thus, the subject under consideration, which is historically fluid mechanics, may equally be termed momentum transfer.

The history of fluid mechanics shows the skillful blending of the nineteenth- and twentieth-century analytical work in hydrodynamics with the empirical knowledge in hydraulics that man has collected over the ages. The mating of these separately developed disciplines was started by Ludwig Prandtl in 1904 with his boundary-layer theory, which was verified by experiment. Modern fluid mechanics, or momentum transfer, is both analytical and experimental.

Each area of study has its phraseology and nomenclature. Momentum transfer being typical, the basic definitions and concepts will be introduced in order to provide a basis for communication.

▶ **1.1**

FLUIDS AND THE CONTINUUM

A fluid is defined as a substance that deforms continuously under the action of a shear stress. An important consequence of this definition is that when a fluid is at rest, there can be no shear stresses. Both liquids and gases are fluids. Some substances such as glass are technically classified as fluids. However, the rate of deformation in glass at normal temperatures is so small as to make its consideration as a fluid impractical.

Concept of a Continuum Fluids, like all matter, are composed of molecules whose numbers stagger the imagination. In a cubic inch of air at room conditions there are some 10^{20} molecules. Any theory that would predict the individual motions of these many molecules would be extremely complex, far beyond our present abilities.

Most engineering work is concerned with the macroscopic or bulk behavior of a fluid rather than with the microscopic or molecular behavior. In most cases it is convenient to think of a fluid as a continuous distribution of matter, or a *continuum*. There are, of course, certain instances in which the concept of a continuum is not valid. Consider, for example, the number of molecules in a small volume of a gas at rest. If the volume were taken small enough, the number of molecules per unit volume would be time-dependent for the microscopic volume even though the macroscopic volume had a constant number of molecules in it. The concept of a continuum would be valid only for the latter case. The validity of the continuum approach is seen to be dependent upon the type of information desired rather than the nature of the fluid. The treatment of fluids as continua is valid whenever the smallest fluid volume of interest contains a sufficient number of molecules to make statistical averages meaningful. The macroscopic properties of a continuum are considered to vary smoothly (continuously) from point to point in the fluid. Our immediate task is to define these properties at a point.

▶ **1.2**

PROPERTIES AT A POINT

When a fluid is in motion, the quantities associated with the state and the motion of the fluid will vary from point to point. The definition of some fluid variables at a point is presented below.

Density at a Point   The density of a fluid is defined as the mass per unit volume. Under flow conditions, particularly in gases, the density may vary greatly throughout the fluid. The density, ρ, at a particular point in the fluid is defined as

$$\rho = \lim_{\Delta V \to \delta V} \frac{\Delta m}{\Delta V}$$

where Δm is the mass contained in a volume ΔV, and δV is the smallest volume surrounding the point for which statistical averages are meaningful. The limit is shown in Figure 1.1.

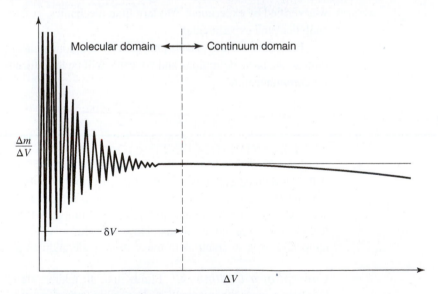

Figure 1.1 Density at a point.

The concept of the density at a mathematical point—that is, at $\Delta V = 0$, is seen to be fictitious; however, taking $\rho = \lim_{\Delta V \to \delta V}(\Delta m/\Delta V)$ is extremely useful, as it allows us to describe fluid flow in terms of continuous functions. The density, in general, may vary from point to point in a fluid and may also vary with respect to time, as in a punctured automobile tire. A table of densities of common fluids is given in Table 1.1.

Fluid Properties and Flow Properties Some fluids, particularly liquids, have densities that remain almost constant over wide ranges of pressure and temperature. Fluids that exhibit this quality are usually treated as being incompressible. The effects of compressibility, however, are more a property of the situation than of the fluid itself. For example, the flow of air at low velocities is described by the same equations that describe the flow of water. From a static viewpoint, air is a compressible fluid and water incompressible. Instead of being classified according to the fluid, compressibility effects are considered a property of the flow. A distinction, often subtle, is made between the properties of the fluid and the properties of the flow, and the student is hereby alerted to the importance of this concept.

Table 1.1 Densities of various fluids (at 20 C) unless otherwise noted

Fluid	ρ (g/cm^3)
Acetone	0.792
Ethanol	0.791
Benzene	0.899
Gasoline	0.670
Glycerin	1.260
Mercury	13.6
Sea water	1.025
Water	1.00
Soap solution	0.900
Blood	1.060 (37 C)

Source: Handbook of Chemistry and Physics, 62nd ed., Chemical Rubber Publishing Co., Cleveland OH, 1980.

Stress at a Point Consider the force $\Delta\mathbf{F}$ acting on an element ΔA of the body shown in Figure 1.2. The force $\Delta\mathbf{F}$ is resolved into components normal and parallel to the surface of the element. The force per unit area or stress at a point is defined as the limit of $\Delta\mathbf{F}/\Delta A$ as $\Delta A \rightarrow \delta A$ where δA is the smallest area for which statistical averages are meaningful:

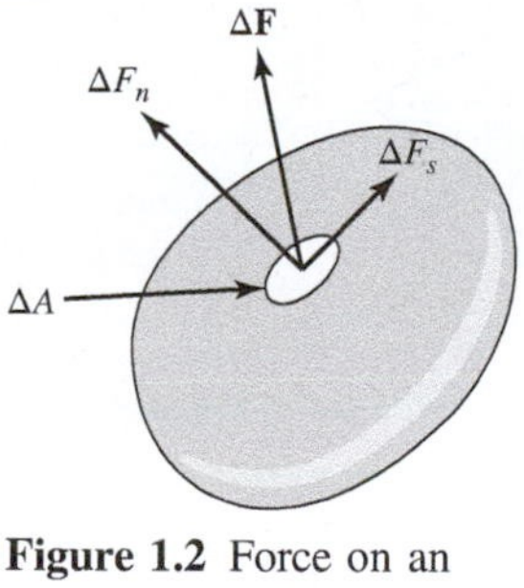

$$\lim_{\Delta A \to \delta A} \frac{\Delta F_n}{\Delta A} = \sigma_{ii} \qquad \lim_{\Delta A \to \delta A} \frac{\Delta F_s}{\Delta A} = \tau_{ij}$$

Here σ_{ii} is the normal stress and τ_{ij} the shear stress. In this text, the double-subscript stress notation as used in solid mechanics will be employed. The student will recall that normal stress is positive in tension. The limiting process for the normal stress is illustrated in Figure 1.3.

Forces acting on a fluid are divided into two general groups: body forces and surface forces. Body forces are those that act without physical contact—for example, gravity and electrostatic forces. On the contrary, pressure and frictional forces require physical contact for transmission. As a surface is required for the action of these forces, they are called surface forces. Stress is therefore a surface force per unit area.[1]

Figure 1.2 Force on an element of fluid.

Pressure at a point in a Static Fluid For a static fluid, the normal stress at a point may be determined from the application of Newton's laws to a fluid element as the fluid element

---

[1] Mathematically, stress is classed as a tensor of second order, as it requires magnitude, direction, and orientation with respect to a plane for its determination.

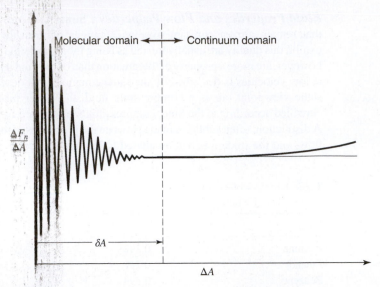

Figure 1.3 Normal stress at a point.

approaches zero size. It may be recalled that there can be no shearing stress in a static fluid. Thus, the only surface forces present will be those due to normal stresses. Consider the element shown in Figure 1.4. This element, while at rest, is acted upon by gravity and normal stresses. The weight of the fluid element is $\rho \mathbf{g}\,(\Delta x \Delta y \Delta z/2)$.

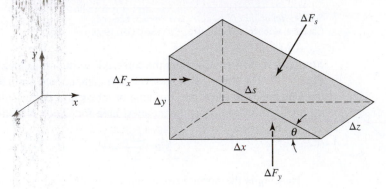

Figure 1.4 Element in a static fluid.

For a body at rest, $\Sigma \mathbf{F} = 0$. In the x direction,

$$\Delta F_x - \Delta F_s \sin \theta = 0$$

Since $\sin \theta = \Delta y/\Delta s$, the above equation becomes

$$\Delta F_x - \Delta F_s \frac{\Delta y}{\Delta s} = 0$$

Dividing through by $\Delta y \Delta z$ and taking the limit as the volume of the element approaches zero, we obtain

$$\lim_{\Delta V \to 0} \left[\frac{\Delta F_x}{\Delta y \Delta z} - \frac{\Delta F_s}{\Delta s \Delta z} \right] = 0$$

Recalling that normal stress is positive in tension, we obtain, by evaluating the above equation,

$$\sigma_{xx} = \sigma_{ss} \tag{1-1}$$

In the y direction, applying $\Sigma\mathbf{F} = 0$ yields

$$\Delta F_y - \Delta F_s \cos\theta - \rho g \frac{\Delta x \Delta y \Delta z}{2} = 0$$

Since $\cos\theta = \Delta x/\Delta s$, one has

$$\Delta F_y - \Delta F_s \frac{\Delta x}{\Delta s} - \rho g \frac{\Delta x \Delta y \Delta z}{2} = 0$$

Dividing through by $\Delta x \Delta z$ and taking the limit as before, we obtain

$$\lim_{\Delta V \to 0} \left[\frac{\Delta F_y}{\Delta x \Delta z} - \frac{\Delta F_s}{\Delta s \Delta z} - \frac{\rho g \Delta y}{2} \right] = 0$$

which becomes

$$-\sigma_{yy} + \sigma_{ss} - \frac{\rho g}{2}(0) = 0$$

or

$$\sigma_{yy} = \sigma_{ss} \tag{1-2}$$

It may be noted that the angle θ does not appear in equation (1-1) or (1-2); thus, the normal stress at a point in a static fluid is independent of direction, and is therefore a scalar quantity.

As the element is at rest, the only surface forces acting are those due to the normal stress. If we were to measure the force per unit area acting on a submerged element, we would observe that it acts inward or places the element in compression. The quantity measured is, of course, pressure, which, in light of the preceding development, must be the negative of the normal stress. This important simplification—the reduction of stress, a tensor, to pressure, a scalar—may also be shown for the case of zero shear stress in a flowing fluid. When shearing stresses are present, the normal stress components at a point may not be equal; however, the pressure is still equal to the average normal stress—that is,

$$P = -\frac{1}{3}\left(\sigma_{xx} + \sigma_{yy} + \sigma_{zz}\right)$$

with very few exceptions, one being flow in shock waves.

Now that certain properties at a point have been discussed, let us investigate the manner in which fluid properties vary from point to point.

► **1.3**

POINT-TO-POINT VARIATION OF PROPERTIES IN A FLUID

In the continuum approach to momentum transfer, use will be made of pressure, temperature, density, velocity, and stress fields. In previous studies, the concept of a gravitational field has been introduced. Gravity, of course, is a vector, and thus a gravitational field is a vector field. In this book, vectors will be written in boldfaced type. Weather maps illustrating the pressure variation over this country are published in newspapers and on websites. As pressure is a scalar quantity, such maps are an illustration of a scalar field. Scalars in this book will be set in regular type.

In Figure 1.5, the lines drawn are the loci of points of equal pressure. The pressure varies continuously throughout the region, and one may observe the pressure levels and infer the manner in which the pressure varies by examining such a map.

Figure 1.5 Weather map—an example of a scalar field.

Of specific interest in momentum transfer is the description of the point-to-point variation in the pressure. Denoting the directions east and north in Figure 1.5 by x and y, respectively, we may represent the pressure throughout the region by the general function $P(x,y)$.

The change in P, written as dP, between two points in the region separated by the distances dx and dy is given by the total differential

$$dP = \frac{\partial P}{\partial x} dx + \frac{\partial P}{\partial y} dy \tag{1-3}$$

In equation (1-3), the partial derivatives represent the manner in which P changes along the x and y axes, respectively.

Along an arbitrary path s in the xy plane the total derivative is

$$\frac{dP}{ds} = \frac{\partial P}{\partial x}\frac{dx}{ds} + \frac{\partial P}{\partial y}\frac{dy}{ds} \tag{1-4}$$

In equation (1-4), the term dP/ds is the directional derivative, and its functional relation describes the rate of change of P in the s direction.

A small portion of the pressure field depicted in Figure 1.5 is shown in Figure 1.6. The arbitrary path s is shown, and it is easily seen that the terms dx/ds and dy/ds are the cosine and sine of the path angle, α, with respect to the x axis. The directional derivative, there may be written as

$$\frac{dP}{ds} = \frac{\partial P}{\partial x}\cos\alpha + \frac{\partial P}{\partial y}\sin\alpha$$

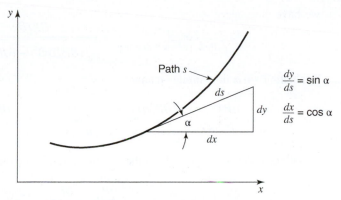

Figure 1.6 Path s in the xy plane.

There are an infinite number of paths to choose from in the xy plane; however, two particular paths are of special interest: the path for which dP/ds is zero and that for which dP/ds is maximum.

The path for which the directional derivative is zero is quite simple to find. Setting dP/ds equal to zero, we have

$$\frac{\sin \alpha}{\cos \alpha}\bigg|_{dP/ds=0} = \tan \alpha\bigg|_{dP/ds=0} = -\frac{\partial P/\partial x}{\partial P/\partial y}$$

or, since $\tan \alpha = dy/dx$, we have

$$\frac{dy}{dx}\bigg|_{dP/ds=0} = -\frac{\partial P/\partial x}{\partial P/\partial y} \tag{1-6}$$

Along the path whose slope is defined by equation (1-6), we have $dP = 0$, and thus P is constant. Paths along which a scalar is constant are called *isolines*.

In order to find the direction for which dP/ds is a maximum, we must have the derivative $(d/d\alpha)(dP/ds)$ equal to zero, or

$$\frac{d}{d\alpha}\frac{dP}{ds} = -\sin \alpha\frac{\partial P}{\partial x} + \cos \alpha\frac{\partial P}{\partial y} = 0$$

or

$$\tan \alpha\bigg|_{dP/ds \text{ is max}} = \frac{\partial P/\partial y}{\partial P/\partial x} \tag{1-7}$$

Comparing equations (1-6) and (1-7), we see that the two directions defined by these equations are perpendicular. The magnitude of the directional derivative when the directional derivative is maximum is

$$\frac{dP}{ds}\bigg|_{max} = \frac{\partial P}{\partial x}\cos \alpha + \frac{\partial P}{\partial y}\sin \alpha$$

where $\cos \alpha$ and $\sin \alpha$ are evaluated along the path given by equation (1-7). As the cosine is related to the tangent by

$$\cos \alpha = \frac{1}{\sqrt{1 + \tan^2 \alpha}}$$

we have

$$\cos \alpha \Big|_{dP/ds \text{ is max}} = \frac{\partial P/\partial x}{\sqrt{(\partial P/\partial x)^2 + (\partial P/\partial y)^2}}$$

Evaluating $\sin \alpha$ in a similar manner gives

$$\frac{dP}{ds}\Big|_{max} = \frac{(\partial P/\partial x)^2 + (\partial P/\partial y)^2}{\sqrt{(\partial P/\partial x)^2 + (\partial P/\partial y)^2}} = \sqrt{\left(\frac{\partial P}{\partial x}\right)^2 + \left(\frac{\partial P}{\partial y}\right)^2} \qquad (1\text{-}8)$$

Equations (1-7) and (1-8) suggest that the maximum directional derivative is a vector of the form

$$\frac{\partial P}{\partial x}\mathbf{e}_x + \frac{\partial P}{\partial y}\mathbf{e}_y$$

where $\mathbf{e}_x$ and $\mathbf{e}_y$ are unit vectors in the x and y directions, respectively.

The directional derivative along the path of maximum value is frequently encountered in the analysis of transfer processes and is given a special name, the *gradient*. Thus, the gradient of P, grad P, is

$$\text{grad } P = \frac{\partial P}{\partial x}\mathbf{e}_x + \frac{\partial P}{\partial y}\mathbf{e}_y$$

where $P = P(x, y)$. This concept can be extended to cases in which $P = P(x, y, z)$. For this more general case,

$$\text{grad } P = \frac{\partial P}{\partial x}\mathbf{e}_x + \frac{\partial P}{\partial y}\mathbf{e}_y + \frac{\partial P}{\partial z}\mathbf{e}_z \qquad (1\text{-}9)$$

Equation (1-9) may be written in more compact form by use of the operation ∇ (pronounced "del"), giving

$$\nabla P = \frac{\partial P}{\partial x}\mathbf{e}_x + \frac{\partial P}{\partial y}\mathbf{e}_y + \frac{\partial P}{\partial z}\mathbf{e}_z$$

where

$$\nabla = \frac{\partial}{\partial x}\mathbf{e}_x + \frac{\partial}{\partial y}\mathbf{e}_y + \frac{\partial}{\partial z}\mathbf{e}_z \qquad (1\text{-}10)$$

Equation (1-10) is the defining relationship for the ∇ operator in Cartesian coordinates. This symbol indicates that differentiation is to be performed in a prescribed manner. In other coordinate systems, such as cylindrical and spherical coordinates, the gradient takes on a different form.[2] However, the geometric meaning of the gradient remains the same; it is a vector having the direction and magnitude of the maximum rate of change of the dependent variable with respect to distance.

▶ **1.4**

UNITS

In addition to the International Standard (SI) system of units, there are two different English systems of units commonly used in engineering. These systems have their roots in Newton's second law of motion: force is equal to the time rate of change of momentum. In defining

---

[2] The forms of the gradient operator in rectangular, cylindrical, and spherical coordinate systems are listed in Appendix B.

each term of this law, a direct relationship has been established between the four basic physical quantities used in mechanics: force, mass, length, and time. Through the arbitrary choice of fundamental dimensions, some confusion has occurred in the use of the English systems of units. Using the SI system of units has greatly reduced these difficulties.

The relationship between force and mass may be expressed by the following statement of Newton's second law of motion:

$$\mathbf{F} = \frac{m\mathbf{a}}{g_c}$$

where g_c is a conversion factor that is included to make the equation dimensionally consistent.

In the SI system, mass, length, and time are taken as basic units. The basic units are mass in kilograms (kg), length in meters (m), and time in seconds (s). The corresponding unit of force is the newton (N). One newton is the force required to accelerate a mass of one kilogram at a rate of one meter per second per second (1 m/s^2). The conversion factor, g_c, is then equal to one kilogram meter per newton per second per second (1 kg · m/N · s^2).

In engineering practice, force, length, and time have been frequently chosen as defining fundamental units. With this system, force is expressed in pounds force (lb$_f$), length in feet, and time in seconds. The corresponding unit of mass will be that which will be accelerated at the rate of 1 ft/(s)^2 by 1 lb$_f$.

This unit of mass having the dimensions of (lb$_f$)(s)^2/(ft) is called the *slug*. The conversion factor, g_c, is then a multiplying factor to convert slugs into (lb$_f$)(s)^2/(ft), and its value is 1 (slug)(ft)/(lb$_f$)(s)^2.

A third system encountered in engineering practice involves all four fundamental units. The unit of force is 1 lb$_f$, the unit of mass is 1 lb$_m$; length and time are given in units of feet and seconds, respectively. When 1 lb$_m$ at sea level is allowed to fall under the influence of gravity, its acceleration will be 32.174 (ft)/(s)^2. The force exerted by gravity on 1 lb$_m$ at sea level is defined as 1 lb$_f$. Therefore, the conversion factor, g_c, for this system is 32.174 (lb$_m$)(ft)/(lb$_f$)(s)^2.[3]

A summary of the values of g_c is given in Table 1.2 for these three English systems of engineering units, along with the units of length, time, force, and mass.

Table 1.2

System	Length	Time	Force	Mass	g_c
1	meter	second	newton	kilogram	$1\dfrac{\text{kg} \cdot \text{m}}{\text{N} \cdot \text{s}^2}$
2	foot	second	lb$_f$	slug	$\dfrac{1\,(\text{slug})(\text{ft})}{(\text{lb}_f)(\text{s})^2}$
3	foot	second	lb$_f$	lb$_m$	$\dfrac{32.174\,(\text{lb}_m)(\text{ft})}{(\text{lb}_f)(\text{s})^2}$

As all three systems are in current use in the technical literature, the student should be able to use formulas given in any particular situation. Careful checking for dimensional consistency will be required in *all* calculations. The conversion factor, g_c, will correctly relate the units corresponding to a system. There will be no attempt by the authors to

---

[3] In subsequent calculations in this book, g_c will be rounded off to a value of 32.2 lb$_m$ft/lb$_f$s^2.

incorporate the conversion factor in any equations; instead, it will be the reader's responsibility to use units that are consistent with every term in the equation.

▶ **1.5**

COMPRESSIBILITY

A fluid is considered *compressible* or *incompressible* depending on whether its density is variable or constant. Liquids are generally considered to be incompressible, whereas gases are certainly compressible.

The *bulk modulus of elasticity*, often referred to as simply the *bulk modulus*, is a fluid property that characterizes compressibility. It is defined according to

$$\beta \equiv \frac{dP}{dV/V} \tag{1-11a}$$

or as

$$\beta \equiv -\frac{dP}{d\rho/\rho} \tag{1-11b}$$

and has the dimensions N/m^2.

Disturbances introduced at some location in a fluid continuum will be propagated at a finite velocity. The velocity is designated the *acoustic velocity*—that is, the speed of sound in the fluid is symbolized C.

It can be shown that the acoustic velocity is related to changes in pressure and density according to

$$C = \left(\frac{dP}{d\rho}\right)^{1/2} \tag{1-12}$$

Introducing equation (1-11b) into this relationship yields

$$C = \left(-\frac{\beta}{\rho}\right)^{1/2} \tag{1-13}$$

For a gas, undergoing an isentropic process where $PV^k = C$, a constant, we have

$$C = \left(\frac{kP}{\rho}\right)^{1/2}, \quad k = C_p/C_v \tag{1-14}$$

or

$$C = (kRT)^{1/2} \tag{1-15}$$

The question arises concerning when a gas, which is compressible, may be treated in a flow situation as incompressible—that is, when density variations are negligibly small. A common criterion for such a consideration involves the *Mach number*. The Mach number, a dimensionless parameter, is defined as the ratio of the fluid velocity, v, to the speed of sound, C, in the fluid:

$$M = \frac{v}{C} \tag{1-16}$$

A general rule of thumb is that when $M < 0.2$, the flow may be treated as incompressible with negligible error.

Example 1

A jet aircraft is flying at an altitude of 15,500 m, where the air temperature is 239 K. Determine whether compressibility effects are significant at airspeeds of (a) 220 km/h and (b) 650 km/h.

The test for compressibility effects requires calculating the Mach number, M, which, in turn, requires that the acoustic velocity at each airspeed be evaluated.

For air, $k = 1.4$, $R = 0.287$ kJ/kg · K, and

$$C = (kRT)^{1/2}$$
$$= [1.4(0.287 \text{ kJ/kg} \cdot \text{K})(239 \text{ K})(1000 \text{ N} \cdot \text{m/kJ})(\text{kg} \cdot \text{m/N} \cdot \text{s}^2)]^{1/2}$$
$$= 310 \text{ m/s}$$

(a) At $v = 220$ km/hr (61.1 m/s),

$$M = \frac{v}{C} = \frac{61.1 \text{ m/s}}{310 \text{ m/s}} = 0.197$$

The flow may be treated as incompressible.

(b) At $v = 650$ km/hr (180.5 m/s),

$$M = \frac{v}{C} = \frac{180.5 \text{ m/s}}{310 \text{ m/s}} = 0.582$$

Compressible effects must be accounted for.

▶ **1.6**

SURFACE TENSION

The situation where a small amount of unconfined liquid forms a spherical drop is familiar to most of us. The phenomenon is the consequence of the attraction that exists between liquid molecules. Within a drop a molecule of liquid is completely surrounded by many others. Particles near the surface, on the contrary, will experience an imbalance of net force because of the nonuniformity in the numbers of adjacent particles. The extreme condition is the density discontinuity at the surface. Particles at the surface experience a relatively strong inwardly directed attractive force.

Given this behavior, it is evident that some work must be done when a liquid particle moves toward the surface. As more fluid is added, the drop will expand, creating additional surface. The work associated with creating this new surface is the *surface tension*, symbolized, σ. Quantitatively, σ is the work per unit area, $N \cdot m/m^2$ or force per unit length of interface in N/m.

A surface is, in reality, an interface between two phases. Thus, both phases will have the property of surface tension. The most common materials involving phase interfaces are water and air, but many others are also possible. For a given interfacial composition, the surface tension property is a function of both pressure and temperature, but a much stronger function of temperature. Table 1.3 lists values of σ for several fluids in air at 1 atm and 20 C. For water in air, the surface tension is expressed as a function of temperature according to

$$\sigma = 0.123(1 - 0.00139\,T)\text{N/m} \tag{1-17}$$

where T is in Kelvin.

Table 1.3 Surface tension of some fluids in air at 1 atm and 20 C

Fluid	σ (N/m)
Ammonia	0.021
Ethyl alcohol	0.028
Gasoline	0.022
Glycerin	0.063
Kerosene	0.028
Mercury	0.440
Soap solution	0.025
SAE 30 oil	0.035

Source: Handbook of Chemistry and Physics, 62nd ed., Chemical Rubber Publishing Co., Cleveland, OH, 1980.

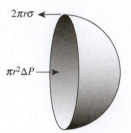

$2\pi r \sigma$

$\pi r^2 \Delta P$

Figure 1.7 A free body diagram of a hemispherical liquid droplet.

In Figure 1.7, we show a free body diagram of a hemispherical drop of liquid with the pressure and surface tension forces in balance. The condition examined is typically used for this analysis as a sphere represents the minimum surface area for a prescribed volume. The pressure difference, ΔP, between the inside and outside of the hemisphere produces a net pressure force that is balanced by the surface tension force. This force balance can be expressed as

$$\pi r^2 \Delta P = 2\pi r \sigma$$

and the pressure difference is given by

$$\Delta P = \frac{2\sigma}{r} \qquad (1\text{-}18)$$

The Young–Laplace equation is a general condition for the equilibrium of normal stresses across a static interface separating a pair of immiscible fluids,

$$\Delta P = \sigma \left(\frac{1}{R_1} + \frac{1}{R_2} \right)$$

where R_1 and R_2 are the radii of curvature at the surface of the body. This equation states that the surface tension causes an increased pressure on the inside of a surface whose magnitude depends on the radii of curvature of the surface. In most systems, $R_1 = R_2$.

For the case of a soap bubble, which has a very thin wall, there are two interfaces, and the pressure difference will be

$$\Delta P = \frac{4\sigma}{r} \qquad (1\text{-}19)$$

Equations (1-18) and (1-19) indicate that the pressure difference is inversely proportional to r. The limit of this relationship is the case of a fully wetted surface where $r \cong \infty$, and the pressure difference due to surface tension is zero.

A consequence of the pressure difference resulting from surface tension is the phenomenon *of capillary action*. This effect is related to how well a liquid *wets* a solid boundary. The indicator for wetting or nonwetting is the *contact angle*, θ, defined as illustrated in Figure 1.8. With θ measured through the liquid, a nonwetting case, as shown in the figure, is associated with $\theta > 90°$. For a wetting case $\theta < 90°$. For mercury in contact

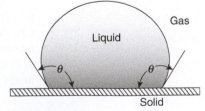

Figure 1.8 Contact angle for a nonwetting gas–liquid–solid interface.

with a clean glass tube, $\theta \cong 130°$. Water in contact with a clean glass surface will completely wet the surface, and, for this case, $\theta \cong 0$.

Illustrated in Figure 1.9 is the case of a small glass tube inserted into a pool of (a) water and (b) mercury. Note that water will rise in the tube and that in mercury the level in the tube is depressed.

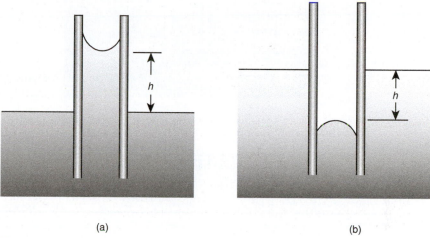

(a)   (b)

Figure 1.9 Capillary effects with a tube inserted in (a) water and (b) mercury.

$2\pi r\sigma$

$\pi r^2 h$

Figure 1.10 Free body diagram of a wetting liquid in a tube.

For the water case, the liquid rises a distance h above the level in the pool. This is the result of attraction between the liquid molecules and the tube wall being greater than the attraction between water molecules at the liquid surface. For the mercury case, the intermolecular forces at the liquid surface are greater than the attractive forces between liquid mercury and the glass surface. The mercury is depressed a distance h below the level of the pool.

A free body diagram of the wetting liquid is shown in Figure 1.10. The upward force, due to surface tension,

$$2\pi r\sigma \cos\theta$$

will be equal to the downward force due to the weight of liquid having volume $V = \pi r^2 h$.

Equating these forces, we obtain

$$2\pi r\,\sigma \cos\theta = \rho g\pi r^2 h$$

and the value of h becomes

$$h = \frac{2\sigma \cos\theta}{\rho gr} \qquad (1\text{-}20)$$

Example 2

Determine the distance h that mercury will be depressed with a 4-mm-diameter glass tube inserted into a pool of mercury at 20 C (Figure 1.11).

Equation (1-20) applies, so we have

$$h = \frac{2\sigma \cos\theta}{\rho gr}$$

Recall that, for mercury and glass, $\theta = 130°$.

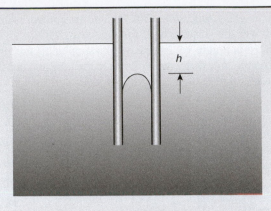

Figure 1.11 Capillary depression of mercury in a glass tube.

For mercury at 20 C, $\rho = 13{,}580 \, \text{kg/m}^3$, and for mercury in air $\sigma = 0.44 \, \text{N/m}$ (Table 1.3), giving

$$h = \frac{2(0.44 \, \text{N/m})(\cos 130°)}{(13{,}580 \, \text{kg/m}^3)(9.81 \, \text{m/s}^2)(2 \times 10^{-3} \, \text{m})}$$

$$= 2.12 \times 10^{-3} \, \text{m} \quad (2.12 \, \text{mm})$$

CHAPTER 2

Fluid Statics

The definition of a fluid variable at a point was considered in Chapter 1. In this chapter, the point-to-point variation of a particular variable, pressure, will be considered for the special case of a fluid at rest.

A frequently encountered static situation exists for a fluid that is stationary on Earth's surface. Although Earth has some motion of its own, we are well within normal limits of accuracy to neglect the absolute acceleration of the coordinate system that, in this situation, would be fixed with reference to Earth. Such a coordinate system is said to be an *inertial reference*. If, on the contrary, a fluid is stationary with respect to a coordinate system that has some significant absolute acceleration of its own, the reference is said to be *noninertial*. An example of this latter situation would be the fluid in a railroad tank car as it travels around a curved portion of track.

The application of Newton's second law of motion to a fixed mass of fluid reduces to the expression that the sum of the external forces is equal to the product of the mass and its acceleration. In the case of an inertial reference, we would naturally have the relation $\Sigma \mathbf{F} = 0$, whereas the more general statement $\Sigma \mathbf{F} = m\mathbf{a}$ must be used for the noninertial case.

▶ 2.1

PRESSURE VARIATION IN A STATIC FLUID

From the definition of a fluid, it is known that there can be no shear stress in a fluid at rest. This means that the only forces acting on the fluid are those due to gravity and pressure. As the sum of the forces must equal zero throughout the fluid, Newton's law may be satisfied by applying it to an arbitrary free body of fluid of differential size. The free body selected, shown in Figure 2.1, is the element of fluid $\Delta x \, \Delta y \, \Delta z$ with a corner at the point *xyz*. The coordinate system *xyz* is an inertial coordinate system.

The pressures that act on the various faces of the element are numbered 1 through 6. To find the sum of the forces on the element, the pressure on each face must first be evaluated.

We shall designate the pressure according to the face of the element upon which the pressure acts. For example, $P_1 = P|_x$, $P_2 = P|_{x+\Delta x}$, and so on. Evaluating the forces acting on each face, along with the force due to gravity acting on the element $\rho \mathbf{g} \, \Delta x \, \Delta y \, \Delta z$, we find that the sum of the forces is

$$\rho \mathbf{g}(\Delta x \Delta y \Delta z) + (P|_x - P|_{x+\Delta x})\Delta y \Delta z \mathbf{e}_x$$
$$+ (P|_y - P|_{y+\Delta y})\Delta x \Delta z \mathbf{e}_y + (P|_z - P|_{z+\Delta z})\Delta x \Delta y \mathbf{e}_z = 0$$

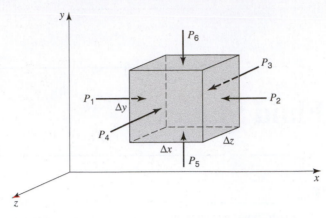

Figure 2.1 Pressure forces on a static fluid element.

Dividing by the volume of the element $\Delta x\,\Delta y\,\Delta z$, we see that the above equation becomes

$$\rho \mathbf{g} - \frac{P|_{x+\Delta x} - P|_x}{\Delta x}\mathbf{e}_x - \frac{P|_{y+\Delta y} - P|_y}{\Delta y}\mathbf{e}_y - \frac{P|_{z+\Delta z} - P|_z}{\Delta z}\mathbf{e}_z = 0$$

where the order of the pressure terms has been reversed. As the size of the element approaches zero, Δx, Δy, and Δz approach zero and the element approaches the point (x, y, z). In the limit

$$\rho \mathbf{g} = \lim_{\Delta x, \Delta y, \Delta z \to 0}\left[\frac{P|_{x+\Delta x} - P|_x}{\Delta x}\mathbf{e}_x + \frac{P|_{y+\Delta y} - P|_y}{\Delta y}\mathbf{e}_y + \frac{P|_{z+\Delta z} - P|_z}{\Delta z}\mathbf{e}_z\right]$$

or

$$\rho \mathbf{g} = \frac{\partial P}{\partial x}\mathbf{e}_x + \frac{\partial P}{\partial y}\mathbf{e}_y + \frac{\partial P}{\partial z}\mathbf{e}_z \tag{2-1}$$

Recalling the form of the gradient, we may write equation (2-1) as

$$\rho \mathbf{g} = \nabla P \tag{2-2}$$

Equation (2-2) is the basic equation of fluid statics and states that the maximum rate of change of pressure occurs in the direction of the gravitational vector. In addition, as isolines are perpendicular to the gradient, constant pressure lines are perpendicular to the gravitational vector. The point-to-point variation in pressure may be obtained by integrating equation (2-2).

Example 1

The manometer, a pressure measuring device, may be analyzed from the previous discussion. The simplest type of manometer is the U-tube manometer shown in Figure 2.2. The pressure in the tank at point A is to be measured. The fluid in the tank extends into the manometer to point B.

Choosing the y-axis in the direction shown, we see that equation (2-2) becomes

$$\frac{dP}{dy}\mathbf{e}_y = -\rho g \mathbf{e}_y$$

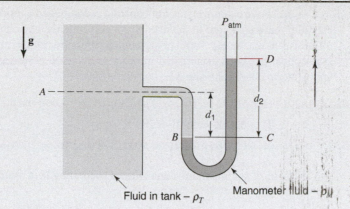

Figure 2.2 A U-tube manometer.

Integrating between C and D in the manometer fluid, we have

$$P_{\text{atm}} - P_C = -\rho_m g d_2$$

and then, integrating between B and A in the tank fluid, we obtain

$$P_A - P_B = -\rho_T g d_1$$

As elevations B and C are equal, the pressures, P_B and P_C, must be the same. We may thus combine the above equations to obtain

$$P_A - P_{\text{atm}} = \rho_m g d_2 - \rho_T g d_1$$

The U-tube manometer measures the difference between the absolute pressure and the atmospheric pressure. This difference is called the *gage pressure* and is frequently used in pressure measurement.

Example 2

In the fluid statics of gases, a relation between the pressure and density is required to integrate equation (2-2). The simplest case is that of the isothermal perfect gas, where $P = \rho RT/M$. Here R is the universal gas constant, M is the molecular weight of the gas, and T is the temperature, which is constant for this case. Selecting the y-axis parallel to g, we see that equation (2-2) becomes

$$\frac{dP}{dy} = -\rho g = -\frac{PMg}{RT}$$

Separating variables, we see that the above differential equation becomes

$$\frac{dP}{P} = -\frac{Mg}{RT} dy$$

Integration between $y = 0$ (where $P = P_{\text{atm}}$) and $y = y$ (where the pressure is P) yields

$$\ln \frac{P}{P_{\text{atm}}} = -\frac{Mgy}{RT}$$

or

$$\frac{P}{P_{\text{atm}}} = \exp\left\{-\frac{Mgy}{RT}\right\}$$

In the above examples, the atmospheric pressure and a model of pressure variation with elevation have appeared in the results. As performance of aircraft, rockets, and many types of industrial machinery varies with ambient pressure, temperature, and density, a standard atmosphere has been established in order to evaluate performance. At sea level, standard atmospheric conditions are

$$P = 29.92 \text{ in. Hg} = 2116.2 \text{ lb}_f/\text{ft}^2 = 14.696 \text{ lb}_f/\text{in.}^2 = 101{,}325 \text{ N/m}^2$$

$$T = 519 \text{ R} = 59 \text{ F} = 288 \text{ K}$$

$$\rho = 0.07651 \text{ lb}_m/\text{ft}^3 = 0.002378 \text{ slug/ft}^3 = 1.226 \text{ kg/m}^3$$

A table of the standard atmospheric properties as a function of altitude is given in Appendix G.[1]

▶ **2.2**

UNIFORM RECTILINEAR ACCELERATION

For the case in which the coordinate system xyz in Figure 2.1 is not an inertial coordinate system, equation (2-2) does not apply. In the case of uniform rectilinear acceleration, however, the fluid will be at rest with respect to the accelerating coordinate system. With a constant acceleration, we may apply the same analysis as in the case of the inertial coordinate system except that $\Sigma \mathbf{F} = m\mathbf{a} = \rho \, \Delta x \, \Delta y \, \Delta z \mathbf{a}$, as required by Newton's second law of motion. The result is

$$\nabla P = \rho(\mathbf{g} - \mathbf{a}) \tag{2-3}$$

The maximum rate of change of pressure is now in the $\mathbf{g} - \mathbf{a}$ direction, and lines of constant pressure are perpendicular to $\mathbf{g} - \mathbf{a}$.

The point-to-point variation in pressure is obtained from integration of equation (2-3).

Example 3

A fuel tank is shown in Figure 2.3. If the tank is given a uniform acceleration to the right, what will be the pressure at point B?

From equation (2-3), the pressure gradient is in the $\mathbf{g} - \mathbf{a}$ direction, so the surface of the fluid will be perpendicular to this direction. Choosing the y-axis parallel to $\mathbf{g} - \mathbf{a}$, we find that equation (2-3) may be integrated between point B and the surface. The pressure gradient becomes $dP/dy\,\mathbf{e}_y$ with the selection of the y-axis parallel to $\mathbf{g} - \mathbf{a}$ as shown in Figure 2.4. Thus,

$$\frac{dP}{dy}\mathbf{e}_y = -\rho|\mathbf{g} - \mathbf{a}|\mathbf{e}_y = -\rho\sqrt{g^2 + a^2}\mathbf{e}_y$$

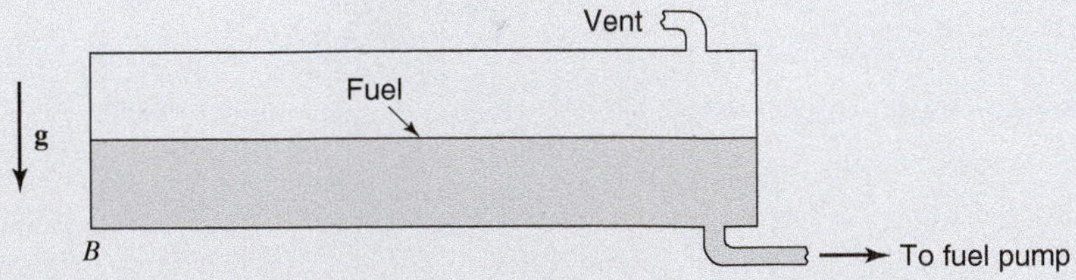

Figure 2.3 Fuel tank at rest.

---

[1] These performance standard sea-level conditions should not be confused with gas-law standard conditions of $P = 29.92$ in Hg $= 14.696$ lb/in^2 $= 101{,}325$ Pa, $T = 492°$ R $= 32$ F $= 273$ K.

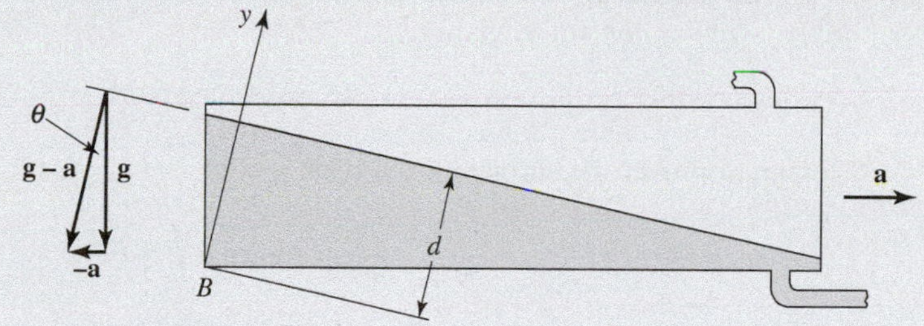

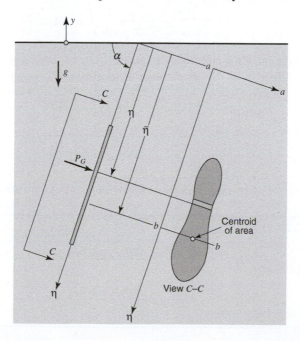

Figure 2.4 Uniformly accelerated fuel tank.

Integrating between $y = 0$ and $y = d$ yields

$$P_{atm} - P_B = \rho\sqrt{g^2 + a^2}(-d)$$

or

$$P_B - P_{atm} = \rho\sqrt{g^2 + a^2}(d)$$

The depth of the fluid, d, at point B is determined from the tank geometry and the angle θ.

▶ **2.3**

FORCES ON SUBMERGED SURFACES

Determination of the force on submerged surfaces is done frequently in fluid statics. As these forces are due to pressure, use will be made of the relations describing the point-to-point variation in pressure that have been developed in the previous sections. The plane surface illustrated in Figure 2.5 is inclined at an angle α to the surface of the fluid. The area of the inclined plane is A and the density of the fluid is ρ.

Figure 2.5 A submerged plane surface.

The magnitude of the force on the element dA is $P_G\,dA$, where P_G is the gage pressure, $P_G = -\rho g y = \rho g \eta \sin \alpha$, giving

$$dF = \rho g \eta \sin \alpha \, dA$$

Integration over the surface of the plate yields

$$F = \rho g \sin \alpha \int_A \eta \, dA$$

The definition or the centroid of the area is

$$\bar{\eta} \equiv \frac{1}{A} \int_A \eta \, dA$$

and thus

$$F = \rho g \sin \alpha \, \bar{\eta} A \qquad (2\text{-}4)$$

Thus, the force due to the pressure is equal to the pressure evaluated at the centroid of the submerged area multiplied by the submerged area. The point at which this force acts (the center of pressure) is not the centroid of the area. In order to find the center of pressure, we must find the point at which the total force on the plate must be concentrated in order to produce the same moment as the distributed pressure, or

$$F \eta_{\text{c.p}} = \int_A \eta P_G \, dA$$

Substitution for the pressure yields

$$F \eta_{\text{c.p.}} = \int_A \rho g \sin \alpha \, \eta^2 \, dA$$

and since $F = \rho g \sin \alpha \, \bar{\eta} A$, we have

$$\eta_{\text{c.p.}} = \frac{1}{A\bar{\eta}} \int_A \eta^2 \, dA = \frac{I_{aa}}{A\bar{\eta}} \qquad (2\text{-}5)$$

The moment of the area about the surface may be translated from an axis aa located at the fluid surface to an axis bb through the centroid by

$$I_{aa} = I_{bb} + \bar{\eta}^2 A$$

and thus

$$\eta_{\text{c.p.}} - \bar{\eta} = \frac{I_{bb}}{A\bar{\eta}} \qquad (2\text{-}6)$$

The center of pressure is located below the centroid a distance $I_{bb}/A\bar{\eta}$.

Example 4

A circular viewing port is to be located 1.5 ft below the surface of a tank as shown in Figure 2.6. Find the magnitude and location of the force acting on the window.

The force on the window is

$$F = \rho g \sin \alpha \, A\bar{\eta}$$

where

$$\alpha = \pi/2 \quad \text{and} \quad \bar{\eta} = 1.5\,\text{ft}$$

The force is

$$F = \rho g A \bar{\eta} = \frac{(62.4\,\text{lb}_m/\text{ft}^3)(32.2\,\text{ft/s}^2)(\pi/4\,\text{ft}^2)(1.5\,\text{ft})}{32.2\,\text{lb}_m\text{ft/s}^2\,\text{lb}_f}$$

$$= 73.5\,\text{lb}_f\,(327\,\text{N})$$

The force F acts at $\bar{\eta} + \dfrac{I_{\text{centroid}}}{A\bar{\eta}}$. For a circular area, $I_{\text{centroid}} = \pi R^4/4$, so we obtain

$$\eta_{\text{c.p.}} = 1.5 + \frac{\pi R^4}{4\pi R^2 1.5}\,\text{ft} = 1.542\,\text{ft}$$

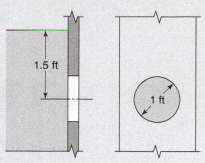

Figure 2.6 Submerged window.

Example 5

Rainwater collects behind the concrete retaining wall shown in Figure 2.7. If the water-saturated soil (specific gravity = 2.2) acts as a fluid, determine the force and center of pressure on a 1 m width of the wall.

    The force on a unit width of the wall is obtained by integrating the pressure difference between the right and left sides of the wall. Taking the origin at the top of the wall and measuring y downward, the force due to pressure is

$$F = \int (P - P_{\text{atm}})(1)\,dy$$

The pressure difference in the region in contact with the water is

$$P - P_{\text{atm}} = \rho_{\text{H}_2\text{O}}\,g\,y$$

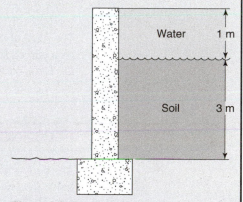

Figure 2.7 Retaining wall.

and the pressure difference in the region in contact with the soil is

$$P - P_{\text{atm}} = \rho_{\text{H}_2\text{O}}\,g(1) + 2.2\,\rho_{\text{H}_2\text{O}}\,g(y-1)$$

The force F is

$$F = \rho_{\text{H}_2\text{O}}\,g \int_0^1 y\,dy + \rho_{\text{H}_2\text{O}}\,g \int_0^4 [1 + 2.2(y-1)]\,dy$$

$$F = (1000\,\text{kg/m}^3)(9.807\,\text{m/s}^2)(1\,\text{m})(13.4\,\text{m}^2) = 131{,}414\,\text{N}\,(29{,}546\,\text{lb}_f)$$

The center of pressure of the force on the wall is obtained by taking moments about the top of the wall.

$$F y_{\text{c.p.}} = \rho_{\text{H}_2\text{O}}\,g \left\{ \int_0^1 y^2\,dy + \int_1^4 y[1 + 2.2(y-1)]\,dy \right\}$$

$$y_{\text{c.p.}} = \frac{1}{(131{,}414\,\text{N})}\,(1000\,\text{kg/m}^3)(9.807\,\text{m/s}^2)(1\,\text{m})(37.53\,\text{m}^3) = 2.80\,\text{m}\,(9.19\,\text{ft})$$

The force on a submerged curved surface may be obtained from knowledge of the force on a plane surface and the laws of statics. Consider the curved surface BC illustrated in Figure 2.8.

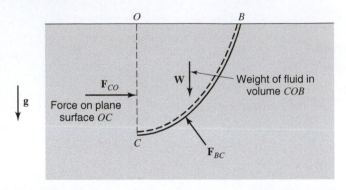

Figure 2.8 Submerged curved surface.

By consideration of the equilibrium of the fictitious body BCO, the force of the curved plate on body BCO may be evaluated. Since $\Sigma \mathbf{F} = 0$, we have

$$\mathbf{F}_{CB} = -\mathbf{W} - \mathbf{F}_{CO} \tag{2-7}$$

The force of the fluid on the curved plate is the negative of this or $\mathbf{W} + \mathbf{F}_{CO}$. Thus, the force on a curved submerged surface may be obtained from the weight on the volume BCO and the force on a submerged plane surface.

▶ **2.4**

BUOYANCY

The body shown in Figure 2.9 is submerged in a fluid with density ρ. The resultant force $\mathbf{F}$ holds the body in equilibrium.

The element of volume $h\,dA$ has gravity and pressure forces acting on it. The component of the force due to the pressure on the top of the element is $-P_2\,dS_2 \cos\alpha\,\mathbf{e}_y$, where α is the angle between the plane of the element dS_2 and the xz plane. The product

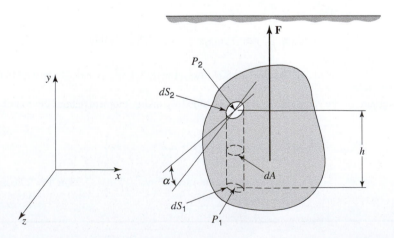

Figure 2.9 Forces on submerged volume.

$dS_2 \cos \alpha$ then is the projection of dS_2 onto the xz plane, or simply dA. The net pressure force on the element is $(P_1 - P_2)\, dA\, \mathbf{e}_y$, and the resultant force on the element is

$$d\mathbf{F} = (P_1 - P_2)\, dA\, \mathbf{e}_y - \rho_B g h\, dA\, \mathbf{e}_y$$

where ρ_B is the density of the body. The difference in pressure $P_1 - P_2$ may be expressed as $\rho g h$, so

$$d\mathbf{F} = (\rho - \rho_B)g h\, dA\, \mathbf{e}_y$$

Integration over the volume of the body, assuming constant densities, yields

$$\mathbf{F} = (\rho - \rho_B)g V \mathbf{e}_y \qquad (2\text{-}8)$$

where V is the volume of the body. The resultant force $\mathbf{F}$ is composed of two parts: the weight $-\rho_B g V \mathbf{e}_y$ and the buoyant force $\rho g V \mathbf{e}_y$. The body experiences an upward force equal to the weight of the displaced fluid. This is the well-known principle of Archimedes. When $\rho > \rho_B$, the resultant force will cause the body to float on the surface. In the case of a floating body, the buoyant force is $\rho g V_s \mathbf{e}_y$, where V_s is the submerged volume.

Example 6

A cube measuring 1 ft on a side is submerged so that its top face is 10 ft below the free surface of water. Determine the magnitude and direction of the applied force necessary to hold the cube in this position if it is made of

(a) cork ($\rho = 10\ \text{lb}_m/\text{ft}^3$)
(b) steel ($\rho = 490\ \text{lb}_m/\text{ft}^3$).

The pressure forces on all lateral surfaces of the cube cancel. Those on the top and bottom do not, as they are at different depths.
    Summing forces on the vertical direction, we obtain

$$\Sigma F_y = -W + P(1)\big|_{\text{bottom}} - P(1)\big|_{\text{top}} + F_y = 0$$

where F_y is the additional force required to hold the cube in position.
    Expressing each of the pressures as $P_{\text{atm}} + \rho_w g h$ and W as $\rho_c g V$, we obtain, for our force balance

$$-\rho_c g V + \rho_w g (11\ \text{ft})(1\ \text{ft}^2) - \rho_w g (10\ \text{ft})(1\ \text{ft}^2) + F_y = 0$$

Solving for F_y, we have

$$F_y = -\rho_w g[(11)(1) - 10(1)] + \rho_c g V = -\rho_w g V + \rho_c g V$$

The first term is seen to be a buoyant force, equal to the weight of displaced water.
    Finally, solving for F_y, we obtain

(a) $\rho_c = 10\ \text{lb}_m/\text{ft}^3$

$$F_y = -\frac{(62.4\ \text{lb}_m/\text{ft}^3)(32.2\ \text{ft/s}^2)(1\ \text{ft}^3)}{32.2\ \text{lb}_m\,\text{ft/s}^2\ \text{lb}_f} + \frac{(10\ \text{lb}_m\,\text{ft}^3)(32.2\ \text{ft/s}^2)(1\ \text{ft}^3)}{32.2\ \text{lb}_m\,\text{ft/s}^2\ \text{lb}_f}$$

$$= -52.4\ \text{lb}_f\,(\text{downward})(-233\ \text{N})$$

(b) $\rho_c = 490 \text{ lb}_m/\text{ft}^3$

$$F_y = -\frac{(62.4 \text{ lb}_m/\text{ft}^3)(32.2 \text{ ft/s}^2)(1 \text{ ft}^3)}{32.2 \text{ lb}_m \text{ft/s}^2 \text{ lb}_f} + \frac{(490 \text{ lb}_m \text{ft}^3)(32.2 \text{ ft/s}^2)(1 \text{ ft}^3)}{32.2 \text{ lb}_m \text{ft/s}^2 \text{lb}_f}$$

$$= +427.6 \text{ lb}_f (\text{upward})(1902 \text{ N})$$

In case (a), the buoyant force exceeded the weight of the cube, thus to keep it submerged 10 ft below the surface, a downward force of over 52 lb was required. In the second case, the weight exceeded the buoyant force, and an upward force was required.

▶ **2.5**

CLOSURE

In this chapter, the behavior of static fluids has been examined. The application of Newton's laws of motion led to the description of the point-to-point variation in fluid pressure, from which force relations were developed. Specific applications have been considered, including manometry, forces on plane and curved submerged surfaces, and the buoyancy of floating objects.

The static analyses that have been considered will later be seen as special cases of more general relations governing fluid behavior. Our next task is to examine the behavior of fluids in motion to describe the effect of that motion. Fundamental laws other than Newton's laws of motion will be necessary for this analysis.

CHAPTER 3

Description of a Fluid in Motion

The development of an analytical description of fluid flow is based upon the expression of the physical laws related to fluid flow in a suitable mathematical form. Accordingly, we shall present the pertinent physical laws and discuss the methods used to describe a fluid in motion.

▶ 3.1

FUNDAMENTAL PHYSICAL LAWS

There are three fundamental physical laws that, with the exception of relativistic and nuclear phenomena, apply to each and every flow independently of the nature of the fluid under consideration. These laws are listed below with the designations of their mathematical formulations.

Law	Equation
1. The law of conservation of mass	Continuity equation
2. Newton's second law of motion	Momentum theorem
3. The first law of thermodynamics	Energy equation

The next three chapters will be devoted exclusively to the development of a suitable working form of these laws.[1]

In addition to the above laws, certain auxiliary or subsidiary relations are employed in describing a fluid. These relations depend upon the nature of the fluid under consideration. Unfortunately, most of these auxiliary relations have also been termed "laws." Already in our previous studies, Hooke's law, the perfect gas law, and others have been encountered. However accurate these "laws" may be over a restricted range, their validity is entirely dependent upon the nature of the material under consideration. Thus, while some of the auxiliary relations that will be used will be called laws, the student will be responsible for noting the difference in scope between the fundamental physical laws and the auxiliary relations.

▶ 3.2

FLUID-FLOW FIELDS: LAGRANGIAN AND EULERIAN REPRESENTATIONS

The term *field* refers to a quantity defined as a function of position and time throughout a given region. There are two different forms of representation for the fields in fluid

---

[1] The second law of thermodynamics is also fundamental to fluid-flow analysis. An analytic consideration of the second law is beyond the scope of the present treatment.

mechanics: Lagrange's form and Euler's form. The difference between these approaches lies in the manner in which the position in the field is identified.

In the Lagrangian approach, the physical variables are described for a particular element of fluid as it traverses the flow. This is the familiar approach of particle and rigid-body dynamics. The coordinates (x, y, z) are the coordinates of the element of fluid and, as such, are functions of time. The coordinates (x, y, z) are therefore dependent variables in the Lagrangian form. The fluid element is identified by its position in the field at some arbitrary time, usually $t = 0$. The velocity field in this case is written in functional form as

$$\mathbf{v} = \mathbf{v}(a, b, c, t) \tag{3-1}$$

where the coordinates (a, b, c) refer to the *initial* position of the fluid element. The other fluid-flow variables, being functions of the same coordinates, may be represented in a similar manner. The Lagrangian approach is seldom used in fluid mechanics, as the type of information desired is usually the value of a particular fluid variable at a fixed point in the flow rather than the value of a fluid variable experienced by an element of fluid along its trajectory. For example, the determination of the force on a stationary body in a flow field requires that we know the pressure and shear stress at every point on the body. The Eulerian form provides us with this type of information.

The Eulerian approach gives the value of a fluid variable at a given point and at a given time. In functional form the velocity field is written as

$$\mathbf{v} = \mathbf{v}(x, y, z, t) \tag{3-2}$$

where x, y, z, and t are *all* independent variables. For a particular point (x_1, y_1, z_1) and t_1, equation (3-2) gives the velocity of the fluid at that location at time t_1. In this text the Eulerian approach will be used exclusively.

▶ 3.3

STEADY AND UNSTEADY FLOWS

In adopting the Eulerian approach, we note that the fluid flow will, in general, be a function of the four independent variables (x, y, z, t). If the flow at every point in the fluid is independent of time, the flow is termed *steady*. If the flow at a point varies with time, the flow is termed *unsteady*. It is possible in certain cases to reduce an unsteady flow to a steady flow by changing the frame of reference. Consider an airplane flying through the air at constant speed v_0, as shown in Figure 3.1. When observed from the stationary x, y, z coordinate system, the flow pattern is unsteady. The flow at the point P illustrated, for example, will vary as the vehicle approaches it.

Now consider the same situation when observed from the x', y', z' coordinate system, which is moving at constant velocity v_0 as illustrated in Figure 3.2.

The flow conditions are now independent of time at every point in the flow field, and thus the flow is steady when viewed from the moving coordinate system. Whenever a body moves through a fluid with a constant velocity, the flow field may be transformed from an unsteady flow to a steady flow by selecting a coordinate system that is fixed with respect to the moving body.

In the wind-tunnel testing of models, use is made of this concept. Data obtained for a static model in a moving stream will be the same as the data obtained for a model moving through a static stream. The physical as well as the analytical simplifications afforded by this transformation are considerable. We shall make use of this transformation whenever applicable.

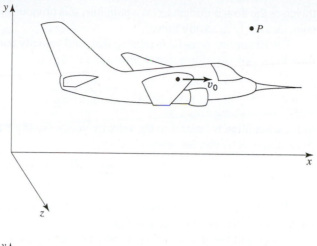

Figure 3.1 Unsteady flow with respect to a fixed coordinate system.

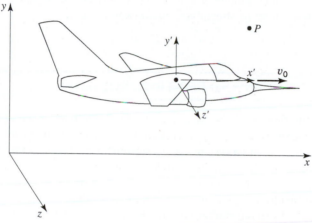

Figure 3.2 Steady flow with respect to a moving coordinate system.

STREAMLINES

A useful concept in the description of a flowing fluid is that of a *streamline*. A streamline is defined as the line drawn tangent to the velocity vector at each point in a flow field. Figure 3.3 shows the streamline pattern for ideal flow past a football-like object. In steady flow, as all velocity vectors are invariant with time, the path of a fluid particle follows a streamline; hence, a streamline is the trajectory of an element of fluid in such a situation. In unsteady flow, streamline patterns change from instant to instant. Thus, the trajectory of a fluid element will be different from a streamline at any particular time. The actual trajectory of a fluid element as it

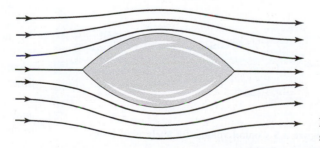

Figure 3.3 Illustration of streamlines.

traverses the flow is designated as a *path line*. It is obvious that path lines and streamlines are coincident only in steady flow.

The streamline is useful in relating that fluid velocity components to the geometry of the flow field. For two-dimensional flow this relation is

$$\frac{v_y}{v_x} = \frac{dy}{dx} \qquad (3\text{-}3)$$

as the streamline is tangent to the velocity vector having x and y components v_x and v_y. In three dimensions this becomes

$$\frac{dx}{v_x} = \frac{dy}{v_y} = \frac{dz}{v_z} \qquad (3\text{-}4)$$

The utility of the above relations is in obtaining an analytical relation between velocity components and the streamline pattern.

Some additional discussion is provided in Chapter 10 regarding the mathematical description of streamlines around solid objects.

▶ 3.5

SYSTEMS AND CONTROL VOLUMES

The three basic physical laws listed in Section 3.1 are all stated in terms of a *system*. A system is defined as a collection of matter of fixed identity. The basic laws give the interaction of a system with its surroundings. The selection of the system for the application of these laws is quite flexible and is, in many cases, a complex problem. Any analysis utilizing a fundamental law must follow the designation of a specific system, and the difficulty of solution varies greatly depending on the choice made.

As an illustration, consider Newton's second law, $\mathbf{F} = m\mathbf{a}$. The terms represented are as follows:

$\mathbf{F}$ = the resultant force exerted by the surroundings on the system.

m = the mass of the system.

$\mathbf{a}$ = the acceleration of the center of mass of the system.

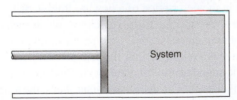

Figure 3.4 An easily identifiable system.

In the piston-and-cylinder arrangement shown in Figure 3.4, a convenient system to analyze, readily identified by virtue of its confinement, is the mass of material enclosed within the cylinder by the piston.

In the case of the nozzle shown in Figure 3.5, the fluid occupying the nozzle changes from instant to instant. Thus, different systems occupy the nozzle at different times.

A more convenient method of analysis of the nozzle would be to consider the region bounded by

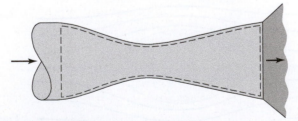

Figure 3.5 Control volume for analysis of flow through a nozzle.

the dotted line. Such a region is a *control volume*. A control volume is a region in space through which fluid flows.[2]

The extreme mobility of a fluid makes the identification of a particular system a tedious task. By developing the fundamental physical laws in a form that applies to a control volume (where the system changes from instant to instant), the analysis of fluid flow is greatly simplified. The control-volume approach circumvents the difficulty in system identification. Succeeding chapters will convert the fundamental physical laws from the system approach to a control-volume approach. The control volume selected may be either finite or infinitesimal.

---

[2] A control volume may be fixed or moving uniformly (inertial), or it may be accelerating (noninertial). Primary consideration here will be given to inertial control volumes.

CHAPTER 4

Conservation of Mass: Control-Volume Approach

The initial application of the fundamental laws of fluid mechanics involves the law of conservation of mass. In this chapter, we shall develop an integral relationship that expresses the law of conservation of mass for a general control volume. The integral relation thus developed will be applied to some often-encountered fluid-flow situations.

▶ 4.1

INTEGRAL RELATION

The law of conservation of mass states that mass may be neither created nor destroyed. With respect to a control volume, the law of conservation of mass may be simply stated as

$$
\left\{\begin{array}{c}\text{rate of mass}\\\text{efflux from}\\\text{control}\\\text{volume}\end{array}\right\} - \left\{\begin{array}{c}\text{rate of mass}\\\text{flow into}\\\text{control}\\\text{volume}\end{array}\right\} + \left\{\begin{array}{c}\text{rate of}\\\text{accumulation}\\\text{of mass within}\\\text{control volume}\end{array}\right\} = 0
$$

Consider now the general control volume located in a fluid-flow field, as shown in Figure 4.1.

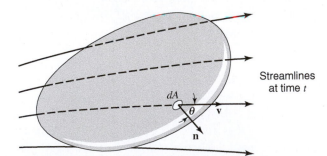

Streamlines at time t

Figure 4.1 Fluid flow through a control volume.

For the small element of area dA on the control surface, the rate of mass efflux $= (\rho v)(dA \cos \theta)$, where $dA \cos \theta$ is the projection of the area dA in a plane normal to the velocity vector, $\mathbf{v}$, and θ is the angle between the velocity vector, $\mathbf{v}$, and the *outward* directed unit normal vector, $\mathbf{n}$, to dA. From vector algebra, we recognize the product

$$\rho v \, dA \, \cos \theta = \rho \, dA |\mathbf{v}| \, |\mathbf{n}| \cos \theta$$

as the "scalar" or "dot" product

$$\rho(\mathbf{v} \cdot \mathbf{n}) \, dA$$

which is the form we shall use to designate the rate of mass efflux through dA. The product ρv is the mass flux, often called the mass velocity, G. Physically, this product represents the amount of mass flowing through a unit cross-sectional area per unit time.

If we now integrate this quantity over the entire control surface, we have

$$\iint_{\text{c.s.}} \rho(\mathbf{v} \cdot \mathbf{n}) \, dA$$

which is the net outward flow of mass across the control surface, or the *net mass efflux* from the control volume.

Note that if mass is entering the control volume—that is, flowing inward across the control surface—the product $\mathbf{v} \cdot \mathbf{n} = |\mathbf{v}| |\mathbf{n}| \cos \theta$ is negative, since $\theta > 90°$, and $\cos \theta$ is therefore negative. Thus, if the integral is

positive, there is a net efflux of mass

negative, there is a net influx of mass

zero, the mass within the control volume is constant

The rate of accumulation of mass within the control volume may be expressed as

$$\frac{\partial}{\partial t} \iiint_{\text{c.v.}} \rho \, dV$$

and the integral expression for the mass balance over a general control volume becomes

$$\iint_{\text{c.s.}} \rho(\mathbf{v} \cdot \mathbf{n}) \, dA + \frac{\partial}{\partial t} \iiint_{\text{c.v.}} \rho \, dV = 0 \qquad (4\text{-}1)$$

► 4.2

SPECIFIC FORMS OF THE INTEGRAL EXPRESSION

Equation (4-1) gives the mass balance in its most general form. We now consider some frequently encountered situations where equation (4-1) may be applied.

If flow is steady relative to coordinates fixed to the control volume, the accumulation term, $\partial/\partial t \iiint_{\text{c.v.}} \rho \, dV$, will be zero. This is readily seen when one recalls that, by the definition of steady flow, the properties of a flow field are invariant with time; hence, the partial derivative with respect to time is zero. Thus, for this situation the applicable form of the continuity expression is

$$\iint_{\text{c.s.}} \rho(\mathbf{v} \cdot \mathbf{n}) \, dA = 0 \qquad (4\text{-}2)$$

Another important case is that of an incompressible flow with fluid filling the control volume. For incompressible flow the density, ρ, is constant, and so the accumulation term involving the partial derivative with respect to time is again zero. Additionally, the density term in the surface integral may be canceled. The conservation-of-mass expression for incompressible flow of this nature thus becomes

$$\iint_{\text{c.s.}} (\mathbf{v} \cdot \mathbf{n}) \, dA = 0 \qquad (4\text{-}3)$$

The following examples illustrate the application of equation (4-1) to some cases that recur frequently in momentum transfer.

Example 1

As our first example, let us consider the common situation of a control volume for which mass efflux and influx are steady and one-dimensional. Specifically, consider the control volume indicated by dashed lines in Figure 4.2.

Figure 4.2 Steady one-dimensional flow into and out of a control volume.

Equation (4-2) applies. As mass crosses the control surface at positions (1) and (2) only, our expression is

$$\iint_{c.s.} \rho(\mathbf{v} \cdot \mathbf{n})\, dA = \iint_{A_1} \rho(\mathbf{v} \cdot \mathbf{n})\, dA + \iint_{A_2} \rho(\mathbf{v} \cdot \mathbf{n})\, dA = 0$$

The absolute value of the scalar product $(\mathbf{v} \cdot \mathbf{n})$ is equal to the magnitude of the velocity in each integral, as the velocity vectors and outwardly directed normal vectors are collinear both at (1) and (2). At (2), these vectors have the same sense, so this product is positive, as it should be for an efflux of mass. At (1), where mass flows into the control volume, the two vectors are opposite in sense—hence, the sign is negative. We may now express the continuity equation in scalar form

$$\iint_{c.s.} \rho(\mathbf{v} \cdot \mathbf{n})\, dA = -\iint_{A_1} \rho v\, dA + \iint_{A_2} \rho v\, dA = 0$$

Integration gives the familiar result

$$\rho_1 v_1 A_1 = \rho_2 v_2 A_2 \tag{4-4}$$

In obtaining equation (4-4), it is noted that the flow situation inside the control volume was unspecified. In fact, this is the beauty of the control-volume approach; the flow inside the control volume can be analyzed from information (measurements) obtained on the surface of the control volume. The box-shaped control volume illustrated in Figure 4.2 is defined for analytical purposes; the actual system contained in this box could be as simple as a pipe or as complex as a propulsion system or a distillation tower.

In solving Example 1, we assumed a constant velocity at Sections (1) and (2). This situation may be approached physically, but a more general case is one in which the velocity varies over the cross-sectional area.

Example 2

Let us consider the case of an incompressible flow, for which the flow area is circular and the velocity profile is parabolic (see Figure 4.3), varying according to the expression

$$v = v_{max}\left[1 - \left(\frac{r}{R}\right)^2\right]$$

where v_{max} is the maximum velocity, which exists at the center of the circular passage (i.e., at $r = 0$), and R is the radial distance to the inside surface of the circular area considered.

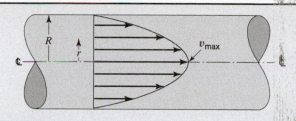

Figure 4.3 A parabolic velocity profile in a circular flow passage.

The above velocity-profile expression may be obtained experimentally. It will also be derived theoretically in Chapter 8 for the case of laminar flow in a circular conduit. This expression represents the velocity at a radial distance, r, from the center of the flow section. As the average velocity is of particular interest in engineering problems, we will now consider the means of obtaining the average velocity from this expression.

At the station where this velocity profile exists, the mass rate of flow is

$$(\rho v)_{avg}A = \iint_A \rho v \, dA$$

For the present case of incompressible flow, the density is constant. Solving for the average velocity, we have

$$v_{avg} = \frac{1}{A} \iint_A v \, dA$$

$$= \frac{1}{\pi R^2} \int_0^{2\pi} \int_0^R v_{max} \left[1 - \left(\frac{r}{R}\right)^2\right] r \, dr \, d\theta$$

$$= \frac{v_{max}}{2}$$

In the previous examples, we were not concerned with the composition of the fluid streams. Equation (4-1) applies to fluid streams containing more than one constituent as well as to the individual constituents alone. This type of application is common to chemical processes in particular. Our final example will use the law of conservation of mass for both the total mass and for a particular species, in this case, salt.

Example 3

Let us now examine the situation illustrated in Figure 4.4. A tank initially contains 1000 kg of brine containing 10% salt by mass. An inlet stream of brine containing 20% salt by mass flows into the tank at a rate of 20 kg/min. The mixture in the tank is kept uniform by stirring. Brine is removed from the tank via an outlet pipe at a rate of 10 kg/min. Find the amount of salt in the tank at any time t, and the elapsed time when the amount of salt in the tank is 200 kg.

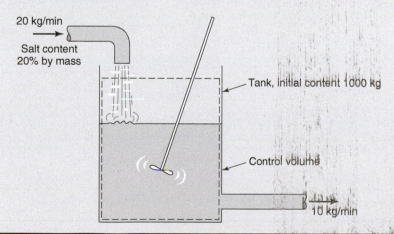

Figure 4.4 A mixing process.

We first apply equation (4-1) to express the total amount of brine in the tank as a function of time. For the control volume shown

$$\iint_{c.s.} \rho(\mathbf{v} \cdot \mathbf{n}) \, dA = 10 - 20 = -10 \, \text{kg/min}$$

$$\frac{\partial}{\partial t} \iiint_{c.v.} \rho \, dV = \frac{d}{dt} \int_{1000}^{M} dM = \frac{d}{dt}(M - 1000)$$

where M is the total mass of brine in the tank at any time. Writing the complete expression, we have

$$\iint_{c.s.} \rho(\mathbf{v} \cdot \mathbf{n}) \, dA + \frac{\partial}{\partial t} \iiint_{c.v.} \rho \, dV = -10 + \frac{d}{dt}(M - 1000) = 0$$

Separating variables and solving for M gives

$$M = 1000 + 10t \quad (\text{kg})$$

We now let S be the amount of salt in the tank at any time. The concentration by weight of salt may be expressed as

$$\frac{S}{M} = \frac{S}{1000 + 10t} \quad \frac{\text{kg salt}}{\text{kg brine}}$$

Using this definition, we may now apply equation (4-1) to the salt, obtaining

$$\iint_{c.s.} \rho(\mathbf{v} \cdot \mathbf{n}) \, dA = \frac{10S}{1000 + 10t} - (0.2)(20) \quad \frac{\text{kg salt}}{\text{min}}$$

and

$$\frac{\partial}{\partial t} \iiint_{c.v.} \rho \, dV = \frac{d}{dt} \int_{S_0}^{S} dS = \frac{dS}{dt} \quad \frac{\text{kg salt}}{\text{min}}$$

The complete expression is now

$$\iint_{c.s.} \rho(\mathbf{v} \cdot \mathbf{n}) \, dA + \frac{\partial}{\partial t} \iiint_{c.v.} \rho \, dV = \frac{S}{100 + t} - 4 + \frac{dS}{dt} = 0$$

This equation may be written in the form

$$\frac{dS}{dt} + \frac{S}{100 + t} = 4$$

which we observe to be a first-order linear differential equation. The general solution is

$$S = \frac{2t(200 + t)}{100 + t} + \frac{C}{100 + t}$$

The constant of integration may be evaluated, using the initial condition that $S = 100$ at $t = 0$ to give $C = 10,000$. Thus, the first part of the answer, expressing the amount of salt present as a function of time, is

$$S = \frac{10,000 + 400t + 2t^2}{100 + t}$$

The elapsed time necessary for S to equal 200 kg may be evaluated to give $t = 36.6 \, \text{min}$.

Example 4

A large well-mixed tank of unknown volume, open to the atmosphere initially, contains pure water. The initial height of the solution in the tank is unknown. At the start of the experiment a potassium chloride solution in water starts flowing into the tank from two separate inlets. The first inlet has a diameter of 1 cm and delivers a solution with a specific gravity of 1.07 and a velocity of 0.2 m/s. The second inlet with a diameter of 2 cm delivers a solution with a velocity of 0.01 m/s and a density of 1053 kg/m^3. The single outflow from this tank has a diameter of 3 cm. A colleague helps you by taking samples of the tank and outflow in your absence. He samples the tank and determines that the tank contains 19.7 kg of potassium chloride. At the same moment he measures the flow rate at the outlet to be 0.5 L/s, and the concentration of potassium chloride to be 13 g/L. Please calculate the exact time when your colleague took the samples. The tank and all inlet solutions are maintained at a constant temperature of 80°C.

First, let's make a rough figure with all the information we are given so we can visualize the problem.

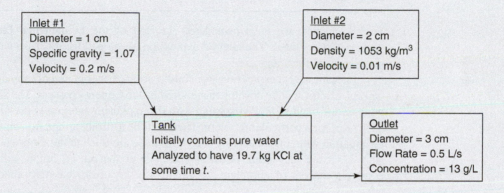

Inlet #1
Diameter = 1 cm
Specific gravity = 1.07
Velocity = 0.2 m/s

Inlet #2
Diameter = 2 cm
Density = 1053 kg/m^3
Velocity = 0.01 m/s

Tank
Initially contains pure water
Analyzed to have 19.7 kg KCl at some time t.

Outlet
Diameter = 3 cm
Flow Rate = 0.5 L/s
Concentration = 13 g/L

Beginning with equation (4.1),

$$\iint \rho(\mathbf{v}\cdot\mathbf{n})dA + \frac{\partial}{\partial t}\iiint \rho\, dV = 0$$

$$\dot{m}_{out} - \dot{m}_{in} + \frac{dM}{dt} = 0$$

Since we have two inputs to the tank,

$$\dot{m}_{out} - [\dot{m}_{in1} + \dot{m}_{in2}] + \frac{dM}{dt} = 0$$

$$\dot{m}_{out} = (0.5\,\text{L/s})(13\,\text{g/L}) = 6.5\,\text{g/s}$$

$$\dot{m}_{in} = \dot{m}_{in1} + \dot{m}_{in2} = \begin{array}{l}(1.07)\left(971.8\,\dfrac{\text{kg}}{\text{m}^3}\right)(0.2\,\text{m/s})\left(\dfrac{\pi(0.01\,\text{m})^2}{4}\right)\left(1000\,\dfrac{\text{g}}{\text{kg}}\right)+\\[2ex](1053\,\text{kg/m}^3)(0.01\,\text{m/s})\left(\dfrac{\pi(0.02\,\text{m})^2}{4}\right)\left(1000\,\dfrac{\text{g}}{\text{kg}}\right)=19.64\,\text{g/s}\end{array}$$

$$6.5\,\text{g/s} - 19.64\,\text{g/s} + \frac{dS}{dt} = 0$$

$$\frac{dS}{dt} = 13.14 \text{ g/s}$$

$$\int_0^{19,700} dS = 13.14 \int_0^t dt$$

$$t = 1500 \text{ s} = 25 \text{ min}$$

So your colleague took the sample 25 minutes after the start of the experiment.

► 4.3

CLOSURE

In this chapter, we have considered the first of the fundamental laws of fluid flow: conservation of mass. The integral expression developed for this case was seen to be quite general in its form and use.

Similar integral expressions for conservation of energy and of momentum for a general control volume will be developed and used in subsequent chapters. The student should now develop the habit of *always* starting with the applicable integral expression and evaluating each term for a particular problem. There will be a strong temptation simply to write down an equation without considering each term in detail. Such temptations should be overcome. This approach may seem needlessly tedious at the outset, but it will always ensure a complete analysis of a problem and circumvent any errors that may otherwise result from a too-hasty consideration.

CHAPTER 5

Newton's Second Law of Motion: Control-Volume Approach

The second of the fundamental physical laws upon which fluid-flow analyses are based is Newton's second law of motion. Starting with Newton's second law, we shall develop integral relations for linear and angular momentum. Applications of these expressions to physical situations will be considered.

▶ **5.1**

INTEGRAL RELATION FOR LINEAR MOMENTUM

Newton's second law of motion may be stated as follows:

> *The time rate of change of momentum of a system is equal to the net force acting on the system and takes place in the direction of the net force.*

We note at the outset two very important parts of this statement: first, this law refers to a specific system, and second, it includes direction as well as magnitude and is therefore a vector expression. In order to use this law, it will be necessary to recast its statement into a form applicable to control volume that contains different fluid particles (i.e., a different system) when examined at different times.

In Figure 5.1, observe the control volume located in a fluid-flow field. The system considered is the material occupying the control volume at time t, and its position is shown both at time t and at time $t + \Delta t$.

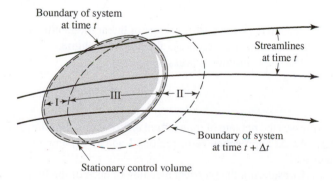

Figure 5.1 Relation between a system and a control volume in a fluid-flow field.

Referring to the figure, we see that

Region I is occupied by the system only at time t.

Region II is occupied by the system at $t + \Delta t$.

Region III is common to the system both at t and at $t + \Delta t$.

Writing Newton's second law for such a situation, we have

$$\Sigma \mathbf{F} = \frac{d}{dt}(m\mathbf{v}) = \frac{d}{dt}\mathbf{P} \tag{5-1}$$

where the symbols $\mathbf{F}$, m, and $\mathbf{v}$ have their usual meanings and $\mathbf{P}$ represents the total linear momentum of the system.

At time $t + \Delta t$, the linear momentum of the system that now occupies regions II and III may be expressed as

$$\mathbf{P}|_{t+\Delta t} = \mathbf{P}_{\text{II}}|_{t+\Delta t} + \mathbf{P}_{\text{III}}|_{t+\Delta t}$$

and at time t we have

$$\mathbf{P}|_t = \mathbf{P}_{\text{I}}|_t + \mathbf{P}_{\text{III}}|_t$$

Subtracting the second of these expressions from the first and dividing by the time interval Δt gives

$$\frac{\mathbf{P}|_{t+\Delta t} - \mathbf{P}|_t}{\Delta t} = \frac{\mathbf{P}_{\text{II}}|_{t+\Delta t} + \mathbf{P}_{\text{III}}|_{t+\Delta t} - \mathbf{P}_{\text{I}}|_t - \mathbf{P}_{\text{III}}|_t}{\Delta t}$$

We may rearrange the right-hand side of this expression and take the limit of the resulting equation to get

$$\lim_{\Delta t \to 0} \frac{\mathbf{P}|_{t+\Delta t} - \mathbf{P}|_t}{\Delta t} = \lim_{\Delta t \to 0} \frac{\mathbf{P}_{\text{III}}|_{t+\Delta t} - \mathbf{P}_{\text{III}}|_t}{\Delta t} + \lim_{\Delta t \to 0} \frac{\mathbf{P}_{\text{II}}|_{t+\Delta t} - \mathbf{P}_{\text{I}}|_t}{\Delta t} \tag{5-2}$$

Considering each of the limiting processes separately, we have, for the left-hand side,

$$\lim_{\Delta t \to 0} \frac{\mathbf{P}|_{t+\Delta t} - \mathbf{P}|_t}{\Delta t} = \frac{d}{dt}\mathbf{P}$$

which is the form specified in the statement of Newton's second law, equation (5-1).

The first limit on the right-hand side of equation (5-2) may be evaluated as

$$\lim_{\Delta t \to 0} \frac{\mathbf{P}_{\text{III}}|_{t+\Delta t} - \mathbf{P}_{\text{III}}|_t}{\Delta t} = \frac{d}{dt}\mathbf{P}_{\text{III}}$$

This we see to be the rate of change of linear momentum of the control volume itself, since, as $\Delta t \to 0$, region III becomes the control volume.

The next limiting process

$$\lim_{\Delta t \to 0} \frac{\mathbf{P}_{\text{II}}|_{t+\Delta t} - \mathbf{P}_{\text{I}}|_t}{\Delta t}$$

expresses the net rate of momentum efflux across the control surface during the time interval Δt. As Δt approaches zero, regions I and II become coincident with the control-volume surface.

Considering the physical meaning of each of the limits in equation (5-2) and Newton's second law, equation (5-1), we may write the following word equation for the conservation

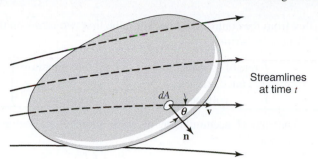

Figure 5.2 Fluid flow through a control volume.

of linear momentum with respect to a control volume:

$$
\left\{\begin{array}{c} \text{sum of} \\ \text{forces acting} \\ \text{on control} \\ \text{volume} \end{array}\right\} = \underbrace{\left\{\begin{array}{c} \text{rate of} \\ \text{momentum} \\ \text{out of control} \\ \text{volume} \end{array}\right\} - \left\{\begin{array}{c} \text{rate of} \\ \text{momentum} \\ \text{into control} \\ \text{volume} \end{array}\right\}}_{\substack{\text{net rate of momentum efflux from} \\ \text{control volume}}} + \left\{\begin{array}{c} \text{rate of} \\ \text{accumulation} \\ \text{of momentum} \\ \text{within control} \\ \text{volume} \end{array}\right\} \quad (5\text{-}3)
$$

We shall now apply equation (5-3) to a general control volume located in a fluid-flow field as shown in Figure 5.2 and evaluate the various terms.

The total force acting on the control volume consists both of surface forces due to interactions between the control-volume fluid, and its surroundings through direct contact, and of body forces resulting from the location of the control volume in a force field. The gravitational field and its resultant force are the most common examples of this latter type. We will designate the total force acting on the control volume as $\Sigma \mathbf{F}$.

If the small area dA on the control surface is considered, we may write

$$\text{rate of momentum efflux} = \mathbf{v}(\rho v)(dA \cos \theta)$$

Observe that the product $(\rho v)(dA \cos \theta)$ is the rate of mass efflux from the control volume through dA, as discussed in Chapter 4. Recall further that $dA \cos \theta$ is the area, dA projected in a direction normal to the velocity vector, $\mathbf{v}$, where θ is the angle between $\mathbf{v}$ and the outwardly directed normal vector, $\mathbf{n}$. We may then multiply the rate of mass efflux by $\mathbf{v}$ to give the rate of momentum efflux through dA. From vector algebra this product may be written as

$$\mathbf{v}(\rho v)(dA \cos \theta) = \mathbf{v}(\rho dA)[|\mathbf{v}|\,|\mathbf{n}| \cos \theta]$$

The term in square brackets is the scalar or dot product, $\mathbf{v} \cdot \mathbf{n}$, and the momentum efflux term becomes

$$\rho \mathbf{v}(\mathbf{v} \cdot \mathbf{n})\, dA$$

Integrating this quantity over the entire control surface, we have

$$\iint_{c.s.} \mathbf{v}\rho(\mathbf{v} \cdot \mathbf{n})\, dA$$

which is the *net momentum efflux* from the control volume.

In its integral form the momentum flux term stated above includes the rate of momentum entering the control volume as well as that leaving. If mass is entering the control volume, the sign of the product $\mathbf{v} \cdot \mathbf{n}$ is negative, and the associated momentum flux is an input. Conversely, a positive sign of the product $\mathbf{v} \cdot \mathbf{n}$ is associated with a momentum efflux from the control volume. Thus, the first two terms on the right-hand side of equation (5-3) may be written

$$\left\{ \begin{array}{c} \text{rate of momentum} \\ \text{out of control} \\ \text{volume} \end{array} \right\} - \left\{ \begin{array}{c} \text{rate of momentum} \\ \text{into control} \\ \text{volume} \end{array} \right\} = \iint_{\text{c.s.}} \mathbf{v}\rho(\mathbf{v}\cdot\mathbf{n}) \, dA$$

The rate of accumulation of linear momentum within the control volume may be expressed as

$$\frac{\partial}{\partial t} \iiint_{\text{c.v.}} \mathbf{v}\rho \, dV$$

and the overall linear-momentum balance for a control volume becomes

$$\Sigma \mathbf{F} = \iint_{\text{c.s.}} \mathbf{v}\rho(\mathbf{v}\cdot\mathbf{n}) \, dA + \frac{\partial}{\partial t} \iiint_{\text{c.v.}} \rho\mathbf{v} \, dV \tag{5-4}$$

This extremely important relation is often referred to in fluid mechanics as the *momentum theorem*. Note the great similarity between equations (5-4) and (4-1) in the form of the integral terms; observe, however, that equation (5-4) is a vector expression opposed to the scalar form of the overall mass balance considered in Chapter 4. In rectangular coordinates the single-vector equation, (5-4), may be written as three scalar equations:

$$\Sigma F_x = \iint_{\text{c.s.}} v_x\rho(\mathbf{v}\cdot\mathbf{n}) \, dA + \frac{\partial}{\partial t} \iiint_{\text{c.v.}} \rho v_x \, dV \tag{5-5a}$$

$$\Sigma F_y = \iint_{\text{c.s.}} v_y\rho(\mathbf{v}\cdot\mathbf{n}) \, dA + \frac{\partial}{\partial t} \iiint_{\text{c.v.}} \rho v_y \, dV \tag{5-5b}$$

$$\Sigma F_z = \iint_{\text{c.s.}} v_z\rho(\mathbf{v}\cdot\mathbf{n}) \, dA + \frac{\partial}{\partial t} \iiint_{\text{c.v.}} \rho v_z \, dV \tag{5-5c}$$

When applying any or all of the above equations, it must be remembered that each term has a sign with respect to the positively defined x, y, and z directions. The determination of the sign of the surface integral should be considered with special care, as both the velocity component (v_x) and the scalar product $(\mathbf{v}\cdot\mathbf{n})$ have signs. The combination of the proper sign associated with each of these terms will give the correct sense to the integral. It should also be remembered that as equations (5-5a–c) are written for the fluid in the control volume, *the forces to be employed in these equations are those acting on the fluid.*

A detailed study of the example problems to follow should aid in the understanding of, and afford facility in using, the overall momentum balance.

▶ **5.2**

APPLICATIONS OF THE INTEGRAL EXPRESSION
FOR LINEAR MOMENTUM

In applying equation (5-4), it is first necessary to define the control volume that will make possible the simplest and most direct solution to the problem at hand. There are no general rules to aid in this definition, but experience in handling problems of this type will enable such a choice to be made readily.

Example 1

Consider first the problem of finding the force exerted on a reducing pipe bend resulting from a steady flow of fluid in it. A diagram of the pipe bend and the quantities significant to its analysis are shown in Figure 5.3.

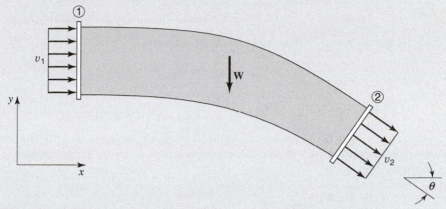

Figure 5.3 Flow in a reducing pipe bend.

The first step is the definition of the control volume. One choice for the control volume, of the several available, is all fluid in the pipe at a given time. The control volume chosen in this manner is designated in Figure 5.4, showing the external forces imposed upon it. The external forces imposed on the fluid include the pressure forces at sections (1) and (2), the body force due to the weight of fluid in the control volume, and the forces due to pressure and shear stress, P_w and τ_w, exerted on the fluid by the pipe wall. The resultant force on the fluid (due to P_w and τ_w) by the pipe is symbolized as **B**, and its x and y components as B_x and B_y, respectively.

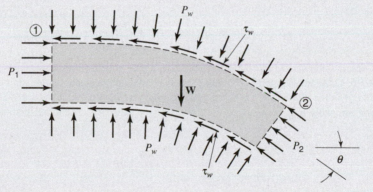

Figure 5.4 Control volume defined by pipe surface.

Considering the x- and y-directional component equations, (5-5a) and (5-5b), of the overall momentum balance, the external forces acting on the fluid in the control volume are

$$\Sigma F_x = P_1 A_1 - P_2 A_2 \cos\theta + B_x$$

and

$$\Sigma F_y = P_2 A_2 \sin\theta - W + B_y$$

Each component of the unknown force **B** is assumed to have a positive sense. The actual signs for these components, when a solution is obtained, will indicate whether or not this assumption is correct.

Evaluating the surface integral in both the x and y directions, we have

$$\iint_{\text{c.s.}} v_x \rho(\mathbf{v} \cdot \mathbf{n})\, dA = (v_2 \cos\theta)(\rho_2 v_2 A_2) + (v_1)(-\rho_1 v_1 A_1)$$

$$\iint_{\text{c.s.}} v_y \rho(\mathbf{v} \cdot \mathbf{n})\, dA = (-v_2 \sin\theta)(\rho_2 v_2 A_2)$$

The complete momentum expressions in the x and y directions are

$$B_x + P_1A_1 - P_2A_2 \cos\theta = (v_2 \cos\theta)(\rho_2 v_2 A_2) + v_1(-\rho_1 v_1 A_1)$$

and

$$B_y + P_2A_2 \sin\theta - W = (-v_2 \sin\theta)(\rho_2 v_2 A_2)$$

Solving for the unknown force components B_x and B_y, we have

$$B_x = v_2^2 \rho_2 A_2 \cos\theta - v_1^2 \rho_1 A_1 - P_1A_1 + P_2A_2 \cos\theta$$

and

$$B_y = -v_2^2 \rho_2 A_2 \sin\theta - P_2A_2 \sin\theta + W$$

Recall that we were to evaluate the force exerted on the pipe rather than that on the fluid. The force sought is the reaction to **B** and has components equal in magnitude and opposite in sense to B_x and B_y. The components of the reaction force, **R**, exerted on the pipe are

$$R_x = -v_2^2 \rho_2 A_2 \cos\theta + v_1^2 \rho_1 A_1 + P_1A_1 - P_2A_2 \cos\theta$$

and

$$R_y = v_2^2 \rho_2 A_2 \sin\theta + P_2A_2 \sin\theta - W$$

Some simplification in form may be achieved if the flow is steady. Applying equation (4-2), we have

$$\rho_1 v_1 A_1 = \rho_2 v_2 A_2 = \dot{m}$$

where $\dot{m}$ is the mass flow rate.

The final solution for the components of **R** may now be written as

$$R_x = \dot{m}(v_1 - v_2 \cos\theta) + P_1A_1 - P_2A_2 \cos\theta$$
$$R_y = \dot{m}v_2 \sin\theta + P_2A_2 \sin\theta - W$$

The control volume shown in Figure 5.4 for which the above solution was obtained is one possible choice. Another is depicted in Figure 5.5. This control volume is bounded simply by the straight planes cutting through the pipe at sections (1) and (2). The fact that a control volume such as this can be used indicates the versatility of this approach, that is, that the results of complicated processes occurring internally may be analyzed quite simply by considering only those quantities of transfer across the control surface.

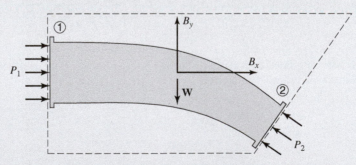

Figure 5.5 Control volume including fluid and pipe.

For this control volume, the x- and y-directional momentum equations are

$$B_x + P_1A_1 - P_2A_2 \cos\theta = (v_2 \cos\theta)(\rho_2 v_2 A_2) + v_1(-v_1 \rho_1 A_1)$$

and

$$B_y + P_2A_2 \sin\theta - W = (-v_2 \sin\theta)(\rho_2 v_2 A_2)$$

where the force having components B_x and B_y is that exerted on the control volume by the section of pipe cut through at sections (1) and (2). The pressures at (1) and (2) in the above equations are gage pressures, as the atmospheric pressures acting on all surfaces cancel.

Note that the resulting equations for this control volume are identical to those obtained for the one defined previously. Thus, a correct solution may be obtained for each of several chosen control volumes so long as they are analyzed carefully and completely.

Example 2

As our second example of the application of the control-volume expression for linear momentum (the momentum theorem), consider the steam locomotive tender schematically illustrated in Figure 5.6, which obtains water from a trough by means of a scoop. The force on the train due to the water is to be obtained.

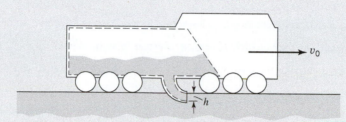

Figure 5.6 Schematic of locomotive tender scooping water from a trough.

The logical choice for a control volume in this case is the water-tank/scoop combination. Our control-volume boundary will be selected as the *interior* of the tank and scoop. As the train is moving with a uniform velocity, there are two possible choices of coordinate systems. We may select a coordinate system either fixed in space or moving[1] with the velocity of the train, v_0. Let us first analyze the system by using a moving coordinate system.

The moving control volume is shown in Figure 5.7 with the xy coordinate system moving at velocity v_0. All velocities are determined with respect to the x and y axes.

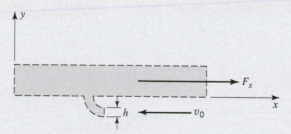

Figure 5.7 Moving coordinate system and control volume.

The applicable expression is equation (5-5a)

$$\Sigma F_x = \iint_{\text{c.s.}} v_x \rho(\mathbf{v} \cdot \mathbf{n})\, dA + \frac{\partial}{\partial t} \iiint_{\text{c.v.}} v_x \rho\, dV$$

In Figure 5.7, ΣF_x is represented as F_x and is shown in the positive sense. As the forces due to pressure and shear are to be neglected, F_x is the total force exerted on the fluid by the train and scoop. The momentum flux term is

$$\iint_{\text{c.s.}} v_x \rho(\mathbf{v} \cdot \mathbf{n})\, dA = \rho(-v_0)(-1)(v_0)(h) \quad \text{(per unit length)}$$

---

[1] Recall that a uniformly translating coordinate system is an inertial coordinate system, so Newton's second law and the momentum theorem may be employed directly.

and the rate of change of momentum within the control volume is zero, as the fluid in the control volume has zero velocity in the x direction.

Thus,

$$F_x = \rho v_0^2 h$$

This is the force exerted by the train on the fluid. The force exerted by the fluid on the train is the opposite of this, or $-\rho v_0^2 h$.

Now let us consider the same problem with a stationary coordinate system (see Figure 5.8). Employing once again the control-volume relation for linear momentum

$$\sum F_x = \iint_{c.s.} v_x \rho(\mathbf{v} \cdot \mathbf{n})\, dA + \frac{\partial}{\partial t} \iiint_{c.v.} v_x \rho\, dV$$

we obtain

$$F_x = 0 + \frac{\partial}{\partial t} \iiint_{c.v.} v_x \rho\, dV$$

where the momentum flux is zero, as the entering fluid has zero velocity. There is, of course, no fluid leaving the control volume. The terms $\frac{\partial}{\partial t} \iiint_{c.v.} v_x \rho\, dV$, as the velocity, $v_x = v_0 = $ constant, may be written as $v_0 \frac{\partial}{\partial t} \iiint_{c.v.} \rho\, dV$ or $v_0(\partial m/\partial t)$, where $\dot{m}$ is the mass of fluid entering the control volume at the rate $\partial m/\partial t = \rho v_0 h$ so that $F_x = \rho v_0^2 h$ as before.

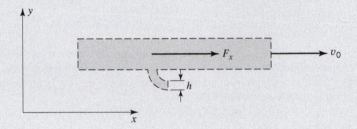

Figure 5.8 Stationary coordinate system and moving control volume.

The student should note that in the case of a stationary coordinate system and a moving control volume, care must be exercised in the interpretation of the momentum flux

$$\iint_{c.s.} \mathbf{v}\rho(\mathbf{v} \cdot \mathbf{n})\, dA$$

Regrouping the terms, we obtain

$$\iint_{c.s.} \mathbf{v}\rho(\mathbf{v} \cdot \mathbf{n})\, dA \equiv \iint_{c.s.} \mathbf{v}\, d\dot{m}$$

Thus, it is obvious that while $\mathbf{v}$ is the velocity relative to fixed coordinates, $\mathbf{v} \cdot \mathbf{n}$ is the velocity relative to the control-volume boundary.

Example 3

A jet of fluid exits a nozzle and strikes a vertical plane surface as shown in Figure 5.9.

    (a) Determine the force required to hold the plate stationary if the jet is composed of

        i. water
        ii. air

(b) Determine the magnitude of the restraining force for a water jet when the plate is moving to the right with a uniform velocity of 4 m/s.

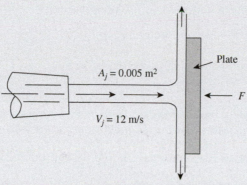

$A_j = 0.005 \text{ m}^2$

$V_j = 12$ m/s

Plate

F

Figure 5.9 A fluid jet striking a vertical plate.

The control volume to be used in this analysis is shown Figure 5.10.

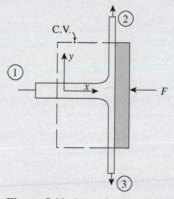

C.V.

F

Figure 5.10 Control volume for Example 3.

The coordinates are fixed with the control volume, which, for parts (a) and (b) of this example, is stationary. Writing the x-directional scalar form of the momentum theorem, we have

$$\Sigma F_x = \iint_{\text{c.s.}} v_x \rho (\mathbf{v} \cdot \mathbf{n})\, dA + \frac{\partial}{\partial t} \iiint_{\text{c.v.}} v_x \rho\, dV$$

Evaluation of each term in this expression yields

$$\Sigma F_x = -F$$

$$\iint_{\text{c.s.}} v_x \rho (\mathbf{v} \cdot \mathbf{n})\, dA = v_j \rho (-v_j A_j)$$

$$\frac{\partial}{\partial t} \iiint_{\text{c.v.}} v_x \rho\, dV = 0$$

and the governing equation is

$$F = \rho A_j v_j^2$$

We may now introduce the appropriate numerical values and solve for F. For case (a),

i. $\rho_w = 1000 \text{ kg/m}^3$

$$F = (1000 \text{ kg/m}^3)(0.005 \text{ m}^2)(12 \text{ m/s})^2$$
$$= 720 \text{ N}$$

ii. $\rho_w = 1.206 \text{ kg/m}^3$

$$F = (1.206 \text{ kg/m}^3)(0.005 \text{ m}^2)(12 \text{ m/s})^2$$
$$= 0.868 \text{ N}$$

For case (b), the same control volume will be used. In this case, however, the control volume and *the coordinate system* are moving to the right at a velocity of 4 m/s. From the perspective of an observer moving with the control volume, the velocity of the incoming water jet is $(v_j - v_0) = 8$ m/s.

The x-directional component form of the momemtum theorem will yield the expression

$$F = \rho A_j (v_j - v_0)^2$$

Substitution of appropriate numerical values yields

$$F = (1000 \text{ kg/m}^3)(0.005 \text{ m}^2)(12 - 4 \text{ m/s})^2$$
$$= 320 \text{ N}$$

▶ **5.3**

INTEGRAL RELATION FOR MOMENT OF MOMENTUM

The integral relation for the moment of momentum of a control volume is an extension of the considerations just made for linear momentum.

Starting with equation (5-1), which is a mathematical expression of Newton's second law of motion applied to a system of particles (Figure 5.11),

$$\sum \mathbf{F} = \frac{d}{dt}(m\mathbf{v}) = \frac{d}{dt}\mathbf{P} \tag{5-1}$$

we take the vector or "cross" product of a position vector, $\mathbf{r}$, with each term and get

$$\mathbf{r} \times \sum \mathbf{F} = \mathbf{r} \times \frac{d}{dt}(m\mathbf{v}) = \mathbf{r} \times \frac{d}{dt}\mathbf{P} \tag{5-6}$$

The quantity on the left-hand side of equation (5-6), $\mathbf{r} \times \sum \mathbf{F}$, is the resultant moment, $\sum \mathbf{M}$, about the origin as shown in Figure 5.11, due to all forces applied to the system. Clearly,

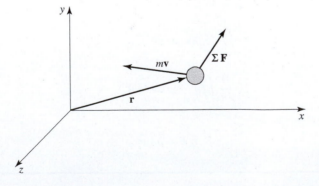

Figure 5.11 A system and its displacement vector $\mathbf{r}$.

we may write

$$\mathbf{r} \times \Sigma\mathbf{F} = \Sigma\mathbf{r} \times \mathbf{F} = \Sigma\mathbf{M}$$

where $\Sigma\mathbf{M}$ is, again, the total moment about the origin of all forces acting on the system.

The right-hand side of equation (5-6) is the moment of the time rate of change of linear momentum. This we can write as

$$\mathbf{r} \times \frac{d}{dt}m\mathbf{v} = \frac{d}{dt}(\mathbf{r} \times m\mathbf{v}) = \frac{d}{dt}(\mathbf{r} \times \mathbf{P}) = \frac{d}{dt}\mathbf{H}$$

Thus, this term is also the time rate of change of the moment of momentum of the system. We shall use the symbol $\mathbf{H}$ to designate moment of momentum. The complete expression is now

$$\Sigma\mathbf{M} = \frac{d}{dt}\mathbf{H} \tag{5-7}$$

As with its analogous expression for linear momentum, equation (5-1), equation (5-7) applies to a specific system. By the same limit process as that used for linear momentum, we may recast this expression into a form applicable to a control volume and achieve a word equation

$$\left\{\begin{array}{c} \text{sum of} \\ \text{moments} \\ \text{acting on} \\ \text{control} \\ \text{volume} \end{array}\right\} = \underbrace{\left\{\begin{array}{c} \text{rate of} \\ \text{moment of} \\ \text{momentum} \\ \text{out of control} \\ \text{volume} \end{array}\right\} - \left\{\begin{array}{c} \text{rate of} \\ \text{moment of} \\ \text{momentum into} \\ \text{control volume} \end{array}\right\}}_{\substack{\text{net rate of efflux of moment of} \\ \text{momentum from control volume}}} + \left\{\begin{array}{c} \text{rate of} \\ \text{accumulation} \\ \text{of moment} \\ \text{of momentum} \\ \text{within} \\ \text{control} \\ \text{volume} \end{array}\right\}$$

$$\tag{5-8}$$

Equation (5-8) may be applied to a general control volume to yield the following equation:

$$\Sigma\mathbf{M} = \iint_{\text{c.s.}} (\mathbf{r} \times \mathbf{v})\rho(\mathbf{v} \cdot \mathbf{n})dA + \frac{\partial}{\partial t}\iiint_{\text{c.v.}} (\mathbf{r} \times \mathbf{v})\rho\, dV \tag{5-9}$$

The term on the left-hand side of equation (5-9) is the total moment of all forces acting on the control volume. The terms on the right-hand side represent the net rate of efflux of moment of momentum through the control surface and the rate of accumulation of moment of momentum within the control volume, respectively.

This single-vector equation may be expressed as three scalar equations for the orthogonal inertial coordinate directions x, y, and z as

$$\Sigma M_x = \iint_{\text{c.s.}} (\mathbf{r} \times \mathbf{v})_x\rho(\mathbf{v} \cdot \mathbf{n})\, dA + \frac{\partial}{\partial t}\iiint_{\text{c.v.}} (\mathbf{r} \times \mathbf{v})_x\rho\, dV \tag{5-10a}$$

$$\Sigma M_y = \iint_{\text{c.s.}} (\mathbf{r} \times \mathbf{v})_y\rho(\mathbf{v} \cdot \mathbf{n})\, dA + \frac{\partial}{\partial t}\iiint_{\text{c.v.}} (\mathbf{r} \times \mathbf{v})_y\rho\, dV \tag{5-10b}$$

and

$$\Sigma M_z = \iint_{\text{c.s.}} (\mathbf{r} \times \mathbf{v})_z\rho(\mathbf{v} \cdot \mathbf{n})\, dA + \frac{\partial}{\partial t}\iiint_{\text{c.v.}} (\mathbf{r} \times \mathbf{v})_z\rho\, dV \tag{5-10c}$$

The directions associated with M_x and $(\mathbf{r} \times \mathbf{v})$ are those considered in mechanics in which the right-hand rule is used to determine the orientation of quantities having rotational sense.

▶ 5.4

APPLICATIONS TO PUMPS AND TURBINES

The moment-of-momentum expression is particularly applicable to two types of devices, generally classified as pumps and turbines. We shall, in this section, consider those having rotary motion only. If energy is derived from a fluid acting on a rotating device, it is designated a turbine, whereas a pump adds energy to a fluid. The rotating part of a turbine is called a runner and that of a pump an impeller.

The following two examples illustrate how moment-of-momentum analysis is used to generate expressions for evaluating turbine performance. Similar approaches will be used in Chapter 14 to evaluate operating characteristics of fans and pumps.

Example 4

Let us first direct our attention to a type of turbine known as the Pelton wheel. Such a device is represented in Figure 5.12. In this turbine, a jet of fluid, usually water, is directed from a nozzle striking a system of buckets on the periphery of the runner. The buckets are shaped so that the water is diverted in such a way as to exert a force on the runner that will, in turn, cause rotation. Using the moment-of-momentum relation, we may determine the torque resulting from such a situation.

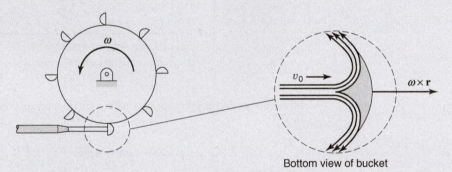

Bottom view of bucket

Figure 5.12 Pelton wheel.

We must initially define our control volume. The dashed line in Figure 5.13 illustrates the control volume chosen. It encloses the entire runner and cuts the jet of water with velocity v_0 as shown. The control surface also cuts through the shaft on both sides of the runner.

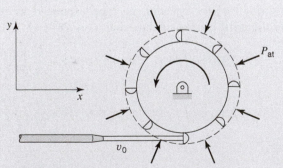

Figure 5.13 Control volume for analysis of Pelton wheel.

The applicable scalar form of the general moment-of-momentum expression is equation (5-10c) written for the z direction. All rotation is in the xy plane, and—according to the right-hand rule—the vector representation of a quantity having angular motion, or a tendency to produce angular motion, has a sense normal to the xy plane—that is, the z direction. Recall that a positive angular sense is that conforming to the direction in which the thumb on the right hand will point when the fingers of the right hand are aligned with the direction of counterclockwise angular motion.

$$\sum M_z = \iint_{\text{c.s.}} (\mathbf{r} \times \mathbf{v})_z \rho(\mathbf{v} \cdot \mathbf{n})\, dA + \frac{\partial}{\partial t} \iiint_{\text{c.v.}} \rho(\mathbf{r} \times \mathbf{v})_z \, dV$$

Evaluating each term separately, we have, for the external moment,

$$\sum M_z = M_{\text{shaft}}$$

where M_{shaft}, the moment applied to the runner by the shaft, is the only such moment acting on the control volume.

The surface integral

$$\iint_{\text{c.s.}} (\mathbf{r} \times \mathbf{v})_z \rho(\mathbf{v} \cdot \mathbf{n})\, dA$$

is the net rate of efflux of moment of momentum. The fluid leaving the control volume is illustrated in Figure 5.14. The x-direction component of the fluid leaving the control volume is

$$\{r\omega - (v_0 - r\omega)\cos\theta\}\mathbf{e}_x$$

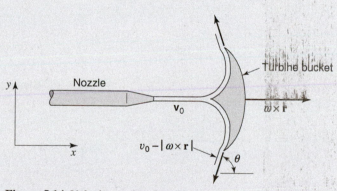

Figure 5.14 Velocity vectors for turbine bucket.

Here it is assumed that the z components of the velocity are equal and opposite. The leaving velocity is the vector sum of the velocity of the turbine bucket, $r\omega$, and that of the exiting fluid relative to the bucket and leaving at an angle θ to the direction of motion of the bucket, $(v_0 - r\omega)\cos\theta$. These velocity vectors are shown in the figure. The final expression for the surface integral is now

$$\iint_{\text{c.s.}} (\mathbf{r} \times \mathbf{v})_z \rho(\mathbf{v} \cdot \mathbf{n})\, dA = r[r\omega - (v_0 - r\omega)\cos\theta]\rho Q - rv_0\rho Q$$

The last term, $rv_0\rho Q$, is the moment of momentum of the incoming fluid stream of velocity v_0 and density ρ, with a volumetric flow rate Q.

As the problem under consideration is one in which the angular velocity, ω, of the wheel is constant, the term expressing the time derivative of moment of momentum of the control volume, $\partial/\partial t \iiint_{\text{c.v.}} (\mathbf{r} \times \mathbf{v})_z \rho \, dV = 0$. Replacing each term in the complete expression by its equivalent, we have

$$\sum M_z = M_{shaft} = \iint_{c.s.} (\mathbf{r} \times \mathbf{v})_z \rho(\mathbf{v} \cdot \mathbf{n}) \, dA + \frac{\partial}{\partial t} \iiint_{c.v.} \rho(\mathbf{r} \times \mathbf{v})_z \, dV$$

$$= r[r\omega - (v_0 - r\omega)\cos\theta]\rho Q - rv_0\rho Q = -r(v_0 - r\omega)(1 + \cos\theta)\rho Q$$

The torque applied to the shaft is equal in magnitude and opposite in sense to M_{shaft}. Thus, our final result is

$$\text{Torque} = -M_{shaft} = r(v_0 - r\omega)(1 + \cos\theta)\rho Q$$

Example 5

The radial-flow turbine illustrated in Figure 5.15 may be analyzed with the aid of the moment-of-momentum expression. In this device, the fluid (usually water) enters the guide vanes, which impart a tangential velocity and hence angular momentum to the fluid before it enters the revolving runner, which reduces the angular momentum of the fluid while delivering torque to the runner.

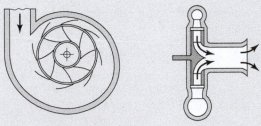

Figure 5.15 Radial-flow turbine.

The control volume to be used is illustrated below in Figure 5.16. The outer boundary of the control volume is at radius r_1, and the inner boundary is at r_2. The width of the runner is h.

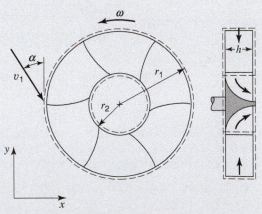

Figure 5.16 Radial-flow turbine-runner control volume.

We will use equation (5-9) in order to determine the torque. For steady flow, this equation becomes

$$\sum \mathbf{M} = \iint_{c.s.} (\mathbf{r} \times \mathbf{v})\rho(\mathbf{v} \cdot \mathbf{n}) \, dA$$

Evaluating each term separately, we have, for the external moment of the runner on the fluid,

$$\Sigma \mathbf{M} = M_{\text{fluid}}\mathbf{e}_z = -T\mathbf{e}_z$$

where T is the shaft torque. The surface integral requires the evaluation of the vector product $(\mathbf{r} \times \mathbf{v})$ at the outer boundary r_1 and at the inner boundary r_2. If we express the velocity of the water in polar coordinates $v = v_r\mathbf{e}_r + v_\theta\mathbf{e}_\theta$, so that $(\mathbf{r} \times \mathbf{v}) = r\mathbf{e}_r \times (v_r\mathbf{e}_r + v_\theta\mathbf{e}_\theta) = rv_\theta\mathbf{e}_z$. Thus the surface integral, assuming uniform velocity distribution, is given by

$$\iint_{\text{c.s.}} (\mathbf{r} \times \mathbf{v})\rho(\mathbf{v} \cdot \mathbf{n})dA = \{r_1 v_{\theta_1}\rho(-v_{r_1})2\pi r_1 h + r_2 v_{\theta_2}\rho v_{r_2}2\pi r_2 h\}\mathbf{e}_z$$

The general result is

$$-T\mathbf{e}_z = (-\rho v_{r_1} v_{\theta_1} 2\pi r_1^2 h + \rho v_{r_2} v_{\theta_2} 2\pi r_2^2 h)\mathbf{e}_z$$

The law of conservation of mass may be used

$$\rho v_{r_1} 2\pi r_1 h = \dot{m} = \rho v_{r_2} 2\pi r_2 h$$

so that the torque is given by

$$T = \dot{m}(r_1 v_{\theta_1} - r_2 v_{\theta_2})$$

The velocity at r_1 is seen from Figures 5.15 and 5.16 to be determined by the flow rate and the guide vane angle α. The velocity at r_2, however, requires knowledge of flow conditions on the runner.

The velocity at r_2 may be determined by the following analysis. In Figure 5.17, the flow conditions at the outlet of the runner are sketched. The velocity of the water v_2 is the vector sum of the velocity with respect to the runner v_2' and the runner velocity $r_2\omega$.

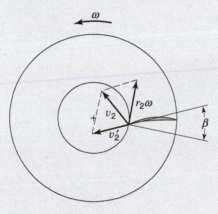

Figure 5.17 Velocity at runner exit (only one blade is shown).

The velocity v_{θ_2}, the tangential velocity of the water leaving the runner, is given by

$$v_{\theta_2} = r_2\omega - v_0' \sin\beta$$

where β is the blade angle as shown. The fluid is assumed to flow in the same direction as the blade. The radial component of the flow may be determined from conservation of mass

$$v_{r_2} = v_2' \cos\beta = \frac{\dot{m}}{2\pi\rho r_2 h}$$

Thus,

$$T = \dot{m}\left(r_1 v_{\theta_1} - r_2\left[r_2\omega - \frac{\dot{m}\tan\beta}{2\pi\rho r_2 h}\right]\right)$$

In practice, the guide vanes are adjustable to make the relative velocity at the runner entrance tangent to the blades.

▶ **5.5**

CLOSURE

In this chapter, the basic relation involved has been Newton's second law of motion. This law, as written for a system, was recast so that it could apply to a control volume. The result of a consideration of a general control volume led to the integral equations for linear momentum, equation (5-4), and moment of momentum, equation (5-9). Equation (5-4) is often referred to as the momentum theorem of fluid mechanics. This equation is one of the most powerful and often-used expressions in this field.

The student is again urged to start always with the complete integral expression when working a problem. A term-by-term analysis from this basis will allow a correct solution, whereas in a hasty consideration certain terms might be evaluated incorrectly or neglected completely. As a final remark, it should be noted that the momentum theorem expression, as developed, applies to an inertial control volume only.

CHAPTER 6

Conservation of Energy: Control-Volume Approach

The third fundamental law to be applied to fluid-flow analyses is the first law of thermodynamics. An integral expression for the conservation of energy applied to a control volume will be developed from the first law of thermodynamics, and examples of the application of the integral expression will be shown.

▶ 6.1

INTEGRAL RELATION FOR THE CONSERVATION OF ENERGY

The first law of thermodynamics may be stated as follows:

If a system is carried through a cycle, the total heat added to the system from its surroundings is proportional to the work done by the system on its surroundings.

Note that this law is written for a specific group of particles—those comprising the defined system. The procedure will then be similar to that used in Chapter 5—that is, recasting this statement into a form applicable to a control volume that contains different fluid particles at different times. The statement of the first law of thermodynamics involves only scalar quantities, however, and thus, unlike the momentum equations considered in Chapter 5, the equations resulting from the first law of thermodynamics will be scalar in form.

The statement of the first law given above may be written in equation form as

$$\oint \delta Q = \frac{1}{J} \oint \delta W \tag{6-1}$$

where the symbol $\oint$ refers to a "cyclic integral" or the integral of the quantity evaluated over a cycle. The symbols δQ and δW represent differential heat transfer and work done, respectively. The differential operator, δ, is used as both heat transfer and work are path functions and the evaluation of integrals of this type requires a knowledge of the path. The more familiar differential operator, d, is used with a "point" function. Thermodynamic properties are, by definition, point functions, and the integrals of such functions may be evaluated without a knowledge of the path by which the change in the property occurs between the initial and final states.[1] The quantity J is the so-called "mechanical equivalent of heat," numerically equal to 778.17 ft lb/Btu in engineering units. In the SI

---

[1] For a more complete discussion of properties, point functions and path fuctions, the reader is referred to G. N. Hatsopoulos and J. H. Keenan, *Principles of General Thermodynamics*. Wiley, New York, 1965, p. 14.

system, $J = 1$ N m/J. This factor will not be written henceforth, and the student is reminded that all equations must be dimensionally homogeneous.

We now consider a general thermodynamic cycle, as shown in Figure 6.1. The cycle a occurs between points 1 and 2 by the paths indicated. Utilizing equation (6-1), we may write, for cycle a,

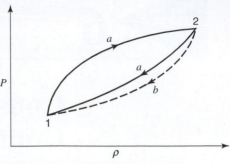

Figure 6.1 Reversible and irreversible thermodynamic cycles.

$$\int_{1a}^{2} \delta Q + \int_{2a}^{1} \delta Q = \int_{1a}^{2} \delta W + \int_{2a}^{1} \delta W$$

$$(6\text{-}2a)$$

A new cycle between points 1 and 2 is postulated as follows: the path between points 1 and 2 is identical to that considered previously; however, the cycle is completed by path b between points 2 and 1, which is any path other than a between these points. Again equation (6-1) allows us to write

$$\int_{1a}^{2} \delta Q + \int_{2b}^{1} \delta Q = \int_{1a}^{2} \delta W + \int_{2b}^{1} \delta W \qquad (6\text{-}2b)$$

Subtracting equation (6-2b) from equation (6-2a) gives

$$\int_{2a}^{1} \delta Q - \int_{2b}^{1} \delta Q = \int_{2a}^{1} \delta W - \int_{2b}^{1} \delta W$$

which may be written

$$\int_{2a}^{1} (\delta Q - \delta W) = \int_{2b}^{1} (\delta Q - \delta W) \qquad (6\text{-}3)$$

As each side of equation (6-3) represents the integrand evaluated between the same two points but along different paths, it follows that the quantity, $\delta Q - \delta W$, is equal to a point function or a property. This property is designated dE, the total energy of the system. An alternate expression for the first law of thermodynamics may be written

$$\delta Q - \delta W = dE \qquad (6\text{-}4)$$

The signs of δQ and δW were specified in the original statement of the first law; δQ is positive when heat is added to the system, δW is positive when work is done by the system.

For a system undergoing a process occurring in time interval dt, equation (6-4) may be written as

$$\frac{\delta Q}{dt} - \frac{\delta W}{dt} = \frac{dE}{dt} \qquad (6\text{-}5)$$

Consider now, as in Chapter 5, a general control volume fixed in inertial space located in a fluid-flow field, as shown in Figure 6.2. The system under consideration, designated by dashed lines, occupies the control volume at time t, and its position is also shown after a period of time Δt has elapsed.

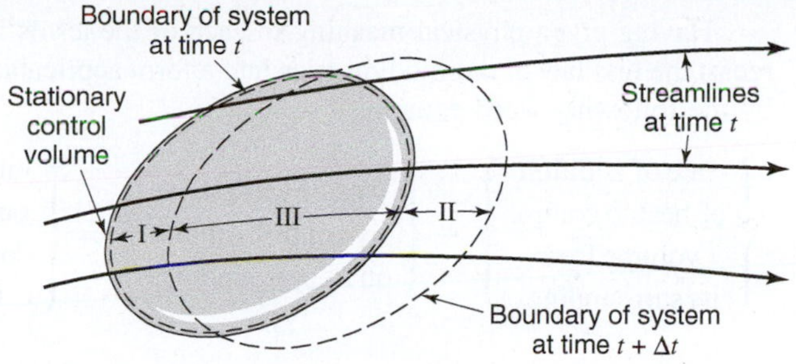

Figure 6.2 Relation between a system and a control volume in a fluid-flow field.

In this figure, region I is occupied by the system at time t, region II is occupied by the system at $t + \Delta t$, and region III is common to the system both at t and at $t + \Delta t$.

At time $t + \Delta t$ the total energy of the system may be expressed as

$$E|_{t+\Delta t} = E_{II}|_{t+\Delta t} + E_{III}|_{t+\Delta t}$$

and at time t

$$E|_t = E_I|_t + E_{III}|_t$$

Subtracting the second expression from the first and dividing by the elapsed time interval, Δt, we have

$$\frac{E|_{t+\Delta t} - E|_t}{\Delta t} = \frac{E_{III}|_{t+\Delta t} + E_{II}|_{t+\Delta t} - E_{III}|_t - E_I|_t}{\Delta t}$$

Rearranging and taking the limit as $\Delta t \to 0$ gives

$$\lim_{\Delta t \to 0} \frac{E|_{t+\Delta t} - E|_t}{\Delta t} = \lim_{\Delta t \to 0} \frac{E_{III}|_{t+\Delta t} - E_{III}|_t}{\Delta t} + \lim_{\Delta t \to 0} \frac{E_{II}|_{t+\Delta t} - E_I|_t}{\Delta t} \tag{6-6}$$

Evaluating the limit of the left-hand side, we have

$$\lim_{\Delta t \to 0} \frac{E|_{t+\Delta t} - E|_t}{\Delta t} = \frac{dE}{dt}$$

which corresponds to the right-hand side of the first-law expression, equation (6-5).

On the right-hand side of equation (6-6) the first limit becomes

$$\lim_{\Delta t \to 0} \frac{E_{III}|_{t+\Delta t} - E_{III}|_t}{\Delta t} = \frac{dE_{III}}{dt}$$

which is the rate of change of the total energy of the system, as the volume occupied by the system as $\Delta t \to 0$ is the control volume under consideration.

The second limit on the right of equation (6-6)

$$\lim_{\Delta t \to 0} \frac{E_{II}|_{t+\Delta t} - E_I|_t}{\Delta t}$$

represents the net rate of energy leaving across the control surface in the time interval Δt.

Having given physical meaning to each of the terms in equation (6-6), we may now recast the first law of thermodynamics into a form applicable to a control volume expressed by the following word equation:

$$
\left\{
\begin{array}{c}
\text{rate of addition} \\
\text{of heat to control} \\
\text{volume from} \\
\text{its surroundings}
\end{array}
\right\}
-
\left\{
\begin{array}{c}
\text{rate of work done} \\
\text{by control volume} \\
\text{on its surroundings}
\end{array}
\right\}
=
\left\{
\begin{array}{c}
\text{rate of energy} \\
\text{out of control} \\
\text{volume due to} \\
\text{fluid flow}
\end{array}
\right\}
$$

(6-7)

$$
-
\left\{
\begin{array}{c}
\text{rate of energy into} \\
\text{control volume due} \\
\text{to fluid flow}
\end{array}
\right\}
+
\left\{
\begin{array}{c}
\text{rate of accumulation} \\
\text{of energy within} \\
\text{control volume}
\end{array}
\right\}
$$

Equation (6-7) will now be applied to the general control volume shown in Figure 6.3.

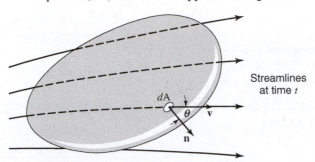

Figure 6.3 Fluid flow through a control volume.

The rates of heat addition to and work done by the control volume will be expressed as $\delta Q/dt$ and $\delta W/dt$.

Consider now the small area dA on the control surface. The rate of energy leaving the control volume through dA may be expressed as

$$\text{rate of energy efflux} = e(\rho v)(dA \cos \theta)$$

The product $(\rho v)(dA \cos \theta)$ is the rate of mass efflux from the control volume through dA, as discussed in the previous chapters. The quantity e is the specific energy or the energy per unit mass. The specific energy includes the potential energy, gy, due to the position of the fluid continuum in the gravitational field; the kinetic energy of the fluid, $v^2/2$, due to its velocity; and the internal energy, u, of the fluid due to its thermal state.

The quantity $dA \cos \theta$ represents the area, dA, projected normal to the velocity vector, $\mathbf{v}$. Theta (θ) is the angle between $\mathbf{v}$ and the outwardly directed normal vector, $\mathbf{n}$. We may now write

$$e(\rho v)(dA \cos \theta) = e\rho \, dA[\|\mathbf{v}\| \, \|\mathbf{n}\|] \cos \theta = e\rho(\mathbf{v} \cdot \mathbf{n}) \, dA$$

which we observe to be similar in form to the expressions previously obtained for mass and momentum. The integral of this quantity over the control surface

$$\iint_{\text{c.s.}} e\rho(\mathbf{v} \cdot \mathbf{n}) \, dA$$

represents the net *efflux of energy* from the control volume. The sign of the scalar product, $\mathbf{v} \cdot \mathbf{n}$, accounts both for efflux and for influx of mass across the control surface as considered previously. Thus, the first two terms on the right-hand side of equation (6-7) may be evaluated as

$$\left\{ \begin{array}{c} \text{rate of energy} \\ \text{out of control} \\ \text{volume} \end{array} \right\} - \left\{ \begin{array}{c} \text{rate of energy} \\ \text{into control} \\ \text{volume} \end{array} \right\} = \iint_{\text{c.s.}} e\rho(\mathbf{v} \cdot \mathbf{n})\,dA$$

The rate of accumulation of energy within the control volume may be expressed as

$$\frac{\partial}{\partial t} \iiint_{\text{c.v.}} e\rho\,dV$$

Equation (6-7) may now be written as

$$\frac{\delta Q}{dt} - \frac{\delta W}{dt} = \iint_{\text{c.s.}} e\rho(\mathbf{v} \cdot \mathbf{n})\,dA + \frac{\partial}{\partial t} \iiint_{\text{c.v.}} e\rho\,dV \tag{6-8}$$

A final form for the first-law expression may be obtained after further consideration of the work-rate or power term, $\delta W/dt$.

There are three types of work included in the work-rate term. The first is the shaft work, W_s, which is that done by the control volume on its surroundings that could cause a shaft to rotate or accomplish the raising of a weight through a distance. A second kind of work done is flow work, W_σ, which is that done on the surroundings to overcome normal stresses on the control surface where there is fluid flow. The third type of work is designated shear work, W_τ, which is performed on the surroundings to overcome shear stresses at the control surface.

Examining our control volume for flow and shear work rates, we have, as shown in Figure 6.4, another effect on the elemental portion of control surface, dA. Vector $\mathbf{S}$ is the force intensity (stress) having components σ_{ii} and τ_{ij} in the directions normal and tangential to the surface, respectively. In terms of $\mathbf{S}$, the force on dA is $\mathbf{S}\,dA$, and the rate of work done by the fluid flowing through dA is $\mathbf{S}\,dA \cdot \mathbf{v}$.

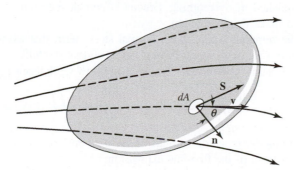

Figure 6.4 Flow and shear work for a general control volume.

The net rate of work done by the control volume on its surroundings due to the presence of $\mathbf{S}$ is

$$-\iint_{\text{c.s.}} \mathbf{v} \cdot \mathbf{S}\,dA$$

where the negative sign arises from the fact that the force per unit area *on the surroundings* is $-\mathbf{S}$.

The first-law expression, equation (6-8), may now be written as

$$\frac{\delta Q}{dt} - \frac{\delta W_s}{dt} + \iint_{\text{c.s.}} \mathbf{v} \cdot \mathbf{S} \, dA = \iint_{\text{c.s.}} e\rho(\mathbf{v} \cdot \mathbf{n}) \, dA + \frac{\partial}{\partial t} \iiint_{\text{c.v.}} e\rho \, dV \qquad (6\text{-}9)$$

where $\delta W_s/dt$ is the shaft work rate.

Writing the normal stress components of $\mathbf{S}$ as $\sigma_{ii}\mathbf{n}$, we obtain, for the net rate of work done in overcoming normal stress,

$$\left(\iint_{\text{c.s.}} \mathbf{v} \cdot \mathbf{S} \, dA \right)_{\text{normal}} = \iint_{\text{c.s.}} \mathbf{v} \cdot \sigma_{ii}\mathbf{n} \, dA = \iint_{\text{c.s.}} \sigma_{ii}(\mathbf{v} \cdot \mathbf{n}) \, dA$$

The remaining part of the work to be evaluated is the part necessary to overcome shearing stresses. This portion of the required work rate, $\delta W_\tau/dt$, is transformed into a form that is unavailable to do mechanical work. This term, representing a loss of mechanical energy, is included in the derivative form given above, and its analysis is included in Example 3, to follow. The work rate now becomes

$$\frac{\delta W}{dt} = \frac{\delta W_s}{dt} + \frac{\delta W_\sigma}{dt} + \frac{\delta W_\tau}{dt} = \frac{\delta W_s}{dt} - \iint_{\text{c.s.}} \sigma_{ii}(\mathbf{v} \cdot \mathbf{n}) \, dA + \frac{\delta W_\tau}{dt}$$

Substituting into equation (6-9), we have

$$\frac{\delta Q}{dt} - \frac{\delta W_s}{dt} + \iint_{\text{c.s.}} \sigma_{ii}(\mathbf{v} \cdot \mathbf{n}) \, dA + \frac{\delta W_\tau}{dt} = \iint_{\text{c.s.}} e\rho(\mathbf{v} \cdot \mathbf{n}) \, dA + \frac{\partial}{\partial t} \iiint_{\text{c.v.}} e\rho \, dV$$

The term involving normal stress must now be presented in a more usable form. A complete expression for σ_{ii} is stated in Chapter 9. For the present, we may say simply that the normal stress term is the sum of pressure effects and viscous effects. Just as with shear work, the work done to overcome the viscous portion of the normal stress is unavailable to do mechanical work. We shall thus combine the work associated with the viscous portion of the normal stress with the shear work to give a single term, $\delta W_\mu/dt$, the work rate accomplished in overcoming viscous effects at the control surface. The subscript, μ, is used to make this distinction.

The remaining part of the normal stress term that associated with pressure may be written in slightly different form if we recall that the bulk stress, σ_{ii}, is the negative of the thermodynamic pressure, P. The shear and flow work terms may now be written as follows:

$$\iint_{\text{c.s.}} \sigma_{ii}(\mathbf{v} \cdot \mathbf{n}) \, dA - \frac{\delta W_\tau}{dt} = -\iint_{\text{c.s.}} P(\mathbf{v} \cdot \mathbf{n}) \, dA - \frac{\delta W_\mu}{dt}$$

Combining this equation with the one written previously and rearranging slightly will yield the final form of the first-law expression:

$$\frac{\delta Q}{dt} - \frac{\delta W_s}{dt} = \iint_{\text{c.s.}} \left(e + \frac{P}{\rho} \right) \rho(\mathbf{v} \cdot \mathbf{n}) \, dA + \frac{\partial}{\partial t} \iiint_{\text{c.v.}} e\rho \, dV + \frac{\delta W_\mu}{dt} \qquad (6\text{-}10)$$

Equations (6-10), (4-1), and (5-4) constitute the basic relations for the analysis of fluid flow via the control-volume approach. A thorough understanding of these three equations and a mastery of their application places at the disposal of the student very powerful means of analyzing many commonly encountered problems in fluid flow.

The use of the overall energy balance will be illustrated in the following example problems.

▶ **6.2**

APPLICATIONS OF THE INTEGRAL EXPRESSION

Example 1

As a first example, let us choose a control volume as shown in Figure 6.5 under the conditions of steady fluid flow and no frictional losses.

For the specified conditions the overall energy expression, equation (6-10), becomes

$$\frac{\delta Q}{dt} - \frac{\delta W_S}{dt} = \iint_{\text{c.s.}} \rho\left(e + \frac{P}{\rho}\right)(\mathbf{v} \cdot \mathbf{n})\,dA + \underbrace{\frac{\partial}{\partial t}\iiint_{\text{c.v.}} e\rho\,dV}_{0-\text{steady flow}} + \underbrace{\frac{\delta W_\mu}{dt}}_{0}$$

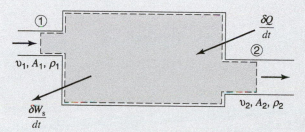

Figure 6.5 Control volume with one-dimensional flow across boundaries.

Considering now the surface integral, we recognize the product $\rho(\mathbf{v} \cdot \mathbf{n})\,dA$ to be the mass flow rate with the sign of this product indicating whether mass flow is into or out of the control volume, dependent upon the sense of $(\mathbf{v} \cdot \mathbf{n})$. The factor by which the mass flow rate is multiplied, $e + P/\rho$, represents the types of energy that may enter or leave the control volume per mass of fluid. The specific total energy, e, may be expanded to include the kinetic, potential, and internal energy contributions, so that

$$e + \frac{P}{\rho} = gy + \frac{v^2}{2} + u + \frac{P}{\rho}$$

As mass enters the control volume only at section (1) and leaves at section (2), the surface integral becomes

$$\iint_{\text{c.s.}} \rho\left(e + \frac{P}{\rho}\right)(\mathbf{v} \cdot \mathbf{n})\,dA = \left[\frac{v_2^2}{2} + gy_2 + u_2 + \frac{P_2}{\rho_2}\right](\rho_2 v_2 A_2) - \left[\frac{v_1^2}{2} + gy_1 + u_1 + \frac{P_1}{\rho_1}\right](\rho_1 v_1 A_1)$$

The energy expression for this example now becomes

$$\frac{\delta Q}{dt} - \frac{\delta W_S}{dt} = \left[\frac{v_2^2}{2} + gy_2 + u_2 + \frac{P_2}{\rho_2}\right](\rho_2 v_2 A_2) - \left[\frac{v_1^2}{2} + gy_1 + u_1 + \frac{P_1}{\rho_1}\right](\rho_1 v_1 A_1)$$

In Chapter 4, the mass balance for this same situation was found to be

$$\dot{m} = \rho_1 v_1 A_1 = \rho_1 v_2 A_2$$

If each term in the above expression is now divided by the mass flow rate, we have

$$\frac{q - \dot{W}_s}{\dot{m}} = \left[\frac{v_2^2}{2} + gy_2 + u_2 + \frac{P_2}{\rho_2}\right] - \left[\frac{v_1^2}{2} + gy_1 + u_1 + \frac{P_1}{\rho_1}\right]$$

or, in more familiar form

$$\frac{v_1^2}{2} + gy_1 + h_1 + \frac{q}{\dot{m}} = \frac{v_2^2}{2} + gy_2 + h_2 + \frac{\dot{W}_s}{\dot{m}}$$

where the sum of the internal energy and flow energy, $u + P/\rho$, has been replaced by the enthalpy, h, which is equal to the sum of these quantities by definition $h \equiv u + P/\rho$.

Example 2

As a second example, consider the situation shown in Figure 6.6. If water flows under steady conditions in which the pump delivers 3 horsepower to the fluid, find the mass flow rate if frictional losses may be neglected.

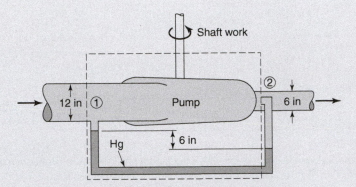

Figure 6.6 A control volume for pump analysis.

Defining the control volume as shown by the dashed lines, we may evaluate equation (6-10) term by term as follows:

$$\frac{\delta Q}{dt} = 0$$

$$-\frac{\delta W_s}{dt} = (3\,hp)(2545\,\text{Btu}/hp-h)(778\,\text{ft}-\text{lb}_f/\text{Btu})(h/3600\,\text{s})$$

$$= 1650\,\text{ft lb}_f/\text{s}$$

$$\iint_{c.s.} \left(e + \frac{P}{\rho}\right)\rho(\mathbf{v}\cdot\mathbf{n})\,dA = \iint_{A_2}\left(e + \frac{P}{\rho}\right)\rho(\mathbf{v}\cdot\mathbf{n})\,dA - \iint_{A_1}\left(e + \frac{P}{\rho}\right)\rho(\mathbf{v}\cdot\mathbf{n})\,dA$$

$$= \left(\frac{v_2^2}{2} + gy_2 + u_2 + \frac{P_2}{\rho_2}\right)(\rho_2 v_2 A_2) - \left(\frac{v_1^2}{2} + gy_1 + u_1 + \frac{P_1}{\rho_1}\right)(\rho_1 v_1 A_1)$$

$$= \left[\frac{v_2^2 - v_1^2}{2} + g(y_2 - y_1) + (u_2 - u_1) + \left(\frac{P_2}{\rho_2} - \frac{P_1}{\rho_1}\right)\right](\rho v A)$$

Here it may be noted that the pressure measured at station (1) is the static pressure while the pressure measured at station (2) is measured using a pressure port that is oriented normal to the oncoming flow—that is, where the velocity has been reduced to zero. Such a pressure is designated the *stagnation pressure*, which is greater than the static pressure by an amount equivalent to the change in kinetic energy of the flow. The stagnation pressure is thus expressed as

$$P_{\text{stagnation}} = P_0 = P_{\text{static}} + \frac{1}{2}\rho v^2$$

for incompressible flow; hence, the energy flux term may be rewritten as

$$\iint_{c.s.} \left(e + \frac{P}{\rho}\right)\rho(\mathbf{v}\cdot\mathbf{n})\,dA = \left(\frac{P_{0_2} - P_1}{\rho} - \frac{v_1^2}{2}\right)(\rho v A)$$

$$= \left\{ \frac{6(1 - 1/13.6)\ \text{in. Hg}(14.7\ \text{lb/in}^2)(144\ \text{in}^2/\text{ft}^2)}{(62.4\ \text{lb}_\text{m}/\text{ft}^3)(29.92\ \text{in. Hg})} \right.$$

$$\left. - \frac{v_1^2}{64.4(\text{lb}_\text{m}\,\text{ft/s}^2\text{lb}_\text{f})} \right\}\{(62.4\ \text{lb}_\text{m}/\text{ft}^3)(v_1)(\pi/4\ \text{ft}^2)\}$$

$$= \left(6.30 - \frac{v_1^2}{64.4}\right)(49\,v_1)\ \text{ft lb}_\text{f}/\text{s}$$

$$\frac{\partial}{\partial t}\iiint_{c.v.} e\rho\,dV = 0$$

$$\frac{\delta W_\mu}{dt} = 0$$

In the evaluation of the surface integral, the choice of the control volume coincided with the location of the pressure taps at sections (1) and (2). The pressure sensed at section (1) is the static pressure, as the manometer opening is parallel to the fluid-flow direction. At section (2), however, the manometer opening is normal to the flowing fluid stream. The pressure measured by such an arrangement includes both the static fluid pressure and the pressure resulting as a fluid flowing with velocity v_2 is brought to rest. The sum of these two quantities is known as the impact or stagnation pressure.

The potential energy change is zero between sections (1) and (2) and as we consider the flow to be isothermal, the variation in internal energy is also zero. Hence, the surface integral reduces to the simple form indicated.

The flow rate of water necessary for the stated conditions to exist is achieved by solving the resulting cubic equation. The solution is

$$v_1 = 16.59\ \text{ft/s}(5.057\ \text{m/s})$$
$$\dot{m} = \rho A v = 813\ \text{lb}_\text{m}/\text{s}(370\ \text{kg/s})$$

Example 3

A shaft is rotating at constant angular velocity ω in the bearing shown in Figure 6.7. The shaft diameter is d and the shear stress acting on the shaft is τ. Find the rate at which energy must be removed from the bearing in order for the lubricating oil between the rotating shaft and the stationary bearing surface to remain at constant temperature.

The shaft is assumed to be lightly loaded and concentric with the journal (the part of the shaft in contact with the bearing). The control volume selected consists of a unit length of the fluid surrounding the shaft as shown in Figure 6.7. The first law of thermodynamics for the control volume is

$$\frac{\delta Q}{dt} - \frac{\delta W_\text{s}}{dt} = \iint_{c.s.} \rho\left(e + \frac{P}{\rho}\right)(\mathbf{v}\cdot\mathbf{n})\,dA$$

$$+ \frac{\partial}{\partial t}\iiint_{c.v.} \rho e\,dV + \frac{\delta W_\mu}{dt}$$

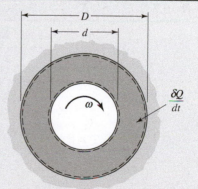

Figure 6.7 Bearing and control volume for bearing analysis.

From the figure we may observe the following:

1. No fluid crosses the control surface.
2. No shaft work crosses the control surface.
3. The flow is steady.

Thus $\delta Q/dt = \delta W_\mu/dt = \delta W_\tau/dt$. The viscous work rate must be determined. In this case all of the viscous work is done to overcome shearing stresses; thus, the viscous work is $\int\int_{c.s.} \tau(\mathbf{v} \cdot \mathbf{e}_t)\,dA$. At the outer boundary, $v = 0$, and at the inner boundary, $\int\int_{c.s.} \tau(\mathbf{v} \cdot \mathbf{e}_t)\,dA = -\tau(\omega d/2)A$, where $\mathbf{e}_t$ indicates the sense of the shear stress, τ, on the surroundings. The resulting sign is consistent with the concept of work being positive when done by a system on its surroundings. Thus,

$$\frac{\delta Q}{dt} = -\tau\frac{\omega\,d^2\pi}{2}$$

which is the heat transfer rate required to maintain the oil at a constant temperature.

If energy is not removed from the system, then $\delta Q/dt = 0$, and

$$\frac{\partial}{\partial t}\iiint_{c.v.} e\rho\,dV = -\frac{\delta W_\mu}{dt}$$

As only the internal energy of the oil will increase with respect to time,

$$\frac{\partial}{\partial t}\iiint_{c.v.} e\rho\,dV = \rho\pi\left(\frac{D^2 - d^2}{4}\right)\frac{d\mu}{dt} - \frac{\delta W_\mu}{dt} = \omega\frac{d^2\pi}{2}\tau$$

or, with constant specific heat c

$$c\frac{dT}{dt} = \frac{2\tau\omega\,d^2}{\rho(D^2 - d^2)}$$

where D is the outer bearing diameter.

In this example, the use of the viscous-work term has been illustrated. Note that

1. The viscous-work term involves only quantities on the surface of the control volume.
2. When the velocity on the surface of the control volume is zero, the viscous-work term is zero.

▶ **6.3**

THE BERNOULLI EQUATION

Under certain flow conditions, the expression of the first law of thermodynamics applied to a control volume reduces to an extremely useful relation known as the Bernoulli equation.

If equation (6-10) is applied to a control volume as shown in Figure 6.8, in which flow is steady, incompressible, and inviscid, and in which no heat transfer or change in internal energy occurs, a term-by-term evaluation of equation (6-10) gives the following:

$$\frac{\delta Q}{dt} = 0$$

$$\frac{\delta W_s}{dt} = 0$$

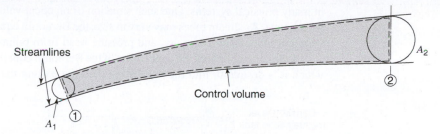

Figure 6.8 Control volume for steady, incompressible, inviscid, isothermal flow.

$$\iint_{c.s.} \rho\left(e + \frac{P}{\rho}\right)(\mathbf{v} \cdot \mathbf{n})\, dA = \iint_{A_1} \rho\left(e + \frac{P}{\rho}\right)(\mathbf{v} \cdot \mathbf{n})\, dA$$

$$+ \iint_{A_2} \rho\left(e + \frac{P}{\rho}\right)(\mathbf{v} \cdot \mathbf{n})\, dA$$

$$= \left(gy_1 + \frac{v_1^2}{2} + \frac{P_1}{\rho_1}\right)(-\rho_1 v_1 A_1)$$

$$+ \left(gy_2 + \frac{v_2^2}{2} + \frac{P_2}{\rho_2}\right)(\rho_2 v_2 A_2)$$

$$\frac{\partial}{\partial t}\iiint_{c.v.} e\rho\, dV = 0$$

The first-law expression now becomes

$$0 = \left(gy_2 + \frac{v_2^2}{2} + \frac{P_2}{\rho}\right)(\rho v_2 A_2) - \left(gy_1 + \frac{v_1^2}{2} + \frac{P_1}{\rho_1}\right)(\rho v_1 A_1)$$

As flow is steady, the continuity equation gives

$$\rho_1 v_1 A_1 = \rho_2 v_2 A_2$$

which may be divided through to give

$$gy_1 + \frac{v_1^2}{2} + \frac{P_1}{\rho} = gy_2 + \frac{v_2^2}{2} + \frac{P_2}{\rho} \qquad (6\text{-}11a)$$

Dividing through by g, we have

$$y_1 + \frac{v_1^2}{2g} + \frac{P_1}{\rho g} = y_2 + \frac{v_2^2}{2g} + \frac{P_2}{\rho g} \qquad (6\text{-}11b)$$

Either of the above expressions is designated the Bernoulli equation.

Note that each term in equation (6-11b) has the unit of length. The quantities are often designated "heads" due to elevation, velocity, and pressure, respectively. These terms, both individually and collectively, indicate the quantities that may be directly converted to produce mechanical energy.

Equation (6-11) may be interpreted physically to mean that the total mechanical energy is conserved for a control volume satisfying the conditions upon which this relation is based—that is, steady, incompressible, inviscid, isothermal flow, with no heat transfer or work done. These conditions may seem overly restrictive, but they are met, or approached, in many physical systems. One such situation of practical value is for flow into and out of a stream tube. As stream tubes may vary in size, the Bernoulli equation can actually describe the variation in elevation, velocity, and pressure head from point-to-point in a fluid-flow field.

A classic example of the application of the Bernoulli equation is depicted in Figure 6.9, in which it is desired to find the velocity of the fluid exiting the tank as shown.

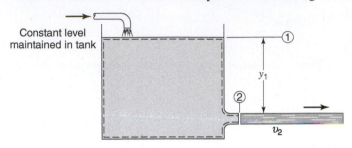

Figure 6.9 Control volume for Bernoulli equation analysis.

The control volume is defined as shown by dashed lines in the figure. The upper boundary of the control volume is just below the fluid surface, and thus can be considered to be at the same height as the fluid. There is fluid flow across this surface, but the surface area is large enough that the velocity of this flowing fluid may be considered negligible.

Under these conditions, the proper form of the first law of thermodynamics is equation (6-11), the Bernoulli equation. Applying equation (6-11), we have

$$y_1 + \frac{P_{\text{atm}}}{\rho g} = \frac{v_2^2}{2g} + \frac{P_{\text{atm}}}{\rho g}$$

from which the exiting velocity may be expressed in the familiar form

$$v_2 = \sqrt{2gy}$$

which is known as Torricelli's equation.

The next illustration of the use of the control-volume relations, an example using all three expressions is presented below.

Example 4

In the sudden enlargement shown below in Figure 6.10, the pressure acting at section (1) is considered uniform with value P_1. Find the change in internal energy between stations (1) and (2) for steady, incompressible flow. Neglect shear stress at the walls and express $u_2 - u_1$ in terms of v_1, A_1, and A_2. The control volume selected is indicated by the dotted line.

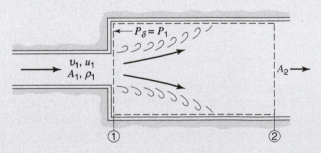

Figure 6.10 Flow through a sudden enlargement.

Conservation of Mass

$$\iint_{\text{c.s.}} \rho(\mathbf{v} \cdot \mathbf{n})\, dA + \frac{\partial}{\partial t} \iiint_{\text{c.v.}} \rho\, dV = 0$$

If we select station (2), a considerable distance downstream from the sudden enlargement, the continuity expression, for steady, incompressible flow, becomes

$$\rho_1 v_1 A_1 = \rho_2 v_2 A_2$$

or

$$v_2 = v_1 \frac{A_1}{A_2} \tag{6-12}$$

Momentum

$$\Sigma \mathbf{F} = \iint_{c.s.} \rho \mathbf{v}(\mathbf{v} \cdot \mathbf{n})\, dA + \frac{\partial}{\partial t} \iiint_{c.v.} \rho \mathbf{v}\, dV$$

and thus

$$P_1 A_2 - P_2 A_2 = \rho v_2^2 A_2 - \rho v_1^2 A_1$$

or

$$\frac{P_1 - P_2}{\rho} = v_2^2 - v_1^2 \left(\frac{A_1}{A_2}\right) \tag{6-13}$$

Energy

$$\frac{\delta Q}{\partial t} - \frac{\delta W_s}{dt} = \iint_{c.s.} \rho \left(e + \frac{P}{\rho}\right)(\mathbf{v} \cdot \mathbf{n})\, dA + \frac{\partial}{\partial t} \iiint_{c.v.} \rho e\, dV + \frac{\delta W_\mu}{dt}$$

Thus,

$$\left(e_1 + \frac{P_1}{\rho}\right)(\rho v_1 A_1) = \left(e_2 + \frac{P_2}{\rho}\right)(\rho v_2 A_2)$$

or, since $\rho v_1 A_1 = \rho v_2 A_2$,

$$e_1 + \frac{P_1}{\rho} = e_2 + \frac{P_2}{\rho}$$

The specific energy is

$$e = \frac{v^2}{2} + gy + u$$

Thus, our energy expression becomes

$$\frac{v_1^2}{2} + gy_1 + u_1 + \frac{P_1}{\rho} = \frac{v_2^2}{2} + gy_2 + u_2 + \frac{P_2}{\rho} \tag{6-14}$$

The three control-volume expressions may now be combined to evaluate $u_2 - u_1$. From equation (6-14), we have

$$u_2 - u_1 = \frac{P_1 - P_2}{\rho} + \frac{v_1^2 - v_2^2}{2} + g(y_1 - y_2) \tag{6-14a}$$

Substituting equation (6-13) for $(P_1 - P_2)/\rho$ and equation (6-12) for v_2 and noting that $y_1 = y_2$, we obtain

$$u_2 - u_1 = v_1^2 \left(\frac{A_1}{A_2}\right)^2 - v_1^2 \frac{A_1}{A_2} + \frac{v_1^2}{2} - \frac{v_1^2}{2}\left(\frac{A_1}{A_2}\right)^2 \tag{6-15}$$

$$= \frac{v_1^2}{2}\left[1 - 2\frac{A_1}{A_2} + \left(\frac{A_1}{A_2}\right)^2\right] = \frac{v_1^2}{2}\left[1 - \frac{A_1}{A_2}\right]^2$$

Equation (6-15) shows that the internal energy increases in a sudden enlargement. The temperature change corresponding to this change in internal energy is insignificant, but from equation (6-14a) it can be seen that the change in total head,

$$\left(\frac{P_1}{\rho} + \frac{v_1^2}{2} + gy_1\right) - \left(\frac{P_2}{\rho} + \frac{v_2^2}{2} + gy_2\right)$$

is equal to the internal energy change. Accordingly, the internal energy change in an incompressible flow is designated as the head loss, h_L, and the energy equation for steady, adiabatic, incompressible flow in a stream tube is written as

$$\frac{P_1}{\rho g} + \frac{v_1^2}{2g} + y_1 = h_L + \frac{P_2}{\rho g} + \frac{v_2^2}{2g} + y_2 \qquad (6\text{-}16)$$

Note the similarity to equation (6–11).

Example 5

The human heart is a four-chambered pump where valves allow the flow of blood while minimizing backflow. The valves open when the inlet pressure is greater than the outlet pressure, and close when the inlet pressure is less than the outlet pressure. Heart valve problems are relatively common, with the most prevalent issues being valve regurgitation or valve stenosis. In valve regurgitation, the valve fails to close, resulting in the unwanted backflow of blood. A stenosis is the narrowing of the opening of the valve. The most common noninvasive method to understand pressure drops across normal and stenoic heart valves is ultrasound, commonly referred to as echocardiography. Bernoulli's equation is used to understand velocity and pressure drop in these patients.

The general form of Bernoulli's equation (6-11) is given below, where the subscript 1 refers to a position upstream in the atrium or ventricle and subscript 2 refers to the position at the valve opening.

$$gy_1 + \frac{v_1^2}{2} + \frac{P_1}{\rho_1} = gy_2 + \frac{v_2^2}{2} + \frac{P_2}{\rho_2} \qquad (6\text{-}11)$$

Presuming the patient is lying down on a table, $z_1 = z_2$:

$$P_1 + \frac{\rho_1 v_1^2}{2} = P_2 + \frac{\rho_2 v_2^2}{2} \qquad (6\text{-}17)$$

Rearranging,

$$P_1 - P_2 = \frac{\rho}{2}(v_2^2 - v_1^2) \qquad (6\text{-}18)$$

Blood has a density of 1070 kg/m^3, which is equal to 8 mmHg(s^2/m^2). Inserting this value into equation (6-18) results in the final equation used to understand pressure drop and velocity in heart valves, an equation that is used daily in hospitals around the world:

$$P_1 - P_2 = 4(v_2^2 - v_1^2) \qquad (6\text{-}19)$$

As always, it is instructive to keep in mind the units of each parameter in the equation. The units of pressure are mmHg (common pressure units used in medical applications), and the number 4 and velocity, have units of (s^2/m^2)mmHg and m/s, respectively.

A patient is being examined for possible mitral valve stenosis by echocardiography. The mitral valve is located between the left atrium and left ventricle of the heart and maintains flow in one direction. In mitral valve stenosis, the valve does not open fully, resulting in decreased blood flow. A patient was examined by echocardiography and was found to have a blood flow velocity in the left atrium of 0.45 m/s, where the area is measured to be 0.0013 m and the maximum velocity through the valve was 1.82 m/s.[2]

---

[2] H. Baumgartner, et. al. *J. Am. Soc. Echocardio.*, **22**, 1 (2009).

We wish to determine the pressure drop between the left ventricle and the valve as well as the valve diameter. The pressure is determined by equation (6-19) as

$$P_1 - P_2 = 4(v_2^2 - v_1^2) = 4\frac{s^2}{m^2}\,\text{mmHg}\left[(1.82\ \text{m/s})^2 - (0.45\ \text{m/s})^2\right] = 12.44\ \text{mmHg}$$

The cross-sectional area of the open mitral valve is

$$\rho_1 v_1 A_1 = \rho_2 v_2 A_2$$

Since blood is at a constant density in the problem, the equation reduces to

$$v_{\text{atrium}}A_{\text{atrium}} = v_{\text{valve}}A_{\text{valve}}$$

Solving for the valve area,

$$A_{\text{valve}} = \frac{v_{\text{atrium}}A_{\text{atrium}}}{v_{\text{valve}}} = \frac{(0.45\ \text{m/s})(0.0013\ \text{m})}{(1.82\ \text{m/s})} = 3.2 \times 10^{-4}\ \text{m}^2$$

As a result, this patient would be further evaluated for possible mild mitral valve stenosis.

▶ **6.4**

CLOSURE

In this chapter the first law of thermodynamics, the third of the fundamental relations upon which fluid-flow analyses are based, has been used to develop an integral expression for the conservation of energy with respect to a control volume. The resulting expression, equation (6-10), is, in conjunction with equations (4-1) and (5-4), one of the fundamental expressions for the control-volume analysis of fluid-flow problems.

A special case of the integral expression for the conservation of energy is the Bernoulli equation, equation (6-11). Although simple in form and use, this expression has broad application to physical situations.

Shear Stress in Laminar Flow

In the analysis of fluid flow thus far, shear stress has been mentioned, but it has not been related to the fluid or flow properties. We shall now investigate this relation for laminar flow. The shear stress acting on a fluid depends upon the type of flow that exists. In the so-called laminar flow, the fluid flows in smooth layers or lamina, and the shear stress is the result of the (nonobservable) microscopic action of the molecules. Turbulent flow is characterized by the large-scale, observable fluctuations in fluid and flow properties, and the shear stress is the result of these fluctuations. The criteria for laminar and turbulent flows and the shear stress in turbulent flow will be discussed in Chapter 12.

▶ **7.1**

NEWTON'S VISCOSITY RELATION

In a solid, the resistance to *deformation* is the modulus of elasticity. The shear modulus of an elastic solid is given by

$$\text{shear modulus} = \frac{\text{shear stress}}{\text{shear strain}} \tag{7-1}$$

Just as the shear modulus of an elastic solid is a property of the solid relating shear stress and shear strain, there exists a relation similar to (7-1), which relates the shear stress in a parallel, laminar flow to a property of the fluid. This relation is Newton's law of viscosity:

$$\text{viscosity} = \frac{\text{shear stress}}{\text{rate of shear strain}} \tag{7-2}$$

Thus, the viscosity is the property of a fluid to resist the *rate* at which deformation takes place when the fluid is acted upon by shear forces. As a property of the fluid, the viscosity depends upon the temperature, composition, and pressure of the fluid, but is independent of the rate of shear strain.

The rate of deformation in a simple flow is illustrated in Figure 7.1. The flow parallel to the x-axis will deform the element if the velocity at the top of the element is different than the velocity at the bottom.

The rate of shear strain at a point is defined as $-d\delta/dt$. From Figure 7.1, it may be seen that

$$
\begin{aligned}
-\frac{d\delta}{dt} &= \lim_{\Delta x, \Delta y, \Delta t \to 0} \frac{\delta|_{t+\Delta t} - \delta|_t}{\Delta t} \\
&= \lim_{\Delta x, \Delta y, \Delta t \to 0} \left(\frac{\{\pi/2 - \arctan[(v|_{y+\Delta y} - v|_y)\Delta t/\Delta y]\} - \pi/2}{\Delta t} \right)
\end{aligned} \tag{7-3}
$$

In the limit, $-d\delta/dt = dv/dy =$ rate of shear strain.

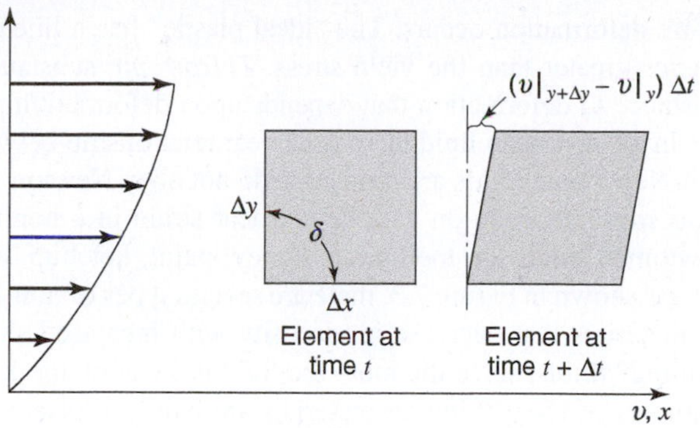

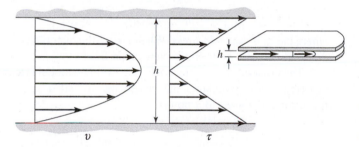

Figure 7.1 Deformation of a fluid element.

Combining equations (7-2) and (7-3) and denoting the viscosity by μ, we may write Newton's law of viscosity as

$$\tau = \mu \frac{dv}{dy}$$

(7-4)

The velocity profile and shear stress variation in a fluid flowing between two parallel plates is illustrated in Figure 7.2. The velocity profile[1] in this case is parabolic; as the shear stress is proportional to the derivative of the velocity, the shear stress varies in a linear manner.

Figure 7.2 Velocity and shear stress profiles for flow between two parallel plates.

▶ **7.2**

NON-NEWTONIAN FLUIDS

Newton's law of viscosity does not predict the shear stress in all fluids. Fluids are classified as Newtonian or non-Newtonian, depending upon the relation between shear stress and the rate of shearing strain. In Newtonian fluids, the relation is linear, as shown in Figure 7.3.

In non-Newtonian fluids, the shear stress depends upon the rate of shear strain. While fluids deform continuously under the action of shear stress, plastics will sustain a shear stress

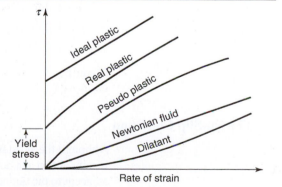

Figure 7.3 Stress rate-of-strain relation for Newtonian and non-Newtonian fluids.

[1] The derivation of velocity profiles is discussed in Chapter 8.

before deformation occurs. The "ideal plastic" has a linear stress rate-of-strain relation for stresses greater than the yield stress. *Thixotropic* substances such as printer's ink have a resistance to deformation that depends upon deformation rate and time.

In a Newtonian fluid there is a linear relationship between shear stress and shear strain. Non-Newtonian fluids are fluids that do not obey Newton's law of viscosity, and as a result shear stress depends on the rate of shear strain in a nonlinear fashion. Examples of non-Newtonian fluids are toothpaste, honey, paint, ketchup, and blood.

As shown in Figure 7.3, there are several types of non-Newtonian fluids. Pseudoplastics are materials that decrease in viscosity with increased shear strain and are called "shear-thinning" fluids, since the more the fluid is sheared, the less viscous it becomes. Common examples of shear-thinning materials are hair gel, plasma, syrup, and latex paint.

Dilatants are non-Newtonian fluids that exhibit an increase in viscosity with shear stress and are called "shear-thickening" fluids, since the more the fluid is sheared, the more viscous it becomes. Common examples of shear-thickening fluids include silly putty, quicksand, and the mixture of cornstarch and water.

A third common type of non-Newtonian fluids are designated viscoelastic fluids that return, either partially or fully, to their original shape after the applied shear stress is released. There are numerous models used to describe and characterize these non-Newtonian fluids.

A Bingham plastic is a material, like toothpaste, mayonnaise, and ketchup, that requires a finite yield stress before it begins to flow and can be described by the following equation:

$$\tau = \mu \frac{dv}{dy} \pm \tau_0 \tag{7-5}$$

where the τ_0 is the yield stress. When $\tau < \tau_0$ the material is a rigid plastic, and when $\tau > \tau_0$, the fluid behaves more like a Newtonian fluid.

The Ostwald–De Waele model or power law model is another commonly used model to describe non-Newtonian fluids where the so-called apparent viscosity is a function of the shear rate raised to a power,

$$\tau = m \left| \frac{dv}{dy} \right|^{n-1} \frac{dv}{dy} \tag{7-6}$$

where m and n are constants that are characteristic of the fluid. Power law fluids are classified based on the value of n:

$n = 1$ fluid is Newtonian and $m = \mu$

$n < 1$ fluid is pseudoplastic or shear thinning

$n > 1$ fluid is a dilatant or shear thickening

It can be seen that when $n = 1$ that the power law model reduces to Newton's law of viscosity (7-4).

The No-Slip Condition

Although the substances above differ in their stress rate-of-strain relations, they are similar in their action at a boundary. In both Newtonian and non-Newtonian fluids, the layer of fluid adjacent to the boundary has zero velocity relative to the boundary. When the boundary is a stationary wall, the layer of fluid next to the wall is at rest. If the boundary or wall is moving, the layer of fluid moves at the velocity of the boundary, hence, the name no-slip (boundary) condition. The no-slip condition is the result of experimental observation and fails when the fluid no longer can be treated as a continuum.

The no-slip condition is a result of the viscous nature of the fluid. In flow situations in which the viscous effects are neglected—the so-called inviscid flows—only the component of the velocity normal to the boundary is zero.

VISCOSITY

The viscosity of a fluid is a measure of its resistance to deformation rate. Tar and molasses are examples of highly viscous fluids; air and water, which are the subject of frequent engineering interest, are examples of fluids with relatively low viscosities. An understanding of the existence of the viscosity requires an examination of the motion of fluid on a molecular basis.

The molecular motion of gases can be described more simply than that of liquids. The mechanism by which a gas resists deformation may be illustrated by examination of the motion of the molecules on a microscopic basis. Consider the control volume shown in Figure 7.4.

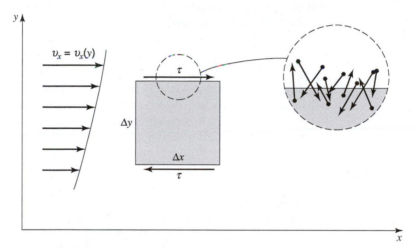

Figure 7.4 Molecular motion at the surface of a control volume.

The top of the control volume is enlarged to show that even though the top of the element is a streamline of the flow, individual molecules cross this plane. The paths of the molecules between collisions are represented by the random arrows. Because the top of the control volume is a streamline, the net molecular flux across this surface must be zero; hence, the upward molecular flux must equal the downward molecular flux. The molecules that cross the control surface in an upward direction have average velocities in the x direction corresponding to their points of origin. Denoting the y coordinate of the top of the control surface as y_0, we shall write the x-directional average velocity of the upward molecular flux as $v_x|_{y-}$, where the minus sign signifies that the average velocity is evaluated at some point below y_0. The x-directional momentum carried across the top of the control surface is then $mv_x|_{y-}$ per molecule, where m is the mass of the molecule. If Z molecules cross the plane per unit time, then the net x-directional momentum flux will be

$$\sum_{n=1}^{Z} m_n \left(v_x|_{y-} - v_x|_{y+} \right)$$

(7-7)

The flux of x-directional momentum on a molecular scale appears as a shear stress when the fluid is observed on a macroscopic scale. The relation between the molecular momentum flux and the shear stress may be seen from the control-volume expression for linear momentum

$$\Sigma \mathbf{F} = \iint_{c.s.} \rho \mathbf{v}(\mathbf{v} \cdot \mathbf{n})dA + \frac{\partial}{\partial t} \iiint_{c.v.} \rho \mathbf{v}\, dV \tag{5-4}$$

The first term on the right-hand side of equation (5-4) is the momentum flux. When a control volume is analyzed on a molecular basis, this term includes both the macroscopic and molecular momentum fluxes. If the molecular portion of the total momentum flux is to be treated as a force, it must be placed on the left-hand side of equation (5-4). Thus, the molecular momentum flux term changes sign. Denoting the negative of the molecular momentum flux as τ, we have

$$\tau = -\sum_{n=1}^{Z} m_n(v_x|_{y-} - v_x|_{y+}) \tag{7-8}$$

We shall treat shear stress exclusively as a force per unit area.

The bracketed term, $(v_x|_{y-} - v_x|_{y+})$ in equation (7-8), may be evaluated by noting that $v_x|_{y-} = v_x|_{y_0} - (dv_x/dy|_{y_0})\delta$, where $y- = y_0 - \delta$. Using a similar expression for $y+$, we obtain, for the shear stress,

$$\tau = 2\sum_{n=1}^{Z} m_n \frac{dv_x}{dy}\bigg|_{y_0} \delta_n$$

In the above expression δ is the y component of the distance between molecular collisions. Borrowing from the kinetic theory of gases, the concept of the mean free path, λ, as the average distance between collisions, and also noting from the same source that $\delta = 2/3\lambda$, we obtain, for a pure gas,

$$\tau = \frac{4}{3} m\lambda Z \frac{dv_x}{dy}\bigg|_{y_0} \tag{7-9}$$

as the shear stress.

Comparing equation (7-9) with Newton's law of viscosity, we see that

$$\mu = \frac{4}{3} m\lambda Z \tag{7-10}$$

The kinetic theory gives $Z = N\overline{C}/4$, where

$$N = \text{molecules per unit volume}$$
$$\overline{C} = \text{average random molecular velocity}$$

and thus

$$\mu = \frac{1}{3} Nm\lambda\overline{C} = \frac{\rho\lambda\overline{C}}{3}$$

or, using[2]

$$\lambda = \frac{1}{\sqrt{2}\pi Nd^2} \quad \text{and} \quad \overline{C} = \sqrt{\frac{8\kappa T}{\pi m}}$$

---

[2] In order of increasing complexity, the expressions for mean free path are presented in R. Resnick and D. Halliday, *Physics*, Part I, Wiley, New York, 1966, Chapter 24, and E. H. Kennard, *Kinetic Theory of Gases*, McGraw-Hill Book Company, New York, 1938, Chapter 2.

where d is the molecular diameter and κ is the Boltzmann constant, we have

$$\mu = \frac{2}{3\pi^{3/2}} \frac{\sqrt{m\kappa T}}{d^2} \tag{7-11}$$

Equation (7-11) indicates that μ is independent of pressure for a gas. This has been shown, experimentally, to be essentially true for pressures up to approximately 10 atmospheres. Experimental evidence indicates that at low temperatures the viscosity varies more rapidly than $\sqrt{T}$. The constant-diameter rigid-sphere model for the gas molecule is responsible for the less-than-adequate viscosity–temperature relation. Even though the preceding development was somewhat crude in that an indefinite property, the molecular diameter, was introduced, the interpretation of the viscosity of a gas being due to the microscopic momentum flux is a valuable result and should not be overlooked. It is also important to note that equation (7-11) expresses the viscosity entirely in terms of fluid properties.

A more realistic molecular model utilizing a force field rather than the rigid-sphere approach will yield a viscosity–temperature relationship much more consistent with experimental data than the $\sqrt{T}$ result. The most acceptable expression for nonpolar molecules is based upon the Lennard–Jones potential energy function. This function and the development leading to the viscosity expression will not be included here. The interested reader may refer to Hirschfelder, Curtiss, and Bird[3] for the details of this approach. The expression for viscosity of a pure gas that results is

$$\mu = 2.6693 \times 10^{-6} \frac{\sqrt{MT}}{\sigma^2 \Omega_\mu} \tag{7-12}$$

where μ is the viscosity, in pascal-seconds; T is absolute temperature, in K; M is the molecular weight; σ is the "collision diameter," a Lennard–Jones parameter, in Å (Angstroms); Ω_μ is the "collision integral," a Lennard–Jones parameter that varies in a relatively slow manner with the dimensionless temperature $\kappa T/\epsilon$; κ is the Boltzmann constant, 1.38×10^{-16} ergs/K; and ϵ is the characteristic energy of interaction between molecules. Values of σ and ϵ for various gases are given in Appendix K, and a table of Ω_μ versus kT/ϵ is also included in Appendix K.

For multicomponent gas mixtures at low density, Wilke[4] has proposed this empirical formula for the viscosity of the mixture:

$$\mu_{\text{mixture}} = \sum_{i=1}^{n} \frac{x_i \mu_i}{\sum x_j \phi_{ij}} \tag{7-13}$$

where x_i, x_j are mole fractions of species i and j in the mixture, and

$$\phi_{ij} = \frac{1}{\sqrt{8}} \left(1 + \frac{M_i}{M_j}\right)^{-1/2} \left[1 + \left(\frac{\mu_i}{\mu_j}\right)^{1/2} \left(\frac{M_j}{M_i}\right)^{1/4}\right]^2 \tag{7-14}$$

where M_i, M_j are the molecular weights of species i and j, and μ_i, μ_j are the viscosities of species i and j. Note that when $i = j$, we have $\phi_{ij} = 1$.

Equations (7-12), (7-13), and (7-14) are for nonpolar gases and gas mixtures at low density. For polar molecules, the preceding relation must be modified.[5]

---

[3] J. O. Hirschfelder, C. F. Curtiss, and R. B. Bird, *Molecular Theory of Gases and Liquids*, Wiley, New York, 1954.

[4] C. R. Wilke, *J. Chem. Phys.*, **18**, 517–519 (1950).

[5] J. O. Hirschfelder, C. F. Curtiss, and R. B. Bird, *Molecular Theory of Gases and Liquids*, Wiley, New York, 1954.

Although the kinetic theory of gases is well developed, and the more sophisticated models of molecular interaction accurately predict viscosity in a gas, the molecular theory of liquids is much less advanced. Hence, the major source of knowledge concerning the viscosity of liquids is experiment. The difficulties in the analytical treatment of a liquid are largely inherent in nature of the liquid itself. Whereas in gases the distance between molecules is so great that we consider gas molecules as interacting or colliding in pairs, the close spacing of molecules in a liquid results in the interaction of several molecules simultaneously. This situation is somewhat akin to an N-body gravitational problem. In spite of these difficulties, an approximate theory has been developed by Eyring, which illustrates the relation of the intermolecular forces to viscosity.[6] The viscosity of a liquid can be considered due to the restraint caused by intermolecular forces. As a liquid heats up, the molecules become more mobile. This results in less restraint from intermolecular forces. Experimental evidence for the viscosity of liquids shows that the viscosity decreases with temperature in agreement with the concept of intermolecular adhesive forces being the controlling factor.

Units of Viscosity

The dimensions of viscosity may be obtained from Newton's viscosity relation,

$$\mu = \frac{\tau}{dv/dy}$$

or, in dimensional form,

$$\frac{F/L^2}{(L/t)(1/L)} = \frac{Ft}{L^2}$$

where F = force, L = length, t = time.

Using Newton's second law of motion to relate force and mass ($F = ML/t^2$), we find that the dimensions of viscosity in the mass–length–time system become M/Lt.

The ratio of the viscosity to the density occurs frequently in engineering problems. This ratio, μ/ρ, is given the name kinematic viscosity and is denoted by the symbol ν. The origin of the name kinematic viscosity may be seen from the dimensions of ν:

$$\nu \equiv \frac{\mu}{\rho} \sim \frac{M/Lt}{M/L^3} = \frac{L^2}{t}$$

The dimensions of ν are those of kinematics: length and time. Either of the two names, absolute viscosity or dynamic viscosity, is frequently employed to distinguish μ from the kinematic viscosity, ν.

In the SI system, dynamic viscosity is expressed in pascal-seconds (1 pascal-second = 1 N · s/m^2 = 10 poise = 0.02089 slugs/ft · s = 0.02089 lb$_f$ · s/ft^2 = 0.6720 lb$_m$/ft · s). Kinematic viscosity in the metric system is expressed in (meters)^2 per second (1 m^2/s = 10^4 stokes = 10.76 ft^2/s).

Absolute and kinematic viscosities are shown in Figure 7.5 for three common gases and two liquids as functions of temperature. A more extensive listing is contained in Appendix I.

---

[6] For a description of Eyring's theory, see R. B. Bird, W. E. Stewart, and E. N. Lightfoot, *Transport Phenomena*, Wiley, New York, 2007, Chapter 1.

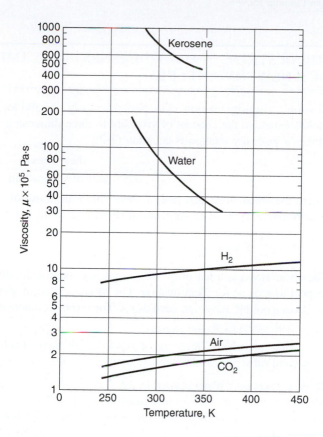

Figure 7.5 Viscosity–temperature variation for some liquids and gases.

Table 7.1 gives viscosities of common fluids.

Table 7.1   Viscosities of common fluids (at 20 C unless otherwise noted)

Fluid	Viscosity (cP) at 20 C
Ethanol	1.194
Mercury	15.47
H_2SO_4	19.15
Water	1.0019
Air	0.018
CO_2	0.015
Blood	2.5 (at 37 C)
SAE 40 motor oil	290
Corn oil	72
Ketchup	50,000
Peanut butter	250,000
Honey	10,000

1 centipoise (cP) = 0.001 kilogram/meter second.
1 centipoise (cP) = 0.001 Pascal second.

SHEAR STRESS IN MULTIDIMENSIONAL LAMINAR FLOWS OF A NEWTONIAN FLUID

Newton's viscosity relation, discussed previously, is valid for only parallel, laminar flows. Stokes extended the concept of viscosity to three-dimensional laminar flow. The basis of Stokes's viscosity relation is equation (7-2),

$$\text{viscosity} = \frac{\text{shear stress}}{\text{rate of shear strain}} \qquad (7\text{-}2)$$

where the shear stress and rate of shear strain are those of a three-dimensional element. Accordingly, we must examine shear stress and strain rate for a three-dimensional body.

Shear Stress

The shear stress is a tensor quantity requiring magnitude, direction, and orientation with respect to a plane for identification. The usual method of identification of the shear stress involves a double subscript, such as τ_{xy}. The tensor component, τ_{ij}, is identified as follows:

$$\tau = \text{magnitude}$$
$$\text{first subscript} = \text{direction of axis to which plane of action of shear stress is normal}$$
$$\text{second subscript} = \text{direction of action of the shear stress}$$

Thus, τ_{xy} acts on a plane normal to the x-axis (the yz plane) and acts in the y direction. In addition to the double subscript, a sense is required. The shear stresses acting on an element $\Delta x \, \Delta y \, \Delta z$, illustrated in Figure 7.6, are indicated in the positive sense. The definition of positive shear stress can be generalized for use in other coordinate systems. A shear stress component is positive when both the vector normal to the surface of action and the shear stress act in the same direction (both positive or both negative).

For example, in Figure 7.6(a), the shear stress τ_{yx} at the top of the element acts on surface $\Delta x \, \Delta z$. The vector normal to this area is in the positive y direction. The stress τ_{yx} acts in the positive x direction—hence, τ_{yx}, as illustrated in Figure 7.6(a), is positive. The student may apply similar reasoning to τ_{yx} acting on the bottom of the element and conclude that τ_{yx} is also positive as illustrated.

As in the mechanics of solids, $\tau_{ij} = \tau_{ji}$ (see Appendix C).

Rate of Shear Strain

The rate of shear strain for a three-dimensional element may be evaluated by determining the shear strain rate in the xy, yz, and xz planes. In the xy plane illustrated in Figure 7.7, the shear strain rate is again $-d\delta/dt$; however, the element may deform in both the x and the y directions. Hence, as the element moves from position 1 to position 2 in time Δt,

$$-\frac{d\delta}{dt} = \lim_{\Delta x, \Delta y, \Delta t \to 0} \frac{\delta|_{t+\Delta t} - \delta|_t}{\Delta t}$$

$$= \lim_{\Delta x, \Delta y, \Delta t \to 0} \left\{ \frac{\pi/2 - \arctan\{[(v_x|_{y+\Delta y} - v_x|_y)\Delta t]/\Delta y\}}{\Delta t} \right.$$

$$\left. - \frac{\arctan\{[(v_y|_{x+\Delta x} - v_y|_x)\Delta t]/\Delta x\} - \pi/2}{\Delta t} \right\}$$

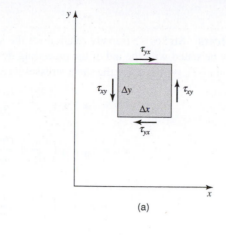

(a)

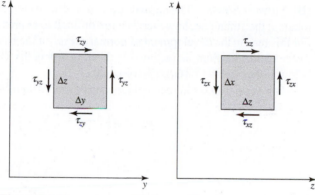

Figure 7.6 Shear stress acting in a positive sense.

As the shear strain evaluated above is in the xy plane, it will be subscripted xy. In the limit, $-d\delta_{xy}/dt = \partial v_x/\partial y + \partial v_y/\partial x$. In a similar manner, the shear strain rates in the yz and xz planes may be evaluated as

$$-\frac{d\delta_{yz}}{dt} = \frac{\partial v_y}{\partial z} + \frac{\partial v_z}{\partial y}$$

$$-\frac{d\delta_{xz}}{dt} = \frac{\partial v_y}{\partial z} + \frac{\partial v_z}{\partial y}$$

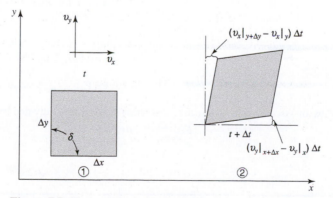

Figure 7.7 Shear strain in the xy plane.

Stokes's Viscosity Relation

(A) *Shear Stress*   Stokes's viscosity relation for the shear-stress components in laminar flow may now be stated with the aid of the preceding developments for rate of shear strain. Using equation (7-2), we have, for the shear stresses written in rectangular coordinate form,

$$\tau_{xy} = \tau_{yx} = \mu\left(\frac{\partial v_x}{\partial y} + \frac{\partial v_y}{\partial x}\right) \tag{7-15a}$$

$$\tau_{yz} = \tau_{zy} = \mu\left(\frac{\partial v_y}{\partial z} + \frac{\partial v_z}{\partial y}\right) \tag{7-15b}$$

and

$$\tau_{zx} = \tau_{xz} = \mu\left(\frac{\partial v_z}{\partial x} + \frac{\partial v_x}{\partial z}\right) \tag{7-15c}$$

(B) *Normal Stress*   The normal stress may also be determined from a stress rate-of-strain relation; the strain rate, however, is more difficult to express than in the case of shear strain. For this reason the development of normal stress, on the basis of a generalized Hooke's law for an elastic medium, is included in detail in Appendix D, with only the result expressed below in equations (7-16a), (7-16b), and (7-16c).

The normal stress in rectangular coordinates written for a Newtonian fluid is given by

$$\sigma_{xx} = \mu\left(2\frac{\partial v_x}{\partial x} - \frac{2}{3}\nabla\cdot\mathbf{v}\right) - P \tag{7-16a}$$

$$\sigma_{yy} = \mu\left(2\frac{\partial v_y}{\partial y} - \frac{2}{3}\nabla\cdot\mathbf{v}\right) - P \tag{7-16b}$$

and

$$\sigma_{zz} = \mu\left(2\frac{\partial v_z}{\partial z} - \frac{2}{3}\nabla\cdot\mathbf{v}\right) - P \tag{7-16c}$$

It is to be noted that the sum of these three equations yields the previously mentioned result: the bulk stress, $\bar{\sigma} = (\sigma_{xx} + \sigma_{yy} + \sigma_{zz})/3$, is the negative of the pressure, P.

Example 1

A Newtonian oil undergoes steady shear between two horizontal parallel plates. The lower plate is fixed, and the upper plate, weighing 0.5 lb$_\text{f}$, moves with a constant velocity of 15 ft/s. The distance between the plates is constant at 0.03 in and the area of the upper plate in contact with the fluid is 0.95 ft^2. What is the viscosity of this fluid?

0.03 in

→ Moving with v = 15 m/s

Fixed bottom plate

Figure 7.8 Flow between two parallel plates where the top plate is moving and the bottom plate is stationary.

We begin with Newton's law of viscosity and also realize that shear stress is force divided by area:

$$\frac{F}{A} = \tau_{yx} = \mu\frac{dv_x}{dy}$$

$$F\,dy = \mu A\,dv_x$$

$$F \int_0^{0.0025 \text{ ft}} dy = \mu A \int_0^{15 \text{ ft/s}} dv_x$$

$$\mu = \frac{F(0.0025 \text{ ft})}{A(15 \text{ ft/s})} = \frac{(0.5 \text{ lb}_f)(0.0025 \text{ ft})}{(0.95 \text{ ft}^2)(15 \text{ ft/s})} \left(32.174 \frac{\text{lb}_m \text{ft}}{\text{lb}_f \text{s}^2}\right) = 2.8 \times 10^{-3} \frac{\text{lb}_m}{\text{ft} \cdot \text{s}}$$

Example 2

Mayonnaise is a non-Newtonian fluid that is classified as a Bingham plastic and is common to daily life. Figure 7.9 illustrates the shear-stress versus shear-strain response to mayonnaise being sheared between two parallel plates where the upper plate is moving and the bottom plate is stationary. In this experiment the gap between the plates is 0.75 in and the moving plate has an area of 1.75 ft^2 and is exerting a force on the fluid of 4.75 lb$_f$. We want to first find the viscosity of mayonnaise and then find the velocity of the moving plate.

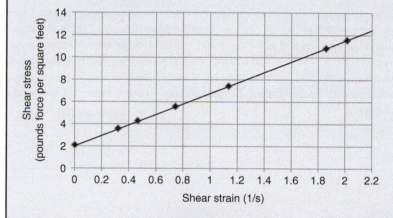

Figure 7.9 Shear stress versus shear strain for mayonnaise.

The viscosity is found from the slope of the shear-stress versus shear-strain curve:

$$\mu = \frac{\Delta y}{\Delta x} = \frac{10 - 4 \text{ lb}_f/\text{ft}^2}{1.7 - 0.4 \text{ s}^{-1}} \left(32.174 \frac{\text{lb}_m \text{ft}}{\text{lb}_f \text{s}^2}\right) = 148.5 \frac{\text{lb}_m}{\text{ft} \cdot \text{s}}$$

Next we want to calculate the velocity of the mayonnaise at the moving plate. Since mayonnaise is a Bingham plastic, we use the equation (notice that the yield stress is additive in this example)

$$\tau = \mu \frac{dv}{dy} + \tau_0 \tag{7-5}$$

From Figure 7.9 we note that the yield stress is approximately 2.1 lb$_f$/ft^2, so the equation for the shear stress is

$$\frac{F}{A} = \left(148.5 \frac{\text{lb}_m}{\text{ft} \cdot \text{s}}\right) \frac{dv}{dy} + (2.1 \text{ lb}_f/\text{ft}^2)$$

Inserting the values for force and area of plate,

$$\frac{4.75 \text{ lb}_f}{1.75 \text{ ft}^2} = \left(148.5 \frac{\text{lb}_m}{\text{ft} \cdot \text{s}}\right) \frac{dv}{dy} + (2.1 \text{ lb}_f/\text{ft}^2)$$

Rearranging and solving for the velocity,

$$\frac{dv}{dy} = 0.00414 \frac{lb_f}{lb_m ft} \left(32.174 \frac{lb_m ft}{lb_f s^2} \right) = 0.1331 \; s^{-1}$$

$$\int_0^{v_{plate}} dv = 0.1331 \int_0^{0.0625 \; ft} dy$$

$$v_{plate} = 0.0083 \frac{ft}{s} = 0.5 \frac{ft}{m}$$

▶ **7.5**

CLOSURE

The shear stress in laminar flow and its dependence upon the viscosity and kinematic derivatives has been presented for a Cartesian coordinate system. The shear stress in other coordinate systems, of course, will occur frequently, and it is to be noted that equation (7-2) forms the general relation between shear stress, viscosity, and the rate of shear strain. The shear stress in other coordinate systems may be obtained from evaluating the shear-strain rate in the associated coordinate systems. Several problems of this nature are included at the end of this chapter.

CHAPTER 8

Analysis of a Differential Fluid Element in Laminar Flow

The analysis of a fluid-flow situation may follow two different paths. One type of analysis has been discussed at length in Chapters 4–6 in which the region of interest has been a definite volume, the macroscopic control volume. In analyzing a problem from the standpoint of a macroscopic control volume, one is concerned only with gross quantities of mass, momentum, and energy crossing the control surface and the total change in these quantities exhibited by the material under consideration. Changes occurring within the control volume by each differential element of fluid cannot be obtained from this type of overall analysis.

In this chapter, we shall direct our attention to elements of fluid as they approach differential size. Our goal is the estimation and description of fluid behavior from a differential point of view; the resulting expressions from such analyses will be differential equations. The solution to these differential equations will give flow information of a different nature than that achieved from a macroscopic examination. Such information may be of less interest to the engineer needing overall design information, but it can give much greater insight into the mechanisms of mass, momentum, and energy transfer.

It is possible to change from one form of analysis to the other—that is, from a differential analysis to an integral analysis by integration and vice versa, rather easily.[1]

A complete solution to the differential equations of fluid flow is possible only if the flow is laminar; for this reason only laminar-flow situations will be examined in this chapter. A more general differential approach will be discussed in Chapter 9.

▶ 8.1

FULLY DEVELOPED LAMINAR FLOW IN A CIRCULAR CONDUIT OF CONSTANT CROSS SECTION

Engineers are often confronted with flow of fluids inside circular conduits or pipes. We shall now analyze this situation for the case of incompressible laminar flow. In Figure 8.1, we have a section of pipe in which the flow is laminar and fully developed—that is, it is not influenced by entrance effects and represents a steady-flow situation. *Fully developed flow* is defined as that for which the velocity profile does not vary along the axis of flow.

---

[1] This transformation may be accomplished by a variety of methods, among which are the methods of vector calculus. We shall use a limiting process in this text.

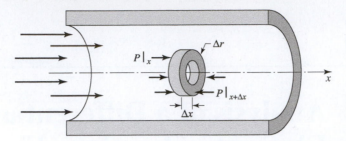

Figure 8.1 Control volume for flow in a circular conduit.

We now consider the cylindrical control volume of fluid having an inside radius, r, thickness Δr, and length Δx. Applying Newton's second law to this control volume, we may evaluate the appropriate force and momentum terms for the x direction. Starting with the control-volume expression for linear momentum in the x direction,

$$\Sigma F_x = \iint_{c.s.} \rho v_x(\mathbf{v} \cdot \mathbf{n})dA + \frac{\partial}{\partial t}\iiint_{c.v.}\rho v_x dV \tag{5-5a}$$

and evaluating each term as it applies to the control volume shown, we have

$$\Sigma F_x = P(2\pi r\,\Delta r)|_x - P(2\pi r\,\Delta r)|_{x+\Delta x} + \tau_{rx}(2\pi r\,\Delta x)|_{r+\Delta r} - \tau_{rx}(2\pi r\,\Delta x)|_r$$

$$\iint_{c.s.} v_x\rho(\mathbf{v}\cdot\mathbf{n})dA = (\rho v_x)(2\pi r\,\Delta r v_x)|_{x+\Delta x} - (\rho v_x)(2\pi r\,\Delta r v_x)|_x$$

and

$$\frac{\partial}{\partial t}\iiint_{c.v.} v_x\rho\,dV = 0$$

in steady flow.

The convective momentum flux

$$(\rho v_x)(2\pi r\,\Delta r v_x)|_{x+\Delta x} - (\rho v_x)(2\pi\,\Delta r v_x)|_x$$

is equal to zero as, by the original stipulation that flow is fully developed, all terms are independent of x. Substitution of the remaining terms into equation (5-5a) gives

$$-[P(2\pi r\,\Delta r)|_{x+\Delta x} - P(2\pi r\,\Delta r)|_x] + \tau_{rx}(2\pi r\,\Delta x)|_{r+\Delta r} - \tau_{rx}(2\pi r\,\Delta x)|_r = 0$$

Canceling terms where possible and rearranging, we find that this expression reduces to the form

$$-r\frac{P|_{x+\Delta x} - P|_x}{\Delta x} + \frac{(r\tau_{rx})|_{r+\Delta r} - (r\tau_{rx})|_r}{\Delta r} = 0$$

Evaluating this expression in the limit as the control volume approaches differential size— that is, as Δx and Δr approach zero, we have

$$-r\frac{dP}{dx} + \frac{d}{dr}(r\tau_{rx}) = 0 \tag{8-1}$$

Note that the pressure and shear stress are functions only of x and r, respectively, and thus, the derivatives formed are total rather than partial derivatives. In a region of fully developed flow, the pressure gradient, dP/dx, is constant.

The variables in equation (8-1) may be separated and integrated to give

$$\tau_{rx} = \left(\frac{dP}{dx}\right)\frac{r}{2} + \frac{C_1}{r}$$

The constant of integration C_1 may be evaluated by knowing a value of τ_{rx} at some r. Such a condition is known at the center of the conduit, $r=0$, where for any finite value of C_1, the shear stress, τ_{rx}, will be infinite. As this is physically impossible, the only realistic value for C_1 is zero. Thus, the shear-stress distribution for the conditions and geometry specified is

$$\tau_{rx} = \left(\frac{dP}{dx}\right)\frac{r}{2} \tag{8-2}$$

We observe that the shear stress varies linearly across the conduit from a value of 0 at $r=0$, to a maximum at $r=R$, the inside surface of the conduit.

Further information may be obtained if we substitute the Newtonian viscosity relationship—that is, assuming the fluid to be Newtonian and recalling that the flow is laminar:

$$\tau_{rx} = \mu\frac{dv_x}{dr} \tag{8-3}$$

Substituting this relation into equation (8-2) gives

$$\mu\frac{dv_x}{dr} = \left(\frac{dP}{dx}\right)\frac{r}{2}$$

which becomes, upon integration,

$$v_x = \left(\frac{dP}{dx}\right)\frac{r^2}{4\mu} + C_2$$

The second constant of integration, C_2, may be evaluated, using the boundary condition that the velocity, v_x, is zero at the conduit surface (the no-slip condition), $r=R$. Thus,

$$C_2 = -\left(\frac{dP}{dx}\right)\frac{R^2}{4\mu}$$

and the velocity distribution becomes

$$v_x = -\left(\frac{dP}{dx}\right)\frac{1}{4\mu}(R^2 - r^2) \tag{8-4}$$

or

$$v_x = -\left(\frac{dP}{dx}\right)\frac{R^2}{4\mu}\left[1 - \left(\frac{r}{R}\right)^2\right] \tag{8-5}$$

Equations (8-4) and (8-5) indicate that the velocity profile is parabolic and that the maximum velocity occurs at the center of the circular conduit where $r=0$. Thus,

$$v_{max} = -\left(\frac{dP}{dx}\right)\frac{R^2}{4\mu} \tag{8-6}$$

and equation (8-5) may be written in the form

$$v_x = v_{max}\left[1 - \left(\frac{r}{R}\right)^2\right] \tag{8-7}$$

Note that the velocity profile written in the form of equation (8-7) is identical to that used in Chapter 4, Example 2. We may, therefore, use the result obtained in Chapter 4, Example 2:

$$v_{avg} = \frac{v_{max}}{2} = -\left(\frac{dP}{dx}\right)\frac{R^2}{8\mu} \tag{8-8}$$

Equation (8-8) may be rearranged to express the pressure gradient, $-dP/dx$, in terms of v_{avg}:

$$-\frac{dP}{dx} = \frac{8\mu v_{\text{avg}}}{R^2} = \frac{32\mu v_{\text{avg}}}{D^2} \tag{8-9}$$

Equation (8-9) is known as the Hagen–Poiseuille equation, in honor of the two men credited with its original derivation. This expression may be integrated over a given length of conduit to find the pressure drop and associated drag force on the conduit resulting from the flow of a viscous fluid.

The conditions for which the preceding equations were derived and apply should be remembered and understood. They are as follows:

1. The fluid
   a. is Newtonian
   b. behaves as a continuum
2. The flow is
   a. laminar
   b. steady
   c. fully developed
   d. incompressible

▶ 8.2

LAMINAR FLOW OF A NEWTONIAN FLUID DOWN AN INCLINED-PLANE SURFACE

The approach used in Section 8.1 will now be applied to a slightly different situation—that of a Newtonian fluid in laminar flow down an inclined-plane surface. This configuration and associated nomenclature are depicted in Figure 8.2. We will examine the two-dimensional case—that is, we consider no significant variation in the z direction.

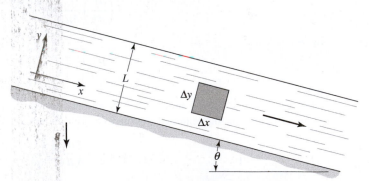

Figure 8.2 Laminar flow down an inclined-plane surface.

The analysis again involves the application of the control-volume expression for linear momentum in the x direction, which is

$$\Sigma F_x = \iint_{\text{c.s.}} v_x \rho (\mathbf{v} \cdot \mathbf{n}) dA + \frac{\partial}{\partial t} \iiint_{\text{c.v.}} \rho v_x dV \tag{5-5a}$$

Evaluating each term in this expression for the fluid element of volume $(\Delta x)(\Delta y)(1)$ as shown in the figure, we have

$$\Sigma F_x = P\Delta y|_x - P\Delta y|_{x+\Delta x} + \tau_{yx}\Delta x|_{y+\Delta y} - \tau_{yx}\Delta x|_y + \rho g\,\Delta x\,\Delta y\sin\theta$$

$$\iint_{c.s.} \rho v_x(\mathbf{v}\cdot\mathbf{n})dA = \rho v_x^2\Delta y|_{x+\Delta x} - \rho v_x^2\Delta y|_x$$

and

$$\frac{\partial}{\partial t}\iiint_{c.v.}\rho v_x\,dV = 0$$

Noting that the convective-momentum terms cancel for fully developed flow and that the pressure–force terms also cancel because of the presence of a free liquid surface, we see that the equation resulting from the substitution of these terms into equation (5-5a) becomes

$$\tau_{yx}\Delta x|_{y+\Delta y} - \tau_{yx}\Delta x|_y + \rho g\,\Delta x\,\Delta y\sin\theta = 0$$

Dividing by $(\Delta x)(\Delta y)(1)$, the volume of the element considered, gives

$$\frac{\tau_{yx}|_{y+\Delta y} - \tau_{yx}|_y}{\Delta y} + \rho g\sin\theta = 0$$

In the limit as $\Delta y \to 0$, we get the applicable differential equation

$$\frac{d}{dy}\tau_{yx} + \rho g\sin\theta = 0 \tag{8-10}$$

Separating the variables in this simple equation and integrating, we obtain for the shear stress

$$\tau_{yx} = -\rho g\sin\theta\,y + C_1$$

The integration constant, C_1, may be evaluated by using the boundary condition that the shear stress, τ_{yx}, is zero at the free surface, $y=L$. Thus, the shear-stress variation becomes

$$\tau_{yx} = \rho g L\sin\theta\left[1 - \frac{y}{L}\right] \tag{8-11}$$

The consideration of a Newtonian fluid in laminar flow enables the substitution of $\mu\,(dv_x/dy)$ to be made for τ_{yx}, yielding

$$\frac{dv_x}{dy} = \frac{\rho g L\sin\theta}{\mu}\left[1 - \frac{y}{L}\right]$$

which, upon separation of variables and integration, becomes

$$v_x = \frac{\rho g L\sin\theta}{\mu}\left[y - \frac{y^2}{2L}\right] + C_2$$

Using the no-slip boundary condition-that is, $v_x = 0$ at $y = 0$-the constant of integration, C_2, is seen to be zero. The final expression for the velocity profile may now be written as

$$v_x = \frac{\rho g L^2\sin\theta}{\mu}\left[\frac{y}{L} - \frac{1}{2}\left(\frac{y}{L}\right)^2\right] \tag{8-12}$$

The form of this solution indicates the velocity variation to be parabolic, reaching the maximum value

$$v_{max} = \frac{\rho g L^2 \sin\theta}{2\mu} \qquad (8\text{-}13)$$

at the free surface, $y = L$.

Additional calculations may be performed to determine the average velocity, as was indicated in Section 8.1. Note that there will be no counterpart in this case to the Hagen–Poiseuille relation, equation (8-9), for the pressure gradient. The reason for this is the presence of a free liquid surface along which the pressure is constant. Thus, for our present case, flow is not the result of a pressure gradient, but rather the manifestation of the gravitational acceleration upon a fluid.

Example 1

Your lab picked a capillary tube without recording the diameter from the literature in the package. It is very difficult to accurately measure the diameter of long capillary tubes, but it can be calculated. The capillary in question is 12 cm long, which is long enough to achieve fully developed flow. Water is flowing through the capillary in steady, continuous laminar flow at a temperature of 313 K and a velocity of 0.05 cm/s. The pressure drop across the capillary is measured to be 6 Pa. You can assume that the density of the fluid is unchanging for this analysis, and that the no-slip boundary condition applies. The diameter of the capillary can be calculated as follows:

The assumptions of the problem allow us to use the Hagen–Poiseuille equation:

$$-\frac{dP}{dz} = \frac{8\,\mu v_{avg}}{R^2} = \frac{32\,\mu v_{avg}}{D^2} \qquad (8\text{-}9)$$

We can take dz to be the length of the capillary tube and dP to be the change in pressure or ΔP, thus,

$$\frac{\Delta P}{L} = \frac{32\,\mu v_{avg}}{D^2}$$

Solving for the diameter,

$$D = \sqrt{\frac{32\,\mu v_{avg} L}{\Delta P}} = \sqrt{\frac{32(6.58 \times 10^{-4}\,\text{Pa} \cdot \text{s})(0.5\,\text{cm/s})(12\,\text{cm})}{6\,\text{Pa}}} = 0.145\,\text{cm} = 1.45\,\text{mm}$$

► **8.3**

CLOSURE

The method of analysis employed in this chapter, that of applying the basic relation for linear momentum to a small control volume and allowing the control volume to shrink to differential size, enables one to find information of a sort different from that obtained previously. Velocity and shear-stress profiles are examples of this type of information. The behavior of a fluid element of differential size can give considerable insight into a given transfer process and provide an understanding available in no other type of analysis.

This method has direct counterparts in heat and mass transfer, where the element may be subjected to an energy or a mass balance.

In Chapter 9, the methods introduced in this chapter will be used to derive differential equations of fluid flow for a general control volume.

Differential Equations of Fluid Flow

The fundamental laws of fluid flow, which have been expressed in mathematical form for an arbitrary control volume in Chapters 4–6, may also be expressed in mathematical form for a special type of control volume, the differential element. These differential equations of fluid flow provide a means of determining the point-to-point variation of fluid properties. Chapter 8 involved the differential equations associated with some one-dimensional, steady, incompressible laminar flows. In Chapter 9, we shall express the law of conservation of mass and Newton's second law of motion in differential form for more general cases. The basic tools used to derive these differential equations will be the control-volume developments of Chapters 4 and 5.

▶ **9.1**

THE DIFFERENTIAL CONTINUITY EQUATION

The continuity equation to be developed in this section is the law of conservation of mass expressed in differential form. Consider the control volume $\Delta x\,\Delta y\,\Delta z$ shown in Figure 9.1.

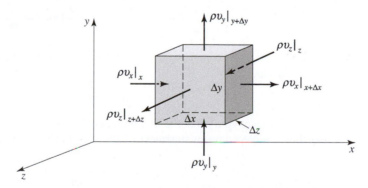

Figure 9.1 Mass flux through a differential control volume.

The control-volume expression for the conservation of mass is

$$\iint \rho(\mathbf{v} \cdot \mathbf{n})dA + \frac{\partial}{\partial t}\iiint \rho\,dV = 0 \tag{4-1}$$

which states that

$$\left\{\begin{array}{c} \text{net rate of mass} \\ \text{flux out of} \\ \text{control volume} \end{array}\right\} + \left\{\begin{array}{c} \text{rate of accumulation} \\ \text{of mass within} \\ \text{control volume} \end{array}\right\} = 0$$

The mass flux $\rho(\mathbf{v} \cdot \mathbf{n})$ at each face of the control volume is illustrated in Figure 9.1. The mass within the control volume is $\rho \, \Delta x \, \Delta y \, \Delta z$, and thus the time rate of change of mass within the control volume is

$$\frac{\partial}{\partial t}(\rho \, \Delta x \, \Delta y \, \Delta z)$$

The student is reminded that the density in general can vary from point to point—that is, $\rho = \rho(x, \ y, \ z, \ t)$.

The net mass flux out of the control volume in the x-direction is

$$(\rho v_x|_{x+\Delta x} - \rho v_x|_x)\Delta y \, \Delta z$$

in the y-direction

$$(\rho v_y|_{y+\Delta y} - \rho v_y|_y)\Delta x \, \Delta z$$

and in the z-direction

$$(\rho v_z|_{z+\Delta z} - \rho v_z|_z)\Delta x \, \Delta y$$

The total net mass flux is the sum of the above three terms. Substituting into equation (4-1) yields

$$(\rho v_x|_{x+\Delta x} - \rho v_x|_x)\Delta y \, \Delta z + (\rho v_y|_{y+\Delta y} - \rho v_y|_y)\Delta x \, \Delta z$$
$$+ (\rho v_z|_{z+\Delta z} - \rho v_z|_z)\Delta x \, \Delta y + \frac{\partial}{\partial t}(\rho \, \Delta x \, \Delta y \, \Delta z) = 0$$

The volume does not change with time, so we may divide the above equation by $\Delta x \, \Delta y \, \Delta z$. In the limit as Δx, Δy, and Δz approach zero, we obtain

$$\frac{\partial}{\partial x}(\rho v_x) + \frac{\partial}{\partial y}(\rho v_y) + \frac{\partial}{\partial z}(\rho v_z) + \frac{\partial \rho}{\partial t} = 0 \qquad (9\text{-}1)$$

The first three terms comprise the divergence of the vector $\rho \mathbf{v}$. The divergence of a vector is the dot product with ∇:

$$\operatorname{div} \mathbf{A} \equiv \nabla \cdot \mathbf{A}$$

The student may verify that the first three terms in equation (9-1) may be written as $\nabla \cdot \rho \mathbf{v}$ and thus a more compact statement of the continuity equation becomes

$$\nabla \cdot \rho \mathbf{v} + \frac{\partial \rho}{\partial t} = 0 \qquad (9\text{-}2)$$

The continuity equation above applies to unsteady, three-dimensional flow. It is apparent that when flow is incompressible, this equation reduces to

$$\nabla \cdot \mathbf{v} = 0 \qquad (9\text{-}3)$$

whether the flow is unsteady or not.

Equation (9-2) may be arranged in a slightly different form to illustrate the use of the *substantial derivative*. Carrying out the differentiation indicated in (9-1), we have

$$\frac{\partial \rho}{\partial t} + v_x \frac{\partial \rho}{\partial x} + v_y \frac{\partial \rho}{\partial y} + v_z \frac{\partial \rho}{\partial z} + \rho\left(\frac{\partial v_x}{\partial x} + \frac{\partial v_y}{\partial y} + \frac{\partial v_z}{\partial z}\right) = 0$$

The first four terms of the above equation comprise the substantial derivative of the density, symbolized as $D\rho/Dt$, where

$$\frac{D}{Dt} = \frac{\partial}{\partial t} + v_x \frac{\partial}{\partial x} + v_y \frac{\partial}{\partial y} + v_z \frac{\partial}{\partial z} \tag{9-4}$$

in Cartesian coordinates. The continuity equation may, thus, be written as

$$\frac{D\rho}{Dt} + \rho\nabla \cdot \mathbf{v} = 0 \tag{9-5}$$

When considering the total differential of a quantity, three different approaches may be taken. If, for instance, we wish to evaluate the change in atmospheric pressure, P, the total differential written in rectangular coordinates is

$$dP = \frac{\partial P}{\partial t}dt + \frac{\partial P}{\partial x}dx + \frac{\partial P}{\partial y}dy + \frac{\partial P}{\partial z}dz$$

where dx, dy, and dz are arbitrary displacements in the x-, y-, and z-directions. The rate of pressure change is obtained by dividing through by dt, giving

$$\frac{dP}{dt} = \frac{\partial P}{\partial t} + \frac{dx}{dt}\frac{\partial P}{\partial x} + \frac{dy}{dt}\frac{\partial P}{\partial y} + \frac{dz}{dt}\frac{\partial P}{\partial z} \tag{9-6}$$

As a first approach, the instrument to measure pressure is located in a weather station, which is, of course, fixed on Earth's surface. Thus, the coefficients dx/dt, dy/dt, dz/dt are all zero, and for a fixed point of observation the total derivative, dP/dt, is equal to the local derivative with respect to time $\partial P/\partial t$.

A second approach involves the pressure-measuring instrument housed in an aircraft, which, at the pilot's discretion, can be made to climb or descend, or fly in any chosen x, y, z directions. In this case, the coefficients dx/dt, dy/dt, dz/dt are the x, y, and z velocities of the aircraft, and they are arbitrarily chosen, bearing only coincidental relationship to the air currents.

The third situation is one in which the pressure indicator is in a balloon that rises, falls, and drifts as influenced by the flow of air in which it is suspended. Here the coefficients dx/dt, dy/dt, dz/dt are those of the *flow* and they may be designated v_x, v_y, and v_z, respectively. This latter situation corresponds to the substantial derivative, and the terms may be grouped as designated below:

$$\frac{dP}{dt} = \frac{DP}{Dt} = \underbrace{\frac{\partial P}{\partial t}}_{\substack{\text{local} \\ \text{rate of} \\ \text{change of} \\ \text{pressure}}} \underbrace{+ v_x \frac{\partial P}{\partial t} + v_y \frac{\partial P}{\partial y} + v_z \frac{\partial P}{\partial z}}_{\substack{\text{rate of change} \\ \text{of pressure} \\ \text{due to motion}}} \tag{9-7}$$

The derivative D/Dt may be interpreted as the time rate of change of a fluid or flow variable along the path of a fluid element. The substantial derivative will be applied to both scalar and vector variables in subsequent sections.

▶ **9.2**

NAVIER–STOKES EQUATIONS

The Navier–Stokes equations are the differential form of Newton's second law of motion. Consider the differential control volume illustrated in Figure 9.1.

The basic tool we shall use in developing the Navier–Stokes equations is Newton's second law of motion for an arbitrary control volume as given in Chapter 5

$$\Sigma \mathbf{F} = \iint_{c.s.} \rho \mathbf{v}(\mathbf{v} \cdot \mathbf{n}) dA + \frac{\partial}{\partial t} \iiint_{c.v.} \rho \mathbf{v} dV \tag{5-4}$$

which states that

$$\left\{ \begin{array}{c} \text{sum of the external} \\ \text{forces acting on the} \\ \text{c.v.} \end{array} \right\} = \left\{ \begin{array}{c} \text{net rate of linear} \\ \text{momentum efflux} \end{array} \right\} + \left\{ \begin{array}{c} \text{time rate of change} \\ \text{of linear momentum} \\ \text{within the c.v.} \end{array} \right\}$$

As the mathematical expression for each of the above terms is rather lengthy, each will be evaluated separately and then substituted into equation (5-4).

The development may be further simplified by recalling that we have, in the prior case, divided by the volume of the control volume and taken the limit as the dimensions approach zero. Equation (5-4) can also be written

$$\lim_{\Delta x, \Delta y, \Delta z \to 0} \frac{\Sigma \mathbf{F}}{\Delta x \, \Delta y \, \Delta z} = \lim_{\Delta x, \Delta y, \Delta z \to 0} \frac{\iint \rho \mathbf{v}(\mathbf{v} \cdot \mathbf{n}) dA}{\Delta x \, \Delta y \, \Delta z} + \lim_{\Delta x, \Delta y, \Delta z \to 0} \frac{\partial/\partial t \iiint \rho \mathbf{v} \, dV}{\Delta x \, \Delta y \, \Delta z}$$
$$\underset{①}{} \qquad \qquad \underset{②}{} \qquad \qquad \underset{③}{}$$

$$\tag{9-8}$$

① ***Sum of the external forces.***   The forces acting on the control volume are those due to the normal stress and to the shear stress, and body forces such as that due to gravity. Figure 9.2 illustrates the various forces acting on the control volume. Summing the forces in the x-direction, we obtain

$$\Sigma F_x = (\sigma_{xx}|_{x+\Delta x} - \sigma_{xx}|_x)\Delta y \, \Delta z + (\tau_{yx}|_{y+\Delta y} - \tau_{yx}|_y)\Delta x \, \Delta z$$
$$+ (\tau_{zx}|_{z+\Delta z} - \tau_{zx}|_z)\Delta x \, \Delta y + g_x \rho \, \Delta x \, \Delta y \, \Delta z$$

where g_x is the component of the gravitational acceleration in the x-direction. In the limit as the dimensions of the element approach zero, this becomes

$$\lim_{\Delta x, \Delta y, \Delta z \to 0} \frac{\Sigma F_x}{\Delta x \, \Delta y \, \Delta z} = \frac{\partial \sigma_{xx}}{\partial x} + \frac{\partial \tau_{yx}}{\partial y} + \frac{\partial \tau_{zx}}{\partial z} + \rho g_x \tag{9-9}$$

Similar expressions are obtained for the force summations in the y- and z-directions:

$$\lim_{\Delta x, \Delta y, \Delta z \to 0} \frac{\Sigma F_y}{\Delta x \, \Delta y \, \Delta z} = \frac{\partial \tau_{xy}}{\partial x} + \frac{\partial \sigma_{yy}}{\partial y} + \frac{\partial \tau_{zy}}{\partial z} + \rho g_y \tag{9-10}$$

$$\lim_{\Delta x, \Delta y, \Delta z \to 0} \frac{\Sigma F_z}{\Delta x \, \Delta y \, \Delta z} = \frac{\partial \tau_{xz}}{\partial x} + \frac{\partial \tau_{yz}}{\partial y} + \frac{\partial \sigma_{zz}}{\partial z} + \rho g_z \tag{9-11}$$

② ***Net momentum flux through the control volume.***   The net momentum flux through the control volume illustrated in Figure 9.3 is

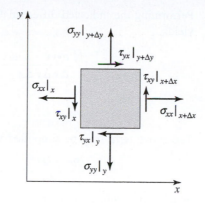

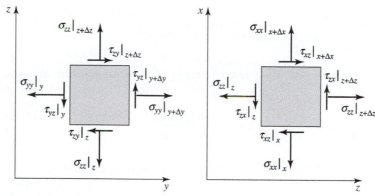

Figure 9.2 Forces acting on a differential control volume.

$$\lim_{\Delta x, \Delta y, \Delta z \to 0} \frac{\iint \rho \mathbf{v}(\mathbf{v} \cdot \mathbf{n}) dA}{\Delta x \, \Delta y \, \Delta z} = \lim_{\Delta x, \Delta y, \Delta z \to 0} \left[\frac{(\rho \mathbf{v} v_x|_{x+\Delta x} - \rho \mathbf{v} v_x|_x) \Delta y \, \Delta z}{\Delta x \, \Delta y \, \Delta z} \right.$$

$$+ \frac{(\rho \mathbf{v} v_y|_{y+\Delta y} - \rho \mathbf{v} v_y|_y) \Delta x \, \Delta z}{\Delta x \, \Delta y \, \Delta z}$$

$$\left. + \frac{(\rho \mathbf{v} v_z|_{z+\Delta z} - \rho \mathbf{v} v_z|_z) \Delta x \, \Delta y}{\Delta x \, \Delta y \, \Delta z} \right]$$

$$= \frac{\partial}{\partial x}(\rho \mathbf{v} v_x) + \frac{\partial}{\partial y}(\rho \mathbf{v} v_y) + \frac{\partial}{\partial z}(\rho \mathbf{v} v_z)$$

(9-12)

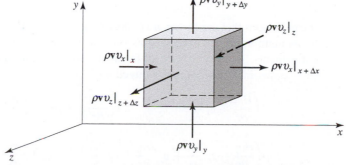

Figure 9.3 Momentum flux through a differential control volume.

Performing the indicated differentiation of the right-hand side of equation (9-12) yields

$$\lim_{\Delta x,\Delta y,\Delta z \to 0} \frac{\iint \rho \mathbf{v}(\mathbf{v} \cdot \mathbf{n})dA}{\Delta x\, \Delta y\, \Delta z} = \mathbf{v}\left[\frac{\partial}{\partial x}(\rho v_x) + \frac{\partial}{\partial y}(\rho v_y) + \frac{\partial}{\partial z}(\rho v_z)\right]$$
$$+ \rho\left[v_x \frac{\partial \mathbf{v}}{\partial x} + v_y \frac{\partial \mathbf{v}}{\partial y} + v_z \frac{\partial \mathbf{v}}{\partial z}\right]$$

The above term may be simplified with the aid of the continuity equation:

$$\frac{\partial \rho}{\partial t} + \frac{\partial}{\partial x}(\rho v_x) + \frac{\partial}{\partial y}(\rho v_y) + \frac{\partial}{\partial z}(\rho v_z) = 0 \qquad (9\text{-}1)$$

which, upon substitution, yields

$$\lim_{\Delta x,\Delta y,\Delta z \to 0} \frac{\iint \rho \mathbf{v}(\mathbf{v} \cdot \mathbf{n})dA}{\Delta x\, \Delta y\, \Delta z} = -\mathbf{v}\frac{\partial \rho}{\partial t} + \rho\left[v_x \frac{\partial \mathbf{v}}{\partial x} + v_y \frac{\partial \mathbf{v}}{\partial y} + v_z \frac{\partial \mathbf{v}}{\partial z}\right] \qquad (9\text{-}13)$$

③ *Time rate of change of momentum within the control volume.* The time rate of change of momentum within the control volume may be evaluated directly:

$$\lim_{\Delta x,\Delta y,\Delta z \to 0} \frac{\partial/\partial t \iiint \mathbf{v}\rho\, dV}{\Delta x\, \Delta y\, \Delta z} = \frac{(\partial/\partial t)\rho \mathbf{v}\, \Delta x\, \Delta y\, \Delta z}{\Delta x\, \Delta y\, \Delta z} = \frac{\partial}{\partial t}\rho \mathbf{v} = \rho\frac{\partial \mathbf{v}}{\partial t} + \mathbf{v}\frac{\partial \rho}{\partial t} \qquad (9\text{-}14)$$

We have now evaluated all terms in equation (9-8):

①

$$\lim_{\Delta x,\, \Delta y,\, \Delta z \to 0} \frac{\Sigma F}{\Delta x\, \Delta y\, \Delta z} = \begin{cases} \left(\dfrac{\partial \sigma_{xx}}{\partial x} + \dfrac{\partial \tau_{yx}}{\partial y} + \dfrac{\partial \tau_{zx}}{\partial z} + \rho g_x\right)\mathbf{e}_x & (9\text{-}9) \\[2mm] \left(\dfrac{\partial \tau_{xy}}{\partial x} + \dfrac{\partial \sigma_{yy}}{\partial y} + \dfrac{\partial \tau_{zy}}{\partial z} + \rho g_y\right)\mathbf{e}_y & (9\text{-}10) \\[2mm] \left(\dfrac{\partial \tau_{xz}}{\partial x} + \dfrac{\partial \tau_{yz}}{\partial y} + \dfrac{\partial \sigma_{zz}}{\partial z} + \rho g_z\right)\mathbf{e}_z & (9\text{-}11) \end{cases}$$

②

$$\lim_{\Delta x,\Delta y,\Delta z \to 0} \frac{\iint \rho \mathbf{v}(\mathbf{v} \cdot \mathbf{n})dA}{\Delta x\, \Delta y\, \Delta z} = -\mathbf{v}\frac{\partial \rho}{\partial t} + \rho\left(v_x \frac{\partial \mathbf{v}}{\partial t} + v_y \frac{\partial \mathbf{v}}{\partial y} + v_z \frac{\partial \mathbf{v}}{\partial z}\right) \qquad (9\text{-}13)$$

③

$$\lim_{\Delta x,\Delta y,\Delta z \to 0} \frac{\partial/\partial t \iiint \rho \mathbf{v} dV}{\Delta x\, \Delta y\, \Delta z} = \rho\frac{\partial \mathbf{v}}{\partial t} + \mathbf{v}\frac{\partial \rho}{\partial t} \qquad (9\text{-}14)$$

It can be seen that the forces are expressed in components, whereas the rate-of-change-of-momentum terms are expressed as vectors. When the momentum terms are expressed as components, we obtain three differential equations that are the statements of Newton's second law in the x-, y-, and z-directions:

$$\rho\left(\frac{\partial v_x}{\partial t} + v_x \frac{\partial v_x}{\partial x} + v_y \frac{\partial v_x}{\partial y} + v_z \frac{\partial v_x}{\partial z}\right) = \rho g_x + \frac{\partial \sigma_{xx}}{\partial x} + \frac{\partial \tau_{yx}}{\partial y} + \frac{\partial \tau_{zx}}{\partial z} \qquad (9\text{-}15a)$$

$$\rho\left(\frac{\partial v_y}{\partial t} + v_x \frac{\partial v_y}{\partial x} + v_y \frac{\partial v_y}{\partial x} + v_z \frac{\partial v_y}{\partial z}\right) = \rho g_y + \frac{\partial \tau_{xy}}{\partial x} + \frac{\partial \sigma_{yy}}{\partial y} + \frac{\partial \tau_{zy}}{\partial z} \qquad (9\text{-}15b)$$

$$\rho\left(\frac{\partial v_z}{\partial t} + v_x\frac{\partial v_z}{\partial x} + v_y\frac{\partial v_z}{\partial x} + v_z\frac{\partial v_z}{\partial z}\right) = \rho g_z + \frac{\partial \tau_{xz}}{\partial x} + \frac{\partial \tau_{yz}}{\partial y} + \frac{\partial \sigma_{zz}}{\partial z} \tag{9-15c}$$

It will be noted that in equations (9-15) above, the terms on the left-hand side represent the time rate of change of momentum, and the terms on the right-hand side represent the forces. Focusing our attention on the left-hand terms in equation (9-15a), we see that

$$\underbrace{\frac{\partial v_x}{\partial t}}_{\substack{\text{local}\\\text{of change}\\\text{of } v_x}} + \underbrace{v_x\frac{\partial v_x}{\partial x} + v_y\frac{\partial v_x}{\partial y} + v_z\frac{\partial v_x}{\partial x}}_{\substack{\text{rate of change in}\\ v_x \text{ due to motion}}} = \left(\frac{\partial}{\partial t} + v_x\frac{\partial}{\partial x} + v_y\frac{\partial}{\partial y} + v_z\frac{\partial}{\partial z}\right)v_x$$

The first term, $\partial v_x/\partial t$, involves the time rate of change of v_x at a point and is called the *local acceleration*. The remaining terms involve the velocity change from point to point—that is, the *convective acceleration*. The sum of these two bracketed terms is the total acceleration. The reader may verify that the terms on the left-hand side of equations (9-15) are all of the form

$$\left(\frac{\partial}{\partial t} + v_x\frac{\partial}{\partial x} + v_y\frac{\partial}{\partial y} + v_z\frac{\partial}{\partial z}\right)v_i$$

where $v_i = v_x$, v_y, or v_z. The above term is the substantial derivative of v_i.

When the substantial derivative notation is used, equations (9-15) become

$$\rho\frac{Dv_x}{Dt} = \rho g_x + \frac{\partial \sigma_{xx}}{\partial x} + \frac{\partial \tau_{yx}}{\partial y} + \frac{\partial \tau_{zx}}{\partial z} \tag{9-16a}$$

$$\rho\frac{Dv_y}{Dt} = \rho g_y + \frac{\partial \tau_{xy}}{\partial x} + \frac{\partial \sigma_{yy}}{\partial y} + \frac{\partial \tau_{zy}}{\partial z} \tag{9-16b}$$

and

$$\rho\frac{Dv_z}{Dt} = \rho g_z + \frac{\partial \tau_{xz}}{\partial x} + \frac{\partial \tau_{yz}}{\partial y} + \frac{\partial \sigma_{zz}}{\partial z} \tag{9-16c}$$

Equations (9-16) are valid for any type of fluid, regardless of the nature of the stress rate-of-strain relation. If Stokes's viscosity relations, equations (7-15) and (7-16), are used for the stress components, equations (9-16) become

$$\rho\frac{Dv_x}{Dt} = \rho g_x - \frac{\partial P}{\partial x} - \frac{\partial}{\partial x}\left(\frac{2}{3}\mu\nabla\cdot\mathbf{v}\right) + \nabla\cdot\left(\mu\frac{\partial \mathbf{v}}{\partial x}\right) + \nabla\cdot(\mu\nabla v_x) \tag{9-17a}$$

$$\rho\frac{Dv_y}{Dt} = \rho g_y - \frac{\partial P}{\partial x} - \frac{\partial}{\partial y}\left(\frac{2}{3}\mu\nabla\cdot\mathbf{v}\right) + \nabla\cdot\left(\mu\frac{\partial \mathbf{v}}{\partial y}\right) + \nabla\cdot(\mu\nabla v_y) \tag{9-17b}$$

and

$$\rho\frac{Dv_z}{Dt} = \rho g_z - \frac{\partial P}{\partial z} - \frac{\partial}{\partial z}\left(\frac{2}{3}\mu\nabla\cdot\mathbf{v}\right) + \nabla\cdot\left(\mu\frac{\partial \mathbf{v}}{\partial z}\right) + \nabla\cdot(\mu\nabla v_z) \tag{9-17c}$$

The above equations are called the Navier–Stokes[1] equations and are the differential expressions of Newton's second law of motion for a Newtonian fluid. As no assumptions relating to the compressibility of the fluid have been made, these equations are valid for both compressible and incompressible flows. In our study of momentum transfer we shall restrict our attention to incompressible flow with constant viscosity. In an incompressible flow, $\nabla \cdot v = 0$. Equations (9-17) thus become

$$\rho \frac{Dv_x}{Dt} = \rho g_x - \frac{\partial P}{\partial x} + \mu \left(\frac{\partial^2 v_x}{\partial x^2} + \frac{\partial^2 v_x}{\partial y^2} + \frac{\partial^2 v_x}{\partial z^2} \right) \tag{9-18a}$$

$$\rho \frac{Dv_y}{Dt} = \rho g_y - \frac{\partial P}{\partial y} + \mu \left(\frac{\partial^2 v_y}{\partial x^2} + \frac{\partial^2 v_y}{\partial y^2} + \frac{\partial^2 v_y}{\partial z^2} \right) \tag{9-18b}$$

$$\rho \frac{Dv_z}{Dt} = \rho g_z - \frac{\partial P}{\partial z} + \mu \left(\frac{\partial^2 v_z}{\partial x^2} + \frac{\partial^2 v_z}{\partial y^2} + \frac{\partial^2 v_z}{\partial z^2} \right) \tag{9-18c}$$

These equations may be expressed in a more compact form in the single vector equation:

$$\rho \frac{D\mathbf{v}}{Dt} = \rho \mathbf{g} - \nabla P + \mu \nabla^2 \mathbf{v} \tag{9-19}$$

The above equation is the Navier–Stokes equation for an incompressible flow. The Navier–Stokes equations are written in cartesian, cylindrical, and spherical coordinate forms in Appendix E. As the development has been lengthy, let us review the assumptions and, therefore, the limitations of equation (9-19). The assumptions are

1. incompressible flow
2. constant viscosity
3. laminar flow[2]

All of the above assumptions are associated with the use of the Stokes viscosity relation. If the flow is inviscid ($\mu = 0$), the Navier–Stokes equation becomes

$$\rho \frac{D\mathbf{v}}{Dt} = \rho \, \mathbf{g} - \nabla P \tag{9-20}$$

which is known as Euler's equation. Euler's equation has only one limitation, that being inviscid flow.

Example 1

Equation (9-19) may be applied to numerous flow systems to provide information regarding velocity variation, pressure gradients, and other information of the type achieved in Chapter 8. Many situations are of sufficient complexity to make the solution extremely difficult and are beyond the scope of this text. A situation for which a solution can be obtained is illustrated in Figure 9.4.

---

[1] L. M. H. Navier, Mémoire sur les Lois du Mouvements des Fluides, *Mem. de l'Acad. d. Sci.*, **6**, 398 (1822); C. G. Stokes, On the Theories of the Internal Friction of Fluids in Motion, *Trans. Cambridge Phys. Soc.*, **8** (1845).
[2] Strictly speaking, equation (9-19) is valid for turbulent flow, as the turbulent stress is included in the momentum flux term. This will be illustrated in Chapter 12.

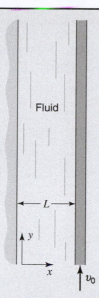

Figure 9.4 Fluid between two vertical plates with the one on the left stationary and the other on the right moving vertically upward with velocity v_0.

Figure 9.4 shows the situation of an incompressible fluid confined between two parallel, vertical surfaces. One surface, shown to the left, is stationary, whereas the other is moving upward at a constant velocity v_0. If we consider the fluid Newtonian and the flow laminar, the governing equation of motion is the Navier–Stokes equation in the form given by equation (9-19). The reduction of each term in the vector equation into its applicable form is shown below:

$$\rho \frac{D\mathbf{v}}{Dt} = \rho \left\{ \frac{\partial \mathbf{v}}{\partial t} + v_x \frac{\partial \mathbf{v}}{\partial x} + v_y \frac{\partial \mathbf{v}}{\partial y} + v_z \frac{\partial \mathbf{v}}{\partial z} \right\} = 0$$

$$\rho \mathbf{g} = -\rho\, g \mathbf{e}_y$$

$$\nabla P = \frac{dP}{dy} \mathbf{e}_y$$

where dP/dy is constant, and

$$\mu \nabla^2 \mathbf{v} = \mu \frac{d^2 v_y}{dx^2} \mathbf{e}_y$$

The resulting equation to be solved is

$$0 = -\rho g - \frac{dP}{dy} + \mu \frac{d^2 v_y}{dx^2}$$

This differential equation is separable. The first integration yields

$$\frac{dv_y}{dx} + \frac{x}{\mu} \left\{ -\rho g - \frac{dP}{dy} \right\} = C_1$$

Integrating once more, we obtain

$$v_y + \frac{x^2}{2\mu} \left\{ -\rho g - \frac{dP}{dy} \right\} = C_1 x + C_2$$

The integration constants may be evaluated, using the boundary conditions that $v_y = 0$ at $x = 0$, and $v_y = v_0$ at $x = L$. The constants thus become

$$C_1 = \frac{v_0}{L} + \frac{L}{2\mu} \left\{ -\rho g - \frac{dP}{dy} \right\} \quad \text{and} \quad C_2 = 0$$

The velocity profile may now be expressed as

$$v_y = \underbrace{\frac{1}{2\mu} \left\{ -\rho g - \frac{dP}{dy} \right\} \{ Lx - x^2 \}}_{①} + \underbrace{v_0 \frac{x}{L}}_{②} \qquad (9\text{-}21)$$

It is interesting to note, in equation (9-21), the effect of the terms labeled ① and ②, which are added. The first term is the equation for a symmetric parabola, the second for a straight line. Equation (9-21) is valid whether v_0 is upward, downward, or zero. In each case, the terms may be added to yield the complete velocity profile. These results are indicated in Figure 9.5. The resulting velocity profile obtained by superposing the two parts is shown in each case.

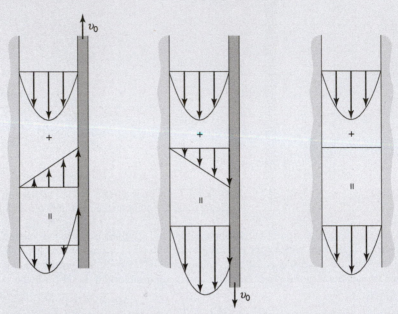

Figure 9.5 Velocity profiles for one surface moving upward, downward, or stationary.

Euler's equation may also be solved to determine velocity profiles, as will be shown in Chapter 10. The vector properties of Euler's equation are illustrated by the example below, in which the form of the velocity profile is given.

Example 2

A rotating shaft, as illustrated in Figure 9.6, causes the fluid to move in circular streamlines with a velocity that is inversely proportional to the distance from the shaft. Find the shape of the free surface if the fluid can be considered inviscid.

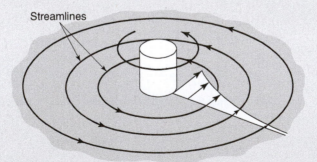

Streamlines

Figure 9.6 Rotating shaft in a fluid.

As the pressure along the free surface will be constant, we may observe that the free surface is perpendicular to the pressure gradient. Determination of the pressure gradient, therefore, will enable us to evaluate the slope of the free surface.

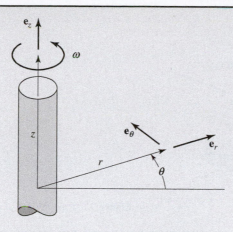

Figure 9.7 Cylindrical coordinate system for rotating shaft and fluid.

Rearranging equation (9-20), we have

$$\nabla P = \rho \mathbf{g} - \rho \frac{D\mathbf{v}}{Dt} \tag{9-20}$$

The velocity $\mathbf{v} = A\mathbf{e}_\theta / r$, where A is a constant, when using the coordinate system shown in Figure 9.7. Assuming that there is no slip between the fluid and the shaft at the surface of the shaft, we have

$$v(R) = \omega R = \frac{A}{R}$$

and thus $A = \omega R^2$ and

$$\mathbf{v} = \frac{\omega R^2}{r} \mathbf{e}_\theta$$

The substantial derivative $D\mathbf{v}/Dt$ may be evaluated by taking the total derivative

$$\frac{d\mathbf{v}}{dt} = -\frac{\omega R^2}{r^2} \mathbf{e}_\theta \dot{r} + \frac{\omega R^2}{r} \frac{d\mathbf{e}_\theta}{dt}$$

where $d\mathbf{e}_\theta / dt = -\dot{\theta} \mathbf{e}_r$. The total derivative becomes

$$\frac{d\mathbf{v}}{dt} = -\frac{\omega R^2}{r^2} \dot{r}\mathbf{e}_\theta - \frac{\omega R^2}{r} \dot{\theta}\mathbf{e}_r$$

Now the fluid velocity in the r-direction is zero, and $\dot{\theta}$ for the fluid is v/r, so

$$\left(\frac{d\mathbf{v}}{dt}\right)_{\text{fluid}} = \frac{D\mathbf{v}}{Dt} = -\frac{\omega R^2}{r^2} v\mathbf{e}_r = -\frac{\omega^2 R^4}{r^3} \mathbf{e}_r$$

This result could have been obtained in a more direct manner by observing that $D\mathbf{v}/Dt$ is the local fluid acceleration, which for this case is $-v^2 \mathbf{e}_r / r$. The pressure gradient becomes

$$\nabla P = -\rho g \mathbf{e}_z + \rho \frac{\omega^2 R^4 \mathbf{e}_r}{r^3}$$

From Figure 9.8, it can be seen that the free surface makes an angle β with the r axis so that

$$\tan \beta = \frac{\rho \omega^2 R^4}{r^3 \rho g}$$

$$= \frac{\omega^2 R^4}{g r^3}$$

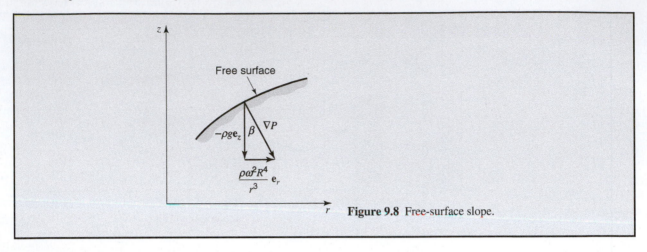

Figure 9.8 Free-surface slope.

▶ **9.3**

BERNOULLI'S EQUATION

Euler's equation may be integrated directly for a particular case, flow along a streamline. In integrating Euler's equation, the use of streamline coordinates is extremely helpful. Streamline coordinates s and n are illustrated in Figure 9.9. The s-direction is parallel to the streamline and the n direction is perpendicular to the streamline, directed away from the instantaneous center of curvature. The flow and fluid properties are functions of position and time. Thus, $\mathbf{v} = \mathbf{v}(s, n, t)$, and $P = P(s, n, t)$. The substantial derivatives of the velocity and pressure gradients in equation (9-20) must be expressed in terms of streamline coordinates so that equation (9-20) may be integrated.

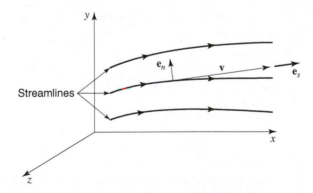

Figure 9.9 Streamline coordinates.

Following the form used in equations (9-6) to obtain the substantial derivative, we have

$$\frac{d\mathbf{v}}{dt} = \frac{d\mathbf{v}}{dt} + \dot{s}\frac{\partial \mathbf{v}}{\partial s} + \dot{n}\frac{\partial \mathbf{v}}{\partial n}$$

As the velocity of the fluid element has components $\dot{s} = v$, $\dot{n} = 0$, the substantial derivative of the velocity in streamline coordinates is

$$\frac{D\mathbf{v}}{Dt} = \frac{\partial \mathbf{v}}{\partial t} + v\frac{\partial \mathbf{v}}{\partial s} \tag{9-22}$$

The pressure gradient in streamline coordinates may be written as

$$\nabla P = \frac{\partial P}{\partial s}\mathbf{e}_s + \frac{\partial P}{\partial n}\mathbf{e}_n \tag{9-23}$$

Taking the dot product of equation (9-20) with $\mathbf{e}_s\,ds$, and using equations (9-22) and (9-23), we obtain

$$\rho\left(\frac{\partial \mathbf{v}}{\partial t}\cdot\mathbf{e}_s\,ds + v\frac{\partial \mathbf{v}}{\partial s}\cdot\mathbf{e}_s\,ds\right) = \rho\mathbf{g}\cdot\mathbf{e}_s\,ds - \left(\frac{\partial P}{\partial s}\mathbf{e}_s + \frac{\partial P}{\partial n}\mathbf{e}_n\right)\cdot\mathbf{e}_s\,ds$$

or, as $\partial\mathbf{v}/\partial s\cdot\mathbf{e}_s = \partial/\partial s(\mathbf{v}\cdot\mathbf{e}_s) = \partial v/\partial s$, we have

$$\rho\left(\frac{\partial \mathbf{v}}{\partial t}\cdot\mathbf{e}_s\,ds + \frac{\partial}{\partial s}\left\{\frac{v^2}{2}\right\}ds\right) = \rho\mathbf{g}\cdot\mathbf{e}_s\,ds - \frac{\partial P}{\partial n}\,ds \tag{9-24}$$

Selecting $\mathbf{g}$ to act in the $-\mathbf{y}$ direction, we have $\mathbf{g}\cdot\mathbf{e}_s\,ds = -g\,dy$. For *steady incompressible flow*, equation (9-24) may be integrated to yield

$$\frac{v^2}{2} + gy + \frac{P}{\rho} = \text{constant} \tag{9-25}$$

which is known as Bernoulli's equation. The limitations are

1. inviscid flow
2. steady flow
3. incompressible flow
4. the equation applies along a streamline

Limitation 4 will be relaxed for certain conditions to be investigated in Chapter 10.

Bernoulli's equation was also developed in Chapter 6 from energy considerations for steady incompressible flow with constant internal energy. It is interesting to note that the constant internal energy assumption and the inviscid flow assumption must be equivalent, as the other assumptions were the same. We may note, therefore, that the viscosity in some way will effect a change in internal energy.

▶ **9.4**

SPHERICAL COORDINATE FORMS OF THE NAVIER–STOKES EQUATIONS[3]

Examples 1 and 2 gave examples of the use of the Navier–Stokes equations in rectangular and cylindrical coordinates, respectively. The use of spherical coordinates is significantly more complex, but is extremely useful.

Consider a solid sphere of radius R rotating in a large body of stagnant fluid as shown in Figure 9.10. This is an example of "creeping flow" and will be revisited in Chapter 12. The sphere rotates steadily about its vertical axis with an angular velocity Ω (rad/s) in an infinite Newtonian liquid of viscosity μ. Our goal is to calculate the velocity in the ϕ-direction. To begin the problem, we need to use the Navier–Stokes equation in spherical coordinates.

---

[3] R. B. Bird, W. E. Stewart, and E. N. Lightfoot, *Transport Phenomena*, Wiley, New York, 2007; W. Deen, *Analysis of Transport Phenomena*, Oxford University Press, 2012.

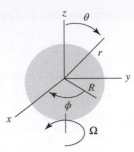

Figure 9.10 A sphere rotating in a stagnant fluid.

The r and θ components of the Navier–Stokes equations are both zero, so we only require the ϕ-direction equation (from Appendix E):

$$\rho\left(\frac{\partial v_\phi}{\partial t} + v_r\frac{\partial v_\phi}{\partial r} + \frac{v_\theta}{r}\frac{\partial v_\phi}{\partial \theta} + \frac{v_\phi}{r\sin(\theta)}\frac{\partial v_\phi}{\partial \phi} + \frac{v_\phi v_r}{r} + \frac{v_\phi v_\theta}{r}\cot(\theta)\right)$$

$$= -\frac{1}{r\sin(\theta)}\frac{\partial P}{\partial \phi} + \rho g_\phi$$

$$+ \mu\left(\frac{1}{r^2}\frac{\partial}{\partial r}\left(r^2\frac{\partial v_\phi}{\partial r}\right) + \frac{1}{r^2\sin\theta}\frac{\partial}{\partial \theta}\left(\sin\theta\frac{\partial v_\phi}{\partial \theta}\right) + \frac{1}{r^2\sin^2\theta}\frac{\partial^2 v_\phi}{\partial \phi^2}\right.$$

$$\left. - \frac{v_\phi}{r^2\sin^2(\theta)} + \frac{2}{r^2\sin(\theta)}\frac{\partial v_r}{\partial \phi} + \frac{2\cos(\theta)}{r^2\sin^2(\theta)}\frac{\partial v_\theta}{\partial \phi}\right)$$

In this analysis we will neglect gravity and pressure, and assume steady, fully developed one-dimensional flow and that there is no velocity in the r and θ directions. We can also assume symmetry about the z-axis (see Appendix B, Figure B.3), which allows simplification by removing all the $\frac{\partial}{\partial \phi}$ terms indicating that there is no dependence on the angle ϕ. The resulting equation is

$$0 = \mu\left(\frac{1}{r^2}\frac{\partial}{\partial r}\left(r^2\frac{\partial v_\phi}{\partial r}\right) + \frac{1}{r^2\sin\theta}\frac{\partial}{\partial \theta}\left(\sin\theta\frac{\partial v_\phi}{\partial \theta}\right) - \frac{v_\phi}{r^2\sin^2(\theta)}\right)$$

This equation is a partial differential equation for v_ϕ, where v_ϕ is a function of r and θ and can be solved using an equation solver or handbook[4] by realizing that it can be rewritten with

$$v_\phi = f(r)\sin\theta$$

as an ordinary differential equation in the form

$$0 = \frac{d}{dr}\left(r^2\frac{df}{dr}\right) - 2f$$

This is an nth-order equidimensional equation (also called a Cauchy-type equation) with a solution of the form $f = r^n$ with $n = 1$ and -2 resulting in a general solution of the form:

$$f(r) = C_1 r + \frac{C_2}{r^2}$$

---

[4] M. Abramowitz and I. A. Stegun, *Handbook of Mathematical Functions*, Dover Publications, 1972, p. 17.

where C_1 and C_2 are integration constants. The resulting equation for v_ϕ is

$$v_\phi = C_1 \, r \sin \theta + \frac{C_2}{r^2} \sin \theta$$

To solve this equation we need two boundary conditions. From Figure 9.10, we obtain

$$\text{B.C. 1} : v_\phi = R \Omega \sin \theta \text{ at } r = R$$

$$\text{B.C. 2} : v_\phi = 0 \text{ at } r = \infty$$

The first boundary condition states that as the sphere rotates at an angular velocity of Ω, and as a result of the no-slip boundary condition, the fluid at the surface is at the same velocity as the surface itself, which is equal to $R \Omega \sin \theta$. The second boundary condition states that far away from the sphere, the fluid is at rest and is not affected by the rotation of the sphere.

Applying B.C. 2 results in $C_1 = 0$, and applying B.C. 1, gives

$$R \Omega = 0 + \frac{C_2}{R^2}$$

$$C_2 = R^3 \Omega$$

This results in the desired velocity equation:

$$v_\phi = \frac{R^3 \Omega}{r^2} \sin \theta$$

▶ **9.5**

CLOSURE

We have developed the differential equations for the conservation of mass and Newton's second law of motion. These equations may be subdivided into two special groups:

$$\frac{\partial \rho}{\partial t} + \nabla \cdot \rho \mathbf{v} = 0 \tag{9-26}$$

(continuity equation)

Inviscid flow

$$\rho \frac{D\mathbf{v}}{Dt} = \rho \mathbf{g} - \nabla P \tag{9-27}$$

(Euler's equation)

Incompressible, viscous flow

$$\nabla \cdot \mathbf{v} = 0 \tag{9-28}$$

(continuity equation)

$$\rho \frac{D\mathbf{v}}{Dt} = \rho \mathbf{g} - \nabla P + \mu \nabla^2 \mathbf{v} \tag{9-29}$$

(Navier–Stokes equation for incompressible flow)

In addition, the student should note the physical meaning of the substantial derivative and appreciate the compactness of the vector representation. In component form, for example, equation (9-29) comprises some 27 terms in Cartesian coordinates.

CHAPTER 10

Inviscid Fluid Flow

$\mathbf{A}$n important area in momentum transfer is inviscid flow, in which, by virtue of the absence of shear stress, analytical solutions to the differential equations of fluid flow are possible.

The subject of inviscid flow has particular application in aerodynamics and hydrodynamics and general application to flow about bodies—the so-called external flows. In this chapter, we shall introduce the fundamentals of inviscid flow analysis.

► **10.1**

FLUID ROTATION AT A POINT

Consider the element of fluid shown in Figure 10.1a. In time Δt the element will move in the xy plane as shown. In addition to translation, the element may also deform and rotate. We have discussed the deformation previously in Chapter 7. Now let us focus our attention on the rotation of the element. Although the element may deform, the orientation will be given by the average rotation of the line segments OB and OA or by denoting the rotation by

$$\omega_z = \frac{d}{dt}\left(\frac{\alpha + \beta}{2}\right)$$

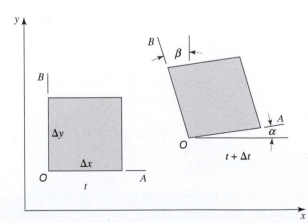

Figure 10.1a Rotation of a fluid element.

where the counterclockwise sense is positive. From Figure 10.1a, we see that

$$\omega_z = \lim_{\Delta x, \Delta y, \Delta z, \Delta t \to 0} \frac{1}{2} \left(\frac{\arctan\left\{[(v_y|_{x+\Delta x} - v_y|_x)\Delta t]/\Delta x\right\}}{\Delta t} \right.$$
$$\left. + \frac{\arctan\{-[(v_x|_{y+\Delta y} - v_x|_y)\Delta t]/\Delta y\}}{\Delta t} \right)$$

which becomes, in the limit,

$$\omega_z = \frac{1}{2}\left(\frac{\partial v_y}{\partial x} - \frac{\partial v_x}{\partial y}\right) \qquad (10\text{-}1)$$

The subscript z indicates that the rotation is about the z-axis.

In the xz and yz planes the rotation at a point is given by

$$\omega_y = \frac{1}{2}\left(\frac{\partial v_x}{\partial z} - \frac{\partial v_z}{\partial x}\right) \qquad (10\text{-}2)$$

and

$$\omega_x = \frac{1}{2}\left(\frac{\partial v_z}{\partial y} - \frac{\partial v_y}{\partial z}\right) \qquad (10\text{-}3)$$

The rotation at a point is related to the vector cross product of the velocity. As the student may verify

$$\nabla \times \mathbf{v} = \left(\frac{\partial v_z}{\partial y} - \frac{\partial v_y}{\partial z}\right)\mathbf{e}_x + \left(\frac{\partial v_x}{\partial z} - \frac{\partial v_z}{\partial x}\right)\mathbf{e}_y + \left(\frac{\partial v_y}{\partial x} - \frac{\partial v_x}{\partial y}\right)\mathbf{e}_z$$

and thus

$$\nabla \times \mathbf{v} = 2\omega \qquad (10\text{-}4)$$

The vector $\nabla \times \mathbf{v}$ is also known as the *vorticity*. When the rotation at a point is zero the flow is said to be *irrotational*. For irrotational flow $\nabla \times \mathbf{v} = 0$, as can be seen from equation (10-4). The significance of fluid rotation at a point may be examined by a different approach. The Navier–Stokes equation for incompressible flow, equation (9-29), may also be written in the form

$$\rho \frac{D\mathbf{v}}{Dt} = -\nabla P + \rho\mathbf{g} - \mu[\nabla \times (\nabla \times \mathbf{v})] \qquad (9\text{-}29)$$

It may be observed from the above equation that if viscous forces act on a fluid, the flow must be rotational.

Figure 10.1b illustrates the effect of vorticity, and whether the fluid element is rotational or irrotational. The vorticity is often described as the measure of the moment of momentum of a small differential area about its own center of mass. The vorticity vector is characterized

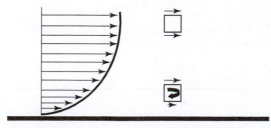

Figure 10.1b Rotation of a fluid element as the result of a velocity gradient.

mathematically as the curl of the velocity vector. In Figure 10.1b, a velocity gradient is moving along a flat plate. Near the plate, as a result of the no-slip boundary condition, the velocity of the fluid is zero at that wall and increases into the bulk fluid. Near the wall, a small differential area of fluid is acted upon by the velocity gradient such that at the bottom of the fluid element, the velocity will be less than at the top of the element resulting in a net rotation or spin of the element in the clockwise direction as shown by the arrow in the center of the differential area. The result of this is the rotation of the fluid element and rotational flow. Out in the bulk, the velocity is the same at the top and bottom of the fluid element, resulting in the differential element having no net rotation—thus the term irrotational flow. Vorticity has many applications in weather patterns, such as in tropical cyclones[1] and in flow of blood in the body.

The kinematic condition $\nabla \times \mathbf{v} = 0$ is not the first time we have encountered a kinematic relation that satisfies one of the fundamental physical laws of fluid mechanics. The law of conservation of mass for an incompressible flow, $\nabla \cdot \mathbf{v} = 0$, is also expressed as a kinematic relation. The use of this relation is the subject of the next section.

Example 1

A two-dimensional velocity is given by the equation

$$v = (6y)e_x + (6x)e_y$$

Is the flow rotational, or irrotational?

To be irrotational, the flow must satisfy the equation $\nabla \times \mathbf{v} = 0$, and as a result, equation (10-1) must be equal to zero.

$$\omega_z = \frac{1}{2}\left(\frac{\partial v_y}{\partial x} - \frac{\partial v_x}{\partial y}\right) = 0$$

First, we must find values for $\partial v_y/\partial x$ and $\partial v_x/\partial y$ to see if the condition of equation (10-1) is satisfied.

From the given equation, we find the velocity components:

$$v_x = 6y \quad \text{and} \quad v_y = 6x$$

Next we take the necessary partial derivatives dictated by equation (10-1):

$$\frac{\partial v_x}{\partial y} = 6$$

$$\frac{\partial v_y}{\partial x} = 6$$

And thus, plugging in the values,

$$\omega_z = \frac{1}{2}(6 - 6) = 0$$

So the condition of irrotational flow is satisfied, and this particle will not rotate.

Let's also examine another different, but similar, condition. A two-dimensional velocity is given by the equation

$$v = (6y)e_x - (6x)e_y$$

Is this flow rotational, or irrotational? We must again see if equation (10-1) is satisfied.

Finding values for $\partial v_y/\partial x$ and $\partial v_x/\partial y$ as we did previously,

$$v_x = 6y \quad \text{and} \quad v_y = -6x$$

---

[1] J. C. L. Chan, *Ann. Rev. Fluid Mech.*, **37**, 99 (2005).

Next we take the necessary partial derivatives dictated by equation (10-1):

$$\frac{\partial v_x}{\partial y} = 6$$

$$\frac{\partial v_y}{\partial x} = -6$$

giving

$$\omega_z = \frac{1}{2}(6 - (-6)) = 6 \neq 0$$

So the condition is not satisfied, and this is an example of rotational flow.

▶ **10.2**

THE STREAM FUNCTION

For a two-dimensional incompressible flow, the continuity equation is

$$\nabla \cdot \mathbf{v} = \frac{\partial v_x}{\partial x} + \frac{\partial v_y}{\partial y} = 0 \qquad (9\text{-}3)$$

Equation (9-3) indicates that v_x and v_y are related in some way so that $\partial v_x/\partial x = -(\partial v_y/\partial y)$. Perhaps the easiest way to express this relation is by having v_x and v_y both related to the same function. Consider the function $F(x, y)$; if $v_x = F(x, y)$, then

$$\frac{\partial v_y}{\partial y} = -\frac{\partial F}{\partial x} \quad \text{or} \quad v_y = -\int \frac{\partial F}{\partial x} dy$$

Unfortunately, the selection of $v_x = F(x, y)$ results in an integral for v_y. We can easily remove the integral sign if we make the original $F(x, y)$ equal to the derivative of some function with respect to y. For example, if $F(x, y) = (\partial \Psi(x, y)/\partial y]$, then

$$v_x = \frac{\partial \Psi}{\partial y}$$

As $\partial v_x/\partial x = -(\partial v_y/\partial y)$, we may write

$$\frac{\partial v_y}{\partial y} = -\frac{\partial}{\partial x}\left(\frac{\partial \Psi}{\partial y}\right) \quad \text{or} \quad \frac{\partial}{\partial y}\left(v_y + \frac{\partial \Psi}{\partial x}\right) = 0$$

for this to be true in general:

$$v_y = -\frac{\partial \Psi}{\partial x}$$

Instead of having two unknowns, v_x and v_y, we now have only one unknown, Ψ. The unknown, Ψ, is called the *stream function*. The physical significance of Ψ can be seen from the following considerations. As $\Psi = \Psi(x, y)$, the total derivative is

$$d\Psi = \frac{\partial \Psi}{\partial x} dx + \frac{\partial \Psi}{\partial y} dy$$

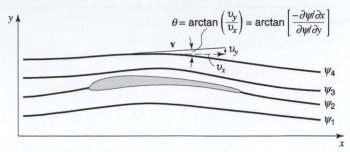

Figure 10.2 Streamlines and the stream function.

Also

$$\frac{\partial \Psi}{\partial x} = -v_y \qquad \text{and} \qquad \frac{\partial \Psi}{\partial y} = v_x$$

and thus

$$d\Psi = -v_y dx + v_x dy \tag{10-5}$$

Consider a path in the xy plane such that $\Psi = $ constant. Along this path, $d\Psi = 0$, and thus equation (10-5) becomes

$$\left. \frac{dy}{dx} \right|_{\Psi = \text{constant}} = \frac{v_y}{v_x} \tag{10-6}$$

The slope of the path $\Psi = $ constant is seen to be the same as the slope of a streamline as discussed in Chapter 3. The function $\Psi(x, y)$ thus represents the streamlines. Figure 10.2 illustrates the streamlines and velocity components for flow about an airfoil.

The differential equation that governs Ψ is obtained by consideration of the fluid rotation, ω, at a point. In a two-dimensional flow, $\omega_z = \frac{1}{2}[(\partial v_y/\partial x) - (\partial v_x/\partial y)]$, and thus, if the velocity components v_y and v_x are expressed in terms of the stream function Ψ, we obtain, for an incompressible, steady flow,

$$-2\omega_z = \frac{\partial^2 \Psi}{\partial x^2} + \frac{\partial^2 \Psi}{\partial y^2} \tag{10-7}$$

When the flow is irrotational, equation (10-7) becomes Laplace's equation:

$$\nabla^2 \Psi = \frac{\partial^2 \Psi}{\partial x^2} + \frac{\partial^2 \Psi}{\partial y^2} = 0 \tag{10-8}$$

Example 2

The stream function for a particular flow is given by the equation $\Psi = 6x^2 - 6y^2$. We wish to determine the velocity components for this flow, and find out whether the flow is rotational or irrotational.

We defined the stream function as

$$\frac{\partial \Psi}{\partial x} = -v_y$$

and

$$\frac{\partial \Psi}{\partial y} = v_x$$

Thus,

$$v_x = \frac{\partial \Psi}{\partial y} = -12y$$

$$v_y = -\frac{\partial \Psi}{\partial x} = -12x$$

The equations for the velocity components are $v_x = -12y$ and $v_y = -12x$.

Next, we want to determine whether the flow is rotational or irrotational. To do this, we must satisfy equation (10-1).

$$\omega_z = \frac{1}{2}\left(\frac{\partial v_y}{\partial x} - \frac{\partial v_x}{\partial y}\right) = 0$$

Solving for the necessary partial derivatives,

$$\frac{\partial v_y}{\partial x} = -12$$

$$\frac{\partial v_x}{\partial y} = -12$$

Thus,

$$\omega_z = \frac{1}{2}(-12 - (-12)) = 0$$

and this flow is seen to be irrotational.

► **10.3**

INVISCID, IRROTATIONAL FLOW ABOUT AN INFINITE CYLINDER

In order to illustrate the use of the stream function, the inviscid, irrotational flow pattern about a cylinder of infinite length will be examined. The physical situation is illustrated in Figure 10.3. A stationary circular cylinder of radius a is situated in uniform, parallel flow in the x-direction.

As there is cylindrical symmetry, polar coordinates are employed. In polar coordinates,[2] equation (10-8) becomes

$$\frac{\partial^2 \Psi}{\partial r^2} + \frac{1}{r}\frac{\partial \Psi}{\partial r} + \frac{1}{r^2}\frac{\partial^2 \Psi}{\partial \theta^2} = 0 \tag{10-9}$$

where the velocity components v_r and v_θ are given by

$$v_r = \frac{1}{r}\frac{\partial \Psi}{\partial \theta} \qquad v_\theta = -\frac{\partial \Psi}{\partial r} \tag{10-10}$$

---

[2] The operator ∇^2 in cylindrical coordinates is developed in Appendix A.

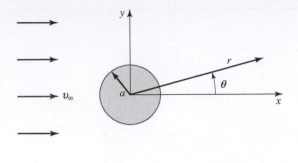

Figure 10.3 Cylinder in a uniform flow.

The solution for this case must meet four boundary conditions. These are as follows:

1. The circle $r = a$ must be a streamline. As the velocity normal to a streamline is zero, $v_r|_{r=a} = 0$ or $\partial \Psi / \partial \theta|_{r=a} = 0$.
2. From symmetry, the line $\theta = 0$ must also be a streamline. Hence, $v_\theta|_{\theta=0} = 0$ or $\partial \Psi / \partial r|_{\theta=0} = 0$.
3. As $r \to \infty$ the velocity must be finite.
4. The magnitude of the velocity as $r \to \infty$ is v_∞, a constant.

The solution to equation (10-9) for this case is

$$\Psi(r, \theta) = v_\infty r \sin \theta \left[1 - \frac{a^2}{r^2}\right] \tag{10-11}$$

The velocity components v_r and v_θ are obtained from equation (10-10):

$$v_r = \frac{1}{r}\frac{\partial \Psi}{\partial \theta} = v_\infty \cos \theta \left[1 - \frac{a^2}{r^2}\right] \tag{10-12}$$

and

$$v_\theta = -\frac{\partial \Psi}{\partial r} = -v_\infty \sin \theta \left[1 + \frac{a^2}{r^2}\right] \tag{10-13}$$

By setting $r = a$ in the above equations, the velocity at the surface of the cylinder may be determined. This results in

$$v_r = 0$$

and

$$v_\theta = -2v_\infty \sin \theta \tag{10-14}$$

The velocity in the radial direction is, of course, zero, as the cylinder surface is a streamline. The velocity along the surface is seen to be zero at $\theta = 0$ and $\theta = 180°$. These points of zero velocity are known as *stagnation points*. The forward stagnation point is at $\theta = 180°$, and the aft or rearward stagnation point is at $\theta = 0°$. The student may verify that each of the boundary conditions for this case are satisfied.

IRROTATIONAL FLOW, THE VELOCITY POTENTIAL

In a two-dimensional irrotational flow $\nabla \times \mathbf{v} = 0$, and thus $\partial v_x/\partial y = \partial v_y/\partial x$. The similarity of this equation to the continuity equation suggests that the type of relation used to obtain the stream function may be used again. Note, however, that the order of differentiation is reversed from the continuity equation. If we let $v_x = \partial \phi(x, y)/\partial x$, we observe that

$$\frac{\partial v_x}{\partial y} = \frac{\partial^2 \phi}{\partial x \partial y} = \frac{\partial v_y}{\partial x}$$

or

$$\frac{\partial}{\partial x}\left(\frac{\partial \phi}{\partial y} - v_y\right) = 0$$

and for the general case

$$v_y = \frac{\partial \phi}{\partial y}$$

The function ϕ is called the *velocity potential*. In order for ϕ to exist, the flow must be irrotational. As the condition of irrotationality is the only condition required, the velocity potential can also exist for compressible, unsteady flows. The velocity potential is commonly used in compressible flow analysis. Additionally, the velocity potential, ϕ, exists for three-dimensional flows, whereas the stream function does not.

The velocity vector is given by

$$\mathbf{v} = v_x\mathbf{e}_x + v_y\mathbf{e}_y + v_z\mathbf{e}_z = \frac{\partial \phi}{\partial x}\mathbf{e}_x + \frac{\partial \phi}{\partial y}\mathbf{e}_y + \frac{\partial \phi}{\partial z}\mathbf{e}_z$$

and thus, in vector notation,

$$\mathbf{v} = \nabla \phi \tag{10-15}$$

The differential equation defining ϕ is obtained from the continuity equation. Considering a steady incompressible flow, we have $\nabla \cdot \mathbf{v} = 0$; thus, using equation (10-15) for $\mathbf{v}$, we obtain

$$\nabla \cdot \nabla \phi = \nabla^2 \phi = 0 \tag{10-16}$$

which is again Laplace's equation; this time the dependent variable is ϕ. Clearly, Ψ and ϕ must be related. This relation may be illustrated by a consideration of isolines of Ψ and ϕ. An isoline of Ψ is, of course, a streamline. Along the isolines

$$d\Psi = \frac{\partial \Psi}{\partial x}dx + \frac{\partial \Psi}{\partial y}dy$$

or

$$\left.\frac{dy}{dx}\right|_{\Psi=\text{constant}} = \frac{v_y}{v_x}$$

and

$$d\phi = \frac{\partial\phi}{\partial x}dx + \frac{\partial\phi}{\partial y}dy \qquad \frac{dy}{dx}\bigg|_{d\phi=0} = -\frac{v_x}{v_y}$$

Accordingly,

$$\frac{dy}{dx}\bigg|_{\phi=\text{constant}} = -\frac{1}{dy/dx}\bigg|_{\Psi=\text{constant}} \tag{10-17}$$

and thus Ψ and ϕ are orthogonal. The orthogonality of the stream function and the velocity potential is a useful property, particularly when graphical solutions to equations (10-8) and (10-16) are employed.

Figure 10.4 illustrates the inviscid, irrotational, steady incompressible flow about an infinite circular cylinder. Both the streamlines and constant-velocity potential lines are shown.

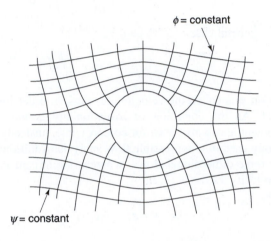

$\phi = \text{constant}$

$\psi = \text{constant}$

Figure 10.4 Streamlines and constant-velocity potential lines for steady, incompressible, irrotational, inviscid flow about a cylinder.

Example 3

The steady, incompressible flow field for two-dimensional flow is given by the following velocity components: $v_x = 16y - x$ and $v_y = 16x + y$. Determine the equation for the stream function and the velocity potential.

First, let's check to make sure continuity is satisfied:

$$\frac{\partial v_x}{\partial x} + \frac{\partial v_y}{\partial y} = \frac{\partial}{\partial x}(16y - x) + \frac{\partial}{\partial y}(16x + y) = -1 + 1 = 0 \tag{1}$$

So continuity is satisfied, which is a necessary condition for us to proceed.

We defined the stream function as

$$\frac{\partial\Psi}{\partial y} = v_x \tag{2}$$

and

$$-\frac{\partial\Psi}{\partial x} = v_y \tag{3}$$

Thus,

$$v_x = \frac{\partial \Psi}{\partial y} = 16y - x \tag{4}$$

$$v_y = -\frac{\partial \Psi}{\partial x} = 16x + y \tag{5}$$

We can begin by integrating equation (4) or equation (5). Either will result in the same answer. (Problem 10.24 will let you verify this.) We will choose to begin by integrating equation (4) partially with respect to y:

$$\Psi = 8y^2 - xy + f_1(x) \tag{6}$$

where $f_1(x)$ is an arbitrary function of x.

Next, we take the other part of the definition of the stream function, equation (3), and differentiate equation (6) with respect to x:

$$v_y = -\frac{\partial \Psi}{\partial x} = y - f_2(x) \tag{7}$$

Here, $f_2(x)$ is $\frac{df}{dx}$, since f is a function of the variable x.

The result is that we now have two equations for v_y, equations (5) and (7). We can now equate these and solve for $f_2(x)$:

$$v_y = y - f_2(x) = 16x + y$$

Solving for $f_2(x)$,

$$f_2(x) = -16x$$

So

$$f_1(x) = -16\frac{x^2}{2} = -8x^2 + C$$

The integration constant C is added to the above equation since f is a function of x only. The final equation for the stream function is

$$\Psi = 8y^2 - xy - 8x^2 + C \tag{8}$$

The constant C is generally dropped from the equation because the value of a constant in this equation is of no significance. The final equation for the stream function is

$$\Psi = 8y^2 - xy - 8x^2$$

One interesting point is that the difference in the value of one stream line in the flow to another is the volume flow rate per unit width between the two streamlines.

Next we want to find the equation for the velocity potential. Since a condition for the velocity potential to exist is irrotational flow, we must first determine whether the flow in this example is irrotational.

To do this, we must satisfy equation (10-1).

$$\omega_z = \frac{1}{2}\left(\frac{\partial v_y}{\partial x} - \frac{\partial v_x}{\partial y}\right) = \frac{1}{2}(16 - 16) = 0$$

So the flow is irrotational, as required.

We now want to determine the equation for the velocity potential. The velocity potential is defined by equation (10-15):

$$v = \nabla \phi$$

or

$$\frac{\partial \phi}{\partial x} = v_x = 16y - x$$

$$\phi = 16xy - \frac{x^2}{2} + f(y)$$

Differentiating with respect to y and equating to v_x,

$$\frac{\partial \phi}{\partial y} = 16x + \frac{d}{dy}f(y) = 16x + y$$

Thus,

$$\frac{d}{dy}f(y) = y$$

and

$$f(y) = \frac{y^2}{2}$$

so that the final equation for the velocity potential is

$$\phi = 16xy - \frac{x^2}{2} + \frac{y^2}{2}$$

▶ **10.5**

TOTAL HEAD IN IRROTATIONAL FLOW

The condition of irrotationality has been shown to be of aid in obtaining analytical solutions in fluid flow. The physical meaning of irrotational flow can be illustrated by the relation between the rotation or vorticity, $\nabla \times \mathbf{v}$, and the total head, $P/\rho + v^2/2 + gy$. For an inviscid flow, we may write

$$\frac{D\mathbf{v}}{Dt} = \mathbf{g} - \frac{\nabla P}{\rho} \qquad \text{(Euler's equation)}$$

and

$$\frac{D\mathbf{v}}{Dt} = \frac{\partial \mathbf{v}}{\partial t} + \nabla \left(\frac{v^2}{2}\right) - \mathbf{v} \times (\nabla \times \mathbf{v}) \qquad \text{(Vector identity)}$$

As the gradient of the potential energy is $-\mathbf{g}$, Euler's equation becomes, for incompressible flow,

$$\nabla \left\{ \frac{P}{\rho} + \frac{v^2}{2} + gy \right\} = \mathbf{v} \times (\nabla \times \mathbf{v}) - \frac{\partial \mathbf{v}}{\partial t}. \qquad (10\text{-}18)$$

If the flow is steady, it is seen from equation (10-18) that the gradient of the total head depends upon the vorticity, $\nabla \times v$. The vector $(\nabla \times v)$ is perpendicular to the velocity vector; hence, the gradient of the total head has no component along a streamline. Thus, along a streamline in an incompressible, inviscid, steady flow,

$$\frac{P}{\rho} + \frac{v^2}{2} + gy = \text{constant} \qquad (10\text{-}19)$$

This is, of course, Bernoulli's equation, which was discussed in Chapters 6 and 9. If the flow is irrotational and steady, equation (10-18) yields the result that Bernoulli's equation is valid throughout the flow field. An irrotational, steady, incompressible flow, therefore, has a constant total head throughout the flow field.[3]

► 10.6

UTILIZATION OF POTENTIAL FLOW

Potential flow has great utility in engineering for the prediction of pressure fields, forces, and flow rates. In the field of aerodynamics, for example, potential flow solutions are used to predict force and moment distributions on wings and other bodies.

An illustration of the determination of the pressure distribution from a potential flow solution may be obtained from the solution for the flow about a circular cylinder presented in Section 10.3. From the Bernoulli equation

$$\frac{P}{\rho} + \frac{v^2}{2} = \text{constant} \tag{10-20}$$

We have deleted the potential energy term in accordance with the original assumption of uniform velocity in the x-direction. At a great distance from the cylinder the pressure is P_∞, and the velocity is v_∞, so equation (10-20) becomes[4]

$$P + \frac{\rho v^2}{2} = P_\infty + \frac{\rho v_\infty^2}{2} = P_0 \tag{10-21}$$

where P_0 is designated the *stagnation pressure* (i.e., the pressure at which the velocity is zero). In accordance with equation (10-19), the stagnation pressure is constant throughout the field in an irrotational flow. The velocity at the surface of the body is $v_\theta = -2v_\infty \sin\theta$, thus the surface pressure is

$$P = P_0 - 2\rho v_\infty^2 \sin^2\theta \tag{10-22}$$

A plot of the potential flow pressure distribution about a cylinder is shown in Figure 10.5.

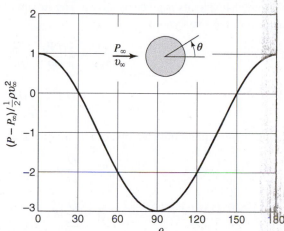

Figure 10.5 Pressure distribution on a cylinder in an inviscid, incompressible, steady flow.

[3] A more general result, Crocco's theorem, relates the vorticity to the entropy. Thus, it can be shown that a steady, inviscid, irrotational flow, either compressible or incompressible, is isentropic.

[4] The stagnation pressure as given in equation (10-21) applies to incompressible flow only.

► **10.7**

POTENTIAL FLOW ANALYSIS—SIMPLE PLANE FLOW CASES

In this section, a number of cases will be considered in which solutions are achieved for two-dimensional, incompressible irrotational flow. We begin with some very straightforward flow situations.

Case 1. Uniform flow in the *x*-direction.

For a uniform flow parallel to the *x*-axis, with velocity v_∞ = constant, the stream function and velocity potential relationships are

$$v_x = v_\infty = \frac{\partial \Psi}{\partial y} = \frac{\partial \phi}{\partial x}$$

$$v_y = 0 = \frac{\partial \Psi}{\partial x} = \frac{\partial \phi}{\partial y}$$

which integrate to yield

$$\Psi = v_\infty y$$

$$\phi = v_\infty x$$

Case 2. A line source or sink.

A line source, in two dimensions, is a flow that is radially outward from the source, which is the origin in this example. The reverse, or sink flow, has the flow directed inward. The source strength is the volume flow rate per unit depth, $Q = 2\pi r v_r$. The radial velocity associated with a source is

$$v_r = \frac{Q}{2\pi r}$$

and the azimuthal velocity is given by $v_\theta = 0$. The stream function and velocity potential are evaluated from the expressions

$$v_r = \frac{Q}{2\pi r} = \frac{1}{r}\frac{\partial \Psi}{\partial \theta} = \frac{\partial \phi}{\partial r}$$

$$v_\theta = 0 = -\frac{\partial \Psi}{\partial r} = \frac{1}{r}\frac{\partial \phi}{\partial \theta}$$

Integrating these expressions, we obtain for the line source

$$\Psi = \frac{Q}{2\pi}\theta$$

$$\phi = \frac{Q}{2\pi}\ln r$$

For sink flow, the sign of the radial velocity is negative (inward), and thus, Q is negative.

The expressions for a line source or sink present a problem at $r = 0$, the origin, which is a singular point. At $r = 0$, the radial velocity approaches infinity. Physically this is unrealistic, and we use only the concept of line source or sink flow under conditions where the singularity is excluded from consideration.

Case 3. A line vortex.

Vortex flow is that which occurs in a circular fashion around a central point, such as a whirlpool. A *free vortex* is one where fluid particles are irrotational, i.e., they do not rotate as they move in concentric circles about the axis of the vortex. This would be analogous to people sitting in cabins on a ferris wheel. For an irrotational flow in polar coordinates (see Appendix B), the product rv_θ must be constant. The stream function and velocity potential can be written directly,

$$v_r = 0 = \frac{1}{r}\frac{\partial \Psi}{\partial \theta} = \frac{\partial \phi}{\partial r}$$

$$v_\theta = \frac{K}{2\pi r} = -\frac{\partial \Psi}{\partial r} = \frac{1}{r}\frac{\partial \phi}{\partial \theta}$$

which, upon integration, become

$$\Psi = -\frac{K}{2\pi}\ln r$$

$$\phi = \frac{K}{2\pi}\theta$$

where K is referred to as the *vortex strength*. When K is positive, the flow is observed to be counterclockwise about the vortex center.

▶ 10.8

POTENTIAL FLOW ANALYSIS—SUPERPOSITION

It was shown earlier that both the stream function and the velocity potential satisfy Laplace's equation for two-dimensional, irrotational, incompressible flow. As Laplace's equation is linear we can use known solutions to achieve expressions for both Ψ and ϕ for more complex situations using the principle of *superposition*. Superposition, simply put, is the process of adding known solutions to achieve another—i.e., if Ψ_1 and Ψ_2 are solutions to $\nabla^2 \Psi = 0$, then so is $\Psi_3 = \Psi_1 + \Psi_2$ a solution.

The reader is reminded that the solutions obtained for these very specialized flow conditions are idealizations. They apply for inviscid flow, which is a reasonable approximation for conditions *outside* the region, near a solid body, where viscous effects are manifested. This region, the *boundary layer,* will be considered in some depth in Chapter 12.

Some cases will now be considered where the elementary plane flows of the previous section give some interesting and useful results through the process of superposition.

Case 4. The doublet.

A useful case is achieved from considering a source-sink pair on the *x*-axis as the separation distance, *2a*, approaches zero. Geometrically, we can note that the streamlines and velocity potential lines are circles with centers on the *y*- and *x*-axes, but with all circles passing through the origin that is a singular point.

The strength of a doublet, designated λ, is defined as the finite limit of the quantity $2aQ$ as $a \to 0$. For our case, the source is placed on the *x*-axis at $-a$ and the sink is placed on the x-axis at $+a$. The resulting expressions for Ψ and ϕ in polar coordinates are

$$\Psi = -\frac{\lambda \sin \theta}{r}$$

$$\phi = \frac{\lambda \cos \theta}{r}$$

Case 5. Flow past a half body—superposition of uniform flow and a source.

The stream function and velocity potentials for uniform flow in the x-direction and for a line source are added together, yielding

$$\Psi = \Psi_{\text{uniform flow}} + \Psi_{\text{source}}$$

$$= v_\infty y + \frac{Q}{2\pi}\theta = v_\infty r \sin \theta + \frac{Q}{2\pi}\theta$$

$$\phi = \phi_{\text{uniform flow}} + \phi_{\text{source}}$$

$$= v_\infty x + \frac{Q}{2\pi}\ln r = v_\infty r \cos \theta + \frac{Q}{2\pi}\ln r$$

Case 6. Flow past a cylinder—superposition of uniform flow and a doublet.

As a final illustration of the superposition method, we will consider a case of considerable utility. When the solutions for uniform flow and the doublet are superposed, the result, similar to the past case, defines a streamline pattern inside and around the outside surface of a body. In this case the body is closed and the exterior flow pattern is that of ideal flow over a cylinder. The expressions for Ψ and φ are

$$\Psi = \Psi_{\text{uniform flow}} + \Psi_{\text{doublet}}$$

$$= v_\infty y - \frac{\lambda \sin \theta}{r} = v_\infty r \sin \theta - \frac{\lambda \sin \theta}{r}$$

$$= \left[v_\infty r - \frac{\lambda}{r}\right]\sin \theta$$

$$\phi = \phi_{\text{uniform flow}} + \phi_{\text{doublet}}$$

$$= v_\infty x + \frac{\lambda \cos \theta}{r} = v_\infty r \cos \theta + \frac{\lambda \cos \theta}{r}$$

$$= \left[v_\infty r + \frac{\lambda}{r}\right]\cos \theta$$

It is useful, at this point, to examine the above expressions in more detail. First, for the stream function

$$\Psi = \left[v_\infty r - \frac{\lambda}{r}\right]\sin \theta$$

$$= v_\infty r \left[1 - \frac{\lambda/v_\infty}{r^2}\right]\sin \theta$$

where, as we recall, λ is the doublet strength. If we choose λ such that

$$\frac{\lambda}{v_\infty} = a^2$$

where a is the radius of our cylinder, we obtain

$$\Psi(r, \theta) = v_\infty r \sin \theta \left[1 - \frac{a^2}{r^2}\right]$$

which is the expression used earlier, designated as equation (10-11).

CLOSURE

In this chapter, we have examined potential flow. A short summary of the properties of the stream function and the velocity potential is given below.

Stream function

1. A stream function $\Psi(x, y)$ exists for each and every two-dimensional, steady, incompressible flow, whether viscous or inviscid.

2. Lines for which $\Psi(x, y) = $ constant are streamlines.

3. In Cartesian coordinates,

$$v_x = \frac{\partial \Psi}{\partial y} \qquad v_y = -\frac{\partial \Psi}{\partial x} \tag{10-23a}$$

and in general,

$$v_s = \frac{\partial \Psi}{\partial n} \tag{10-23b}$$

where n is 90° counterclockwise from s.

4. The stream function identically satisfies the continuity equation.

5. For an irrotational, steady incompressible flow,

$$\nabla^2 \Psi = 0 \tag{10-24}$$

Velocity potential

1. The velocity potential exists if and only if the flow is irrotational. No other restrictions are required.

2. $\nabla \phi = \mathbf{v}$.

3. For irrotational, incompressible flow, $\nabla^2 \phi = 0$.

4. For steady, incompressible two-dimensional flows, lines of constant velocity potential are perpendicular to the streamlines.

CHAPTER 11

Dimensional Analysis and Similitude

An important consideration in all equations written thus far has been dimensional homogeneity. At times it has been necessary to use proper conversion factors in order that an answer be correct numerically and have the proper units. The idea of dimensional consistency can be used in another way, by a procedure known as dimensional analysis, to group the variables in a given situation into dimensionless parameters that are less numerous than the original variables. Such a procedure is very helpful in experimental work, in which the very number of significant variables presents an imposing task of correlation. By combining the variables into a smaller number of dimensionless parameters, the work of experimental data reduction is considerably reduced.

This chapter will include means of evaluating dimensionless parameters both in situations in which the governing equation is known, and in those in which no equation is available. Certain dimensionless groups emerging from this analysis will be familiar, and some others will be encountered for the first time. Finally, certain aspects of similarity will be used to predict the flow behavior of equipment on the basis of experiments with scale models.

▶ **11.1**

DIMENSIONS

In dimensional analysis, certain dimensions must be established as fundamental, with all others expressible in terms of these. One of these fundamental dimensions is length, symbolized L. Thus, area and volume may dimensionally be expressed as L^2 and L^3, respectively. A second fundamental dimension is time, symbolized t. The kinematic quantities, velocity and acceleration, may now be expressed as L/t and L/t^2, respectively.

Another fundamental dimension is mass, symbolized M. An example of a quantity whose dimensional expression involves mass is the density that would be expressed as M/L^3. Newton's second law of motion gives a relation between force and mass and allows force to be expressed dimensionally as $F = Ma = ML/t^2$. Some texts reverse this procedure and consider force fundamental, with mass expressed in terms of F, L, and t according to Newton's second law of motion. Here, mass will be considered a fundamental unit.

The significant quantities in momentum transfer can all be expressed dimensionally in terms of M, L, and t; thus, these comprise the fundamental dimensions we shall be concerned with presently. The dimensional analysis of energy problems in Chapter 19 will require the addition of two more fundamental dimensions, heat and temperature.

Some of the more important variables in momentum transfer and their dimensional representations in terms of M, L, and t are given in Table 11.1.

Table 11.1 Important variables in momentum transfer

Variable	Symbol	Dimension
Mass	M	M
Length	L	L
Time	t	t
Velocity	v	L/t
Gravitational acceleration	g	L/t^2
Force	F	ML/t^2
Pressure	P	ML/t^2
Density	ρ	M/L^3
Viscosity	μ	M/Lt
Surface tension	σ	M/t^2
Sonic velocity	a	L/t

▶ **11.2**

DIMENSIONAL ANALYSIS OF GOVERNING DIFFERENTIAL EQUATIONS

The differential equations that describe fluid behavior as developed in Chapter 9 are powerful tools for analyzing and predicting fluid phenomena and their effects. The Navier–Stokes equations have been solved analytically for a few simple situations. For more complex applications, these relationships provide the basis for a number of sophisticated and powerful numerical codes.

In this section, we will use the differential forms of the continuity and momentum (Navier–Stokes) equations to develop some useful dimensionless parameters that will be valuable tools for subsequent analysis. This process will now be illustrated as we examine two-dimensional incompressible flow.

The governing differential equations are the following.

Continuity:

$$\frac{\partial v_x}{\partial x} + \frac{\partial v_y}{\partial y} = 0 \tag{9-3}$$

Momentum:

$$\rho\left(\frac{\partial \mathbf{v}}{\partial t} + v_x \frac{\partial \mathbf{v}}{\partial x} + v_y \frac{\partial \mathbf{v}}{\partial y}\right) = \rho\mathbf{g} - \nabla P + \mu\left(\frac{\partial^2 \mathbf{v}}{\partial x^2} + \frac{\partial^2 \mathbf{v}}{\partial y^2}\right) \tag{9-19}$$

We now stipulate the reference values for length and velocity:

- reference length L
- reference velocity v_∞

and, accordingly, specify nondimensional quantities for the variables in equations (9-3) and (9-19) as

$$\begin{aligned}
x^* &= x/L & v_x^* &= v_x/v_\infty \\
y^* &= y/L & v_y^* &= v_y/v_\infty \\
t^* &= \frac{tv_\infty}{L} & \mathbf{v}^* &= \mathbf{v}/v_\infty \\
& & \nabla^* &= L\nabla
\end{aligned}$$

The last quantity in this list, ∇^*, is the dimensionless gradient operator. As ∇ is composed of first derivatives with respect to space coordinates, the product $L\nabla$ is seen to be dimensionless.

The next step is to nondimensionalize our governing equations by introducing the specified dimensionless variables. This process involves the chain rule for differentiation; for example, the two terms in equation (9-3) are transformed as follows:

$$\frac{\partial v_x}{\partial x} = \frac{\partial v_x^*}{\partial x^*}\frac{\partial v_x}{\partial v_x^*}\frac{\partial x^*}{\partial x} = \frac{\partial v_x^*}{\partial x^*}(v_\infty)(1/L) = \frac{v_\infty}{L}\frac{\partial v_x^*}{\partial x^*}$$

$$\frac{\partial v_y}{\partial y} = \frac{\partial v_y^*}{\partial y^*}\frac{\partial v_y}{\partial v_y^*}\frac{\partial y_*}{\partial y} = \frac{v_\infty}{L}\frac{v_x^*}{\partial x^*}$$

Substitution into equation (9-3) gives

$$\frac{\partial v_x^*}{\partial x^*} + \frac{\partial v_y^*}{\partial y^*} = 0 \tag{11-1}$$

and we see that the continuity equation has the same form in terms of dimensionless variables as it had originally.

Utilizing the chain rule in the same manner as just discussed, the equation of motion becomes

$$\frac{\rho v_\infty^2}{L}\left(\frac{\partial \mathbf{v}^*}{\partial t^*} + v_x^*\frac{\partial \mathbf{v}^*}{\partial x^*} + v_y^*\frac{\partial \mathbf{v}^*}{\partial y^*}\right) = \rho\mathbf{g} + \frac{1}{L}\nabla^* P + \frac{\mu v_\infty}{L^2}\left(\frac{\partial^2 \mathbf{v}^*}{\partial x^{*2}} + \frac{\partial^2 \mathbf{v}^*}{\partial y^{*2}}\right) \tag{11-2}$$

In equation (11-2), we note that each term has the units $M/L^2 t^2$ or F/L^3. Also, it should be observed that each term represents a certain kind of force—that is,

- $\dfrac{\rho v_\infty^2}{L}$ is an inertial force
- $\dfrac{\mu v_\infty^2}{L}$ is a viscous force
- ρg is a gravitational force
- P/L is a pressure force

If we next divide through by the quantity, $\rho v_\infty^2/L$, our dimensionless equation becomes

$$\frac{\partial \mathbf{v}^*}{\partial t^*} + v_x^*\frac{\partial \mathbf{v}^*}{\partial x^*} + v_y^*\frac{\partial \mathbf{v}^*}{\partial y^*} = \mathbf{g}\frac{L}{v_\infty^2} - \frac{\nabla^* P}{\rho v_\infty^2} + \frac{\mu}{L v_\infty \rho}\left(\frac{\partial^2 \mathbf{v}^*}{\partial x^{*2}} + \frac{\partial^2 \mathbf{v}^*}{\partial y^{*2}}\right) \tag{11-3}$$

This resulting dimensionless equation has the same general characteristics as its original, except that, as a result of its transformation into dimensionless form, each of the original force terms (those on the right-hand side) has a coefficient composed of a combination of variables. An examination of these coefficients reveals that each is dimensionless. Additionally, because of the manner in which they were formed, the parameters can be interpreted as a ratio of forces.

Consideration of the first term, gL/v_∞^2, reveals that it is, indeed, dimensionless. The choice of gL/v_∞^2 or v_∞^2/gL is arbitrary; clearly both forms are dimensionless.

The conventional choice is the latter form. The *Froude number* is defined as

$$\text{Fr} \equiv v_\infty^2/gL \tag{11-4}$$

This parameter can be interpreted as a measure of the ratio of inertial to gravitational forces. The Froude number arises in analyzing flows involving a free liquid surface. It is an important parameter when dealing with open-channel flows.

The next parameter, $P/\rho v_\infty^2$, is observed to be the ratio of pressure forces to inertial forces. In this form it is designated the *Euler number*,

$$\text{Eu} \equiv P/\rho v_\infty^2 \qquad (11\text{-}5)$$

A modified form of equation (11-5), also clearly dimensionless, is the coefficient of drag

$$C_\text{D} = \frac{F/A}{\rho v_\infty^2/2} \qquad (11\text{-}6)$$

which, we will see directly, has application to both internal and external flows.

The third dimensionless ratio that has been generated is the *Reynolds number*, which is conventionally expressed as

$$\text{Re} \equiv L v_\infty \rho/\mu \qquad (11\text{-}7)$$

In this form, the Reynolds number is observed to represent the ratio of inertial forces to viscous forces. The Reynolds number is generally considered the most important dimensionless parameter in the field of fluid mechanics. It is ubiquitous in all of the transport processes. We will encounter it frequently throughout the remainder of this text.

If equation (11-3) can be solved, the results will provide the functional relationships between applicable dimensionless parameters. If direct solution is not possible, then one must resort to numerical modeling or experimental determination of these functional relationships.

▶ 11.3

THE BUCKINGHAM METHOD

The procedure introduced in the previous section is, obviously, quite powerful when one knows the differential equation that pertains to a specific fluid-flow process. There are, however, many situations of interest in which the governing equation is not known. In these cases, we need an alternative method for dimensional analysis. In this section, we discuss a more general approach for generating dimensionless groups of variables. This procedure was proposed by Buckingham[1] in the early part of the twentieth century. It is generally referred to as the *Buckingham method*.

The initial step in applying the Buckingham method requires the listing of the variables significant to a given problem. It is then necessary to determine the number of dimensionless parameters into which the variables may be combined. This number may be determined using the *Buckingham pi* theorem, which states

> *The number of dimensionless groups used to describe a situation involving* n *variables is equal to* n − r, *where* r *is the rank of the dimensional matrix of the variables.*

Thus,

$$i = n - r \qquad (11\text{-}8)$$

where

$i =$ the number of independent dimensionless groups
$n =$ the number of variables involved

---

[1] E. Buckingham, *Phys. Rev.* **2**, 345 (1914).

and

$$r = \text{the rank of the dimensional matrix}$$

The dimensional matrix is simply the matrix formed by tabulating the exponents of the fundamental dimensions M, L, and t, which appear in each of the variables involved.

An example of the evaluation of r and i, as well as the application of the Buckingham method, follows.

Example 1

Determine the dimensionless groups formed from the variables involved in the flow of fluid external to a solid body. The force exerted on the body is a function of v, ρ, μ, and L (a significant dimension of the body).

A usual first step is to construct a table of the variables and their dimensions.

Variable	Symbol	Dimension
Force	F	ML/t^2
Velocity	v	L/t^2
Density	ρ	M/L^3
Viscosity	μ	M/Lt
Length	L	L

Before determining the number of dimensionless parameters to be formed, we must know r. The dimensional matrix that applies is formed from the following tabulation:

	F	v	ρ	μ	L
M	1	0	1	1	0
L	1	1	−3	−1	1
t	−2	−1	0	−1	0

The numbers in the table represent the exponents of M, L, and t in the dimensional expression for each variable involved. For example, the dimensional expression of F is ML/t^2; hence, the exponents 1, 1, and −2 are tabulated versus M, L, and t, respectively, the quantities with which they are associated. The matrix is then the array of numbers shown below:

$$\begin{pmatrix} 1 & 0 & 1 & 1 & 0 \\ 1 & 1 & -3 & -1 & 1 \\ -2 & -1 & -0 & -1 & 0 \end{pmatrix}$$

The rank, r, of a matrix is the number of rows (columns) in the largest nonzero determinant that can be formed from it. The rank is 3 in this case. Thus, the number of dimensionless parameters to be formed may be found by applying equation (11-4). In this example, $i = 5 - 3 = 2$.

The two dimensionless parameters will be symbolized π_1 and π_2 and may be formed in several ways. Initially, a *core group* of r variables must be chosen, which will consist of those variables that will appear in each pi group and, among them, contain all of the fundamental dimensions. One way to choose a core is to exclude from it those variables whose effect one desires to isolate. In the present problem it would be desirable to have the drag force in only one dimensionless group; hence, it will not be in the core. Let us arbitrarily let the viscosity be the other exclusion from the core. Our core group now consists of the remaining variables v, ρ, and L, which, we observe, include M, L, and t among them.

We now know that π_1 and π_2 both include ρ, L, and v; that one of them includes F and the other μ; and that they are both dimensionless. In order that each be dimensionless, the variables must be raised to certain exponents. Writing

$$\pi_1 = v^a \rho^b L^c F \quad \text{and} \quad \pi_2 = v^d \rho^e L^f \mu$$

we shall evaluate the exponents as follows. Considering each π group independently, we write

$$\pi_1 = v^a \rho^b L^c F$$

and dimensionally

$$M^0 L^0 t^0 = 1 = \left(\frac{L}{t}\right)^a \left(\frac{M}{L^3}\right)^b (L)^c \frac{ML}{t^2}$$

Equating exponents of M, L, and t on both sides of this expression, we have, for M,

$$0 = b + 1$$

for L,

$$0 = a - 3b + c + 1$$

and for t,

$$0 = -a - 2$$

From these, we find that $a = -2$, $b = -1$, and $c = -2$, giving

$$\pi_1 = \frac{F}{L^2 \rho v^2} = \frac{F/L^2}{\rho v^2} = \mathrm{Eu}$$

Similarly for π_2 we have, in dimensional form,

$$1 = \left(\frac{L}{t}\right)^d \left(\frac{M}{L^3}\right)^e (L)^f \frac{M}{Lt}$$

and for exponents of M,

$$0 = e + 1$$

for L,

$$0 = d - 3e + f - 1$$

and for t,

$$0 = -d - 1$$

giving $d = -1$, $e = -1$ and $f = -1$. Thus, for our second dimensionless group we have

$$\pi_2 = \mu/\rho v L = 1/\mathrm{Re}$$

Dimensional analysis has enabled us to relate the original five variables in terms of only two dimensionless parameters in the form

$$\mathrm{Eu} = \phi(\mathrm{Re}) \tag{11-9}$$

$$C_\mathrm{D} = f(\mathrm{Re}) \tag{11-10}$$

The two parameters, Eu and C_D, were also generated in the previous section by an alternate method. The functions $\phi(\text{Re})$ and $f(\text{Re})$ must be determined by experiment.

Table 11.2 lists several dimensionless groups that pertain to fluid flow. Similar tables will be included in later chapters that list dimensionless parameters common to heat transfer and to mass transfer.

Table 11.2 Common dimensionless parameters in momentum transfer

Name/symbol	Dimensionless group	Physical meaning	Area of application
Reynolds number, Re	$\dfrac{Lv\rho}{\mu}$	$\dfrac{\text{Inertial force}}{\text{Viscous force}}$	Widely applicable in a host of fluid flow situations
Euler number, Eu	$\dfrac{P}{\rho v^2}$	$\dfrac{\text{Pressure force}}{\text{Inertial force}}$	Flows involving pressure differences due to frictional effects
Coefficient of skin friction, C_f	$\dfrac{F/A}{\rho v^2/2}$	$\dfrac{\text{Drag force}}{\text{Dynamic force}}$	Flows in aerodynamics and hydrodynamics
Froude number, Fr	$\dfrac{v^2}{gL}$	$\dfrac{\text{Inertial force}}{\text{Gravitational force}}$	Flows involving free liquid surfaces
Weber number, We	$\dfrac{\rho v^2 L}{\sigma}$	$\dfrac{\text{Inertial force}}{\text{Surface tension force}}$	Flows with significant surface tension effects
Mach number, M	$\dfrac{v}{C}$	$\dfrac{\text{Inertial force}}{\text{Compressibility force}}$	Flows with significant compressibility effects

► **11.4**

GEOMETRIC, KINEMATIC, AND DYNAMIC SIMILARITY

An important application and use of the dimensionless parameters listed in Table 11.2 is in using experimental results obtained using models to predict the performance of full-size prototypical systems. The validity of such *scaling* requires that the models and prototypes possess *similarity*. Three types of similarity are important in this regard; they are geometric, kinematic, and dynamic similarity.

Geometric similarity exists between two systems if the ratio of all significant dimensions is the same for each system. For example, if the ratio a/b for the diamond-shaped section in Figure 11.1 is equal in magnitude to the ratio a/b for the larger section, they are

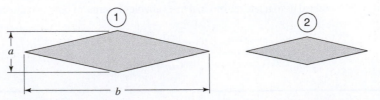

Figure 11.1 Two geometrically similar objects.

geometrically similar. In this example, there are only two significant dimensions. For more complex geometries, geometric similarity would be achieved when all geometric ratios between model and prototype are equal.

Kinematic similarity similarly exists when, in geometrically similar systems ① and ②, the velocities at the same locations are related according to

$$\left(\frac{v_x}{v_y}\right)_1 = \left(\frac{v_x}{v_y}\right)_2 \quad \left(\frac{v_x}{v_z}\right)_1 = \left(\frac{v_x}{v_z}\right)_2$$

The third type of similarity, *dynamic similarity*, exists when, in geometrically and kinematically similar systems, the ratios of significant forces are equal between model and prototype. These force ratios that are important in fluid-flow applications include the dimensionless parameters listed in Table 11.2.

The process of scaling using these similarity requirements will be presented in Section 11.5.

▶ 11.5

MODEL THEORY

In the design and testing of large equipment involving fluid flow, it is customary to build small models geometrically similar to the larger prototypes. Experimental data achieved for the models are then scaled to predict the performance of full-sized prototypes according to the requirements of geometric, kinematic, and dynamic similarity. The following examples will illustrate the manner of utilizing model data to evaluate the conditions for a full-scale device.

Example 2

A cylindrical mixing tank is to be scaled up to a larger size such that the volume of the larger tank is five times that of the smaller one. What will be the ratios of diameter and height between the two?

Geometric similarity between tanks A and B in Figure 11.2 requires that

$$\frac{D_a}{h_a} = \frac{D_b}{h_b}$$

or

$$\frac{h_b}{h_a} = \frac{D_b}{D_a}$$

The volumes of the two tanks are

$$V_a = \frac{\pi}{4} D_a^2 h_a \quad \text{and} \quad V_b = \frac{\pi}{4} D_b^2 h_b$$

The scaling ratio between the two is stipulated as $\dfrac{V_b}{V_a} = 5$, thus,

$$\frac{V_b}{V_a} = \frac{(\pi/4)D_b^2 h_b}{(\pi/4)D_a^2 h_a} = 5$$

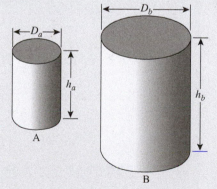

Figure 11.2 Cylindrical mixing tanks for Example 2.

and we get

$$\left(\frac{D_b}{D_a}\right)^2 \frac{h_b}{h_a} = 5$$

We now substitute the geometric similarity requirement that gives

$$\left(\frac{D_b}{D_a}\right)^3 = \left(\frac{L_b}{L_a}\right)^3 = 5$$

and the two ratios of interest become

$$\frac{D_b}{D_a} = \frac{L_b}{L_a} = 5^{1/3} = 1.71$$

Example 3

Dynamic similarity may be obtained by using a cryogenic wind tunnel in which nitrogen at low temperature and high pressure is employed as the working fluid. If nitrogen at 5 atm and 183 K is used to test the low-speed aerodynamics of a prototype that has a 24.38 m wing span and is to fly at standard sea-level conditions at a speed of 60 m/s, determine

1. The scale of the model to be tested.
2. The ratio of forces between the model and the full-scale aircraft.

Conditions of dynamic similarity should prevail. The speed of sound in nitrogen at 183 K is 275 m/s.

For dynamic similarity to exist, we know that both model and prototype must be geometrically similar and that the Reynolds number and the Mach number must be the same. A table such as the following is helpful.

	Model	Prototype
Characteristic length	L	24.38 m
Velocity	v	60 m/s
Viscosity	μ	$1.789 \cdot 10^{-5}$ Pa $\cdot$ s
Density	ρ	1.225 kg/m^3
Speed of sound	275 m/s	340 m/s

The conditions listed for the prototype have been obtained from Appendix I. Equating Mach numbers, we obtain

$$M_m = M_p$$

$$v = \frac{275}{340} 60 = 48.5 \text{ m/s}$$

Equating the Reynolds numbers of the model and the prototype, we obtain

$$\text{Re}_m = \text{Re}_p$$

$$\frac{\rho \, 48.5 L}{\mu} = \frac{1.225 \cdot 60 \cdot 24.38}{1.789 \cdot 10^{-5}} = 1.002 \times 10^8$$

Using equation (7-10), we may evaluate μ for nitrogen. From Appendix K, $\varepsilon/\kappa = 91.5\,\mathrm{K}$ and $\sigma = 3.681\,\text{Å}$ for nitrogen so that $\kappa T/\varepsilon = 2$ and $\Omega_\mu = 1.175$ (Appendix K). Thus,

$$\mu = 2.6693 \cdot 10^{-6} \frac{\sqrt{28 \cdot 183}}{(3.681)^2 (1.175)} = 1.200 \cdot 10^{-5}\,\mathrm{Pa \cdot s}$$

The density may be approximated from the perfect gas law

$$\rho = \frac{P}{P_1} \frac{M}{M_1} \frac{T_1}{T} \rho_1$$

so that

$$\rho = 5 \left(\frac{28}{28.96} \right) \left(\frac{288}{183} \right) 1.225 = 9.32\,\mathrm{kg/m^3}$$

Solving for the wing span of the model, we obtain

$$L = 3.26\,\mathrm{m}\ (10.7\,\mathrm{ft})$$

The ratio of the forces on the model to the forces experienced by the prototype may be determined equating values of Eu between the model and the prototype. Hence

$$\left(\frac{F}{\rho V^2 A_R} \right)_{\text{model}} = \left(\frac{F}{\rho V^2 A_R} \right)_{\text{prototype}}$$

where A_R is a suitable reference area. For an aircraft, this reference area is the projected wing area. The ratio of model force to prototype force is then given by

$$\frac{F_m}{F_p} = \frac{\rho_m}{\rho_p} \frac{V_m^2}{V_p^2} \frac{A_{R,m}}{A_{R,p}} = \frac{(\rho V^2)_m}{(\rho V^2)_p} \left(\frac{l_m}{l_p} \right)^2$$

where the ratio of reference areas can be expressed in terms of the scale ratio. Substituting numbers,

$$\frac{F_m}{F_p} = \frac{9.32}{1.225} \left(\frac{48.5}{60.0} \right)^2 \left(\frac{3.26}{24.38} \right)^2 = 0.089$$

The forces on the model are seen to be 8.9% the prototype forces.

▶ **11.6**

CLOSURE

The dimensional analysis of a momentum-transfer problem is simply an application of the requirement of dimensional homogeneity to a given situation. By dimensional analysis, the work and time required to reduce and correlate experimental data are decreased substantially by the combination of individual variables into dimensionless π groups, which are fewer in number than the original variables. The indicated relations between dimensionless parameters are then useful in expressing the performance of the systems to which they apply.

It should be kept in mind that dimensional analysis *cannot* predict which variables are important in a given situation, nor does it give any insight into the physical transfer mechanism involved. Even with these limitations, dimensional analysis techniques are a valuable aid to the engineer.

If the equation describing a given process is known, the number of dimensionless groups is automatically determined by taking ratios of the various terms in the expression to one another. This method also gives physical meaning to the groups thus obtained.

If, on the contrary, no equation applies, an empirical method, the *Buckingham* method, may be used. This is a very general approach, but gives no physical meaning to the dimensionless parameters obtained from such an analysis.

The requirements of geometric, kinematic, and dynamic similarity enable one to use model date to predict the behavior of a prototype or full-size piece of equipment. *Model theory* is thus an important application of the parameters obtained in a dimensional analysis.

Viscous Flow

The concept of fluid viscosity was developed and defined in Chapter 7. Clearly, all fluids are viscous, but in certain situations and under certain conditions, a fluid may be considered ideal or inviscid, making possible an analysis by the methods of Chapter 10.

Our task in this chapter is to consider viscous fluids and the role of viscosity as it affects the flow. Of particular interest is the case of flow past solid surfaces and the interrelations between the surfaces and the flowing fluid.

▶ **12.1**

REYNOLDS'S EXPERIMENT

The existence of two distinct types of viscous flow is a universally accepted phenomenon. The smoke emanating from a lighted cigarette is seen to flow smoothly and uniformly for a short distance from its source and then change abruptly into a very irregular, unstable pattern. Similar behavior may be observed for water flowing slowly from a faucet.

The well-ordered type of flow occurs when adjacent fluid layers slide smoothly over one another with mixing between layers or lamina occurring only on a molecular level. It was for this type of flow that Newton's viscosity relation was derived, and in order for us to measure the viscosity, μ, this *laminar* flow must exist.

The second flow regime, in which small packets of fluid particles are transferred between layers, giving it a fluctuating nature, is called the *turbulent* flow regime.

The existence of laminar and turbulent flow, although recognized earlier, was first described quantitatively by Reynolds in 1883. His classic experiment is illustrated in Figure 12.1. Water was allowed to flow through a transparent pipe, as shown, at a rate controlled by a valve. A dye having the same specific gravity as water was introduced at the pipe opening and its pattern observed for progressively larger flow rates of water. At low rates of flow, the dye pattern was regular and formed a single line of color as shown in Figure 12.1(a). At high flow rates, however, the dye became dispersed throughout the pipe

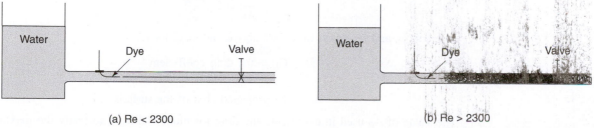

(a) Re < 2300   (b) Re > 2300

Figure 12.1 Reynolds's experiment.

cross section because of the very irregular fluid motion. The difference in the appearance of the dye streak was, of course, due to the orderly nature of laminar flow in the first case and to the fluctuating character of turbulent flow in the latter case.

The transition from laminar to turbulent flow in pipes is thus a function of the fluid velocity. Actually, Reynolds found that fluid velocity was the only one variable determining the nature of pipe flow, the others being pipe diameter, fluid density, and fluid viscosity. These four variables, combined into the single dimensionless parameter

$$\mathrm{Re} \equiv \frac{D\rho v}{\mu} \qquad (12\text{-}1)$$

form the Reynolds number, symbolized Re, in honor of Osborne Reynolds and his important contributions to fluid mechanics.

For flow in circular pipes, it is found that below a value for Reynolds number of 2300, the flow is *laminar*. Above this value the flow may be laminar as well, and indeed, laminar flow has been observed for Reynolds numbers as high as 40,000 in experiments wherein external disturbances were minimized. Above a Reynolds number of 2300, small disturbances will cause a transition to *turbulent* flow, whereas below this value disturbances are damped out and laminar flow prevails. The *critical Reynolds number for pipe flow* thus is 2300.

▶ 12.2

DRAG

Reynolds's experiment clearly demonstrated the two different regimes of flow: laminar and turbulent. Another manner of illustrating these different flow regimes and their dependence upon Reynolds number is through the consideration of drag. A particularly illustrative case is that of external flow (i.e., flow around a body as opposed to flow inside a conduit).

The drag force due to friction is caused by the shear stresses at the surface of a solid object moving through a viscous fluid. Frictional drag is evaluated by using the expression

$$\frac{F}{A} \equiv C_f \frac{\rho v_\infty^2}{2} \qquad (12\text{-}2)$$

where F is the force, A is the area of contact between the solid body and the fluid, C_f is the coefficient of skin friction, ρ is the fluid density, and v_∞ is the free-stream fluid velocity.

The coefficient of skin friction, C_f, which is defined by equation (12-2), is dimensionless.

The total drag on an object may be due to pressure as well as frictional effects. In such a situation another coefficient, C_D, is defined as

$$\frac{F}{A_P} \equiv C_D \frac{\rho v_\infty^2}{2} \qquad (12\text{-}3)$$

where F, ρ, and v_∞ are as described above and, additionally,

$$C_D = \text{the drag coefficient}$$

and

$$A_P = \text{the projected area of the surface}$$

The value of A_P used in expressing the drag for blunt bodies is normally the maximum projected area for the body.

The quantity $\rho v_\infty^2 / 2$ appearing in equations (12-2) and (12-3) is frequently called the *dynamic pressure*.

Pressure drag arises from two principal sources.[1] One is induced drag, or drag due to lift. The other source is wake drag, which arises from the fact that the shear stress causes the streamlines to deviate from their inviscid flow paths, and in some cases to separate from the body altogether. This deviation in streamline pattern prevents the pressure over the rest of a body from reaching the level it would attain otherwise. As the pressure at the front of the body is now greater than that at the rear, a net rearward force develops.

In an incompressible flow, the drag coefficient depends upon the Reynolds number and the geometry of a body. A simple geometric shape that illustrates the drag dependence upon the Reynolds number is the circular cylinder. The inviscid flow about a circular cylinder was examined in Chapter 10. The inviscid flow about a cylinder, of course, produced no drag, as there existed neither frictional nor pressure drag. The variation in the drag coefficient with the Reynolds number for a smooth cylinder is shown in Figure 12.2. The flow pattern about the cylinder is illustrated for several different values of Re. The flow pattern and general shape of the curve suggest that the drag variation, and hence the effects of shear stress on the flow, may be subdivided into four regimes. The features of each regime will be examined.

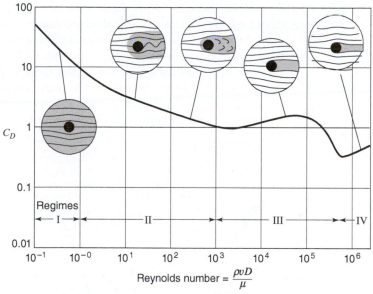

Figure 12.2 Drag coefficient for circular cylinders as a function of Reynolds number. Shaded regions indicate areas influenced by shear stress.

Regime 1

In this regime the entire flow is laminar and the Reynolds number small, being less than 1. Recalling the physical significance of the Reynolds number from Chapter 11 as the ratio of the inertia forces to the viscous forces, we may say that in regime 1 the viscous forces predominate. The flow pattern in this case is almost symmetric, the flow adheres to the body, and the wake is free from oscillations. In this regime of the so-called *creeping flow*, viscous effects predominate and extend throughout the flow field.

---

[1] A third source of pressure drag, wave drag, is associated with shock waves.

Regime 2

Two illustrations of the flow pattern are shown in the second regime. As the Reynolds number is increased, small eddies form at the rear stagnation point of the cylinder. At higher values of the Reynolds number, these eddies grow to the point at which they separate from the body and are swept downstream into the wake. The pattern of eddies shown in regime 2 is called a von Kármán vortex trail. This change in the character of the wake from a steady to an unsteady nature is accompanied by a change in the slope of the drag curve. The paramount features of this regime are (a) the unsteady nature of the wake and (b) flow separation from the body.

Regime 3

In the third regime the point of flow separation stabilizes at a point about 80 degrees from the forward stagnation point. The wake is no longer characterized by large eddies, although it remains unsteady. The flow on the surface of the body from the stagnation point to the point of separation is laminar, and the shear stress in this interval is appreciable only in a thin layer near the body. The drag coefficient levels out at a near-constant value of approximately 1.

Regime 4

At a Reynolds number near 5×10^5, the drag coefficient suddenly decreases to 0.3. When the flow about the body is examined, it is observed that the point of separation has moved past 90 degrees. In addition, the pressure distribution about the cylinder (shown in Figure 12.3) up to the point of separation is fairly close to the inviscid flow pressure distribution depicted in Figure 10.5. In the figure it will be noticed that the pressure variation about the surface is a changing function of Reynolds number. The minimum point on the curves for Reynolds numbers of 10^5 and 6×10^5 are both at the point of flow separation. From this figure it is seen that separation occurs at a larger value of θ for $\mathrm{Re} = 6 \times 10^5$ than it does for $\mathrm{Re} = 10^5$.

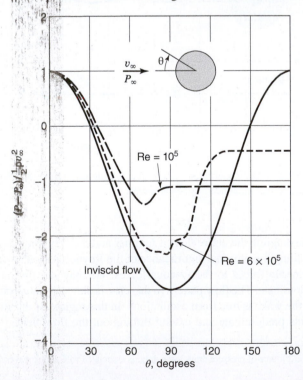

Figure 12.3 Pressure distribution on a circular cylinder at various Reynolds numbers.

The layer of flow near the surface of the cylinder is turbulent in this regime, undergoing transition from laminar flow close to the forward stagnation point. The marked decrease in drag is due to the change in the point of separation. In general, a turbulent flow resists flow separation better than a laminar flow. As the Reynolds number is large in this regime, it may be said that the inertial forces predominate over the viscous forces.

The four regimes of flow about a cylinder illustrate the decreasing realm of influence of viscous forces as the Reynolds number is increased. In regimes 3 and 4, the flow pattern over the forward part of the cylinder agrees well with the inviscid flow theory. For other geometries, a similar variation in the realm of influence of viscous forces is observed and, as might be expected, agreement with inviscid-flow predictions at a given Reynolds number increases as the slenderness of the body increases. The majority of cases of engineering interest involving external flows have flow fields similar to those of regimes 3 and 4.

Figure 12.4 shows the variation in the drag coefficient with the Reynolds number for a sphere, for infinite plates, and for circular disks and square plates. Note the similarity in form of the curve of C_D for the sphere to that for a cylinder in Figure 12.2. Specifically, one may observe the same sharp decrease in C_D to a minimum value near a Reynolds number value of 5×10^5. This is again due to the change from laminar to turbulent flow in the boundary layer.

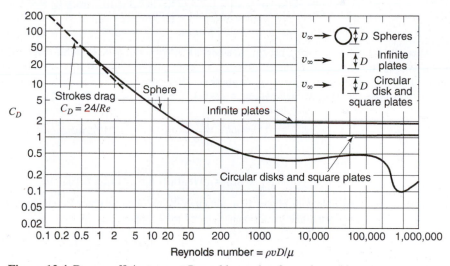

Figure 12.4 Drag coefficient versus Reynolds number for various objects.

Example 1

Evaluate the terminal velocity of a 7.5-mm-diameter glass sphere falling freely through (a) air at 300 K, (b) water at 300 K, and (c) glycerin at 300 K. The density of glass is 2250 kg/m³.

The terminal (steady-state) velocity of a falling body is reached when the force due to fluid drag matches the body's weight. In this case the weight of the glass sphere can be expressed as

$$\rho_s \frac{\pi d^3}{6} g$$

The fluid drag force is given by

$$C_D \frac{\rho_f v_\infty^2}{2} \frac{\pi d^2}{4}$$

and a force balance yields

$$C_D v_\infty^2 = \frac{4}{3} \frac{\rho_s}{\rho_f} dg$$

The drag coefficient, C_D, is plotted as a function of Reynolds number, Re_d, in Figure 12.4. As C_D is a function of v_∞, we are unable to solve explicitly for v_∞ unless $Re_d < 1$, which would permit the use of Stokes's law in expressing C_D. A trial-and-error solution is, thus, required. The conditions to be satisfied are our force balance expression and the graphical relation between C_D and Re_d in Figure 12.4.

For air at 300 K

$$\nu = 1.569 \times 10^{-5}\,\text{m}^2/\text{s}$$

$$\rho = 1.177\,\text{kg/m}^3$$

$$Re_d = \frac{d v_\infty}{\nu} = \frac{(7.5 \times 10^{-3}\,\text{m}) v_\infty}{1.569 \times 10^{-5}\,\text{m}^2/\text{s}} \tag{A}$$

$$= 478.0\, v_\infty$$

Inserting known values into our force balance expression, we have

$$C_D v_\infty^2 = \left(\frac{4}{3}\right) \frac{2250\,\text{kg/m}^3}{1.177\,\text{kg/m}^3} (7.5 \times 10^{-3}\,\text{m})(9.81\,\text{m/s}^2) \tag{B}$$

$$= 187.5\,\text{m}^2/\text{s}^2$$

Normally the trial-and-error procedure to achieve a solution would be straightforward. In this case, however, the shape of the C_D vs. Re_d curve, given in Figure 12.4, poses a bit of a problem. Specifically, the value of C_D remains nearly uniform—that is, $0.4 < C_D < 0.5$—over a range in Re_d between $500 < Re_d < 10^5$: over three orders of magnitude!

In such a specific case, we will assume $C_D \cong 0.4$, and solve equation B for v_∞:

$$v_\infty = \left[\frac{187.5}{0.4}\,\text{m}^2/\text{s}^2\right]^{1/2} = 21.65\,\text{m/s}$$

Equation (B) then yields

$$Re_d = (478.0)(21.65) = 1.035 \times 10^4$$

These results are compatible with Figure 12.4, although the absolute accuracy is obviously not great.

Finally, for air, we determine the terminal velocity to be, approximately,

$$v_\infty \cong 21.6\,\text{m/s} \tag{a}$$

For water at 300 K,

$$\nu = 0.879 \times 10^{-6}\,\text{m}^2/\text{s}$$

$$\rho = 996.1\,\text{kg/m}^3$$

$$Re_d = \frac{(7.5 \times 10^{-3}\,\text{m}) v_\infty}{0.879 \times 10^{-6}\,\text{m}^2/\text{s}} = 8530\, v_\infty$$

$$C_D v_\infty^2 = \left(\frac{4}{3}\right) \frac{2250\,\text{kg/m}^3}{996\,\text{kg/m}^3} (7.5 \times 10^{-3}\,\text{m})(9.81\,\text{m/s}^2)$$

$$= 0.2216\,\text{m}^2/\text{s}^2$$

As in part (a), we will initially assume $C_D \cong 0.4$, and achieve the result that

$$v_\infty = 0.744\,\text{m/s}$$

$$Re_d = 6350$$

These results, again, satisfy Figure 12.4. Thus, in water

$$v_\infty = 0.744 \text{ m/s} \qquad \text{(b)}$$

Finally, for glycerin at 300 K,

$$\nu = 7.08 \times 10^{-4} \text{ m}^2/\text{s}$$

$$\rho = 1260 \text{ kg/m}^3$$

$$\text{Re}_d = \frac{(7.5 \times 10^{-3} \text{ m})v_\infty}{7.08 \times 10^{-4} \text{ m}^2/\text{s}} = 10.59 \, v_\infty$$

$$C_D v_\infty^2 = \left(\frac{4}{3}\right) \frac{2250 \text{ kg/m}^3}{1260 \text{ kg/m}^3} (7.5 \times 10^{-3} \text{ m})(9.81 \text{ m/s}^2)$$

$$= 0.1752 \text{ m}^2/\text{s}^2$$

In this case we suspect the Reynolds number will be quite small. As an initial guess we will assume Stokes's law applies, thus $C_D = 24/\text{Re}$.

Solving for v_∞ for this case, we have

$$C_D v_\infty^2 = \frac{24\nu}{d v_\infty} v_\infty^2 = 0.1752 \text{ m}^2/\text{s}^2$$

$$v_\infty = \frac{(0.1752 \text{ m}^2/\text{s}^2)(7.5 \times 10^{-3} \text{ m})}{24(7.08 \times 10^{-4} \text{ m}^2/\text{s})}$$

$$= 0.0773 \text{ m/s}$$

To validate the use of Stokes's law, we check the value of Reynolds number and get

$$\text{Re}_d = \frac{(7.5 \times 10^{-3} \text{ m})(0.0773 \text{ m/s})}{7.08 \times 10^{-4} \text{ m}^2/\text{s}}$$

$$= 0.819$$

which is in the allowable range. The terminal velocity in glycerin thus

$$v_\infty = 0.0773 \text{ m/s} \qquad \text{(c)}$$

▶ **12.3**

THE BOUNDARY-LAYER CONCEPT

The observation of a decreasing region of influence of shear stress as the Reynolds number is increased led Ludwig Prandtl to the boundary-layer concept in 1904. According to Prandtl's hypothesis, the effects of fluid friction at high Reynolds numbers are limited to a thin layer near the boundary of a body, hence, the term *boundary layer*. Further, there is no significant pressure change across the boundary layer. This means that the pressure in the boundary layer is the same as the pressure in the inviscid flow outside the boundary layer. The significance of the Prandtl theory lies in the simplification that it allows in the analytical treatment of viscous flows. The pressure, for example, may be obtained from experiment or inviscid flow theory. Thus, the only unknowns are the velocity components.

The boundary layer on a flat plate is shown in Figure 12.5. The thickness of the boundary layer, δ, is arbitrarily taken as the distance away from the surface where the velocity reaches 99% of the free-stream velocity. The thickness is exaggerated for clarity.

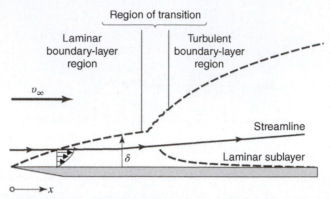

Figure 12.5 Boundary layer on a flat plate. (The thickness is exaggerated for clarity.)

Figure 12.5 illustrates how the thickness of the boundary layer increases with distance x from the leading edge. At relatively small values of x, flow within the boundary layer is laminar, and this is designated as the laminar boundary-layer region. At larger values of x the transition region is shown where fluctuations between laminar and turbulent flows occur within the boundary layer. Finally, for a certain value of x and above, the boundary layer will always be turbulent. In the region in which the boundary layer is turbulent, there exists, as shown, a very thin film of fluid called the *laminar sublayer*, wherein flow is still laminar and large velocity gradients exist.

The criterion for the type of boundary layer present is the magnitude of Reynolds number, Re_x, known as the *local Reynolds number*, based on the distance x from the leading edge. The local Reynolds number is defined as

$$\text{Re}_x \equiv \frac{x v \rho}{\mu} \qquad (12\text{-}4)$$

For flow past a flat plate, as shown in Figure 12.5, experimental data indicate that for

(a) $\text{Re}_x < 2 \times 10^5$ the boundary layer is laminar
(b) $2 \times 10^5 < \text{Re}_x < 3 \times 10^6$ the boundary layer may be either laminar or turbulent
(c) $3 \times 10^6 < \text{Re}_x$ the boundary layer is turbulent

► **12.4**

THE BOUNDARY-LAYER EQUATIONS

The concept of a relatively thin boundary layer at high Reynolds numbers leads to some important simplifications of the Navier–Stokes equations. For incompressible, two-dimensional flow over a flat plate, the Navier–Stokes equations are

$$\rho \left\{ \frac{\partial v_x}{\partial t} + v_x \frac{\partial v_x}{\partial x} + v_y \frac{\partial v_x}{\partial y} \right\} = \frac{\partial \sigma_{xx}}{\partial x} + \frac{\partial \tau_{yx}}{\partial y} \qquad (12\text{-}5)$$

and

$$\rho \left\{ \frac{\partial v_y}{\partial t} + v_x \frac{\partial v_y}{\partial x} + v_y \frac{\partial v_y}{\partial y} \right\} = \frac{\partial \tau_{xy}}{\partial x} + \frac{\partial \sigma_{yy}}{\partial y} \tag{12-6}$$

where $\tau_{xy} = \tau_{yx} = \mu \left(\partial v_x / \partial y + \partial v_y / \partial x \right)$, $\sigma_{xx} = -P + 2\mu(\partial v_x / \partial x)$, and $\sigma_{yy} = -P + 2\mu(\partial v_y / \partial y)$. The shear stress in a thin boundary layer is closely approximated by $\mu \left(\partial v_x / \partial y \right)$. This can be seen by considering the relative magnitudes of $\partial v_x / \partial y$ and $\partial v_y / \partial x$. From Figure 12.5, we may write $v_x|_\delta / v_y|_\delta \sim \mathbb{O}(x/\delta)$, where $\mathbb{O}$ signifies the order of magnitude. Then

$$\frac{\partial v_x}{\partial y} \sim \mathbb{O} \left(\frac{v_x|_\delta}{\delta} \right) \quad \frac{\partial v_y}{\partial x} \sim \mathbb{O} \left(\frac{v_y|_\delta}{x} \right)$$

so

$$\frac{\partial v_x / \partial y}{\partial v_y / \partial x} \sim \mathbb{O} \left(\frac{x}{\delta} \right)^2$$

which, for a relatively thin boundary layer, is a large number, and thus $\partial v_x / \partial y \gg \partial v_y / \partial x$. The normal stress at a large Reynolds number is closely approximated by the negative of the pressure as $\mu(\partial v_x / \partial x) \sim \mathbb{O}(\mu v_\infty / x) = \mathbb{O}(\rho v_\infty^2 / \mathrm{Re}_x)$; therefore, $\sigma_{xx} \simeq \sigma_{yy} \simeq -P$. When these simplifications in the stresses are incorporated, the equations for flow over a flat plate become

$$\rho \left(\frac{\partial v_x}{\partial t} + v_x \frac{\partial v_x}{\partial x} + v_y \frac{\partial v_x}{\partial y} \right) = -\frac{\partial P}{\partial x} + \mu \frac{\partial^2 v_x}{\partial y^2} \tag{12-7}$$

and

$$\rho \left(\frac{\partial v_y}{\partial t} + v_x \frac{\partial v_y}{\partial x} + v_y \frac{\partial v_y}{\partial y} \right) = -\frac{\partial P}{\partial y} + \mu \frac{\partial^2 v_y}{\partial x^2} \tag{12-8}$$

Furthermore,[2] the terms in the second equation are much smaller than those in the first equation, and thus $\partial P / \partial y \simeq 0$; hence $\partial P / \partial x = dP/dx$, which according to Bernoulli's equation is equal to $-\rho v_\infty dv_\infty / dx$.

The final form of equation (12-7) becomes

$$\frac{\partial v_x}{\partial t} + v_x \frac{\partial v_x}{\partial x} + v_y \frac{\partial v_x}{\partial y} = v_\infty \frac{dv_\infty}{dx} + \nu \frac{\partial^2 v_x}{\partial y^2} \tag{12-9}$$

The above equation, and the continuity equation

$$\frac{\partial v_x}{\partial x} + \frac{\partial v_y}{\partial y} = 0 \tag{12-10}$$

are known as the boundary-layer equations.

---

[2] The order of magnitude of each term may be considered as above. For example,
$$v_x \left(\frac{\partial v_y}{\partial x} \right) \sim \mathbb{O} \left(v_\infty \left(\frac{v_\infty}{x} \right) \left(\frac{\delta}{x} \right) \right) = \mathbb{O} \left(\frac{v_\infty^2 \delta}{x^2} \right).$$

BLASIUS'S SOLUTION FOR THE LAMINAR BOUNDARY LAYER ON A FLAT PLATE

One very important case in which an analytical solution of the equations of motion has been achieved is that for the laminar boundary layer on a flat plate in steady flow.

For flow parallel to a flat surface, $v_\infty(x) = v_\infty$ and $dP/dx = 0$, according to the Bernoulli equation. The equations to be solved are now

$$v_x \frac{\partial v_x}{\partial x} + v_y \frac{\partial v_x}{\partial y} = \nu \frac{\partial^2 v_x}{\partial y^2} \tag{12-11a}$$

and

$$\frac{\partial v_x}{\partial x} + \frac{\partial v_y}{\partial y} = 0 \tag{12-11b}$$

with boundary conditions $v_x = v_y = 0$ at $y = 0$, and $v_x = v_\infty$ at $y = \infty$.

Blasius[3] obtained a solution to the set of equations (12-11) by first introducing the stream function, Ψ, as described in Chapter 10, which automatically satisfies the two-dimensional continuity equation, equation (12-11b). This set of equations may be reduced to a single ordinary differential equation by transforming the independent variables x, y, to η and the dependent variables from $\Psi(x, y)$ to $f(\eta)$ where

$$\eta(x, y) = \frac{y}{2} \left(\frac{v_\infty}{\nu x} \right)^{1/2} \tag{12-12}$$

and

$$f(\eta) = \frac{\Psi(x, y)}{(\nu x v_\infty)^{1/2}} \tag{12-13}$$

The appropriate terms in equation (12-11a) may be determined from equations (12-12) and (12-13). The following expressions will result. The reader may wish to verify the mathematics involved.

$$v_x = \frac{\partial \Psi}{\partial y} = \frac{v_\infty}{2} f'(\eta) \tag{12-14}$$

$$v_y = -\frac{\partial \Psi}{\partial x} = \frac{1}{2} \left(\frac{\nu v_\infty}{x} \right)^{1/2} (\eta f' - f) \tag{12-15}$$

$$\frac{\partial v_x}{\partial x} = -\frac{v_\infty \eta}{4x} f'' \tag{12-16}$$

$$\frac{\partial v_x}{\partial y} = \frac{v_\infty}{4} \left(\frac{v_\infty}{\nu x} \right)^{1/2} f'' \tag{12-17}$$

$$\frac{\partial^2 v_x}{\partial y^2} = \frac{v_\infty}{8} \frac{v_\infty}{\nu x} f''' \tag{12-18}$$

Substitution of (12-14) through (12-18) into equation (12-11a) and cancellation gives, as a single ordinary differential equation,

$$f''' + ff'' = 0 \tag{12-19}$$

---

[3] H. Blasius, Grenzshichten in Flüssigkeiten mit kleiner Reibung, *Z. Math. U. Phys. Sci.*, **1**, 1908.

with the appropriate boundary conditions

$$f = f' = 0 \qquad \text{at } \eta = 0$$
$$f' = 2 \qquad \text{at } \eta = \infty$$

Observe that this differential equation, although ordinary, is nonlinear and that, of the end conditions on the variable $f(\eta)$, two are initial values and the third is a boundary value. This equation was solved first by Blasius, using a series expansion to express the function, $f(\eta)$, at the origin and an asymptotic solution to match the boundary condition at $\eta = \infty$. Howarth[4] later performed essentially the same work but obtained more accurate results. Table 12.1 presents the significant numerical results of Howarth. A plot of these values is included in Figure 12.6.

Table 12.1   Values of f, f', f'', and v_x/v_∞ for laminar flow parallel to a flat plate (after Howarth)

$\eta = \dfrac{y}{2}\sqrt{\dfrac{v_\infty}{\nu x}}$	f	f'	f''	$\dfrac{v_x}{v_\infty}$
0	0	0	1.32824	0
0.2	0.0266	0.2655	1.3260	0.1328
0.4	0.1061	0.5294	1.3096	0.2647
0.6	0.2380	0.7876	1.2664	0.3938
0.8	0.4203	1.0336	1.1867	0.5168
1.0	0.6500	1.2596	1.0670	0.6298
1.2	0.9223	1.4580	0.9124	0.7290
1.4	1.2310	1.6230	0.7360	0.8115
1.6	1.5691	1.7522	0.5565	0.8761
1.8	1.9295	1.8466	0.3924	0.9233
2.0	2.3058	1.9110	0.2570	0.9555
2.2	2.6924	1.9518	0.1558	0.9759
2.4	3.0853	1.9756	0.0875	0.9878
2.6	3.4819	1.9885	0.0454	0.9943
2.8	3.8803	1.9950	0.0217	0.9915
3.0	4.2796	1.9980	0.0096	0.9990
3.2	4.6794	1.9992	0.0039	0.9996
3.4	5.0793	1.9998	0.0015	0.9999
3.6	5.4793	2.0000	0.0005	1.0000
3.8	5.8792	2.0000	0.0002	1.0000
4.0	6.2792	2.0000	0.0000	1.0000
5.0	8.2792	2.0000	0.0000	1.0000

A simpler way of solving equation (12-19) has been suggested in Goldstein,[5] who presented a scheme whereby the boundary conditions on the function f are initial values.

If we define two new variables in terms of the constant, C, so that

$$\phi = f/C \tag{12-20}$$

and

$$\xi = C\eta \tag{12-21}$$

[4] L. Howarth, "On the solution of the laminar boundary layer equations," *Proc. Roy. Soc. London*, **A164** 547 (1938).

[5] S. Goldstein, *Modern Developments in Fluid Dynamics*, Oxford University Press, London, 1938, p. 135.

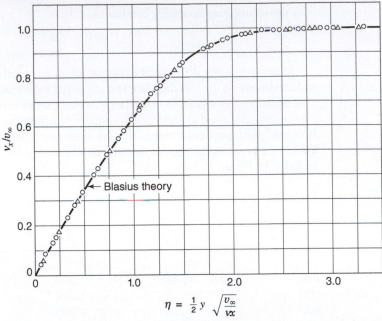

Figure 12.6 Velocity distribution in the laminar boundary layer over a flat plate. Experimental data by J. Nikuradse (monograph, Zentrale F. wiss. Berichtswesen, Berlin, 1942) for the Reynolds number range from 1.08×10^5 to 7.28×10^5.

then the terms in equation (12-19) become

$$f(\eta) = C\phi(\xi) \tag{12-22}$$

$$f' = C^2\phi' \tag{12-23}$$

$$f'' = C^3\phi'' \tag{12-24}$$

and

$$f''' = C^4\phi''' \tag{12-25}$$

The resulting differential equation in $\phi(\xi)$ becomes

$$\phi''' + \phi\phi'' = 0 \tag{12-26}$$

and the initial conditions on ϕ are

$$\phi = 0 \quad \phi' = 0 \quad \phi'' = ? \quad \text{at } \xi = 0$$

The other boundary condition may be expressed as follows:

$$\phi'(\xi) = \frac{f'(\eta)}{C^2} = \frac{2}{C^2} \quad \text{at } \xi = \infty$$

An initial condition may be matched to this boundary condition if we let $f''(\eta = 0)$ equal some constant A; then $\phi''(\xi = 0) = A/C^3$. The constant A must have a certain value to satisfy the original boundary condition on f'. As an estimate we let $\phi''(\xi = 0) = 2$, giving $A = 2C^3$. Thus, initial values of ϕ, ϕ', and ϕ'' are now specified. The estimate on $\phi''(0)$ requires that

$$\phi'(\infty) = \frac{2}{C^2} = 2\left(\frac{2}{A}\right)^{2/3} \tag{12-27}$$

Thus, equation (12-26) may be solved as an initial-value problem with the answer scaled according to equation (12-27) to match the boundary condition at $\eta = \infty$.

The significant results of Blasius's work are the following:

(a) The boundary thickness, δ, is obtained from Table 12.1. When $\eta = 2.5$, we have $v_x/v_\infty \cong 0.99$, thus, designating $y = \delta$ at this point, we have

$$\eta = \frac{y}{2}\sqrt{\frac{v_\infty}{\nu x}} = \frac{\delta}{2}\sqrt{\frac{v_\infty}{\nu x}} = 2.5$$

and thus

$$\delta = 5\sqrt{\frac{\nu x}{v_\infty}}$$

or

$$\frac{\delta}{x} = \frac{5}{\sqrt{\dfrac{v_\infty x}{\nu}}} = \frac{5}{\sqrt{\mathrm{Re}_x}} \tag{12-28}$$

(b) The velocity gradient at the surface is given by equation (12-27):

$$\left.\frac{\partial v_x}{\partial y}\right|_{y=0} = \frac{v_\infty}{4}\left(\frac{v_\infty}{\nu x}\right)^{1/2} f''(0) = 0.332\, v_\infty\sqrt{\frac{v_\infty}{\nu x}} \tag{12-29}$$

As the pressure does not contribute to the drag for flow over a flat plate, all the drag is viscous. The shear stress at the surface may be calculated as

$$\tau_0 = \mu\left.\frac{\partial v_x}{\partial y}\right|_{y=0}$$

Substituting equation (12-29) into this expression, we have

$$\tau_0 = \mu\, 0.332\, v_\infty\sqrt{\frac{v_\infty}{\nu x}} \tag{12-30}$$

The coefficient of skin friction may be determined by employing equation (12-2) as follows:

$$C_{fx} \equiv \frac{\tau}{\rho v_\infty^2/2} = \frac{F_d/A}{\rho v_\infty^2/2} = \frac{0.332\mu v_\infty\sqrt{\dfrac{v_\infty}{\nu x}}}{\rho v_\infty^2/2}$$

$$= 0.664\sqrt{\frac{\nu}{x v_\infty}}$$

$$C_{fx} = \frac{0.664}{\sqrt{\mathrm{Re}_x}} \tag{12-31}$$

Equation (12-31) is a simple expression for the coefficient of skin friction at a particular value of x. For this reason the symbol C_{fx} is used, the x subscript indicating a *local coefficient*.

While it is of interest to know values of C_{fx}, it is seldom that a local value is useful; most often one wishes to calculate the total drag resulting from viscous flow over some surface of finite size. The mean coefficient of skin friction that is helpful in this regard may be determined quite simply from C_{fx} according to

$$F_d = AC_{fL}\frac{\rho v_\infty^2}{2} = \frac{\rho v_\infty^2}{2}\int_A C_{fx}dA$$

or the mean coefficient, designated C_{fL}, is related to C_{fx} by

$$C_{fL} = \frac{1}{A}\int_A C_{fx}dA$$

For the case solved by Blasius, consider a plate of uniform width W, and length L, for which

$$C_{fL} = \frac{1}{L}\int_0^L C_{fx}dx = \frac{1}{L}\int_0^L 0.664\sqrt{\frac{\nu}{v_\infty}}x^{-1/2}dx$$

$$= 1.328\sqrt{\frac{\nu}{Lv_\infty}}$$

$$C_{fL} = \frac{1.328}{\sqrt{\mathrm{Re}_L}} \tag{12-32}$$

▶ **12.6**

FLOW WITH A PRESSURE GRADIENT

In Blasius's solution for laminar flow over a flat plate, the pressure gradient was zero. A much more common flow situation involves flow with a pressure gradient. The pressure gradient plays a major role in flow separation, as can be seen with the aid of the boundary-layer equation (12-7). If we make use of the boundary conditions at the wall $v_x = v_y = 0$, at $y = 0$, equation (12-7) becomes

$$\mu\frac{\partial^2 v_x}{\partial y^2}\bigg|_{y=0} = \frac{dP}{dx} \tag{12-33}$$

which relates the curvature of the velocity profile at the surface to the pressure gradient. Figure 12.7 illustrates the variation in v_x, $\partial v_x/\partial y$, and $\partial^2 v_x/\partial y^2$ across the boundary layer for the case of a zero-pressure gradient.

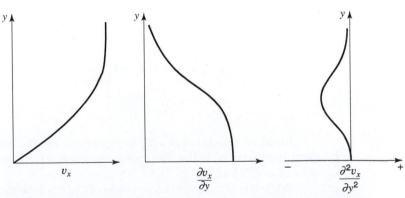

Figure 12.7 Variation in velocity and velocity derivatives across the laminar boundary layer when $dP/dx = 0$.

When $dP/dx = 0$, the second derivative of the velocity at the wall must also be zero; hence, the velocity profile is linear near the wall. Further, out in the boundary layer, the velocity gradient becomes smaller and gradually approaches zero. The decrease in the velocity gradient means that the second derivative of the velocity must be negative. The derivative $\partial^2 v_x / \partial y^2$ is shown as being zero at the wall, negative within the boundary layer, and approaching zero at the outer edge of the boundary layer. It is important to note that the second derivative must approach zero from the negative side as $y \rightarrow \delta$. For values of $dP/dx \neq 0$, the variation in v_x and its derivatives is shown in Figure 12.8.

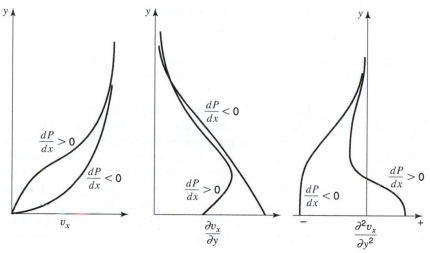

Figure 12.8 Variation in v_x and its derivatives across the boundary layer for various pressure gradients.

A negative pressure gradient is seen to produce a velocity variation somewhat similar to that of the zero-pressure-gradient case. A positive value of dP/dx, however, requires a positive value of $\partial^2 v_x / \partial y^2$ at the wall. As this derivative must approach zero from the negative side, at some point within the boundary layer the second derivative must equal zero. A zero-second derivative, it will be recalled, is associated with an inflection point. The inflection point is shown in the velocity profile of Figure 12.8. We may now turn our attention to the subject of flow separation.

In order for flow separation to occur, the velocity in the layer of fluid adjacent to the wall must be zero or negative, as shown in Figure 12.9. This type of velocity profile is seen to require a point of inflection. As the only type of boundary-layer flow that has an

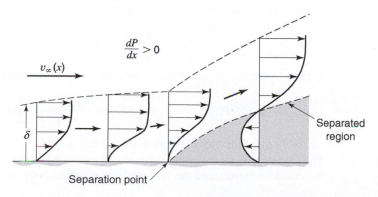

Figure 12.9 Velocity profiles in separated-flow region.

inflection point is flow with a positive pressure gradient, it may be concluded that a positive pressure gradient is necessary for flow separation. For this reason a positive pressure gradient is called an *adverse pressure gradient*. Flow can remain unseparated with an adverse pressure gradient; thus, $dP/dx > 0$ is a necessary but not a sufficient condition for separation. In contrast a negative pressure gradient, in the absence of sharp corners, cannot cause flow separation. Therefore, a negative pressure gradient is called a favorable pressure gradient.

The presence of a pressure gradient also affects the magnitude of the skin friction coefficient, as can be inferred from Figure 12.8. The velocity gradient at the wall increases as the pressure gradient becomes more favorable.

▶ **12.7**

VON KÁRMÁN MOMENTUM INTEGRAL ANALYSIS

The Blasius solution is obviously quite restrictive in application, applying only to the case of a laminar boundary layer over a flat surface. Any situation of practical interest more complex than this involves analytical procedures that have, to the present time, proved inferior to experiment. An approximate method providing information for systems involving other types of flow and having other geometries will now be considered.

Consider the control volume in Figure 12.10. The control volume to be analyzed is of unit depth and is bounded in the *xy* plane by the *x*-axis, here drawn tangent to the surface at point 0; the *y*-axis; the edge of the boundary layer; and a line parallel to the *y*-axis a distance Δx away. We shall consider the case of two-dimensional, incompressible steady flow.

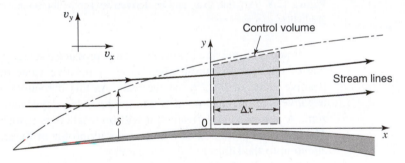

Figure 12.10 Control volume for integral analysis of the boundary layer.

A momentum analysis of the defined control volume involves the application of the *x*-directional scalar form of the momentum theorem

$$\Sigma F_x = \iint_{\text{c.s.}} v_x \rho(\mathbf{v} \cdot \mathbf{n})dA + \frac{\partial}{\partial t} \iiint_{\text{c.v.}} v_x \rho \, dV \tag{5-5a}$$

A term-by-term analysis of the present problem yields the following:

$$\Sigma F_x = P\delta|_x - P\delta|_{x+\Delta x} + \left(P|_x + \frac{P|_{x+\Delta x} - P|_x}{2} \right)(\delta|_{x+\Delta x} - \delta|_x) - \tau_0 \Delta x$$

where δ represents the boundary-layer thickness, and body forces are assumed negligible. The above terms represent the *x*-directional pressure forces on the left, right, and top sides of the control volume, and the frictional force on the bottom, respectively.

The surface integral term becomes

$$\iint_{c.s.} v_x \rho (\mathbf{v} \cdot \mathbf{n}) dA = \int_0^\delta \rho v_x^2 dy \bigg|_{x+\Delta x} - \int_0^\delta \rho v_x^2 dy \bigg|_x - v_\infty \dot{m}_{top}$$

and the accumulation term is

$$\frac{\partial}{\partial t} \iiint_{c.v.} v_x \rho dV = 0$$

as this is a steady-flow situation.

An application of the integral equation for conservation of mass will give

$$\iint_{c.s.} \rho(\mathbf{v.n}) dA + \frac{\partial}{\partial t} \iiint_{c.v.} \rho dV = 0 \tag{4-1}$$

$$\iint_{c.s.} \rho(\mathbf{v.n}) dA = \int_0^\delta \rho v_x dy \bigg|_{x+\Delta x} - \int_0^\delta \rho v_x dy \bigg|_x - \dot{m}_{top}$$

$$\frac{\partial}{\partial t} \iiint_{c.v.} \rho \, dV = 0$$

and the mass-flow rate into the top of the control volume, $\dot{m}_{top}$, may be evaluated as

$$\dot{m}_{top} = \int_0^\delta \rho v_x dy \bigg|_{x+\Delta x} - \int_0^\delta \rho v_x dy \bigg|_x \tag{12-34}$$

The momentum expression, including equation (12-34), now becomes

$$-(P\delta|_{x+\Delta x} - P\delta|_x) + \left(\frac{P|_{x+\Delta x} - P|_x}{2} + P|_x\right)(\delta|_{x+\Delta x} - \delta|_x) - \tau_0 \Delta x$$

$$= \int_0^\delta \rho v_x^2 dy \bigg|_{x+\Delta x} - \int_0^\delta \rho v_x^2 dy \bigg|_x - v_\infty \left(\int_0^\delta \rho v_x dy \bigg|_{x+\Delta x} - \int_0^\delta \rho v_x dy \bigg|_x\right)$$

Rearranging this expression and dividing through by Δx, we get

$$-\left(\frac{P|_{x+\Delta x} - P|_x}{\Delta x}\right)\delta|_{x+\Delta x} + \left(\frac{P|_{x+\Delta x} - P|_x}{2}\right)\left(\frac{\delta|_{x+\Delta x} - \delta|_x}{\Delta x}\right) + \left(\frac{P\delta|_x - P\delta|_x}{\Delta x}\right)$$

$$= \left(\frac{\int_0^\delta \rho v_x^2 dy|_{x+\Delta x} - \int_0^\delta \rho v_x^2 dy|_x}{\Delta x}\right) - v_\infty\left(\frac{\int_0^\delta \rho v_x dy|_{x+\Delta x} - \int_0^\delta \rho v_x dy|_x}{\Delta x}\right) + \tau_0$$

Taking the limit as $\Delta x \to 0$, we obtain

$$-\delta \frac{dP}{dx} = \tau_0 + \frac{d}{dx}\int_0^\delta \rho v_x^2 dy - v_\infty \frac{d}{dx}\int_0^\delta \rho v_x dy \tag{12-35}$$

The boundary-layer concept assumes inviscid flow outside the boundary layer, for which we may write Bernoulli's equation,

$$\frac{dP}{dx} + \rho v_\infty \frac{dv_\infty}{dx} = 0$$

which may be rearranged to the form

$$\frac{\delta}{\rho}\frac{dP}{dx} = \frac{d}{dx}(\delta v_\infty^2) - v_\infty \frac{d}{dx}(\delta v_\infty) \tag{12-36}$$

Notice that the left-hand sides of equations (12-35) and (12-36) are similar. We may thus relate the right-hand sides and, with proper rearrangement, get the result

$$\frac{\tau_0}{\rho} = \left(\frac{d}{dx} v_\infty\right) \int_0^\delta (v_\infty - v_x)\, dy + \frac{d}{dx} \int_0^\delta v_x(v_\infty - v_x)\, dy \tag{12-37}$$

Equation (12-37) is the von Kármán momentum integral expression, named in honor of Theodore von Kármán, who first developed it.

Equation (12-37) is a general expression whose solution requires a knowledge of the velocity, v_x, as a function of distance from the surface, y. The accuracy of the final result will depend on how closely the assumed velocity profile approaches the real one.

As an example of the application of equation (12-37), let us consider the case of laminar flow over a flat plate, a situation for which an exact answer is known. In this case the free-stream velocity is constant; therefore, $(d/dx)\,v_\infty = 0$ and equation (12-36) simplifies to

$$\frac{\tau_0}{\rho} = \frac{d}{dx} \int_0^\delta v_x(v_\infty - v_x)\, dy \tag{12-38}$$

An early solution to equation (12-38) was achieved by Pohlhausen, who assumed for the velocity profile a cubic function

$$v_x = a + by + cy^2 + dy^3 \tag{12-39}$$

The constants a, b, c, and d may be evaluated if we know certain boundary conditions that must be satisfied in the boundary layer. These are

$$(1) \quad v_x = 0 \quad \text{at } y = 0$$

$$(2) \quad v_x = v_\infty \quad \text{at } y = \delta$$

$$(3) \quad \frac{\partial v_x}{\partial y} = 0 \quad \text{at } y = \delta$$

and

$$(4) \quad \frac{\partial^2 v_x}{\partial y^2} = 0 \quad \text{at } y = 0$$

Boundary condition (4) results from equation (12-33), which states that the second derivative at the wall is equal to the pressure gradient. As the pressure is constant in this case, $\partial^2 v_x/\partial y^2 = 0$. Solving for a, b, c, and d from these conditions, we get

$$a = 0 \quad b = \frac{3}{2\delta} v_\infty \quad c = 0 \quad d = -\frac{v_\infty}{2\delta^3}$$

which, when substituted in equation (12-39), give the form of the velocity profile

$$\frac{v_x}{v_\infty} = \frac{3}{2}\left(\frac{y}{\delta}\right) - \frac{1}{2}\left(\frac{y}{\delta}\right)^3 \tag{12-40}$$

Upon substitution, equation (12-38) becomes

$$\frac{3\nu}{2}\frac{v_\infty}{\delta_\infty} = \frac{d}{dx}\int_0^\delta v_\infty^2 \left(\frac{3}{2}\frac{y}{\delta} - \frac{1}{2}\left(\frac{y}{\delta}\right)^3\right)\left(1 - \frac{3}{2}\frac{y}{\delta} + \frac{1}{2}\left(\frac{y}{\delta}\right)^3\right) dy$$

or, after integrating

$$\frac{3}{2}\nu\frac{v_\infty}{\delta} = \frac{39}{280}\frac{d}{dx}(v_\infty^2 \delta)$$

As the free-stream velocity is constant, a simple ordinary differential equation in δ results

$$\delta \, d\delta = \frac{140}{13} \frac{\nu \, dx}{v_\infty}$$

This, upon integration, yields

$$\frac{\delta}{x} = \frac{4.64}{\sqrt{Re_x}} \tag{12-41}$$

The local skin-friction coefficient, C_{fx}, is given by

$$C_{fx} \equiv \frac{\tau_0}{\frac{1}{2}\rho v_\infty^2} = \frac{2\nu}{v_\infty^2}\frac{3}{2}\frac{v_\infty}{\delta} = \frac{0.646}{\sqrt{Re_x}} \tag{12-42}$$

Integration of the local skin-friction coefficient between $x=0$ and $x=L$ as in equation (12-32) yields

$$C_{fL} = \frac{1.292}{\sqrt{Re_L}} \tag{12-43}$$

Comparing equations (12-41), (12-42), and (12-43) with exact results obtained by Blasius for the same situation, equations (12-28), (12-31), and (12-32), we observe a difference of about 7% in δ and 3% in C_f. This difference could, of course, have been smaller had the assumed velocity profile been a more accurate representation of the actual profile.

This comparison has shown the utility of the momentum integral method for the solution of the boundary layer and indicates a procedure that may be used with reasonable accuracy to obtain values for boundary-layer thickness and the coefficient of skin friction where an exact analysis is not feasible. The momentum integral method may also be used to determine the shear stress from the velocity profile.

▶ **12.8**

DESCRIPTION OF TURBULENCE

Turbulent flow is the most frequently encountered type of viscous flow, yet the analytical treatment of turbulent flow is not nearly so well developed as that of laminar flow. In this section, we examine the phenomenon of turbulence, particularly with respect to the mechanism of turbulent contributions to momentum transfer.

In a turbulent flow the fluid and flow variables vary with time. The instantaneous velocity vector, for example, will differ from the average velocity vector in both magnitude and direction. Figure 12.11 illustrates the type of time dependence experienced by the axial component of the velocity for turbulent flow in a tube. While the velocity in Figure 12.11(a) is seen to be steady in its mean value, small random fluctuations in velocity occur about the mean value. Accordingly, we may express the fluid and flow variables in terms of a mean value and a fluctuating value. For example, the x-directional velocity is expressed as

$$v_x = \bar{v}_x(x, y, z) + v'_x(x, y, z, t) \tag{12-44}$$

Here $\bar{v}_x(x, y, z)$ represents the time-averaged velocity at the point (x, y, z)

$$\bar{v}_x = \frac{1}{t_1}\int_0^{t_1} v_x(x, y, z, t)\,dt \tag{12-45}$$

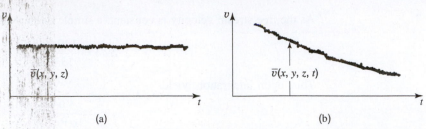

Figure 12.11 Time dependence of velocity in a turbulent flow: (a) steady mean flow; (b) unsteady mean flow.

where t_1 is a time that is very long in comparison with the duration of any fluctuation. The mean value of $v'_x(x, y, z, t)$ is zero, as expressed by

$$\bar{v}'_x = \frac{1}{t_1} \int_0^{t_1} v'_x(x, y, z, t)\,dt = 0 \qquad (12\text{-}46)$$

Hereafter, $\bar{Q}$ will be used to designate the time average of the general property, Q, according to $\bar{Q} = 1/t_1 \int_0^{t_1} Q(x, y, z, t)\,dt$. While the mean value of the turbulent fluctuations is zero, these fluctuations contribute to the mean value of certain flow quantities. For example, the mean kinetic energy per unit volume is

$$\overline{KE} = \frac{1}{2}\rho\overline{[(\bar{v}_x + v'_x)^2 + (\bar{v}_y + v'_y)^2 + (\bar{v}_z + v'_z)^2]}$$

The average of a sum is the sum of the averages; hence, the kinetic energy becomes

$$\overline{KE} = \frac{1}{2}\rho\left\{\overline{(\bar{v}_x^2 + 2\bar{v}_x v'_x + v'^2_x)} + \overline{(\bar{v}_y^2 + 2\bar{v}_y v'_y + v'^2_y)} + \overline{(\bar{v}_z^2 + 2\bar{v}_z v'_z + v'^2_z)}\right\}$$

or, since, $\overline{\bar{v}_x v'_x} = \bar{v}_x \overline{v'_x} = 0$,

$$\overline{KE} = \frac{1}{2}\rho(\bar{v}_x^2 + \bar{v}_y^2 + \bar{v}_z^2 + \overline{v'^2_x} + \overline{v'^2_y} + \overline{v'^2_z}) \qquad (12\text{-}47)$$

A fraction of the total kinetic energy of a turbulent flow is seen to be associated with the magnitude of the turbulent fluctuations. It can be shown that the rms (root mean square) value of the fluctuations, $(\overline{v'^2_x} + \overline{v'^2_y} + \overline{v'^2_z})^{1/2}$, is a significant quantity. The level or *intensity of turbulence* is defined as

$$I \equiv \frac{\sqrt{(\overline{v'^2_x} + \overline{v'^2_y} + \overline{v'^2_z})/3}}{v_\infty} \qquad (12\text{-}48)$$

where v_∞ is the mean velocity of the flow. The intensity of turbulence is a measure of the kinetic energy of the turbulence and is an important parameter in flow simulation. In model testing, simulation of turbulent flows requires not only duplication of Reynolds number but also duplication of the turbulent kinetic energy. Thus, the measurement of turbulence is seen to be a necessity in many applications.

The general discussion so far has indicated the fluctuating nature of turbulence. The random nature of turbulence lends itself to statistical analysis. We shall now turn our attention to the effect of the turbulent fluctuations on momentum transfer.

TURBULENT SHEARING STRESSES

In Chapter 7, the random molecular motion of the molecules was shown to result in a net momentum transfer between two adjacent layers of fluid. If the (molecular) random motions give rise to momentum transfer, it seems reasonable to expect that large-scale fluctuations, such as those present in a turbulent flow, will also result in a net transfer of momentum. Using an approach similar to that of Section 7.3, let us consider the transfer of momentum in the turbulent flow illustrated in Figure 12.12.

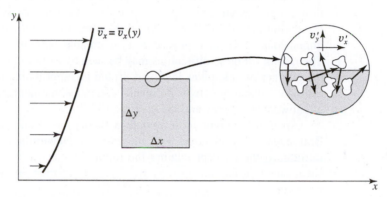

Figure 12.12 Turbulent motion at the surface of a control volume.

The relation between the macroscopic momentum flux due to the turbulent fluctuations and the shear stress may be seen from the control-volume expression for linear momentum:

$$\Sigma \mathbf{F} = \iint_{\text{c.s.}} \mathbf{v}\rho(\mathbf{v} \cdot \mathbf{n})dA + \frac{\partial}{\partial t}\iiint_{\text{c.v.}} \mathbf{v}\rho\, dV \qquad (5\text{-}4)$$

The flux of x-directional momentum across the top of the control surface is

$$\iint_{\text{top}} \mathbf{v}\rho(\mathbf{v} \cdot \mathbf{n})\, dA = \iint_{\text{top}} v_y'\rho(\bar{v}_x + v_x')dA \qquad (12\text{-}49)$$

If the mean value of the momentum flux over a period of time is evaluated for the case of steady mean flow, the time derivative in equation (5-4) is zero; thus,

$$\overline{\Sigma F_x} = \iint \overline{v_y'\,\rho(\bar{v}_x + v_x')dA} = \iint \overline{v_y'\rho\bar{v}_x}^{\,0}\, dA + \iint \overline{\rho v_y'v_x'}\, dA \qquad (12\text{-}50)$$

The presence of the turbulent fluctuations is seen to contribute a mean x directional momentum flux of $\rho v_x'v_y's$ per unit area. Although the turbulent fluctuations are functions of position and time, their analytical description has not been achieved, even for the simplest case. The close analogy between the molecular exchange of momentum in laminar flow and the macroscopic exchange of momentum in turbulent flow suggests that the term $\rho v_x'v_y'$ be regarded as a shear stress. Transposing this term to the left-hand side of equation (5-4) and incorporating it with the shear stress due to molecular momentum transfer, we see that the total shear stress becomes

$$\tau_{yx} = \mu \frac{d\bar{v}_x}{dy} - \overline{\rho v'_x v'_y} \tag{12-51}$$

The turbulent contribution to the shear stress is called the *Reynolds stress*. In turbulent flows it is found that the magnitude of the Reynolds stress is much greater than the molecular contribution except near the walls.

An important difference between the molecular and turbulent contributions to the shear stress is to be noted. Whereas the molecular contribution is expressed in terms of a property of the fluid and a derivative of the mean flow, the turbulent contribution is expressed solely in terms of the fluctuating properties of the flow. Further, these flow properties are not expressible in analytical terms. While Reynolds stresses exist for multidimensional flows,[6] the difficulties in analytically predicting even the one-dimensional case have proved insurmountable without the aid of experimental data. The reason for these difficulties may be seen by examining the number of equations and the number of unknowns involved. In the incompressible turbulent boundary layer, for example, there are two pertinent equations–momentum and continuity–and four unknowns: $\bar{v}_x$, $\bar{v}_y$, v'_x, and v'_y.

An early attempt to formulate a theory of turbulent shear stress was made by Boussinesq.[7] By analogy with the form of Newton's viscosity relation, Boussinesq introduced the concept relating the turbulent shear stress to the shear-strain rate. The shear stress in laminar flow is $\tau_{yx} = \mu (dv_x/dy)$; thus, by analogy, the Reynolds stress becomes

$$(\tau_{yx})_{\text{turb}} = A_t \frac{d\bar{v}_x}{dy}$$

where A_t is the *eddy viscosity*. Subsequent refinements have led to the introduction of the *eddy diffusivity of momentum*, $\epsilon_M \equiv A_t/\rho$, and thus

$$(\tau_{yx})_{\text{turb}} = \rho \epsilon_M \frac{d\bar{v}_x}{dy} \tag{12-52}$$

The difficulties in analytical treatment still exist, however, as the eddy diffusivity, ϵ_M, is a property of the flow and not of the fluid. By analogy with the kinematic viscosity in a laminar flow, it may be observed that the units of the eddy diffusivity are L^2/t.

▶ **12.10**

THE MIXING-LENGTH HYPOTHESIS

A general similarity between the mechanism of transfer of momentum in turbulent flow and that in laminar flow permits an analog to be made for turbulent shear stress. The analog to the mean free path in molecular momentum exchange for the turbulent case is the mixing length proposed by Prandtl[8] in 1925. Consider the simple turbulent flow shown in Figure 12.13.

---

[6] The existence of the Reynolds stresses may also be shown by taking the time average of the Navier–Stokes equations.

[7] J. Boussinesq, *Mem. Pre. par div. Sav.*, XXIII (1877).

[8] L. Prandtl, *ZAMM*, **5**, 136 (1925).

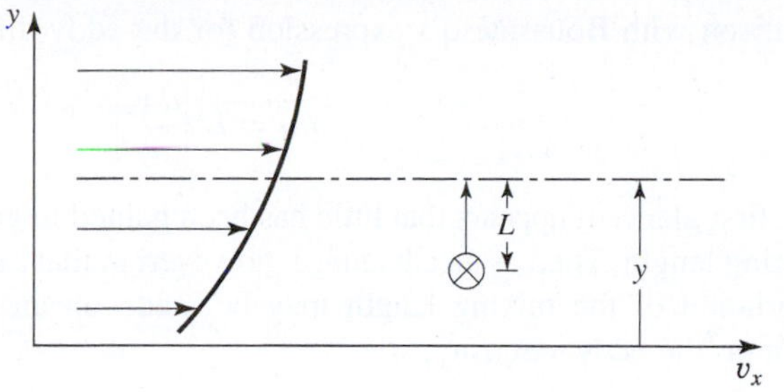

Figure 12.13 The Prandtl mixing length.

The velocity fluctuation v_x' is hypothesized as being due to the y-directional motion of a "lump" of fluid through a distance L. In undergoing translation the lump of fluid retains the mean velocity from its point of origin. Upon reaching a destination, a distance L from the point of origin, the lump of fluid will differ in mean velocity from that of the adjacent fluid by an amount $\bar{v}_x|_{y-L} - \bar{v}_x|_y$. If the lump of fluid originated at $y+L$, the velocity difference would be $\bar{v}_x|_{y+L} - \bar{v}_x|_y$. The instantaneous value of $v_x'|_y$ is then $\bar{v}_x|_{y\pm L} - \bar{v}_x|_y$, the sign of L, of course, depending on the point of origin with respect to y. Further, the mixing length, although finite, is assumed to be small enough to permit the velocity difference to be written as

$$\bar{v}_x|_{y\pm L} - \bar{v}_x|_y = \pm L \frac{d\bar{v}_x}{dy}$$

and thus

$$v_x' = \pm L \frac{d\bar{v}_x}{dy} \tag{12-52}$$

The concept of the mixing length is somewhat akin to that of the mean free path of a gas molecule. The important differences are its magnitude and dependence upon flow properties rather than fluid properties. With an expression for v_x' at hand, an expression for v_y' is necessary to determine the turbulent shear stress, $-\rho \overline{v_x' v_y'}$.

Prandtl assumed that v_x' must be proportional to v_y'. If v_x' and v_y' were completely independent, then the time average of their product would be zero. Both the continuity equation and experimental data show that there is some degree of proportionality between v_x' and v_y'. Using the fact that $v_y' \sim v_x'$, Prandtl expressed the time average, $\overline{v_x' v_y'}$, as

$$\overline{v_x' v_y'} = -(\text{constant})L^2 \left| \frac{d\bar{v}_x}{dy} \right| \frac{d\bar{v}_x}{dy} \tag{12-53}$$

The constant represents the unknown proportionality between v_x' and v_y' as well as their correlation in taking the time average. The minus sign and the absolute value were introduced to make the quantity $\overline{v_x' v_y'}$ agree with experimental observations. The constant in (12-53), which is unknown, may be incorporated into the mixing length, which is also unknown, giving

$$\overline{v_x' v_y'} = -L^2 \left| \frac{d\bar{v}_x}{dy} \right| \frac{d\bar{v}_x}{dy} \tag{12-54}$$

Comparison with Boussinesq's expression for the eddy diffusivity yields

$$\epsilon_M = L^2 \left| \frac{d\bar{v}_x}{dy} \right| \tag{12-55}$$

At first glance it appears that little has been gained in going from the eddy viscosity to the mixing length. There is an advantage, however, in that assumptions regarding the nature and variation of the mixing length may be made on an easier basis than assumptions concerning the eddy viscosity.

▶ **12.11**

VELOCITY DISTRIBUTION FROM THE MIXING-LENGTH THEORY

One of the important contributions of the mixing-length theory is its use in correlating velocity profiles at large Reynolds numbers. Consider a turbulent flow as illustrated in Figure 12.13. In the neighborhood of the wall the mixing length is assumed to vary directly with y, and thus $L = Ky$, where K remains a dimensionless constant to be determined via experiment. The shear stress is assumed to be entirely due to turbulence and to remain constant over the region of interest. The velocity $\bar{v}_x$ is assumed to increase in the y direction, and thus $d\bar{v}_x/dy = |d\bar{v}_x/dy|$. Using these assumptions, we may write the turbulent shear stress as

$$\tau_{yx} = \rho K^2 y^2 \left(\frac{d\bar{v}_x}{dy} \right)^2 = \tau_0 \text{(a constant)}$$

or

$$\frac{d\bar{v}_x}{dy} = \frac{\sqrt{\tau_0/\rho}}{Ky}$$

The quantity $\sqrt{\tau_0/\rho}$ is observed to have units of velocity. Integration of the above equation yields

$$\bar{v}_x = \frac{\sqrt{\tau_0/\rho}}{K} \ln y + C \tag{12-56}$$

where C is a constant of integration. This constant may be evaluated by setting $\bar{v}_x = \bar{v}_{x\,\text{max}}$ at $y = h$, whereby

$$\frac{\bar{v}_{x\,\text{max}} - \bar{v}_x}{\sqrt{\tau_0/\rho}} = -\frac{1}{K} \left[\ln \frac{y}{h} \right] \tag{12-57}$$

The constant K was evaluated by Prandtl[9] and Nikuradse[10] from data on turbulent flow in tubes and found to have a value of 0.4. The agreement of experimental data for turbulent flow in smooth tubes with equation (12-57) is quite good, as can be seen from Figure 12.14.

The empirical nature of the preceding discussion cannot be overlooked. Several assumptions regarding the flow are known to be incorrect for flow in tubes, namely that the shear stress is not constant and that the geometry was treated from a two-dimensional viewpoint rather than an axisymmetric viewpoint. In view of these obvious difficulties, it is remarkable that equation (12-57) describes the velocity profile so well.

[9] L. Prandtl, *Proc. Intern. Congr. Appl. Mech.*, 2nd Congr., Zurich (1927), 62.
[10] J. Nikuradse, *VDI-Forschungsheft*, **356** (1932).

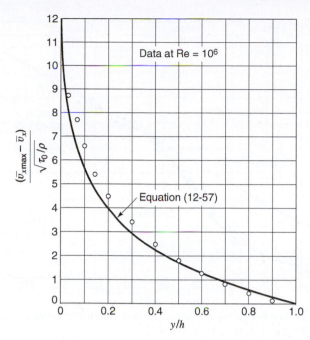

Figure 12.14 Comparison of data for flow in smooth tube with equation (12-57).

▶ **12.12**

THE UNIVERSAL VELOCITY DISTRIBUTION

For turbulent flow in smooth tubes, equation (12-57) may be taken as a basis for a more general development. Recalling that the term $\sqrt{\tau_0/\rho}$ has the units of velocity, we may introduce a dimensionless velocity $\bar{v}_x/\sqrt{\tau_0/\rho}$. Defining

$$v^+ \equiv \frac{\bar{v}_x}{\sqrt{\tau_0/\rho}} \tag{12-58}$$

we may write equation (12-56) as

$$v^+ = \frac{1}{K}[\ln y] + C \tag{12-59}$$

The left-hand side of (12-59) is, of course, dimensionless; therefore, the right-hand side of this equation must also be dimensionless. A pseudo-Reynolds number is found useful in this regard. Defining

$$y^+ \equiv \frac{\sqrt{\tau_0/\rho}}{\nu} y \tag{12-60}$$

we find that equation (12-59) becomes

$$v^+ = \frac{1}{K}\ln \frac{\nu y^+}{\sqrt{\tau_0/\rho}} + C = \frac{1}{K}(\ln y^+ + \ln \beta) \tag{12-61}$$

where the constant β is dimensionless.

Equation (12-61) indicates that for flow in smooth tubes $v^+ = f(y^+)$ or

$$v^+ \equiv \frac{\bar{v}_x}{\sqrt{\tau_0/\rho}} = f\left\{\ln \frac{y\sqrt{\tau_0/\rho}}{\nu}\right\} \tag{12-62}$$

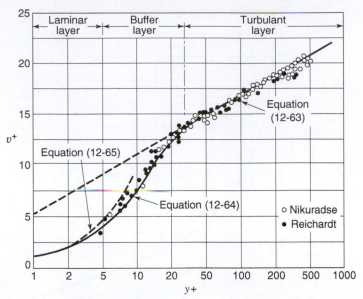

Figure 12.15 Velocity correlation for flow in circular smooth tubes at high Reynolds number (H. Reichardt, NACA TM1047, 1943).

The range of validity of equation (12-61) may be observed from a plot (see Figure 12.15) of v^+ versus $\ln y^+$, using the data of Nikuradse and Reichardt.

Three distinct regions are apparent: a turbulent core, a buffer layer, and a laminar sublayer. The velocity is correlated as follows:

for turbulent core, $y^+ \geq 30$:

$$v^+ = 5.5 + 2.5 \ln y^+ \tag{12-63}$$

for the buffer layer, $30 \geq y^+ \geq 5$:

$$v^+ = -3.05 + 5 \ln y^+ \tag{12-64}$$

for the laminar sublayer, $5 > y^+ > 0$:

$$v^+ = y^+ \tag{12-65}$$

Equations (12-63) through (12-65) define the *universal velocity distribution*. Because of the empirical nature of these equations, there are, of course, inconsistencies. The velocity gradient, for example, at the center of the tube predicted by (12-63) is not zero. In spite of this and other inconsistencies, these equations are extremely useful for describing flow in smooth tubes.

In rough tubes, the scale of the roughness e is found to affect the flow in the turbulent core, but not in the laminar sublayer. The constant β in equation (12-61) becomes $\ln \beta = 3.4 - \ln[(e\sqrt{\tau_0/\rho})/\nu]$ for rough tubes. As the wall shear stress appears in the revised expression for $\ln \beta$, it is important to note that wall roughness affects the magnitude of the shear stress in a turbulent flow.

▶ **12.13**

FURTHER EMPIRICAL RELATIONS FOR TURBULENT FLOW

Two important experimental results that are helpful in studying turbulent flows are the power-law relation for velocity profiles and a turbulent-flow shear-stress relation due to Blasius. Both of these relations are valid for flow adjacent to smooth surfaces.

For flow in smooth circular tubes, it is found that over much of the cross section the velocity profile may be correlated by

$$\frac{\overline{v}_x}{\overline{v}_{x\,max}} = \left(\frac{y}{R}\right)^{1/n} \tag{12-66}$$

where R is the radius of the tube and n is a slowly varying function of Reynolds number. The exponent n is found to vary from a value of 6 at Re = 4000 to 10 at Re = 3,200,000. At Reynolds numbers of 10^5 the value of n is 7. This leads to the frequently used one-seventh-power law, $\overline{v}_x/\overline{v}_{x\,max} = (y/R)^{1/7}$. The power-law profile has also been found to represent the velocity distribution in boundary layers. For boundary layers of thickness δ, the power law is written

$$\frac{\overline{v}_x}{\overline{v}_{x\,max}} = \left(\frac{y}{\delta}\right)^{1/n} \tag{12-67}$$

The power-law profile has two obvious difficulties: the velocity gradients at the wall and those at δ are incorrect. This expression indicates that the velocity gradient at the wall is infinite and that the velocity gradient at δ is nonzero.

In spite of these inconsistencies, the power law is extremely useful in connection with the von Kármán integral relation, as we shall see in Section 12.14.

Another useful relation is Blasius's correlation for shear stress. For pipe-flow Reynolds numbers up to 10^5 and flat-plate Reynolds numbers up to 10^7, the wall shear stress in a turbulent flow is given by

$$\tau_0 = 0.0225 \rho \overline{v}_{x\,max}^2 \left(\frac{\nu}{\overline{v}_{x\,max}\, y_{max}}\right)^{1/4} \tag{12-68}$$

where $y_{max} = R$ in pipes and $y_{max} = \delta$ for flat surfaces.

▶ **12.14**

THE TURBULENT BOUNDARY LAYER ON A FLAT PLATE

The variation in boundary-layer thickness for turbulent flow over a smooth flat plate may be obtained from the von Kármán momentum integral. The manner of approximation involved in a turbulent analysis differs from that used previously. In a laminar flow, a simple polynomial was assumed to represent the velocity profile. In a turbulent flow, we have seen that the velocity profile depends upon the wall shear stress and that no single function adequately represents the velocity profile over the entire region. The procedure we shall follow in using the von Kármán integral relation in a turbulent flow is to utilize a simple profile for the integration with the Blasius correlation for the shear stress. For a zero-pressure gradient, the von Kármán integral relation is

$$\frac{\tau_0}{\rho} = \frac{d}{dx} \int_0^\delta v_x (v_\infty - v_x)\, dy \tag{12-38}$$

Employing the one-seventh-power law for v_x and the Blasius relation, equation (12-68), for τ_0, we see that equation (12-38), becomes

$$0.0225 v_\infty^2 \left(\frac{\nu}{v_\infty \delta}\right)^{1/4} = \frac{d}{dx} \int_0^\delta v_\infty^2 \left\{ \left(\frac{y}{\delta}\right)^{1/7} - \left(\frac{y}{\delta}\right)^{2/7} \right\} dy \tag{12-69}$$

where the free-stream velocity, v_∞, is written in place of $\overline{v}_{x\,max}$. Performing the indicated integration and differentiation, we obtain

$$0.0225\left(\frac{\nu}{v_\infty\delta}\right)^{1/4} = \frac{7}{72}\frac{d\delta}{dx} \tag{12-70}$$

which becomes, upon integration,

$$\left(\frac{\nu}{v_\infty}\right)^{1/4} x = 3.45\,\delta^{5/4} + C \tag{12-71}$$

If the boundary layer is assumed to be turbulent from the leading edge, $x = 0$ (a poor assumption), the above equation may be rearranged to give

$$\frac{\delta}{x} = \frac{0.376}{\mathrm{Re}_x^{1/5}} \tag{12-72}$$

The local skin-friction coefficient may be computed from the Blasius relation for shear stress, equation (12-67), to give

$$C_{fx} = \frac{0.0576}{\mathrm{Re}_x^{1/5}} \tag{12-73}$$

Several things are to be noted about these expressions. First, they are limited to values of $\mathrm{Re}_x < 10^7$, by virtue of the Blasius relation. Second, they apply only to smooth flat plates. Last, a major assumption has been made in assuming the boundary layer to be turbulent from the leading edge. The boundary layer is known to be laminar initially and to undergo transition to turbulent flow at a value of Re_x of about 2×10^5. We shall retain the assumption of a completely turbulent boundary layer for the simplicity it affords; it is recognized, however, that this assumption introduces some error in the case of a boundary layer that is not completely turbulent.

A comparison of a laminar and a turbulent boundary layer can be made from Blasius's laminar-flow solution and equations (12-28), (12-72), and (12-73). At the same Reynolds number, the turbulent boundary layer is observed to be thicker, and is associated with a larger skin-friction coefficient. While it would appear then that a laminar boundary layer is more desirable, the reverse is generally true. In most cases of engineering interest, a turbulent boundary layer is desired because it resists separation better than a laminar boundary layer. The velocity profiles in laminar and turbulent boundary layers are compared qualitatively in Figure 12.16.

It can be seen that the turbulent boundary layer has a greater mean velocity, and hence both greater momentum and energy, than the laminar boundary layer. The greater momentum and energy permit the turbulent boundary layer to remain unseparated for a greater distance in the presence of an adverse pressure gradient than would be the case for a laminar boundary layer.

Consider a flat plate with transition from laminar flow to turbulent flow occurring on the plate. If transition from laminar flow to turbulent flow is assumed to occur abruptly (for computational purposes), a problem arises in how to join the laminar boundary layer to the turbulent layer at the point of transition. The prevailing procedure is to equate the momentum thicknesses, equation (12-44), at the transition point. That is, at the start of the turbulent portion of the boundary layer, the momentum thickness, θ, is equal to the momentum thickness at the end of the laminar portion of the boundary layer.

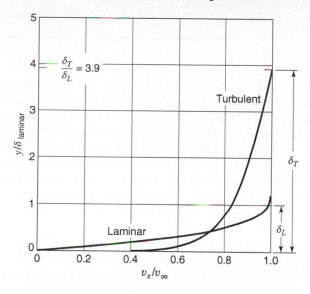

Figure 12.16 Comparison of velocity profiles in laminar and turbulent boundary layers. The Reynolds number is 500,000.

The general approach for turbulent boundary layers with a pressure gradient involves the use of the von Kármán momentum integral as given in equation (12-46). Numerical integration is required.

► **12.15**

FACTORS AFFECTING THE TRANSITION FROM LAMINAR TO TURBULENT FLOW

The velocity profiles and momentum-transfer mechanisms have been examined for both laminar and turbulent flow regimes and found to be quite different. Laminar flow has also been seen to undergo transition to turbulent flow at certain Reynolds numbers.

So far the occurrence of transition has been expressed in terms of the Reynolds number alone, while a variety of factors other than Re actually influence transition. The Reynolds number remains, however, the principal parameter for predicting transition.

Table 12.2 indicates the influence of some of these factors on the transition Reynolds number.

Table 12.2   Factors affecting the Reynolds number of transition from laminar to turbulent flow

Factor	Influence
Pressure gradient	Favorable pressure gradient retards transition; unfavorable pressure gradient hastens it.
Free-stream turbulence	Free-stream turbulence decreases transition Reynolds number.
Roughness	No effect in pipes; decreases transition in external flow.
Suction	Suction greatly increases transition Re.
Wall curvatures	Convex curvature increases transition Re. Concave curvature decreases it.
Wall temperature	Cool walls increase transition Re. Hot walls decrease it.

► **12.16**

CLOSURE

Viscous flow has been examined in this chapter for both internal and external geometries. Two approaches were employed for analyzing laminar boundary-layer flows—exact analysis using boundary-layer equations and approximate integral methods. For turbulent boundary-layer analysis along a plane surface, an integral method was employed.

Concepts of skin friction and drag coefficients were introduced and quantitative relationships were developed for both internal and external flows.

Approaches to modeling turbulent flows were introduced, culminating in expressions for the "universal" velocity distribution. This approach considers turbulent flows to be described in three parts: the laminar sublayer, the transition or buffer layer, and the turbulent core.

The concepts developed in this chapter will be used to develop important expressions for momentum flow application in the next chapter. Similar applications will be developed in the sections related to both heat and mass transfer in later sections of this text.

CHAPTER 13

Flow in Closed Conduits

Many of the theoretical relations that have been developed in the previous chapters apply to special situations such as inviscid flow, incompressible flow, and the like. Some experimental correlations were introduced in Chapter 12 for turbulent flow in or past surfaces of simple geometry. In this chapter, an application of the material that has been developed thus far will be considered with respect to a situation of considerable engineering importance—namely, fluid flow, both laminar and turbulent, through closed conduits.

▶ **13.1**

DIMENSIONAL ANALYSIS OF CONDUIT FLOW

As an initial approach to conduit flow, we shall utilize dimensional analysis to obtain the significant parameters for the flow of an incompressible fluid in a straight, horizontal, circular pipe of constant cross section.

The significant variables and their dimensional expressions are represented in the following table:

Variable	Symbol	Dimension
Pressure drop	ΔP	M/Lt^2
Velocity	v	L/t
Pipe diameter	D	L
Pipe length	L	L
Pipe roughness	e	L
Fluid viscosity	μ	M/Lt
Fluid density	ρ	M/L^3

Each of the variables is familiar, with the exception of the pipe roughness, symbolized e. The roughness is included to represent the condition of the pipe surface and may be thought of as the characteristic height of projections from the pipe wall, hence, the dimension of length.

According to the Buckingham π theorem, the number of independent dimensionless groups to be formed with these variables is four. If the core group consists of the variables v, D, and ρ, then the groups to be formed are as follows:

$$\pi_1 = v^a D^b \rho^c \Delta P$$
$$\pi_2 = v^d D^e \rho^f L$$
$$\pi_3 = v^g D^h \rho^i e$$
$$\pi_4 = v^j D^k \rho^l \mu$$

Carrying out the procedure outlined in Chapter 11 to solve for the unknown exponents in each group, we see that the dimensionless parameters become

$$\pi_1 = \frac{\Delta P}{\rho v^2}$$

$$\pi_2 = \frac{L}{D}$$

$$\pi_3 = \frac{e}{D}$$

and

$$\pi_4 = \frac{vD\rho}{\mu}$$

The first π group is the Euler number. As the pressure drop is due to fluid friction, this parameter is often written with $\Delta P/\rho$ replaced by gh_L, where h_L is the "head loss"; thus, π_1 becomes

$$\frac{h_L}{v^2/g}$$

The third π group, the ratio of pipe roughness to diameter, is the so-called relative roughness. The fourth π group is the Reynolds number, Re.

A functional expression resulting from dimensional analysis may be written as

$$\frac{h_L}{v^2/g} = \phi_1\left(\frac{L}{D}, \frac{e}{D}, \text{Re}\right) \qquad (13\text{-}1)$$

Experimental data have shown that the head loss in fully developed flow is directly proportional to the ratio L/D. This ratio may, then, be removed from the functional expression, giving

$$\frac{h_L}{v^2/g} = \frac{L}{D}\phi_2\left(\frac{e}{D}, \text{Re}\right) \qquad (13\text{-}2)$$

The function ϕ_2, which varies with the relative roughness and Reynolds number, is designated f, the friction factor. Expressing the head loss from equation (13-2) in terms of f, we have

$$h_L = 2f_f\frac{L}{D}\frac{v^2}{g} \qquad (13\text{-}3)$$

With the factor 2 inserted in the right-hand side, equation (13-3) is the defining relation for f_f, the *Fanning friction factor*. Another friction factor in common use is the *Darcy friction factor*, f_D, defined by equation (13-4).

$$h_L = f_D\frac{L}{D}\frac{v^2}{2g} \qquad (13\text{-}4)$$

Quite obviously, $f_D = 4\,f_f$. The student should be careful to note which friction factor is being used to properly calculate frictional head loss by either equation (13-3) or (13-4). The Fanning friction factor, f_f, will be used exclusively in this text. The student may easily verify that the Fanning friction factor is the same as the skin-friction coefficient C_f.

Our task now becomes that of determining suitable relations for f_f from that theory and experimental data.

► **13.2**

FRICTION FACTORS FOR FULLY DEVELOPED LAMINAR, TURBULENT, AND TRANSITION FLOW IN CIRCULAR CONDUITS

Laminar Flow

Some analysis has been performed already for incompressible laminar flow. As fluid behavior can be described quite well in this regime according to Newton's viscosity relation, we should expect no difficulty in obtaining a functional relationship for f_f in the case of laminar flow. Recall that, for closed conduits, the flow may be considered laminar for values of the Reynolds number less than 2300.

From Chapter 8, the Hagen–Poiseuille equation was derived for incompressible, laminar, conduit flow

$$-\frac{dP}{dx} = 32\frac{\mu v_{avg}}{D^2} \tag{8-9}$$

Separating variables and integrating this expression along a length, L, of the passage, we get

$$-\int_{P_0}^{P} dP = 32\frac{\mu v_{avg}}{D^2}\int_0^L dx$$

and

$$\Delta P = 32\frac{\mu v_{avg}L}{D^2} \tag{13-5}$$

Recall that equation (8-9) held for the case of fully developed flow; thus, v_{avg} does not vary along the length of the passage.

Forming an expression for frictional head loss from equation (13-5), we have

$$h_L = \frac{\Delta P}{\rho g} = 32\frac{\mu v_{avg}L}{g\rho D^2} \tag{13-6}$$

Combining this equation with equation (13-3), the defining relation for f_f

$$h_L = 32\frac{\mu v_{avg}L}{g\rho D^2} = 2f_f\frac{L}{D}\frac{v^2}{g}$$

and solving for f_f, we obtain

$$f_f = 16\frac{\mu}{D v_{avg}\rho} = \frac{16}{Re} \tag{13-7}$$

This very simple result indicates that f_f is inversely proportional to Re in the laminar flow range; the friction factor is *not* a function of pipe roughness for values of Re < 2300, but varies only with the Reynolds number.

This result has been experimentally verified and is the manifestation of the viscous effects in the fluid, damping out any irregularities in the flow caused by protrusions from a rough surface.

Turbulent Flow

In the case of turbulent flow in closed conduits or pipes, the relation for f_f is not so simply obtained or expressed as in the laminar case. No easily derived relation such as the Hagen–Poiseuille law applies; however, some use can be made of the velocity profiles expressed in Chapter 12 for turbulent flow. All development will be based on circular conduits; thus, we are primarily concerned with pipes or tubes. In turbulent flow a distinction must be made between smooth- and rough-surfaced tubes.

Smooth Tubes The velocity profile in the turbulent core has been expressed as

$$v^+ = 5.5 + 2.5 \ln y^+ \tag{12-63}$$

where the variables v^+ and y^+ are defined according to the relations

$$v^+ \equiv \frac{\bar{v}}{\sqrt{\tau_0/\rho}} \tag{12-58}$$

and

$$y^+ \equiv \frac{\sqrt{\tau_0/\rho}}{\nu} y \tag{12-60}$$

The average velocity in the turbulent core for flow in a tube of radius R can be evaluated from equation (12-63) as follows:

$$v_{avg} = \frac{\int_0^A \bar{v}\, dA}{A}$$

$$= \frac{\sqrt{\tau_0/\rho} \int_0^R \left(2.5 \ln\left\{ \frac{\sqrt{\tau_0/\rho}\, y}{\nu} \right\} + 5.5 \right) 2\pi r\, dr}{\pi R^2}$$

Letting $y = R - r$, we obtain

$$v_{avg} = 2.5 \sqrt{\tau_0/\rho} \ln\left\{ \frac{\sqrt{\tau_0/\rho}\, R}{\nu} \right\} + 1.75 \sqrt{\tau_0/\rho} \tag{13-8}$$

The functions $\sqrt{\tau_0/\rho}$ and C_f are related according to equation (12-2). As C_f and f_f are equivalent, we may write

$$\frac{v_{avg}}{\sqrt{\tau_0/\rho}} = \frac{1}{\sqrt{f_f/2}} \tag{13-9}$$

The substitution of equation (13-9) into equation (13-8) yields

$$\frac{1}{\sqrt{f_f/2}} = 2.5 \ln\left\{ \frac{R}{\nu}\, v_{avg} \sqrt{f_f/2} \right\} + 1.75 \tag{13-10}$$

Rearranging the argument of the logarithm into Reynolds number form, and changing to $\log_{10}$, we see that equation (13-10) reduces to

$$\frac{1}{\sqrt{f_f}} = 4.06 \log_{10}\left\{ \mathrm{Re}\sqrt{f_f} \right\} - 0.60 \tag{13-11}$$

This expression gives the relation for the friction factor as a function of Reynolds number for turbulent flow in smooth circular tubes. The preceding development was first performed by von Kármán.[1] Nikuradse,[2] from experimental data, obtained the equation

$$\frac{1}{\sqrt{f_f}} = 4.0 \log_{10}\left\{\mathrm{Re}\sqrt{f_f}\right\} - 0.40 \tag{13-12}$$

which is very similar to equation (13-11).

Rough Tubes   By an analysis similar to that used for smooth tubes, von Kármán developed equation (13-13) for turbulent flow in rough tubes:

$$\frac{1}{\sqrt{f_f}} = 4.06 \log_{10}\frac{D}{e} + 2.16 \tag{13-13}$$

which compares very well with the equation obtained by Nikuradse from experimental data:

$$\frac{1}{\sqrt{f_f}} = 4.0 \log_{10}\frac{D}{e} + 2.28 \tag{13-14}$$

Nikuradse's results for fully developed pipe flow indicated that the surface condition—that is, roughness—had nothing to do with the transition from laminar to turbulent flow. Once the Reynolds number becomes large enough that flow is fully turbulent, then either equation (13-12) or (13-14) must be used to obtain the proper value for f_f. These two equations are quite different in that equation (13-12) expresses f_f as a function of Re only and equation (13-14) gives f_f as a function only of the relative roughness. The difference is, of course, that the former equation is for smooth tubes and the latter for rough tubes. The question that naturally arises at this point is "what is 'rough'?"

It has been observed from experiment that equation (13-12) describes the variation in f_f for a range in Re, even for rough tubes. Beyond some value of Re, this variation deviates from the smooth-tube equation and achieves a constant value dictated by the tube roughness as expressed by equation (13-14). The region wherein f_f varies both with Re and e/D is called the *transition region*. An empirical equation describing the variation of f_f in the transition region has been proposed by Colebrook.[3]

$$\frac{1}{\sqrt{f_f}} = 4 \log_{10}\frac{D}{e} + 2.28 - 4 \log_{10}\left(4.67\frac{D/e}{\mathrm{Re}\sqrt{f_f}} + 1\right) \tag{13-15}$$

Equation (13-15) is applicable to the transition region above a value of $(D/e)/(\mathrm{Re}\sqrt{f_f}) = 0.01$. Below this value, the friction factor is independent of the Reynolds number, and the flow is said to be *fully turbulent*.

To summarize the development of this section, the following equations express the friction-factor variation for the surface and flow conditions specified:

For laminar flow (Re < 2300),

$$f_f = \frac{16}{\mathrm{Re}} \tag{13-7}$$

[1] T. von Kármán, NACA TM 611, 1931.

[2] J. Nikuradse, *VDI-Forschungsheft*, **356**, 1932.

[3] C. F. Colebrook, *J. Inst. Civil Engr.* (London) II, **133** (1938–39).

For turbulent flow (smooth pipe, Re > 3000),

$$\frac{1}{\sqrt{f_f}} = 4.0 \log_{10}\left\{\text{Re}\sqrt{f_f}\right\} - 0.40 \tag{13-12}$$

For turbulent flow (rough pipe (Re > 3000, $D/e)/(\text{Re}\sqrt{f_f}) < 0.01$),

$$\frac{1}{\sqrt{f_f}} = 4.0 \log_{10}\frac{D}{e} + 2.28 \tag{13-14}$$

And for transition flow,

$$\frac{1}{\sqrt{f_f}} = 4 \log_{10}\frac{D}{e} + 2.28 - 4 \log_{10}\left(4.67\frac{D/e}{\text{Re}\sqrt{f_f}} + 1\right) \tag{13-15}$$

► 13.3

FRICTION FACTOR AND HEAD-LOSS DETERMINATION FOR PIPE FLOW

Friction Factor

A single friction-factor plot based upon equations (13-7), (13-13), (13-14), and (13-15) has been presented by Moody.[4] Figure 13.1 is a plot of the Fanning friction factor vs. the Reynolds number for a range of values of the roughness parameter e/D.

When using the friction-factor plot, Figure 13.1, it is necessary to know the value of the roughness parameter that applies to a pipe of given size and material. After a pipe or tube has been in service for some time, its roughness may change considerably, making the determination of e/D quite difficult. Moody has presented a chart, reproduced in Figure 13.2, by which a value of e/D can be determined for a given size tube or pipe constructed of a particular material.

The combination of these two plots enables the frictional head loss for a length, L, of pipe having diameter D to be evaluated, using the relation

$$h_L = 2 f_f \frac{L}{D}\frac{v^2}{g} \tag{13-3}$$

Recently Haaland[5] has shown that over the range $10^8 \geq \text{Re} \geq 4 \times 10^4$, $0.05 \geq e/D \geq 0$, the friction factor may be expressed (within ±1.5%) as

$$\frac{1}{\sqrt{f_f}} = -3.6 \log_{10}\left[\frac{6.9}{\text{Re}} + \left(\frac{e}{3.7D}\right)^{10/9}\right] \tag{13-15a}$$

This expression allows explicit calculation of the friction factor.

Head Losses Due to Fittings

The frictional head loss calculated from equation (13-3) is only a part of the total head loss that must be overcome in pipe lines and other fluid-flow circuits. Other losses may occur due

---

[4] L. F. Moody, *Trans. ASME*, **66**, 671 (1944).
[5] S. E. Haaland, *Trans. ASME, JFE*, **105**, 89 (1983).

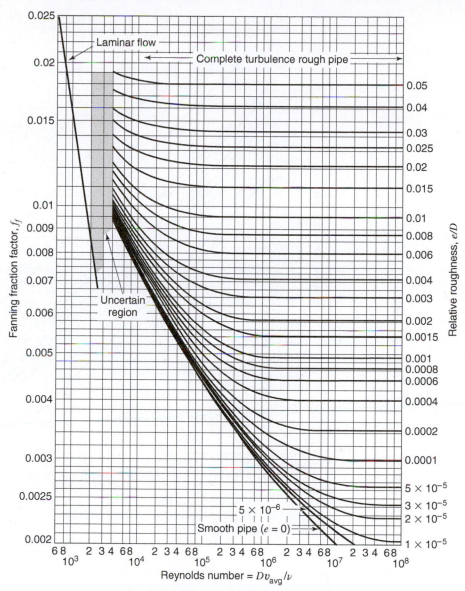

Figure 13.1 The Fanning friction factor as a function of Re and D/e.

to the presence of valves, elbows, and any other fittings that involve a change in the direction of flow or in the size of the flow passage. The head losses resulting from such fittings are functions of the geometry of the fitting, the Reynolds number, and the roughness. As the losses in fittings, to a first approximation, have been found to be independent of the Reynolds number, the head loss may be evaluated as

$$h_L = \frac{\Delta P}{\rho g} = K \frac{v^2}{2g} \tag{13-16}$$

where K is a coefficient depending upon the fitting.

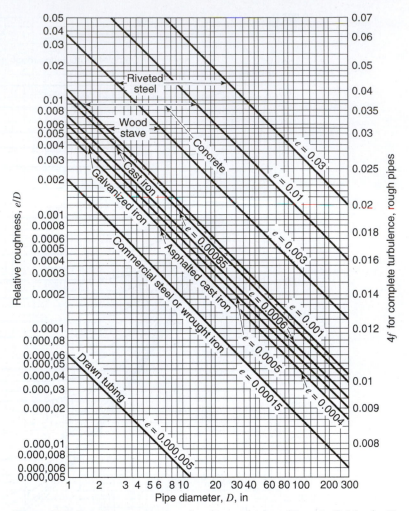

Figure 13.2 Roughness parameters for pipes and tubes. (From L. F. Moody, *Trans. ASME* (1944).) Values of *e* are given in feet.

An equivalent method of determining the head loss in fittings is to introduce an *equivalent length*, L_{eq}, such that

$$h_L = 2f_f \frac{L_{eq}}{D} \frac{v^2}{g}$$ (13-17)

where L_{eq} is the length of pipe that produces a head loss equivalent to the head loss in a particular fitting. Equation (13-17) is seen to be in the same form as equation (13-3), and thus the total head loss for a piping system may be determined by adding the equivalent lengths for the fittings to the pipe length to obtain the total effective length of the pipe.

Comparison of equations (13-16) and (13-17) shows that the constant K must be equal to $4f_f L_{eq}/D$. Although equation (13-17) appears to be dependent upon the Reynolds number because of the appearance of the Fanning friction factor, it is not. The assumption made in both equations (13-16) and (13-17) is that the Reynolds number is large enough that the flow is fully turbulent. The friction coefficient for a given fitting, then, is

Table 13.1 Friction loss factors for various pipe fittings

Fitting	K	L_{eq}/D
Globe valve, wide open	7.5	350
Angle valve, wide open	3.8	170
Gate valve, wide open	0.15	7
Gate valve, $\frac{3}{4}$ open	0.85	40
Gate valve, $\frac{1}{2}$ open	4.4	200
Gate valve, $\frac{1}{4}$ open	20	900
Standard 90° elbow	0.7	32
Short-radius 90° elbow	0.9	41
Long-radius 90° elbow	0.4	20
Standard 45° elbow	0.35	15
Tee, through side outlet	1.5	67
Tee, straight through	0.4	20
180° bend	1.6	75

dependent only upon the roughness of the fitting. Typical values for K and L_{eq}/D are given in Table 13.1.

Recall that the head loss due to a sudden expansion is calculated in Chapter 6, with the result given in equation (6-16).

Equivalent Diameter

Equations (13-16) and (13-17) are based upon a circular flow passage. These equations may be used to estimate the head loss in a closed conduit of any configuration if an "equivalent diameter" for a noncircular flow passage is used. An equivalent diameter is calculated according to

$$D_{eq} = 4 \frac{\text{cross-sectional area of flow}}{\text{wetted perimeter}} \qquad (13\text{-}18)$$

The ratio of the cross-sectional area of flow to the wetted perimeter is called the hydraulic radius.

The reader may verify that D_{eq} corresponds to D for a circular flow passage. One type of noncircular flow passage often encountered in transfer processes is the annular area between two concentric pipes. The equivalent diameter for this configuration is determined as follows:

$$\text{Cross-sectional area} = \frac{\pi}{4}(D_0^2 - D_i^2)$$

$$\text{Wetted perimeter} = \pi(D_0 + D_i)$$

yielding

$$D_{eq} = 4 \frac{\pi/4}{\pi} \frac{(D_0^2 - D_i^2)}{(D_0 + D_i)} = D_0 - D_i \qquad (13\text{-}19)$$

This value of D_{eq} may now be used to evaluate the Reynolds number, the friction factor, and the frictional head loss, using the relations and methods developed previously for circular conduits.

► **13.4**

PIPE-FLOW ANALYSIS

Application of the equations and methods developed in the previous sections is common in engineering systems involving pipe networks. Such analyses are always straightforward but may vary as to the complexity of calculation. The following four example problems are typical, but by no means all-inclusive, of the types of problems found in engineering practice.

Example 1

Water at 59 F flows through a straight section of a 6-in-ID cast-iron pipe with an average velocity of 4 fps. The pipe is 120 ft long, and there is an increase in elevation of 2 ft from the inlet of the pipe to its exit.

Find the power required to produce this flow rate for the specified conditions.

The control volume in this case is the pipe and the water it encloses. Applying the energy equation to this control volume, we obtain

$$\frac{\delta Q}{dt} - \frac{\delta W_s}{dt} - \frac{\delta W_\mu}{dt} = \iint_{\text{c.s.}} \rho\left(e + \frac{P}{\rho}\right)(\mathbf{v} \cdot \mathbf{n}) \, dA + \frac{\partial}{\partial t} \iiint_{\text{c.v.}} \rho e \, dV \tag{6-10}$$

An evaluation of each term yields

$$\frac{\delta Q}{dt} = 0 \qquad \frac{\delta W_s}{dt} = \dot{W}$$

$$\iint_{\text{c.s.}} \rho\left(e + \frac{P}{\rho}\right)(\mathbf{v} \cdot \mathbf{n}) \, dA = \rho A v_{\text{avg}}\left(\frac{v_2^2}{2} + g y_2 + \frac{P_2}{\rho} + u_2 - \frac{v_1^2}{2} - g y_1 - \frac{P_1}{\rho} - u_1\right)$$

$$\frac{\partial}{\partial t} \iiint_{\text{c·v}} \rho e \, dV = 0$$

and

$$\frac{\delta W_\mu}{dt} = 0$$

The applicable form of the energy equation written on a unit mass basis is now

$$\dot{W}/\dot{m} = \frac{v_1^2 - v_2^2}{2} + g(y_1 - y_2) + \frac{P_1 - P_2}{\rho} + u_1 - u_2$$

and with the internal energy change written as $g h_L$, the frictional head loss, the expression for w becomes

$$\dot{W}/\dot{m} = \frac{v_1^2 - v_2^2}{2} + g(y_1 - y_2) + \frac{P_1 - P_2}{\rho} - g h_L$$

Assuming the fluid at both ends of the control volume to be at atmospheric pressure, $(P_1 - P_2)/\rho = 0$, and for a pipe of constant cross section $(v_1^2 - v_2^2)/2 = 0$, giving for $\dot{W}/\dot{m}$

$$\dot{W}/\dot{m} = g(y_1 - y_2) - g h_L$$

Evaluating h_L, we have

$$\text{Re} = \frac{(\frac{1}{2})(4)}{1.22 \times 10^{-5}} = 164{,}000$$

$$\frac{e}{D} = 0.0017 \quad \text{(from Figure 13.2)}$$

$$f_f = 0.0059 \quad \text{(from equation (13-15a))}$$

yielding

$$h_L = \frac{2(0.0059)(120 \text{ ft})(16 \text{ ft}^2/\text{s}^2)}{(0.5 \text{ ft})(32.2 \text{ ft/s}^2)} = 1.401 \text{ ft}$$

The power required to produce the specified flow conditions thus becomes

$$\dot{W} = \frac{-g((-2 \text{ ft}) - 1.401 \text{ ft})}{550 \text{ ft lb}_f/\text{hp-s}} \left[\frac{62.3 \text{ lb}_m/\text{ft}^3}{32.2 \text{ lb}_m \text{ft/s}^2 \text{ lb}_f} \left(\frac{\pi}{4}\right) \left(\frac{1}{2}\text{ft}\right)^2 \left(4\frac{\text{ft}}{\text{s}}\right) \right]$$
$$= 0.300 \text{ hp}$$

Example 2

A heat exchanger is required, which will be able to handle 0.0567 m³/s of water through a smooth pipe with an equivalent length of 122 m. The total pressure drop is 103,000 Pa. What size pipe is required for this application?

Once again, applying equation (6-10), we see that a term-by-term evaluation gives

$$\frac{\delta Q}{dt} = 0 \qquad \frac{\delta W_s}{dt} = 0 \qquad \frac{\delta W_\mu}{dt} = 0$$

$$\iint_{\text{c.s.}} \rho\left(e + \frac{P}{\rho}\right)(\mathbf{v} \cdot \mathbf{n}) \, dA = \rho A \, v_{\text{avg}} \left(\frac{v_2^2}{2} + gy_2 + \frac{P_2}{\rho} + u_2 - \frac{v_1^2}{2} - gy_1 - \frac{P_1}{\rho} - u_1 \right)$$

$$\frac{\partial}{\partial t} \iiint_{\text{c.v.}} \rho e \, dV = 0$$

and the applicable equation for the present problem is

$$0 = \frac{P_2 - P_1}{\rho} + gh_L$$

The quantity desired, the diameter, is included in the head-loss term but cannot be solved for directly, as the friction factor also depends on D. Inserting numerical values into the above equation and solving, we obtain

$$0 = -\frac{103,000 \text{ Pa}}{1000 \text{ kg/m}^3} + 2f_f \left(\frac{0.0567}{\pi D^2/4}\right)^2 \frac{\text{m}^2}{\text{s}^2} \cdot \frac{122 \text{ m}}{D} \frac{g}{\text{m} g}$$

or

$$0 = -103 + 1.27 \frac{f_f}{D^5}$$

The solution to this problem must now be obtained by trial and error. A possible procedure is the following:

1. Assume a value for f_f.
2. Using this f_f, solve the above equation for D.
3. Calculate Re with this D.
4. Using e/D and the calculated Re, check the assumed value of f_f.
5. Repeat this procedure until the assumed and calculated friction-factor values agree.

Carrying out these steps for the present problem, the required pipe diameter is 0.132 m (5.2 in).

Example 3

An existing heat exchanger has a cross section as shown in Figure 13.3 with nine 1-in-OD tubes inside a 5-in-ID pipe. For a 5-ft length of heat exchanger, what flow rate of water at 60 F can be achieved in the shell side of this unit for a pressure drop of 3 psi?

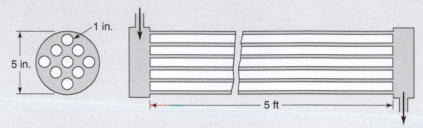

Figure 13.3 Shell-and-tube heat-exchanger configuration.

An energy-equation analysis using equation (6-10) will follow the same steps as in Example 2 in this chapter, yielding, as the governing equation,

$$0 = \frac{P_2 - P_1}{\rho} + gh_L$$

The equivalent diameter for the shell is evaluated as follows:

$$\text{Flow area} = \frac{\pi}{4}(25 - 9) = 4\pi \text{ in.}^2$$

$$\text{Wetted perimeter} = \pi(5 + 9) = 14\pi \text{ in.}$$

thus

$$D_{eq} = 4\frac{4\pi}{14\pi} = 1.142 \text{ in.}$$

Substituting the proper numerical values into the energy equation for this problem reduces it to

$$0 = -\frac{3 \text{ lb}_f/\text{in.}^2(144 \text{ in.}^2/\text{ft}^2)}{1.94 \text{ slugs/ft}^3} + 2f_f v_{avg}^2 \text{ ft}^2/\text{s}^2 \frac{5 \text{ ft}}{(1.142/12) \text{ ft}} \frac{g}{g}$$

or

$$0 = -223 + 105 f_f v_{avg}^2$$

As f_f cannot be determined without a value of Re, which is a function of v_{avg}, a simple trial-and-error procedure such as the following might be employed:

1. Assume a value for f_f.
2. Calculate v_{avg} from the above expression.
3. Evaluate Re from this value of v_{avg}.
4. Check the assumed value of f_f using equation (13-15a).
5. If the assumed and calculated values for f_f do not agree, repeat this procedure until they do.

Employing this method, we find the velocity to be 23.6 fps, giving a flow rate for this problem of 2.06 ft^3/min (0.058 m^3/s).

Notice that in each of the last two examples in which a trial-and-error approach was used, the assumption of f_f was made initially. This was not, of course, the only way to approach these problems; however, in both cases a value for f_f could be assumed within a much closer range than either D or v_{avg}.

Example 4

Water at 80 F flows through a 0.21-ft-diameter 50-ft-long cast-iron pipe at 53 ft/min. Using Figures 13.1 and 13.2, determine the fanning friction factor and then the pressure drop in this system.

The first thing to do is calculate the Reynolds number. The density and viscosity of water at the given temperature can be found in Appendix I. The Reynolds number is

$$\text{Re} = \frac{\rho v D}{\mu} = \frac{(62.2 \text{ lb}_m/\text{ft}^3)(53 \text{ ft/min})(\text{min}/60 \text{ sec})(0.21 \text{ ft})}{0.578 \times 10^{-3} \frac{\text{lb}_m}{\text{ft s}}} = 2 \times 10^4$$

Figure 13.2 contains roughness parameters for various materials, including cast iron, for which the figure is $e = 0.00085$. We next calculate the relative roughness:

$$\frac{e}{D} = \frac{0.00085}{0.21 \text{ ft}} = 0.004$$

We now turn to Figure 13.1. Along the right-side y-axis are values for the relative roughness. We find 0.004 on that axis. Next, follow the line for 0.004 until we reach the value for the Reynolds number from the x-axis. The calculated Reynolds number from above is 2×10^4; where these lines intersect, we read the value for the fanning friction factor on the left-side y-axis to be 0.00825.

Since the flow is turbulent, we use equations (13-3) and (13-16) in combination to calculate the pressure drop:

$$h_L = \frac{\Delta P}{\rho g} = 2 f_f \frac{L_{eq}}{D} \frac{v^2}{g}$$

Rearrange and solve for the pressure change:

$$\Delta P = 2 \rho f_f \frac{L_{eq}}{D} v^2 = 2 \frac{(62.2 \text{ lb}_m/\text{ft}^3)}{32.174 \text{ lb}_m \text{ft}/\text{lb}_f \text{s}^2}(0.00825)\frac{50 \text{ ft}}{0.21 \text{ ft}}(0.88 \text{ ft/s})^2 = 5.88 \text{ lb}_f/\text{ft}^2$$

▶ **13.5**

FRICTION FACTORS FOR FLOW IN THE ENTRANCE TO A CIRCULAR CONDUIT

The development and problems in the preceding section have involved flow conditions that did not change along the axis of flow. This condition is often met, and the methods just described will be adequate to evaluate and predict the significant flow parameters.

In many real flow systems this condition is never realized. A boundary layer forms on the surface of a pipe, and its thickness increases in a similar manner to that of the boundary layer on a flat plate as described in Chapter 12. The buildup of the boundary layer in pipe flow is depicted in Figure 13.4.

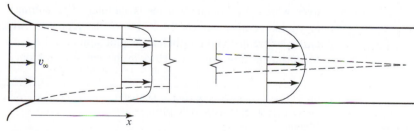

Figure 13.4 Boundary-layer buildup in a pipe.

A boundary layer forms on the inside surface and occupies a larger amount of the flow area for increasing values of x, the distance downstream from the pipe entrance. At some value of x, the boundary layer fills the flow area. The velocity profile will not change the downstream from this point, and the flow is said to be *fully developed*. The distance downstream from the pipe entrance to where flow becomes fully developed is called the entrance length, symbolized as L_e. Observe that the fluid velocity outside the boundary layer increases with x, as is required to satisfy continuity. The velocity at the center of the pipe finally reaches a value of $2v_\infty$ for fully developed laminar flow.

The entrance length required for a fully developed velocity profile to form in laminar flow has been expressed by Langhaar[6] according to

$$\frac{L_e}{D} = 0.0575 \, \text{Re} \tag{13-20}$$

where D represents the inside diameter of the pipe. This relation, derived analytically, has been found to agree well with experiment.

There is no relation available to predict the entrance length for a fully developed turbulent velocity profile. An additional factor that affects the entrance length in turbulent flow is the nature of the entrance itself. The reader is referred to the work of Deissler[7] for experimentally obtained turbulent velocity profiles in the entrance region of the circular pipes. A general conclusion of the results of Deissler and others is that the turbulent velocity profile becomes fully developed after a minimum distance of 50 diameters downstream from the entrance.

The reader should realize that the entrance length for the velocity profile differs considerably from the entrance length for the velocity gradient at the wall. As the friction factor is a function of dv/dy at the pipe surface, we are also interested in this starting length.

Two conditions exist in the entrance region, which cause the friction factor to be greater than in fully developed flow. The first of these is the extremely large wall velocity gradient right at the entrance. The gradient decreases in the downstream direction, becoming constant before the velocity profile becomes fully developed. The other factor is the existence of a "core" of fluid outside the viscous layer whose velocity must increase as dictated by continuity. The fluid in the core is thus being accelerated, thereby producing an additional drag force whose effect is incorporated in the friction factor.

The friction factor for laminar flow in the entrance to a pipe has been studied by Langhaar.[8] His results indicated the friction factor to be highest in the vicinity of the entrance, then to decrease smoothly to the fully developed flow value. Figure 13.5 is a qualitative representation of this variation. Table 13.2 gives the results of Langhaar for the average friction factor between the entrance and a location, a distance x from the entrance.

For turbulent flow in the entrance region, the friction factor as well as the velocity profile is difficult to express. Deissler[9] has analyzed this situation and presented his results graphically.

Even for very high free-stream velocities, there will be some portion of the entrance over which the boundary layer is laminar. The entrance configuration, as well as the Reynolds number, affects the length of the pipe over which the laminar boundary layer exists before becoming turbulent. A plot similar to Figure 13.5 is presented in Figure 13.6 for turbulent-flow friction factors in the entrance region.

---

[6] H. L. Langhaar, *Trans. ASME*, **64**, A-55 (1942).
[7] R. G. Deissler, NACA TN 2138 (1950).
[8] *Op cit.*
[9] R. G. Deissler, NACA TN 3016 (1953).

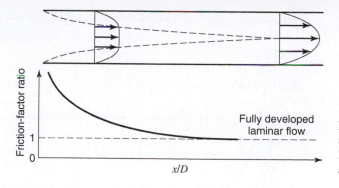

Figure 13.5 Velocity profile and friction-factor variation for laminar flow in the region near a pipe entrance.

Table 13.2   Average friction factor for laminar flow in the entrance to a circular pipe

$\dfrac{x/D}{\text{Re}}$	$f_f\left(\dfrac{x}{D}\right)$
0.000205	0.0530
0.000830	0.0965
0.001805	0.1413
0.003575	0.2075
0.00535	0.2605
0.00838	0.340
0.01373	0.461
0.01788	0.547
0.02368	0.659
0.0341	0.845
0.0449	1.028
0.0620	1.308
0.0760	1.538

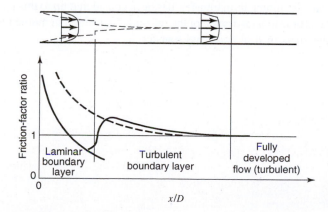

Figure 13.6 Velocity profile and friction-factor variation in turbulent flow in the region near a pipe entrance.

The foregoing description of the entrance region has been qualitative. For an accurate analytical consideration of a system involving entrance-length phenomena, Deissler's results portrayed in Figure 13.7 may be utilized.

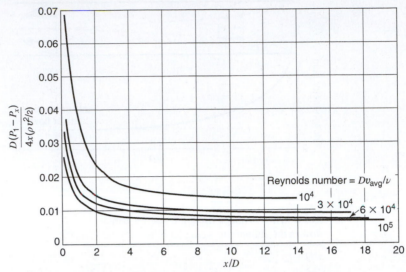

Figure 13.7 Static pressure drop due to friction and momentum change in the entrance to a smooth, horizontal, circular tube (Deissler).

It is important to realize that in many situations flow is never fully developed; thus, the friction factor will be higher than that predicted from the equations for fully developed flow or the friction-factor plot.

▶ **13.6**

CLOSURE

The information and techniques presented in this chapter have included applications of the theory developed in earlier chapters supported by correlations of experimental data.

The chapters to follow will be devoted to heat and mass transfer. One specific type of transfer, momentum transfer, has been considered up to this point. The student will find that he is able to apply much of the information learned in momentum transfer to counterparts in the areas of heat and mass transfer.

CHAPTER 14

Fluid Machinery

In this chapter we will examine the operating principles of mechanical devices that exchange fluid energy and mechanical work. A *pump* is a machine whose purpose is to apply mechanical energy to a fluid, thereby generating flow, or producing a higher pressure, or both. A *turbine* does just the opposite—producing work through the application of fluid energy.

There are two principal types of fluid machines—*positive-displacement machines* and *turbomachines.* In positive-displacement machines, a fluid is confined in a chamber whose volume is varied. Examples of positive-displacement-type machines are shown in Figure 14.1.

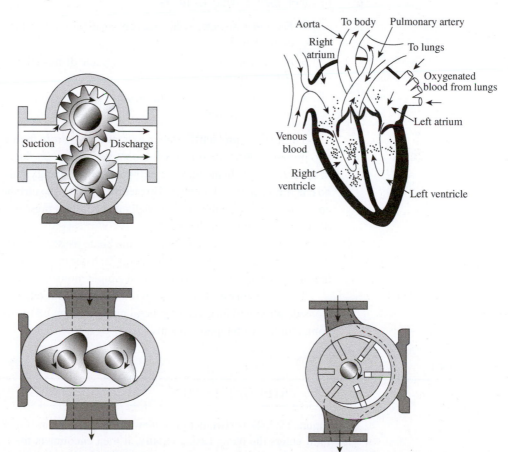

Figure 14.1 Some examples of positive-displacement configurations.

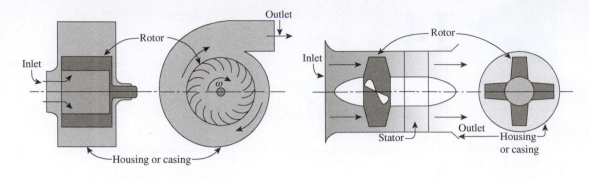

Radial or centrifugal flow   Axial flow

Figure 14.2 Turbomachines.

Turbomachines, as the name implies, involve rotary motion. Window fans and aircraft propellers are examples of *unshrouded* turbomachines. Pumps used with liquids generally have *shrouds* that confine and direct the flow. The two general types of pumps in this category are shown in Figure 14.2. The designations *radial flow* and *axial flow* refer to the direction of fluid flow relative to the axis of rotation of the rotating element.

The term *pump* is generally used when the working fluid is a liquid. If the fluid is a gas or vapor, the following terms are used:

- *Fans* are associated with relatively small pressure changes, on the order of $\Delta P \sim 35$ cm of H_2O (0.5 psi).
- *Blowers* are of both positive and variable displacement types, with ΔP up to 2.8 m of H_2O (40 psi).
- *Compressors* are of both positive and variable configurations having delivery pressures as high as 69 MPa (10^3 psi).

Turbines, as previously stated, extract energy from high-pressure fluids. They are of two primary types, *impulse* and *reaction*, which convert fluid energy into mechanical work in different ways. In the impulse turbine, the high-energy fluid is converted, by means of a nozzle, into a high-velocity jet. This jet then strikes the turbine blades as they pass. In this configuration, the jet flow is essentially at a constant pressure. The basic analysis of these devices is examined in Chapter 5.

In reaction turbines the fluid fills the blade passages and a pressure decrease occurs as it flows through the impeller. The energy transfer in such devices involves some thermodynamic considerations beyond simple momentum analysis.

The remainder of this chapter will be devoted entirely to pumps and fans. Consideration will be given to general pump and fan performances, scaling laws, and their compatibility with piping systems.

▶ 14.1

CENTRIFUGAL PUMPS

Figure 14.3 shows two cutaway views of a typical centrifugal pump. In this configuration, fluid enters the pump casing axially. It then encounters the impeller blades that direct the flow tangentially and radially outward into the outer part of the casing and is then

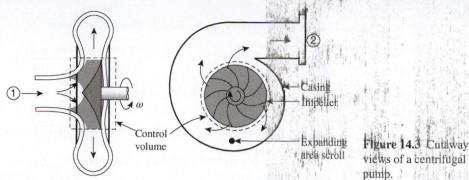

Figure 14.3 Cutaway views of a centrifugal pump.

discharged. The fluid experiences an increase in velocity and pressure as it passes through the impeller. The discharge section, which is doughnut shaped, causes the flow to decelerate and the pressure to increase further.

The impeller blades shown have a *backward-curved* shape, which is the most common configuration.

Pump Performance Parameters

We now focus on the control volume designated in Figure 14.3 by dashed lines. Note that flow enters at section one and leaves at two.

Applying the first law of thermodynamics to this control volume we have

$$\frac{\delta Q}{dt} - \frac{\delta W_s}{dt} = \iint_{c.s.} \left(e + \frac{P}{\rho}\right)\rho(\mathbf{v}\cdot\mathbf{n})\,dA + \frac{\partial}{\partial t}\iiint_{c.v.} e\rho\,dV + \frac{\delta W_\mu}{dt} \qquad (6\text{-}10)$$

which for steady, adiabatic flow with no viscous work, becomes

$$-\frac{\delta W_s}{dt} = \dot{m}\left[h_2 - h_1 + \frac{v_2^2 - v_1^2}{2} + g(y_2 - y_1)\right]$$

It is customary to neglect the small differences in velocity and elevation between sections one and two, thus

$$v_2^2 - v_1^2 \approx 0 \quad \text{and} \quad y_2 + y_1 \approx 0$$

and the remaining expression is

$$-\frac{\delta W_s}{dt} = \dot{m}(h_2 - h_1) = \dot{m}\left(u_2 - u_1 + \frac{P_2 - P_1}{\rho}\right)$$

Recalling that the term $u_2 - u_1$ represents the loss due to friction and other irreversible effects, we write

$$u_2 - u_1 = h_L$$

The net pressure head produced in the pump is

$$\frac{P_2 - P_1}{\rho} = \frac{1}{\dot{m}}\frac{\delta W_s}{\delta t} - h_L \qquad (14\text{-}1)$$

An important performance parameter, the *efficiency,* can now be expressed in broad terms as the ratio of actual output to required input. For a centrifugal pump the efficiency, designated η, is

$$\eta = \frac{\text{power added to the fluid}}{\text{shaft power to the impeller}}$$

The power added to the fluid is given by equation (14-1)

$$\frac{\delta W}{dt}\bigg|_{\text{fluid}} = \dot{m}\left(\frac{P_2 - P_1}{\rho}\right) \tag{14-2}$$

and the efficiency can be expressed as

$$\eta = \frac{\dot{m}(P_2 - P_1)}{\rho(\delta W_s/\delta t)_{\text{c.v.}}} \tag{14-3}$$

The difference between $\delta W_s/dt|_{\text{c.v.}}$ and $\delta W_s/dt|_{\text{liquid}}$ is clearly the head loss, h_L.

Equations (14-1), (14-2), and (14-3) provide general relationships for important pump performance parameters. To develop actual performance information for centrifugal pumps, we must examine our control volume once again from a moment of momentum perspective.

The governing equation for this analysis is

$$\Sigma M_z = \iint_{\text{c.s.}} (\mathbf{r} \times \mathbf{v})_z \, \rho(\mathbf{v} \cdot \mathbf{n}) \, dA + \frac{\partial}{\partial t} \iiint_{\text{c.v.}} (\mathbf{r} \times \mathbf{v})_z \, \rho \, dV \tag{5-10c}$$

The axis of rotation of the rotor depicted in Figure 14.3 has been chosen as the z-direction, hence, our choice of equation (5-10c).

We now wish to solve for M_z by applying equation (5-10c) to the control volume in Figure 14.3 for one-dimensional steady flow. The coordinate system will be fixed with the z-direction along the axis of rotation. Recall that the rotor contains *backward-curved* blades. In Figure 14.4, we show a detailed view of a single rotor blade. The blade is attached to the rotor hub at distance r_1 from the z-axis; the outer dimension of the blade has the value r_2.

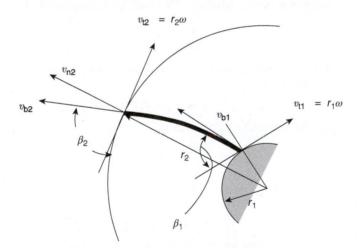

Figure 14.4 Velocity diagram for flow exiting a centrifugal pump impeller.

In this figure

v_{b1}, v_{b2} represent velocities *along the blade* at r_1 and r_2, respectively.

v_{n2} is the normal velocity of the flow at r_2.

v_{t2} is the tangential velocity of the flow at r_2.

β_1, β_2 are the angles made between the blade and tangent directions at r_1 and r_2, respectively.

Equation (5-10c) can now be written as

$$M_z = \dot{m}\left[\begin{vmatrix} \mathbf{e}_r & \mathbf{e}_\theta & \mathbf{e}_z \\ r & 0 & 0 \\ v_r & v_\theta & v_z \end{vmatrix}_2 - \begin{vmatrix} \mathbf{e}_r & \mathbf{e}_\theta & \mathbf{e}_z \\ r & 0 & 0 \\ v_r & v_\theta & v_z \end{vmatrix}_1\right]_z$$

which becomes

$$M_z = \dot{m}[(rv_\theta)_2 - 0]$$
$$= \rho \dot{V} r_2\, v_{\theta 2} \tag{14-4}$$

The velocity, $v_{\theta 2}$, is the tangential component of the fluid stream exiting the rotor relative to the fixed-coordinate system. The quantities shown in Figure 14.4 will be useful in evaluating $v_{\theta 2}$.

The absolute velocity of the existing flow v_2 is the vector sum of the velocity relative to the impeller blade and the velocity of the blade tip relative to our coordinate system. For blade length, L, normal to the plane of Figure 14.4, we evaluate the following:

- The normal velocity of flow at r_2:

$$v_{n2} = \frac{\dot{V}}{2\pi r_2 L} \tag{14-5}$$

- The velocity of flow along the blade at r_2:

$$v_{b2} = \frac{v_{n2}}{\sin\beta_2} \tag{14-6}$$

- The blade tip velocity:

$$v_{t2} = r_2\omega \tag{14-7}$$

The velocity we want, $v_{\theta 2}$, can now be evaluated as

$$v_{\theta 2} = v_{r2} - v_{b2}\cos\beta_2$$

Substitution from equations (14-6) and (14-7) yields

$$v_{\theta 2} = r_2\omega - \frac{v_{n2}}{\sin\beta_2}\cos\beta_2$$
$$= r_2\omega - v_{n2}\cot\beta_2$$

Finally, introducing the expression for v_{n2} from equation (14-5) we have

$$v_\theta = r_2\omega - \frac{\dot{V}}{2\pi r_2 L}\cot\beta_2 \tag{14-8}$$

and the desired moment is

$$M_z = \rho\dot{V}r_2\left[r_2\omega - \frac{\dot{V}}{2\pi r_2 L}\cot\beta_2\right] \tag{14-9}$$

The power delivered to the fluid is, by definition, $M_z\omega$; thus

$$\dot{W} = \frac{\delta W_s}{dt} = M_z\omega = \rho\dot{V}r_2\omega\left[r_2\omega - \frac{\dot{V}}{2\pi r_2 L}\cot\beta_2\right] \qquad (14\text{-}10)$$

Equation (14-10) expresses the power imparted to the fluid for an impeller with dimensions r_2, β_2, and L, operating at angular velocity, ω, with mass flow rate $\rho\dot{V}$.

This expression may be related to equations (14-2) and (14-3) to evaluate the imparted pressure head and the pump efficiency.

It is a standard practice to minimize friction loss at r_1, the radial location at which flow enters the impeller. This is accomplished by configuring the angle, β_1, such that inlet flow is along the blade surface. Referring to Figure 14.4, the design point for minimum losses is achieved when

$$v_{b1}\cos\beta_1 = r_1\omega$$

or, equivalently, when

$$v_{r1} = v_{b1}\sin\beta_1 = r_1\omega\frac{\sin\beta_1}{\cos\beta_1}$$

and, finally, when

$$v_{r1} = r_1\omega\tan\beta_1 \qquad (14\text{-}11)$$

Typical performance curves, for a centrifugal pump, are shown in Figure 14.5. Pressure head, brake horsepower, and efficiency are all shown as functions of volumetric flow rate. It is reasonable to choose operating conditions at or near the flow rate where maximum efficiency is achieved.

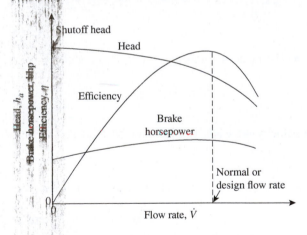

Figure 14.5 Centrifugal pump performance curves.

Example 1 illustrates how the analysis presented above relates to centrifugal pump performance.

Example 1

Water flow is produced by a centrifugal pump with the following dimensions:

$$r_1 = 6\,\text{cm} \qquad \beta_1 = 33°$$
$$r_2 = 10.5\,\text{cm} \qquad \beta_2 = 21°$$
$$L = 4.75\,\text{cm}$$

At a rotational speed of 1200 rpm determine

   (a) the design flow rate
   (b) the power added to the flow
   (c) the maximum pressure head at the pump discharge

To experience minimum losses, equation (14-11) must be satisfied; thus,

$$v_{r1} = r_1 \omega \, \tan\beta_1$$
$$= (0.06 \text{ m})\left(1200 \frac{\text{rev}}{\text{min}}\right)\left(\frac{2\pi \text{ rad}}{\text{rev}}\right)\left(\frac{\text{min}}{60 \text{ s}}\right)(\tan 33°) \tag{14-11}$$
$$= 4.896 \text{ m/s}$$

The corresponding flow rate is

$$\dot{V} = 2\pi r_1 L v_{r1}$$
$$= 2\pi(0.06 \text{ m})(0.0475 \text{ m})(4.896 \text{ m/s})$$
$$= 0.0877 \text{ m}^3/\text{s} \qquad (1390 \text{ gpm}) \tag{a}$$

The power imparted to the flow is expressed by equation (14-10):

$$\dot{W} = \rho \dot{V} r_2 \omega \left[r_2 \omega - \frac{\dot{V}}{2\pi r L} \cot\beta_2 \right]$$

Evaluate the following:

$$\omega = \left(1200 \frac{\text{rev}}{\text{min}}\right)\left(2\pi \frac{\text{rad}}{\text{rev}}\right)\left(\frac{\text{min}}{60 \text{ s}}\right) = 125.7 \text{ rad/s}$$
$$\rho \dot{V} r_2 \omega = (1000 \text{ kg/m}^3)(0.0877 \text{ m}^3/\text{s})(0.105 \text{ m})(125.7 \text{ rad/s})$$
$$= 1157 \text{ kg} \cdot \text{m/s}$$

$$\frac{\dot{V}}{2\pi r_2 L} = \frac{0.0877 \text{ m}^3/\text{s}}{2\pi(0.105 \text{ m})(0.0475 \text{ m})}$$
$$= 2.80 \text{ m/s}$$

we obtain

$$\dot{W} = (1157 \text{ kg} \cdot \text{m/s})[(0.105 \text{ m})(125.7 \text{ rad/s}) - (2.80 \text{ m/s})(\cot 21)]$$
$$= 6830 \text{ W} = 6.83 \text{ kW} \tag{b}$$

Equation (14-1) expresses the net pressure head as

$$\frac{P_2 - P_1}{\rho g} = -\frac{\dot{W}}{\dot{m}g} - h_L \tag{14-1}$$

The maximum value, with negligible losses, will be

$$\frac{P_2 - P_1}{\rho g} = \frac{6830 \text{ W}}{(1000 \text{ kg/m}^3)(0.0877 \text{ m}^3/\text{s})(9.81 \text{ m/s}^2)}$$
$$= 7.94 \text{ m H}_2\text{O gage}$$

$$\text{For } P_1 = 1 \text{ atm} = 14.7 \text{ psi} = 10.33 \text{ m H}_2\text{O}$$
$$P_2 = (7.94 + 10.33)\text{m H}_2\text{O} = 18.3 \text{ m H}_2\text{O} \ (26 \text{ psi}) \tag{c}$$

The actual discharge pressure will be less than this owing to friction and other irreversible losses.

Net Positive Suction Head

A major concern in pump operation is the presence of *cavitation*. Cavitation occurs when a liquid being pumped vaporizes or boils. If this occurs, the vapor bubbles that have been formed cause a decrease in efficiency and, often, structural damage to the pump that may lead to catastrophic failure. The parameter designated *net positive suction head* (NPSH) characterizes the likelihood for cavitation to occur.

At the suction side of the impeller, where pressure is lowest, thus the location where cavitation will first occur, the NPSH can be expressed as

$$\text{NPSH} + \frac{P_v}{\rho g} = \frac{v_i^2}{2g} + \frac{P_i}{\rho g} \tag{14-12}$$

where v_i and P_i are evaluated at pump inlet and P_v is the liquid vapor pressure. Values of NPSH are, in general, determined experimentally over a range in flow rates, for a given pump. A typical variation of NPSH vs. $\dot{V}$ is shown in Figure 14.6.

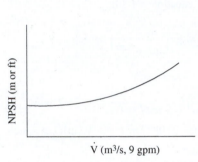

$\dot{V}$ (m³/s, 9 gpm)

Figure 14.6 Typical variation of NPSH with $\dot{V}$.

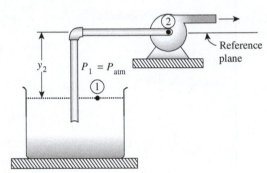

Figure 14.7 Pump installation at a level y above a supply reservoir.

In Figure 14.7, a representative pump installation is shown with the liquid being drawn from a reservoir located a distance, y, below the pump inlet. An energy balance between the pump inlet and the reservoir level yields

$$\frac{P_{\text{atm}}}{\rho g} = y_2 + \frac{P_2}{\rho g} + \frac{v_2^2}{2g} + \Sigma h_L \tag{14-13}$$

where the term Σh_L represents head losses between locations 1 and 2 as discussed in Chapter 13.

Combining this relationship with equation (14-12), we get

$$\begin{aligned}
\text{NPSH} &= \frac{v_2^2}{2g} + \frac{P_2}{\rho g} - \frac{P_v}{\rho g} \\
&= \frac{P_{\text{atm}}}{\rho g} - y_2 - \frac{P_v}{\rho g} - \Sigma h_L
\end{aligned} \tag{14-14}$$

For proper pump installation, the value of NPSH evaluated using equation (14-14) should be greater than the value obtained from a pump performance plot at the same flow rate. The principal use of these ideas is to establish a maximum value for the height, y_2. Example 2 illustrates the use of NPSH.

Example 2

A system like the one shown in Figure 14.7 is to be assembled to pump water. The inlet pipe to the centrifugal pump is 12 cm in diameter and the desired flow rate is 0.025 m³/s. At this flow rate, the specifications for this pump show a value of NPSH of 4.2 m. The minor loss coefficient for the system may be taken as $K = 12$. Water properties are to be evaluated at 300 K. Determine the maximum value of y, the distance between pump inlet and reservoir level.

The quantity desired, y, is given by

$$y = \frac{P_{atm} - P_v}{\rho g} - \Sigma h_L - NPSH \tag{14-14}$$

Water properties required, at 300 K, are

$$\rho = 997 \text{ kg/m}^3$$
$$P_v = 3598 \text{ Pa}$$

and we have

$$v = \frac{\dot{V}}{A} = \frac{0.025 \text{ m}^2/\text{s}}{\frac{\pi}{4}(0.12 \text{ m})^2} = 2.21 \text{ m/s}$$

$$\Sigma h_L = K_L \frac{v^2}{2g} = \frac{12(2.21 \text{ m/s}^2)^2}{2(9.81 \text{ m/s}^2)} = 2.99 \text{ m}$$

We can now complete the solution

$$y = \frac{(101{,}360 - 3598)P_a}{(997 \text{ kg/m}^3)(9.81 \text{ m/s}^2)} - 2.99 \text{ m} - 4.2 \text{ m}$$

$$= 2.805 \text{ m} \quad (9.2 \text{ ft})$$

Combined Pump and System Performance

As depicted in Figure 14.5, a pump has the capability of operating over a range in flow rates with its delivered head, operating efficiency, and NPSH values, all being flow-rate-dependent. An important task of the engineer is to match a given pump, with its known operating characteristics, to the performance of the system in which the pump produces flow. Piping system performance is discussed in Chapter 13.

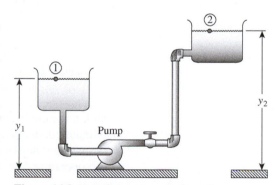

Figure 14.8 Pumping system configuration.

A simple flow system is illustrated in Figure 14.8 where a pump is used to produce flow between two reservoirs at different elevations.

With the two reservoir surfaces designated ① and ② as shown in the figure, an energy balance between these two locations yields

$$-\frac{\dot{W}}{\dot{m}} = g(y_2 - y_1) + \frac{P_2 - P_1}{\rho} + (u_2 - u_1) \tag{14-15}$$

Observing that $P_1 = P_2 = P_{atm}$ and expressing $u_2 - u_1 = \Sigma h_L$, we have

$$-\frac{\dot{W}}{\dot{m}g} = y_2 - y_1 + \Sigma h_L \qquad (14\text{-}16)$$

From Chapter 13, we can write, for the head loss,

$$\Sigma h_L = \Sigma K \frac{v^2}{2g}$$

where the quantity ΣK accounts for frictional pipe loss as well as minor losses due to valves, elbows, and fittings.

The operating line for system performance is now expressed by

$$-\frac{\dot{W}}{\dot{m}g} = y_2 - y_1 + \Sigma K \frac{v^2}{2g} \qquad (14\text{-}17)$$

Plotting the system operating line together with the plot of pump performance yields the combined performance diagram, as shown in Figure 14.9.

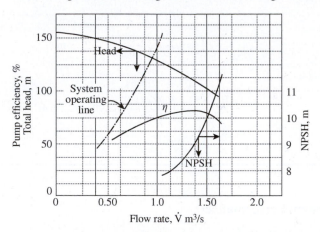

Figure 14.9 Combined pump and system performance.

We note that the two operating lines intersect at a flow rate where the required head for system operation matches that which the particular pump can produce. At this operating flow rate, one can read the corresponding efficiency from the chart. A system designer would, naturally, want the system to operate at, or as near as possible to, the flow rate of maximum pump efficiency. If the operating point corresponds to an undesirable efficiency value, changes must be made either to the system, which is generally a difficult process, or to pump-operating conditions.

▶ **14.2**

SCALING LAWS FOR PUMPS AND FANS

The concepts of similarity and scaling are introduced in Chapter 11. The requirements of geometric, kinematic, and dynamic similarity find important applications in the scaling of rotating fluid mechanics. In this section we will develop the "fan laws" that are used to predict the effect of changing the fluid, size, or speed of rotating machines, which are in a geometrically similar family.

Dimensional Analysis of Rotating Machines

The Buckingham method of dimensional analysis, which is introduced in Chapter 11, is a useful tool in generating the dimensionless parameters that apply to rotating fluid machines. As discussed earlier, the first step to be undertaken is to develop a table of variables that are important to our application. Table 14.1 lists the variables of interest along with their symbols and dimensional representation in the MLt system.

Table 14.1 Pump performance variables

Variable	Symbol	Dimension
Total head	gh	L^2/t^2
Flow rate	$\dot{V}$	L^3/t
Impeller diameter	D	L
Shaft speed	ω	$1/t$
Fluid density	ρ	M/L^3
Fluid viscosity	μ	M/Lt
Power	$\dot{W}$	ML^2/t^3

Without repeating all details regarding the Buckingham method, we can establish the following:

- $i = n - r = 7 - 3 = 4$
- With a core group including the variables D, ω, ρ, the dimensionless pi groups become

$$\pi_1 = gh/D^2\omega^2$$
$$\pi_2 = \dot{V}/\omega D^3$$
$$\pi_3 = \dot{W}/\rho\omega^3 D^5$$
$$\pi_4 = \mu/D^2\omega\rho$$

The group $\pi_4 = \mu/D^2\omega\rho$ is a form of the Reynolds number. The other three groups here are designated by the pump community as

$$\pi_1 = gh/D^2\omega^2 = C_H \qquad \text{—the head coefficient} \qquad (14\text{-}18)$$

$$\pi_2 = \dot{V}/\omega D^3 = C_Q \qquad \text{—the flow coefficient} \qquad (14\text{-}19)$$

$$\pi_3 = \dot{W}/\rho\omega^3 D^5 = C_P \qquad \text{—the power coefficient} \qquad (14\text{-}20)$$

Figure 14.10 is a plot of the dimensionless parameters C_H and C_P vs. the flow coefficient, C_Q, for a representative centrifugal pump family.

There is, of course, one additional dimensionless performance parameter, efficiency. The efficiency is related to the other parameters defined above according to

$$\eta = \frac{C_H C_Q}{C_P} \qquad (14\text{-}21)$$

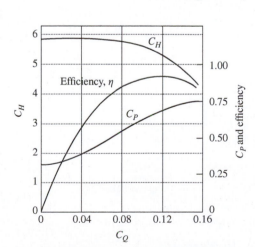

Figure 14.10 Dimensionless performance curves for a typical centrifugal pump.

As the parameters on the right-hand side of the equation are functionally related to C_Q, the efficiency, η, is also a function of C_Q, and is included as one of the dependent variables in Figure 14.10.

The three coefficients C_H, C_Q, and C_P provide the basis for the fan laws. For similar pumps, designated 1 and 2, we may write, for C_H

$$C_{H1} = C_{H2}$$

or

$$\frac{gh_1}{\omega_1^2 D_1^2} = \frac{gh_2}{\omega_2^2 D_2^2}$$

thus

$$\frac{h_2}{h_1} = \left(\frac{\omega_2}{\omega_1}\right)^2 \left(\frac{D_2}{D_1}\right)^2 \tag{14-22}$$

Performing the same equations on C_Q and C_P, we obtain

$$\frac{\dot{V}_2}{\dot{V}_1} = \frac{\omega_2}{\omega_1} \left(\frac{D_2}{D_1}\right)^3 \tag{14-23}$$

$$\frac{P_2}{P_1} = \frac{\rho_2}{\rho_1} \left(\frac{\omega_2}{\omega_1}\right)^3 \left(\frac{D_2}{D_1}\right)^5 \tag{14-24}$$

These three equations comprise the "fan laws" or "pump laws" that are used extensively for scaling rotating machines as well as for predicting their performance.

Example 3 illustrates the use of these expressions.

Example 3

A centrifugal pump operating at 1100 rpm against a head of 120 m H_2O produces a flow of 0.85 m³/s.

(a) For a geometrically similar pump, operating at the same speed but with an impeller diameter 30% greater than the original, what flow rate will be achieved?

(b) If the new larger pump described in part (a) is also operated at 1300 rpm, what will be the new values of flow rate and total head?

Specifying for pump 1, $D = D_1$, then for the larger pump, $D_2 = 1.3 D_1$; thus, the new flow rate will be, using equation (14-23),

$$\frac{\dot{V}_2}{\dot{V}_1} = \frac{\omega_2}{\omega_1} \left(\frac{D_2}{D_1}\right)^3 \tag{14-23}$$

$$\dot{V}_2 = 0.85 \text{ m}^3/\text{s} \left(\frac{1.3 D_1}{D_1}\right)^3$$

$$= 1.867 \text{ m}^3/\text{s}$$

For the case with $D_2 = 1.3 D_1$ and $\omega_2 = 1300$ rpm, we have, from equation (14-23),

$$\dot{V}_2 = 0.85 \text{ m}^3/\text{s} \left(\frac{1300 \text{ rpm}}{1100 \text{ rpm}}\right) \left(\frac{1.3 D_1}{D_1}\right)^3$$

$$= 2.207 \text{ m}^3/\text{s}$$

The new head is determined, using equation (14-22):

$$\frac{h_2}{h_1} = \left(\frac{\omega_2}{\omega_1}\right)^2 \left(\frac{D_2}{D_1}\right)^2 \tag{14-22}$$

$$= 120 \text{ m } H_2O \left(\frac{1300 \text{ rpm}}{1100 \text{ rpm}}\right)^2 \left(\frac{1.3 D_1}{D_1}\right)^2$$

$$= 283 \text{ m } H_2O$$

► **14.3**

AXIAL- AND MIXED-FLOW PUMP CONFIGURATIONS

Our examination of pumps thus far has focused on centrifugal pumps. The other basic configuration is axial flow. The designation centrifugal flow or axial flow relates to the direction of fluid flow in the pump. In the centrifugal case, flow is turned 90° to the axis of rotation; in the axial-flow case, flow is in the direction of the axis of rotation. There is an intermediate case, designated *mixed flow*, where the flow has both normal and axial components.

The choice of centrifugal-, axial-, or mixed-flow configurations depends on the desired values of flow rate and head needed in a specific application. The single parameter that includes both head and flow rate effects is designated N_S, the *specific speed*. It is defined as

$$N_S = \frac{C_Q^{1/2}}{C_H^{3/4}} \tag{14-25}$$

Figure 14.11 is a plot of optimum efficiencies of the three pump types as functions of N_S. The values of N_S shown in this plot correspond to the somewhat unusual units shown.

The basic message conveyed by Figure 14.11 is that higher delivery head and lower flow rate combinations dictate the use of centrifugal pumps, whereas lower heads and higher flow rates require mixed-flow or axial-flow pumps.

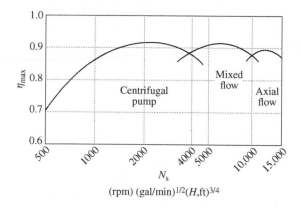

Figure 14.11 Optimum pump efficiency as a function of specific speed.

► **14.4**

TURBINES

Analysis of turbines follows the same general steps as has been done for pumps. The reader is referred to Section 5.4 in Chapter 5 for a review of the analysis of an impulse turbine.

Turbine operation uses the energy of a fluid, emanating from a nozzle, to interact with blades attached to the rotating unit, designated the *rotor*. The momentum exchange produced as the fluid changes direction generates power at the rotor shaft.

A detailed discussion of turbine operation, other than that presented in Chapter 5, is beyond the scope of this book. Numerous treatises are available to the interested reader. A good introductory discussion, along with extensive references, is included in the text by Munson et al. (1998).

▶ **14.5**

CLOSURE

This chapter has been devoted to the examination of rotating fluid machines. External power applied to pumps and fans produces higher pressure, increased flow, or both. Turbines operate in the reverse, producing power from a high-energy fluid.

Types of pumps or fans are characterized by the direction of flow through the rotor. In centrifugal pumps, the flow is turned 90 degrees to the flow axis; flow is parallel to the flow axis in axial-flow pumps. Machines with both centrifugal and axial-flow components are designated mixed-flow pumps.

Standard performance plots for a family of geometrically similar pumps or fans show the head, power, efficiency, and NPSH as functions of flow rate for a designated speed of rotation.

Scaling laws were developed using parameters generated from dimensional analysis. The resulting "fan laws" that relate two similar systems are

$$\frac{h_2}{h_1} = \left(\frac{\omega_2}{\omega_1}\right)^2 \left(\frac{D_2}{D_1}\right)^2 \tag{14-22}$$

$$\frac{\dot{V}_2}{\dot{V}_1} = \frac{\omega_2}{\omega_1} \left(\frac{D_2}{D_1}\right)^3 \tag{14-23}$$

$$\frac{P_2}{P_1} = \frac{\rho_2}{\rho_1} \left(\frac{\omega_2}{\omega_1}\right)^3 \left(\frac{D_2}{D_1}\right)^5 \tag{14-24}$$

CHAPTER 15

Fundamentals of Heat Transfer

The next nine chapters deal with the transfer of energy. Gross quantities of heat added to or rejected from a system may be evaluated by applying the control-volume expression for the first law of thermodynamics, as discussed in Chapter 6. The result of a first-law analysis is only a part of the required information necessary for the complete evaluation of a process or situation that involves energy transfer. The overriding consideration is, in many instances, the rate at which energy transfer takes place. Certainly, in designing a plant in which heat must be exchanged with the surroundings, the size of heat-transfer equipment, the materials of which it is to be constructed, and the auxiliary equipment required for its utilization are all important considerations for the engineer. Not only must the equipment accomplish its required mission, but it must also be economical to purchase and to operate.

Considerations of an engineering nature such as these require both a familiarity with the basic mechanisms of energy transfer and an ability to evaluate quantitatively these rates as well as the important associated quantities. Our immediate goal is to examine the basic mechanisms of energy transfer and to consider the fundamental equations for evaluating the rate of energy transfer.

There are three modes of energy transfer: conduction, convection, and radiation. All heat-transfer processes involve one or more of these modes. The remainder of this chapter will be devoted to an introductory description and discussion of these types of transfer.

▶ **15.1**

CONDUCTION

Energy transfer by conduction is accomplished in two ways. The first mechanism is that of molecular interaction, in which the greater motion of a molecule at a higher energy level (temperature) imparts energy to adjacent molecules at lower energy levels. This type of transfer is present, to some degree, in all systems in which a temperature gradient exists and in which molecules of a solid, liquid, or gas are present.

The second mechanism of conduction heat transfer is by "free" electrons. The free-electron mechanism is significant primarily in pure-metallic solids; the concentration of free electrons varies considerably for alloys and becomes very low for nonmetallic solids. The ability of solids to conduct heat varies directly with the concentration of free electrons; thus, it is not surprising that pure metals are the best heat conductors, as our experience has indicated.

As heat conduction is primarily a molecular phenomenon, we might expect the basic equation used to describe this process to be similar to the expression used in the molecular transfer of momentum, equation (7-4). Such an equation was first stated in 1822 by Fourier in the form

$$\frac{q_x}{A} = -k\frac{dT}{dx} \tag{15-1}$$

where q_x is the heat-transfer rate in the x-direction, in Watts or Btu/h; A is the area *normal* to the direction of heat flow, in m^2 or ft^2; dT/dx is the temperature gradient in the x-direction, in K/m or F/ft; and k is the thermal conductivity, in W/(m · K) or Btu/h ft F. The ratio q_x/A, having the dimensions of W/m^2 or Btu/h ft^2, is referred to as the heat flux in the x-direction. A more general relation for the heat flux is equation (15-2),

$$\frac{\mathbf{q}}{A} = -k\nabla T \tag{15-2}$$

which expresses the heat flux as proportional to the temperature gradient. The proportionality constant is seen to be the thermal conductivity, which plays a role similar to that of the viscosity in momentum transfer. The negative sign in equation (15-2) indicates that heat flow is in the direction of a negative temperature gradient. Equation (15-2) is the vector form of the *Fourier rate equation*, often referred to as Fourier's first law of heat conduction.

The thermal conductivity, k, which is defined by equation (15-1), is assumed independent of direction in equation (15-2); thus, this expression applies to an *isotropic* medium only. Most materials of engineering interest are isotropic. Wood is a good example of an *anisotropic* material where the thermal conductivity parallel to the grain may be greater than that normal to the grain by a factor of 2 or more. The thermal conductivity is a property of a conducting medium and, like the viscosity, is primarily a function of temperature, varying significantly with pressure only in the case of gases subjected to high pressures.

▶ **15.2**

THERMAL CONDUCTIVITY

As the mechanism of conduction heat transfer is one of the molecular interaction, it will be illustrative to examine the motion of gas molecules from a standpoint similar to that in Section 7.3.

Considering the control volume shown in Figure 15.1, in which energy transfer in the y-direction is on a molecular scale only, we may utilize the first-law analysis of Chapter 6 as follows. Mass transfer across the top of this control volume is considered to occur only on the molecular scale. This criterion is met for a gas in laminar flow.

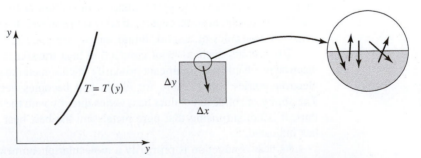

Figure 15.1 Molecular motion at the surface of a control volume.

Applying equation (6-10) and considering transfer only across the top face of the element considered,

$$\frac{\delta Q}{dt} - \frac{\delta W_s}{dt} - \frac{\delta W_\mu}{dt} = \iint_{c.s.} \left(e + \frac{P}{\rho}\right)\rho(\mathbf{v} \cdot \mathbf{n}) \, dA + \frac{\partial}{\partial t}\iiint_{c.v.} e\rho \, dV \qquad (6\text{-}10)$$

For Z molecules crossing the plane $\Delta x \, \Delta z$ per unit time, this equation reduces to

$$q_y = \sum_{n=1}^{Z} m_n C_p (T|_{y-} - T|_{y+})\Delta x \, \Delta z \qquad (15\text{-}3)$$

where m_n is the mass per molecule; C_p is the molecular heat capacity of the gas; Z is the frequency with which molecules will cross area $\Delta x \, \Delta z$; and $T|_{y-}$, $-T|_{y+}$ are the temperatures of the gas slightly below and slightly above the plane considered, respectively. The right-hand term is the summation of the energy flux associated with the molecules crossing the control surface. Noting now that $T|_{y-} = T - T/y|_{y_0}\delta$, where $y- = y_0 - \delta$, and that a similar expression may be written for $T|_{y+}$, we may rewrite equation (15-3) in the form

$$\frac{q_y}{A} = -2\sum_{n=1}^{Z} m_n C_p \delta \frac{dT}{dy}\bigg|_{y_0} \qquad (15\text{-}4)$$

where δ represents the y component of the distance between collisions. We note, as previously in Chapter 7, that $\delta = (\frac{2}{3})\lambda$, where λ is the mean free path of a molecule. Using this relation and summing over Z molecules, we have

$$\frac{q_y}{A} = -\frac{4}{3}\rho c_p Z\lambda \frac{dT}{dy}\bigg|_{y_0} \qquad (15\text{-}5)$$

Comparing equation (15-5) with the y component of equation (15-2),

$$\frac{q_y}{A} = -k\frac{dT}{dy}$$

it is apparent that the thermal conductivity, k, can be written as

$$k = \frac{4}{3}\rho c_p Z\lambda$$

Utilizing further the results of the kinetic theory of gases, we may make the following substitutions:

$$Z = \frac{N\overline{C}}{4}$$

where $\overline{C}$ is the average random molecular velocity, $\overline{C} = \sqrt{8\kappa T/\pi m}$ (κ being the Boltzmann constant),

$$\lambda = \frac{1}{\sqrt{2}\pi N d^2}$$

where d is the molecular diameter, and

$$c_p = \frac{3}{2}\frac{\kappa}{N}$$

giving, finally,

$$k = \frac{1}{\pi^{3/2}d^2}\sqrt{\kappa^3 T/m} \qquad (15\text{-}6)$$

This development, applying specifically to monatomic gases, is significant in that it shows the thermal conductivity of a gas to be independent of pressure, and to vary as the 1/2 power of the absolute temperature. Some relations for thermal conductivity of gases, based upon more sophisticated molecular models, may be found in Bird, Stewart, and Lightfoot.[1]

The Chapman–Enskog theory used in Chapter 7 to predict gas viscosities at low pressures has a heat-transfer counterpart. For a monatomic gas, the recommended equation is

$$k = 0.0829\sqrt{(T/M)}/\sigma^2\Omega_k \tag{15-7}$$

where k is in W/m · K, σ is in Angstroms, M is the molecular weight, and Ω_k is the Lennard–Jones collision integral, identical with Ω, as discussed in Section 7.3. Both σ and Ω_k may be evaluated from Appendices J and K.

The thermal conductivity of a liquid is not amenable to any simplified kinetic-theory development, as the molecular behavior of the liquid phase is not clearly understood and no universally accurate mathematical model presently exists. Some empirical correlations have met with reasonable success, but these are so specialized that they will not be included in this book. For a discussion of molecular theories related to the liquid phase and some empirical correlations of thermal conductivities of liquids, the reader is referred to Reid and Sherwood.[2] A general observation about liquid thermal conductivities is that they vary only slightly with temperature and are relatively independent of pressure. One problem in experimentally determining values of the thermal conductivity in a liquid is making sure the liquid is free of convection currents.

In the solid phase, thermal conductivity is attributed both to molecular interaction, as in other phases, and to free electrons, which are present primarily in pure metals. The solid phase is amenable to quite precise measurements of thermal conductivity, as there is no problem with convection currents. The thermal properties of most solids of engineering interest have been evaluated, and extensive tables and charts of these properties, including thermal conductivity, are available.

The free-electron mechanism of heat conduction is directly analogous to the mechanism of electrical conduction. This realization led Wiedemann and Franz, in 1853, to relate the two conductivities in a crude way; and in 1872, Lorenz[3] presented the following relation, known as the Wiedemann, Franz, Lorenz equation:

$$L = \frac{k}{k_e T} = \text{constant} \tag{15-8}$$

where k is the thermal conductivity, k_e is the electrical conductivity, T is the absolute temperature, and L is the Lorenz number.

The numerical values of the quantities in equation (15-8) are of secondary importance at this time. The significant point to note here is the simple relation between electrical and thermal conductivities and, specifically, that those materials that are good conductors of electricity are likewise good heat conductors, and vice versa.

Figure 15.2 illustrates the thermal conductivity variation with temperature of several important materials in gas, liquid, and solid phases. A more complete tabulation of thermal conductivity may be found in Appendices H and I.

The following two examples illustrate the use of the Fourier rate equation in solving simple heat-conduction problems.

---

[1] R. B. Bird, W. E. Stewart, and E. N. Lightfoot, *Transport Phenomena,* Wiley, New York, 2007, Chapter 9.

[2] Reid and Sherwood, *The Properties of Gases and Liquids,* McGraw-Hill Book Company, New York, 1958, Chapter 7.

[3] L. Lorenz, *Ann. Physik und Chemie* (Poggendorffs), **147**, 429 (1872).

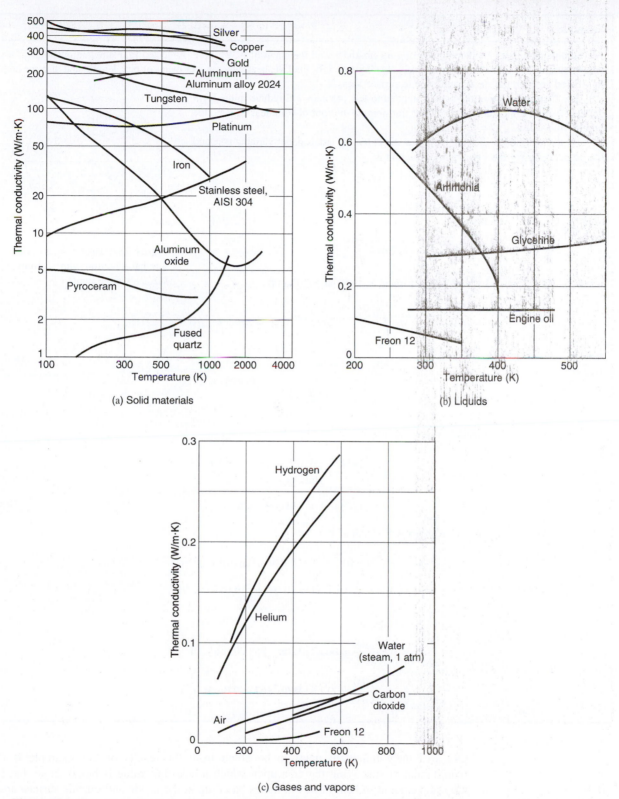

(a) Solid materials

(b) Liquids

(c) Gases and vapors

Figure 15.2 Thermal conductivity of several materials as functions of temperature.

Example 1

A steel pipe having an inside diameter of 1.88 cm and a wall thickness of 0.391 cm is subjected to inside and outside surface temperature of 367 and 344 K, respectively (see Figure 15.3). Find the heat flow rate per meter of pipe length, and also the heat flux based on both the inside and outside surface areas.

The first law of thermodynamics applied to this problem will reduce to the form $\delta Q/dt = 0$, indicating that the rate of heat transfer into the control volume is equal to the rate leaving—that is, $Q = q = $ constant.

As the heat flow will be in the radial direction, the independent variable is r, and the proper form for the Fourier rate equation is

$$q_r = -kA\frac{dT}{dr}$$

Writing $A = 2\pi rL$, we have

$$q_r = -k(2\pi rL)\frac{dT}{dr}$$

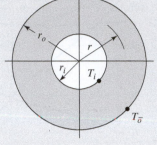

Figure 15.3 Heat conduction in a radial direction with uniform surface temperatures.

where q_r is constant, which may be separated and solved as follows:

$$q_r \int_{r_i}^{r_o} \frac{dr}{r} = -2\pi kL \int_{T_i}^{T_o} dT = 2\pi kL \int_{T_o}^{T_i} dT$$

$$q_r \ln\frac{r_o}{r_i} = 2\pi kL(T_i - T_o) \qquad (15\text{-}9)$$

$$q_r = \frac{2\pi kL}{\ln r_o/r_i}(T_i - T_o)$$

Substituting the given numerical values, we obtain

$$q_r = \frac{2\pi(42.90\,\text{W/m}\cdot\text{K})(367 - 344)\text{K}}{\ln(2.66/1.88)}$$

$$= 17{,}860\,\text{W/m}\,(18{,}600\,\text{Btu/h}\cdot\text{ft})$$

The inside and outside surface areas per unit length of pipe are

$$A_i = \pi(1.88)(10^{-2})(1) = 0.059\,\text{m}^2/\text{m}\,(0.194\,\text{ft}^2/\text{ft})$$

$$A_o = \pi(2.662)(10^{-2})(1) = 0.084\,\text{m}^2/\text{m}\,(0.275\,\text{ft}^2/\text{ft})$$

giving

$$\frac{q_r}{A_i} = \frac{17{,}860}{0.059} = 302.7\,\text{kW/m}^2\,(95{,}900\,\text{Btu/h}\cdot\text{ft}^2)$$

$$\frac{q_o}{A_i} = \frac{17{,}860}{0.084} = 212.6\,\text{kW/m}^2\,(67{,}400\,\text{Btu/h}\cdot\text{ft}^2)$$

One extremely important point to be noted from the results of this example is the requirement of specifying the area upon which a heat-flux value is based. Note that for the same amount of heat flow, the fluxes based upon the inside and outside surface areas differ by approximately 42%.

Example 2

Consider a hollow cylindrical heat-transfer medium having inside and outside radii of r_i and r_o with the corresponding surface temperatures T_i and T_o. If the thermal-conductivity variation may be described as a linear function of temperature according to

$$k = k_o(1 + \beta T)$$

then calculate the steady-state heat-transfer rate in the radial direction, using the above relation for the thermal conductivity, and compare the result with that using a k value calculated at the arithmetic mean temperature.

Figure 15.3 applies. The equation to be solved is now

$$q_r = -[k_o(1 + \beta T)](2\pi r L)\frac{dT}{dr}$$

which, upon separation and integration, becomes

$$q_r \int_{r_i}^{r_o} \frac{dr}{r} = -2\pi k_o L \int_{T_i}^{T_o} (1 + \beta T)dT$$

$$= 2\pi k_o L \int_{T_o}^{T_i} (1 + \beta T)dT$$

$$q_r = \frac{2\pi k_o L}{\ln r_o/r_i}\left[T + \frac{\beta T^2}{2}\right]_{T_o}^{T_i}$$

$$q_r = \frac{2\pi k_o L}{\ln r_o/r_i}\left[1 + \frac{\beta}{2}(T_i + T_o)\right](T_i - T_o) \tag{15-10}$$

Noting that the arithmetic average value of k would be

$$k_{avg} = k_o\left[1 + \frac{\beta}{2}(T_i + T_o)\right]$$

we see that equation (15-10) could also be written as

$$q_r = \frac{2\pi k_{avg} L}{\ln r_o/r_i}(T_i - T_o)$$

Thus, the two methods give identical results.

▶ **15.3**

CONVECTION

Heat transfer due to convection involves the energy exchange between a surface and an adjacent fluid. A distinction must be made between *forced convection*, wherein a fluid is made to flow past a solid surface by an external agent such as a fan or pump, and *free* or *natural convection* wherein warmer (or cooler) fluid next to the solid boundary causes circulation because of the density difference resulting from the temperature variation throughout a region of the fluid.

The rate equation for convective heat transfer was first expressed by Newton in 1701, and is referred to as the *Newton rate equation*, or Newton's "law" of cooling. This equation is

$$q/A = h\Delta T \tag{15-11}$$

where q is the rate of convective heat transfer, in W or Btu/h; A is the area normal to direction of heat flow, in m^2 or ft^2; ΔT is the temperature difference between surface and fluid, in K or F; and h is the convective heat-transfer coefficient, in $W/m^2 \cdot K$ or $Btu/h\,ft^2\,F$. Equation (15-11) defined the coefficient h. A substantial portion of our work in the chapters to follow will involve the determination of this coefficient. It is, in general, a function of system geometry, fluid and flow properties, and the magnitude of ΔT.

As flow properties are so important in the evaluation of the convective heat-transfer coefficient, we may expect many of the concepts and methods of analysis introduced in the preceding chapters to be of continuing importance in convective heat-transfer analysis; this is indeed the case.

From our previous experience we should also recall that even when a fluid is flowing in a turbulent manner past a surface, there is still a layer, sometimes extremely thin, close to the surface, where flow is laminar; also, the fluid particles next to the solid boundary are at rest. As this is always true, the mechanism of heat transfer between a solid surface and a fluid must involve conduction through the fluid layers close to the surface. This "film" of fluid often presents the controlling resistance to convective heat transfer, and the coefficient h is often referred to as the *film coefficient*.

Two types of heat transfer that differ somewhat from free or forced convection but that are still treated quantitatively by equation (15-11) are the phenomena of boiling and condensation. The film coefficients associated with these two kinds of transfer are quite high. Table 15.1 represents some order-of-magnitude values of h for different convective mechanisms.

Table 15.1 Approximate values of the convective heat-transfer coefficient

Mechanism	h, Btu/h ft^2 F	h, W/(m^2 · K)
Free convection, air	1–10	5–50
Forced convection, air	5–50	25–250
Forced convection, water	50–3000	250–15,000
Boiling water	500–5000	2500–25,000
Condensing water vapor	1000–20,000	5000–100,000

It will also be necessary to distinguish between local heat-transfer coefficients—that is, those that apply at a point, and total or average values of h that apply over a given surface area. We will designate the local coefficient h_x, according to equation (15-11):

$$dq = h_x \Delta T \, dA$$

Thus, the average coefficient, h, is related to h_x according to the relation

$$q = \int_A h_x \Delta T \, dA = hA \, \Delta T \tag{15-12}$$

The values given in Table 15.1 are average convective heat-transfer coefficients.

▶ **15.4**

RADIATION

Radiant heat transfer between surfaces differs from conduction and convection in that no medium is required for its propagation; indeed, energy transfer by radiation is maximum when the two surfaces that are exchanging energy are separated by a perfect vacuum.

The rate of energy emission from a perfect radiator or *blackbody* is given by

$$\frac{q}{A} = \sigma T^4 \tag{15-13}$$

where q is the rate of radiant-energy emission, in W or Btu/h; A is the area of the emitting surface, in m^2 or ft^2; T is the absolute temperature, in K or R; and σ is the Stefan–Boltzmann constant, which is equal to 5.676×10^{-8} W/$m^2 \cdot K^4$ or 0.1714×10^{-8} Btu/h ft^2 R^4. The proportionally constant relating radiant-energy flux to the fourth power of the absolute temperature is named after Stefan, who, from experimental observations, proposed equation (15-13) in 1879, and Boltzmann, who derived this relation theoretically in 1884. Equation (15-13) is most often referred to as the Stefan–Boltzmann law of thermal radiation.

Certain modifications will be made in equation (15-13) to account for the *net* energy transfer between two surfaces, the degree of deviation of the emitting and receiving surfaces from blackbody behavior, and geometrical factors associated with radiant exchange between a surface and its surroundings. These considerations are discussed at length in Chapter 23.

▶ **15.5**

COMBINED MECHANISMS OF HEAT TRANSFER

The three modes of heat transfer have been considered separately in Section 15.4. It is rare, in actual situations, for only one mechanism to be involved in the transfer of energy. It will be instructive to look at some situations in which heat transfer is accomplished by a combination of these mechanisms.

Consider the case depicted in Figure 15.4, that of steady-state conduction through a plane wall with its surfaces held at constant temperatures T_1 and T_2.

Writing the Fourier rate equation for the x-direction, we have

$$\frac{q_x}{A} = -k\frac{dT}{dx} \tag{15-1}$$

Solving this equation for q_x subject to the boundary conditions $T = T_1$ at $x = 0$ and $T = T_2$ at $x = L$, we obtain

$$\frac{q_x}{A}\int_0^L dx = -k\int_{T_1}^{T_2} dT = k\int_{T_2}^{T_1} dT$$

or

$$q_x = \frac{kA}{L}(T_1 - T_2) \tag{15-14}$$

Equation (15-14) bears an obvious resemblance to the Newton rate equation

$$q_x = hA\,\Delta T \tag{15-11}$$

We may utilize this similarity in form in a problem in which both types of energy transfer are involved.

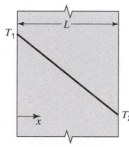

Figure 15.4 Steady-state conduction through a plane wall.

Consider the composite plane wall constructed of three materials in layers with dimensions as shown in Figure 15.5. We wish to express the steady-state heat-transfer rate per unit area between a hot gas at temperature T_h on one side of this wall and a cool gas at T_c on the other side. Temperature designations and dimensions are as shown in the figure. The following relations for q_x arise from the application of equations (15-11) and (15-14):

$$q_x = h_h A(T_h - T_1) = \frac{k_1 A}{L_1}(T_1 - T_2) = \frac{k_2 A}{L_2}(T_2 - T_3)$$
$$= \frac{k_3 A}{L_3}(T_3 - T_4) = h_c A(T_4 - T_c)$$

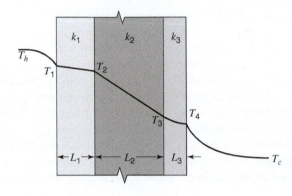

Figure 15.5 Steady-state heat transfer through a composite wall.

Each temperature difference is expressed in terms of q_x as follows:

$$T_h - T_1 = q_x(1/h_h A)$$
$$T_1 - T_2 = q_x(L_1/k_1 A)$$
$$T_2 - T_3 = q_x(L_2/k_2 A)$$
$$T_3 - T_4 = q_x(L_3/k_3 A)$$
$$T_4 - T_c = q_x(1/h_c A)$$

Adding these equations, we obtain

$$T_h - T_c = q_x\left(\frac{1}{h_h A} + \frac{L_1}{k_1 A} + \frac{L_2}{k_2 A} + \frac{L_3}{k_3 A} + \frac{1}{h_c A}\right)$$

and finally, solving for q_x, we have

$$q_x = \frac{T_h - T_c}{1/h_h A + L_1/k_1 A + L_2/k_2 A + L_3/k_3 A + 1/h_c A} \tag{15-15}$$

Note that the heat-transfer rate is expressed in terms of the *overall* temperature difference. If a series electrical circuit

is considered, we may write

$$I = \frac{\Delta V}{R_1 + R_2 + R_3 + R_4 + R_5} = \frac{\Delta V}{\sum R_i}$$

The analogous quantities in the expressions for heat flow and electrical current are apparent:

$$\Delta V \rightarrow \Delta T$$

$$I \rightarrow q_x$$

$$R_i \rightarrow 1/hA, \ L/kA$$

and each term in the denominator of equation (15-15) may be thought of as a thermal resistance due to convection or conduction. Equation (15-15) thus becomes a heat-transfer analog to Ohm's law, relating heat flow to the overall temperature difference divided by the total thermal resistance between the points of known temperature. Equation (15-15) may now be written simply as

$$q = \frac{\Delta T}{\sum R_{\text{thermal}}} \qquad (15\text{-}16)$$

This relation applies to steady-state heat transfer in systems of other geometries as well. The thermal-resistance terms will change in form for cylindrical or spherical systems, but once evaluated, they can be utilized in the form indicated by equation (15-16). With specific reference to equation (15-9), it may be noted that the thermal resistance of a cylindrical conductor is

$$\frac{\ln(r_o/r_i)}{2\pi kL}$$

Another common way of expressing the heat-transfer rate for a situation involving a composite material or combination of mechanisms is with the *overall heat-transfer coefficient* defined as

$$U \equiv \frac{q_x}{A \, \Delta T} \qquad (15\text{-}17)$$

where U is the overall heat-transfer coefficient having the same units as h, in $W/m^2 \cdot K$ or $Btu/h \, ft^2 \, F$.

Example 3

Saturated steam at 0.276 MPa flows inside a steel pipe having an inside diameter of 2.09 cm and an outside diameter of 2.67 cm. The convective coefficients on the inner and outer pipe surfaces are 5680 and 22.7 $W/m^2 \cdot K$, respectively. The surrounding air is at 294 K. Find the heat loss per meter of bare pipe and for a pipe having a 3.8 cm thickness of 85% magnesia insulation on its outer surface.

In the case of the bare pipe there are three thermal resistances to evaluate:

$$R_1 = R_{\text{convection inside}} = 1/h_i A_i$$

$$R_2 = R_{\text{convection outside}} = 1/h_o A_o$$

$$R_3 = R_{\text{conduction}} = \ln(r_o/r_i)/2\pi kL$$

For conditions of this problem, these resistances have the values

$$R_1 = 1/[(5680 \ W/m^2 \cdot K)(\pi)(0.0209 \ m)(1 \ m)]$$

$$= 0.00268 \ K/W \qquad \left(0.00141 \ \frac{h \, R}{Btu}\right)$$

$$R_2 = 1/[(22.7 \ W/m^2 \cdot K)(\pi)(0.0267 \ m)(1 \ m)]$$

$$= 0.525 \ K/W \qquad \left(0.277 \ \frac{h \, R}{Btu}\right)$$

and

$$R_3 = \frac{\ln(2.67/2.09)}{2\pi(42.9 \text{ W/m} \cdot \text{K})(1 \text{ m})}$$

$$= 0.00091 \text{ K/W} \qquad \left(0.00048 \frac{\text{h R}}{\text{Btu}}\right)$$

The inside temperature is that of 0.276 MPa saturated steam, 404 K or 267 F. The heat-transfer rate per meter of pipe may now be calculated as

$$q = \frac{\Delta T}{\sum R} = \frac{404 - 294 \text{ K}}{0.528 \text{ K/W}}$$

$$= 208 \text{ W} \qquad \left(710 \frac{\text{Btu}}{\text{h}}\right)$$

In the case of an insulated pipe, the total thermal resistance would include R_1 and R_3 evaluated above, plus additional resistances to account for the insulation. For the insulation,

$$R_4 = \frac{\ln(10.27/2.67)}{2\pi(0.0675 \text{ W/m} \cdot \text{K})(1 \text{ m})}$$

$$= 3.176 \text{ K/W} \qquad \left(1.675 \frac{\text{h R}}{\text{Btu}}\right)$$

and for the outside surface of the insulation,

$$R_5 = 1/[(22.7 \text{ W/m}^2 \cdot \text{K})(\pi)(0.1027 \text{ m})(1 \text{ m})]$$

$$= 0.1365 \text{ K/W} \qquad \left(0.0720 \frac{\text{h R}}{\text{Btu}}\right)$$

Thus, the heat loss for the insulated pipe becomes

$$q = \frac{\Delta T}{\sum R} = \frac{\Delta T}{R_1 + R_3 + R_4 + R_5} = \frac{404 - 294 \text{ K}}{3.316 \text{ K/W}}$$

$$= 33.2 \text{ W} \qquad \left(113 \frac{\text{Btu}}{\text{h}}\right)$$

This is a reduction of approximately 85%!

It is apparent from this example that certain parts of the heat-transfer path offer a negligible resistance. If, for instance, in the case of the bare pipe, an increased rate of heat transfer were desired, the obvious approach would be to alter the outside convective resistance, which is almost 200 times the magnitude of the next-highest thermal-resistance value.

Example 3 could also have been worked by using an *overall heat-transfer coefficient*, which would be, in general,

$$U = \frac{q_x}{A \Delta T} = \frac{\Delta T / \sum R}{A \Delta T} = \frac{1}{A \sum R}$$

or, for the specific case considered,

$$U = \frac{1}{A\{1/A_i h_i + [\ln(r_o/r_i)]/2\pi k L + 1/A_o h_o\}} \qquad (15\text{-}18)$$

Equation (15-18) indicates that the overall heat-transfer coefficient, U, may have a different numerical value, depending on which area it is based upon. If, for instance, U is based upon the outside surface area of the pipe, A_o, we have

$$U_o = \frac{1}{A_o/A_i h_i + [A_o \ln(r_o/r_i)]/2\pi k L + 1/h_o}$$

Thus, it is necessary, when specifying an overall coefficient, to relate it to a specific area.

One other means of evaluating heat-transfer rates is by means of the *shape factor*, symbolized as S. Considering the steady-state relations developed for plane and cylindrical shapes

$$q = \frac{kA}{L} \Delta T \tag{15-14}$$

and

$$q = \frac{2\pi k L}{\ln(r_o/r_i)} \Delta T \tag{15-9}$$

if that part of each expression having to do with the geometry is separated from the remaining terms, we have, for a plane wall,

$$q = k\left(\frac{A}{L}\right) \Delta T$$

and for a cylinder

$$q = k\left(\frac{2\pi L}{\ln(r_o/r_i)}\right) \Delta T$$

Each of the bracketed terms is the shape factor for the applicable geometry. A general relation utilizing this form is

$$q = kS \, \Delta T \tag{15-19}$$

Equation (15-19) offers some advantages when a given geometry is required because of space and configuration limitations. If this is the case, then the shape factor may be calculated and q determined for various materials displaying a range of values of k.

▶ **15.6**

CLOSURE

In this chapter, the basic modes of energy transfer—conduction, convection, and radiation—have been introduced, along with the simple relations expressing the rates of energy transfer associated therewith. The transport property, thermal conductivity, has been discussed and some consideration given to energy transfer in a monatomic gas at low pressure.

The rate equations for heat transfer are as follows:

Conduction: the Fourier rate equation

$$\frac{\mathbf{q}}{A} = -k \, \nabla T$$

Convection: the Newton rate equation

$$\frac{q_x}{A} = h \, \Delta T$$

Radiation: the Stefan–Boltzmann law for energy emitted from a black surface

$$\frac{q}{A} = \sigma T^4$$

Combined modes of heat transfer were considered, specifically with respect to the means of calculating heat-transfer rates when several transfer modes were involved. The three ways of calculating steady-state heat-transfer rates are represented by the equations

$$q_x = \frac{\Delta T}{\sum R_T} \tag{15-16}$$

where $\sum R_T$ is the total thermal resistance along the transfer path,

$$q_x = UA \, \Delta T \tag{15-17}$$

where U is the overall heat transfer coefficient, and

$$q_x = kS \, \Delta T \tag{15-19}$$

where S is the shape factor.

The equations presented will be used throughout the remaining chapters dealing with energy transfer. A primary object of the chapters to follow will be the evaluation of the heat-transfer rates for special geometries or conditions of flow, or both.

Note: Effects of thermal radiation are included, along with convection, in values of surface coefficients.

CHAPTER 16

Differential Equations of Heat Transfer

$\mathbf{P}$aralleling the treatment of momentum transfer undertaken in Chapter 9, we shall now generate the fundamental equations for a differential control volume from a first-law-of-thermodynamics approach. The control-volume expression for the first law will provide our basic analytical tool. Additionally, certain differential equations already developed in previous sections will be applicable.

▶ **16.1**

THE GENERAL DIFFERENTIAL EQUATION FOR ENERGY TRANSFER

Consider the control volume having dimensions Δx, Δy, and Δz as depicted in Figure 16.1. Refer to the control-volume expression for the first law of thermodynamics:

$$\frac{\delta Q}{dt} - \frac{\delta W_s}{dt} - \frac{\delta W_\mu}{dt} = \iint_{\text{c.s.}} \left(e + \frac{P}{\rho} \right) \rho(\mathbf{v} \cdot \mathbf{n}) dA + \frac{\partial}{\partial t} \iiint_{\text{c.v.}} e\rho\, dV \qquad (6\text{-}10)$$

The individual terms are evaluated and their meanings are discussed below.

The net rate of heat added to the control volume will include all conduction effects, the net release of thermal energy within the control volume due to volumetric effects such as a chemical reaction or induction heating, and the dissipation of electrical or nuclear energy. The generation effects will be included in the single term, $\dot{q}$, which is the volumetric rate of thermal energy generation having units W/m^3 or Btu/h ft^3. Thus, the first term may be expressed as

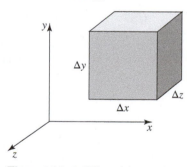

Figure 16.1 A differential control volume.

$$\frac{\delta Q}{dt} = \left[k\frac{\partial T}{\partial x}\bigg|_{x+\Delta x} - k\frac{\partial T}{\partial x}\bigg|_{x} \right] \Delta y\, \Delta z + \left[k\frac{\partial T}{\partial y}\bigg|_{y+\Delta y} - k\frac{\partial T}{\partial y}\bigg|_{y} \right] \Delta x\, \Delta z$$

$$+ \left[k\frac{\partial T}{\partial z}\bigg|_{z+\Delta z} - k\frac{\partial T}{\partial z}\bigg|_{z} \right] \Delta x\, \Delta y + \dot{q}\, \Delta x\, \Delta y\, \Delta z \qquad (16\text{-}1)$$

The shaft work rate or power term will be taken as zero for our present purposes. This term is specifically related to work done by some effect within the control volume that, for the differential case, is not present. The power term is thus evaluated as

$$\frac{\delta W_s}{dt} = 0 \tag{16-2}$$

The viscous work rate, occurring at the control surface, is formally evaluated by integrating the dot product of the viscous stress and the velocity over the control surface. As this operation is tedious, we shall express the viscous work rate as $\Lambda\,\Delta x\,\Delta y\,\Delta z$, where Λ is the viscous work rate per unit volume. The third term in equation (6-10) is thus written as

$$\frac{\delta W_\mu}{dt} = \Lambda\,\Delta x\,\Delta y\,\Delta z \tag{16-3}$$

The surface integral includes all energy transfer across the control surface due to fluid flow. All terms associated with the surface integral have been defined previously. The surface integral is

$$\iint_{c.s.} \left(e + \frac{P}{\rho}\right)\rho(\mathbf{v}\cdot\mathbf{n})dA$$

$$= \left[\rho v_x\left(\frac{v^2}{2} + gy + u + \frac{P}{\rho}\right)\Big|_{x+\Delta x} - \rho v_x\left(\frac{v^2}{2} + gy + u + \frac{P}{\rho}\right)\Big|_x\right]\Delta y\,\Delta z$$

$$+ \left[\rho v_y\left(\frac{v^2}{2} + gy + u + \frac{P}{\rho}\right)\Big|_{y+\Delta y} - \rho v_y\left(\frac{v^2}{2} + gy + u + \frac{P}{\rho}\right)\Big|_y\right]\Delta x\,\Delta z$$

$$+ \left[\rho v_z\left(\frac{v^2}{2} + gy + u + \frac{P}{\rho}\right)\Big|_{z+\Delta z} - \rho v_z\left(\frac{v^2}{2} + gy + u + \frac{P}{\rho}\right)\Big|_z\right]\Delta x\,\Delta y \tag{16-4}$$

The energy accumulation term, relating the variation in total energy within the control volume as a function of time, is

$$\frac{\partial}{\partial t}\iiint_{c.v.} e\rho\,dV = \frac{\partial}{\partial t}\left[\frac{v^2}{2} + gy + u\right]\rho\,\Delta x\,\Delta y\,\Delta z \tag{16-5}$$

Equations (16-1) through (16-5) may now be combined as indicated by the general first-law expression, equation (6-10). Performing this combination and dividing through by the volume of the element, we have

$$\frac{k(\partial T/\partial x)|_{x+\Delta x} - k(\partial T/\partial x)|_x}{\Delta x} + \frac{k(\partial T/\partial y)|_{y+\Delta y} - k(\partial T/\partial y)|_y}{\Delta y}$$

$$+ \frac{k(\partial T/\partial z)|_{z+\Delta z} - k(\partial T/\partial z)|_z}{\Delta z} + \dot{q} + \Lambda$$

$$= \frac{\{\rho v_x[(v^2/2) + gy + u + (P/\rho)]|_{x+\Delta x} - \rho v_x[(v^2/2) + gy + u + (P/\rho)]|_x\}}{\Delta x}$$

$$+ \frac{\{\rho v_y[(v^2/2) + gy + u + (P/\rho)]|_{y+\Delta y} - \rho v_y[(v^2/2) + gy + u + (P/\rho)]|_y\}}{\Delta y}$$

$$+ \frac{\{\rho v_z[(v^2/2) + gy + u + (P/\rho)]|_{z+\Delta z} - \rho v_z[(v^2/2) + gy + u + (P/\rho)]|_z\}}{\Delta z}$$

$$+ \frac{\partial}{\partial t}\rho\left(\frac{v^2}{2} + gy + u\right)$$

Evaluated in the limit as Δx, Δy, and Δz approach zero, this equation becomes

$$\frac{\partial}{\partial x}\left(k\frac{\partial T}{\partial x}\right) + \frac{\partial}{\partial y}\left(k\frac{\partial T}{\partial y}\right) + \frac{\partial}{\partial z}\left(k\frac{\partial T}{\partial z}\right) + \dot{q} + \Lambda$$

$$= \frac{\partial}{\partial x}\left[\rho v_x\left(\frac{v^2}{2} + gy + u + \frac{P}{\rho}\right)\right] + \frac{\partial}{\partial y}\left[\rho v_y\left(\frac{v^2}{2} + gy + u + \frac{P}{\rho}\right)\right]$$

$$+ \frac{\partial}{\partial z}\left[\rho v_z\left(\frac{v^2}{2} + gy + u + \frac{P}{\rho}\right)\right] + \frac{\partial}{\partial t}\left[\rho\left(\frac{v^2}{2} + gy + u\right)\right] \tag{16-6}$$

Equation (16-6) is completely general in application. Introducing the substantial derivative, we may write equation (16-6) as

$$\frac{\partial}{\partial x}\left(k\frac{\partial T}{\partial x}\right) + \frac{\partial}{\partial y}\left(k\frac{\partial T}{\partial y}\right) + \frac{\partial}{\partial z}\left(k\frac{\partial T}{\partial z}\right) + \dot{q} + \Lambda$$

$$= \nabla \cdot (P\mathbf{v}) + \left(\frac{v^2}{2} + u + gy\right)\left(\nabla \cdot \rho\mathbf{v} + \frac{\partial \rho}{\partial t}\right) + \frac{\rho}{2}\frac{Dv^2}{Dt} + \rho\frac{Du}{Dt} + \rho\frac{D(gy)}{Dt}$$

Utilizing the continuity equation, equation (9-2), we reduce this to

$$\frac{\partial}{\partial x}\left(k\frac{\partial T}{\partial x}\right) + \frac{\partial}{\partial y}\left(k\frac{\partial T}{\partial y}\right) + \frac{\partial}{\partial z}\left(k\frac{\partial T}{\partial z}\right) + \dot{q} + \Lambda$$

$$= \nabla \cdot P\mathbf{v} + \frac{\rho}{2}\frac{Dv^2}{Dt} + \rho\frac{Du}{Dt} + \rho\frac{D(gy)}{Dt} \tag{16-7}$$

With the aid of equation (9-19), which is valid for incompressible flow of a fluid with constant μ, the second term on the right-hand side of equation (16-7) becomes

$$\frac{\rho}{2}\frac{Dv^2}{Dt} = \mathbf{v} \cdot \nabla P + \mathbf{v} \cdot \rho g + \mathbf{v} \cdot \mu\nabla^2\mathbf{v} \tag{16-8}$$

Also, for incompressible flow, the first term on the right-hand side of equation (16-7) becomes

$$\nabla \cdot P\mathbf{v} = \mathbf{v} \cdot \nabla P \tag{16-9}$$

Substituting equations (16-8) and (16-9) into equation (16-7), and writing the conduction terms as $\nabla \cdot k\nabla T$, we have

$$\nabla \cdot k\nabla T + \dot{q} + \Lambda = \rho\frac{Du}{Dt} + \rho\frac{D(gy)}{Dt} + \mathbf{v} \cdot \rho g + \mathbf{v} \cdot \mu\nabla^2\mathbf{v} \tag{16-10}$$

It will be left as an exercise for the reader to verify that equation (16-10) reduces further to the form

$$\nabla \cdot k\nabla T + \dot{q} + \Lambda = \rho c_v\frac{DT}{Dt} + \mathbf{v} \cdot \mu\nabla^2\mathbf{v} \tag{16-11}$$

The function Λ may be expressed in terms of the viscous portion of the normal- and shear-stress terms in equations (7-15) and (7-16). For the case of incompressible flow, it is written as

$$\Lambda = \mathbf{v} \cdot \mu\nabla^2\mathbf{v} + \Phi \tag{16-12}$$

where the "dissipation function," Φ, is given by

$$\Phi = 2\mu\left[\left(\frac{\partial v_x}{\partial x}\right)^2 + \left(\frac{\partial v_y}{\partial y}\right)^2 + \left(\frac{\partial v_z}{\partial z}\right)^2\right]$$
$$+ \mu\left[\left(\frac{\partial v_x}{\partial y} + \frac{\partial v_y}{\partial x}\right)^2 + \left(\frac{\partial v_y}{\partial z} + \frac{\partial v_z}{\partial y}\right)^2 + \left(\frac{\partial v_z}{\partial x} + \frac{\partial v_x}{\partial z}\right)^2\right]$$

Substituting for Λ in equation (16-11), we see that the energy equation becomes

$$\nabla \cdot k\nabla T + \dot{q} + \Phi = \rho c_v \frac{DT}{Dt} \tag{16-13}$$

From equation (16-13), Φ is seen to be a function of fluid viscosity and shear-strain rates, and is positive-definite. The effect of viscous dissipation is always to increase internal energy at the expense of potential energy or stagnation pressure. The dissipation function is negligible in all cases that we will consider; its effect becomes significant in supersonic boundary layers.

▶ **16.2**

SPECIAL FORMS OF THE DIFFERENTIAL ENERGY EQUATION

The applicable forms of the energy equation for some commonly encountered situations follow. In every case the dissipation term is considered negligibly small.

I. For an incompressible fluid without energy sources and with constant k,

$$\rho c_v \frac{DT}{Dt} = k\,\nabla^2 T \tag{16-14}$$

II. For isobaric flow without energy sources and with constant k, the energy equation is

$$\rho c_v \frac{DT}{Dt} = k\,\nabla^2 T \tag{16-15}$$

Note that equations (16-14) and (16-15) are identical, yet apply to completely different physical situations.

III. In a situation where there is no fluid motion, all heat transfer is by conduction. If this situation exists, as it most certainly does in solids where $c_v \simeq c_p$, the energy equation becomes

$$\rho c_p \frac{\partial T}{\partial t} = \nabla \cdot k\,\nabla T + \dot{q} \tag{16-16}$$

Equation (16-16) applies in general to heat conduction. No assumption has been made concerning constant k. If the thermal conductivity is constant, the energy equation is

$$\frac{\partial T}{\partial t} = \alpha\,\nabla^2 T + \frac{\dot{q}}{\rho c_p} \tag{16-17}$$

where the ratio $k/\rho c_p$ has been symbolized by α and is designated the *thermal diffusivity*. It is easily seen that α has the units, L^2/t; in the SI system α is expressed in m^2/s, and as ft^2/h in the English system.

If the conducting medium contains no heat sources, equation (16-17) reduces to the *Fourier field equation*:

$$\frac{\partial T}{\partial t} = \alpha \nabla^2 T \tag{16-18}$$

which is often referred to as Fourier's second law of heat conduction.

For a system in which heat sources are present but there is no time variation, equation (16-17) reduces to the *Poisson equation*:

$$\nabla^2 T + \frac{\dot{q}}{k} = 0 \tag{16-19}$$

The final form of the heat-conduction equation to be presented applies to a steady-state situation without heat sources. For this case, the temperature distribution must satisfy the *Laplace equation*:

$$\nabla^2 T = 0 \tag{16-20}$$

Each of equations (16-17) through (16-20) has been written in general form; thus, each applies to any orthogonal coordinate system. Writing the Laplacian operator, ∇^2, in the appropriate form will accomplish the transformation to the desired coordinate system. The Fourier field equation written in rectangular coordinates is

$$\frac{\partial T}{\partial t} = \alpha \left[\frac{\partial^2 T}{\partial x^2} + \frac{\partial^2 T}{\partial y^2} + \frac{\partial^2 T}{\partial z^2} \right] \tag{16-21}$$

in cylindrical coordinates,

$$\frac{\partial T}{\partial t} = \alpha \left[\frac{\partial^2 T}{\partial r^2} + \frac{1}{r}\frac{\partial T}{\partial r} + \frac{1}{r^2}\frac{\partial^2 T}{\partial \theta^2} + \frac{\partial^2 T}{\partial z^2} \right] \tag{16-22}$$

and in spherical coordinates,

$$\frac{\partial T}{\partial t} = \alpha \left[\frac{1}{r^2}\frac{\partial}{\partial r}\left(r^2 \frac{\partial T}{\partial r} \right) + \frac{1}{r^2 \sin\theta}\frac{\partial}{\partial \theta}\left(\sin\theta \frac{\partial T}{\partial \theta} \right) + \frac{1}{r^2 \sin^2\theta}\frac{\partial^2 T}{\partial \phi^2} \right] \tag{16-23}$$

The reader is referred to Appendix B for an illustration of the variables in cylindrical and spherical coordinate systems.

▶ **16.3**

COMMONLY ENCOUNTERED BOUNDARY CONDITIONS

In solving one of the differential equations developed thus far, the existing physical situation will dictate the appropriate initial or boundary conditions, or both, which the final solutions must satisfy.

Initial conditions refer specifically to the values of T and $\mathbf{v}$ at the start of the time interval of interest. Initial conditions may be as simply specified as stating that $T|_{t=0} = T_0$ (a constant), or more complex if the temperature distribution at the start of time measurement is some function of the space variables.

Boundary conditions refer to the values of T and $\mathbf{v}$ existing at specific positions on the boundaries of a system—that is, for given values of the significant space variables. Frequently encountered boundary conditions for temperature are the case of *isothermal boundaries*, along which the temperature is constant, and *insulated boundaries*, across which no heat conduction occurs, where, according to the Fourier rate equation, the

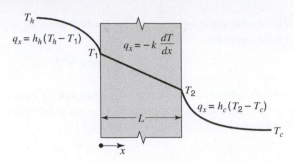

T_h

$q_x = h_h(T_h - T_1)$

T_1

$q_x = -k\dfrac{dT}{dx}$

L

x

T_2

$q_x = h_c(T_2 - T_c)$

T_c

Figure 16.2 Conduction and convention at a system boundary.

temperature derivative normal to the boundary is zero. More complicated temperature functions often exist at system boundaries, and the surface temperature may also vary with time. Combinations of heat-transfer mechanisms may dictate boundary conditions as well. One situation often existing at a solid boundary is the equality between heat transfer to the surface by conduction and that leaving the surface by convection. This condition is illustrated in Figure 16.2. At the left-hand surface, the boundary condition is

$$h_h\left(T_h - T|_{x=0}\right) = -k\frac{\partial T}{\partial x}\bigg|_{x=0} \tag{16-24}$$

and at the right-hand surface,

$$h_c\left(T|_{x=L} - T_c\right) = -k\frac{\partial T}{\partial x}\bigg|_{x=L} \tag{16-25}$$

It is impossible at this time to foresee all the initial and boundary conditions that will be needed. The student should be aware, however, that these conditions are dictated by the physical situation. The differential equations of energy transfer are not numerous, and a specific form applying to a given situation may be found easily. It remains for the user of these equations to choose the appropriate initial and boundary conditions to make the solution meaningful.

Example 1

Radioactive waste is stored in the spherical stainless-steel container as shown schematically in the figure below. Heat generated within the waste material must be dissipated by conduction through the steel shell and then by convection to the cooling water.

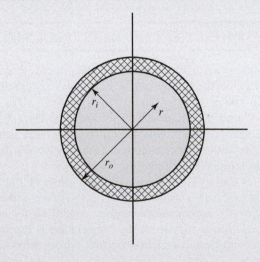

Pertinent dimensions and system operating variables are

Inside radius of steel shell $(r = r_i) = 0.4\,\mathrm{m}$
Outside radius of steel shell $(r = r_o) = 0.5\,\mathrm{m}$
Volumetric rate of heat generated within waste material $= 10^5\,\mathrm{W/m^3 \cdot K}$
Thermal conductivity of steel $= 15\,\mathrm{W/m \cdot K}$
Thermal conductivity of waste $= 20\,\mathrm{W/m \cdot K}$
Conductive heat-transfer coefficient between steel shell and water $= 800\,\mathrm{W/m^2 \cdot K}$
Water temperature $= 25\,\mathrm{C}$

We wish to evaluate the following:

(a) The location and magnitude of the maximum temperature in the waste material.
(b) The temperature at the interface between the waste material and the stainless-steel container
(c) The outside surface temperature of the stainless steel

Initially, we will consider the waste material within the region $0 \leq r \leq r_1$. Equation (16-9) applies:

$$\nabla^2 T + \frac{\dot{q}}{k} = 0 \qquad\qquad (16\text{-}19)$$

For temperature variation and heat flow in the radial direction, only equation (16-19) reduces to the form

$$\frac{1}{r^2} \frac{d}{dr}\left(r^2 \frac{dT}{dr} \right) + \frac{\dot{q}}{k} = 0$$

Separating variables gives

$$\frac{d}{dr}\left(r^2 \frac{dT}{dr} \right) + \frac{\dot{q}}{k} r^2 = 0$$

and integrating, we obtain

$$r^2 \frac{dT}{dr} + \frac{\dot{q}}{k} \frac{r^3}{3} = C_1$$

The constant, C_1, can be evaluated using the requirement that the temperature gradient at the center cannot be infinite:

$$\frac{dT}{dr} \text{ at } r = 0 \neq \infty, \text{ thus } C_1 = 0.$$

Our expression now reduces to

$$\frac{dT}{dr} + \frac{\dot{q}}{k} \frac{r}{3} = 0$$

which is integrated once more to yield

$$T + \frac{\dot{q}}{k} \frac{r^2}{6} = C_2$$

Noting that T has a maximum value at $r = 0$, we can write

$$C_2 = T_{w,\max}$$

and the temperature distribution in the waste material is

$$T = T_{w,\max} - \frac{\dot{q}}{k} \frac{r^2}{6} \qquad (1)$$

Directing our attention to the stainless-steel shell, the governing expression is the one-dimensional form of equation (16-23):

$$\frac{d}{dr}\left(r^2\frac{dT}{dr}\right) = 0$$

Integrating this expressional yields

$$r^2\frac{dT}{dr} = C_3 \quad \text{or} \quad \frac{dT}{dr} = \frac{C_3}{r^2}$$

and integrating once more, we have

$$T = -\frac{C_3}{r} + C_4$$

Applicable boundary conditions are the temperatures at the inside and outside steel surfaces:

$$T(r_i) = T_i \quad T(r_o) = T_o$$

and the constants, C_3 and C_4, are evaluated by solving the two expressions

$$T_i = -\frac{C_3}{r_i} + C_4$$

$$T_o = -\frac{C_3}{r_o} + C_4$$

Performing the required algebra the temperature variation in the steel shell, $T(r)$ has the form

$$\frac{T_i - T}{T_i - T_o} = \frac{1/r_i - 1/r}{1/r_i - 1/r_o} \qquad (2)$$

We can now solve for the desired quantities.

The rate of heat generation within the radioactive core is

$$q = \dot{q}'''V = \dot{q}\left(\frac{4}{3}\pi r_i^3\right)$$

$$= \left(10^5\,\frac{W}{m^3\cdot h}\right)\left(\frac{4}{3}\pi\right)(0.4\,m)^3$$

$$= 26{,}800\,W\ (26.8\,kW)$$

At r_o, the steel–water interface,

$$q = hA(T_o - T_\infty) = h(4\pi r_0^2)(T_o - T_\infty)$$

$$26{,}800\,W = \left(800\,\frac{W}{m^2\cdot k}\right)(4\pi)(0.5\,m)^2(T_0 - 25\,C)$$

and, solving for T_0,

$$T_o = 25\,C + 10.7\,C = 35.7\,C$$

The heat-transfer rate at $r = r_o$ can also be expressed as

$$q = -kA\frac{dT}{dr}\bigg|_{r_0} = -k\left(4\pi r_0^2\right)\frac{dT}{dr}\bigg|_{r_0}$$

Utilizing equation (2), we get

$$r_0^2\frac{dT}{dr}\bigg|_{r_0} = -(T_i - T_o)\left[\frac{1}{1/r_i - 1/r_o}\right]$$

and thus

$$q = k_{ss}(4\pi)(T_i - T_o)\left[\frac{1}{1/r_i - 1/r_o}\right]$$

Now, inserting numbers, we get

$$T_i - T_o = 284 \text{ C}$$

or

$$T_i = 35.7 \text{ C} + 284 \text{ C} = 319.7 \text{ C}$$

Equation (1) can now be used to get $T_{w,\text{max}}$:

$$T_{w,\text{max}} = T_i + \frac{\dot{q}''' r_i^2}{6k_w}$$

$$= 319.7 \text{ C} + \frac{\left(10^5 \frac{\text{W}}{\text{m}^3}\right)(0.4 \text{ m})^2}{6\left(20 \frac{\text{W}}{\text{m} \cdot \text{K}}\right)}$$

$$= 319.7 \text{ C} + 133.3 \text{ C}$$

$$= 453 \text{ C}$$

▶ **16.4**

CLOSURE

The general differential equations of energy transfer have been developed in this chapter, and some forms applying to more specific situations were presented. Some remarks concerning initial and boundary conditions have been made as well.

In the chapters to follow, analyses of energy transfer will start with the applicable differential equation. Numerous solutions will be presented, and still more assigned as student exercises. The tools for heat-transfer analysis have now been developed and examined. Our remaining task is to develop a familiarity with and facility in their use.

CHAPTER 17

Steady-State Conduction

In most equipment used in transferring heat, energy flows from one fluid to another through a solid wall. As the energy transfer through each medium is one step in the overall process, a clear understanding of the conduction mechanism of energy transfer through homogeneous solids is essential to the solutions of most heat-transfer problems.

In this chapter, we shall direct our attention to steady-state heat conduction. Steady state implies that the conditions, temperature, density, and the like at all points in the conduction region are independent of time. Our analyses will parallel the approaches used for analyzing a differential fluid element in laminar flow and those that will be used in analyzing steady-state molecular diffusion. During our discussions, two types of presentations will be used: (1) The governing differential equation will be generated by means of the control-volume concept and (2) the governing differential equation will be obtained by eliminating all irrelevant terms in the general differential equation for energy transfer.

▶ 17.1 ONE-DIMENSIONAL CONDUCTION

For steady-state conduction independent of any internal generation of energy, the general differential equation reduces to the Laplace equation:

$$\nabla^2 T = 0 \tag{16-20}$$

Although this equation implies that more than one space coordinate is necessary to describe the temperature field, many problems are simpler because of the geometry of the conduction region or because of symmetries in the temperature distribution. One-dimensional cases often arise.

The one-dimensional, steady-state transfer of energy by conduction is the simplest process to describe as the condition imposed upon the temperature field is an ordinary differential equation. For one-dimensional conduction, equation (16-20) reduces to

$$\frac{d}{dx}\left(x^i \frac{dT}{dx}\right) = 0 \tag{17-1}$$

where $i = 0$ for rectangular coordinates, $i = 1$ for cylindrical coordinates, and $i = 2$ for spherical coordinates.

One-dimensional processes occur in flat planes, such as furnace walls; in cylindrical elements, such as steam pipes; and in spherical elements, such as nuclear-reactor pressure vessels. In this section, we shall consider steady-state conduction through simple systems in which the temperature and the energy flux are functions of a single space coordinate.

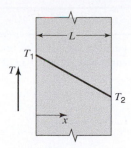

Figure 17.1 Plane wall with a one-dimensional temperature distribution.

Plane Wall Consider the conduction of energy through a plane wall as illustrated in Figure 17.1. The one-dimensional Laplace equation is easily solved, yielding

$$T = C_1 x + C_2 \qquad (17\text{-}2)$$

The two constants are obtained by applying the boundary conditions

$$\text{at } x = 0 \quad T = T_1$$

and

$$\text{at } x = L \quad T = T_2$$

These constants are

$$C_2 = T_1$$

and

$$C_1 = \frac{T_2 - T_1}{L}$$

The temperature profile becomes

$$T = \frac{T_2 - T_1}{L} x + T_1$$

or

$$T = T_1 - \frac{T_1 - T_2}{L} x \qquad (17\text{-}3)$$

and is linear, as illustrated in Figure 17.1.

The energy flux is evaluated, using the Fourier rate equation

$$\frac{q_x}{A} = -k \frac{dT}{dx} \qquad (15\text{-}1)$$

The temperature gradient, dT/dx, is obtained by differentiating equation (17-3), yielding

$$\frac{dT}{dx} = -\frac{T_1 - T_2}{L}$$

Substituting this term into the rate equation, we obtain for a flat wall with constant thermal conductivity

$$q_x = \frac{kA}{L}(T_1 - T_2) \qquad (17\text{-}4)$$

The quantity kA/L is characteristic of a flat wall or a flat plate and is designated the *thermal conductance*. The reciprocal of the thermal conductance, L/kA, is the *thermal resistance*.

Composite Walls The steady flow of energy through several walls in series is often encountered. A typical furnace design might include one wall for strength, an intermediate wall for insulation, and the third outer wall for appearance. This composite plane wall is illustrated in Figure 17.2.

For a solution to the system shown in this figure, the reader is referred to Section 15.5.

The following example illustrates the use of the composite-wall energy-rate equation for predicting the temperature distribution in walls.

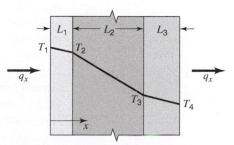

Figure 17.2 Temperature distribution for steady-state conduction of energy through a composite plane wall.

Example 1

A furnace wall is composed of three layers, 10 cm of firebrick ($k = 1.560$ W/m · K), followed by 23 cm of kaolin insulating brick ($k = 0.073$ W/m · K), and finally 5 cm of masonry brick ($k = 1.0$ W/m · K). The temperature of the inner wall surface is 1370 K and the outer surface is at 360 K. What are the temperatures at the contacting surfaces?

The individual material thermal resistances per m^2 of area are

$$R_1, \text{ firebrick} = \frac{L_1}{k_1 A_1} = \frac{0.10 \text{ m}}{(1.560 \text{ W/m} \cdot \text{K})(1 \text{ m}^2)} = 0.0641 \text{ K/W}$$

$$R_2, \text{ kaolin} = \frac{L_2}{k_2 A_2} = \frac{0.23}{(0.073)(1)} = 3.15 \text{ K/W}$$

$$R_3, \text{ masonry} = \frac{L_3}{k_3 A_3} = \frac{0.05}{(1.0)(1)} = 0.05 \text{ K/W}$$

The total resistance of the composite wall is equal to $0.0641 + 3.15 + 0.05 = 3.26$ K/W. The total temperature drop is equal to $(T_1 - T_4) = 1370 - 360 = 1010$ K.

Using equation (15-16), the energy-transfer rate is

$$q = \frac{T_1 - T_4}{\Sigma R} = \frac{1010 \text{ K}}{3.26 \text{ K/W}} = 309.8 \text{ W}$$

As this is a steady-state situation, the energy-transfer rate is the same for each part of the transfer path (i.e., through each wall section). The temperature at the firebrick-kaolin interface, T_2, is given by

$$T_1 - T_2 = q(R_1)$$
$$= (309.8 \text{ W})(0.0641 \text{ K/W}) = 19.9 \text{ K}$$

giving

$$T_2 = 1350.1$$

Similarly,

$$T_3 - T_4 = q(R_3)$$
$$= (309.8 \text{ W})(0.05 \text{ K/W}) = 15.5 \text{ K}$$

giving

$$T_3 = 375.5 \text{ K}$$

There are numerous situations in which a composite wall involves a combination of series and parallel energy-flow paths. An example of such a wall is illustrated in Figure 17.3, where steel is used as reinforcement for a concrete wall. The composite wall can be divided into three sections of length L_1, L_2, and L_3, and the thermal resistance for each of these lengths may be evaluated.

The intermediate layer between planes 2 and 3 consists of two separate thermal paths in parallel; the effective thermal conductance is the sum of the conductances for the two materials. For the section of the wall of height $y_1 + y_2$ and unit depth, the resistance is

$$R_2 = \frac{1}{\dfrac{k_1 y_1}{L_2} + \dfrac{k_2 y_2}{L_2}} = L_2 \left(\frac{1}{k_1 y_1 + k_2 y_2} \right)$$

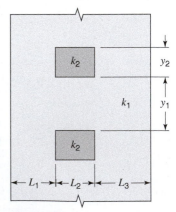

Figure 17.3 A series-parallel composite wall.

The total resistance for this wall is

$$\Sigma R_T = R_1 + R_2 + R_3$$

or

$$\Sigma R_T = \frac{L_1}{k_1(y_1 + y_2)} + L_2\left(\frac{1}{k_1 y_1 + k_2 y_2}\right) + \frac{L_3}{k_1(y_1 + y_2)}$$

The electrical circuit —/\/\/— is an analog to the composite wall.

The rate of energy transferred from plane 1 to plane 4 is obtained by a modified form of equation (15-16).

$$q = \frac{T_1 - T_4}{\Sigma R_T} = \frac{T_1 - T_4}{\dfrac{L_1}{k_1(y_1 + y_2)} + L_2\left(\dfrac{1}{k_1 y_1 + k_2 y_2}\right) + \dfrac{L_3}{k_1(y_1 + y_2)}} \tag{17-5}$$

It important to recognize that this equation is only an approximation. Actually, there is a significant temperature distribution in the y-direction close to the material that has the higher thermal conductivity.

In our discussions of composite walls, no allowance was made for a temperature drop at the contact face between two different solids. This assumption is not always valid, as there will often be vapor spaces caused by rough surfaces, or even oxide films on the surfaces of metals. These additional contact resistances must be accounted for in a precise energy-transfer equation.

Long, Hollow Cylinder Radial energy flow by conduction through a long, hollow cylinder is another example of one-dimensional conduction. The radial heat flow for this configuration is evaluated in Example 1 of Chapter 15 as

$$\frac{q_r}{L} = \frac{2\pi k}{\ln(r_o/r_i)}(T_i - T_o) \tag{17-6}$$

where r_i is the inside radius, r_o is the outside radius, T_i is the temperature on the inside surface, and T_o is the temperature on the outside surface. The resistance concept may again be used; the thermal resistance of the hollow cylinder is

$$R = \frac{\ln(r_o/r_i)}{2\pi k L} \tag{17-7}$$

The radial temperature distribution in a long, hollow cylinder may be evaluated by using equation (17-1) in cylindrical form

$$\frac{d}{dr}\left(r\frac{dT}{dr}\right) = 0 \tag{17-8}$$

Solving this equation subject to the boundary conditions

$$\text{at } r = r_i \quad T = T_i$$

and

$$\text{at } r = r_o \quad T = T_o$$

we see that temperature profile is

$$T(r) = T_i - \frac{T_i - T_o}{\ln(r_o/r_i)}\ln\frac{r}{r_i} \tag{17-9}$$

Thus, the temperature in a long, hollow cylinder is a logarithmic function of radius r, whereas for the plane wall the temperature distribution is linear.

The following example illustrates the analysis of radial energy conduction through a long, hollow cylinder.

Example 2

A long steam pipe of outside radius r_2 is covered with thermal insulation having an outside radius of r_3. The temperature of the outer surface of the pipe, T_2, and the temperature of the surrounding air, T_∞, are fixed. The energy loss per unit area of outside surface of the insulation is described by the Newton rate equation

$$\frac{q_r}{A} = h(T_3 - T_\infty) \qquad (15\text{-}11)$$

Can the energy loss increase with an increase in the thickness of insulation? If possible, under what conditions will this situation arise? Figure 17.4 may be used to illustrate this composite cylinder.

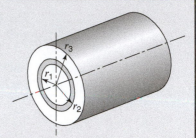

Figure 17.4 A series composite hollow cylinder.

In Example 3 of Chapter 15, the thermal resistance of a hollow cylindrical element was shown to be

$$R = \frac{\ln(r_o/r_i)}{2\pi k L} \qquad (17\text{-}10)$$

In the present example, the total difference in temperature is $T_2 - T_\infty$ and the two resistances, due to the insulation and the surrounding air film, are

$$R_2 = \frac{\ln(r_3/r_2)}{2\pi k_2 L}$$

for the insulation and

$$R_3 = \frac{1}{hA} = \frac{1}{h 2\pi r_3 L}$$

for the air film.

Substituting these terms into the radial heat flow equation and rearranging, we obtain

$$q_r = \frac{2\pi L (T_2 - T_\infty)}{[\ln(r_3/r_2)]/k_2 + 1/hr_3} \qquad (17\text{-}11)$$

The dual effect of increasing the resistance to energy transfer by conduction and simultaneously increasing the surface area as r_3 is increased suggests that for a pipe of given size, a particular outer radius exists for which the heat loss is maximum. As the ratio r_3/r_2 increases logarithmically, and the term $1/r_3$ decreases as r_3 increases, the relative importance of each resistance term will change as the insulation thickness is varied. In this example, L, T_2, T_∞, k_2, h, and r_2 are considered constant. Differentiating equation (17-11) with respect to r_3, we obtain

$$\frac{dq_r}{dr_3} = -\frac{2\pi L (T_2 - T_\infty)\left(\dfrac{1}{k_2 r_3} - \dfrac{1}{h r_3^2}\right)}{\left[\dfrac{1}{k_2}\ln\left(\dfrac{r_3}{r_2}\right) + \dfrac{1}{h r_3}\right]^2} \qquad (17\text{-}12)$$

The radius of insulation associated with the maximum energy transfer, the *critical radius*, found by setting $dq_r/dr_3 = 0$; equation (17-12) reduces to

$$(r_3)_{\text{critical}} = \frac{k_2}{h} \tag{17-13}$$

In the case of 85% magnesia insulation ($k = 0.0692$ W/m · K) and a typical value for the heat-transfer coefficient in natural convection ($h = 34$ W/m^2 · K), the critical radius is calculated as

$$r_{\text{crit}} = \frac{k}{h} = \frac{0.0692 \text{ W/m} \cdot \text{K}}{34 \text{ W/m}^2 \cdot \text{K}} = 0.0020 \text{ m} \quad (0.0067 \text{ ft})$$
$$= 0.20 \text{ cm} \quad (0.0787 \text{ in})$$

These very small numbers indicate that the critical radius will be exceeded in any practical problem. The question then is whether the critical radius given by equation (17-13) represents a maximum or a minimum condition for q. The evaluation of the second derivative, d^2q_r/dr_3^2, when $r_3 = k/h$ yields a negative result; thus, r_{crit} is a maximum condition. It now follows that q_r will be decreased for any value of r_3 greater than 0.0020 m.

Hollow Sphere   Radial heat flow through a hollow sphere is another example of one-dimensional conduction. For constant thermal conductivity, the modified Fourier rate equation

$$q_r = -k\frac{dT}{dr}A$$

applies, where $A =$ surface area of a sphere $= 4\pi r^2$, giving

$$q_r = -4\pi k r^2 \frac{dT}{dr} \tag{17-14}$$

This relation, when integrated between the boundary conditions

$$\text{at } T = T_i \quad r = r_i$$

and

$$\text{at } T = T_o \quad r = r_o$$

yields

$$q = \frac{4\pi k(T_i - T_o)}{\dfrac{1}{r_i} - \dfrac{1}{r_o}} \tag{17-15}$$

The hyperbolic temperature distribution

$$T = T_i - \left(\frac{T_i - T_o}{1/r_i - 1/r_o}\right)\left(\frac{1}{r_i} - \frac{1}{r}\right) \tag{17-16}$$

is obtained by using the same procedure that was followed to obtain equation (17-9).

Variable Thermal Conductivity If the thermal conductivity of the medium through which the energy is transferred varies significantly, the preceding equations in this section do not apply. As Laplace's equation involves the assumption of constant thermal conductivity, a new differential equation must be determined from the general equation for heat transfer. For steady-state conduction in the x-direction without internal generation of energy, the equation that applies is

$$\frac{d}{dx}\left(k\frac{dT}{dx}\right) = 0 \tag{17-17}$$

where k may be a function of T.

In many cases, the thermal conductivity may be a linear function of temperature over a considerable range. The equation of such a straight-line function may be expressed by

$$k = k_o(1 + \beta T)$$

where k_o and β are constants for a particular material. In general, for materials satisfying this relation, β is negative for good conductors and positive for good insulators. Other relations for varying k have been experimentally determined for specific materials. The evaluation of the rate of energy transfer when the material has a varying thermal conductivity is illustrated in Example 2 of Chapter 15.

▶ 17.2

ONE-DIMENSIONAL CONDUCTION WITH INTERNAL GENERATION OF ENERGY

In certain systems, such as electric resistance heaters and nuclear fuel rods, heat is generated within the conducting medium. As one might expect, the generation of energy within the conducting medium produces temperature profiles different than those for simple conduction.

In this section, we shall consider two simple example cases: steady-state conduction in a circular cylinder with uniform or homogeneous energy generation and steady-state conduction in a plane wall with variable energy generation. Carslaw and Jaeger[1] and Jakob[2] have written excellent treatises dealing with more complicated problems.

Cylindrical Solid with Homogeneous Energy Generation Consider a cylindrical solid with internal energy generation as shown in Figure 17.5. The cylinder will be considered long enough that only radial conduction occurs. The density, ρ, the heat capacity, c_p, and the thermal conductivity of the material will be considered constant.

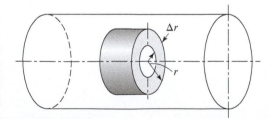

Figure 17.5 Annular element in a long, circular cylinder with internal heat generation.

[1] H. S. Carslaw and J. C. Jaeger, *Conduction of Heat in Solids,* 2nd edition, Oxford University Press, New York, 1959.

The energy balance for the element shown is

$$\left\{\begin{array}{c} \text{rate of energy} \\ \text{conduction into} \\ \text{the element} \end{array}\right\} + \left\{\begin{array}{c} \text{rate of energy} \\ \text{generation within} \\ \text{the element} \end{array}\right\} - \left\{\begin{array}{c} \text{rate of energy} \\ \text{conduction out} \\ \text{of the element} \end{array}\right\}$$

$$= \left\{\begin{array}{c} \text{rate of accumulation} \\ \text{of energy} \\ \text{within the element} \end{array}\right\}$$

(17-18)

Applying the Fourier rate equation and letting $\dot{q}$ represent the rate of energy generated per unit volume, we may express equation (17-18) by the algebraic expression

$$-k(2\pi rL)\frac{\partial T}{\partial r}\Big|_r + \dot{q}(2\pi rL\,\Delta r) - \left[-k(2\pi rL)\frac{\partial T}{\partial r}\Big|_{r+\Delta r}\right] = \rho c_p \frac{\partial T}{\partial t}(2\pi rL\Delta r)$$

Dividing each term by $2\pi rL\,\Delta r$, we obtain

$$\dot{q} + \frac{k[r(\partial T/\partial r)|_{r+\Delta r} - r(\partial T/\partial r)|_r]}{r\Delta r} = \rho c_p \frac{\partial T}{\partial t}$$

In the limit as Δr approaches zero, the following differential equation is generated:

$$\dot{q} + \frac{k}{r}\frac{\partial}{\partial r}\left(r\frac{\partial T}{\partial r}\right) = \rho c_p \frac{\partial T}{\partial t} \tag{17-19}$$

For steady-state conditions, the accumulation term is zero. When we eliminate this term from the above expression, the differential equation for a solid cylinder with homogeneous energy generation becomes

$$\dot{q} + \frac{k}{r}\frac{d}{dr}\left(r\frac{dT}{dr}\right) = 0 \tag{17-20}$$

The variables in this equation may be separated and integrated to yield

$$rk\frac{dT}{dr} + \dot{q}\frac{r^2}{2} = C_1$$

or

$$k\frac{dT}{dr} + \dot{q}\frac{r}{2} = \frac{C_1}{r}$$

Because of the symmetry of the solid cylinder, a boundary condition that must be satisfied stipulates that the temperature gradient must be finite at the center of the cylinder, where $r = 0$. This can be true only if $C_1 = 0$. Accordingly, the above relation reduces to

$$k\frac{dT}{dr} + \dot{q}\frac{r}{2} = 0 \tag{17-21}$$

A second integration will now yield

$$T = -\frac{\dot{q}r^2}{4k} + C_2 \tag{17-22}$$

If the temperature T is known at any radial value, such as a surface, the second constant, C_2, may be evaluated. This, of course, provides the completed expression for the temperature profile. The energy flux in the radial direction may be obtained from

$$\frac{q_r}{A} = -k\frac{dT}{dr}$$

by substituting equation (17-21), yielding

$$\frac{q_r}{A} = \dot{q}\frac{r}{2}$$

or

$$q_r = (2\pi rL)\dot{q}\frac{r}{2} = \pi r^2 L\dot{q} \qquad (17\text{-}23)$$

Plane Wall with Variable Energy Generation   The second case associated with energy generation involves a temperature-dependent, energy-generating process. This situation develops when an electric current is passed through a conducting medium possessing an electrical resistivity that varies with temperature. In our discussion, we shall assume that the energy-generation term varies linearly with temperature, and that the conducting medium is a flat plate with temperature T_L at both surfaces. The internal energy generation is described by

$$\dot{q} = \dot{q}_L[1 + \beta(T - T_L)] \qquad (17\text{-}24)$$

where $\dot{q}_L$ is the generation rate at the surface and β is a constant.

With this model for the generation function, and as both surfaces have the same temperature, the temperature distribution within the flat plate is symmetric about the midplane. The plane wall and its coordinate system are illustrated in Figure 17.6. The symmetry of the temperature distribution requires a zero temperature gradient at $x = 0$. With steady-state conditions, the differential equation may be obtained by eliminating the irrelevant terms in the general differential equation for heat transfer. Equation (16-19) for the case of steady-state conduction in the x-direction in a stationary solid with constant thermal conductivity becomes

$$\frac{d^2T}{dx^2} + \frac{\dot{q}_L}{k}[1 + \beta(T - T_L)] = 0$$

The boundary conditions are

$$\text{at } x = 0 \quad \frac{dT}{dx} = 0$$

and

$$\text{at } x = \pm L \quad T = T_L$$

These relations may be expressed in terms of a new variable, $\theta = T - T_L$, by

$$\frac{d^2\theta}{dx^2} + \frac{\dot{q}_L}{k}(1 + \beta\theta) = 0$$

or

$$\frac{d^2\theta}{dx^2} + C + s\theta = 0$$

where $C = \dot{q}_L/k$ and $s = \beta\dot{q}_L/k$. The boundary conditions are

$$\text{at } x = 0 \quad \frac{d\theta}{dx} = 0$$

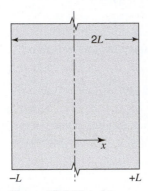

Figure 17.6 Flat plate with temperature-dependent energy generation.

and

$$\text{at } x = \pm L \quad \theta = 0$$

The integration of this differential equation is simplified by a second change in variables; inserting ϕ for $C + s\theta$ into the differential equation and the boundary conditions, we obtain

$$\frac{d^2\phi}{dx^2} + s\phi = 0$$

for

$$x = 0, \quad \frac{d\phi}{dx} = 0$$

and

$$x = \pm L, \quad \phi = C$$

The solution is

$$\phi = C + s\theta = A\cos(x\sqrt{s}) + B\sin(x\sqrt{s})$$

or

$$\theta = A_1\cos(x\sqrt{s}) + A_2\sin(x\sqrt{s}) - \frac{C}{s}$$

The temperature distribution becomes

$$T - T_L = \frac{1}{\beta}\left[\frac{\cos(x\sqrt{s})}{\cos(L\sqrt{s})} - 1\right] \tag{17-25}$$

where $s = \beta\dot{q}_L/k$ is obtained by applying the two boundary conditions.

The cylindrical and spherical examples of one-dimensional temperature-dependent generation are more complex; solutions to these may be found in the technical literature.

▶ **17.3**

HEAT TRANSFER FROM EXTENDED SURFACES

A very useful application of one-dimensional heat-conduction analysis is that of describing the effect of extended surfaces. It is possible to increase the energy transfer between a surface and an adjacent fluid by increasing the amount of surface area in contact with the fluid. This increase in area is accomplished by adding extended surfaces that may be in the forms of fins or spines of various cross sections.

The one-dimensional analysis of extended surfaces may be formulated in general terms by considering the situation depicted in Figure 17.7.

The shaded area represents a portion of the extended surface that has variable cross-sectional area, $A(x)$, and surface area, $S(x)$, which are functions of x alone. For steady-state conditions, the first law of thermodynamics, equation (6-10), reduces to the simple expression

$$\frac{\delta Q}{dt} = 0$$

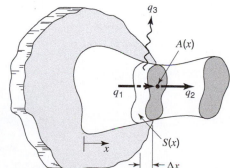

Figure 17.7 An extended surface of general configuration.

Thus, in terms of the heat flow rates designated in the figure, we may write

$$q_1 = q_2 + q_3 \tag{17-26}$$

The quantities q_1 and q_2 are conduction terms, while q_3 is a convective heat-flow rate. Evaluating each of these in the appropriate way and substituting into equation (17-26), we obtain

$$kA\frac{dT}{dx}\Big|_{x+\Delta x} - kA\frac{dT}{dx}\Big|_x - hS(T - T_\infty) = 0 \tag{17-27}$$

where T_∞ is the surrounding fluid temperature. Expressing the surface area, $S(x)$, in terms of the width, Δx, times the perimeter, $P(x)$, and dividing through by Δx, we obtain

$$\frac{kA(dT/dx)|_{x+\Delta x} - kA(dT/dx)|_x}{\Delta x} - hP(T - T_\infty) = 0$$

Evaluating this equation in the limit as $\Delta x \to 0$, we obtain the differential equation

$$\frac{d}{dx}\left(kA\frac{dT}{dx}\right) - hP(T - T_\infty) = 0 \tag{17-28}$$

One should note, at this point, that the temperature gradient, dT/dx, and the surface temperature, T, are expressed such that T is a function of x only. This treatment assumes the temperature to be "lumped" in the transverse direction. This is physically realistic when the cross section is thin or when the material thermal conductivity is large. Both of these conditions apply in the case of fins. More will be said about the "lumped parameter" approach in Chapter 18. This approximation in the present case leads to equation (17-28), an ordinary differential equation. If we did not make this simplifying analysis, we would have a distributed parameter problem that would require solving a partial differential equation.

A wide range of possible forms exist when equation (17-28) is applied to specific geometries. Three possible applications and the resulting equations are described in the following paragraphs.

(1) Fins or Spines of Uniform Cross Section   For either of the cases shown in Figure 17.8, the following are true: $A(x) = A$, and $P(x) = P$, both constants. If, additionally, both k and h are taken to be constant, equation (17-28) reduces to

$$\frac{d^2T}{dx^2} - \frac{hP}{kA}(T - T_\infty) = 0 \tag{17-29}$$

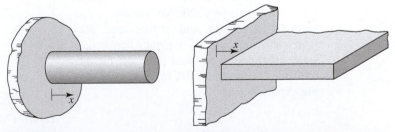

Figure 17.8 Two examples of extended surfaces with constant cross section.

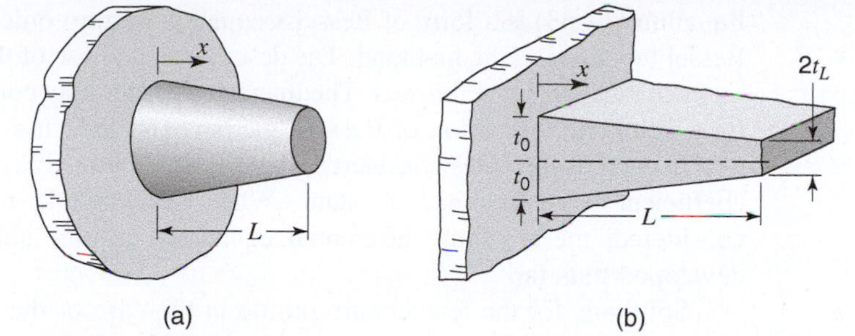

Figure 17.9 Two examples of straight extended surfaces with variable cross section.

(2) Straight Surfaces with Linearly Varying Cross Section   Two configurations for which A and P are not constant are shown in Figure 17.9. If the area and perimeter both vary in a linear manner from the primary surface, $x=0$, to some lesser value at the end, $x=L$, both A and P may be expressed as

$$A = A_0 - (A_0 - A_L)\frac{x}{L} \tag{17-30}$$

and

$$P = P_0 - (P_0 - P_L)\frac{x}{L} \tag{17-31}$$

In the case of the rectangular fin, shown in Figure 17.9(b), the appropriate values of A and P are

$$A_0 = 2t_0 W \qquad A_L = 2t_L W$$

$$P_0 = 2[2t_0 + W] \quad P_L = 2[2t_L + W]$$

where t_0 and t_L represent the semi-thickness of the fin evaluated at $x=0$ and $x=L$, respectively, and W is the total depth of the fin.

For constant h and k, equation (17-28) applied to extended surfaces with cross-sectional area varying linearly becomes

$$\left[A_0 - (A_0 - A_L)\frac{x}{L}\right]\frac{d^2 T}{dx^2} - \frac{A_0 - A_L}{L}\frac{dT}{dx} - \frac{h}{k}\left[P_0 - (P_0 - P_L)\frac{x}{L}\right](T - T_\infty) = 0 \tag{17-32}$$

(3) Curved Surfaces of Uniform Thickness   A common type of extended surface is that of the circular fin of constant thickness as depicted in Figure 17.10. The appropriate expressions for A and P, in this case, are

and
$$\left.\begin{array}{c} A = 4\pi rt \\[4pt] P = 4\pi r \end{array}\right\} r_0 \leq r \leq r_L$$

When these expressions are substituted into equation (17-28), the applicable differential equation, considering k and h constant, is

$$\frac{d^2 T}{dr^2} + \frac{1}{r}\frac{dT}{dr} - \frac{h}{kt}(T - T_\infty) = 0 \tag{17-33}$$

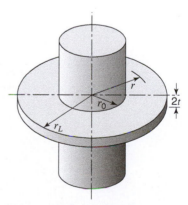

Figure 17.10 A curved fin of constant thickness.

Equation (17-33) is a form of Bessel's equation of zero order. The solution is in terms of Bessel functions of the first kind. The description and use of these functions are beyond the mathematical scope of this text. The interested reader may consult the work of Kraus et al.[3] for a complete discussion of Bessel functions and their use.

In each of the cases considered, the thermal conductivity and convective heat-transfer coefficient were assumed constant. When the variable nature of these quantities is considered, the resulting differential equations become still more complex than those developed thus far.

Solutions for the temperature profile in the case of the straight fin of constant cross section will now be considered; equation (17-29) applies.

The general solution to equation (17-29) may be written

$$\theta = c_1 e^{mx} + c_2 e^{-mx} \qquad (17\text{-}34)$$

or

$$\theta = A \cosh mx + B \sinh mx \qquad (17\text{-}35)$$

where $m^2 = hP/kA$ and $\theta = T - T_\infty$. The evaluation of the constants of integration requires that two boundary conditions be known. The three sets of boundary conditions that we shall consider are as follows:

(a) $\qquad T = T_0 \qquad$ at $\quad x = 0$
$\qquad\qquad T = T_L \qquad$ at $\quad x = L$

(b) $\qquad T = T_0 \qquad$ at $\quad x = 0$
$\qquad\qquad \dfrac{dT}{dx} = 0 \qquad$ at $\quad x = L$

and

(c) $\qquad T = T_0 \qquad\qquad\qquad$ at $\quad x = 0$
$\qquad\qquad -k\dfrac{dT}{dx} = h(T - T_\infty) \qquad$ at $\quad x = L$

The first boundary condition of each set is the same and stipulates that the temperature at the base of the extended surface is equal to that of the primary surface. The second boundary condition relates the situation at a distance L from the base. In set (a) the condition is that of a known temperature at $x = L$. In set (b) the temperature gradient is zero at $x = L$. In set (c) the requirement is that heat flow to the end of an extended surface by conduction be equal to that leaving this position by convection.

The temperature profile, associated with the first set of boundary conditions, is

$$\frac{\theta}{\theta_0} = \frac{T - T_\infty}{T_0 - T_\infty} = \left(\frac{\theta_L}{\theta_0} - e^{-mL}\right)\left(\frac{e^{mx} - e^{-mx}}{e^{mL} - e^{-mL}}\right) + e^{-mx} \qquad (17\text{-}36)$$

A special case of this solution applies when L becomes very large—that is, $L \to \infty$, for which equation (17-36) reduces to

$$\frac{\theta}{\theta_0} = \frac{T - T_\infty}{T_0 - T_\infty} = e^{-mx} \qquad (17\text{-}37)$$

The constants, c_1 and c_2, obtained by applying set (b), yield, for the temperature profile,

$$\frac{\theta}{\theta_0} = \frac{T - T_\infty}{T_0 - T_\infty} = \frac{e^{mx}}{1 + e^{2mL}} + \frac{e^{-mx}}{1 + e^{-2mL}} \qquad (17\text{-}38)$$

---

[3] A. D. Kraus, A. Aziz, and J. R. Welty, *Extended Surface Heat Transfer*, Wiley-Interscience, New York, 2001.

An equivalent expression to equation (17-38) but in a more compact form is

$$\frac{\theta}{\theta_0} = \frac{T - T_\infty}{T_0 - T_\infty} = \frac{\cosh[m(L - x)]}{\cosh mL} \tag{17-39}$$

Note that in either equation (17-38) or (17-39), as $L \to \infty$, the temperature profile approaches that expressed in equation (17-37).

The application of set (c) of the boundary conditions yields, for the temperature profile,

$$\frac{\theta}{\theta_0} = \frac{T - T_\infty}{T_0 - T_\infty} = \frac{\cosh[m(L - x)] + (h/mk)\sinh[m(L - x)]}{\cosh mL + (h/mk)\sinh mL} \tag{17-40}$$

It may be noted that this expression reduces to equation (17-39) if $d\theta/dx = 0$ at $x = L$ and to equation (17-37) if $T = T_\infty$ at $L = \infty$.

The expressions for $T(x)$ that have been obtained are particularly useful in evaluating the total heat transfer from an extended surface. This total heat transfer may be determined by either of two approaches. The first is to integrate the convective heat-transfer expression over the surface according to

$$q = \int_S h[T(x) - T_\infty]dS = \int_S h\theta \, dS \tag{17-41}$$

The second method involves evaluating the energy conducted into the extended surface at the base as expressed by

$$q = -kA \frac{dT}{dx}\bigg|_{x=0} \tag{17-42}$$

The latter of these two expressions is easier to evaluate; accordingly, we will use this equation in the following development.

Using equation (17-36), we find that the heat-transfer rate, when set (a) of the boundary conditions applies, is

$$q = kAm\theta_0 \left[1 - 2\frac{\theta_L/\theta_0 - e^{-mL}}{e^{mL} - e^{-mL}} \right] \tag{17-43}$$

If the length L is very long, this expression becomes

$$q = kAm\theta_0 = kAm(T_0 - T_\infty) \tag{17-44}$$

Substituting equation (17-39) [obtained by using set (b) of the boundary conditions] into equation (17-42), we obtain

$$q = kAm\theta_0 \tanh mL \tag{17-45}$$

Equation (17-40), utilized in equation (17-42), yields for q the expression

$$q = kAm\theta_0 \frac{\sinh mL + (h/mk)\cosh mL}{\cosh mL + (h/mk)\sinh mL} \tag{17-46}$$

The equations for the temperature profile and total heat transfer for extended surfaces of more involved configuration have not been considered. Certain of these cases will be left as exercises for the reader.

A question that is logically asked at this point is, "What benefit is accrued by the addition of extended surfaces?" A term that aids in answering this question is the *fin efficiency*, symbolized as η_f, defined as the ratio of the actual heat transfer from an extended surface to the maximum possible heat transfer from the surface. The maximum heat transfer would occur if the temperature of the extended surface were equal to the base temperature, T_0, at all points.

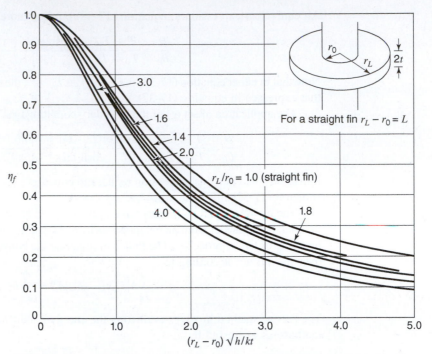

Figure 17.11 Fin efficiency for straight and circular fins of constant thickness.

Figure 17.11 is a plot of η_f as a function of a significant parameter for both straight and circular fins of constant thickness (when fin thickness is small, $t \ll r_L - r_0$).

The total heat transfer from a finned surface is

$$q_{\text{total}} = q_{\text{primary surface}} + q_{\text{fin}}$$
$$= A_0 h(T_0 - T_\infty) + A_f h(T - T_\infty) \qquad (17\text{-}47)$$

The second term in equation (17-47) is the actual heat transfer from the fin surface in terms of the variable surface temperature. This may be written in terms of the fin efficiency, yielding

$$q_{\text{total}} = A_0 h(T_0 - T_\infty) + A_f h\eta_f(T_0 - T_\infty)$$

or

$$q_{\text{total}} = h(A_0 + A_f\eta_f)(T_0 - T_\infty) \qquad (17\text{-}48)$$

In this expression A_0 represents the exposed area of the primary surface, A_f is the total fin surface area, and the heat-transfer coefficient, h, is assumed constant.

The application of equation (17-48) as well as an idea of the effectiveness of fins is illustrated in Example 3.

Example 3

Water and air are separated by a mild-steel plane wall. It is proposed to increase the heat-transfer rate between these fluids by adding straight rectangular fins of 1.27-mm thickness and 2.5-cm length, spaced 1.27 cm apart. The air-side and water-side heat-transfer coefficients may be assumed constant with values of 11.4 and 256 W/m² · K, respectively. Determine the percent change in total heat transfer when fins are placed on (a) the water side, (b) the air side, and (c) both sides.

For a $1\,m^2$ section of the wall, the areas of the primary surface and of the fins are

$$A_o = 1\,m^2 - 79\,\text{fins}\,(1\,m)\left[\frac{0.00127\,m}{fin}\right]$$
$$= 0.90\,m^2$$

$$A_f = 79\,\text{fins}\,(1\,m)[(2)(0.025\,m)] + 0.10\,m^2$$
$$= 4.05\,m^2$$

Values of fin efficiency can now be determined from Figure 17.11. For the air side,

$$L\sqrt{h/kt} = 0.025\,m\left[\frac{11.4\,W/m^2 \cdot K}{(42.9\,W/m \cdot K)(0.00127\,m)}\right]^{1/2}$$
$$= 0.362$$

and for the water side,

$$L\sqrt{h/kt} = 0.025\,m\left[\frac{256\,W/m^2 \cdot K}{(42.9\,W/m \cdot K)(0.00127\,m)}\right]^{1/2}$$
$$= 1.71$$

The fin efficiencies are then read from the figure as

$$\eta_{\text{air}} \cong 0.95$$
$$\eta_{\text{water}} \cong 0.55$$

The total heat-transfer rates can now be evaluated. For fins on the air side,

$$q = h_a \Delta T_a [A_o + \eta_{fa}\,A_f]$$
$$= 11.4\,\Delta T_a[0.90 + 0.95(4.05)]$$
$$= 54.1\Delta T_a$$

and on the water side,

$$q = h_w \Delta T_w [A_o + \eta_{fw}\,A_f]$$
$$= 256\,\Delta T_w[0.90 + 0.55(4.05)]$$
$$= 801\,\Delta T_w$$

The quantities ΔT_a and ΔT_w represent the temperature differences between the steel surface at temperature T_o and each of the two fluids.

The reciprocals of the coefficients are the thermal resistances of the finned surfaces.

Without fins the heat-transfer rate in terms of the overall temperature difference, $\Delta T = T_w - T_a$, neglecting the conductive resistance of the steel wall, is

$$q = \frac{\Delta T}{\dfrac{1}{11.4} + \dfrac{1}{256}} = 10.91\,\Delta T$$

With fins on the air side alone,

$$q = \frac{\Delta T}{\dfrac{1}{54.1} + \dfrac{1}{256}} = 44.67\,\Delta T$$

an increase of 310% compared with the bare-wall case.

With fins on the water side alone,

$$q = \frac{\Delta T}{\dfrac{1}{11.4} + \dfrac{1}{801}} = 11.24\,\Delta T$$

an increase of 3.0%.

With fins on both sides the heat-flow rate is

$$q = \frac{\Delta T}{\dfrac{1}{54.1} + \dfrac{1}{801}} = 50.68\,\Delta T$$

an increase of 365%.

This result indicates that adding fins is particularly beneficial where the convection coefficient has a relatively small value.

▶ **17.4**

TWO- AND THREE-DIMENSIONAL SYSTEMS

In Sections 17.2 and 17.3, we discussed systems in which the temperature and the energy transfer were functions of a single-space variable. Although many problems fall into this category, there are many other systems involving complicated geometry or temperature boundary conditions, or both, for which two or even three spatial coordinates are necessary to describe the temperature field.

In this section, we shall review some of the methods for analyzing heat transfer by conduction in two- and three-dimensional systems. The problems will mainly involve two-dimensional systems, as they are less cumbersome to solve, yet illustrate the techniques of analysis.

Analytical Solution An analytical solution to any transfer problem must satisfy the differential equation describing the process as well as the prescribed boundary conditions. Many mathematical techniques have been used to obtain solutions for particular energy conduction situations in which a partial differential equation describes the temperature field. Carslaw and Jaeger[4] and Boelter et al.[5] have written excellent treatises that deal with the mathematical solutions for many of the more complex conduction problems. As most of this material is too specialized for an introductory course, a solution will be obtained to one of the first cases analyzed by Fourier[6] in the classical treatise that established the theory of energy transfer by conduction. This solution of a two-dimensional conduction medium employs the mathematical method of separation of variables.

Consider a thin, infinitely long rectangular plate that is free of heat sources, as illustrated in Figure 17.12. For a thin plate $\partial T/\partial z$ is negligible, and the temperature is a function of x and y only. The solution will be obtained for the case in which the two edges of the plate are maintained at zero temperature and the bottom is maintained at T_1 as shown. The steady-state temperature distribution in the plate of constant thermal conductivity must satisfy the differential equation

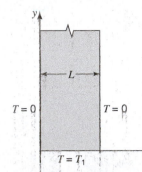

Figure 17.12 Model for two-dimensional conduction analysis.

$$\frac{\partial^2 T}{\partial x^2} + \frac{\partial^2 T}{\partial y^2} = 0 \tag{17-49}$$

[4] H. S. Carslaw and J. C. Jaeger, *Conduction of Heat in Solids, 2nd* edition, Oxford University Press, New York, 1959.

[5] L. M. K. Boelter, V. H. Cherry, H. A. Johnson, and R. C. Martinelli, *Heat Transfer Notes,* McGraw-Hill Book Company, New York, 1965.

[6] J. B. J. Fourier, *Theorie Analytique de la Chaleur,* Gauthier-Villars, Paris, 1822.

and the boundary conditions

$$T = 0 \quad \text{at} \quad x = 0 \quad \text{for all values of } y$$

$$T = 0 \quad \text{at} \quad x = L \quad \text{for all values of } y$$

$$T = T_1 \quad \text{at} \quad y = 0 \quad \text{for } 0 \leq x \leq L$$

and

$$T = 0 \quad \text{at} \quad y = \infty \quad \text{for } 0 \leq x \leq L$$

Equation (17-49) is a linear, homogeneous partial differential equation. This type of equation usually can be solved by assuming that the temperature distribution, $T(x, y)$, is of the form

$$T(x, y) = X(x)Y(y) \tag{17-50}$$

where $X(x)$ is a function of x only and $Y(y)$ is a function of y only. Substituting this equation into equation (17-49), we obtain an expression in which the variables are separated

$$-\frac{1}{X}\frac{d^2X}{dx^2} = \frac{1}{Y}\frac{d^2Y}{dy^2} \tag{17-51}$$

As the left-hand side of equation (17-51) is independent of y and the equivalent right-hand side is independent of x, it follows that both must be independent of x and y, and hence must be equal to a constant. If we designate this constant λ^2, two ordinary differential equations result:

$$\frac{d^2X}{dx^2} + \lambda^2 X = 0 \tag{17-52}$$

and

$$\frac{d^2Y}{dy^2} - \lambda^2 Y = 0 \tag{17-53}$$

These differential equations may be integrated, yielding

$$X = A \cos \lambda x + B \sin \lambda x$$

and

$$Y = C e^{\lambda y} + D e^{-\lambda y}$$

According to equation (17-50), the temperature distribution is defined by the relation

$$T(x, y) = XY = (A \cos \lambda x + B \sin \lambda x)(C e^{\lambda y} + D e^{-\lambda y}) \tag{17-54}$$

where A, B, C, and D are constants to be evaluated from the four boundary conditions. The condition that $T = 0$ at $x = 0$ requires that $A = 0$. Similarly, $\sin \lambda x$ must be zero at $x = L$; accordingly, λL must be an integral multiple of π or $\lambda = n\pi/L$. Equation (17-54) is now reduced to

$$T(x, y) = B \sin\left(\frac{n\pi x}{L}\right)\left(C e^{n\pi y/L} + D e^{-n\pi y/L}\right) \tag{17-55}$$

The requirement that $T = 0$ at $y = \infty$ requires that C must be zero. A combination of B and D into the single constant E reduces equation (17-55) to

$$T(x, y) = E e^{-n\pi y/L} \sin\left(\frac{n\pi x}{L}\right)$$

This expression satisfies the differential equation for any integer n greater than or equal to zero. The general solution is obtained by summing all possible solutions, giving

$$T = \sum_{n=1}^{\infty} E_n e^{-n\pi y/L} \sin\left(\frac{n\pi x}{L}\right) \tag{17-56}$$

The last boundary condition, $T = T_1$ at $y = 0$, is used to evaluate E_n according to the expression

$$T_1 = \sum_{n=1}^{\infty} E_n \sin\left(\frac{n\pi x}{L}\right) \quad \text{for } 0 \le x \le L$$

The constants E_n are the Fourier coefficients for such an expansion and are given by

$$E_n = \frac{4T_1}{n\pi} \quad \text{for} \quad n = 1, 3, 5, \ldots$$

and

$$E_n = 0 \quad \text{for} \quad n = 2, 4, 6, \ldots$$

The solution to this two-dimensional conduction problem is

$$T = \frac{4T_1}{\pi} \sum_{n=0}^{\infty} \frac{e^{[-(2n+1)\pi y]/L}}{2n+1} \sin \frac{(2n+1)\pi x}{L} \tag{17-57}$$

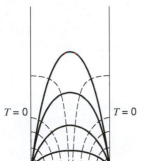

Figure 17.13 Isotherms and energy flow lines for the rectangular plate in Figure 17.12.

The isotherms and energy flow lines are plotted in Figure 17.13. The isotherms are shown in the figure as solid lines, and the dashed lines, which are orthogonal to the isotherms, are energy-flow lines. Note the similarity to the lines of constant velocity potential and stream function, as discussed in momentum transfer.

The separation of variables method can be extended to three-dimensional cases by assuming T to be equal to the product $X(x)Y(y)Z(z)$ and substituting this expression for T into the applicable differential equation. When the variables are separated, three second-order ordinary differential equations are obtained, which may be integrated subject to the given boundary conditions.

Analytical solutions are useful when they can be obtained. There are, however, practical problems with complicated geometry and boundary conditions, which cannot be solved analytically. As an alternative approach, one must turn to numerical methods.

Shape Factors for Common Configurations

The shape factor, **S**, is defined and discussed briefly in Chapter 15. When a geometric case of interest involves conduction between a source and a sink, both with isothermal boundaries, a knowledge of the shape factor makes the determination of heat flow a simple calculation.

Table 17.1 lists expressions for shape factors of five configurations. In every case depicted, it is presumed that the heat-transfer problem is two-dimensional—that is, the dimension normal to the plane shown is very large.

Numerical Solutions

Each of the solution techniques discussed thus far for multidimensional conduction has considerable utility when conditions permit its use. Analytical solutions require relatively simple functions and geometries; the use of shape factors requires isothermal boundaries. When the situation of interest becomes sufficiently complex or when

Table 17.1   Conduction shape factors

Shape	Shape factor, S $q/L = kS(T_i - T_o)$
Concentric circular cylinders.	$\dfrac{2\pi}{\ln(r_o/r_i)}$
Eccentric circular cylinders.	$\dfrac{2\pi}{\cosh^{-1}\left(\dfrac{1+\rho^2-e^2}{2\rho}\right)}$ $\rho \equiv r_i/r_o,\, e \equiv e/r_o$
Circular cylinder in a hexagonal cylinder.	$\dfrac{2\pi}{\ln(r_o/r_i)-0.10669}$
Circular cylinder in a square cylinder.	$\dfrac{2\pi}{\ln(r_o/r_i)-0.27079}$
Infinite cylinder buried in semi-infinite medium.	$\dfrac{2\pi}{\cosh^{-1}(\rho/r)}$

boundary conditions preclude the use of simple solution techniques, one must turn to numerical solutions.

With the presence of digital computers to accomplish the large number of manipulations inherent in numerical solutions rapidly and accurately, this approach is now very common. In this section, we shall introduce the concepts of numerical problem formulation and solution. A more complete and detailed discussion of numerical solutions to heat conduction problems may be found in Carnahan et al.[7] and in Welty.[8]

Shown in Figure 17.14 is a two-dimensional representation of an element within a conducting medium. The element

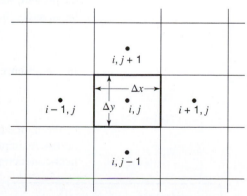

Figure 17.14 Two-dimensional volume element in a conducting medium.

[7] B. Carnahan, H. A. Luther, and J. O. Wilkes, *Applied Numerical Methods,* Wiley, New York, 1969.
[8] J. R. Welty, *Engineering Heat Transfer,* Wiley, New York, 1974.

or "node" i, j is centered in the figure along with its adjacent nodes. The designation, i, j, implies a general location in a two-dimensional system where i is a general index in the x direction and j is the y index. Adjacent node indices are shown in Figure 17.14. The grid is set up with constant node width, Δx, and constant height, Δy. It may be convenient to make the grid "square"—that is, $\Delta x = \Delta y$—but for now we will allow these dimensions to be different.

A direct application of equation (6-10) to node i, j yields

$$\frac{\delta Q}{dt} = \frac{\partial}{\partial t} \iiint_{\text{c.v.}} e\rho dV \tag{17-58}$$

The heat input term, $\delta Q/dt$, may be evaluated allowing for conduction into node i, j from the adjacent nodes and by energy generation within the medium. Evaluating $\delta Q/dt$ in this manner, we obtain

$$\frac{\delta Q}{dt} = k\frac{\Delta y}{\Delta x}(T_{i-1,j} - T_{i,j}) + k\frac{\Delta y}{\Delta x}(T_{i+1,j} - T_{i,j})$$

$$+ k\frac{\Delta x}{\Delta y}(T_{i,j-1} - T_{i,j}) + k\frac{\Delta x}{\Delta y}(T_{i,j+1} - T_{i,j}) + \dot{q}\,\Delta x\Delta y \tag{17-59}$$

The first two terms in this expression relate conduction in the x-direction, the third and fourth express y-directional conduction, and the last is the generation term. All of these terms are positive; heat transfer is assumed positive.

The rate of energy increase within node i, j may be written simply as

$$\frac{\partial}{\partial t} \iiint_{\text{c.v.}} e\rho dV = \left[\frac{\rho c T|_{t+\Delta t} - \rho c T|_t}{\Delta t}\right]\Delta x\Delta y \tag{17-60}$$

Equation (17-58) indicates that the expressions given by equations (17-59) and (17-60) may be equated. Setting these expressions equal to each other and simplifying, we have

$$k\frac{\Delta y}{\Delta x}[T_{i-1,j} + T_{i+1,j} - 2T_{i,j}] + k\frac{\Delta x}{\Delta y}[T_{i,j-1} + T_{i,j+1} - 2T_{i,j}]$$

$$+ \dot{q}\,\Delta x\Delta y = \left[\frac{\rho c T_{i,j}|_{t+\Delta t} - \rho c T_{i,j}|_t}{\Delta t}\right]\Delta x\Delta y \tag{17-61}$$

This expression has been considered in a more complete form in the next chapter. For the present, we will not consider time-variant terms; moreover, we will consider the nodes to be square—that is, $\Delta x = \Delta y$. With these simplifications, equation (17-61) becomes

$$T_{i-1,j} + T_{i+1,j} + T_{i,j-1} + T_{i,j+1} - 4T_{i,j} + \dot{q}\frac{\Delta x^2}{k} = 0 \tag{17-62}$$

In the absence of internal generation, equation (17-62) may be solved for T_{ij} to yield

$$T_{i,j} = \frac{T_{i-1,j} + T_{i+1,j} + T_{i,j-1} + T_{i,j+1}}{4} \tag{17-63}$$

or the temperature of node i, j is the arithmetic mean of the temperatures of its adjacent nodes. A simple example showing the use of equation (17-63) in solving a two-dimensional heat-conduction problem follows.

Example 4

A hollow square duct of the configuration shown (left) has its surfaces maintained at 200 and 100 K, respectively. Determine the steady-state heat-transfer rate between the hot and cold surfaces of this duct. The wall material has a thermal conductivity of 1.21 W/m · K. We may take advantage of the eightfold symmetry of this figure to lay out the simple square grid shown in Figure 17.15.

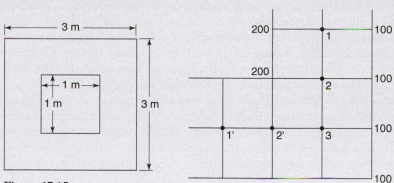

Figure 17.15

The grid chosen is square with $\Delta x = \Delta y = 1/2$ m. Three interior node points are thus identified; their temperatures may be determined by proper application of equation (17-63). Writing the proper expressions for T_1, T_2, and T_3 using equation (17-68) as a guide, we have

$$T_1 = \frac{200 + 100 + 2T_2}{4}$$

$$T_2 = \frac{200 + 100 + T_1 + T_3}{4}$$

$$T_3 = \frac{100 + 100 + 2T_2}{4}$$

This set of three equations and three unknowns may be solved quite easily to yield $T_1 = 145.83$ K, $T_2 = 141.67$ K, $T_3 = 120.83$ K.

The temperatures just obtained may now be used to find heat transfer. Implicit in the procedure of laying out a grid of the sort we have specified is the assumption that heat flows in the x- and y-directions between nodes. On this basis heat transfer occurs from the hot surface to the interior only to nodes 1 and 2; heat transfer occurs to the cooler surface from nodes 1, 2, and 3. We should also recall that the section of duct that has been analyzed is one-eighth of the total; thus, of the heat transfer to and from node 1, only one-half should be properly considered part of the element analyzed.

We now solve for the heat-transfer rate from the hotter surface, and write

$$q = \frac{k(200 - T_1)}{2} + k(200 - T_2)$$

$$= k\left[\left(\frac{200 - 145.83}{2}\right) + (200 - 141.67)\right]$$

$$= 85.41k \quad (q \text{ in W/m}, k \text{ in W/m} \cdot \text{K})$$

A similar accounting for the heat flow from nodes 1, 2, and 3 to the cooler surface is written as

$$q = \frac{k(T_1 - 100)}{2} + k(T_2 - 100) + k(T_3 - 100)$$

$$= k\left[\left(\frac{145.83 - 100}{2}\right) + (141.67 - 100) + (120.83 - 100)\right]$$

$$= 85.41\,k \quad (q \text{ in W/m}, k \text{ in W/m} \cdot \text{K})$$

Observe that these two different means of solving for q yield identical results. This is obviously a requirement of the analysis and serves as a check on the formulation and numerical work.

The example may now be concluded. The total heat transfer per meter of duct is calculated as

$$q = 8(8.415 \text{ K})(1.21 \text{ W/m} \cdot \text{K})$$

$$= 81.45 \text{ W/m}$$

Example 4 has illustrated, in simple fashion, the numerical approach to solving two-dimensional steady-state conduction problems. It is apparent that any added complexity in the form of more involved geometry, other types of boundary conditions such as convection, radiation, specified heat flux, among others, or simply a greater number of interior nodes, will render a problem too complex for hand calculation. Techniques for formulating such problems and some solution techniques are described by Welty.

In this section, we have considered techniques for solving two- and three-dimensional steady-state conduction problems. Each of these approaches has certain requirements that limit their use. The analytical solution is recommended for problems of simple geometrical shapes and simple boundary conditions. Numerical techniques may be used to solve complex problems involving nonuniform boundary conditions and variable physical properties.

▶ 17.5

CLOSURE

In this chapter, we have considered solutions to steady-state conduction problems. The defining differential equations were frequently established by generating the equation through the use of the control-volume expression for the conservation of energy as well as by using the general differential equation for energy transfer. It is hoped that this approach will provide the student with an insight into the various terms contained in the general differential equation and thus enable one to decide, for each solution, which terms are relevant.

One-dimensional systems with and without internal generation of energy were considered.

CHAPTER 18

Unsteady-State Conduction

Transient processes, in which the temperature at a given point varies with time, will be considered in this chapter. As the transfer of energy is directly related to the temperature gradient, these processes involve an *unsteady-state* flux of energy.

Transient conduction processes are commonly encountered in engineering design. These design problems generally fall into two categories: the process that ultimately reaches steady-state conditions and the process that is operated a relatively short time in a continually changing temperature environment. Examples of this second category would include metal stock or ingots undergoing heat treatment, missile components during reentry into Earth's atmosphere, or the thermal response of a thin laminate being bonded using a laser source.

In this chapter, we shall consider problems and their solutions that deal with unsteady-state heat transfer within systems both with and without internal energy sources.

▶ **18.1**

ANALYTICAL SOLUTIONS

The solution of an unsteady-state conduction problem is, in general, more difficult than that for a steady-state problem because of the dependence of temperature on both time and position. The solution is approached by establishing the defining differential equation and the boundary conditions. In addition, the initial temperature distribution in the conducting medium must be known. By finding the solution to the partial differential equation that satisfies the initial and boundary conditions, the variation in the temperature distribution with time is established, and the flux of energy at a specific time can then be evaluated.

In heating or cooling a conducting medium, the rate of energy transfer is dependent upon both the internal and surface resistances, the limiting cases being represented either by negligible internal resistance or by negligible surface resistance. Both of these cases will be considered, as well as the more general case in which both resistances are important.

Lumped Parameter Analysis—Systems with Negligible Internal Resistance

Equation (16-17) will be the starting point for transient conduction analysis. It is repeated below for reference.

$$\frac{\partial T}{\partial t} = \alpha \nabla^2 T + \frac{\dot{q}}{\rho c_p} \tag{16-17}$$

Recall that, in the derivation of this expression, thermal properties were taken to be independent of position and time; however, the rate of internal generation, $\dot{q}$, can vary in both.

It is frequently the case that temperature within a medium varies significantly in fewer than all three space variables. A circular cylinder, heated at one end with a fixed boundary condition, will show a temperature variation in the axial and radial directions as well as time. If the cylinder has a length that is large compared to its diameter, or if it is composed of a material with high thermal conductivity, temperature will vary with axial position and time only. If a metallic specimen, initially with uniform temperature, is suddenly exposed to surroundings at a different temperature, it may be that size, shape, and thermal conductivity may combine in such a way that the temperature within the material varies with time only—that is, is not a significant function of position. These conditions are characteristic of a "lumped" system, where the temperature of a body varies only with time; this case is the easiest of all to analyze. Because of this we will consider, as our first transient conduction case, that of a completely lumped-parameter system.

Shown in Figure 18.1, we have a spherical metallic specimen, initially at uniform temperature T_0 after it has been immersed in a hot oil bath at temperature T_∞ for a period of time t. It is presumed that the temperature of the metallic sphere is uniform at any given time. A first-law analysis using equation (6-10), applied to a spherical control volume coinciding with the specimen in question, will reduce to

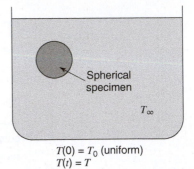

$$T(0) = T_0 \text{ (uniform)}$$
$$T(t) = T$$

Figure 18.1

$$\frac{\delta Q}{dt} = \frac{\partial}{\partial t} \iiint_{\text{c.v.}} e\rho \, dV \qquad (18\text{-}1)$$

The rate of heat addition to the control volume, $\delta Q/dt$, is due to convection from the oil and is written as

$$\frac{\delta Q}{dt} = hA(T_\infty - T) \qquad (18\text{-}2)$$

The rate of energy increase within the specimen, $\partial/\partial t \iiint_{\text{c.v.}} e\rho \, dV$, with constant properties, may be expressed as

$$\frac{\partial}{\partial t} \iiint_{\text{c.v.}} e\rho \, dV = \rho V c_p \frac{dT}{dt} \qquad (18\text{-}3)$$

Equating these expressions as indicated by equation (18-1), we have, with slight rearrangement

$$\frac{dT}{dt} = \frac{hA(T_\infty - T)}{\rho V c_p} \qquad (18\text{-}4)$$

We may now obtain a solution for the temperature variation with time by solving equation (18-4) subject to the initial condition, $T(0) = T_0$, and obtain

$$\frac{T - T_\infty}{T_0 - T_\infty} = e^{-hAt/\rho c_p V} \qquad (18\text{-}5)$$

The exponent is observed to be dimensionless. A rearrangement of terms in the exponent may be accomplished as follows:

$$\frac{hAt}{\rho c_p V} = \left(\frac{hV}{kA}\right)\left(\frac{A^2 k}{\rho V^2 c_p}t\right) = \left[\frac{hV/A}{k}\right]\left[\frac{\alpha t}{(V/A)^2}\right] \qquad (18\text{-}6)$$

Each of the bracketed terms in equation (18-6) is dimensionless. The ratio, V/A, having units of length, is also seen to be a part of each of these new parametric forms. The first of the new nondimensional parameters formed is the *Biot modulus*, abbreviated Bi:

$$\text{Bi} = \frac{hV/A}{k} \tag{18-7}$$

By analogy with the concepts of thermal resistance, discussed at length earlier, the Biot modulus is seen to be the ratio of $(V/A)/k$, the conductive (internal) resistance to heat transfer, to $1/h$, the convective (external) resistance to heat transfer. The magnitude of Bi thus has some physical significance in relating where the greater resistance to heat transfer occurs. A large value of Bi indicates that the conductive resistance controls—that is, there is more capacity for heat to leave the surface by convection than to reach it by conduction. A small value for Bi represents the case where internal resistance is negligibly small and there is more capacity to transfer heat by conduction than there is by convection. In this latter case, the controlling heat-transfer phenomenon is convection, and temperature gradients within the medium are quite small. An extremely small internal temperature gradient is the basic assumption in a lumped-parameter analysis.

A natural conclusion to the foregoing discussion is that the magnitude of the Biot modulus is a reasonable measure of the likely accuracy of a lumped-parameter analysis. A commonly used rule of thumb is that the error inherent in a lumped-parameter analysis will be less than 5% for a value of Bi less than 0.1. The evaluation of the Biot modulus should thus be the first thing done when analyzing an unsteady-state conduction situation.

The other bracketed term in equation (18-6) is the *Fourier modulus*, abbreviated Fo, where

$$\text{Fo} = \frac{\alpha t}{(V/A)^2} \tag{18-8}$$

The Fourier modulus is frequently used as a nondimensional time parameter.

The lumped-parameter solution for transient conduction may now be written as

$$\frac{T - T_\infty}{T_0 - T_\infty} = e^{-\text{BiFo}} \tag{18-9}$$

Equation (18-9) is portrayed graphically in Figure 18.2. The use of equation (18-9) is illustrated in the following example.

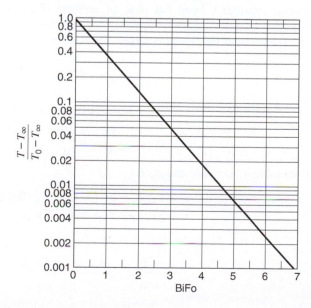

Figure 18.2 Time-temperature history of a body at initial temperature, T_0, exposed to an environment at T_∞; lumped-parameter case.

Example 1

A long copper wire, 0.635 cm in diameter, is exposed to an air stream at a temperature, T_∞, of 310 K.

After 30 s, the average temperature of the wire increased from 280 to 297 K. Using this information, estimate the average surface conductance, h.

In order to determine if equation (18-9) is valid for this problem, the value of Bi must be determined. The Biot number is expressed as

$$\text{Bi} = \frac{hV/A}{k} = \frac{h\dfrac{\pi D^2 L}{4\pi DL}}{386 \text{ W/m}\cdot\text{K}} = \frac{h\dfrac{0.00635 \text{ m}}{4}}{386 \text{ W/m}\cdot\text{K}}$$

$$= 4.11 \times 10^{-6}\, h$$

Setting Bi = 0.1, which is the limiting value of Bi for a lumped-parameter analysis to be valid, and solving for h, we obtain

$$h = 0.1/4.11 \times 10^{-6} = 24300 \text{ W/m}^2\cdot\text{K}$$

We may conclude that a lumped-parameter solution is valid if $h < 24,300 \text{ W/m}^2\cdot\text{K}$, which is a near certainty.

Proceeding, we can apply equation (18-4) to yield

$$h = \frac{\rho c_p V}{tA}\ln\frac{T_0 - T_\infty}{T - T_\infty}$$

$$= \frac{(8890 \text{ kg/m}^3)(385 \text{ J/kg}\cdot\text{K})\left(\dfrac{\pi D^2 L}{4\pi DL}\right)}{(30 \text{ s})}\ln\frac{280 - 310}{297 - 310}$$

$$= 151 \text{ W/m}^2\cdot\text{K}$$

This result is much less than the limiting value of h, indicating that a lumped-parameter solution is probably very accurate.

Heating a Body Under Conditions of Negligible Surface Resistance A second class of time-dependent energy-transfer processes is encountered when the surface resistance is small relative to the overall resistance, that is, Bi is 0.1. For this limiting case, the temperature of the surface, T_s is constant for all time, $t > 0$, and its value is essentially equal to the ambient temperature, T_∞.

To illustrate the analytical method of solving this class of transient heat-conduction problems, consider a large flat plate of uniform thickness L. The initial temperature distribution through the plate will be assumed to be an arbitrary function of x. The solution for the temperature history must satisfy the Fourier field equation

$$\frac{\partial T}{\partial t} = \alpha\nabla^2 T \tag{16-18}$$

For one-directional energy flow

$$\frac{\partial T}{\partial t} = \alpha\frac{\partial^2 T}{\partial x^2} \tag{18-10}$$

with initial and boundary conditions

$$T = T_0(x) \quad \text{at } t = 0 \quad \text{for } 0 \leq x \leq L$$
$$T = T_s \quad \text{at } x = 0 \quad \text{for } t > 0$$

and

$$T = T_s \quad \text{at } x = L \quad \text{for } t > 0$$

For convenience, let $Y = (T - T_s)/(T_0 - T_s)$, where T_0 is an arbitrarily chosen reference temperature; the partial differential equation may be rewritten in terms of the new temperature variable as

$$\frac{\partial Y}{\partial t} = \alpha \frac{\partial^2 Y}{\partial x^2} \tag{18-11}$$

and the initial and boundary conditions become

$$
\begin{array}{lll}
Y = Y_0(x) & \text{at } t = 0 & \text{for } 0 \leq x \leq L \\
Y = 0 & \text{at } x = 0 & \text{for } t > 0
\end{array}
$$

and

$$Y = 0 \qquad \text{at } x = L \qquad \text{for } t > 0$$

Solving equation (18-11) by the method of separation of variables leads to product solutions of the form

$$Y = (C_1 \cos \lambda x + C_2 \sin \lambda x) e^{-\alpha \lambda^2 t}$$

The constants C_1 and C_2 and the parameter λ are obtained by applying the initial and boundary conditions. The complete solution is

$$Y = \frac{2}{L} \sum_{n=1}^{\infty} \sin\left(\frac{n\pi}{L}\right) e^{-(n\pi/2)^2 \text{Fo}} \int_0^L Y_0(x) \sin\frac{n\pi}{L} x \, dx \tag{18-12}$$

where $\text{Fo} = \alpha t/(L/2)^2$. Equation (18-12) points out the necessity for knowing the initial temperature distribution in the conducting medium, $Y_0(x)$, before the complete temperature history may be evaluated. Consider the special case in which the conducting body has a uniform initial temperature, $Y_0(x) = Y_0$. With this temperature distribution, equation (18-12) reduces to

$$\frac{T - T_s}{T_0 - T_s} = \frac{4}{\pi} \sum_{n=1}^{\infty} \frac{1}{n} \sin\left(\frac{n\pi}{L} x\right) e^{-(n\pi/2)^2 \text{Fo}} \qquad n = 1, 3, 5, \ldots \tag{18-13}$$

The temperature history at the center of the infinite plane, as well as the central temperature history in other solids, is illustrated in Figure 18.3. The central temperature history for the plane wall, infinite cylinder, and sphere is presented in Appendix F, in "Heissler charts." These charts cover a much greater range in the Fourier modulus than Figure 18.3.

The heat rate, q, at any plane in the conducting medium may be evaluated by

$$q_x = -kA \frac{\partial T}{\partial x} \tag{18-14}$$

In the case of the infinite flat plate with an initial uniform temperature distribution of T_0, the heat rate at any time t is

$$q_x = 4\left(\frac{kA}{L}\right)(T_s - T_0) \sum_{n=1}^{\infty} \cos\left(\frac{n\pi}{L} x\right) e^{-(n\pi/2)^2 \text{Fo}} \qquad n = 1, 3, 5, \ldots \tag{18-15}$$

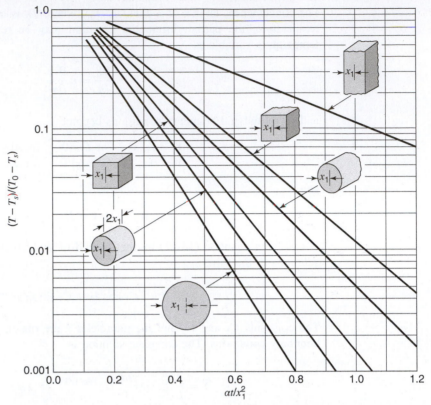

Figure 18.3 Central temperature history of various solids with initial uniform temperature, T_0, and constant surface temperature, T_s. (From P. J. Schneider, *Conduction Heat Transfer,* Addison-Wesley Publishing Co., Inc., Reading Mass., 1955, p. 249. By permission of the publishers.)

In the following example, the use of the central temperature history figure will be illustrated.

Example 2

A concrete cylinder, 0.1 m in length and 0.1 m in diameter, is initially at room temperature, 292 K. It is suspended in a steam environment where water vapor at 373 K condenses on all surfaces with an effective film coefficient, h, of 8500 W/m$^2 \cdot$ K. Determine the time required for the center of this stubby cylinder to reach 310 K. If the cylinder were sufficiently long as to be considered infinite, how long would it take?

For the first case, the finite cylinder, the Biot number is evaluated as

$$\text{Bi} = \frac{h(V/A)}{k} = \frac{h\left(\dfrac{\pi D^2 L}{4}\right)}{k\left(\pi D L + \dfrac{\pi D^2}{2}\right)} = \frac{h(DL/4)}{k(L + D/2)}$$

$$= \frac{(8500 \text{ W/m}^2 \cdot \text{K})(0.1 \text{ m})(0.1 \text{ m})/4}{1.21 \text{ W/m} \cdot \text{K}(0.1 + 0.1/2) \text{ m}}$$

$$= 117$$

For this large value, Figure 18.3 may be used. The second line from the bottom in this figure applies to a cylinder with height equal to diameter, as in this case. The ordinate is

$$\frac{T - T_s}{T_0 - T_s} = \frac{310 - 373}{292 - 373} = 0.778$$

and the corresponding abscissa value is approximately 0.11. The time required may now be determined as

$$\frac{\alpha t}{x_1^2} = 0.11$$

Thus,

$$t = 0.11 \frac{(0.05 \text{ m})^2}{5.95 \times 10^{-7} \text{ m}^2/\text{s}} = 462 \text{ s}$$

$$= 7.7 \text{ min}$$

In the case of an infinitely long cylinder, the fourth line from the bottom applies. The Biot number in this case is

$$\text{Bi} = \frac{h(V/A)}{k} = \frac{h\left(\dfrac{\pi D^2 L}{4}\right)}{k(\pi D L)} = \frac{h\dfrac{D}{4}}{k}$$

$$= \frac{(8500 \text{ W/m}^2 \cdot \text{K})(0.1 \text{ m})/4}{1.21 \text{ W/m} \cdot \text{K}} = 176$$

which is even larger than the finite cylinder case. Figure 18.3 will again be used. The ordinate value of 0.778 yields, for the abscissa, a value of approximately 0.13. The required time, in this case, is

$$t = \frac{0.13(0.05 \text{ m})^2}{5.95 \times 10^{-7} \text{ m}^2/\text{s}} = 546 \text{ s}$$

$$= 9.1 \text{ min}$$

Heating a Body with Finite Surface and Internal Resistances The most general cases of transient heat-conduction processes involve significant values of both internal and surface resistances. The solution for the temperature history without internal generation must satisfy the Fourier field equation, which may be expressed for one-dimensional heat flow by

$$\frac{\partial T}{\partial t} = \alpha \frac{\partial^2 T}{\partial x^2} \tag{18-7}$$

A case of considerable practical interest is one in which a body having a uniform temperature is placed in a new fluid environment with its surfaces suddenly and simultaneously exposed to the fluid at temperature T_∞. In this case, the temperature history must satisfy the initial, symmetry, and convective boundary conditions

$$T = T_0 \qquad\qquad \text{at } t = 0$$

$$\frac{\partial T}{\partial x} = 0 \qquad\qquad \text{at the center of the body}$$

and

$$-\frac{\partial T}{\partial x} = \frac{h}{k}(T - T_\infty) \quad \text{at the surface}$$

One method of solution for this class of problems involves separation of variables, which results in product solutions as previously encountered when only the internal resistance was involved.

Solutions to this case of time-dependent energy-transfer processes have been obtained for many geometries. Excellent treatises discussing these solutions have been written by Carslaw and Jaeger[1] and by Ingersoll, Zobel, and Ingersoll.[2] If we reconsider the infinite flat plate of thickness, $2x_1$, when inserted into a medium at constant temperature, T_∞, but now include a constant surface conductance, h, the following solution is obtained

$$\frac{T - T_\infty}{T_0 - T_\infty} = 2 \sum_{n=1}^{\infty} \frac{\sin \delta_n \cos (\delta_n x/x_1)}{\delta_n + \sin \delta_n \cos \delta_n} e^{-\delta_n^2 \text{Fo}} \qquad (18\text{-}16)$$

where δ_n is defined by the relation

$$\delta_n \tan \delta_n = \frac{hx_1}{k} \qquad (18\text{-}17)$$

The temperature history for this relatively simple geometrical shape is a function of three dimensionless quantities: $\alpha t/x_1^2$, hx_1/k, and the relative distance, x/x_1.

The complex nature of equation (18-17) has led to a number of graphical solutions for the case of one-dimensional transient conduction. The resulting plots, with dimensionless temperature as a function of other dimensionless parameters as listed above, are discussed in Section 18.2.

Heat Transfer to a Semi-Infinite Wall

An analytical solution to the one-dimensional heat-conduction equation for the case of the semi-infinite wall has some utility as a limiting case in engineering computations. Consider the situation illustrated in Figure 18.4. A large plane wall initially at a constant temperature T_0 is subjected to a surface temperature T_s, where $T_s > T_0$. The differential equation to be solved is

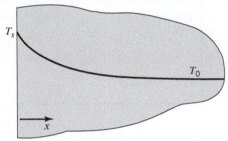

Figure 18.4 Temperature distribution in a semi-infinite wall at time t.

$$\frac{\partial T}{\partial t} = \alpha \frac{\partial^2 T}{\partial x^2} \qquad (18\text{-}10)$$

and the initial and boundary conditions are

$$T = T_0 \quad \text{at} \quad t = 0 \quad \text{for all } x$$
$$T = T_s \quad \text{at} \quad x = 0 \quad \text{for all } t$$

and

$$T \rightarrow T_0 \quad \text{as} \quad x \rightarrow \infty \quad \text{for all } t$$

The solution to this problem may be accomplished in a variety of ways, among which are the Laplace transformation and the Fourier transformation. We shall use an alternative procedure, which is less involved mathematically. The variables in equation (18-10) may be expressed in dimensionless form by analogy with the previous case. Thus, we may write

---

[1] H. S. Carslaw and J. C. Jaeger, *Conduction of Heat in Solids,* Oxford University Press, 1947.
[2] L. R. Ingersoll, O. J. Zobel, and A. C. Ingersoll, *Heat Conduction (with Engineering and Geological Applications),* McGraw-Hill Book Company, New York, 1948.

$$\frac{T - T_0}{T_s - T_0} = f\left(\frac{x}{x_1}, \frac{\alpha t}{x_1^2}\right)$$

However, in this problem there is no finite characteristic dimension, x_1, and thus $(T - T_0)/(T_s - T_0) = f(\alpha t/x^2)$, or, with equal validity, $(T - T_0)/(T_s - T_0) = f(x/\sqrt{\alpha t})$. If $\eta = x/2\sqrt{\alpha t}$ is selected as the independent variable and the dependent variable $Y = (T - T_0)/(T_s - T_0)$ is used, substitution into equation (18-10) yields the ordinary differential equation

$$d^2Y/d\eta^2 + 2\eta \, dY/d\eta = 0 \tag{18-18}$$

with the transformed boundary and initial conditions

$$Y \to 0 \quad \text{as} \quad \eta \to \infty$$

and

$$Y = 1 \quad \text{at} \quad \eta = 0$$

The first condition above is the same as the initial condition $T = T_0$ at $t = 0$, and the boundary condition $T \to T_0$ as $x \to \infty$. Equation (18-18) may be integrated once to yield

$$\ln\frac{dY}{d\eta} = c_1 - \eta^2$$

or

$$\frac{dY}{d\eta} = c_2 e^{-\eta^2}$$

and integrated once more to yield

$$Y = c_3 + c_2 \int e^{-\eta^2} \, d\eta \tag{18-19}$$

The integral is related to a frequently encountered form, the *error function*, designated "erf," where

$$\text{erf } \phi \equiv \frac{2}{\sqrt{\pi}} \int_0^\phi e^{-\eta^2} \, d\eta$$

and erf $(0) = 0$, erf $(\infty) = 1$, A short table of erf ϕ is given in Appendix L. Applying the boundary conditions to equation (18-19), we obtain

$$Y = 1 - \text{erf}\left(\frac{x}{2\sqrt{\alpha t}}\right)$$

or

$$\frac{T - T_0}{T_s - T_0} = 1 - \text{erf}\left(\frac{x}{2\sqrt{\alpha t}}\right)$$

or

$$\frac{T_s - T}{T_s - T_0} = \text{erf}\left(\frac{x}{2\sqrt{\alpha t}}\right) \tag{18-20}$$

This equation is extremely simple to use and is quite valuable.

Consider a finite wall of thickness L subject to the surface temperature T_s. Until the temperature change at $x = L$ exceeds some nominal amount, say $(T - T_0)/(T_s - T_0)$ equal to

0.5%, the solution for the finite and infinite walls will be the same. The value of $L/(2\sqrt{\alpha t})$ corresponding to a 0.5% change in $(T - T_0)/(T_s - T_0)$ is $L/(2\sqrt{\alpha t}) \simeq 2$, so for $L/(2\sqrt{\alpha t}) > 2$, equation (18-20) may be used for finite geometry with little or no error.

For the case of finite surface resistance, the solution to equation (18-10) for a semi-infinite wall is

$$\frac{T_\infty - T}{T_\infty - T_0} = \operatorname{erf}\frac{x}{2\sqrt{\alpha t}} + \exp\left(\frac{hx}{k} + \frac{h^2 \alpha t}{k^2}\right)\left[1 - \operatorname{erf}\left(\frac{h\sqrt{\alpha t}}{k} + \frac{x}{2\sqrt{\alpha t}}\right)\right] \quad (18-21)$$

This equation may be used to determine the temperature distribution in finite bodies for small times in the same manner as equation (18-20). The surface temperature is particularly easy to obtain from the above equation, if we let $x = 0$, and the heat-transfer rate may be determined from

$$\frac{q}{A} = h(T_s - T_\infty)$$

▶ **18.2**

TEMPERATURE-TIME CHARTS FOR SIMPLE GEOMETRIC SHAPES

For unsteady-state energy transfer in several simple shapes with certain restrictive boundary conditions, the equations describing temperature profiles have been solved,[3] and the results have been presented in a wide variety of charts to facilitate their use. Two forms of these charts are available in Appendix F.

Solutions are presented in Appendix F for the flat plate, sphere, and long cylinder in terms of four dimensionless ratios:

$$Y, \text{ unaccomplished temperature change} = \frac{T_\infty - T}{T_\infty - T_0}$$

$$X, \text{ relative time} = \frac{\alpha t}{x_1^2}$$

$$n, \text{ relative position} = \frac{x}{x_1}$$

and

$$m, \text{ relative resistance} = \frac{k}{hx_1}$$

where x_1 is the radius or semi-thickness of the conducting medium. These charts may be used to evaluate temperature profiles for cases involving transport of energy into or out of the conducting medium if the following conditions are met:

a. Fourier's field equation describes the process—i.e., constant thermal diffusivity and no internal heat source.
b. The conducting medium has a uniform initial temperature, T_0.
c. The temperature of the boundary or the adjacent fluid is changed to a new value, T_∞, for $t \geq 0$.

For flat plates where the transport takes place from only one of the faces, the relative time, position, and resistance are evaluated as if the thickness were twice the true value.

---

[3] Equation (18-16) pertains to a plane wall of thickness, L, and boundary conditions $T(x, 0) = T_0$ and $dT/dx (0, t) = 0$.

Although the charts were drawn for one-dimensional transport, they may be combined to yield solutions for two- and three-dimensional problems. The following is a summary of these combined solutions:

1. For transport in a rectangular bar with insulated ends,

$$Y_{\text{bar}} = Y_a Y_b \tag{18-22}$$

where Y_a is evaluated with width $x_1 = a$, and Y_b is evaluated with thickness $x_1 = b$.

2. For transport in a rectangular parallelepiped,

$$Y_{\text{parallelepiped}} = Y_a Y_b Y_c \tag{18-23}$$

where Y_a is evaluated with width $x_1 = a$, Y_b is evaluated with thickness $x_1 = b$, and Y_c is evaluated with depth $x_1 = c$.

3. For transport in a cylinder, including both ends,

$$Y_{\substack{\text{cylinder} \\ \text{plus ends}}} = Y_{\text{cylinder}} Y_a \tag{18-24}$$

where Y_a is evaluated by using the flat-plate chart, and thickness $x_1 = a$.

The use of temperature-time charts is demonstrated in the following examples.

Example 3

A flat wall of fire-clay brick, 0.5 m thick and originally at 200 K, has one of its faces suddenly exposed to a hot gas at 1200 K. If the heat-transfer coefficient on the hot side is 7.38 W/m² · K and the other face of the wall is insulated so that no heat passes out of that face, determine (a) the time necessary to raise the center of the wall to 600 K; (b) the temperature of the insulated wall face at the time evaluated in (a).

From the table of physical properties given in Appendix H, the following values are listed:

$$k = 1.125 \text{ W/m} \cdot \text{K}$$

$$c_p = 919 \text{ J/kg} \cdot \text{K}$$

$$\rho = 2310 \text{ kg/m}^3$$

and

$$\alpha = 5.30 \times 10^{-7} \text{ m}^2/\text{s}$$

The insulated face limits the energy transfer into the conducting medium to only one direction. This is equivalent to heat transfer from a 1-m-thick wall, where x is then measured from the line of symmetry, the insulated face. The relative position, x/x_1, is 1/2. The relative resistance, k/hx_1, is $1.125/[(7.38)(0.5)]$ or 0.305. The dimensionless temperature, $Y = (T_\infty - T)/(T_\infty - T_0)$, is equal to $(1200 - 600)/(1200 - 200)$, or 0.6. From Figure F.7, in Appendix F, the abscissa, $\alpha t/x_1^2$, is 0.35 under these conditions. The time required to raise the centerline to 600 F is

$$t = \frac{0.35\, x_1^2}{\alpha} = \frac{0.35(0.5)^2}{5.30 \times 10^{-7}} = 1.651 \times 10^5 \text{ s} \quad \text{or} \quad 45.9 \text{ h}$$

The relative resistance and the relative time for (b) will be the same as in part (a). The relative position, x/x_1, will be 0. Using these values and Figure F.1 in Appendix F, we find the dimensionless temperature, Y, to be 0.74. Using this value, the desired temperature can be evaluated by

$$\frac{T_s - T}{T_s - T_0} = \frac{1200 - T}{1200 - 200} = 0.74$$

or

$$T = 460 \text{ K} \quad (368 \text{ F})$$

Example 4

A billet of steel 30.5 cm in diameter, 61 cm long, initially at 645 K, is immersed in an oil bath that is maintained at 310 K. If the surface conductance is 34 W/m² · K, determine the center temperature of the billet after 1 h.

From Appendix H, the following average physical properties will be used:

$$k = 49.9 \text{ W/m} \cdot \text{K}$$
$$c_p = 473 \text{ J/kg} \cdot \text{K}$$
$$\rho = 7820 \text{ kg/m}^3$$
$$\alpha = 1.16 \times 10^{-5} \text{ m}^2/\text{s}$$

Figure F.7 applies. To evaluate Y_a the following dimensionless parameters apply

$$X = \frac{\alpha t}{x_1^2} = \frac{(1.16 \times 10^{-5} \text{ m}^2/\text{s})(3600 \text{ s})}{(0.305 \text{ m})^2} = 0.449$$

$$n = x/x_1 = 0$$

$$m = k/hx_1 = \frac{42.9 \text{ W/m} \cdot \text{K}}{(34 \text{ W/m}^2 \cdot \text{K})(0.305 \text{ m})} = 4.14$$

Using these values with Figure F.7 in Appendix F, the corresponding values of dimensionless temperature, Y_a, is approximately 0.95.

For the cylindrical surface the appropriate values are

$$X = \frac{\alpha t}{x_1^2} = \frac{(1.16 \times 10^{-5} \text{ m}^2/\text{s})(3600 \text{ s})}{(0.1525 \text{ m})^2}$$

$$= 1.80$$

$$n = \frac{x}{x_1} = 0$$

$$m = k/hx_1 = 42.9/(34)(0.1525) = 8.27$$

and, from Figure F.8 in Appendix F, we obtain

$$Y_{\text{cl}} = \frac{T - T_\infty}{T_0 - T_\infty} \bigg|_{\text{cyl}} \cong 0.7$$

Now, for heat transfer across the cylindrical surface and both ends,

$$Y|_{\text{total}} = \frac{T_{\text{cl}} - T_\infty}{T_0 - T_\infty} = Y_a Y_{\text{cl}} = (0.95)(0.7) = 0.665$$

The desired center temperature is now calculated as

$$T_{\text{cl}} = T_\infty + 0.665(T_0 - T_\infty)$$
$$= 310 \text{ K} + 0.665(645 - 310) \text{ K}$$
$$= 533 \text{ K} \quad (499 \text{ F})$$

▶ **18.3**

NUMERICAL METHODS FOR TRANSIENT CONDUCTION ANALYSIS

In many time-dependent or unsteady-state conduction processes, actual initial and/or boundary conditions do not correspond to those mentioned earlier with regard to analytical solutions. An initial temperature distribution may be nonuniform in nature; ambient temperature, surface conductance, or system geometry may be variable or quite irregular. For such complex cases, numerical techniques offer the best means to achieve solutions.

More recently, with sophisticated computing codes available, numerical solutions are being obtained for heat-transfer problems of all types, and this trend will doubtlessly continue. It is likely that many users of this book will be involved in code development for such analysis.

Some numerical work is introduced in Chapter 17, dealing with two-dimensional, steady-state conduction. In this section, we will consider variation in time as well as position.

To begin our discussion, the reader is referred to equation (17-61) and the development leading up to it. For the case of no internal generation of energy, equation (17-61) reduces to

$$k\frac{\Delta y}{\Delta x}(T_{i-1,j} + T_{i+1,j} - 2T_{i,j}) + k\frac{\Delta x}{\Delta y}(T_{i,j-1} + T_{i,j+1} - 2T_{i,j})$$

$$= \left(\frac{\rho c_p T_{i,j}|_{t+\Delta t} - \rho c_p T_{i,j}|_t}{\Delta t}\right)\Delta x\Delta y \qquad (18\text{-}25)$$

This expression applies to two dimensions; however, it can be extended easily to three dimensions.

The time-dependent term on the right of equation (18-25) is written such that the temperature at node i, j is presumed known at time t; this equation can then be solved to find T_{ij} at the end of time interval Δt. As $T_{i,j}|_{t+\Delta t}$ appears only once in this equation, it can be evaluated quite easily. This means of evaluating T_{ij} at the end of a time increment is designated an "explicit" technique. A more thorough discussion of explicit solutions is given by Carnahan.

Equation (18-25) may be solved to evaluate the temperature at node i, j for all values of i, j that comprise the region of interest. For large numbers of nodes, it is clear that a great number of calculations are needed and that much information must be stored for use in subsequent computation. Digital computers obviously provide the only feasible way to accomplish solutions.

We will next consider the one-dimensional form of equation (18-25). For a space increment Δx, the simplified expression becomes

$$\frac{k}{\Delta x}(T_{i-1}|_t + T_{i+1}|_t - 2T_i|_t) = \left(\frac{\rho c_p T_i|_{t+\Delta t} - \rho c_p T_i|_t}{\Delta t}\right)\Delta x \qquad (18\text{-}26)$$

where the j notation has been dropped. The absence of variation in the y-direction allows several terms to be deleted. We next consider properties to be constant and represent the ratio $k = \rho c_p$ as α. Solving for $T_{i/t+\Delta t}$, we obtain

$$T_i|_{t+\Delta t} = \frac{\alpha\Delta t}{(\Delta x)^2}(T_{i+1}|_t + T_{i-1}|_t) + \left(1 - \frac{2\alpha\Delta t}{(\Delta x)^2}\right)T_i|t \qquad (18\text{-}27)$$

The ratio, $\alpha\Delta t/(\Delta x)^2$, a form resembling the Fourier modulus, is seen to arise naturally in this development. This grouping relates the time step, Δt, to the space increment, Δx. The magnitude of this grouping will, quite obviously, have an effect on the solution. It has been determined that equation (18-27) is numerically "stable" when

$$\frac{\alpha\Delta t}{(\Delta x)^2} \leq \frac{1}{2} \qquad (18\text{-}28)$$

For a discussion of numerical stability the reader is referred to Carnahan et al.[4]

---

[4] B. Carnahan, H. A. Luther, and J. O. Wilkes, *Applied Numerical Methods,* Wiley, New York, 1969.

The choice of a time step involves a trade-off between solution accuracy—a smaller time step will produce greater accuracy—and computation time: a solution will be achieved more rapidly for larger values of Δt. When computing is done by machine, a small time step will likely be used without major difficulty.

An examination of equation (18-27) indicates considerable simplification to be achieved if the equality in equation (18-28) is used. For the case with $\alpha \Delta t/(\Delta x^2) = 1/2$, equation (18-27) becomes

$$T_{i|t+\Delta t} = \frac{T_{i+1|t} + T_{i-1|t}}{2} \tag{18-29}$$

Example 5

A brick wall ($\alpha = 4.72 \times 10^{-7}\,\text{m}^2/\text{s}$) with a thickness of 0.5 m is initially at a uniform temperature of 300 K. Determine the length of time required for its center temperature to reach 425 K if its surfaces are raised to, and maintained at 425 and 600 K, respectively.

Although relatively simple, this one-dimensional problem is not amenable to a solution using the charts because there is no axis of symmetry. Either analytical or numerical methods must therefore be employed.

An analytical solution using Laplace transform or separation-of-variables methodology is relatively straightforward. However, the solution is in terms of infinite series involving eigenvalues and the determination of a final answer is cumbersome. The simplest approach is thus numerical and we will proceed with the ideas introduced in this section.

The illustration below depicts the wall divided into 10 increments. Each of the nodes within the wall is at the center of a subvolume, having a width, Δx. The shaded subvolume at node 4 is considered to have uniform properties averaged at its center—i.e., the location of node 4. This same idea prevails for all 11 nodes; this includes the surface nodes, 0 and 10.

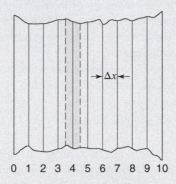

An energy balance for any internal node, having width Δx, will yield equation (18-27) as a result. This relationship includes the dimensionless ratio, $\alpha \Delta t/\Delta x^2$, which relates the time increment, Δt, to the space increment Δx. In this example, we have specified $\Delta x = 0.05\,\text{m}$.

The quantity, $\alpha \Delta t/\Delta x^2$, can have any value equal to or less than 0.5, which is the limit for a stable solution. If the limiting value is chosen, equation (18-27) reduces to a simple algorithmic form

$$T_{i,t+1} = \frac{T_{i-1,t} + T_{i+1,t}}{2} \tag{18-29}$$

This expression is valid for $i = 1$ to 9; however, as nodes 0 and 10 are at constant temperature for all time, the algorithms for nodes 1 and 9 can be written as

$$\begin{aligned} T_{1,t+1} &= \frac{T_{0,t} + T_{2,t}}{2} = \frac{425 + T_{2,t}}{2} \\ T_{9,t+1} &= \frac{T_{8,t} + T_{10,t}}{2} = \frac{T_{8,t} + 600}{2} \end{aligned} \tag{18-30}$$

The problem solution now proceeds as equations (18-29) and (18-30) and are solved at succeeding times to update nodal temperatures until the desired result, $T_s - 425\,\text{K}$, is achieved.

Equations (18-29) and (18-30) are quite simple and easily programmed to achieve a solution. In this case, a spreadsheet approach could also be used. The table below summarizes the form of the results for $T_{i,t}$.

	T_0	T_1	T_2	T_3	T_4	T_5	T_6	T_7	T_8	T_9	T_{10}
$t=0$	425	300	300	300	300	300	300	300	300	300	600
$\vdots$											
$t=10$	425	394.8	372.1	349.4	347.9	346.4	367.1	405.7	466.1	526.4	600
$\vdots$											
$t=20$	425	411.4	403.3	395.3	402.3	409.4	436.2	463.1	506.5	550.0	600
$\vdots$											
$t=22$	425	414.2	408.5	409.8	411.0	428.2	445.3	471.4	512.3	553.3	600
$t=23$	425	416.8	408.5	409.8	411.0	428.2	445.3	478.8	512.3	556.2	600

The desired center temperature is reached between time increments 22 and 23; an interpolated value is $n=22.6$ time increments. As discussed earlier, the increment Δt is related to α and Δx according to the ratio

$$\frac{\alpha \Delta t}{\Delta x^2} = 1/2$$

or

$$\Delta t = \frac{1}{2}\frac{\Delta x^2}{\alpha} = \frac{1}{2}\frac{(0.05\,\text{m})^2}{4.72 \times 10^{-7}\,\text{m}^2/\text{s}}$$
$$= 2648\,\text{s}$$
$$= 0.736\,\text{h}$$

The answer for total time elapsed is thus

$$t = 22.6(0.736) = 16.6\,\text{h}$$

18.4

AN INTEGRAL METHOD FOR ONE-DIMENSIONAL UNSTEADY CONDUCTION

The von Kármán momentum integral approach to the hydrodynamic boundary layer has a counterpart in conduction. Figure 18.5 shows a portion of a semi-infinite wall, originally at uniform temperature T_0, exposed to a fluid at temperature T_∞, with the surface of the wall at any time at temperature T_s.

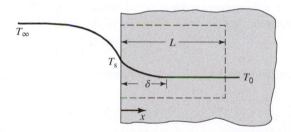

Figure 18.5 A portion of a semi-infinite wall used in integral analysis.

At any time t, heat transfer from the fluid to the wall affects the temperature profile within the wall. The "penetration distance," designated as δ, is the distance from the surface wherein this effect is manifested. At distance δ, the temperature gradient, $\partial T/\partial x$, is taken as zero.

Applying the first law of thermodynamics, equation (6-10), to a control volume extending from $x=0$ to $x=L$, where $L>\delta$, we have

$$\frac{\delta Q}{dt} - \frac{\delta W_s}{dt} - \frac{\delta W_\mu}{dt} = \iint_{c.s.} \left(e + \frac{P}{\rho}\right)\rho(\mathbf{v}\cdot\mathbf{n})dA + \frac{\partial}{\partial t}\iiint_{c.v.} e\rho\, dV \qquad (6\text{-}10)$$

with

$$\frac{\delta W_s}{dt} = \frac{\delta W_\mu}{dt} = \iint_{c.s.} \left(e + \frac{P}{\rho}\right)\rho(\mathbf{v}\cdot\mathbf{n})\, dA = 0$$

The applicable form of the first law is now

$$\frac{\delta Q}{dt} = \frac{\partial}{\partial t}\iiint_{c.v.} e\rho\, dV$$

Considering all variables to be functions of x alone, we may express the heat flux as

$$\frac{q_x}{A} = \frac{d}{dt}\int_0^L \rho u\, dx = \frac{d}{dt}\int_0^L \rho c_p T\, dx \qquad (18\text{-}32)$$

The interval from 0 to L will now be divided into two increments, giving

$$\frac{q_x}{A} = \frac{d}{dt}\left[\int_0^\delta \rho c_p T\, dx + \int_\delta^L \rho c_p T_0\, dx\right]$$

and, since T_0 is constant, this becomes

$$\frac{q_x}{A} = \frac{d}{dt}\left[\int_0^\delta \rho c_p T\, dx + \rho c_p T_0(L - \delta)\right]$$

The integral equation to be solved is now

$$\frac{q_x}{A} = \frac{d}{dt}\int_0^\delta \rho c_p T\, dx + \rho c_p T_0\frac{d\delta}{dt} \qquad (18\text{-}33)$$

If a temperature profile of the form $T = T(x, \delta)$ is assumed, equation (18-33) will produce a differential equation in $\delta(t)$, which may be solved, and one may use this result to express the temperature profile as $T(x, t)$.

The solution of equation (18-33) is subject to three different boundary conditions at the wall, $x=0$, in the sections to follow.

Case 1. Constant wall temperature

The wall, initially at uniform temperature T_0, has its surface maintained at temperature T_s for $t>0$. The temperature profile at two different times is illustrated in Figure 18.6. Assuming the temperature profile to be parabolic of the form

$$T = A + Bx + Cx^2$$

and requiring that the boundary conditions

$$\begin{aligned} T &= T_s &&\text{at } x = 0 \\ T &= T_0 &&\text{at } x = \delta \end{aligned}$$

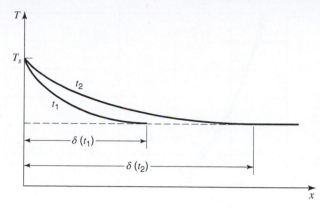

Figure 18.6 Temperature profiles at two times after the surface temperature is raised to T_s.

and

$$\frac{\partial T}{\partial x} = 0 \qquad \text{at } x = \delta$$

be satisfied, we see that the expression for $T(x)$ becomes

$$\frac{T - T_0}{T_s - T_0} = \left(1 - \frac{x}{\delta}\right)^2 \tag{18-34}$$

The heat flux at the wall may now be evaluated as

$$\frac{q_x}{A} = -k\frac{\partial T}{\partial x}\Big|_{x=0} = 2\frac{k}{\delta}(T_s - T_0) \tag{18-35}$$

which may be substituted into the integral expression along with equation (18-33), yielding

$$2\frac{k}{\delta}(T_s - T_0) = \frac{d}{dt}\int_0^\delta \rho c_p\left[T_0 + (T_s - T_0)\left(1 - \frac{x}{\delta}\right)^2\right]dx - \rho c_p T_0\frac{d\delta}{dt}$$

and, after dividing through by ρc_p, both quantities being considered constant, we have

$$2\frac{\alpha}{\delta}(T_s - T_0) = \frac{d}{dt}\int_0^\delta\left[T_0 + (T_s - T_0)\left(1 - \frac{x}{\delta}\right)^2\right]dx - T_0\frac{d\delta}{dt} \tag{18-36}$$

After integration, equation (18-36) becomes

$$\frac{2\alpha}{\delta}(T_s - T_0) = \frac{d}{dt}\left[(T_s - T_0)\frac{\delta}{3}\right]$$

and cancelling $(T_s - T_0)$, we obtain

$$6\alpha = \delta\frac{d\delta}{dt} \tag{18-37}$$

and thus the penetration depth becomes

$$\delta = \sqrt{12\alpha t} \tag{18-38}$$

The corresponding temperature profile may be obtained from equation (18-34) as

$$\frac{T - T_0}{T_s - T_0} = \left[1 - \frac{x}{\sqrt{3}(2\sqrt{\alpha t})}\right]^2 \tag{18-39}$$

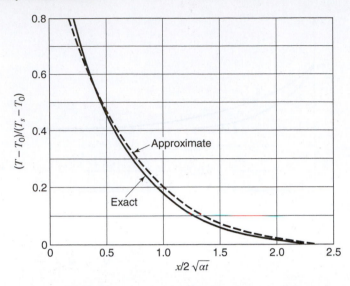

Figure 18.7 A comparison of exact and approximate results for one-dimensional conduction with a constant wall temperature.

which compares reasonably well with the exact result

$$\frac{T - T_0}{T_s - T_0} = 1 - \text{erf}\,\frac{x}{2\sqrt{\alpha t}} \tag{18-40}$$

Figure 18.7 shows a comparison of these two results.

Case 2. A specified heat flux at the wall

In this case, the appropriate boundary conditions are

$$T = T_0 \qquad \text{at } x = \delta$$

$$\frac{\partial T}{\partial x} = 0 \qquad \text{at } x = \delta$$

and

$$\frac{\partial T}{\partial x} = -\frac{F(t)}{k} \qquad \text{at } x = 0$$

where the heat flux at the wall is expressed as the general function $F(t)$.

If the parabolic temperature profile is used, the above boundary conditions yield

$$T - T_0 = \frac{[F(t)](\delta - x)^2}{2k\delta} \tag{18-41}$$

which, when substituted into equation (18-38), yields

$$\frac{d}{dt}\left(\frac{F(t)\delta^2}{6k}\right) = \frac{\alpha F(t)}{k} \tag{18-42}$$

and

$$\delta(t) = \sqrt{6\alpha}\left[\frac{1}{F(t)}\int_0^t F(t)\,dt\right]^{1/2} \tag{18-43}$$

For a constant heat flux of magnitude q_0/A, the resulting expression for T_s is

$$T_s - T_0 = \frac{q_0}{ak}\sqrt{\frac{3}{2}\alpha t} \tag{18-44}$$

which differs by approximately 8% from the exact expression

$$T_s - T_0 = \frac{1.13 q_0}{Ak}\sqrt{\alpha t} \tag{18-45}$$

Case 3. Convection at the surface

The wall temperature is a variable in this case; however, it may be easily determined. If the temperature variation within the medium is expressed generally as

$$\frac{T - T_0}{T_s - T_0} = \phi\left(\frac{x}{\delta}\right) \tag{18-46}$$

we note that the temperature gradient at the surface becomes

$$\left.\frac{\partial T}{\partial x}\right|_{x=0} = -\frac{T_s - T_0}{\delta}N \tag{18-47}$$

where N is a constant depending upon the form of $\phi(x/\delta)$.

At the surface, we may write

$$\left.\frac{q}{A}\right|_{x=0} = -k\left.\frac{\partial T}{\partial x}\right|_{x=0} = h(T_\infty - T_s)$$

which becomes, upon substituting equation (18-47),

$$T_s - T_0 = \frac{h\delta}{Nk}(T_\infty - T_0) \tag{18-48}$$

or

$$T_s = \frac{T_0 + (h\delta/Nk)T_\infty}{1 + h\delta/Nk} \tag{18-49}$$

We may now write

$$\frac{T_s - T_0}{T_\infty - T_0} = \frac{h\delta/Nk}{1 + h\delta/Nk} \tag{18-50}$$

and

$$\frac{T_\infty - T_s}{T_\infty - T_0} = \frac{1}{1 + h\delta/Nk} \tag{18-51}$$

The appropriate substitutions into the integral equation and subsequent solution follow the same procedures as in cases (a) and (b); the details of this solution are left as a student exercise.

The student should recognize the marked utility of the integral solution for solving one-dimensional unsteady-state conduction problems. Temperature profile expressions more complex than a parabolic form may be assumed. However, additional boundary conditions are needed in such cases to evaluate the constants. The similarity between the penetration depth and the boundary-layer thickness from the integral analysis of Chapter 12 should also be noted.

▶ **18.5**

CLOSURE

In this chapter, some of the techniques for solving transient or unsteady-state heat-conduction problems have been presented and discussed. Situations considered included cases of negligible internal resistance, negligible surface resistance, and those for which both resistances were significant.

For flat slabs, cylinders, and spheres, with a uniform initial temperature, whose surfaces are suddenly exposed to surroundings at a different temperature, charts are available for evaluating the temperature at any position and time. Numerical and integral methods were also introduced.

CHAPTER 19

Convective Heat Transfer

$\mathbf{H}$eat transfer by convection is associated with energy exchange between a surface and an adjacent fluid. There are very few energy-transfer situations of practical importance in which fluid motion is not in some way involved. This effect has been eliminated as much as possible in the preceding chapters, but will now be considered in some depth.

The rate equation for convection has been expressed previously as

$$\frac{q}{A} = h\Delta T \tag{15-11}$$

where the heat flux, q/A, occurs by virtue of a temperature difference. This simple equation is the defining relation for h, the convective heat-transfer coefficient. The determination of the coefficient h is, however, not at all a simple undertaking. It is related to the mechanism of fluid flow, the properties of the fluid, and the geometry of the specific system of interest.

In light of the intimate involvement between the convective heat-transfer coefficient and fluid motion, we may expect many of the considerations from our earlier treatment of momentum transfer to be of interest. In the analyses to follow, much use will be made of the developments and concepts of Chapters 4 through 14.

▶ **19.1**

FUNDAMENTAL CONSIDERATIONS IN CONVECTIVE HEAT TRANSFER

As mentioned in Chapter 12, the fluid particles immediately adjacent to a solid boundary are stationary, and a thin layer of fluid close to the surface will be in laminar flow regardless of the nature of the free stream. Thus, molecular energy exchange or conduction effects will always be present, and play a major role in any convection process. If fluid flow is laminar, then all energy transfer between a surface and contacting fluid or between adjacent fluid layers is by molecular means. If, on the other hand, flow is turbulent, then there is bulk mixing of fluid particles between regions at different temperatures, and the heat-transfer rate is increased. The distinction between laminar and turbulent flow will thus be a major consideration in any convective situation.

There are two main classifications of convective heat transfer. These have to do with the driving force causing fluid to flow. *Natural* or *free convection* designates the type of process wherein fluid motion results from the heat transfer. When a fluid is heated or cooled, the associated density change and buoyant effect produce a natural circulation in which the affected fluid moves of its own accord past the solid surface, the fluid that replaces it is similarly affected by the energy transfer, and the process is repeated. *Forced convection* is the classification used to describe those convection situations in which fluid circulation is produced by an external agency such as a fan or a pump.

The hydrodynamic boundary layer, analyzed in Chapter 12, plays a major role in convective heat transfer, as one would expect. Additionally, we shall define and analyze the *thermal boundary layer*, which will also be vital to the analysis of a convective energy-transfer process.

There are four methods of evaluating the convective heat-transfer coefficient that will be discussed in this book. These are as follows:

(a) Dimensional analysis, which to be useful requires experimental results
(b) Exact analysis of the boundary layer
(c) Approximate integral analysis of the boundary layer
(d) Analogy between energy and momentum transfer

► **19.2**

SIGNIFICANT PARAMETERS IN CONVECTIVE HEAT TRANSFER

Certain parameters will be found useful in the correlation of convective data and in the functional relations for the convective heat-transfer coefficients. Some parameters of this type have been encountered earlier; these include the Reynolds and the Euler numbers. Several of the new parameters to be encountered in energy transfer will arise in such a manner that their physical meaning is unclear. For this reason, we shall devote a short section to the physical interpretation of two such terms.

The molecular diffusivities of momentum and energy have been defined previously as

$$\text{momentum diffusivity}: \qquad \nu \equiv \frac{\mu}{\rho}$$

and

$$\text{thermal diffusivity}: \qquad \alpha \equiv \frac{k}{\rho c_p}$$

That these two are designated similarly would indicate that they must also play similar roles in their specific transfer modes. This is indeed the case, as we shall see several times in the developments to follow. For the moment we should note that both have the same dimensions, those of L^2/t; thus, their ratio must be dimensionless. This ratio, that of the molecular diffusivity of momentum to the molecular diffusivity of heat, is designated the *Prandtl number*.

$$\Pr \equiv \frac{\nu}{\alpha} = \frac{\mu c_p}{k} \qquad (19\text{-}1)$$

The Prandtl number is observed to be a combination of fluid properties; thus, Pr itself may be thought of as a property. The Prandtl number is primarily a function of temperature and is tabulated in Appendix I, at various temperatures for each fluid listed.

The temperature profile for a fluid flowing past a surface is depicted in Figure 19.1. In the figure, the surface is at a higher temperature than the fluid. The temperature profile that exists is due to the energy exchange resulting from this temperature difference. For such a case, the heat-transfer rate between the surface and the fluid may be written as

$$q_y = hA(T_s - T_\infty) \qquad (19\text{-}2)$$

and, because heat transfer at the surface is by conduction,

$$q_y = -kA\frac{\partial}{\partial y}(T - T_s)\big|_{y=0} \qquad (19\text{-}3)$$

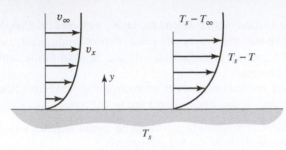

Figure 19.1 Temperature and velocity profiles for a fluid flowing past a heated plate.

These two terms must be equal; thus,

$$h(T_s - T_\infty) = -k\frac{\partial}{\partial y}(T - T_s)\big|_{y=0}$$

which may be rearranged to give

$$\frac{h}{k} = \frac{\partial(T_s - T)/\partial y\big|_{y=0}}{T_s - T_\infty} \tag{19-4}$$

Equation (19-4) may be made dimensionless if a length parameter is introduced. Multiplying both sides by a representative length, L, we have

$$\frac{hL}{k} = \frac{\partial(T_s - T)/\partial y\big|_{y=0}}{(T_s - T_\infty)/L} \tag{19-5}$$

The right-hand side of equation (19-5) is now the ratio of the temperature gradient at the surface to an overall or reference temperature gradient. The left-hand side of this equation is written in a manner similar to that for the Biot modulus encountered in Chapter 18. It may be considered a ratio of conductive thermal resistance to the convective thermal resistance of the fluid. This ratio is referred to as the *Nusselt number*:

$$\mathrm{Nu} \equiv \frac{hL}{k} \tag{19-6}$$

where the thermal conductivity is that of the fluid as opposed to that of the solid, which was the case in the evaluation of the Biot modulus.

These two parameters, Pr and Nu, will be encountered many times in the work to follow.

▶ **19.3**

DIMENSIONAL ANALYSIS OF CONVECTIVE ENERGY TRANSFER

Forced Convection The specific forced-convection situation, which we shall now consider, is that of fluid flowing in a closed conduit at some average velocity, v, with a temperature difference existing between the fluid and the tube wall.

The important variables, their symbols, and dimensional representations are listed below. It is necessary to include two more dimensions—Q, heat, and T, temperature,—to the fundamental group considered in Chapter 11; thus, all variables must be expressed dimensionally as some combination of M, L, t, Q, and T. The above variables include terms descriptive of the system geometry, thermal and flow properties of the fluid, and the quantity of primary interest, h.

Variable	Symbol	Dimension
Tube diameter	D	L
Fluid density	ρ	M/L^3
Fluid viscosity	μ	M/Lt
Fluid heat capacity	c_p	Q/MT
Fluid thermal conductivity	k	Q/tLT
Velocity	v	L/t
Heat-transfer coefficient	h	Q/tL^2T

Utilizing the Buckingham method of grouping the variables as presented in Chapter 11, the required number of dimensionless groups is found to be 3. Note that the rank of the dimensional matrix is 4, one more than the total number of fundamental dimensions.

Choosing D, k, μ, and v as the four variables comprising the core, we find that the three π groups to be formed are

$$\pi_1 = D^a k^b \mu^c v^d \rho$$
$$\pi_2 = D^e k^f \mu^g v^h c_p$$

and

$$\pi_3 = D^i k^j \mu^k v^l h$$

Writing π_1 in dimensional form,

$$1 = (L)^a \left(\frac{Q}{LtT}\right)^b \left(\frac{M}{Lt}\right)^c \left(\frac{L}{t}\right)^d \frac{M}{L^3}$$

and equating the exponents of the fundamental dimensions on both sides of this equation, we have for

$$
\begin{aligned}
L: &\quad 0 = a - b - c + d - 3 \\
Q: &\quad 0 = b \\
t: &\quad 0 = -b - c - d \\
T: &\quad 0 = -b
\end{aligned}
$$

and

$$M: \quad 0 = c + 1$$

Solving these equations for the four unknowns yields

$$
\begin{aligned}
a = 1 &\quad c = -1 \\
b = 0 &\quad d = 1
\end{aligned}
$$

and π_1 becomes

$$\pi_1 = \frac{Dv\rho}{\mu}$$

which is the Reynolds number. Solving for π_2 and π_3 in the same way will give

$$\pi_2 = \frac{\mu c_p}{k} = \text{Pr} \quad \text{and} \quad \pi_3 = \frac{hD}{k} = \text{Nu}$$

The result of a dimensional analysis of forced-convection heat transfer in a circular conduit indicates that a possible relation correlating the important variables is of the form

$$Nu = f_1(Re, Pr) \tag{19-7}$$

If, in the preceding case, the core group had been chosen to include ρ, μ, c_p, and v, the analysis would have yielded the groups $Dv\rho/\mu$, $\mu c_p/k$, and $h/\rho v c_p$. The first two of these we recognize as Re and Pr. The third is the *Stanton number*:

$$St \equiv \frac{h}{\rho v c_p} \tag{19-8}$$

This parameter could also have been formed by taking the ratio Nu/(Re Pr). An alternative correlating relation for forced convection in a closed conduit is thus

$$St = f_2(Re, Pr) \tag{19-9}$$

Natural Convection In the case of natural-convection heat transfer from a vertical plane wall to an adjacent fluid, the variables will differ significantly from those used in the preceding case. The velocity no longer belongs in the group of variables, as it is a result of other effects associated with the energy transfer. New variables to be included in the analysis are those accounting for fluid circulation. They may be found by considering the relation for buoyant force in terms of the density difference due to the energy exchange.

The coefficient of thermal expansion, β, is given by

$$\rho = \rho_0(1 - \beta \Delta T) \tag{19-10}$$

where ρ_0 is the bulk fluid density, ρ is the fluid density inside the heated layer, and ΔT is the temperature difference between the heated fluid and the bulk value. The buoyant force per unit volume, $F_{buoyant}$, is

$$F_{buoyant} = (\rho_0 - \rho)g$$

which becomes, upon substituting equation (19-10),

$$F_{buoyant} = \beta g \rho_0 \Delta T \tag{19-11}$$

Equation (19-11) suggests the inclusion of the variables β, g, and ΔT into the list of those important to the natural-convection situation.

The list of variables for the problem under consideration is given below.

Variable	Symbol	Dimension
Significant length	L	L
Fluid density	ρ	M/L^3
Fluid viscosity	μ	M/Lt
Fluid heat capacity	c_p	Q/MT
Fluid thermal conductivity	k	Q/LtT
Fluid coefficient of thermal expansion	β	$1/T$
Gravitational acceleration	g	L/t^2
Temperature difference	ΔT	T
Heat-transfer coefficient	h	Q/L^2tT

The Buckingham π theorem indicates that the number of independent dimensionless parameters applicable to this problem is $9 - 5 = 4$. Choosing L, μ, k, g, and β as the core group, we see that the π groups to be formed are

$$\pi_1 = L^a \mu^b k^c \beta^d g^e c_p$$

$$\pi_2 = L^f \mu^g k^h \beta^i g^j \rho$$

$$\pi_3 = L^k \mu^l k^m \beta^n g^o \Delta T$$

and

$$\pi_4 = L^p \mu^q k^r \beta^s g^t h$$

Solving for the exponents in the usual way, we obtain

$$\pi_1 = \frac{\mu c_p}{k} = \mathrm{Pr} \qquad \qquad \pi_3 = \beta \Delta T$$

$$\pi_2 = \frac{L^3 g \rho^2}{\mu^2} \qquad \text{and} \qquad \pi_4 = \frac{hL}{k} = \mathrm{Nu}$$

The product of π_2 and π_3, which must be dimensionless, is $(\beta g \rho^2 L^3 \Delta T)/\mu^2$. This parameter, used in correlating natural-convection data, is the *Grashof number*.

$$\mathrm{Gr} \equiv \frac{\beta g \rho^2 L^3 \Delta T}{\mu^2} \qquad (19\text{-}12)$$

From the preceding brief dimensional-analyses considerations, we have obtained the following possible forms for correlating convection data:

(a) *Forced convection*

$$\mathrm{Nu} = f_1(\mathrm{Re},\ \mathrm{Pr}) \qquad (19\text{-}7)$$

or

$$\mathrm{St} = f_2(\mathrm{Re},\ \mathrm{Pr}) \qquad (19\text{-}9)$$

(b) *Natural convection*

$$\mathrm{Nu} = f_3(\mathrm{Gr},\ \mathrm{Pr}) \qquad (19\text{-}13)$$

The similarity between the correlations of equations (19-7) and (19-13) is apparent. In equation (19-13), Gr has replaced Re in the correlation indicated by equation (19-7). It should be noted that the Stanton number can be used only in correlating forced-convection data. This becomes obvious when we observe the velocity, v, contained in the expression for St.

▶ **19.4**

EXACT ANALYSIS OF THE LAMINAR BOUNDARY LAYER

An exact solution for a special case of the hydrodynamic boundary layer is discussed in Section 12.5. Blasius's solution for the laminar boundary layer on a flat plate may be extended to include the convective heat-transfer problem for the same geometry and laminar flow.

The boundary-layer equations considered previously include the two-dimensional, incompressible continuity equation:

$$\frac{\partial v_x}{\partial x} + \frac{\partial v_y}{\partial y} = 0 \tag{12-10}$$

and the equation of motion in the x-direction:

$$\frac{\partial v_x}{\partial t} + v_x \frac{\partial v_x}{\partial x} + v_y \frac{\partial v_x}{\partial y} = v_\infty \frac{dv_\infty}{dx} + \nu \frac{\partial^2 v_x}{\partial y^2} \tag{12-9}$$

Recall that the y-directional equation of motion gave the result of constant pressure through the boundary layer. The proper form of the energy equation will thus be equation (16-14), for isobaric flow, written in two-dimensional form as

$$\frac{\partial T}{\partial t} + v_x \frac{\partial T}{\partial x} + v_y \frac{\partial T}{\partial y} = \alpha \left(\frac{\partial^2 T}{\partial x^2} + \frac{\partial^2 T}{\partial y^2} \right) \tag{19-14}$$

With respect to the thermal boundary layer depicted in Figure 19.2, $\partial^2 T/\partial x^2$ is much smaller in magnitude than $\partial^2 T/\partial y^2$.

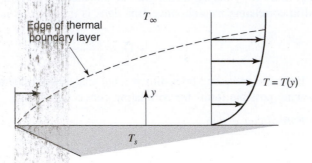

Figure 19.2 The thermal boundary layer for laminar flow past a flat surface.

In steady, incompressible, two-dimensional, isobaric flow, the energy equation that applies is now

$$v_x \frac{\partial T}{\partial x} + v_y \frac{\partial T}{\partial y} = \alpha \frac{\partial^2 T}{\partial y^2} \tag{19-15}$$

From Chapter 12, the applicable equation of motion with uniform free-stream velocity is

$$v_x \frac{\partial v_x}{\partial x} + v_y \frac{\partial v_x}{\partial y} = \nu \frac{\partial^2 v_x}{\partial y^2} \tag{12-11a}$$

and the continuity equation

$$\frac{\partial v_x}{\partial x} + \frac{\partial v_y}{\partial y} = 0 \tag{12-11b}$$

The latter two of the above equations were originally solved by Blasius to give the results discussed in Chapter 12. The solution was based upon the boundary conditions

$$\frac{v_x}{v_\infty} = \frac{v_y}{v_\infty} = 0 \quad \text{at } y = 0$$

and

$$\frac{v_x}{v_\infty} = 1 \quad \text{at } y = \infty$$

The similarity in form between equations (19-15) and (12-11a) is obvious. This situation suggests the possibility of applying the Blasius solution to the energy equation. In order that this be possible, the following conditions must be satisfied:

1. The coefficients of the second-order terms must be equal. This requires that $\nu = \alpha$ or that $\text{Pr} = 1$.

2. The boundary conditions for temperature must be compatible with those for the velocity. This may be accomplished by changing the dependent variable from T to $(T - T_s)/(T_\infty - T_s)$. The boundary conditions now are

$$\frac{v_x}{v_\infty} = \frac{v_y}{v_\infty} = \frac{T - T_s}{T_\infty - T_s} = 0 \qquad \text{at } y = 0$$

$$\frac{v_x}{v_\infty} = \frac{T - T_s}{T_\infty - T_s} = 1 \qquad \text{at } y = \infty$$

Imposing these conditions upon the set of equations (19-15) and (12-11a), we may now write the results obtained by Blasius for the energy-transfer case. Using the nomenclature of Chapter 12,

$$f' = 2\frac{v_x}{v_\infty} = 2\frac{T - T_s}{T_\infty - T_s} \tag{19-16}$$

$$\eta = \frac{y}{2}\sqrt{\frac{v_\infty}{\nu x}} = \frac{y}{2x}\sqrt{\frac{x v_\infty}{\nu}} = \frac{y}{2x}\sqrt{\text{Re}_x} \tag{19-17}$$

and applying the Blasius result, we obtain

$$\begin{aligned}
\left.\frac{df'}{d\eta}\right|_{y=0} &= f''(0) = \left.\frac{d[2(v_x/v_\infty)]}{d[(y/2x)\sqrt{\text{Re}_x}]}\right|_{y=0} \\
&= \left.\frac{d\{2[(T/T_s)/(T_\infty/T_s)]\}}{d[(y/2x)\sqrt{\text{Re}_x}]}\right|_{y=0} = 1.328
\end{aligned} \tag{19-18}$$

It should be noted that according to equation (19-16), the dimensionless velocity profile in the laminar boundary layer is identical with the dimensionless temperature profile. This is a consequence of having $\text{Pr} = 1$. A logical consequence of this situation is that the hydrodynamic and thermal boundary layers are of equal thickness. It is significant that the Prandtl numbers for most gases are sufficiently close to unity that the hydrodynamic and thermal boundary layers are of similar extent.

We may now obtain the temperature gradient at the surface:

$$\left.\frac{\partial T}{\partial y}\right|_{y=0} = (T_\infty - T_s)\left[\frac{0.332}{x}\text{Re}_x^{1/2}\right] \tag{19-19}$$

Application of the Newton and Fourier rate equations now yields

$$\frac{q_y}{A} = h_x(T_s - T_\infty) = -k\left.\frac{\partial T}{\partial y}\right|_{y=0}$$

from which

$$h_x = -\frac{k}{T_s - T_\infty}\left.\frac{\partial T}{\partial y}\right|_{y=0} = \frac{0.332\,k}{x}\text{Re}_x^{1/2} \tag{19-20}$$

or

$$\frac{h_x x}{k} = \text{Nu}_x = 0.332 \, \text{Re}_x^{1/2} \qquad (19\text{-}21)$$

Pohlhausen[1] considered the same problem with the additional effect of a Prandtl number other than unity. He was able to show the relation between the thermal and hydrodynamic boundary layers in laminar flow to be approximately given by

$$\frac{\delta}{\delta_t} = \text{Pr}^{1/3} \qquad (19\text{-}22)$$

The additional factor of $\text{Pr}^{1/3}$ multiplied by η allows the solution to the thermal boundary layer to be extended to Pr values other than unity. A plot of the dimensionless temperature vs. $\eta \, \text{Pr}^{1/3}$ is shown in Figure 19.3. The temperature variation given in this form leads to an expression for the convective heat-transfer coefficient similar to equation (19-20). At $y = 0$, the gradient is

$$\left. \frac{\partial T}{\partial y} \right|_{y=0} = (T_\infty - T_s) \left[\frac{0.332}{x} \text{Re}_x^{1/2} \text{Pr}^{1/3} \right] \qquad (19\text{-}23)$$

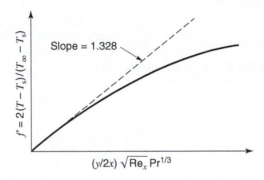

Figure 19.3 Temperature variation for laminar flow over a flat plate.

which, when used with the Fourier and Newton rate equations, yields

$$h_x = 0.332 \frac{k}{x} \text{Re}_x^{1/2} \, \text{Pr}^{1/3} \qquad (19\text{-}24)$$

or

$$\frac{h_x x}{k} = \text{Nu}_x = 0.332 \, \text{Re}_x^{1/2} \, \text{Pr}^{1/3} \qquad (19\text{-}25)$$

The inclusion of the factor $\text{Pr}^{1/3}$ in these equations extends the range of application of equations (19-20) and (19-21) to situations in which the Prandtl number differs considerably from 1.

[1] E. Pohlhausen, *ZAMM*, **1**, 115 (1921).

The mean heat-transfer coefficient applying over a plate of width w and length L may be obtained by integration. For a plate of these dimensions,

$$q_y = hA(T_s - T_\infty) = \int_A h_x(T_s - T_\infty)\,dA$$

$$h(wL)(T_s - T_\infty) = 0.332\, kw\, \mathrm{Pr}^{1/3}(T_s - T_\infty) \int_0^L \frac{\mathrm{Re}_x^{1/2}}{x}\,dx$$

$$hL = 0.332\, k\, \mathrm{Pr}^{1/3}\left(\frac{v_\infty \rho}{\mu}\right)^{1/2} \int_0^L x^{-1/2}\,dx$$

$$= 0.664\, k\, \mathrm{Pr}^{1/3}\left(\frac{v_\infty \rho}{\mu}\right)^{1/2} L^{1/2}$$

$$= 0.664\, k\, \mathrm{Pr}^{1/3}\mathrm{Re}_L^{1/2}$$

The mean Nusselt number becomes

$$\mathrm{Nu}_L = \frac{hL}{k} = 0.664\, \mathrm{Pr}^{1/3}\mathrm{Re}_L^{1/2} \tag{19-26}$$

and it is seen that

$$\mathrm{Nu}_L = 2\, \mathrm{Nu}_x \quad \text{at } x = L \tag{19-27}$$

In applying the results of the foregoing analysis, it is customary to evaluate all fluid properties at the *film temperature*, which is defined as

$$T_f = \frac{T_s + T_\infty}{2} \tag{19-28}$$

which is the arithmetic mean between the wall and bulk fluid temperatures.

▶ **19.5**

APPROXIMATE INTEGRAL ANALYSIS OF THE THERMAL BOUNDARY LAYER

The application of the Blasius solution to the thermal boundary layer in Section 19.4 was convenient, although very limited in scope. For flow other than laminar or for a configuration other than a flat surface, another method must be utilized to estimate the convective heat-transfer coefficient. An approximate method for analysis of the thermal boundary layer employs the integral analysis as used by von Kármán for the hydrodynamic boundary layer. This approach is discussed in Chapter 12.

Consider the control volume designated by the dashed lines in Figure 19.4, applying to flow parallel to a flat surface with no pressure gradient, having width Δx, a height equal to

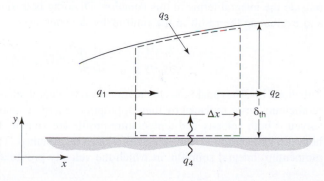

Figure 19.4 Control volume for integral energy analysis.

the thickness of the thermal boundary layer, δ_t, and a unit depth. An application of the first law of thermodynamics in integral form,

$$\frac{\delta Q}{dt} - \frac{\delta W_s}{dt} - \frac{\delta W_\mu}{dt} = \iint_{\text{c.s.}} (e + P/\rho)\rho(\mathbf{v} \cdot \mathbf{n})\, dA + \frac{\partial}{\partial t} \iiint_{\text{c.v.}} e\rho\, dV \qquad (6\text{-}10)$$

yields the following under steady-state conditions:

$$\frac{\delta Q}{dt} = -k\Delta x \frac{\partial T}{\partial y}\bigg|_{y=0}$$

$$\frac{\delta W_s}{dt} = \frac{\partial W_\mu}{dt} = 0$$

$$\iint_{\text{c.s.}} (e + P/\rho)\rho(\mathbf{v}\cdot\mathbf{n})dA = \int_0^{\delta_t} \left(\frac{v_x^2}{2} + gy + u + \frac{P}{\rho}\right)\rho v_x\, dy\bigg|_{x+\Delta x}$$

$$- \int_0^{\delta_t} \left(\frac{v_x^2}{2} + gy + u + \frac{P}{\rho}\right)\rho v_x\, dy\bigg|_x$$

$$- \frac{d}{dx}\int_0^{\delta_t} \left[\rho v_x\left(\frac{v_x^2}{2} + gy + u + \frac{P}{\rho}\right)\bigg|_{\delta_t}\right] dy\, \Delta x$$

and

$$\frac{\partial}{\partial t} \iiint_{\text{c.v.}} e\rho\, dV = 0$$

In the absence of significant gravitational effects, the convective-energy-flux terms become

$$\frac{v_x^2}{2} + u + \frac{P}{\rho} = h_0 \simeq c_p T_0$$

where h_0 is the stagnation enthalpy and c_p is the constant-pressure heat capacity. The stagnation temperature will now be written merely as T (without subscript) to avoid confusion. The complete energy expression is now

$$-k\Delta x \frac{\partial T}{\partial y}\bigg|_{y=0} = \int_0^{\delta_t} \rho v_x c_p T dy\bigg|_{x+\Delta x} - \int_0^{\delta_t} \rho v_x c_p T dy\bigg|_x - \rho c_p \Delta x \frac{d}{dx}\int_0^{\delta_t} v_x T_\infty\, dy \quad (19\text{-}29)$$

Equation (19-29) can also be written as $q_4 = q_2 - q_1 - q_3$ where these quantities are shown in Figure 19.4. In equation (19-29), T_∞ represents the free-stream stagnation temperature. If flow is incompressible, and an average value of c_p is used, the product ρc_p may be taken outside the integral terms in this equation. Dividing both sides of equation (19-29) by Δx and evaluating the result in the limit as by Δx approaches zero, we obtain

$$\frac{k}{\rho c_p}\frac{\partial T}{\partial y}\bigg|_{y=0} = \frac{d}{dx}\int_0^{\delta_t} v_x(T_\infty - T)\, dy \qquad (19\text{-}30)$$

Equation (19-30) is analogous to the momentum integral relation, equation (12-37), with the momentum terms replaced by their appropriate energy counterparts. This equation may be solved if both velocity and temperature profile are known. Thus, for the energy equation both the variation in v_x and in T with y must be assumed. This contrasts slightly with the momentum integral solution, in which the velocity profile alone was assumed.

An assumed temperature profile must satisfy the boundary conditions

1. $T - T_s = 0$ at $y = 0$
2. $T - T_s = T_\infty - T_s$ at $y = \delta_t$
3. $\dfrac{\partial}{\partial y}(T - T_s) = 0$ at $y = \delta_t$
4. $\dfrac{\partial^2}{\partial y^2}(T - T_s) = 0$ at $y = 0$ [see equation (19-15)]

If a power-series expression for the temperature variation is assumed in the form

$$T - T_s = a + by + cy^2 + dy^3$$

the application of the boundary conditions will result in the expression for $T - T_s$

$$\frac{T - T_s}{T_\infty - T_s} = \frac{3}{2}\left(\frac{y}{\delta_t}\right) - \frac{1}{2}\left(\frac{y}{\delta_t}\right)^3 \tag{19-31}$$

If the velocity profile is assumed in the same form, then the resulting expression, as obtained in Chapter 12, is

$$\frac{v}{v_\infty} = \frac{3}{2}\frac{y}{\delta} - \frac{1}{2}\left(\frac{y}{\delta}\right)^3 \tag{12-40}$$

Substituting equations (19-31) and (12-40) into the integral expression and solving, we obtain the result

$$\mathrm{Nu}_x = 0.36\,\mathrm{Re}_x^{1/2}\mathrm{Pr}^{1/3} \tag{19-32}$$

which is approximately 8% larger than the exact result expressed in equation (19-25).

This result, although inexact, is sufficiently close to the known value to indicate that the integral method may be used with confidence in situations in which an exact solution is not known. It is interesting to note that equation (19-32) again involves the parameters predicted from dimensional analysis.

A condition of considerable importance is that of an unheated starting length. Problem 19.22 at the end of the chapter deals with this situation where the wall temperature, T_s, is related to the distance from the leading edge, x, and the *unheated starting, length X*, according to

$$T_s = T_\infty \quad \text{for} \quad 0 < x < X$$
$$\text{and} \quad T_s > T_\infty \quad \text{for} \quad X < x$$

The integral technique, as presented in this section, has proved effective in generating a modified solution for this situation. The result for $T_s = $ constant, and assuming both the hydrodynamic and temperature profiles to be cubic, is

$$\mathrm{Nu}_x \cong 0.33\left[\frac{\mathrm{Pr}}{1 - (X/x)^{3/4}}\right]^{\frac{1}{3}}\mathrm{Re}_x \tag{19-33}$$

Note that this expression reduces to equation (19-25) for $X = 0$.

▶ **19.6**

ENERGY- AND MOMENTUM-TRANSFER ANALOGIES

Many times in our consideration of heat transfer thus far, we have noted the similarities to momentum transfer both in the transfer mechanism itself and in the manner of its quantitative description. This section will deal with these analogies and use them to develop relations to describe energy transfer.

Osborne Reynolds first noted the similarities in mechanism between energy and momentum transfer in 1874.[2] In 1883, he presented[3] the results of his work on frictional resistance to fluid flow in conduits, thus making possible the quantitative analogy between the two transport phenomena.

As we have noted in the previous sections, for flow past a solid surface with a Prandtl number of unity, the dimensionless velocity and temperature gradients are related as follows:

$$\frac{d}{dy} \frac{v_x}{v_\infty}\bigg|_{y=0} = \frac{d}{dy} \left(\frac{T - T_s}{T_\infty - T_s}\right)\bigg|_{y=0} \tag{19-34}$$

For $Pr = \mu c_p / k = 1$, we have $\mu c_p = k$ and we may write equation (19-34) as

$$\mu c_p \frac{d}{dy} \left(\frac{v_x}{v_\infty}\right)\bigg|_{y=0} = k \frac{d}{dy} \left(\frac{T - T_s}{T_\infty - T_s}\right)\bigg|_{y=0}$$

which may be transformed to the form

$$\frac{\mu c_p}{v_\infty} \frac{dv_x}{dy}\bigg|_{y=0} = -\frac{k}{T_s - T_\infty} \frac{d}{dy}(T - T_s)\bigg|_{y=0} \tag{19-35}$$

Recalling a previous relation for the convective heat-transfer coefficient

$$\frac{h}{k} = \frac{d}{dy} \left[\frac{(T_s - T)}{(T_s - T_\infty)}\right]\bigg|_{y=0} \tag{19-4}$$

it is seen that the entire right-hand side of equation (19-34) may be replaced by h, giving

$$h = \frac{\mu c_p}{v_\infty} \frac{dv_x}{dy}\bigg|_{y=0} \tag{19-36}$$

Introducing next the coefficient of skin friction,

$$C_f \cong \frac{\tau_0}{\rho v_\infty^2 / 2} = \frac{2\mu}{\rho v_\infty^2} \frac{dv_x}{dy}\bigg|_{y=0}$$

we may write equation (19-36) as

$$h = \frac{C_f}{2} (\rho v_\infty c_p)$$

which, in dimensionless form, becomes

$$\frac{h}{\rho v_\infty c_p} \equiv St = \frac{C_f}{2} \tag{19-37}$$

Equation (19-37) is the *Reynolds analogy* and is an excellent example of the similar nature of energy and momentum transfer. For those situations satisfying the basis for the development of equation (19-37), a knowledge of the coefficient of frictional drag will enable the convective heat-transfer coefficient to be readily evaluated.

The restrictions on the use of the Reynolds analogy should be kept in mind; they are (1) $Pr = 1$ and (2) no form drag. The former of these was the starting point in the preceding development and obviously must be satisfied. The latter is sensible when one considers that

---

[2] O. Reynolds, *Proc. Manchester Lit. Phil. Soc.*, **14**, 7 (1874).
[3] O. Reynolds, *Trans. Roy. Soc.* (London), **174A**, 935 (1883).

in relating two transfer mechanisms, the manner of expressing them quantitatively must remain consistent. Obviously the description of drag in terms of the coefficient of skin friction requires that the drag be wholly viscous in nature. Thus, equation (19-37) is applicable only for those situations in which form drag is not present. Some possible areas of application would be flow parallel to plane surfaces or flow in conduits. The coefficient of skin friction for conduit flow has already been shown to be equivalent to the Fanning friction factor, which may be evaluated by using Figure 19.1.

The restriction that $Pr = 1$ makes the Reynolds analogy of limited use. Colburn[4] has suggested a simple variation of the Reynolds analogy form that allows its application to situations where the Prandtl number is other than unity. The Colburn analogy expression is

$$St\,Pr^{2/3} = \frac{C_f}{2} \tag{19-38}$$

which obviously reduces to the Reynolds analogy when $Pr = 1$.

Colburn applied this expression to a wide range of data for flow and geometries of different types and found it to be quite accurate for conditions where (1) no form drag exists and (2) $0.5 < Pr < 50$. The Prandtl number range is extended to include gases, water, and several other liquids of interest. The Colburn analogy is particularly helpful for evaluating heat transfer in internal forced flows. It can be easily shown that the exact expression for a laminar boundary layer on a flat plate reduces to equation (19-38).

The Colburn analogy is often written as

$$j_H = \frac{C_f}{2} \tag{19-39}$$

where

$$j_H = St\,Pr^{2/3} \tag{19-40}$$

is designated the Colburn j factor for heat transfer. A mass-transfer j factor is discussed in Chapter 28.

Note that for $Pr = 1$, the Colburn and Reynolds analogies are the same. Equation (19-38) is thus an extension of the Reynolds analogy for fluids having Prandtl numbers other than unity, within the range 0.5–50 as specified above. High and low Prandtl number fluids falling outside this range would be heavy oils at one extreme and liquid metals at the other.

► **19.7**

TURBULENT FLOW CONSIDERATIONS

The effect of the turbulent flow on energy transfer is directly analogous to the similar effects on momentum transfer as discussed in Chapter 12. Consider the temperature profile variation in Figure 19.5 to exist in turbulent flow. The distance moved by a fluid "packet" in the y direction, which is normal to the direction of bulk flow, is denoted by L, the Prandtl mixing length. The packet of fluid moving through the distance L retains the mean temperature from its point of origin, and upon reaching its destination, the packet will differ in temperature from that of the adjacent fluid by an amount $T|_{y\pm L} - T|_y$. The mixing length is assumed small enough to permit the temperature difference to be written as

$$T\big|_{y\pm L} - T\big|_y = \pm L \frac{dT}{dy}\bigg|_y \tag{19-41}$$

---

[4] A. P. Colburn, *Trans. A.I.Ch.E.*, **29**, 174 (1933).

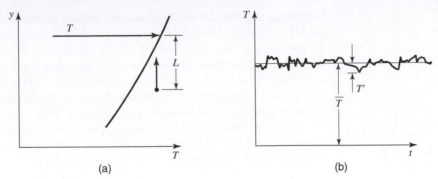

Figure 19.5 Turbulent flow temperature variation.

We now define the quantity T' as the fluctuating temperature, synonymous with the fluctuating velocity component, v'_x, described in Chapter 12. The instantaneous temperature is the sum of the mean and fluctuating values, as indicated in Figure 19.5(b), or, in equation form,

$$T = \overline{T} + T' \tag{19-42}$$

Any significant amount of energy transfer in the y-direction, for bulk flow occurring in the x direction, is accomplished because of the fluctuating temperature, T'; thus, it is apparent from equations (19-41) and (19-42) that

$$T' = \pm L \frac{d\overline{T}}{dy} \tag{19-43}$$

The energy flux in the y-direction may now be written as

$$\left.\frac{q_y}{A}\right|_y = \rho c_p T v'_y \tag{19-44}$$

where v'_y may be either positive or negative. Substituting for T its equivalent, according to equation (19-42)

$$\left.\frac{q_y}{A}\right|_y = \rho c_p v'_y (\overline{T} + T')$$

and taking the time average, we obtain, for the y-directional energy flux due to turbulent effects,

$$\left.\frac{q_y}{A}\right|_{\text{turb}} = \rho c_p \overline{(v'_y T')} \tag{19-45}$$

or, with T' in terms of the mixing length,

$$\left.\frac{q_y}{A}\right|_{\text{turb}} = \rho c_p \overline{v'_y L} \frac{d\overline{T}}{dy} \tag{19-46}$$

The total energy flux due to both microscopic and turbulent contributions may be written as

$$\frac{q_y}{A} = -\rho c_p [\alpha + \overline{|v'_y L|}] \frac{d\overline{T}}{dy} \tag{19-47}$$

As α is the molecular diffusivity of heat, the quantity $\overline{|v'_y L|}$ is the *eddy diffusivity of heat*, designated as ϵ_H. This quantity is exactly analogous to the eddy diffusivity of momentum, ϵ_M, as defined in equation (12-52). In a region of turbulent flow, $\epsilon_H \gg \alpha$ for all fluids except liquid metals.

As the Prandtl number is the ratio of the molecular diffusivities of momentum and heat, an analogous term, the *turbulent Prandtl number*, can be formed by the ratio ϵ_M/ϵ_H. Utilizing equations (19-47) and (12-55), we have

$$\text{Pr}_{\text{turb}} = \frac{\varepsilon_M}{\varepsilon_H} = \frac{L^2|dv_x/dy|}{|Lv'_y|} = \frac{L^2|dv_x/dy|}{L^2|dv_x/dy|} = 1 \tag{19-48}$$

Thus, in a region of fully turbulent flow the effective Prandtl number is unity, and the Reynolds analogy applies in the absence of form drag.

In terms of the eddy diffusivity of heat, the heat flux can be expressed as

$$\left.\frac{q_y}{A}\right|_{\text{turb}} = -\rho c_p \varepsilon_H \frac{D\overline{T}}{dy} \tag{19-49}$$

The total heat flux, including both molecular and turbulent contributions, thus becomes

$$\frac{q_y}{A} = -\rho c_p(\alpha + \varepsilon_H)\frac{d\overline{T}}{dy} \tag{19-50}$$

Equation (19-50) applies both to the region wherein flow is laminar, for which $\alpha \gg \epsilon_H$, and to that for which flow is turbulent and $\alpha \gg \epsilon_H$. It is in this latter region that the Reynolds analogy applies. Prandtl[5] achieved a solution that includes the influences of both the laminar sublayer and the turbulent core. In his analysis, solutions were obtained in each region and then joined at $y = \xi$, the hypothetical distance from the wall that is assumed to be the boundary separating the two regions.

Within the laminar sublayer the momentum and heat flux equations reduce to

$$\tau = \rho\nu\frac{dv_x}{dy}\text{ (a constant)}$$

and

$$\frac{q_y}{A} = -\rho c_p \alpha \frac{dT}{dy}$$

Separating variables and integrating between $y = 0$ and $y = \xi$, we have, for the momentum expression,

$$\int_0^{v_x|_\xi} dv_x = \frac{\tau}{\rho\nu}\int_0^\xi dy$$

and, for the heat flux,

$$\int_{T_s}^{T_\xi} dT = -\frac{q_y}{A\rho c_p \alpha}\int_0^\xi dy$$

Solving for the velocity and temperature profiles in the laminar sublayer yields

$$v_x|_\xi = \frac{\tau\xi}{\rho\nu} \tag{19-51}$$

and

$$T_s - T_\xi = \frac{q_y\xi}{A\rho c_p \alpha} \tag{19-52}$$

---

[5] L. Prandtl, *Zeit. Physik.*, **11**, 1072 (1910).

Eliminating the distance ξ between these two expressions gives

$$\frac{\rho v v_x|_\xi}{\tau} = \frac{\rho A c_p \alpha}{q_y}(T_s - T_\xi) \tag{19-53}$$

Directing our attention now to the turbulent core where the Reynolds analogy applies, we may write equation (19-37)

$$\frac{h}{\rho c_p(v_\infty - v_x|_\xi)} = \frac{C_f}{2} \tag{19-37}$$

and, expressing h and C_f in terms of their defining relations, we obtain

$$\frac{q_y/A}{\rho c_p(v_\infty - v_x|_\xi)(T_\xi - T_\infty)} = \frac{\tau}{\rho(v_\infty - v_x|_\xi)^2}$$

Simplifying and rearranging this expression, we have

$$\frac{\rho(v_\infty - v_x|_\xi)}{\tau} = \rho A c_p \frac{(T_\xi - T_\infty)}{q_y} \tag{19-54}$$

which is a modified form of the Reynolds analogy applying from $y = \xi$ to $y = y_{max}$. Eliminating T_ξ between equations (19-53) and (19-54), we have

$$\frac{\rho}{\tau}\left[v_\infty + v_x|_\xi\left(\frac{v}{\alpha} - 1\right)\right] = \frac{\rho A c_p}{q_y}(T_s - T_\infty) \tag{19-55}$$

Introducing the coefficient of skin friction

$$C_f = \frac{\tau}{\rho v_\infty^2/2}$$

and the convective heat-transfer coefficient

$$h = \frac{q_y}{A(T_s - T_\infty)}$$

we may reduce equation (19-54) to

$$\frac{v_\infty + v_x|_\xi(v/\alpha - 1)}{v_\infty^2 C_f/2} = \frac{\rho c_p}{h}$$

Inverting both sides of this expression and making it dimensionless, we obtain

$$\frac{h}{\rho c_p v_\infty} \equiv St = \frac{C_f/2}{1 + (v_x|_\xi/v_\infty)[(v/\alpha) - 1]} \tag{19-56}$$

This equation involves the ratio v/α, which has been defined previously as the Prandtl number. For a value of $Pr = 1$, equation (19-56) reduces to the Reynolds analogy. For $Pr = 1$, the Stanton number is a function of C_f, Pr, and the ratio $v_x|_\xi/v_\infty$. It would be convenient to eliminate the velocity ratio; this may be accomplished by recalling some results from Chapter 12.

At the edge of the laminar sublayer

$$v^+ = y^+ = 5$$

and by definition $v^+ = v_x/(\sqrt{\tau/\rho})$. Thus, for the case at hand,

$$v^+ = v_x|_\xi/(\sqrt{\tau/\rho}) = 5$$

Again introducing the coefficient of skin friction in the form

$$C_f = \frac{\tau}{\rho v_\infty^2 / 2}$$

we may write

$$\sqrt{\frac{\tau}{\rho}} = v_\infty \sqrt{\frac{C_f}{2}}$$

which, when combined with the previous expression given for the velocity ratio, gives

$$\frac{v_x|_\xi}{v_\infty} = 5 \sqrt{\frac{C_f}{2}} \tag{19-57}$$

Substitution of equation (19-57) into (19-56) gives

$$St = \frac{C_f / 2}{1 + 5\sqrt{C_f / 2}(Pr - 1)} \tag{19-58}$$

which is known as the *Prandtl analogy*. This equation is written entirely in terms of measurable quantities.

Von Kármán[6] extended Prandtl's work to include the effect of the transition or buffer layer in addition to the laminar sublayer and turbulent core. His result, the von Kármán analogy, is expressed as

$$St = \frac{C_f / 2}{1 + 5\sqrt{C_f / 2}\{Pr - 1 + \ln[1 + \frac{5}{6}(Pr - 1)]\}} \tag{19-59}$$

Note that, just as for the Prandtl analogy, equation (19-59) reduces to the Reynolds analogy for a Prandtl number of unity.

The application of the Prandtl and von Kármán analogies is, quite logically, restricted to those cases in which there is negligible form drag. These equations yield the most accurate results for Prandtl numbers greater than unity.

An illustration of the use of the four relations developed in this section is given in the example below.

Example 1

Water at 50 F enters a heat-exchanger tube having an inside diameter of 1 in and a length of 10 ft. The water flows at 20 gal/min. For a constant wall temperature of 210 F, estimate the exit temperature of the water using (a) the Reynolds analogy, (b) the Colburn analogy, (c) the Prandtl analogy, and (d) the von Kármán analogy. Entrance effects are to be neglected, and the properties of water may be evaluated at the arithmetic-mean bulk temperature.

Considering a portion of the heat-exchanger tube shown in Figure 19.6, we see that an application of the first law of thermodynamics to the control volume indicated will yield the result that

$$\left\{ \begin{array}{c} \text{rate of heat} \\ \text{transfer into c.v.} \\ \text{by fluid flow} \end{array} \right\} + \left\{ \begin{array}{c} \text{rate of heat} \\ \text{transfer into c.v.} \\ \text{by convection} \end{array} \right\} = \left\{ \begin{array}{c} \text{rate of heat} \\ \text{transfer out of c.v.} \\ \text{by fluid flow} \end{array} \right\}$$

---

[6] T. von Kármán, *Trans. ASME*, **61**, 705 (1939).

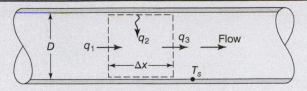

Figure 19.6 Analog analysis of water flowing in a circular tube.

If these heat-transfer rates are designated as q_1, q_2, and q_3, they may be evaluated as follows:

$$q_1 = \rho \frac{\pi D^2}{4} v_x c_p T|_x$$

$$q_2 = h \pi D \Delta_x (T_s - T)$$

and

$$q_3 = \rho \frac{\pi D^2}{4} v_x c_p T|_{x+\Delta x}$$

The substitution of these quantities into the energy balance expression gives

$$\rho \frac{\pi D^2}{4} v_x c_p [T|_{x+\Delta x} - T|_x] - h \pi D \Delta x (T_s - T) = 0$$

which may be simplified and rearranged into the form

$$\frac{D}{4} \frac{T|_{x+\Delta x} - T|_x}{\Delta x} + \frac{h}{\rho v_x c_p} (T - T_s) = 0 \tag{19-60}$$

Evaluated in the limit as $\Delta x \to 0$, equation (19-59) reduces to

$$\frac{dT}{dx} + \frac{h}{\rho v_x c_p} \frac{4}{D} (T - T_s) = 0 \tag{19-61}$$

Separating the variables, we have

$$\frac{dT}{T - T_s} + \frac{h}{\rho v_x c_p} \frac{4}{D} dx = 0$$

and integrating between the limits indicated, we obtain

$$\int_{T_0}^{T_L} \frac{dT}{T - T_s} + \frac{h}{\rho v_x c_p} \frac{4}{D} \int_0^L dx = 0$$

$$\ln \frac{T_L - T_s}{T_0 - T_s} + \frac{h}{\rho v_x c_p} \frac{4L}{D} = 0 \tag{19-62}$$

Equation (19-62) may now be solved for the exit temperature T_L. Observe that the coefficient of the right-hand term, $h/\rho v_x c_p$, is the Stanton number. This parameter has been achieved quite naturally from our analysis.

The coefficient of skin friction may be evaluated with the aid of Figure 13.1. The velocity is calculated as

$$v_x = 20 \text{ gal/min}(\text{ft}^3/7.48 \text{ gal})[144/(\pi/4)(1^2)] \text{ ft}^2(\text{min}/60 \text{ s}) = 8.17 \text{ fps}$$

Initially, we will assume the mean bulk temperature to be 90 F. The film temperature will then be 150 F, at which $\nu = 0.474 \times 10^{-5}$ ft^2/s. The Reynolds number is

$$\text{Re} = \frac{D v_x}{\nu} = \frac{(1/12 \text{ ft})(8.17 \text{ ft/s})}{0.474 \times 10^{-5} \text{ ft}^2/\text{s}} = 144,000$$

At this value of Re, the friction factor, f_f, assuming smooth tubing, is 0.0042. For each of the four analogies, the Stanton number is evaluated as follows:

(a) *Reynolds analogy*

$$St = \frac{C_f}{2} = 0.0021$$

(b) *Colburn analogy*

$$St = \frac{C_f}{2}Pr^{-2/3} = 0.0021(2.72)^{-2/3} = 0.00108$$

(c) *Prandtl analogy*

$$St = \frac{C_f/2}{1 + 5\sqrt{C_f/2}(Pr - 1)}$$

$$= \frac{0.0021}{1 + 5\sqrt{0.0021}(1.72)} = 0.00151$$

(d) *von Kármán analogy*

$$St = \frac{C_f/2}{1 + 5\sqrt{C_f/2}\left\{Pr - 1 + \ln\left[1 + \frac{5}{6}(Pr - 1)\right]\right\}}$$

$$= \frac{0.0021}{1 + 5\sqrt{0.0021}\left\{2.72 - 1 + \ln\left[1 + \frac{5}{6}(2.72 - 1)\right]\right\}}$$

$$= 0.00131$$

Substituting these results into equation (19-62), we obtain, for T_L, the following results:

(a) $T_L = 152$ F
(b) $T_L = 115$ F
(c) $T_L = 132$ F
(d) $T_L = 125$ F

Some fine-tuning of these results may be necessary to adjust the physical property values for the calculated film temperatures. In none of these cases is the assumed film temperature different than the calculated one by more than 6 F, so the results are not going to change much.

The Reynolds analogy value is much different from the other results obtained. This is not surprising, as the Prandtl number was considerably above a value of 1. The last three analogies yielded quite consistent results. The Colburn analogy is the simplest to use and is preferable from that standpoint.

▶ **19.8**

CLOSURE

The fundamental concepts of convection heat transfer have been introduced in this chapter. New parameters pertinent to convection are the Prandtl, Nusselt, Stanton, and Grashof numbers.

Four methods of analyzing a convection heat-transfer process have been discussed. These are as follows:

1. Dimensional analysis coupled with experiment
2. Exact analysis of the boundary layer
3. Integral analysis of the boundary layer
4. Analogy between momentum and energy transfer

Several empirical equations for the prediction of convective heat-transfer coefficients will be given in the chapters to follow.

CHAPTER 20

Convective Heat-Transfer Correlations

Convective heat transfer was treated from an analytical point of view in Chapter 19. Although the analytic approach is very meaningful, it may not offer a practical solution to every problem. There are many situations for which no mathematical models have as yet been successfully applied. Even in those cases for which an analytical solution is possible, it is necessary to verify the results by experiment. In this chapter, we shall present some of the most useful correlations of experimental heat-transfer data available. Most correlations are in the forms indicated by dimensional analysis.

The sections to follow include discussion and correlations for natural convection, forced convection for internal flow, and forced convection for external flow, respectively. In each case, those analytical relations that are available are presented along with the most satisfactory empirical correlations for a particular geometry and flow condition.

▶ **20.1**

NATURAL CONVECTION

The mechanism of energy transfer by natural convection involves the motion of a fluid past a solid boundary, which is the result of the density differences resulting from the energy exchange. Because of this, it is quite natural that the heat-transfer coefficients and their correlating equations will vary with the geometry of a given system.

Vertical Plates   The natural convection system most amenable to analytical treatment is that of a fluid adjacent to a vertical wall.

Standard nomenclature for a two-dimensional consideration of natural convection adjacent to a vertical plane surface is indicated in Figure 20.1. The x-direction is commonly taken along the wall, with y measured normal to the plane surface.

Schmidt and Beckmann[1] measured the temperature and velocity of air at different locations near a vertical plate and found a significant variation in both quantities along the direction parallel to the plate. The variations of velocity and temperature for a 12.5-cm-high vertical plate are shown in Figures 20.2 and 20.3 for the conditions $T_s = 65$ C, $T_\infty = 15$ C.

The two limiting cases for vertical plane walls are those with constant surface temperature and with constant wall heat flux. The former of these cases have been solved by Ostrach[2] and the latter by Sparrow and Gregg.[3]

---

[1] E. Schmidt and W. Beckmann, *Tech. Mech. U. Thermodynamik*, **1**, 341 and 391 (1930).
[2] S. Ostrach, NACA Report 1111, 1953.
[3] E. M. Sparrow and J. L. Gregg, *Trans. A.S.M.E.*, **78**, 435 (1956).

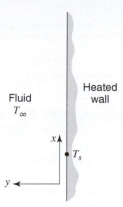

Figure 20.1 Coordinate system for the analysis of natural convection adjacent to a heated vertical wall.

Ostrach, employing a similarity transformation with governing equations of mass conservation, motion, and energy in a free-convection boundary layer, obtained an expression for local Nusselt number of the form

$$\mathrm{Nu}_x = f(\mathrm{Pr})\left(\frac{\mathrm{Gr}_x}{4}\right)^{1/4} \qquad (20\text{-}1)$$

The coefficient, $f(\mathrm{Pr})$, varies with Prandtl number, with values given in Table 20.1.

We usually find the mean Nusselt number, Nu_L, to be of more value than Nu_x. Using an integration procedure, as discussed earlier, an expression for Nu_L may be determined using

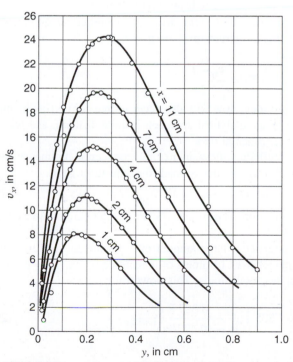

Figure 20.2 Velocity distribution in the vicinity of a vertical heated plate in air.

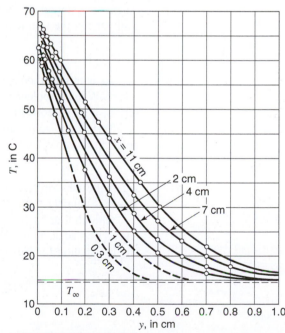

Figure 20.3 Temperature distribution in the vicinity of a vertical heated plate in air.

Table 20.1 Values of the coefficient $f(Pr)$ for use in equation (20-1)

Pr	0.01	0.072	1	2	10	100	1000
$f(Pr)$	0.081	0.505	0.567	0.716	1.169	2.191	3.966

equation (20-1). The mean heat-transfer coefficient for a vertical surface of height, L, is related to the local value according to

$$h_L = \frac{1}{L}\int_0^L h_x \, dx$$

Inserting equation (20-1) appropriately, we proceed to

$$h_L = \frac{k}{L}f(Pr)\left[\frac{\beta g \Delta T}{4\nu^2}\right]^{1/4}\int_0^L x^{-1/4}dx$$

$$= \left(\frac{4}{3}\right)\left(\frac{k}{L}\right)f(Pr)\left[\frac{\beta g L^3 \Delta T}{4\nu^2}\right]^{1/4}$$

and, in dimensionless form, we have

$$Nu_L = \frac{4}{3}f(Pr)\left(\frac{Gr_L}{4}\right)^{1/4} \tag{20-2}$$

Sparrow and Gregg's[4] results for the constant wall heat flux case compared within 5% to those of Ostrach for like values of Pr. Equations (20-1) and (20-2), along with coefficients from Table 20.1, may thus be used, with reasonable accuracy, to evaluate any vertical plane surface regardless of wall conditions, provided boundary-layer flow is laminar.

Fluid properties, being temperature dependent, will have some effect on calculated results. It is important, therefore, that properties involved in equations (20-1) and (20-2) be evaluated at the film temperature.

$$T_f = \frac{T_s + T_\infty}{2}$$

As with forced convection, turbulent flow will also occur in free convection boundary layers. When turbulence is present, an analytical approach is quite difficult, and we must rely heavily on correlations of experimental data.

Transition from laminar to turbulent flow in natural convection boundary layers adjacent to vertical plane surfaces has been determined to occur at, or near,

$$Gr_t \, Pr = Ra_t \cong 10^9 \tag{20-3}$$

where the subscript, t, indicates transition. The product, $Gr_t \, Pr$, is often referred to as Ra, the *Rayleigh* number.

Churchill and Chu[5] have correlated a large amount of experimental data for natural convection adjacent to vertical planes over 13 orders of magnitude of Ra. They propose a single equation for Nu_L that applies to all fluids. This powerful equation is

$$Nu_L = \left\{0.825 + \frac{0.387 \, Ra_L^{1/6}}{[1 + (0.492/Pr)^{9/16}]^{8/27}}\right\}^2 \tag{20-4}$$

[4] E. M. Sparrow and J. L. Gregg, *Trans. A.S.M.E.*, **78**, 435 (1956).
[5] S. W. Churchill and H. H. S. Chu, *Int. J. Heat & Mass Tr.*, **18**, 1323 (1975).

Churchill and Chu show this expression to provide accurate results for both laminar and turbulent flows. Some improvement was found for the laminar range ($Ra_L < 10^9$) by using the following equation:

$$Nu_L = 0.68 + \frac{0.670\,Ra_L^{1/4}}{[1 + (0.492/Pr)^{9/16}]^{4/9}} \qquad (20\text{-}5)$$

Vertical Cylinders   For the case of cylinders with their axes vertical, the expressions presented for plane surfaces can be used provided the curvature effect is not too great. The criterion for this is expressed in equation (20-6); specifically, a vertical cylinder can be evaluated using correlations for vertical plane walls when

$$\frac{D}{L} \geq \frac{35}{Gr_L^{1/4}} \qquad (20\text{-}6)$$

Physically, this represents the limit where boundary-layer thickness is small relative to cylinder diameter, D.

Horizontal Plates   The correlations suggested by McAdams[6] are well accepted for this geometry. A distinction is made regarding whether the fluid is hot or cool, relative to the adjacent surface, and whether the surface faces up or down. It is clear that the induced buoyancy will be much different for a hot surface facing up than down. McAdams's correlations are, for a hot surface facing up or cold surface facing down,

$$10^5 < Ra_L < 2 \times 10^7 \quad Nu_L = 0.54\,Ra_L^{1/4} \qquad (20\text{-}7)$$

$$2 \times 10^7 < Ra_L < 3 \times 10^{10} \quad Nu_L = 0.14\,Ra_L^{1/3} \qquad (20\text{-}8)$$

and, for a hot surface facing down or cold surface facing up,

$$3 \times 10^5 < Ra_L < 10^{10} \quad Nu_L = 0.27\,Ra_L^{1/4} \qquad (20\text{-}9)$$

In each of these correlating equations, the film temperature, T_f, should be used for fluid property evaluation. The length scale, L, is the ratio of the plate-surface area to perimeter.

For plane surfaces inclined at an angle, θ, with the vertical, equations (20-4) and (20-5) may be used, with modification, for values of θ up to 60°. Churchill and Chu[7] suggest replacing g by $g\cos\theta$ in equation (20-5) when boundary-layer flow is laminar. With turbulent flow, equation (20-4) may be used without modification.

Horizontal Cylinders   With cylinders of sufficient length that end effects are insignificant, two correlations are recommended. Churchill and Chu[8] suggest the following correlation

$$Nu_D = \left\{ 0.60 + \frac{0.387\,Ra_D^{1/6}}{[1 + (0.559/Pr)^{9/16}]^{8/27}} \right\}^2 \qquad (20\text{-}10)$$

over the Rayleigh number range $10^{-5} < Ra_D < 10^{12}$.

---

[6] W. H. McAdams, *Heat Transmission*, Third Edition, Chapter 7, McGraw-Hill Book Company, New York, 1957.
[7] S. W. Churchill and H. H. S. Chu, *Int. J. Heat & Mass Tr.*, **18**, 1323 (1975).
[8] S. W. Churchill and H. H. S. Chu, *Int. J. Heat & Mass Tr.*, **18**, 1049 (1975).

A simpler equation has been suggested by Morgan,[9] in terms of variable coefficients

$$\mathrm{Nu}_D = C\,\mathrm{Ra}_D^n \tag{20-11}$$

where values of C and n are specified as functions of Ra_D in Table 20.2.

Table 20.2   Values of constants C and n in equation (20-11)

	C	n
$10^{-10} < \mathrm{Ra}_D < 10^{-2}$	0.675	0.058
$10^{-2} < \mathrm{Ra}_D < 10^{2}$	1.02	0.148
$10^{2} < \mathrm{Ra}_D < 10^{4}$	0.850	0.188
$10^{4} < \mathrm{Ra}_D < 10^{7}$	0.480	0.250
$10^{7} < \mathrm{Ra}_D < 10^{12}$	0.125	0.333

The film temperature should be used in evaluating fluid properties in the above equations.

Spheres   The correlation suggested by Yuge[10] is recommended for the case with $\mathrm{Pr} \simeq 1$, and $1 < \mathrm{Ra}_D < 10^5$.

$$\mathrm{Nu}_D = 2 + 0.43\,\mathrm{Ra}_D^{1/4} \tag{20-12}$$

We may notice that, for the sphere, as Ra approaches zero, heat transfer from the surface to the surrounding medium is by conduction. This problem may be solved to yield a limiting value for Nu_D equal to 2. This result is obviously compatible with equation (20-12).

Rectangular Enclosures   Shown in Figure 20.4 is the configuration and nomenclature pertinent to rectangular enclosures. These cases have become much more important in recent years due to their application in solar collectors. Clearly, heat transfer will be

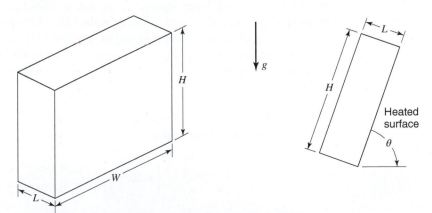

Figure 20.4 The rectangle enclosure.

---

[9] V. T. Morgan, *Advances in Heat Transfer*, Vol. II, T. F. Irvine and J. P. Hartnett, Eds., Academic Press, New York, 1975, pp. 199–264.

[10] T. Yuge, *J. Heat Transfer*, **82**, 214 (1960).

affected by the angle of tilt, θ; by the aspect ratio, H/L; and by the usual dimensionless parameters, Pr and Ra_L.

In each of the correlations to follow, the temperature of the hotter of the two large surfaces is designated as T_1, and the cooler surface is at temperature T_2. Fluid properties are evaluated at the film temperature, $T_f = (T_1 + T_2)/2$. Convective heat flux is expressed as

$$\frac{q}{A} = h(T_1 - T_2) \tag{20-13}$$

Case 1. Horizontal enclosures, $\theta = 0$

With the bottom surface heated, a critical Rayleigh member has been determined by several investigators to exist. For cases where

$$Ra_L = \frac{\beta g L^3 (T_1 - T_2)}{\alpha \nu} > 1700$$

conditions within an enclosure are thermally unstable and natural convection will occur. A correlation for this case has been proposed by Globe and Dropkin[11] in the form

$$Nu_L = 0.069\, Ra_L^{1/3} Pr^{0.074} \tag{20-14}$$

for the range $3 \times 10^5 < Ra_L < 7 \times 10^9$.

When $\theta = 180°$, that is, the upper surface is heated, or when $Ra_L < 1700$, heat transfer is by conduction; thus, $Nu_L = 1$.

Case 2. Vertical enclosures, $\theta = 90°$

For aspect ratios less than 10, Catton[12] suggests the use of the following correlations:

$$Nu_L = 0.18 \left(\frac{Pr}{0.2 + Pr} Ra_L \right)^{0.29} \tag{20-15}$$

when

$$1 < H/L < 2,\ 10^{-3} < Pr < 10^5,\ 10^3 < Ra_L Pr/(0.2 + Pr)$$

and

$$Nu_L = 0.22 \left(\frac{Pr}{0.22 + Pr} Ra_L \right)^{0.28} \left(\frac{H}{L} \right)^{-1/4} \tag{20-16}$$

when $2 < H/L < 10$, $Pr < 10^5$, $Ra_L < 10^{10}$.

For higher values of H/L, the correlations of MacGregor and Emery[13] are recommended. These are

$$Nu_L = 0.42\, Ra_L^{1/4} Pr^{0.012} (H/L)^{-0.3} \tag{20-17}$$

for

$$10 < H/L < 40,\ 1 < Pr < 2 \times 10^4,\ 10^4 < Ra_L < 10^7$$

and

$$Nu_L = 0.046\, Ra_L^{1/3} \tag{20-18}$$

---

[11] S. Globe and D. Dropkin, *J Heat Transfer*, **81C**, 24 (1959).

[12] I. Catton, *Proc. 6th Int. Heat Tr. Conference, Toronto, Canada*, **6**, 13 (1978).

[13] P. K. MacGregor and A. P. Emery, *J. Heat Transfer*, **91**, 391 (1969).

for

$$10 < H/L < 40, \ 1 < \text{Pr} < 20, \ 10^6 < \text{Ra}_L < 10^9$$

Case 3. Tilted vertical enclosures, $0 < \theta = 90°$

Numerous publications have dealt with this configuration. Correlations for this case, when the aspect ratio is large ($H/L > 12$), are the following:

$$\text{Nu}_L = 1 + 1.44\left[1 - \frac{1708}{\text{Ra}_L\cos\theta}\right]\left[1 - \frac{1708(\sin 1.8\,\theta)^{1.6}}{\text{Ra}_L\cos\theta}\right]$$

$$+ \left[\left(\frac{\text{Ra}_L\cos\theta}{5830}\right)^{1/3} - 1\right] \tag{20-19}$$

when $H/L \geq 12$, $0 < \theta < 70°$. In applying this relationship any bracketed term with a negative value should be set equal to zero. Equation (20-19) was suggested by Hollands et al.[14] With enclosures nearing the vertical, Ayyaswamy and Catton[15] suggest the relationship

$$\text{Nu}_L = \text{Nu}_{LV}(\sin\theta)^{1/4} \tag{20-20}$$

for all aspect ratios, and $70° < \theta < 90°$. The value $\theta = 70°$ is termed the "critical" tilt angle for vertical enclosures with $H/L > 12$. For smaller aspect ratios the critical angle of tilt is also smaller. A recommended review article on the subject of inclined rectangular cavities is that of Buchberg, Catton, and Edwards.[16]

Example 1

Determine the surface temperature of a cylindrical tank measuring 0.75 m in diameter and 1.2 m high. The tank contains a transformer immersed in an oil bath that produces a uniform surface temperature condition. All heat loss from the surface may be assumed to be due to natural convection to surrounding air at 295 K. The heat dissipation rate from the transformer is constant at 1.5 KW.

Surface areas that apply are

$$A_{\text{top}} = A_{\text{bottom}} = \frac{\pi}{4}(0.75\,\text{m})^2 = 0.442\,\text{m}^2$$

$$A_{\text{side}} = \pi(1.2\,\text{m})(0.75\,\text{m}) = 2.83\,\text{m}^2$$

The total heat transfer is the sum of the contributions from the three surfaces. This may be written as

$$q_{\text{total}} = [h_t(0.442) + h_b(0.442) + h_s(2.83)](T - 295)$$

where subscripts t (top), b (bottom), and s (sides) apply to the surfaces in question.

Rewriting this expression in terms of Nu, we have

$$q_{\text{total}} = \left[\text{Nu}_t\frac{k}{0.75}(0.442) + \text{Nu}_b\frac{k}{0.75}(0.442) + \text{Nu}_s\frac{k}{1.2}(2.83)\right](T - 295)$$

---

[14] K. G. T. Hollands, S. E. Unny, G. D. Raithby, and L. Konicek, *J Heat Transfer*, **98**, 189 (1976).

[15] P. S. Ayyaswamy and I. Catton, *J Heat Transfer*, **95**, 543 (1973).

[16] H. Buchberg, I. Catton, and D. K. Edwards, *J Heat Transfer*, **98**, 182 (1976).

or

$$q_{\text{total}} = [0.589\,\text{Nu}_t + 0.589\,\text{Nu}_b + 2.36\,\text{Nu}_s]k(T - 295)$$

A complication exists in solving this equation, because the unknown quantity is surface temperature. The procedure to be used is trial and error, where, initially, a surface temperature value is assumed for property evaluation, and then solved for T. This new value for surface temperature will then be used and the procedure continued until the resulting temperature agrees with the value used in finding fluid properties.

To begin the problem, we assume that $T_{\text{surface}} = 38.5\,\text{K}$. Properties will thus be evaluated at $T_f = 340\,\text{K}$. For air at 340 K, $\nu = 1.955 \times 10^{-5}\,\text{m}^2/\text{s}$, $k = 0.0293\,\text{W/m}\cdot\text{K}$, $\alpha = 2.80 \times 10^{-5}\,\text{m}^2/\text{s}$, $\text{Pr} = 0.699$, and $\beta g/\nu^2 = 0.750 \times 10^8\,1/\text{K}\cdot\text{m}^3$.

For the vertical surface,

$$\text{Gr} = \frac{\beta g}{\nu^2}L^3\Delta T$$

$$= (0.750 \times 10^8\,1/\text{K}\cdot\text{m}^3)(1.2\,\text{m})^3(90\,\text{K})$$

$$= 11.7 \times 10^9$$

According to equation (20-6), the effect of curvature may be neglected if

$$\frac{D}{L} \geq \frac{35}{(11.7 \times 10^9)^{1/4}} = 0.106$$

In the present case $D/L = 0.75/1.2 = 0.625$; thus, the vertical surface will be treated using equations for a plane wall.

Values must now be determined for Nu_t, Nu_b, and Nu_s. Equations (20-8), (20-9), and (20-4) will be employed. The three values for Nu are determined as follows:

Nu_t:

$$L = A/p = \frac{\pi D^2/4}{\pi D} = \frac{D}{4} = 0.1875\,\text{m}$$

$$\text{Nu}_t = 0.14[(0.750 \times 10^8)(0.1875)^3(90)(0.699)]^{1/3}$$

$$= 44.0$$

Nu_b:

$$\text{Nu}_b = 0.27[(0.750 \times 10^8)(0.1875)^3(90)(0.699)]^{1/4}$$

$$= 20.2$$

Nu_s:

$$\text{Nu}_s = \left\{ 0.825 + \frac{0.387[(0.750 \times 10^8)(1.2)^3(90)(0.699)]^{1/6}}{\left[1 + \left(\frac{0.492}{0.699}\right)^{9/16}\right]^{8/27}} \right\}^2 = 236$$

The solution for T is now

$$T = 295 + \frac{1500\,\text{W}}{[0.589(44) + 0.589(20.2) + 2.36(236)](0.0293)}$$

$$= 381.1\,\text{K}$$

Using this as the new estimate for T_{surface}, we have a film temperature, $T_f \cong 338\,\text{K}$. Air properties at this temperature are

$$\nu = 1.936 \times 10^{-5}\,\text{m}^2/\text{s} \quad k = 0.0291\,\text{W/m}\cdot\text{K}$$

$$\alpha = 2.77 \times 10^{-5}\,\text{m}^2/\text{s} \quad \text{Pr} = 0.699 \quad \beta g/\nu^2 = 0.775 \times 10^8\,1/\text{K}\cdot\text{m}^3$$

The new values for Nu become

$$\text{Nu}_t = 0.14[(0.775 \times 10^8)(0.1875)^3(86)(0.699)]^{1/3} = 43.8$$

$$\text{Nu}_t = 0.27[(0.775 \times 10^8)(0.1875)^3(86)(0.699)]^{1/4} = 20.10$$

$$\text{Nu}_s = \left\{ 0.825 + \frac{0.387[0.775 \times 10^8(1.2)^3(86)(0.699)]^{1/6}}{[1 + (0.492/0.699)^{9/16}]^{8/27}} \right\}^2$$

$$= 235$$

The revised value for T_s is now

$$T_s = 295 + \frac{1500/0.0293}{[0.589(43.8) + 0.589(20.1) + 2.36(235)]}$$

$$= 381.4 \text{ K}$$

This result is obviously close enough, and the desired result for surface temperature is

$$T_{\text{surface}} \cong 381 \text{ K}$$

▶ **20.2**

FORCED CONVECTION FOR INTERNAL FLOW

Undoubtedly the most important convective heat-transfer process from an industrial point of view is that of heating or cooling a fluid that is flowing inside a closed conduit. The momentum transfer associated with this type of flow is studied in Chapter 13. Many of the concepts and terminology of that chapter will be used in this section without further discussion.

Energy transfer associated with forced convection inside closed conduits will be considered separately for laminar and turbulent flow. The reader will recall that the critical Reynolds number for conduit flow is approximately 2300.

Laminar Flow The first analytical solution for laminar flow forced convection inside tubes was formulated by Graetz[17] in 1885. The assumptions basic to the Graetz solution are as follows:

1. The velocity profile is parabolic and fully developed before any energy exchange between the tube wall and the fluid occurs.

2. All properties of the fluid are constant.

3. The surface temperature of the tube is constant at a value T_s during the energy transfer.

Considering the system as depicted in Figure 20.5, we may write the velocity profile as

$$v_x = v_{\max} \left[1 - \left(\frac{r}{R} \right)^2 \right] \tag{8-7}$$

[17] L. Graetz, Ann. *Phys. u. Chem.*, **25**, 337 (1885).

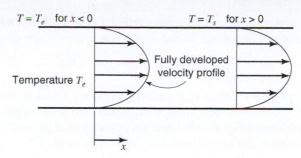

$T = T_e$ for $x < 0$ $T = T_s$ for $x > 0$

Fully developed
velocity profile

Temperature T_e

Figure 20.5 Boundary and flow
conditions for the Graetz solution.

or, recalling that $v_{\max} = 2v_{\text{avg}}$, we may write

$$v_x = 2v_{\text{avg}}\left[1 - \left(\frac{r}{R}\right)^2\right] \tag{20-21}$$

The applicable form of the energy equation written in cylindrical coordinates, assuming radial symmetry, and neglecting $\partial^2 T/\partial x^2$ (axial conduction) in comparison to the radial variation in temperature, is

$$v_x \frac{\partial T}{\partial x} = \alpha\left[\frac{1}{r}\frac{\partial}{\partial r}\left(r\frac{\partial T}{\partial r}\right)\right] \tag{20-22}$$

Substituting equation (20-21) for v_x into equation (20-22) gives

$$2v_{\text{avg}}\left[1 - \left(\frac{r}{R}\right)^2\right]\frac{\partial T}{\partial x} = \alpha\left[\frac{1}{r}\frac{\partial}{\partial r}\left(r\frac{\partial T}{\partial r}\right)\right] \tag{20-23}$$

which is the equation to be solved subject to the boundary conditions

$$
\begin{aligned}
T &= T_e &&\text{at } x = 0 &&\text{for } 0 \leq r \leq R\\
T &= T_s &&\text{at } x > 0, &&\qquad r = R
\end{aligned}
$$

and

$$\frac{\partial T}{\partial r} = 0 \quad \text{at} \quad x > 0, \quad r = 0$$

The solution to equation (20-23) takes the form

$$\frac{T - T_e}{T_s - T_e} = \sum_{n=0}^{\infty} c_n f\left(\frac{r}{R}\right)\exp\left[-\beta_n^2 \frac{\alpha}{Rv_{\text{avg}}}\frac{x}{R}\right] \tag{20-24}$$

The terms c_n, $f(r/R)$, and β_n are all coefficients to be evaluated by using appropriate boundary conditions.

The argument of the exponential, exclusive of β_n—that is, $(\alpha/Rv_{\text{avg}})(x/R)$—may be rewritten as

$$\frac{4}{(2Rv_{\text{avg}}/\alpha)(2R/x)} = \frac{4}{(Dv_{\text{avg}}\rho/\mu)(c_p\mu/k)(D/x)}$$

or, in terms of dimensionless parameters already introduced, this becomes

$$\frac{4}{\text{Re Pr } D/x} = \frac{4x/D}{\text{Pe}}$$

The product of Re and Pr is often referred to as the *Peclet number*, Pe. Another parameter encountered in laminar forced convection is the *Graetz number*, Gz, defined as

$$\text{Gz} \equiv \frac{\pi}{4} \frac{D}{x} \text{Pe}$$

Detailed solutions of equation (20-24) are found in the literature, and Knudsen and Katz[18] summarize these quite well. Figure 20.6 presents the results of the Graetz solution graphically for two different boundary conditions at the wall, these being (1) a constant wall temperature and (2) uniform heat input at the wall.

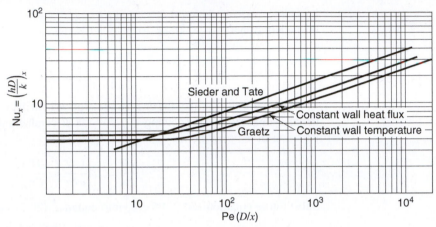

Figure 20.6 Variation in the local Nusselt number for laminar flow in tubes.

Note that, in Figure 20.6, the analytical results approach constant limiting values for large values of x. These limits are

$$\text{Nu}_x = 3.658 \qquad \text{for } T_{\text{wall}} = \text{constant} \tag{20-25}$$

$$\text{Nu}_x = 4.364 \qquad \text{for } q/A_{\text{wall}} = \text{constant} \tag{20-26}$$

Experimental data for laminar flow in tubes have been correlated by Sieder and Tate[19] by the equation

$$\text{Nu}_D = 1.86 \left(\text{Pe} \frac{D}{L} \right)^{1/3} \left(\frac{\mu_b}{\mu_w} \right)^{0.14} \tag{20-27}$$

The Sieder–Tate relation is also shown in Figure 20.6 along with the two Graetz results. These results cannot be compared directly because the Graetz results yield local values of h_x and the Sieder–Tate equation gives mean values of the heat-transfer coefficient. The last part of equation (20-27), the ratio of the fluid viscosity at the arithmetic-mean bulk temperature to that at the temperature of the wall, takes into account the significant effect that variable fluid viscosity has on the heat-transfer rate. All properties other than μ_w are evaluated at the bulk fluid temperature.

Turbulent Flow When considering energy exchange between a conduit surface and a fluid in turbulent flow, we must resort to correlations of experimental data as suggested by

---

[18] J. G. Knudsen and D. L. Katz, *Fluid Dynamics and Heat Transfer*, McGraw-Hill Book Company, New York, 1958, p. 370.
[19] F. N. Sieder and G. E. Tate, *Ind. Eng. Chem.*, **28**, 1429 (1936).

dimensional analysis. The three most-used equations of this nature and the restrictions on their use are as follows.

Dittus and Boelter[20] proposed the following equation of the type suggested earlier by dimensional analysis, equation (19-7):

$$\mathrm{Nu}_D = 0.023\,\mathrm{Re}_D^{0.8}\mathrm{Pr}^n \tag{20-28}$$

where

1. $n = 0.4$ if the fluid is being heated, $n = 0.3$ if the fluid is being cooled
2. All fluid properties are evaluated at the arithmetic-mean bulk temperature
3. The value of Re_D should be $> 10^4$
4. Pr is in the range $0.7 < \mathrm{Pr} < 100$
5. $L/D > 60$

Colburn[21] proposed an equation using the Stanton number, St, in place of Nu_D as related in equation (19-9). His equation is

$$\mathrm{St} = 0.023\,\mathrm{Re}_D^{-0.2}\mathrm{Pr}^{-2/3} \tag{20-29}$$

where

1. Re_D and Pr are evaluated at the *film* temperature, and St is evaluated at the bulk temperature
2. Re_D, Pr, and L/D should have values within the following limits

$$\mathrm{Re}_D > 10^4 \quad 0.7 < \mathrm{Pr} < 160 \quad \text{and} \quad L/D > 60$$

To account for high Prandtl number fluids, such as oils, Sieder and Tate[22] proposed the equation

$$\mathrm{St} = 0.023\,\mathrm{Re}_D^{-0.2}\mathrm{Pr}^{-2/3}\left(\frac{\mu_b}{\mu_w}\right)^{0.14} \tag{20-30}$$

where

1. All fluid properties except μ_w are evaluated at bulk temperature
2. $\mathrm{Re}_D > 10^4$
3. $0.7 < \mathrm{Pr} < 17\,000$

and

4. $L/D > 60$

Of the three equations presented, the first two are most often used for those fluids whose Prandtl numbers are within the specified range. The Dittus–Boelter equation is simpler to use than the Colburn equation because of the fluid property evaluation at the bulk temperature.

[20] F. W. Dittus and L. M. K. Boelter, University of California, *Publ. Eng.*, **2**, 443 (1930).
[21] A. P. Colburn, *Trans. A.I.Ch.E.*, **29**, 174 (1933).
[22] E. N. Sieder and G. E. Tate, *Ind. Eng. Chem.*, **28**, 1429 (1936).

The following examples illustrate the use of some of the expressions presented in this section.

Example 2

Hydraulic fluid (MIL-M-5606), in fully developed flow, flows through a 2.5-cm-diameter copper tube that is 0.61 m long, with a mean velocity of 0.05 m/s. The oil enters at 295 K. Steam condenses on the outside tube surface with an effective heat-transfer coefficient of 11,400 W/m² · K. Find the rate of heat transfer to the oil.

To evaluate oil properties at either the film temperature or the mean bulk temperature, we need to know the existing oil temperature. Equation (19-61) applies in this case:

$$\ln\frac{T_L - T_s}{T_o - T_s} + 4\frac{L}{D}\frac{h}{\rho v c_p} = 0 \tag{19-62}$$

If the thermal resistance of the copper tube wall is negligible, the heat-transfer rate can be expressed as

$$q = \frac{A_{\text{surf}}(T_{\text{stm}} - T_{\text{oil}})}{1/h_i + 1/h_o} = \rho A v c_p(T_L - T_o)$$

To get an indication of whether the flow is laminar or turbulent, we will assume a bulk oil temperature of 300 K. The Reynolds number, at this temperature, is

$$\mathrm{Re}_D = \frac{(0.025\ \text{m})(0.05\ \text{m/s})}{9.94 \times 10^{-6}\text{m}^2/\text{s}} = 126$$

and the flow is clearly laminar. The heat-transfer coefficient on the oil side can then be determined using equation (20-27)

$$h_i = \frac{k}{D}\mathrm{Nu}_D = \frac{k}{D}1.86\left(\mathrm{Pe}\frac{D}{L}\right)^{0.33}\left(\frac{\mu_b}{\mu_w}\right)^{0.14}$$

Initially, the bulk temperature of the oil and the wall temperature will be assumed to be 300 and 372 K, respectively. Using fluid properties at these temperatures, we have

$$h_i = \frac{(0.123\ \text{W/m}\cdot\text{K})(1.86)}{0.025\ \text{m}}\left[(126)(155)\frac{0.025}{0.61}\right]^{0.33}\left(\frac{1.036 \times 10^{-4}}{3.72 \times 10^{-3}}\right)^{0.14}$$

$$= 98.1\ \text{W/m}^2\cdot\text{K}$$

Substituting into equation (19-61), we get

$$\ln\frac{T_s - T_L}{T_s - T_o} = -4\left(\frac{0.61\ \text{m}}{0.025\ \text{m}}\right)\frac{98.1\ \text{W/m}^2\cdot\text{K}}{(843\ \text{kg/m}^3)(0.05\ \text{m/s})(1897\ \text{J/kg}\cdot\text{K})}$$

$$= -0.120$$

$$\frac{T_s - T_L}{T_s - T_o} = e^{-0.120} = 0.887$$

$$T_L = 372 - 0.887(372 - 295)$$

$$= 304\ \text{K}$$

With this value of T_L, the mean bulk temperature of the oil is

$$T_b = \frac{295 + 304}{2} = 299.5\ \text{K}$$

which is sufficiently close to the initial assumption that there is no need to iterate further.

With an exiting temperature of 304 K, the heat-transfer rate to the oil is

$$q = \rho A v c_p (T_L - T_o)$$

$$= (843 \text{ kg/m}^3)\left(\frac{\pi}{4}\right)(0.025 \text{ m})^2(0.05 \text{ m/s})(1897 \text{ J/kg} \cdot \text{K})(9 \text{ K})$$

$$= 353 \text{ W}$$

Example 3

Air at 1 atm and a temperature of 290 K enters a 1.27-cm-ID tube at a velocity of 24 m/s. The wall temperature is maintained at 372 K by condensing steam. Evaluate the convective heat-transfer coefficient for this situation if the tube is 1.52 m long.

As in the previous example, it will be necessary to evaluate the exiting air temperature by

$$\ln \frac{T_L - T_s}{T_o - T_s} + 4 \frac{L}{D} \frac{h}{\rho v c_p} = 0 \tag{19-62}$$

To determine the type of flow, we first evaluate the Reynolds number at the tube entrance.

$$\text{Re} = \frac{Dv}{\nu} = \frac{(0.0127 \text{ m})(24 \text{ m/s})}{1.478 \times 10^{-5} \text{Pa} \cdot \text{s}} = 20{,}600$$

Flow is clearly turbulent and Re is sufficiently large that equation (20-28), (20-29), or (20-30) may be used.

Equation (20-29) will be used. An exit temperature of 360 K will be assumed; the corresponding mean bulk temperature is 325 K and $T_f = 349$. We now have

$$\text{St} = \frac{h}{\rho v c_p} = 0.023 \, \text{Re}^{-0.2} \text{Pr}^{-2/3}$$

$$= 0.023 \left[\frac{(0.0127)(24)}{2.05 \times 10^{-5}}\right]^{-0.2} (0.697)^{-2/3}$$

$$= 0.00428$$

Substituting into equation (19-61), we have

$$\frac{T_L - T_s}{T_o - T_s} = \exp\left[-4\left(\frac{1.52 \text{ m}}{0.0127 \text{ m}}\right)(0.00428)\right]$$

$$= 0.129$$

and the calculated value of T_L is

$$T_L = 372 - (0.129)(372 - 290)$$

$$= 361 \text{ K}$$

This value agrees closely with the initially assumed value for T_L, so there is no need to perform a second calculation. The heat-transfer coefficient is now evaluated as

$$h = \rho v c_p \text{St}$$

$$= (1.012 \text{ kg/m}^3)(24 \text{ m/s})(1009 \text{ J/kg} \cdot \text{K})(0.00428)$$

$$= 105 \text{ W/m}^2 \cdot \text{K}$$

For flow in short passages the correlations presented thus far must be modified to account for variable velocity and temperature profiles along the axis of flow. Deissler[23] has

---

[23] R. G. Deissler, *Trans. A.S.M.E.*, **77**, 1221 (1955).

analyzed this region extensively for the case of turbulent flow. The following equations may be used to modify the heat-transfer coefficients in passages for which $L/D < 60$:

for $2 < L/D < 20$

$$\frac{h_L}{h_\infty} = 1 + (D/L)^{0.7} \tag{20-31}$$

and for $20 < L/D < 60$

$$\frac{h_L}{h_\infty} = 1 + 6\,D/L \tag{20-32}$$

Both of these expressions are approximations relating the appropriate coefficient, h_L, in terms of h_∞, where h_∞ is the value calculated for $L/D > 60$.

▶ 20.3

FORCED CONVECTION FOR EXTERNAL FLOW

Numerous situations exist in practice in which one is interested in analyzing or describing heat transfer associated with the flow of a fluid past the exterior surface of a solid. The sphere and cylinder are the shapes of greatest engineering interest, with heat transfer between these surfaces and a fluid in crossflow frequently encountered.

The reader will recall the nature of momentum-transfer phenomena discussed in Chapter 12 relative to external flow. The analysis of such flow and of heat transfer in these situations is complicated when the phenomenon of boundary-layer separation is encountered. Separation will occur in those cases in which an adverse pressure gradient exists; such a condition will exist for most situations of engineering interest.

Flow Parallel to Plane Surfaces This condition is amenable to analysis and has already been discussed in Chapter 19. The significant results are repeated here for completeness.

We recall that in this case, the boundary-layer flow regimes are laminar for $\mathrm{Re}_x < 2 \times 10^5$ and turbulent for $3 \times 10^6 < \mathrm{Re}_x$. For the laminar range

$$\mathrm{Nu}_x = 0.332\,\mathrm{Re}_x^{1/2}\,\mathrm{Pr}^{1/3} \tag{19-25}$$

and

$$\mathrm{Nu}_L = 0.664\,\mathrm{Re}_L^{1/2}\,\mathrm{Pr}^{1/3} \tag{19-26}$$

With turbulent flow in the boundary layer, an application of the Colburn analogy

$$\mathrm{St}_x\mathrm{Pr}^{2/3} = \frac{C_{fx}}{2} \tag{19-37}$$

along with equation (12-73) yields

$$\mathrm{Nu}_x = 0.0288\,\mathrm{Re}_x^{4/5}\,\mathrm{Pr}^{1/3} \tag{20-33}$$

A mean Nusselt number can be calculated using this expression for Nu_x. The resulting expression is

$$\mathrm{Nu}_L = 0.036\,\mathrm{Re}_L^{4/5}\,\mathrm{Pr}^{1/3} \tag{20-34}$$

Fluid properties should be evaluated at the film temperature when using these equations.

Cylinders in Crossflow　Eckert and Soehngen[24] evaluated local Nusselt numbers at various positions on a cylindrical surface past, which flowed an air stream with a range in Reynolds numbers from 20 to 600. Their results are shown in Figure 20.7. A much higher Reynolds number range was investigated by Giedt,[25] whose results are shown in Figure 20.8.

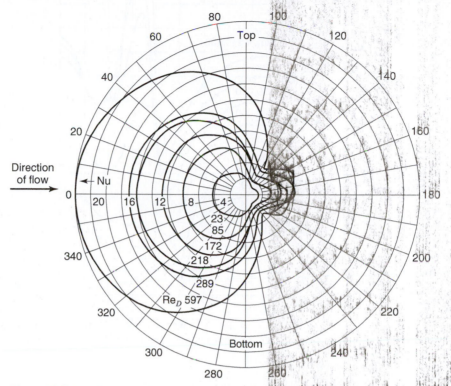

Figure 20.7 Local numbers for crossflow about a circular cylinder at low Reynolds numbers. (From E. R. G. Eckert and E. Soehngen, *Trans. A.S.M.E.*, **74**, 346 (1952). By permission of the publishers.)

Figures 20.7 and 20.8 show a smooth variation in the Nusselt number near the stagnation point. At low Reynolds numbers, the film coefficient decreases almost continuously from the stagnation point, the only departure being a slight rise in the separated-wake region of the cylinder. At higher Reynolds numbers, as illustrated in Figure 20.7, the film coefficient reaches a second maximum, which is greater than the stagnation-point value. The second peak in the Nusselt number at high Reynolds numbers is due to the fact that the boundary layer undergoes transition from laminar to turbulent flow. In the bottom curves of Figure 20.7, the laminar boundary layer separates from the cylinder near 80 degrees from the stagnation point, and no large change in the Nusselt number occurs. The effect of higher Reynolds number is twofold. First, the separation point moves past 90 degrees as the boundary layer becomes turbulent; thus, less of the cylinder is engulfed in the wake. A second effect is that the Nusselt number reaches a value that is higher than the stagnation-point value. The increase is due to the greater conductance of the turbulent boundary layer.

It is quite apparent from the figures that the convective heat-transfer coefficient varies in an irregular, complex manner in external flow about a cylinder. It is likely, in practice,

---

[24] E. R. G. Eckert and E. Soehngen, *Trans. A.S.M.E.*, **74**, 343 (1952).
[25] W. H. Giedt, *Trans. A.S.M.E.*, **71**, 378 (1949).

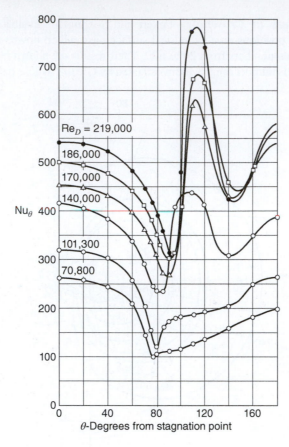

Figure 20.8 Local Nusselt numbers for crossflow about a circular cylinder at high Reynolds numbers. (From W. H. Giedt, *Trans. A.S.M.E.*, **71**, 378 (1949). By permission of the publishers.)

that an average h for the entire cylinder is desired. McAdams[26] has plotted the data of 13 separate investigations for the flow of air normal to single cylinders and found excellent agreement when plotted as Nu_D vs. Re_D. His plot is reproduced in Figure 20.9. Note that values of Nu_D are for $Pr = 1$. For other fluids, a correction factor, $Pr^{1/3}$, should be employed—that is, $Nu_D = Nu_{D(\text{figure})}Pr^{1/3}$.

A widely used correlation for these data is of the form

$$Nu_D = B\,Re^n\,Pr^{1/3} \tag{20-35}$$

where the constants B and n are functions of the Reynolds number. Values for these constants are given in Table 20.3. The film temperature is appropriate for physical property evaluation.

Churchill and Bernstein[27] have recommended a single correlating equation covering conditions for which $Re_D Pr > 0.2$. This correlation is expressed in equation (20-36).

$$Nu_D = 0.3 + \frac{0.62\,Re_D^{1/2}Pr^{1/3}}{[1 + (0.4/Pr)^{2/3}]^{1/4}}\left[1 + \left(\frac{Re_D}{282{,}000}\right)^{5/8}\right]^{4/5} \tag{20-36}$$

[26] W. H. McAdams, *Heat Transmission*, Third Edition, McGraw-Hill Book Company, New York, 1949.

[27] S. W. Churchill and M. Bernstein, *J. Heat Transfer*, **99**, 300 (1977).

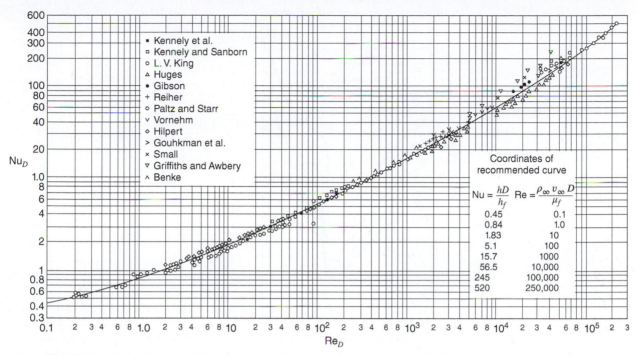

Figure 20.9 Nu vs. Re for flow normal to single cylinders. Note: These values are strictly valid when Pr ≅ 1. (From W. H. McAdams, *Heat Transmission*, 3rd edition, McGraw-Hill Book Company, New York, 1954, p. 259.)

Table 20.3   Values of B and n for use in equation (20-35)

Re_D	B	n
0.4/–4	0.989	0.330
4–40	0.911	0.385
40–4000	0.683	0.466
4000–40,000	0.193	0.618
40,000–400,000	0.027	0.805

Single Spheres   Local convective heat-transfer coefficients at various positions relative to the forward stagnation point for flow past a sphere are plotted in Figure 20.10, following the work of Cary.[28]

McAdams[29] has plotted the data of several investigators relating Nu_D vs. Re_D for air flowing past spheres. His plot is duplicated in Figure 20.11.

A recent correlation proposed by Whitaker[30] is recommended for the following conditions: $0.71 < Pr < 380$, $3.5 < Re_D < 7.6 \times 10^4$, $1.0 \times \mu_\infty/\mu_s < 3.2$. All properties

---

[28] J. R. Cary, *Trans. A.S.M.E.*, **75**, 483 (1953).
[29] W. H. McAdams, *op. cit.*
[30] S. Whitaker, *A.I.Ch.E.J.*, **18**, 361 (1972).

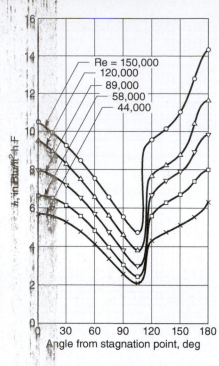

Figure 20.10 Local heat-transfer coefficients for flow past a sphere. (From J. R. Cary, *Trans. A.S.M.E.*, **75**, 485 (1953). By permission of the publishers.)

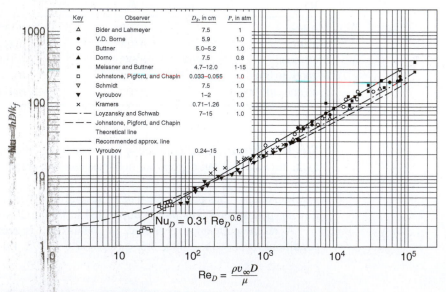

Figure 20.11 Nu_D vs. Re_D for air flow past single spheres. (From McAdams, *Heat Transmission*, 3rd edition, McGraw-Hill Book Company, New York, 1954, p. 266. By permission of the publishers.)

are evaluated at T_∞ except for μ_s, which is the value at the surface temperature. Whitaker's correlation is

$$\text{Nu}_D = 2 + (0.4\,\text{Re}_D^{1/2} + 0.06\,\text{Re}_D^{2/3})\text{Pr}^{0.4}(\mu_\infty/\mu_s)^{1/4} \tag{20-37}$$

An important case is that of falling liquid drops, modeled as spheres. The correlation of Ranz and Marshall,[31] for this case, is

$$\text{Nu}_D = 2 + 0.6\,\text{Re}_D^{1/2}\,\text{Pr}^{1/3} \tag{20-38}$$

Tube Banks in Crossflow   When a number of tubes are placed together in a bank or bundle, as might be encountered in a heat exchanger, the effective heat-transfer coefficient is affected by the tube arrangement and spacing, in addition to those factors already considered for flow past single cylinders. Several investigators have made significant contributions to the analysis of these configurations.

As fluid flow through and past tube bundles involves an irregular flow path, some investigators have chosen significant lengths other than D, the tube diameter, to use in calculating Reynolds numbers. One such term is the equivalent diameter of a tube bundle D_{eq}, defined as

$$D_{eq} = \frac{4(S_L S_T - \pi D^2/4)}{\pi D} \tag{20-39}$$

where S_L is the center-to-center distance between tubes *along* the direction of flow, S_T is the center-to-center distance between tubes *normal* to the flow direction, and D is the OD of a tube.

Bergelin, Colburn, and Hull[32] studied the flow of liquids past tube bundles in the region of laminar flow with $1 < \text{Re} < 1000$. Their results, plotted as $\text{St}\,\text{Pr}^{2/3}(\mu_w/\mu_b)^{0.14}$ vs. Re for various configurations, are presented in Figure 20.12. In that figure all fluid properties except μ_w are evaluated at the average bulk temperature.

For liquids in transition flow across tube bundles, Bergelin, Brown, and Doberstein[33] extended the work just mentioned for five of the tube arrangements to include values of Re up to 10^4. Their results are presented both for energy transfer and friction factor vs. Re in Figure 20.13.

In addition to the greater Reynolds number range, Figure 20.13 involves Re calculated by using the tube diameter, D, as opposed to Figure 20.12, in which D_{eq}, defined by equation (20-37), was used.

Windchill Equivalent Temperature   The reader is undoubtedly familiar with weather reports where, in addition to measured air temperatures, an indication of how cold one actually feels is expressed as the *windchill equivalent temperature*. This temperature indicates how cold the wind actually makes one feel; the stronger the wind blows, the colder the air feels to the human body.

The determination of windchill equivalent temperature is an interesting example of the combined effects of convective and conductive heat transfer between the body and the adjacent air. For a complete explanation of the modeling used in determining this quantity,

[31] W. Ranz and W. Marshall, *Chem. Engr. Progr.*, **48**, 141 (1952).
[32] O. P. Bergelin, A. P. Colburn, and H. L. Hull, Univ. Delaware, Eng. Expt. Sta. Bulletin No. 2 (1950).
[33] O. P. Bergelin, G. A. Brown, and S. C. Doberstein, *Trans. A.S.M.E.*, **74**, 953 (1952).

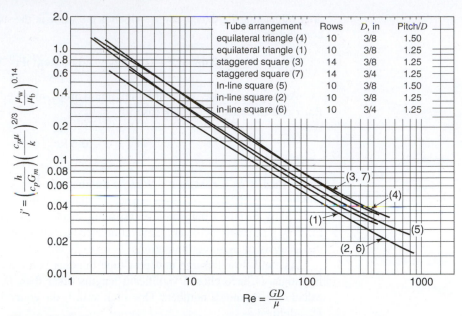

Figure 20.12 Convective heat-transfer exchange between liquids in laminar flow and tube bundles. (From O. P. Bergelin, A. P. Colburn, and H. L. Hull, Univ. of Delaware, Engr. Dept., Station Bulletin 10.2, 1950, p. 8. By permission of the publishers.)

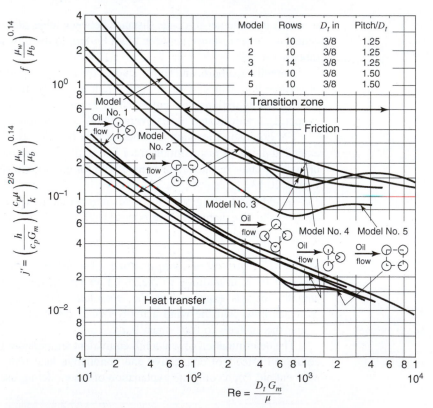

Figure 20.13 Energy transfer and frictional loss for liquids in laminar and transition flow past tube bundles. (From O. P. Bergelin, G. A. Brown, and S. C. Doberstein, *Trans. A.S.M.E.*, **74**, 1958 (1952). By permission of the publishers.)

Table 20.4 Windchill equivalent temperature—English units

		35	30	25	20	15	10	5	0	−5	−10	−15	−20	−25
						Air temperature (F)								
Wind speed (mi/h)	5	32	27	22	16	11	6	0	−5	−10	−15	−21	−26	−31
	10	22	16	10	3	−3	−9	−15	−22	−27	−34	−40	−46	−52
	15	16	9	2	−5	−11	−18	−25	−31	−38	−45	−51	−58	−65
	20	12	4	−3	−10	−17	−24	−31	−39	−46	−53	−60	−67	−74
	25	8	1	−7	−15	−22	−29	−36	−44	−51	−59	−66	−74	−81
	30	6	−2	−10	−18	−25	−33	−41	−49	−56	−64	−71	−79	−86
	35	4	−4	−12	−20	−27	−35	−43	−52	−58	−67	−74	−82	−89
	40	3	−5	−13	−21	−29	−37	−45	−53	−60	−69	−76	−84	−92

Table 20.5 Windchill equivalent temperature—SI units

		8	4	0	−4	−8	−12	−16	−20	−24	−28	−32	−36	−40
						Air temperature (C)								
Wind speed (km/h)	0	8	4	0	−4	−8	−12	−16	−20	−24	−28	−32	−36	−40
	10	5	0	−4	−8	−13	−17	−22	−26	−31	−35	−40	−44	−49
	20	0	−5	−10	−15	−21	−26	−31	−36	−42	−47	−52	−57	−63
	30	−3	−8	−14	−20	−25	−31	−37	−43	−48	−54	−60	−65	−71
	40	−5	−11	−17	−23	−29	−35	−41	−47	−53	−59	−65	−71	−77
	50	−6	−12	−18	−25	−31	−37	−43	−49	−56	−62	−68	−74	−80
	60	−7	−13	−19	−26	−32	−39	−45	−51	−58	−64	−70	−77	−83

the reader is referred to a 1971 paper by Steadman.[34] Tables 20.4 and 20.5 provide values of the windchill equivalent temperature as a function of air temperature and wind speed, in English and SI units, respectively.

▶ 20.4

CLOSURE

Many of the more useful experimentally developed correlations for predicting convective heat-transfer coefficients have been presented in this chapter. Those graphs and equations presented are a small part of the information of this type available in the literature. The information included should, in any case, allow most of the more common convection heat-transfer coefficients to be predicted with some confidence.

The convection phenomena considered have included the following:

1. *Natural convection* past vertical and horizontal surfaces, plus some useful simplified expressions for air

2. *Forced convection for internal flow*, including laminar and turbulent flow correlations

[34] R. G. Steadman, Indices of windchill of clothed persons, *J. App. Meteorol*, **10**, 674–683 (1971).

3. *Forced convection for external flow* with cylinders, spheres, and tube bundles being the types of surfaces so considered

The reader is reminded to observe any special considerations relative to the equations and plots in this chapter. Such considerations include whether to evaluate fluid properties at the bulk or film temperature, what significant length is used in a given correlation, and what is the allowable Prandtl and Reynolds number range for a given set of data.

CHAPTER 21

Boiling and Condensation

$\mathbf{E}$nergy-transfer processes associated with the phenomena of boiling and condensation may achieve relatively high heat-transfer rates, whereas the accompanying temperature differences may be quite small. These phenomena, associated with the change in phase between a liquid and a vapor, are more involved and thus more difficult to describe than the convective heat-transfer processes discussed in the preceding chapters. This is due to the additional considerations of surface tension, latent heat of vaporization, surface characteristics, and other properties of two-phase systems that were not involved in the earlier considerations. The processes of boiling and condensation deal with opposite effects relative to the change in phase between a liquid and its vapor. These phenomena will be considered separately in the following sections.

▶ **21.1**

BOILING

Boiling heat transfer is associated with a change in phase from liquid to vapor. Extremely high heat fluxes may be achieved in conjunction with boiling phenomena, making the application particularly valuable where a small amount of space is available to accomplish a relatively large energy transfer. One such application is the cooling of nuclear reactors. Another is the cooling of electronic devices where space is very critical. The advent of these applications has spurred the interest in boiling, and concentrated research in this area in recent years has shed much light on the mechanism and behavior of the boiling phenomenon.

There are two basic types of boiling: *pool boiling* and *flow boiling*. Pool boiling occurs on a heated surface submerged in a liquid pool that is not agitated. Flow boiling occurs in a flowing stream, and the boiling surface may itself be a portion of the flow passage. The flow of liquid and vapor associated with flow boiling is an important type of two-phase flow.

Regimes of Boiling An electrically heated horizontal wire submerged in a pool of water at its saturation temperature is a convenient system to illustrate the regimes of boiling heat transfer. A plot of the heat flux associated with such a system as the ordinate vs. the temperature difference between the heated surface and saturated water is depicted in Figure 21.1. There are six different regimes of boiling associated with the behavior exhibited in this figure.

In regime I, the wire surface temperature is only a few degrees higher than that of the surrounding saturated liquid. Natural convection currents circulate the superheated liquid, and evaporation occurs at the free liquid surface as the superheated liquid reaches it.

An increase in wire temperature is accompanied by the formation of vapor bubbles on the wire surface. These bubbles form at certain surface sites, where vapor bubble nuclei are

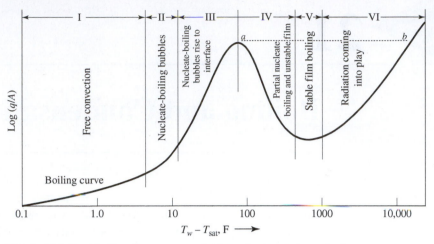

Figure 21.1 Pool boiling in water on a horizontal wire at atmospheric pressure.

present, break off, rise, and condense before reaching the free liquid surface. This is the process occurring in regime II.

At a still higher wire surface temperature, as in regime III, larger and more numerous bubbles form, break away from the wire surface, rise, and reach the free surface. Regimes II and III are associated with *nucleate boiling*.

Beyond the peak of this curve, the transition boiling regime is entered. This is region IV on the curve. In this regime, a vapor film forms around the wire, and portions of this film break off and rise, briefly exposing a portion of the wire surface. This film collapse and reformation, and this unstable nature of the film, is characteristic of the transition regime. When present, the vapor film provides a considerable resistance to heat transfer; thus, the heat flux decreases.

When the surface temperature reaches a value of approximately 400 F above the saturated liquid, the vapor film around the wire becomes stable. This is region V, the *stable film-boiling* regime.

For surface temperatures of 1000 F or greater above that of the saturated liquid, radiant energy transfer comes into play, and the heat flux curve rises once more. This is designated as region VI in Figure 21.1.

The curve in Figure 21.1 can be achieved if the energy source is a condensing vapor. If, however, electrical heating is used, then regime IV will probably not be obtained because of wire "burnout." As the energy flux is increased, ΔT increases through regions I, II, and III. When the peak value of q/A is exceeded slightly, the required amount of energy cannot be transferred by boiling. The result is an increase in ΔT accompanied by a further decrease in the possible q/A. This condition continues until point b is reached. As ΔT at point b is extremely high, the wire will long since have reached its melting point. Point a on the curve is often referred to as the "burnout point" for these reasons.

Note the somewhat anomalous behavior exhibited by the heat flux associated with boiling. One normally considers a flux to be proportional to the driving force; thus, the heat flux might be expected to increase continuously as the temperature difference between the heated surface and the saturated liquid increases. This, of course, is not the case; the very high heat fluxes associated with moderate temperature differences in the nucleate-boiling regime are much higher than the heat fluxes resulting from much higher temperature differences in the film-boiling regime. The reason for this is the presence of the vapor film, which covers and insulates the heating surface in the latter case.

Correlations of Boiling Heat-Transfer Data As the fluid behavior in a boiling situation is very difficult to describe, there is no adequate analytical solution available for boiling transfer. Various correlations of experimental data have been achieved for the different boiling regimes; the most useful of these follow.

In the natural convection regime, *regime I* of Figure 21.1, the correlations presented in Chapter 20 for natural convection may be used.

Regime II, the regime of partial nucleate boiling and partial natural convection, is a combination of regimes I and III, and the results for each of these two regimes may be superposed to describe a process in regime II.

The nucleate-boiling regime, *regime III*, is of great engineering importance because of the very high heat fluxes possible with moderate temperature differences. Data for this regime are correlated by equations of the form

$$\text{Nu}_b = \phi(\text{Re}_b, \text{Pr}_L) \tag{21-1}$$

The parameter Nu_b in equation (21-1) is a Nusselt number defined as

$$\text{Nu}_b \equiv \frac{(q/A)D_b}{(T_s - T_{\text{sat}})k_L} \tag{21-2}$$

where q/A is the total heat flux, D_b is the maximum bubble diameter as it leaves the surface, $T_s - T_{\text{sat}}$ is the *excess temperature* or the difference between the surface and saturated-liquid temperatures, and k_L is the thermal conductivity of the liquid. The quantity, Pr_L, is the Prandtl number for the liquid. The bubble Reynolds number, Re_b, is defined as

$$\text{Re}_b \equiv \frac{D_b G_b}{\mu_L} \tag{21-3}$$

where G_b is the average mass velocity of the vapor leaving the surface and μ_L is the liquid viscosity

The mass velocity, G_b, may be determined from

$$G_b = \frac{q/A}{h_{fg}} \tag{21-4}$$

where h_{fg} is the latent heat of vaporization.

Rohsenow[1] has used equation (21-1) to correlate Addoms's[2] pool-boiling data for a 0.024-in-diameter platinum wire immersed in water. This correlation is shown in Figure 21.2 and is expressed in equation form as

$$\frac{q}{A} = \mu_L h_{fg} \left[\frac{g(\rho_L - \rho_u)}{\sigma} \right]^{1/2} \left[\frac{c_{pL}(T_s - T_{\text{sat}})}{C_{sf} h_{fg} \, \text{Pr}_L^{1.7}} \right]^3 \tag{21-5}$$

where c_{pL} is the heat capacity for the liquid and the other terms have their usual meanings.

The coefficients C_{sf} in equation (21-5) vary with the surface-fluid combination. The curve drawn in Figure 21.2 is for $C_{sf} = 0.013$. A table of C_{sf} for various combinations of fluid and surface is presented by Rohsenow and Choi[3] and duplicated here as Table 21.1.

[1] W. M. Rohsenow, *A.S.M.E. Trans.,* **74**, 969 (1952).

[2] J. N. Addoms, D.Sc. Thesis, Chemical Engineering Department, Massachusetts Institute of Technology, June 1948.

[3] W. M. Rohsenow and H. Y. Choi, *Heat, Mass, and Momentum Transfer,* Prentice-Hall, Inc., Englewood Cliffs, N.J., 1961.

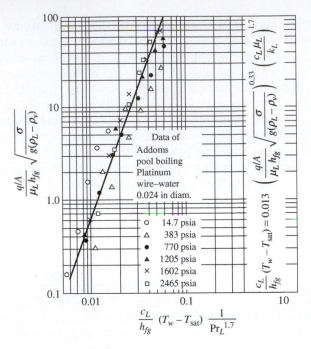

Figure 21.2 Correlation of pool-boiling data. (From W. M. Rohsenow and H. Choi, *Heat, Mass, and Momentum Transfer,* Prentice-Hall, Inc., Englewood Cliffs, N.J., 1961, p. 224. By permission of the publishers.)

Table 21.1 Values of C_{sf} for equation (21-5)

Surface-fluid combination	C_{sf}
Water/nickel	0.006
Water/platinum	0.013
Water/copper	0.013
Water/brass	0.006
CCl$_4$/copper	0.013
Benzene/chromium	0.010
n-Pentane/chromium	0.015
Ethyl alcohol/chromium	0.0027
Isopropyl alcohol/copper	0.0025
35% K$_2$CO$_3$/copper	0.0054
50% K$_2$CO$_3$/copper	0.0027
n-Butyl alcohol/copper	0.0030

From an earlier discussion, it is clear that the burnout point has considerable importance. The "critical heat flux" is the value of q/A represented by point a in Figure 21.1.

An analysis of conditions at burnout modified by experimental results is expressed in equation (21-6) as

$$q/A|_{\text{critical}} = 0.18 h_{fg}\rho_v \left[\frac{\sigma g(\rho_L - \rho_v)}{\rho_v^2} \right]^{1/4} \tag{21-6}$$

The interested reader is referred to the work of Zuber[4] for a discussion of this subject.

---

[4] N. Zuber, *Trans. A.S.M.E.*, **80**, 711 (1958).

Regime IV, that of unstable film boiling, is not of great engineering interest, and no satisfactory correlation has been found for this region as yet.

The stable-film-boiling region, *regime V*, requires high surface temperatures; thus, few experimental data have been reported for this region.

Stable film boiling on the surface of horizontal tubes and vertical plates has been studied both analytically and experimentally by Bromley.[5,6] Considering conduction alone through the film on a horizontal tube, Bromley obtained the expression

$$h = 0.62 \left[\frac{k_v^3 \rho_v (\rho_L - \rho_v) g (h_{fg} + 0.4 c_{pL} \Delta T)}{D_o \mu_v (T_s - T_{sat})} \right]^{1/4} \tag{21-7}$$

where all terms are self-explanatory except D_o, which is the outside diameter of the tube.

A modification in equation (21-7) was proposed by Berenson[7] to provide a similar correlation for stable film boiling on a horizontal surface. In Berenson's correlation, the tube diameter, D_o, is replaced by the term $[\sigma/g(\rho_L - \rho_v)]^{1/2}$, and the recommended expression is

$$h = 0.425 \left[\frac{k_{vf}^3 \rho_{vf} (\rho_L - \rho_v) g (h_{fg} + 0.4 c_{pL} \Delta T)}{\mu_{vf} (T_s - T_{sat}) \sqrt{\sigma/g(\rho_L - \rho_v)}} \right]^{1/4} \tag{21-8}$$

where k_{vf}, ρ_{vf}, and μ_{vf} are to be evaluated at the film temperature as indicated.

Hsu and Westwater[8] considered film boiling for the case of a vertical tube. Their test results were correlated by the equation

$$h \left[\frac{\mu_v^2}{g \rho_v (\rho_L - \rho_v) k_v^3} \right]^{1/3} = 0.0020 \, \mathrm{Re}^{0.6} \tag{21-9}$$

where

$$\mathrm{Re} = \frac{4\dot{m}}{\pi D_v \mu_v} \tag{21-10}$$

$\dot{m}$ being the flow rate of vapor in lb_m/h at the upper end of the tube and the other terms being identical to those in equation (21-7). Hsu[9] states that heat-transfer rates for film boiling are higher for vertical tubes than for horizontal tubes when all other conditions remain the same.

In *regime VI*, the correlations for film boiling still apply; however, the superimposed contribution of radiation is appreciable, becoming dominant at extremely high values of ΔT. Without any appreciable flow of liquid, the two contributions may be combined, as indicated by equation (21-11).

The contribution of radiation to the total heat-transfer coefficient may be expressed as

$$h = h_c \left(\frac{h_c}{h} \right)^{1/3} + h_r \tag{21-11}$$

[5] L. A. Bromley, *Chem. Eng Prog.*, **46**, (5), 221 (1950).
[6] L. A. Bromley, et al., *Ind. Eng Chem.*, **45**, 2639 (1953).
[7] P. Berenson, A.I.Ch.E. Paper No. 18, Heat Transfer Conference, Buffalo, N.Y., August 14–17, 1960.
[8] Y. Y. Hsu and J. W. Westwater, *A.I.Ch.E. J.*, **4**, 59 (1958).
[9] S. T. Hsu, *Engineering Heat Transfer*, Van Nostrand, Princeton, N.J., 1963.

where h is the total heat-transfer coefficient, h_c is the coefficient for the boiling phenomenon, and h_r is an effective radiant heat-transfer coefficient considering exchange between two parallel planes with the liquid between assigned a value of unity for its emissivity. This term is discussed in Chapter 23.

When there is appreciable flow of either the liquid or the vapor, the foregoing correlations are unsatisfactory. The description of *flow boiling* or *two-phase flow* will not be discussed here. The interested reader is referred to the recent literature for pertinent discussion of these phenomena. It is evident that for vertical surfaces or large-diameter horizontal tubes, the density difference between liquid and vapor will produce significant local velocities. Any correlation that neglects flow contributions should, therefore, be used with caution.

▶ **21.2**

CONDENSATION

Condensation occurs when a vapor contacts a surface that is at a temperature below the saturation temperature of the vapor. When the liquid condensate forms on the surface, it will flow under the influence of gravity.

Normally the liquid wets the surface, spreads out, and forms a film. Such a process is called *film condensation*. If the surface is not wetted by the liquid, then droplets form and run down the surface, coalescing as they contact other condensate droplets. This process is designated *dropwise condensation*. After a condensate film has been developed in filmwise condensation, additional condensation will occur at the liquid–vapor interface, and the associated energy transfer must occur by conduction through the condensate film. Dropwise condensation, on the contrary, always has some surface present as the condensate drop forms and runs off. Dropwise condensation is, therefore, associated with the higher heat-transfer rates of the two types of condensation phenomena. Dropwise condensation is very difficult to achieve or maintain commercially; therefore, all equipment is designed on the basis of filmwise condensation.

Film Condensation: The Nusselt Model   In 1916, Nusselt[10] achieved an analytical result for the problem of filmwise condensation of a pure vapor on a vertical wall. The meanings of the various terms in this analysis will be made clear by referring to Figure 21.3.

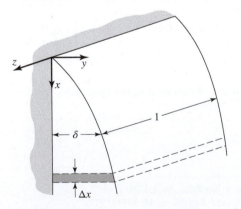

Figure 21.3 Filmwise condensation on a vertical plane wall.

[10] W. Nusselt, *Zeitschr. d. Ver. deutsch. Ing.*, **60**, 514 (1916).

In this figure, the film thickness, δ, is seen to be zero at the top of the vertical wall, $x = 0$, and to increase with increasing values of x.

The initial assumption made by Nusselt was that of wholly laminar flow in the condensate film. Under these conditions, the velocity profile may be easily obtained from equation (21-12):

$$v_x = \frac{\rho g L^2 \sin \theta}{\mu} \left[\frac{y}{L} - \frac{1}{2} \left(\frac{y}{L} \right)^2 \right] \tag{8-12}$$

For the present application, $\sin \theta = 1$ and $L = \delta$. It is also necessary to modify the density for the present case. In the derivation of equation (21-12), the density of the gas or vapor at the liquid surface was neglected. This may be true in many cases of a condensation process; however, the process may occur at a sufficiently high pressure that the vapor density, ρ_v, is significant in comparison to that of the liquid, ρ_L. To account for this possibility, the density function to be used in the present case is $\rho_L - \rho_v$ instead of simply ρ_L. The resulting expression for the velocity profile in the condensate film at a particular distance x from the top of the wall becomes

$$v_x = \frac{(\rho_L - \rho_v) g \delta^2}{\mu} \left[\frac{y}{\delta} - \frac{1}{2} \left(\frac{y}{\delta} \right)^2 \right] \tag{21-12}$$

The flow rate per unit width, Γ, at any value $x > 0$ is

$$\Gamma = \int_0^\delta v_x \, dy$$
$$= \frac{(\rho_L - \rho_v) g \delta^3}{3\mu} \tag{21-13}$$

A differential change, $d\Gamma$, in the flow rate is evaluated from this expression to be

$$d\Gamma = \frac{(\rho_L - \rho_v) g \delta^2 \, d\delta}{\mu} \tag{21-14}$$

This result has been obtained from momentum considerations alone. We shall now, as Nusselt did originally, look at the related energy transfer.

As the flow of condensate is assumed to be laminar, it is not unreasonable to consider energy transfer through the film from the temperature at the vapor–liquid interface, T_{sat}, to the wall–liquid boundary at temperature, T_w, to be purely by conduction. On this basis, the temperature profile is linear and the heat flux to the wall is

$$\frac{q_y}{A} = k \frac{(T_{sat} - T_w)}{\delta} \tag{21-15}$$

This same amount of energy must be transferred from the vapor as it condenses and then cools to the average liquid temperature. Relating these two effects, we may write

$$\frac{q_y}{A} = k \frac{(T_{sat} - T_w)}{\delta} = \rho_L \left[h_{fg} + \frac{1}{\rho_L \Gamma} \int_0^\delta \rho_L v_x c_{pL} (T_{sat} - T) dy \right] \frac{d\Gamma}{dx}$$

which, if a linear temperature variation in y is utilized, becomes

$$\frac{q_y}{A} = \frac{k(T_{sat} - T_w)}{\delta} = \rho_L \left[h_{fg} + \frac{3}{8} c_p L (T_{sat} - T_w) \right] \frac{d\Gamma}{dx} \tag{21-16}$$

Solving equation (21-16) for $d\Gamma$, we have

$$d\Gamma = \frac{k(T_{\text{sat}} - T_w)dx}{\rho_L\delta\left[h_{fg} + \frac{3}{8}c_{pL}(T_{\text{sat}} - T_w)\right]} \qquad (21\text{-}17)$$

which may now be equated to the result in equation (21-14), giving

$$\frac{(\rho_L - \rho_v)}{\mu}\delta^2\,d\delta = \frac{k(T_{\text{sat}} - T_w)}{\rho_L\delta\left[h_{fg} + \frac{3}{8}c_{pL}(T_{\text{sat}} - T_w)\right]}dx$$

Simplifying this result and solving for δ, we obtain

$$\delta = \left[\frac{4k\mu(T_{\text{sat}} - T_w)x}{\rho_L\delta(\rho_L - \rho_v)\left[h_{fg} + \frac{3}{8}c_{pL}(T_{\text{sat}} - T_w)\right]}\right]^{1/4} \qquad (21\text{-}18)$$

We may now solve for the heat-transfer coefficient, h, from the expression

$$h = \frac{q_y/A}{T_{\text{sat}} - T_w} = \frac{k}{\delta}$$

The substitution of equation (21-18) into this expression yields

$$h_x = \left\{\frac{\rho_L g k^3(\rho_L - \rho_v)\left[h_{fg} + \frac{3}{8}c_{pL}(T_{\text{sat}} - T_w)\right]}{4\mu(T_{\text{sat}} - T_w)x}\right\}^{1/4} \qquad (21\text{-}19)$$

The average heat-transfer coefficient for a surface of length L is determined from

$$h = \frac{1}{L}\int_0^L h_x\,dx$$

which, when equation (21-19) is substituted, becomes

$$h = 0.943\left\{\frac{\rho_L g k^3(\rho_L - \rho_v)\left[h_{fg} + \frac{3}{8}c_{pL}(T_{\text{sat}} - T_w)\right]}{L\mu(T_{\text{sat}} - T_w)}\right\}^{1/4} \qquad (21\text{-}20)$$

The latent heat term, h_{fg}, in equation (21-20) and those preceding it, should be evaluated at the saturation temperature. Liquid properties should all be taken at the film temperature.

An expression similar to equation (21-20) may be achieved for a surface inclined at an angle θ from the horizontal if $\sin\theta$ is introduced into the bracketed term. This extension obviously has a limit and should not be used when θ is small—that is, when the surface is near horizontal. For such a condition, the analysis is quite simple; Example 1 illustrates such a case.

Rohsenow[11] performed a modified integral analysis of this same problem, obtaining a result that differs only in that the term $\left[h_{fg} + \frac{3}{8}c_{pL}(T_{\text{sat}} - T_w)\right]$ is replaced by $\left[h_{fg} + 0.68c_{pL}(T_{\text{sat}} - T_w)\right]$. Rohsenow's results agree well with experimental data achieved for values of $\text{Pr} > 0.5$ and $c_{pL}(T_{\text{sat}} - T_w)/h_{fg} < 1.0$.

---

[11] W. M. Rohsenow, *A.S.M.E. Trans.*, **78**, 1645 (1956).

Example 1

A square pan with its bottom surface maintained at 350 K is exposed to water vapor at 1 atm pressure and 373 K. The pan has a lip all around, so the condensate that forms cannot flow away. How deep will the condensate film be after 10 min have elapsed at this condition?

We will employ a "pseudo-steady-state" approach to solve this problem. An energy balance at the vapor–liquid interface will indicate that the heat flux and rate of mass condensed, $\dot{m}_{cond}$, are related as

$$\frac{q}{A}\bigg|_{in} = \frac{\dot{m}_{cond}\, h_{fg}}{A}$$

The condensation rate, $\dot{m}_{cond}$, may be expressed as follows:

$$\dot{m}_{cond} = \rho \dot{V}_{cond} = \rho A \frac{d\delta}{dt}$$

where $d\delta/dt$ is the rate at which the condensate film thickness, δ, grows. The heat flux at the interface may now be expressed as

$$\frac{q}{A}\bigg|_{in} = \rho h_{fg} \frac{d\delta}{dt}$$

This heat flux is now equated to that which must be conducted through the film to the cool pan surface. The heat flux expression that applies is

$$\frac{q}{A}\bigg|_{out} = \frac{k_L}{\delta}(T_{sat} - T_s)$$

This is a steady-state expression; that is, we are assuming δ to be constant. If δ is not rapidly varying, this "pseudo-steady-state" approximation will give satisfactory results. Now, equating the two heat fluxes, we have

$$\rho h_{fg} \frac{d\delta}{dt} = \frac{k_L}{\delta}(T_{sat} - T_s)$$

and, progressing, the condensate film thickness is seen to vary with time according to

$$\delta \frac{d\delta}{dt} = \frac{k_L}{\rho h_{fg}}(T_{sat} - T_s)$$

$$\int_0^\delta \delta\, d\delta = \frac{k_L}{\rho h_{fg}}(T_{sat} - T_s) \int_0^t dt$$

$$\delta = \left[\frac{2k_L}{\rho h_{fg}}(T_{sat} - T_s)\right]^{1/2} t^{1/2}$$

A quantitative answer to our example problem now yields the result

$$\delta = \left[\frac{2(0.674\,\text{W/m}\cdot\text{K})(23\,\text{K})(600\,\text{s})}{(966\,\text{kg/m}^3)(2250\,\text{kJ/kg})}\right]^{1/2}$$

$$= 2.93\,\text{mm}$$

Film Condensation: Turbulent-Flow Analysis It is logical to expect the flow of the condensate film to become turbulent for relatively long surfaces or for high condensation rates. The criterion for turbulent flow is, as we should expect, a Reynolds number for the condensate film. In terms of an equivalent diameter, the applicable Reynolds number is

$$\text{Re} = \frac{4A}{P}\frac{\rho_L v}{\mu_f} \tag{21-21}$$

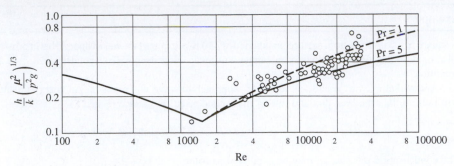

Figure 21.4 Film condensation including the regions of both laminar and turbulent flows.

where A is the condensate flow area, P is the wetted perimeter, and v is the velocity of condensate. The critical value of Re in this case is approximately 2000.

The first attempt to analyze the case of turbulent flow of a condensate film was that of Colburn,[12] who used the same j factor determined for internal pipe flow. On the basis partly of analysis and partly of experiment, Colburn formulated the plot shown in Figure 21.4. The data points shown are those of Kirkbride.[13] The correlating equations for the two regions shown are for $4\Gamma_c/\mu_f < 2000$:

$$h_{\text{avg}} = 1.51 \left(\frac{k^3 \rho^2 g}{\mu^2}\right)_f^{1/3} \left(\frac{4\Gamma_c}{\mu_f}\right)^{-1/3} \tag{21-22}$$

and for $4\Gamma_c/\mu_f > 2000$:

$$h_{\text{avg}} = 0.045 \frac{(k^3 \rho^2 g/\mu^2)_f^{1/3}(4\Gamma_c/\mu_f)\text{Pr}^{1/3}}{[(4\Gamma_c/\mu_f)^{4/5} - 364] + 576\,\text{Pr}^{1/3}} \tag{21-23}$$

In these expressions, Γ_c is the mass flow rate per unit width of surface; that is, $\Gamma_c = \rho_L v_{\text{avg}} \delta$, δ being film thickness and v_{avg} the average velocity. The term $4\Gamma_c/\mu_f$ is thus a Reynolds number for a condensate film on a plane vertical wall. McAdams[14] recommends a simpler expression for the turbulent range, $\text{Re}_\delta > 2000$, as

$$h = 0.0077 \left[\frac{\rho_L g(\rho_L - \rho_v)k_L^3}{\mu_L^2}\right]^{1/3} \text{Re}_\delta^{0.4} \tag{21-24}$$

Film Condensation: Analysis of the Horizontal Cylinder An analysis by Nusselt[15] produced the following expression for the mean heat-transfer coefficient for a horizontal cylinder:

$$h_{\text{avg}} = 0.725 \left\{\frac{\rho_L g(\rho_L - \rho_v)k^3 \left[h_{fg} + \frac{3}{8}c_{pL}(T_{\text{sat}} - T_w)\right]}{\mu D(T_{\text{sat}} - T_w)}\right\}^{1/4} \tag{21-25}$$

[12] A. P. Colburn, *Ind. Eng. Chem.*, **26**, 432 (1934).

[13] C. G. Kirkbride, *Ind. Eng. Chem.*, **26**, 4 (1930).

[14] W. H. McAdams, *Heat Transmission*, 3rd edition, McGraw-Hill Book Company, New York, 1954.

[15] W. Nusselt, *Zeitschr. d. Ver. deutsch. Ing.*, **60**, 569 (1916).

The similarity between equation (21-25) for a horizontal tube and equation (21-20) for a vertical tube is marked. Combining these expressions and canceling similar terms, we obtain the result that

$$\frac{h_{\text{vert}}}{h_{\text{horiz}}} = \frac{0.943}{0.725}\left(\frac{D}{L}\right)^{1/4} = 1.3\left(\frac{D}{L}\right)^{1/4} \tag{21-26}$$

For the case of equal heat-transfer coefficients, the relation between D and L is

$$\frac{L}{D} = 2.86 \tag{21-27}$$

or equal amounts of energy can be transferred from the same tube in either the vertical or the horizontal position if the ratio L/D is 2.86. For L/D values greater than 2.86, the horizontal position has the greater heat-transfer capability.

Film Condensation: Banks of Horizontal Tubes For a bank of horizontal tubes, there is, naturally, a different value of h for each tube, as the condensate film from one tube will drop on the next tube below it in the line. This process is depicted in Figure 21.5.

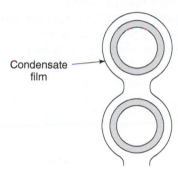

Condensate film

Figure 21.5 Condensation on a horizontal tube bank.

Nusselt also considered this situation analytically and achieved, for a vertical bank of n tubes in line, the expression

$$h_{\text{avg}} = 0.725\left\{\frac{\rho_L g(\rho_L - \rho_v)k^3\left[h_{fg} + \frac{3}{8}c_{pL}(T_{\text{sat}} - T_w)\right]}{nD\mu(T_{\text{sat}} - T_w)}\right\}^{1/4} \tag{21-28}$$

This equation yields a mean heat-transfer coefficient averaged over all n tubes.

Observing that experimental data exceeded those values predicted from equation (21-28), Chen[16] modified this expression to include the effect of condensation on the liquid layer between the tubes. His resulting equation is

$$h_{\text{avg}} = 0.725\left[1 + 0.02\frac{c_{pL}(T_{\text{sat}} - T_w)}{h_{fg}}(n-1)\right]$$
$$\times \left\{\frac{\rho_L g(\rho_L - \rho_v)k^3\left[h_{fg} + \frac{3}{8}c_{pL}(T_{\text{sat}} - T_w)\right]}{nD\mu(T_{\text{sat}} - T_w)}\right\}^{1/4} \tag{21-29}$$

which is valid for values of $c_{pL}(T_{\text{sat}} - T_w)(n-1)/h_{fg} > 2$. Chen's equation agrees reasonably well with experimental data for condensation on vertical banks of horizontal tubes.

---

[16] M. M. Chen, *A.S.M.E. (Trans.) Series C*, **83**, 48 (1961).

Drop Condensation   Dropwise condensation, as mentioned earlier, is associated with higher heat-transfer coefficients than the filmwise condensation phenomenon. For dropwise condensation to occur, the surface must not be "wetted" by the condensate. Normally this requires that metal surfaces be specially treated.

Dropwise condensation is an attractive phenomenon for applications where extremely large heat-transfer rates are desired. At present, it is difficult to maintain this condition for several reasons. Because of its uncertain nature and the conservative approach of a design based on lower heat-transfer coefficients, filmwise condensation is the type predominantly used in design.

▶ 21.3

CLOSURE

The phenomena of boiling and condensation have been examined in this chapter. Both conditions have a prominent place in engineering practice and both are difficult to describe analytically. Several empirical correlations for these phenomena for various surfaces oriented in different ways have been presented.

Boiling is normally described as nucleate type, film type, or a combination of the two. Very high heat-transfer rates are possible in the nucleate-boiling regime with relatively low-temperature differences between the primary surface and the saturation temperature of the liquid. Film boiling is associated with a higher temperature difference, yet a lower rate of heat transfer. This anomalous behavior is peculiar to the boiling phenomenon.

Condensation is categorized as either filmwise or dropwise. Dropwise condensation is associated with much higher heat-transfer coefficients than filmwise; however, it is difficult both to achieve and to maintain. Thus, filmwise condensation is of primary interest. Analytical solutions have been presented, along with empirical results, for filmwise condensation on vertical and horizontal plates and cylinders and for banks of horizontal cylinders.

CHAPTER 22

Heat-Transfer Equipment

$\mathbf{A}$ device whose primary purpose is the transfer of energy between two fluids is called a *heat exchanger*. Heat exchangers are usually classified into three categories:

1. Regenerators
2. Open-type exchangers
3. Closed-type exchangers or recuperators

Regenerators are exchangers in which the hot and cold fluids flow alternately through the same space with as little physical mixing between the two streams as possible. The amount of energy transfer is dependent upon the fluid and flow properties of the fluid stream as well as the geometry and thermal properties of the surface. The required analytical tools for handling this type of heat exchanger have been developed in the preceding chapters.

Open-type heat exchangers are, as implied in their designation, devices wherein physical mixing of the two fluid streams actually occurs. Hot and cold fluids enter open-type heat exchangers and leave as a single stream. The nature of the exit stream is predicted by continuity and the first law of thermodynamics. No rate equations are necessary for the analysis of this type of exchanger.

The third type of heat exchanger, the recuperator, is the one of primary importance and is the one to which we shall direct most of our attention. In the recuperator, the hot and cold fluid streams do not come into direct contact with each other, but are separated by a tube wall or a surface that may be flat or curved in some manner. Energy exchange is thus accomplished from one fluid to an intermediate surface by convection, through the wall or plate by conduction, and then by convection from the surface to the second fluid. Each of these energy-transfer processes has been considered separately in the preceding chapters. We shall, in the following sections, investigate the conditions under which these three energy-transfer processes act in series with one another, resulting in a continuous change in the temperature of at least one of the fluid streams involved.

We shall be concerned with a thermal analysis of these exchangers. A complete design of such equipment involves an analysis of pressure drop, using techniques from Chapter 13, as well as material and structural considerations that are not within the scope of this text.

▶ 22.1 TYPES OF HEAT EXCHANGERS

In addition to being considered a closed-type exchanger, a recuperator is classified according to its configuration and the number of passes made by each fluid stream as it traverses the heat exchanger.

A *single-pass* heat exchanger is one in which each fluid flows through the exchanger only once. An additional descriptive term identifies the relative directions of the two streams, the terms used being *parallel flow* or *cocurrent flow* if the fluids flow in the same direction, *countercurrent flow* or simply *counterflow* if the fluids flow in opposite directions, and *crossflow* if the two fluids flow at right angles to one another. A common single-pass configuration is the double-pipe arrangement shown in Figure 22.1. A crossflow arrangement is shown in Figure 22.2.

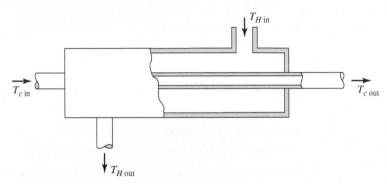

Figure 22.1 A double-pipe heat exchanger.

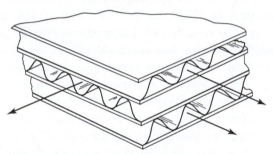

Figure 22.2 A crossflow heat exchanger.

Variations on the crossflow configuration occur when one or the other, or both, fluids are mixed. The arrangement shown in Figure 22.2 is one in which neither fluid is mixed. If the baffles or corrugations were not present, the fluid streams would be unseparated or mixed. In a condition such as that depicted in the figure, the fluid leaving at one end of the sandwich arrangement will have a nonuniform temperature variation from one side to the other, as each section contacts an adjacent fluid stream at a different temperature. It is normally desirable to have one or both fluids unmixed.

In order to accomplish as much transfer of energy in as little space as possible, it is desirable to utilize multiple passes of one or both fluids. A popular configuration is the *shell-and-tube* arrangement shown in Figure 22.3. In this figure, the *tube-side fluid* makes two passes, whereas the *shell-side fluid* makes one pass. Good mixing of the shell-side fluid is accomplished with the baffles shown. Without these baffles, the fluid becomes stagnant in certain parts of the shell, the flow is partially channeled past these stagnant or "dead" regions, and less-than-optimum performance is achieved. Variations on the number of tube-and-shell passes are encountered in numerous applications; seldom are more than two shell-side passes used.

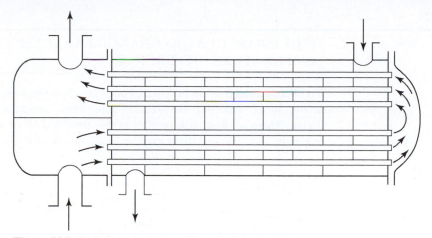

Figure 22.3 Shell-and-tube heat exchanger.

A number of more recent heat-transfer applications require more compact configurations than that afforded by the shell-and-tube arrangement. The subject of "compact heat exchangers" has been investigated and reported both carefully and quite thoroughly by Kays and London.[1] Typical compact arrangements are shown in Figure 22.4.

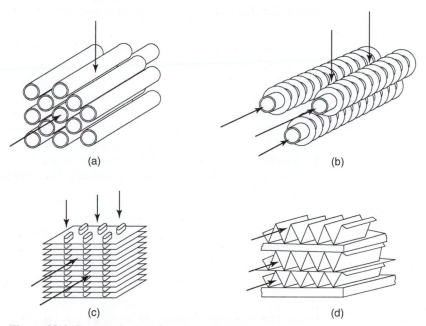

Figure 22.4 Compact heat-exchanger configurations.

The analysis of shell-and-tube, compact, or any multiple-pass heat exchanger is quite involved. As each is a composite of several single-pass arrangements, we shall initially focus our attention on the single-pass heat exchanger.

---

[1] W. M. Kays and A. L. London, *Compact Heat Exchangers*, 2nd edition, McGraw-Hill Book Company, 1964.

SINGLE-PASS HEAT-EXCHANGER ANALYSIS: THE LOG-MEAN TEMPERATURE DIFFERENCE

It is useful, when considering parallel or counterflow single-pass heat exchangers, to draw a simple sketch depicting the general temperature variation experienced by each fluid stream. There are four such profiles in this category, all of which are shown and labeled in Figure 22.5. Each of these may be found in a double-pipe arrangement.

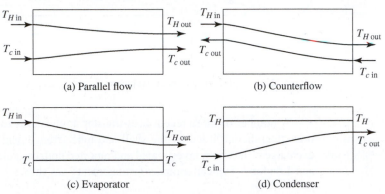

Figure 22.5 Temperature profiles for single-pass, double-pipe heat exchangers.

In Figure 22.5(c) and (d), one of the two fluids remains at constant temperature while exchanging heat with the other fluid whose temperature is changing. This situation occurs when energy transfer results in a change of phase rather than of temperature as in the cases of evaporation and condensation shown. The direction of flow of the fluid undergoing a change in phase is not depicted in the figure, as it is of no consequence to the analysis. If the situation occurs where the complete phase change such as condensation occurs within the exchanger along with some subcooling, then the diagram will appear as in Figure 22.6. In such a case, the direction of flow of the condensate stream is important. For purposes of analysis, this process may be considered the superposition of a condenser and a counterflow exchanger, as depicted in the diagram.

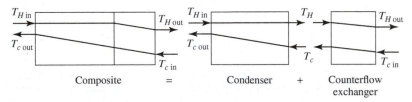

Figure 22.6 Temperature profile in a condenser with subcooling.

Also quite noticeable from Figure 22.5(a) and (b) is the significant difference in temperature profile exhibited by the parallel and counterflow arrangements. It is apparent that the exit temperatures of the hot and cold fluids in the parallel-flow case approach the same value. It is a simple exercise to show that this temperature is the one resulting if the two fluids are mixed in an open-type heat exchanger.

In the counterflow arrangement, it is possible for the hot fluid to leave the exchanger at a temperature below that at which the cold fluid leaves. This situation obviously corresponds to a case of greater total energy transfer per unit area of heat-exchanger surface than would be obtained if the same fluids entered a parallel-flow configuration. The obvious conclusion to this discussion is that the counterflow configuration is the most desirable of the single-pass arrangements. It is thus the single-pass counterflow arrangement to which we shall direct our primary attention.

The detailed analysis of a single-pass counterflow heat exchanger that follows is referred to the diagram and nomenclature of Figure 22.7.

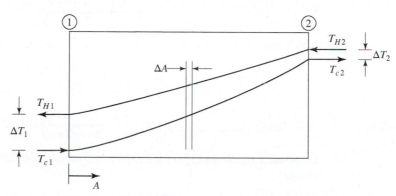

Figure 22.7 Diagram of temperature vs. contact area for single-pass counterflow analysis.

The abscissa of this figure is area. For a double-pipe arrangement, the heat-transfer area varies linearly with distance from one end of the exchanger; in the case shown, the zero reference is the end of the exchanger at which the cold fluid enters.

With reference to a general increment of area, ΔA, between the ends of this unit, a first-law-of-thermodynamics analysis of the two fluid streams will yield

$$\Delta q = (\dot{m}c_p)_c \, \Delta T_c$$

and

$$\Delta q = (\dot{m}c_p)_H \, \Delta T_H$$

As the incremental area approaches differential size, we may write

$$dq = (\dot{m}c_p)_c dT_c = C_c \, dT_c \tag{22-1}$$

and

$$dq = (\dot{m}c_p)_H dT_H = C_H \, dT_H \tag{22-2}$$

where the capacity coefficient, C, is introduced in place of the more cumbersome product, $\dot{m}c_p$.

Writing equation (15-17) for the energy transfer between the two fluids at this location, we have

$$dq = U \, dA(T_H - T_c) \tag{22-3}$$

which utilizes the overall heat-transfer coefficient, U, introduced in Chapter 15. Designating $T_H - T_c$ as ΔT, we have

$$d(\Delta T) = dT_H - dT_c \tag{22-4}$$

and substituting for dT_H and dT_c from equations (22-1) and (22-2), we obtain

$$d(\Delta T) = dq \left(\frac{1}{C_H} - \frac{1}{C_c} \right) = \frac{dq}{C_H} \left(1 - \frac{C_H}{C_c} \right) \tag{22-5}$$

We should also note that dq is the same in each of these expressions; thus, equations (22-1) and (22-2) may be equated and integrated from one end of the exchanger to the other, yielding, for the ratio C_H/C_c,

$$\frac{C_H}{C_c} = \frac{T_{c2} - T_{c1}}{T_{H2} - T_{H1}} \tag{22-6}$$

which may be substituted into equation (22-5) and rearranged as follows:

$$d(\Delta T) = \frac{dq}{C_H} \left(1 - \frac{T_{c2} - T_{c1}}{T_{H2} - T_{H1}} \right) = \frac{dq}{C_H} \left(\frac{T_{H2} - T_{H1} - T_{c2} + T_{c1}}{T_{H2} - T_{H1}} \right)$$

$$= \frac{dq}{C_H} \left(\frac{\Delta T_2 - \Delta T_1}{T_{H2} - T_{H1}} \right) \tag{22-7}$$

Combining equations (22-3) and (22-7), and noting that $C_H(T_{H2} - T_{H1}) = q$, we have, for constant U,

$$\int_{\Delta T_1}^{\Delta T_2} \frac{d(\Delta T)}{\Delta T} = \frac{U}{q} (\Delta T_2 - \Delta T_1) \int_0^A dA \tag{22-8}$$

which, upon integration, becomes

$$\ln \frac{\Delta T_2}{\Delta T_1} = \frac{UA}{q} (\Delta T_2 - \Delta T_1)$$

This result is normally written as

$$q = UA \frac{\Delta T_2 - \Delta T_1}{\ln \dfrac{\Delta T_2}{\Delta T_1}} \tag{22-9}$$

The driving force, on the right-hand side of equation (22-9), is seen to be a particular sort of mean temperature difference between the two fluid streams. This ratio, $(\Delta T_2 - \Delta T_1)/\ln(\Delta T_2/\Delta T_1)$, is designated ΔT_{lm}, the *logarithmic-mean temperature difference*, and the expression for q is written simply as

$$q = UA\, \Delta T_{lm} \tag{22-10}$$

Even though equation (22-10) was developed for the specific case of counterflow, it is equally valid for any of the single-pass operations depicted in Figure 22.5.

It was mentioned earlier, but bears repeating, that equation (22-10) is based upon a constant value of the overall heat-transfer coefficient, U. This coefficient will not, in general, remain constant; however, calculations based upon a value of U taken midway between the ends of the exchanger are usually accurate enough. If there is considerable variation in U from one end of the exchanger to the other, then a step-by-step numerical integration is necessary, equations (22-1) through (22-3) being evaluated repeatedly over a number of small-area increments.

It is also possible that the temperature differences in equation (22-9), evaluated at either end of a counterflow exchanger, are equal. In such a case, the log-mean temperature difference is indeterminate; that is,

$$\frac{\Delta T_2 - \Delta T_1}{\ln(\Delta T_2/\Delta T_1)} = \frac{0}{0}, \qquad \text{if } \Delta T_1 = \Delta T_2$$

In such a case, L'Hôpital's rule may be applied as follows:

$$\lim_{\Delta T_2 \to \Delta T_1} \frac{\Delta T_2 - \Delta T_1}{\ln(\Delta T_2/\Delta T_1)} = \lim_{\Delta T_2/\Delta T_1 \to 1} \left[\frac{\Delta T_1\{(\Delta T_2/\Delta T_1) - 1\}}{\ln(\Delta T_2/\Delta T_1)} \right]$$

when the ratio $\Delta T_2/\Delta T_1$ is designated by the symbol F, we may write

$$= \lim_{F \to 1} \Delta T \left(\frac{F - 1}{\ln F} \right)$$

Differentiating numerator and denominator with respect to F yields the result that

$$\lim_{\Delta T_2 \to \Delta T_1} \frac{\Delta T_2 - \Delta T_1}{\ln(\Delta T_2/\Delta T_1)} = \Delta T$$

or that equation (22-10) may be used in the simple form

$$q = UA \, \Delta T \tag{22-11}$$

From the foregoing simple analysis, it should be apparent that equation (22-11) may be used and achieve reasonable accuracy so long as ΔT_1 and ΔT_2 are not vastly different. It turns out that a simple arithmetic mean is within 1% of the logarithmic-mean temperature difference for values of $(\Delta T_2/\Delta T_1) < 1.5$.

Example 1

Light lubricating oil ($c_p = 2090 \, \text{J/kg} \cdot \text{K}$) is cooled by allowing it to exchange energy with water in a small heat exchanger. The oil enters and leaves the heat exchanger at 375 and 350 K, respectively, and flows at a rate of 0.5 kg/s. Water at 280 K is available in sufficient quantity to allow 0.201 kg/s to be used for cooling purposes. Determine the required heat-transfer area for (a) counterflow and (b) parallel-flow operations (see Figure 22.8). The overall heat-transfer coefficient may be taken as $250 \, \text{W/m}^2 \cdot \text{K}$.

The outlet water temperature is determined by applying equations (22-1) and (22-2):

$$q = (0.5 \, \text{kg/s})(2090 \, \text{J/kg} \cdot \text{K})(25 \, \text{K}) = 26\,125 \, \text{W}$$

$$= (0.201 \, \text{kg/s})(4177 \, \text{J/kg} \cdot \text{K})(T_{w \, \text{out}} - 280 \, \text{K})$$

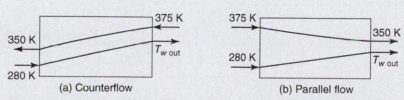

(a) Counterflow   (b) Parallel flow

Figure 22.8 Single-pass temperature profiles for counterflow and parallel flow.

from which we obtain

$$T_{w\,\text{out}} = 280 + \frac{(0.5)(2090)(25)}{(0.201)(4177)} = 311.1\,\text{K} \quad (100\,\text{F})$$

This result applies to both parallel flow and counterflow. For the counterflow configuration, ΔT_{lm} is calculated as

$$\Delta T_{lm} = \frac{70 - 63.9}{\ln \dfrac{70}{63.9}} = 66.9\,\text{K} \quad (120.4\,\text{F})$$

and applying equation (22-10), we see that the area required to accomplish this energy transfer is

$$A = \frac{26125\,\text{W}}{(250\,\text{W/m}^2 \cdot \text{K})(66.9\,\text{K})} = 1.562\,\text{m}^2 \quad (16.81\,\text{ft}^2)$$

Performing similar calculations for the parallel-flow situation, we obtain

$$\Delta T_{lm} = \frac{95 - 38.9}{\ln \dfrac{95}{38.90}} = 62.8\,\text{K} \quad (113\,\text{F})$$

$$A = \frac{26125\,\text{W}}{(250\,\text{W/m}^2 \cdot \text{K})(62.8\,\text{K})} = 1.66\,\text{m}^2 \quad (17.9\,\text{ft}^2)$$

The area required to transfer 26,125 W is seen to be lower for the counterflow arrangement by approximately 7%.

▶ 22.3

CROSSFLOW AND SHELL-AND-TUBE HEAT-EXCHANGER ANALYSIS

More complicated flow arrangements than the ones considered in the previous sections are much more difficult to treat analytically. Correction factors to be used with equation (22-10) have been presented in chart form by Bowman, Mueller, and Nagle[2] and by the Tubular Exchanger Manufacturers Association.[3] Figures 22.9 and 22.10 present correction factors for six types of heat-exchanger configurations. The first three are for different shell-and-tube configurations and the latter three are for different crossflow conditions.

The parameters in Figures 22.9 and 22.10 are evaluated as follows:

$$Y = \frac{T_{t\,\text{out}} - T_{t\,\text{in}}}{T_{s\,\text{in}} - T_{t\,\text{in}}} \tag{22-12}$$

$$Z = \frac{(\dot{m}c_p)_{\text{tube}}}{(\dot{m}c_p)_{\text{shell}}} = \frac{C_t}{C_s} = \frac{T_{s\,\text{in}} - T_{s\,\text{out}}}{T_{t\,\text{out}} - T_{t\,\text{in}}} \tag{22-13}$$

[2] R. A. Bowman, A. C. Mueller, and W. M. Nagle, *Trans. A.S.M.E.*, **62**, 283 (1940).
[3] Tubular Exchanger Manufacturers Association, Standards, 3rd edition, TEMA, New York, 1952.

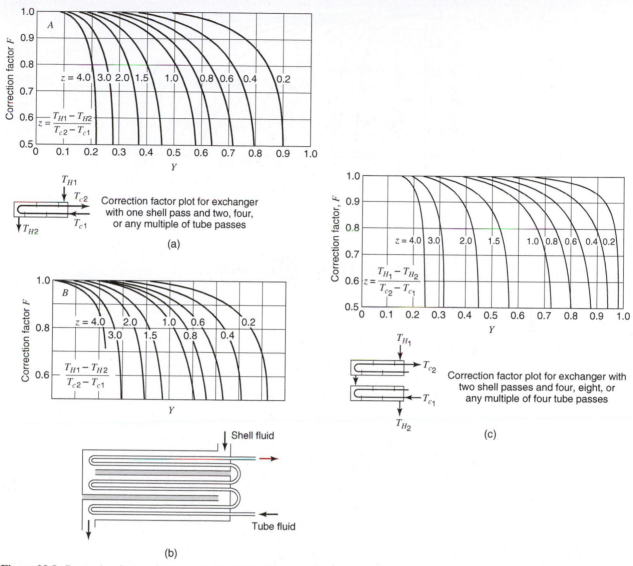

Figure 22.9 Correction factors for three shell-and-tube heat-exchanger configurations. (a) One shell pass and two or a multiple of two tube passes. (b) One shell pass and three or a multiple of three tube passes. (c) Two shell passes and two or a multiple of two tube passes. (From R. A. Bowman, A. C. Mueller, and W. M. Nagle, *Trans. A.S.M.E.*, **62**, 284, 285 (1940). By permission of the publishers.) Correction factors, F, based on counterflow LMTD.

where the subscripts s and t refer to the shell-side and tube-side fluids, respectively. The quantity read on the ordinate of each plot, for given values of Y and Z, is F, the correction factor to be applied to equation (22-10), and thus, these more complicated configurations may be treated in much the same way as the single-pass double-pipe case. The reader is cautioned to apply equation (22-10), using the factor F as in equation (22-14),

$$q = UA(F \, \Delta T_{lm}) \tag{22-14}$$

with the logarithmic-mean temperature difference calculated on the basis of *counterflow*.

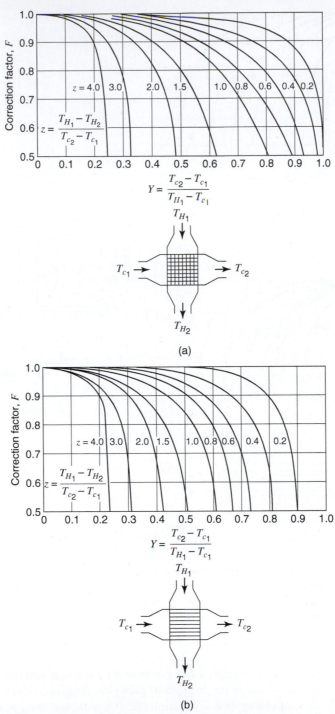

(a)

(b)

Figure 22.10 Correction factors for three crossflow heat-exchanger configurations. (a) Crossflow, single-pass, both fluids unmixed. (b) Crossflow, single-pass, one fluid unmixed. (c) Crossflow, tube passes mixed; fluid flows over first and second passes in series. (From R. A. Bowman, A. C. Mueller, and W. M. Nagle, *Trans. A.S.M.E.*, **62**, 288–289 (1940). By permission of the publishers.)

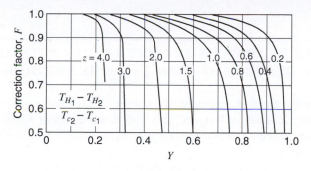

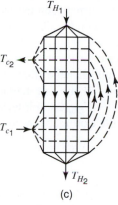

Figure 22.10 (*Continued*)

The manner of using Figures 22.9 and 22.10 may be illustrated by referring to the following example.

Example 2

In the oil–water energy transfer described in Example 1, compare the result obtained with the result that would be obtained if the heat exchanger were

    (a) crossflow, water-mixed
    (b) shell-and-tube with four tube-side passes, oil being the tube-side fluid

For part (a), Figure 22.10(b) must be used. The parameters needed to use this figure are

$$Y = \frac{T_{t\,out} - T_{t\,in}}{T_{s\,in} - T_{t\,in}} = \frac{25}{95} = 0.263$$

and

$$Z = \frac{T_{t\,in} - T_{s\,out}}{T_{s\,out} - T_{t\,in}} = \frac{31.1}{25} = 1.244$$

and from the figure we read $F = 0.96$. The required area for part (a) is thus equal to $(1.562)/(0.96) = 1.63\,\mathrm{m}^2$.

    The values of Y and Z determined above are the same in part (b), yielding a value of F equal to 0.97. The area for part (b) becomes $(1.562)/(0.97) = 1.61\,\mathrm{m}^2$.

▶ **22.4**

THE NUMBER-OF-TRANSFER-UNITS (NTU) METHOD OF HEAT-EXCHANGER ANALYSIS AND DESIGN

Earlier mention was made of the work of Kays and London[1] with particular reference to compact heat exchangers. The book *Compact Heat Exchangers*, by Kays and London, also presents charts useful for heat-exchanger design on a different basis than discussed thus far.

Nusselt[4] in 1930, proposed the method of analysis based upon the heat-exchanger effectiveness $\mathscr{E}$. This term is defined as the ratio of the actual heat transfer in a heat exchanger to the maximum possible heat transfer that would take place if infinite surface area were available. By referring to a temperature profile diagram for counterflow operation, as in Figure 22.11, it is seen that, in general, one fluid undergoes a greater total temperature change than the other. It is apparent that the fluid experiencing the larger change in temperature is the one having the smaller capacity coefficient, which we designate C_{min}. If $C_c = C_{min}$, as in Figure 22.11(a), and if there is infinite area available for energy transfer, the exit temperature of the cold fluid will equal the inlet temperature of the hot fluid.

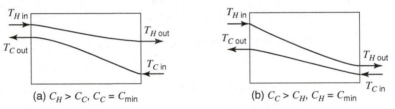

(a) $C_H > C_C$, $C_C = C_{min}$ (b) $C_C > C_H$, $C_H = C_{min}$

Figure 22.11 Temperature profiles for counterflow heat exchangers.

According to the definition of effectiveness, we may write

$$\mathscr{E} = \frac{C_H(T_{H\,in} - T_{H\,out})}{C_c(T_{c\,out} - T_{c\,in})_{max}} = \frac{C_{max}(T_{H\,in} - T_{H\,out})}{C_{min}(T_{H\,in} - T_{c\,in})} \tag{22-15}$$

If the hot fluid is the minimum fluid, as in Figure 22.11(b), the expression for $\mathscr{E}$ becomes

$$\mathscr{E} = \frac{C_c(T_{c\,out} - T_{c\,in})}{C_H(T_{H\,in} - T_{H\,out})_{max}} = \frac{C_{max}(T_{c\,out} - T_{c\,in})}{C_{min}(T_{H\,in} - T_{c\,in})} \tag{22-16}$$

Notice that the denominators in both equations (22-15) and (22-16) are the same, and that, in each case, the numerator represents the actual heat transfer. It is thus possible to write a fifth expression for q as

$$q = \mathscr{E}C_{min}(T_{H\,in} - T_{c\,in}) \tag{22-17}$$

[4] W. Nusselt, *Tech. Mechan. Thermodyn.*, **12** (1930).

which, along with the integrated forms of equations (22-1) and (22-2), as well as equations (22-10) and (22-14), expresses q, the rate of heat transfer, in all of its useful forms as far as heat-exchanger analysis and design are concerned. Equation (22-17) is conspicuous among these others, as the temperature difference appearing is that between the inlet streams alone. This is a definite advantage when a given heat exchanger is to be used under conditions other than those for which it was designed. The exit temperatures of the two streams are then unknown quantities, and equation (22-17) is obviously the easiest means of attaining this knowledge if one can determine the value of $\mathscr{E}$.

To determine $\mathscr{E}$ for a single-pass case, we initially write equation (22-17) in the form

$$\mathscr{E} = \frac{C_H(T_{H\,\text{in}} - T_{H\,\text{out}})}{C_{\min}(T_{H\,\text{in}} - T_{c\,\text{in}})} = \frac{C_c(T_{c\,\text{out}} - T_{c\,\text{in}})}{C_{\min}(T_{H\,\text{in}} - T_{c\,\text{in}})} \tag{22-18}$$

The appropriate form for equation (22-18) depends on which of the two fluids has the smaller value of C. We shall consider the cold fluid to be the minimum fluid and consider the case of counterflow. For these conditions, equation (22-10) may be written as follows (numerical subscripts correspond to the situation shown in Figure 22.7):

$$q = C_c(T_{c2} - T_{c1}) = UA\frac{(T_{H1} - T_{c1}) - (T_{H2} - T_{c2})}{\ln[(T_{H1} - T_{c1})/(T_{H2} - T_{c2})]} \tag{22-19}$$

The entering temperature of the hot fluid, T_{H2}, may be written in terms of $\mathscr{E}$ by use of equation (22-18), yielding

$$T_{H2} = T_{c1} + \frac{1}{\mathscr{E}}(T_{c2} - T_{c1}) \tag{22-20}$$

and also

$$T_{H2} - T_{c2} = T_{c1} - T_{c2} + \frac{1}{\mathscr{E}}(T_{c2} - T_{c1})$$

$$= \left(\frac{1}{\mathscr{E}} - 1\right)(T_{c2} - T_{c1}) \tag{22-21}$$

From the integrated forms of equations (22-1) and (22-2), we have

$$\frac{C_c}{C_H} = \frac{T_{H2} - T_{H1}}{T_{c2} - T_{c1}}$$

which may be rearranged to the form

$$T_{H1} = T_{H2} - \frac{C_{\min}}{C_{\max}}(T_{c2} - T_{c1})$$

or

$$T_{H1} - T_{c1} = T_{H2} - T_{c1} - \frac{C_{\min}}{C_{\max}}(T_{c2} - T_{c1}) \tag{22-22}$$

Combining this expression with equation (22-20), we obtain

$$T_{H1} - T_{c1} = \frac{1}{\mathscr{E}}(T_{c2} - T_{c1}) - \frac{C_{min}}{C_{max}}(T_{c2} - T_{c1})$$

$$= \left(\frac{1}{\mathscr{E}} - \frac{C_{min}}{C_{max}}\right)(T_{c2} - T_{c1}) \tag{22-23}$$

Now substituting equations (22-21) and (22-23) into equation (22-19) and rearranging, we have

$$\ln \frac{1/\mathscr{E} - C_{min}/C_{max}}{1/\mathscr{E} - 1} = \frac{UA}{C_{min}}\left(1 - \frac{C_{min}}{C_{max}}\right)$$

Taking the antilog of both sides of this expression and solving for $\mathscr{E}$ we have, finally,

$$\mathscr{E} = \frac{1 - \exp\left[-\dfrac{UA}{C_{min}}\left(1 - \dfrac{C_{min}}{C_{max}}\right)\right]}{1 - (C_{min}/C_{max})\exp\left[-\dfrac{UA}{C_{min}}\left(1 - \dfrac{C_{min}}{C_{max}}\right)\right]} \tag{22-24}$$

The ratio UA/C_{min} is designated the *number of transfer units*, abbreviated NTU. Equation (22-24) was derived on the basis that $C_c = C_{min}$; if we had initially considered the hot fluid to be minimum, the same result would have been achieved. Thus, equation (22-25),

$$\mathscr{E} = \frac{1 - \exp\left[-\text{NTU}\left(1 - \dfrac{C_{min}}{C_{max}}\right)\right]}{1 - (C_{min}/C_{max})\exp\left[-\text{NTU}\left(1 - \dfrac{C_{min}}{C_{max}}\right)\right]} \tag{22-25}$$

is valid for counterflow operation in general. For parallel flow, an analogous development to the preceding will yield

$$\mathscr{E} = \frac{1 - \exp\left[-\text{NTU}\left(1 + \dfrac{C_{min}}{C_{max}}\right)\right]}{1 + C_{min}/C_{max}} \tag{22-26}$$

Kays and London[1] have put equations (22-25) and (22-26) into chart form, along with comparable expressions for the effectiveness of several shell-and-tube and crossflow arrangements. Figures 22.12 and 22.13 are charts for $\mathscr{E}$ as functions of NTU for various values of the parameter C_{min}/C_{max}.

With the aid of these figures, equation (22-17) may be used both as an original design equation and as a means of evaluating existing equipment when it operates at other than design conditions.

The utility of the NTU approach is illustrated in the following example.

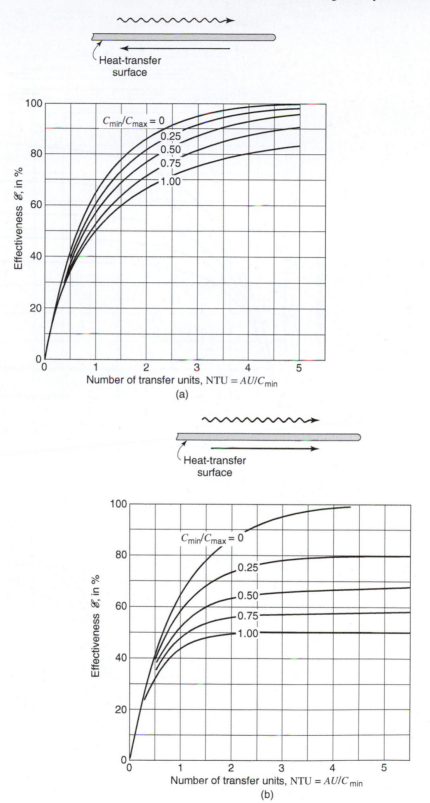

Figure 22.12 Heat-exchanger effectiveness for three shell-and-tube configurations, (a) Counterflow. (b) Parallel flow. (c) One shell pass and two or a multiple of two tube passes.

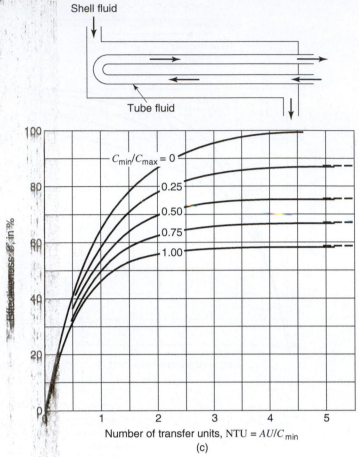

(c)

Figure 22.12 (*Continued*)

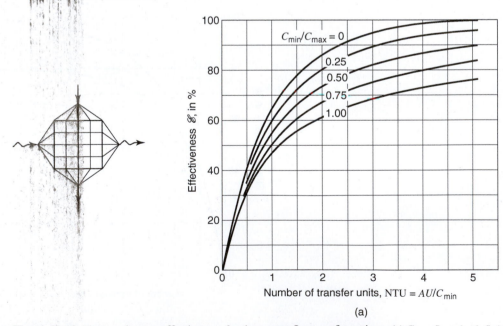

(a)

Figure 22.13 Heat-exchanger effectiveness for three crossflow configurations. (a) Crossflow, both fluids unmixed. (b) Crossflow, one fluid mixed. (c) Crossflow, multiple passes.

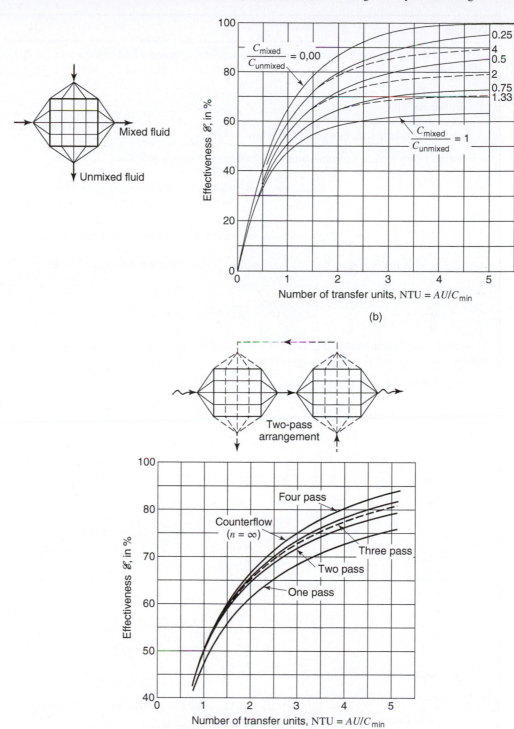

Figure 22.13 (*Continued*)

Example 3

Repeat the calculations for Examples 1 and 2 to determine the required heat-transfer area for the specified conditions if the configurations are

- **(a)** counterflow
- **(b)** parallel flow
- **(c)** crossflow, water-mixed
- **(d)** shell-and-tube with four tube-side passes

It is first necessary to determine the capacity coefficients for the oil and water:

$$C_{oil} = (\dot{m}c_p)_{oil} = (0.5 \text{ kg/s})(2090 \text{ J/kg} \cdot \text{K}) = 1045 \text{ J/s} \cdot \text{K}$$

and

$$C_{water} = (\dot{m}c_p)_w = (0.201 \text{ kg/s})(4177 \text{ J/kg} \cdot \text{K}) = 841.2 \text{ J/s} \cdot \text{K}$$

Thus, the water is the minimum fluid. From equation (22-16), the effectiveness is evaluated as

$$\mathscr{E} = \frac{26125 \text{ W}}{(841.2 \text{ J/kg} \cdot \text{s})(95 \text{ K})} = 0.327$$

By using the appropriate chart in Figures 22.12 and 22.13, the appropriate NTU values and, in turn, the required area may be evaluated for each heat-exchanger configuration.

(a) *Counterflow*

$$NTU = 0.47$$

$$A = \frac{(0.47)(841.2)}{250} = 1.581 \text{ m}^2$$

(b) *Parallel flow*

$$NTU = 0.50$$

$$A = \frac{(0.50)(841.2)}{250} = 1.682 \text{ m}^2$$

(c) *Crossflow, water-mixed*

$$NTU = 0.48$$

$$A = \frac{(0.48)(841.2)}{250} = 1.615 \text{ m}^2$$

(d) *Shell-and-tube, four tube-side passes*

$$NTU = 0.49$$

$$A = \frac{(0.49)(841.2)}{250} = 1.649 \text{ m}^2$$

These results are comparable to those obtained earlier, with some possible inaccuracies involved in reading the chart.

The NTU method offers no distinct advantage over the procedure introduced earlier, using the log-mean temperature difference, when performing calculations of the type involved in the preceding examples. In Example 4, however, the NTU approach is clearly superior.

Example 4

In the energy exchange between water and lubricating oil as considered in the preceding examples, a crossflow heat exchanger with the shell-side fluid (water) mixed is constructed with a heat-transfer area of 1.53 m². A new pump is attached to the water supply line, enabling the water flow rate to be increased to 1000 kg/h. What will be the exit temperatures of the water and oil for the new operating conditions?

If the ΔT_{lm} method were used in this problem, a trial-and-error method would be necessary as ΔT_{lm}, Y, and F are all dependent on one or both exit stream temperatures. The NTU method is thus a bit simpler. Using the NTU method, it is first necessary to calculate the capacity coefficients:

$$C_{oil} = (0.5 \text{ kg/s})(2090 \text{ J/kg} \cdot \text{K}) = 1045 \text{ J/s} \cdot \text{K}$$
$$C_w = (1000 \text{ kg/h})(\text{h}/3600 \text{ s})(4177 \text{ J/kg} \cdot \text{K})$$
$$= 1160 \text{ J/kg} \cdot \text{K}$$

Oil is now the "minimum" fluid. With $C_{oil} = C_{min}$ we have

$$\text{NTU} = \frac{UA}{C_{min}} = \frac{(250 \text{ W/m}^2 \cdot \text{K})(1.53 \text{ m}^2)}{1045 \text{ J/s} \cdot \text{K}}$$
$$= 0.366$$

and, from Figure 22.13, the effectiveness is

$$\mathscr{E} \cong 0.29$$

Using equation (22-17) we may evaluate the heat-transfer rate as

$$q = (0.29)(1045 \text{ J/s} \cdot \text{K})(95 \text{ K})$$
$$= 28.8 \text{ K}$$

which is an increase of over 10%. This value may now be used in equations (22-1) and (22-2) to yield the required answers.

$$T_{oil \text{ out}} = 375 - 28.8 \text{ kW}/(1045 \text{ W/K})$$
$$= 347.4 \text{ K}$$
$$T_{w \text{ out}} = 280 + 28.8 \text{ kW}/(1160 \text{ W/K})$$
$$= 304.8 \text{ K}$$

▶ **22.5**

ADDITIONAL CONSIDERATIONS IN HEAT-EXCHANGER DESIGN

After a heat exchanger has been in service for some time, its performance may change as a result of the buildup of scale on a heat-transfer surface or of the deterioration of the surface by a corrosive fluid. When the nature of the surface is altered in some way so as to affect the heat-transfer capability, the surface is said to be "fouled."

When a fouling resistance exists, the thermal resistance is increased and a heat exchanger will transfer less energy than the design value. It is extremely difficult to predict the rate of scale buildup or the effect such buildup will have upon heat transfer. Some evaluation can be done after a heat exchanger has been in service for some time by comparing its performance with that when the surfaces were clean. The thermal resistance of the scale is determined by

$$R_{sc} = \frac{1}{U_f} - \frac{1}{U_0} \qquad (22\text{-}27)$$

where U_0 is the overall heat-transfer coefficient of the clean exchanger, U_f is the overall heat-transfer coefficient of the fouled exchanger, and R_{sc} is the thermal resistance of the scale.

Fouling resistances that have been obtained from experiments may be used to roughly predict the overall heat-transfer coefficient by incorporation into an expression similar to equation (15-18). The following equation includes the fouling resistances, R_i on the inside tube surface and R_o on the outside tube surface:

$$U_f = \frac{1}{A_0/A_i h_i + R_i + [A_0 \ln(r_o/r_i)]/2\pi k/L + R_o + 1/h_o} \qquad (22\text{-}28)$$

Fouling resistances to be used in equation (22-28) have been compiled by the Tubular Exchanger Manufacturers Association.[5] Some useful values are given in Table 22.1.

Table 22.1 Heat-exchanger fouling resistances

Fluid	Fouling resistances $(\text{m}^2 \cdot \text{K/W} \times 10^5)$
Distilled water	8.8
Seawater, below 325 K	8.8
above 325 K	17.6
Boiler feed water, treated	17.6
City or well water, below 325 K	17.6
above 325 K	35.2
Refrigerating liquids	17.6
Refrigerating vapors	35.2
Liquid gasoline, organic vapors	8.8
Fuel oil	88.1
Quenching oil	70.5
Steam, non-oil-bearing	8.8
Industrial air	35.2

It is often useful to have "ballpark" figures on heat-exchanger size, flow rates, and the like. The most difficult quantity to estimate quickly is the overall heat-transfer coefficient, U. Mueller[6] has prepared the very useful table of approximate U values, which is reproduced here as Table 22.2.

[5] Tubular Exchanger Manufacturers Association, *TEMA Standards*, 3rd edition, New York, 1952.
[6] A. C. Mueller, *Purdue Univ. Eng. Exp. Sta. Eng. Bull. Res. Ser.*, **121** (1954).

Table 22.2 Approximate values for overall heat-transfer coefficients

Fluid combination	U (W/m$^2 \cdot$ K)
Water to compressed air	55–165
Water to water, jacket water coolers	850–1560
Water to brine	570–1140
Water to gasoline	340–480
Water to gas oil or distillate	200–340
Water to organic solvents, alcohol	280–850
Water to condensing alcohol	250–680
Water to lubricating oil	110–340
Water to condensing oil vapors	220–570
Water to condensing or boiling Freon-12	280–850
Water to condensing ammonia	850–1350
Steam to water, instantaneous heater	2280–3400
storage-tank heater	990–1700
Steam to oil, heavy fuel	55–165
light fuel	165–340
light petroleum distillate	280–1140
Steam to aqueous solutions	570–3400
Steam to gases	28–280
Light organics to light organics	220–425
Medium organics to medium organics	110–340
Heavy organics to heavy organics	55–220
Heavy organics to light organics	55–340
Crude oil to gas oil	165–310

▶ **22.6**

CLOSURE

The basic equations and procedures for heat-exchanger design are presented and developed in this chapter. All heat-exchanger design and analysis involve one or more of the following equations:

$$dq = C_c \, dT_c \tag{22-1}$$

$$dq = C_H \, dT_H \tag{22-2}$$

$$dq = U \, dA(T_H - T_c) \tag{22-3}$$

$$q = UA \, \Delta T_{lm} \tag{22-10}$$

and

$$q = \mathscr{E} C_{\min}(T_{H \text{ in}} - T_{C \text{ in}}) \tag{22-17}$$

Charts were presented by which single-pass techniques could be extended to include the design and analysis of crossflow and shell-and-tube configurations.

The two methods for heat-exchanger design utilize either equation (22-10) or equation (22-17). Either is reasonably rapid and straightforward for designing an exchanger. Equation (22-17) is a simpler and more direct approach when analyzing an exchanger that operates at other than design conditions.

CHAPTER 23

Radiation Heat Transfer

The mechanism of radiation heat transfer has no analogy in either momentum or mass transfer. Radiation heat transfer is extremely important in many phases of engineering design, such as boilers, home heating, and spacecraft. In this chapter, we will concern ourselves first with understanding the nature of thermal radiation. Next, we will discuss properties of surfaces and consider how system geometry influences radiant heat transfer. Finally, we will illustrate some techniques for solving relatively simple problems where surfaces and some gases participate in radiant-energy exchange.

▶ **23.1**

NATURE OF RADIATION

The transfer of energy by radiation has several unique characteristics when contrasted with conduction or convection. First, matter is not required for radiant heat transfer; indeed, the presence of a medium will impede radiation transfer between surfaces. Cloud cover is observed to reduce maximum daytime temperatures and to increase minimum evening temperatures, both of which are dependent upon radiant-energy transfer between Earth and space. A second unique aspect of radiation is that both the amount of radiation *and* the quality of the radiation depend upon temperature. In conduction and convection, the amount of heat transfer was found to depend upon the temperature difference; in radiation, the amount of heat transfer depends upon both the temperature difference between two bodies and their absolute temperatures. In addition, radiation from a hot object will be different in quality than radiation from a body at a lower temperature. The color of incandescent objects is observed to change as the temperature is changed. The changing optical properties of radiation with temperature are of paramount importance in determining the radiant-energy exchange between bodies.

Radiation travels at the speed of light, having both wave properties and particle-like properties. The electromagnetic spectrum shown in Figure 23.1 illustrates the tremendous range of frequency and wavelength over which radiation occurs.

The unit of wavelength we shall use in discussing radiation is the micron, symbolized μ. One micron is 10^{-6} m or $3.94(10)^{-5}$ in. The frequency, ν, of radiation is related to the wavelength λ, by $\lambda\nu = c$, where c is the speed of light. Short-wavelength radiation such as gamma rays and X-rays is associated with very high energies. To produce radiation of this type, we must disturb the nucleus or the inner-shell electrons of an atom. Gamma rays and X-rays also have great penetrating ability; surfaces that are opaque to visible radiation are easily traversed by gamma rays and X-rays. Very-long-wavelength radiation, such as radio waves, also may pass through solids; however, the energy associated with these waves is much less than that for short-wavelength radiation.

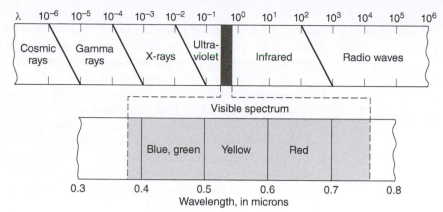

Figure 23.1 The electromagnetic spectrum.

In the range from $\lambda = 0.38$ to 0.76 microns, radiation is sensed by the optical nerve of the eye and is what we call light. Radiation in the visible range is observed to have little penetrating power except in some liquids, plastics, and glasses. The radiation between wavelengths of 0.1 and 100 microns is termed *thermal* radiation. The thermal band of the spectrum includes a portion of the ultraviolet and all of the infrared regions.

▶ **23.2**

THERMAL RADIATION

Thermal radiation incident upon a surface as shown in Figure 23.2 may be either absorbed, reflected, or transmitted.

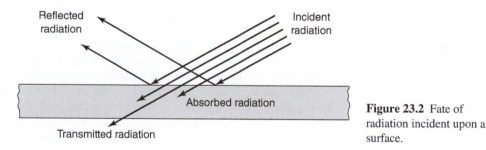

Figure 23.2 Fate of radiation incident upon a surface.

If ρ, α, and τ are the fractions of the incident radiation that are reflected, absorbed, and transmitted, respectively, then

$$\rho + \alpha + \tau = 1 \qquad (23\text{-}1)$$

where ρ is called the *reflectivity*, α is called the *absorptivity*, and τ is called *transmissivity*.

There are two types of reflection that can occur, specular reflection and diffuse reflection. In *specular reflection*, the angle of incidence of the radiation is equal to the angle of reflection. The reflection shown in Figure 23.2 is specular reflection. Most bodies do not reflect in a specular manner; they reflect radiation in all directions. *Diffuse reflection* is sometimes likened to a situation in which the incident thermal radiation is absorbed and then reemitted from the surface, still retaining its initial wavelength.

Absorption of thermal radiation in solids takes place in a very short distance, on the order of 1 μm in electrical conductors and about 0.05 in. in electrical nonconductors, the difference being caused by the different population of energy states in electrical conductors, which can absorb energy at thermal radiation frequencies.

For most solids, the transmissivity is zero, and thus they may be called *opaque* to thermal radiation. Equation (23-1) becomes, for an opaque body, $\rho + \alpha = 1$.

The ideally absorbing body, for which $\alpha = 1$, is called a *black body*. A black body neither reflects nor transmits any thermal radiation. As we see reflected light (radiation), a so-called black body will appear black, no light being reflected from it. A small hole in a large cavity closely approaches a black body, regardless of the nature of the interior surface. Radiation incident to the hole has very little opportunity to be reflected back out of the hole. Black bodies may also be made of bright objects, as can be shown by looking at a stack of razor blades, sharp edge forward.

The *total emissive power*, E, of a surface is defined as the total rate of thermal energy emitted via radiation from a surface in all directions and at all wavelengths per unit surface area. The total emissive power is also referred to elsewhere as the emittance or the total hemispheric intensity. Closely related to the total emissive power is the emissivity. The *emissivity*, ϵ, is defined as the ratio of the total emissive power of a surface to the total emissive power of an ideally radiating surface at the same temperature. The ideal radiating surface is also called a black body, so we may write

$$\epsilon = \frac{E}{E_b} \tag{23-2}$$

where E_b is the total emissive power of a black body. As the total emissive power includes radiant-energy contributions from all wavelengths, the *monochromatic emissive power*, E_λ, may also be defined. The radiant energy E_λ contained between wavelengths λ and $\lambda + d\lambda$ is the monochromatic emissive power; thus,

$$dE = E_\lambda \, d\lambda, \qquad \text{or} \qquad E = \int_0^\infty E_\lambda \, d\lambda$$

The monochromatic emissivity, ϵ_λ, is simply $\epsilon_\lambda = E_\lambda/E_{\lambda,b}$, where $E_{\lambda,b}$ is the monochromatic emissive power of a black body at wavelength λ at the same temperature. A monochromatic absorptivity, α_λ, may be defined in the same manner as the monochromatic emissivity. The *monochromatic absorptivity* is defined as the ratio of the incident radiation of wavelength λ that is absorbed by a surface to the incident radiation absorbed by a black surface.

A relation between the absorptivity and the emissivity is given by Kirchhoff's law. Kirchhoff's law states that for a system in thermodynamic equilibrium, the following equality holds for each surface:

$$\epsilon_\lambda = \alpha_\lambda \tag{23-3}$$

Thermodynamic equilibrium requires that all surfaces be at the same temperature so that there is no net heat transfer. The utility of Kirchhoff's law lies in its use for situations in which the departure from equilibrium is small. In such situations the emissivity and the absorptivity may be assumed to be equal. For radiation between bodies at greatly different temperatures, such as between Earth and the sun, Kirchhoff's law does not apply. A frequent error in using Kirchhoff's law arises from confusing thermal equilibrium with steady-state conditions. Steady state means that time derivatives are zero, whereas equilibrium refers to the equality of temperatures.

▶ **23.3**

THE INTENSITY OF RADIATION

In order to characterize the quantity of radiation that travels from a surface along a specified path, the concept of a single ray is not adequate. The amount of energy traveling in a given direction is determined from I, the *intensity* of radiation. With reference to Figure 23.3, we are interested in knowing the rate at which radiant energy is emitted from a representative portion, dA, of the surface shown in a prescribed direction. Our perspective will be that of an observer at point P looking at dA. Standard spherical coordinates will be used, these being r, the radial coordinate; θ, the zenith angle shown in Figure 23.3; and ϕ, the azimuthal angle, which will be discussed shortly. If a unit area of surface, dA, emits a total energy dq, then the *intensity of radiation* is given by

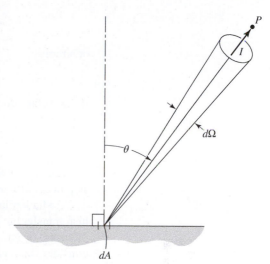

Figure 23.3 The intensity of radiation.

$$I \equiv \frac{d^2 q}{dA \, d\Omega \cos \theta} \tag{23-4}$$

where $d\Omega$ is a differential solid angle—that is, a portion of space. Note that with the eye located at point P, in Figure 23.3, the apparent size of the emitting area is $dA \cos \theta$. It is important to remember that the intensity of radiation is independent of direction for a diffusely radiating surface. Rearranging equation (23-4), we see that the relation between the total emissive power, $E = dq/dA$, and the intensity, I, is

$$\frac{dq}{dA} = E = \int I \cos \theta \, d\Omega = I \int \cos \theta \, d\Omega$$

The relation is seen to be purely geometric for a diffusely radiating ($I \neq I(\theta)$) surface. Consider an imaginary hemisphere of radius r covering the plane surface on which dA is located. The solid angle $d\Omega$ intersects the shaded area on the hemisphere as shown in Figure 23.4. A solid angle is defined by $\Omega = A/r^2$ or $d\Omega = dA/r^2$, and thus

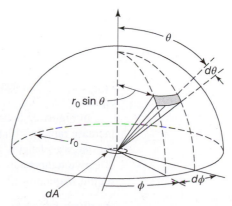

Figure 23.4 Integration of intensity over solid angles.

$$d\Omega = \frac{(r \sin \theta \, d\phi)(r d\theta)}{r^2} = \sin \theta \, d\theta \, d\phi$$

The total emissive power per unit area becomes

$$E = I \int \cos\theta \, d\Omega$$

$$= I \int_0^{2\pi} \int_0^{\pi/2} \cos\theta \sin\theta \, d\theta \, d\phi$$

or simply

$$E = \pi I \tag{23-5}$$

If the surface does not radiate diffusely, then

$$E = \int_0^{2\pi} \int_0^{\pi/2} I \cos\theta \sin\theta \, d\theta \, d\phi \tag{23-6}$$

The relation between the intensity of radiation, I, and the total emissive power is an important step in determining the total emissive power.

Radiation intensity is fundamental in formulating a quantitative description of radiant heat transfer but its definition, as already discussed, is cumbersome. Equation (23-5) relates intensity to emissive power that, potentially, is much easier to describe. We will now consider the means of such a description.

▶ **23.4**

PLANCK'S LAW OF RADIATION

Planck[1] introduced the quantum concept in 1900, and with it the idea that radiation is emitted not in a continuous energy state, but in discrete amounts or quanta. The intensity of radiation emitted by a black body, derived by Planck, is

$$I_{b,\lambda} = \frac{2c^2 h \lambda^{-5}}{\exp\left(\dfrac{ch}{\kappa \lambda T}\right) - 1}$$

where $I_{b,\lambda}$ is the intensity of radiation from a black body between wavelengths λ and $\lambda + d\lambda$, c is the speed of light, h is Planck's constant, κ is the Boltzmann constant, and T is the temperature. The total emissive power between wavelengths λ and $\lambda + d\lambda$ is then

$$E_{b,\lambda} = \frac{2\pi c^2 h \lambda^{-5}}{\exp\left(\dfrac{ch}{\kappa \lambda T}\right) - 1} \tag{23-7}$$

Figure 23.5 illustrates the spectral energy distribution of energy of a black body as given by equation (23-7).

In Figure 23.5 the area under the curve of $E_{b,\lambda}$ vs. λ (the total emitted energy) is seen to increase rapidly with temperature. The peak energy is also observed to occur at shorter and shorter wavelengths as the temperature is increased. For a black body at 5800 K (the effective temperature of solar radiation), a large part of the emitted energy is in the visible region. Equation (23-7) expresses, functionally, $E_{b,\lambda}$ as a function of wavelength and

---

[1] M. Planck, *Verh. d. deut. physik. Gesell.*, **2**, 237 (1900).

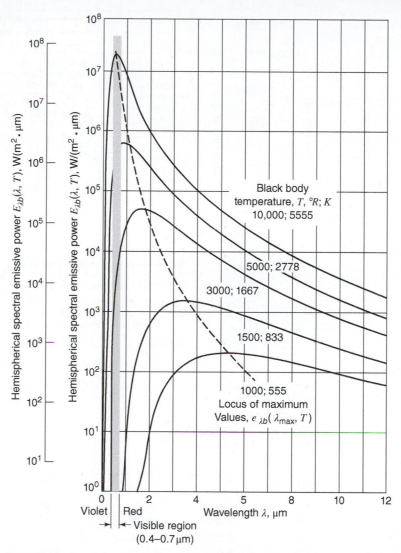

Figure 23.5 Spectral emissive power for a black body for several temperatures. (From R. Siegel and J. R. Howell, *Thermal Radiation Heat Transfer*, 3rd edition, Hemisphere Publishers, Washington, 1992. By permission of the publishers.)

temperature. Dividing both sides of this equation by T^5, we get

$$\frac{E_{b\lambda}}{T^5} = \frac{2\pi^2 h(\lambda T)^{-5}}{\exp\left(\dfrac{ch}{\kappa\lambda T}\right) - 1} \qquad (23\text{-}8)$$

where the quantity $E_{b\lambda}/T^5$ is expressed as a function of the λT product, which can be treated as a single independent variable. This functional relationship is plotted in Figure 23.6, and discrete values of $E_{b\lambda}/\sigma T^5$ are given in Table 23.1. The constant, σ, will be discussed in the next section.

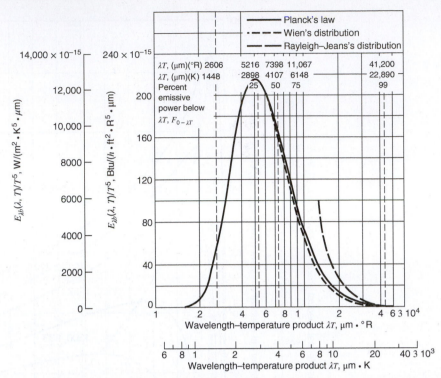

Figure 23.6 Spectral energy distribution for a black body as a function of λT. (From R. Siegel and J. R. Howell, *Thermal Radiation Heat Transfer*, 3rd edition, Hemisphere Publishers, Washington, 1992. By permission of the publishers.)

Table 23.1   Planck radiation functions

λT (μm K)	$F_{0-\lambda_r}$	$\dfrac{E_b}{\sigma T^5}\left(\dfrac{1}{\text{cm K}}\right)$	λT (μm K)	$F_{0-\lambda_r}$	$\dfrac{E_b}{\sigma T^5}\left(\dfrac{1}{\text{cm K}}\right)$
1000	0.0003	0.0372	3000	0.2732	2.2624
1100	0.0009	0.0855	3100	0.2958	2.2447
1200	0.0021	0.1646	3200	0.3181	2.2175
1300	0.0043	0.2774	3300	0.3401	2.1824
1400	0.0078	0.4222	3400	0.3617	2.1408
1500	0.0128	0.5933	3500	0.3829	2.0939
1600	0.0197	0.7825	3600	0.4036	2.0429
1700	0.0285	0.9809	3700	0.4238	1.9888
1800	0.0393	1.1797	3800	0.4434	1.9324
1900	0.0521	1.3713	3900	0.4624	1.8745
2000	0.0667	1.5499	4000	0.4809	1.8157
2100	0.0830	1.7111	4100	0.4987	1.7565
2200	0.1009	1.8521	4200	0.5160	1.6974
2300	0.1200	1.9717	4300	0.5327	1.6387
2400	0.1402	2.0695	4400	0.5488	1.5807
2500	0.1613	2.1462	4500	0.5643	1.5238
2600	0.1831	2.2028	4600	0.5793	1.4679
2700	0.2053	2.2409	4700	0.5937	1.4135
2800	0.2279	2.2623	4800	0.6075	1.3604
2900	0.2505	2.2688	4900	0.6209	1.3089

Table 23.1 (*Continued*)

λT (µm K)	$F_{0-\lambda_r}$	$\frac{E_b}{\sigma T^5}\left(\frac{1}{\text{cm K}}\right)$	λT (µm K)	$F_{0-\lambda_r}$	$\frac{E_b}{\sigma T^5}\left(\frac{1}{\text{cm K}}\right)$
5000	0.6337	1.2590	9500	0.9030	0.2404
5100	0.6461	1.2107	9600	0.9054	0.2328
5200	0.6579	1.1640	9700	0.9077	0.2255
5300	0.6694	1.1190	9800	0.9099	0.2185
5400	0.6803	1.0756	9900	0.9121	0.2117
5500	0.6909	1.0339	10,000	0.9142	0.2052
5600	0.7010	0.9938	11,000	0.9318	0.1518
5700	0.7108	0.9552	12,000	0.9451	0.1146
5800	0.7201	0.9181	13,000	0.9551	0.0878
5900	0.7291	0.8826	14,000	0.9628	0.0684
6000	0.7378	0.8485	15,000	0.9689	0.0540
6100	0.7461	0.8158	16,000	0.9738	0.0432
6200	0.7541	0.7844	17,000	0.9777	0.0349
6300	0.7618	0.7543	18,000	0.9808	0.0285
6400	0.7692	0.7255	19,000	0.9834	0.0235
6500	0.7763	0.6979	20,000	0.9856	0.0196
6600	0.7832	0.6715	21,000	0.9873	0.0164
6700	0.7897	0.6462	22,000	0.9889	0.0139
6800	0.7961	0.6220	23,000	0.9901	0.0118
6900	0.8022	0.5987	24,000	0.9912	0.0101
7000	0.8081	0.5765	25,000	0.9922	0.0087
7100	0.8137	0.5552	26,000	0.9930	0.0075
7200	0.8192	0.5348	27,000	0.9937	0.0065
7300	0.8244	0.5152	28,000	0.9943	0.0057
7400	0.8295	0.4965	29,000	0.9948	0.0050
7500	0.8344	0.4786	30,000	0.9953	0.0044
7600	0.8391	0.4614	31,000	0.9957	0.0039
7700	0.8436	0.4449	32,000	0.9961	0.0035
7800	0.8480	0.4291	33,000	0.9964	0.0031
7900	0.8522	0.4140	34,000	0.9967	0.0028
8000	0.8562	0.3995	35,000	0.9970	0.0025
8100	0.8602	0.3856	36,000	0.9972	0.0022
8200	0.8640	0.3722	37,000	0.9974	0.0020
8300	0.8676	0.3594	38,000	0.9976	0.0018
8400	0.8712	0.3472	39,000	0.9978	0.0016
8500	0.8746	0.3354	40,000	0.9979	0.0015
8600	0.8779	0.3241	41,000	0.9981	0.0014
8700	0.8810	0.3132	42,000	0.9982	0.0012
8800	0.8841	0.3028	43,000	0.9983	0.0011
8900	0.8871	0.2928	44,000	0.9984	0.0010
9000	0.8900	0.2832	45,000	0.9985	0.0009
9100	0.8928	0.2739	46,000	0.9986	0.0009
9200	0.8955	0.2650	47,000	0.9987	0.0008
9300	0.8981	0.2565	48,000	0.9988	0.0007
9400	0.9006	0.2483	49,000	0.9988	0.0007

(From M. Q. Brewster, *Thermal Radiative Transfer and Properties*, John Wiley & Sons, New York, 1992. By permission of the publishers.)

The peak energy is observed to be emitted at $\lambda T = 2897.6 \, \mu m K$ ($5215.6 \, \mu m°R$) as can be determined by maximizing equation (23-8). The relation, $\lambda_{max} T = 2897 \, \mu K$, is called Wien's displacement law. Wien obtained this result in 1893, 7 years prior to Planck's development.

We are often interested in knowing how much emission occurs in a specific portion of the total wavelength spectrum. This is conveniently expressed as a fraction of the total emissive power. The fraction between wavelengths λ_1 and λ_2 is designated $F_{\lambda_1 - \lambda_2}$ and may be expressed as

$$F_{\lambda_1 - \lambda_2} = \frac{\int_{\lambda_1}^{\lambda_2} E_{b\lambda} \, d\lambda}{\int_0^\infty E_{b\lambda} \, d\lambda} = \frac{\int_{\lambda_1}^{\lambda_2} E_{b\lambda} \, d\lambda}{\sigma T^4} \tag{23-9}$$

Equation (23-9) is conveniently broken into two integrals, as follows:

$$F_{\lambda 1 - \lambda 2} = \frac{1}{\sigma T^4} \left(\int_0^{\lambda_2} E_{b\lambda} \, d\lambda - \int_0^{\lambda_1} E_{b\lambda} \, d\lambda \right) \tag{23-10}$$

$$= F_{0 - \lambda_2} - F_{0 - \lambda_1}$$

So, at a given temperature, the fraction of emission between any two wavelengths can be determined by subtraction.

This process can be simplified if the temperature is eliminated as a separate variable. This may be accomplished by using the fraction $E_{b\lambda}/\sigma T^5$. Equation (23-10) may be modified in this manner to yield

$$F_{\lambda_1 T - \lambda_2 T} = \int_0^{\lambda_2 T} \frac{E_{b\lambda}}{\sigma T^5} \, d(\lambda T) - \int_0^{\lambda_1 T} \frac{E_{b\lambda}}{\sigma T^5} \, d(\lambda T)$$

$$= F_{0 - \lambda_2 T} - F_{0 - \lambda_1 T} \tag{23-11}$$

Values of $F_{0 - \lambda T}$ given as functions of the product, λT, in Table 23.1.

▶ **23.5**

STEFAN–BOLTZMANN LAW

Planck's law of radiation may be integrated over wavelengths from zero to infinity to determine the total emissive power. The result is

$$E_b = \int_0^\infty E_{b,\lambda} \, d\lambda = \frac{2\pi^5 \kappa^4 T^4}{15 c^2 h^3} = \sigma T^4 \tag{23-12}$$

where σ is called the Stefan–Boltzmann constant and has the value $\sigma = 5.676 \times 10^{-8} \, W/m^2 \cdot K^4$ ($0.1714 \times 10^{-8} \, Btu/h \, ft^2 \, °R^4$). This constant is observed to be a combination of other physical constants. The Stefan–Boltzmann relation, $E_b = \sigma T^4$, was obtained prior to Planck's law, via experiment by Stefan in 1879 and via a thermodynamic derivation by Boltzmann in 1884. The exact value of the Stefan–Boltzmann constant, σ, and its relation to other physical constants were obtained after the presentation of Planck's law in 1900.

Example 1

For a black, diffusely emitting surface at a temperature of 1600 K, determine the rate of radiant-energy emission, E_b, in the wave length interval $1.5 \, \mu m < \lambda < 3 \, \mu m$ for values of the azimuthal angle, θ, in the range $0 < \theta < 60°$.

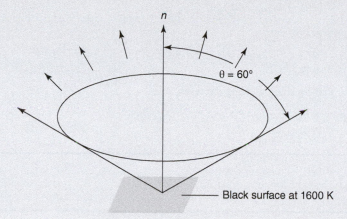

Black surface at 1600 K

Figure 23.7 Black surface at 1600 K.

Solution

For the conditions specified in the problem statement equation (23-6) can be written in the modified form

$$\Delta E_b = \int_{1.5}^{3} \int_{0}^{2\pi} \int_{0}^{60°} I_{b,\lambda} \cos\theta \sin\theta \, d\theta \, d\phi \, d\lambda$$

We proceed as follows:

$$\Delta E_b = \int_{1.5}^{3} I_{b,\lambda} \left[\int_{0}^{\pi/2} \int_{0}^{\pi/3} \cos\theta \sin\theta \, d\theta \, d\phi \right] d\lambda$$

$$= \int_{1.5}^{3} I_{b,\lambda} \left[2\pi \frac{\sin^2\theta}{2} \Big|_{0}^{\pi/3} \right] d\lambda$$

$$= 0.75\pi \int_{1.5}^{3} I_{b,\lambda} \, d\lambda$$

Using equation (23-6), we can write

$$\Delta E_b = 0.75 \int_{1.5}^{3} E_{b,\lambda} \, d\lambda$$

This expression can now be modified to the form allowing the use of Table 23.1 by multiplying and dividing by E_b, giving

$$\Delta E_b = 0.75 \, E_b \int_{1.5}^{3} \frac{E_{b,\lambda}}{E_b} \, d\lambda$$

which may be written as

$$\Delta E_b = 0.75 \, E_b [F_{0-3} - F_{0-1.5}]$$

From Table 23.1,

$$\lambda_2 T = 3(1600) = 4800 \, \mu m \cdot K$$

$$\lambda_1 T = 1.5(1600) = 2400 \, \mu m \cdot K$$

giving

$$F_{0-\lambda_{2T}} = 0.6075$$

$$F_{0-\lambda_{2T}} = 0.1402$$

which yields

$$\Delta E_b = 0.75 \, E_b[0.6075 - 0.1402]$$
$$= 0.350 \, E_b$$

Finally, we calculate E_b using the Stephan rate equation, yielding the result

$$\Delta E_b = 0.350(5.67 \times 10^{-8} \, \text{W/m}^2 \cdot \text{K}^4)(1600 \, \text{K})^4$$

$$= 1.302 \times 10^5 \, \text{W/m}^2$$

▶ 23.6

EMISSIVITY AND ABSORPTIVITY OF SOLID SURFACES

Whereas thermal conductivity, specific heat, density, and viscosity are the important physical properties of matter in heat conduction and convection, emissivity and absorptivity are the controlling properties in heat exchange by radiation.

From preceding sections it is seen that, for black-body radiation, $E_b = \sigma T^4$. For actual surfaces, $E = \epsilon E_b$, following the definition of emissivity. The emissivity of the surface, so defined, is a gross factor, as radiant energy is being sent out from a body not only in all directions but also over various wavelengths. For actual surfaces, the emissivity may vary with wavelength as well as the direction of emission. Consequently, we have to differentiate the monochromatic emissivity ϵ_λ and the directional emissivity ϵ_θ from the total emissivity ϵ.

Monochromatic Emissivity   By definition, the monochromatic emissivity of an actual surface is the ratio of its monochromatic emissive power to that of a black surface at the same temperature. Figure 23.8 represents a typical distribution of the intensity of radiation of two such surfaces at the same temperature over various wavelengths. The monochromatic emissivity at a certain wavelength, λ_1, is seen to be the ratio of two ordinates, such as $\overline{OQ}$ and $\overline{OP}$. That is,

$$\epsilon_{\lambda_1} = \frac{\overline{OQ}}{\overline{OP}}$$

which is equal to the monochromatic absorptivity α_{λ_1} from radiation of a body at the same temperature. This is the direct consequence of Kirchhoff's law. The total emissivity of the surface is given by the ratio of the shaded area shown in Figure 23.8 to that under the curve for the black-body radiation.

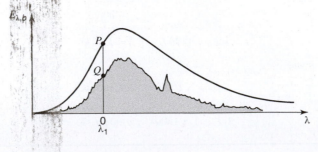

Figure 23.8 Emissivity at various wavelengths.

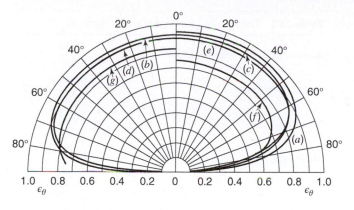

Figure 23.9 Emissivity variation with direction for nonconductors. (a) Wet ice. (b) Wood. (c) Glass. (d) Paper. (e) Clay. (f) Copper oxide. (g) Aluminum oxide.

Directional Emissivity   The cosine variation discussed previously, equation (23-5), is strictly applicable to radiation from a black surface, but fulfilled only approximately by materials present in nature. This is due to the fact that the emissivity (averaged over all wavelengths) of actual surfaces is not a constant in all directions. The variation of emissivity of materials with the direction of emission can be conveniently represented by polar diagrams.

If the cosine law is fulfilled, the distribution curves should take the form of semicircles. Most nonconductors have much smaller emissivities for emission angles in the neighborhood of 90° (see Figure 23.9).

Deviation from the cosine law is even greater for many conductors (see Figure 23.10). The emissivity stays fairly constant in the neighborhood of the normal direction of emission; as the emission angle is increased, it first increases and then decreases as the former approaches 90°.

The average total emissivity may be determined by using the following expression:

$$\epsilon = \int_0^{\pi/2} \epsilon_\theta \sin 2\theta \, d\theta$$

The emissivity, ϵ, is, in general, different from the normal emissivity, ϵ_n (emissivity in the normal direction). It has been found that for most bright metallic surfaces, the total emissivity is approximately 20% higher than ϵ_n. Table 23.2 lists the ratio of ϵ/ϵ_n for a few representative bright metallic surfaces. For nonmetallic or other surfaces, the ratio ϵ/ϵ_n is slightly less than unity. Because of the inconsistency that can often be found among various sources, the normal emissivity values can be used, without appreciable error, for total emissivity (see Table 23.3).

A few generalizations may be made concerning the emissivity of surfaces:

a. In general, emissivity depends on surface conditions.
b. The emissivity of highly polished metallic surfaces is very low.
c. The emissivity of all metallic surfaces increases with temperature.
d. The formation of a thick oxide layer and roughening of the surface increase the emissivity appreciably.

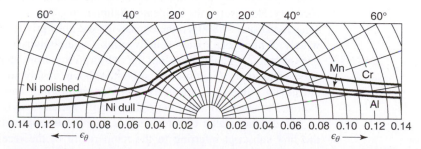

Figure 23.10 Emissivity variation with direction for conductors.

Table 23.2   The ratio ϵ/ϵ_n for bright metallic surfaces

Aluminum, bright rolled (443 K)	$\dfrac{0.049}{0.039} = 1.25$
Nickel, bright matte (373 K)	$\dfrac{0.046}{0.041} = 1.12$
Nickel, polished (373 K)	$\dfrac{0.053}{0.045} = 1.18$
Manganin, bright rolled (392 K)	$\dfrac{0.057}{0.048} = 1.19$
Chromium, polished (423 K)	$\dfrac{0.071}{0.058} = 1.22$
Iron, bright etched (423 K)	$\dfrac{0.158}{0.128} = 1.23$
Bismuth, bright (353 K)	$\dfrac{0.340}{0.336} = 1.08$

Table 23.3   The ratio ϵ/ϵ_n for nonmetallic and other surfaces

Copper oxide (300 F)	0.96
Fire clay (183 F)	0.99
Paper (200 F)	0.97
Plywood (158 F)	0.97
Glass (200 F)	0.93
Ice (32 F)	0.95

e. The ratio ϵ/ϵ_n is always greater than unity for bright metallic surfaces. The value 1.2 can be taken as a good average.

f. The emissivities of nonmetallic surfaces are much higher than for metallic surfaces and show a decrease as temperature increases.

g. The emissivities of colored oxides of heavy metals like Zn, Fe, and Cr are much larger than emissivities of white oxides of light metals like Ca, Mg, and Al.

Absorptivity   The absorptivity of a surface depends on the factors affecting the emissivity and, in addition, on the quality of the incident radiation. It may be remarked once again that Kirchhoff's law holds strictly true under thermal equilibrium. That is, if a body at temperature T_1 is receiving radiation from a black body also at temperature T_1, then $\alpha = \epsilon$. For most materials, in the usual range of temperature encountered in practice (from room temperature up to about 1370 K) the simple relationship $\alpha = \epsilon$ holds with good accuracy. However, if the incident radiation is that from a very-high-temperature source, say solar radiation ($\sim$5800 K), the emissivity and absorptivity of ordinary surfaces may differ widely. White metal oxides usually exhibit an emissivity (and absorptivity) value of about 0.95 at ordinary temperature, but their absorptivity drops sharply to 0.15 if these oxides are exposed to solar radiation. Contrary to the above, freshly polished metallic surfaces have an emissivity value (and absorptivity under equilibrium conditions) of about 0.05. When exposed to solar radiation, their absorptivity increases to 0.2 or even 0.4.

Under these latter circumstances a double-subscript notation, $\alpha_{1,2}$, may be employed, the first subscript referring to the temperature of the receiving surface and the second subscript to the temperature of the incident radiation.

Gray Surfaces   Like emissivity, the monochromatic absorptivity, α_λ, of a surface may vary with wavelength. If α_λ is a constant and thus independent of λ the surface is called *gray*. For a gray surface, the total average absorptivity will be independent of the spectral-energy distribution of the incident radiation. Consequently, the emissivity, ϵ, may be used in place of α, even though the temperatures of the incident radiation and the receiver are not the same. Good approximations of a gray surface are slate, tar board, and dark linoleum. Table 23.4 lists emissivities, at various temperatures, for several materials.

Table 23.4   Normal total emissivity of various surfaces (Compiled by H. C. Hottel)[†]

Surface	T, F[‡]	Emissivity
A. Metals and their oxides		
Aluminum:		
Highly polished plate, 98.3% pure	440–1070	0.039–0.057
Commercial sheet	212	0.09
Oxidized at 1110 F	390–1110	0.11–0.19
Heavily oxidized	200–940	0.20–0.31
Brass:		
Polished	100–600	0.10
Oxidized by heating at 1110 F	390–1110	0.61–0.59
Chromium (see nickel alloys for Ni–Cr steels):		
Polished	100–2000	0.08–0.36
Copper		
Polished	212	0.052
Plate heated at 1110 F	390–1110	0.57
Cuprous oxide	1470–2010	0.66–0.54
Molten copper	1970–2330	0.16–0.13
Gold:		
Pure, highly polished	440–1160	0.018–0.035
Iron and steel (not including stainless):		
Metallic surfaces (or very thin oxide layer)		
Iron, polished	800–1880	0.14–0.38
Cast iron, polished	392	0.21
Wrought iron, highly polished	100–80	0.28
Oxidized surfaces		
Iron plate, completely rusted	67	0.69
Steel plate, rough	100–700	0.94–0.97
Molten surfaces		
Cast iron	2370–2550	0.29
Mild steel	2910–3270	0.28
Lead:		
Pure (99.96%), unoxidized	260–40	0.057–0.075
Gray oxidized	75	0.28
Nickel alloys:		
Chromnickel	125–1894	0.64–0.76
Copper–nickel, polished	212	0.059
Nichrome wire, bright	120–1830	0.65–0.79

(Continued)

Table 23.4 (*Continued*)

Surface	T, F[‡]	Emissivity
Nichrome wire, oxidized	120–930	0.95–0.98
Platinum:		
Pure, polished plate	440–1160	0.054–0.104
Strip	1700–2960	0.12–0.17
Filament	80–2240	0.036–0.192
Wire	440–2510	0.073–0.182
Silver:		
Polished, pure	440–1160	0.020–0.032
Polished	100–700	0.022–0.031
Stainless steels:		
Polished	212	0.074
Type 310 (25 Cr; 20 Ni)		
Brown, splotched, oxidized from furnace service	420–980	0.90–0.97
Tin:		
Bright tinned iron	76	0.043 and 0.064
Bright	122	0.06
Commercial tin-plated sheet iron	212	0.07, 0.08
Tungsten:		
Filament, aged	80–6000	0.032–0.35
Filament	6000	0.39
Polished coat	212	0.066
Zinc:		
Commercial 99.1% pure, polished	440–620	0.045–0.053
Oxidized by heating at 750 F	750	0.11

B. Refractories, building materials, paints, and miscellaneous

Surface	T	Emissivity
Asbestos:		
Board	74	0.96
Paper	100–700	0.93–0.94
Brick		
Red, rough, but no gross irregularities	70	0.93
Brick, glazed	2012	0.75
Building	1832	0.45
Fireclay	1832	0.75
Carbon:		
Filament	1900–2560	0.526
Lampblack-waterglass coating	209–440	0.96–0.95
Thin layer of same on iron plate	69	0.927
Glass:		
Smooth	72	0.94
Pyrex, lead, and soda	500–1000	0.95–0.85
Gypsum, 0.02 in. thick on smooth or blackened plate	70	0.903
Magnesite refractory brick	1832	0.38
Marble, light gray, polished	72	0.93
Oak, planed	70	0.90

Table 23.4 *(Continued)*

Surface	T, F‡	Emissivity
B. Refractories, building materials, paints, and miscellaneous		
Paints, lacquers, varnishes:		
Snow-white enamel varnish on rough iron plate	73	0.906
Black shiny lacquer, sprayed on iron	76	0.875
Black shiny shellac on tinned iron sheet	70	0.821
Black matte shellac	170–295	0.91
Black or white lacquer	100–200	0.80–0.95
Flat black lacquer	100–200	0.96–0.98
Oil paints, 16 different, all colors	212	0.92–0.96
A1 paint, after heating to 620 F	300–600	0.35
Plaster, rough lime	50–190	0.91
Roofing paper	69	0.91
Rubber:		
Hard, glossy plate	74	0.94
Soft, gray, rough (reclaimed)	76	0.86
Water	32–212	0.95–0.963

† By permission from W. H. McAdams (ed.), *Heat Transmission*, 3rd edition, McGraw-Hill Book Company, Table of normal total emissivity compiled by H. C. Hottel.
‡ When temperatures and emissivities appear in pairs separated by dashes, they correspond and linear interpolation is permissible.

The reader should note that the units of temperatures in Table 23.4 are F in contrast to K, as has been used throughout the text thus far. Table 23.4 is presented as originally published by McAdams.

▶ **23.7**

RADIANT HEAT TRANSFER BETWEEN BLACK BODIES

The exchange of energy between black bodies is dependent upon the temperature difference and the geometry with the geometry, in particular, playing a dominant role. Consider the two surfaces illustrated in Figure 23.11. The radiant energy emitted from a black surface at dA_1 and received at dA_2 is

$$dq_{1 \to 2} = I_{b_1} \cos \theta_1 \, d\Omega_{1-2} \, dA_1$$

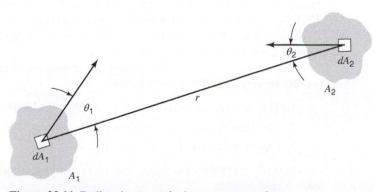

Figure 23.11 Radiant heat transfer between two surfaces.

where $d\Omega_{1-2}$ is the solid angle subtended by dA_2 as seen from dA_1. Thus,

$$d\Omega_{1-2} = \cos\theta_2 \frac{dA_2}{r^2}$$

and as $I_{b1} = E_{b1}/\pi$, the heat transfer from 1 to 2 is

$$dq_{1\to2} = E_{b_1}\, dA_2 \left\{ \frac{\cos\theta_1 \cos\theta_2\, dA_1}{\pi r^2} \right\}$$

The bracketed term is seen to depend solely upon geometry. In exactly the same manner, the energy emitted by dA_2 and captured by dA_1 may be determined. This is

$$dq_{2\to1} = E_{b_2}\, dA_2 \left\{ \frac{\cos\theta_2 \cos\theta_1\, dA_1}{\pi r^2} \right\}$$

The net heat transfer between surfaces dA_1 and dA_2 is then simply

$$dq_{1-2\,\text{net}} = dq_{1\rightleftharpoons2} = dq_{1\to2} - dq_{2\to1}$$

or

$$dq_{1\rightleftharpoons2} = (E_{b_1} - E_{b_2}) \frac{\cos\theta_1 \cos\theta_2\, dA_1\, dA_2}{\pi r^2}$$

Integrating over surfaces 1 and 2, we obtain

$$q_{1\rightleftharpoons2} = (E_{b_1} - E_{b_2}) \int_{A_1}\int_{A_2} \frac{\cos\theta_1 \cos\theta_2\, dA_2\, dA_1}{\pi r^2}$$

The insertion of A_1/A_1 yields

$$q_{1\rightleftharpoons2} = (E_{b_1} - E_{b_2})A_1 \left[\frac{1}{A_1} \int_{A_1}\int_{A_2} \frac{\cos\theta_1 \cos\theta_2\, dA_2\, dA_1}{\pi r^2} \right] \tag{23-13}$$

The bracketed term in the above equation is called the *view factor*, F_{12}. If we had used A_2 as a reference, then the view factor would be F_{21}. Clearly, the net heat transfer is not affected by these operations, and thus $A_1F_{12} = A_2F_{21}$. This simple but extremely important expression is called the *reciprocity* relationship.

A physical interpretation of the view factor may be obtained from the following argument. As the total energy leaving surface A_1 is E_bA_1, the amount of heat that surface A_2 receives is $E_{b_1}A_1F_{12}$. The amount of heat lost by surface A_2 is $E_{b_2}A_2$, whereas the amount that reaches A_1 is $E_{b_2}A_2F_{21}$. The net rate of heat transfer between A_1 and A_2 is the difference or $E_{b_1}A_1F_{12} - E_{b_2}A_2F_{21}$. This may be arranged to yield $(E_{b_1} - E_{b_2})A_1F_{12}$. Thus, the view factor F_{12} can be interpreted as the fraction of black-body energy leaving A_1, which reaches A_2. Clearly the view factor cannot exceed unity.

Before some specific view factors are examined, there are several generalizations worthy of note concerning view factors:

1. The *reciprocity relation*, $A_1F_{12} = A_2F_{21}$, is always valid.

2. The view factor is independent of temperature. It is purely geometric.

3. For an enclosure, $F_{11} + F_{12} + F_{13} + \cdots = 1$.

In many cases the view factor may be determined without integration. An example of such a case follows.

Example 2

Consider the view factor between a hemisphere and a plane as shown in the figure. Determine the view factors F_{11}, F_{12}, and F_{21}.

The view factor F_{21} is unity, as surface 2 sees only surface 1. For surface 1 we may write $F_{11} + F_{12} = 1$ and $A_1 F_{12} = A_2 F_{21}$. As $F_{21} = 1$, $A_2 = \pi r_0^2$, and $A_1 = 2\pi r_0^2$, the above relations give

$$F_{12} = F_{21} \frac{A_2}{A_1} = (1)\left(\frac{\pi r_0^2}{2\pi r_0^2}\right) = \frac{1}{2}$$

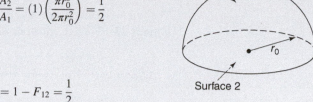

and

$$F_{11} = 1 - F_{12} = \frac{1}{2}$$

The view factor F_{12} can, in general, be determined by integration. As

$$F_{12} \equiv \frac{1}{A_1} \int_{A_1} \int_{A_2} \frac{\cos\theta_1 \cos\theta_2 \, dA_2 \, dA_1}{\pi r^2} \tag{23-14}$$

this integration process becomes quite tedious, and the view factor for a complex geometry is seen to require numerical methods. In order to illustrate the analytical evaluation of view factors, consider the view factor between the differential area dA_1 and the parallel plane A_2 shown in Figure 23.12. The view factor $F_{dA_1 A_2}$ is given by

$$F_{dA_1 A_2} = \frac{1}{dA_1} \int_{dA_1} \int_{A_2} \frac{\cos\theta_1 \cos\theta_2 \, dA_2 \, dA_1}{\pi r^2}$$

and as $A_2 \gg dA_1$ the view of dA_2 from dA_1 is independent of the position on dA_1, hence,

$$F_{dA_1 A_2} = \frac{1}{\pi} \int_{A_2} \frac{\cos\theta_1 \cos\theta_2}{r^2} \, dA_2$$

Also, it may be noted that $\theta_1 = \theta_2$ and $\cos\theta = D/r$, where $r^2 = D^2 + x^2 + y^2$. The resulting integral becomes

$$F_{dA_1 A_2} = \frac{1}{\pi} \int_0^{L_1} \int_0^{L_2} \frac{D^2 \, dx \, dy}{(D^2 + x^2 + y^2)^2}$$

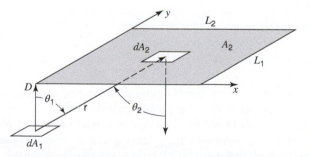

Figure 23.12 Differential area and parallel-finite area.

or

$$F_{dA_1A_2} = \frac{1}{2\pi} \left\{ \frac{L_1}{\sqrt{D^2 + L_1^2}} \tan^{-1} \frac{L_2}{\sqrt{D^2 + L_1^2}} + \frac{L_2}{\sqrt{D^2 + L_2^2}} \tan^{-1} \frac{L_1}{\sqrt{D^2 + L_2^2}} \right\} \quad (23\text{-}15)$$

The view factor given by equation (23-15) is shown graphically in Figure 23.13. Figures 23.14–23.16 illustrate some view factors for some other simple geometries.

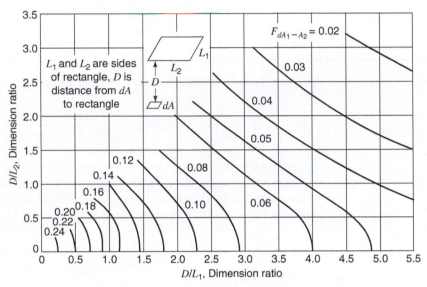

Figure 23.13 View factor for a surface element and a rectangular surface parallel to it. (From H. C. Hottel, "Radiant Heat Transmission," *Mech. Engrg.*, **52** (1930). By permission of the publishers.)

Example 3

Determine the view factor from a 1 m square to a parallel rectangular plane 10 m by 12 m centered 8 m above the 1 m square.

The smaller area may be considered a differential area, and Figure 23.13 may be used. The 10 m by 12 m area may be divided into four 5 m by 6 m rectangles directly over the smaller area. Thus, the total view factor is the sum of the view factors to each subdivided rectangle. Using $D = 8$, $L_1 = 6$, $L_2 = 5$, we find that the view factor from Figure 23.13 is 0.09. The total view factor is the sum of the view factors, or 0.36.

View-Factor Algebra

View factors between combinations of differential and finite-size areas have been expressed in equation form thus far. Some generalizations can be made that will be useful in evaluating radiant-energy exchange in cases that, at first glance, seem quite difficult.

In an enclosure all energy leaving one surface, designated i, will be incident on the other surfaces that it can "see." If there are n surfaces in total, with j designating any surface

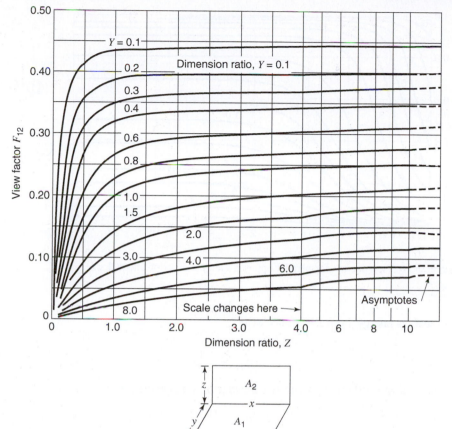

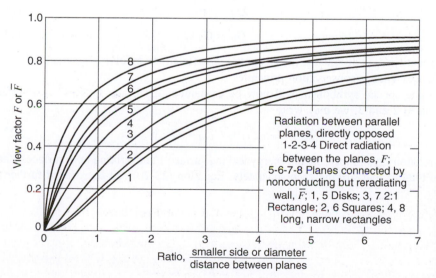

Figure 23.14 View factor for adjacent rectangles in perpendicular planes. (From H. C. Hottel, "Radiant Heat Transmission," *Mech. Engrg.*, **52** (1930). By permission of the publishers.)

Figure 23.15 View factors for equal and parallel squares, rectangles, and disks. The curves labeled 5, 6, 7, and 8 allow for continuous variation in the sidewall temperatures from top to bottom. (From H. C. Hottel, "Radiant Heat Transmission," *Mech. Engrg.*, **52** (1930). By permission of the publishers.)

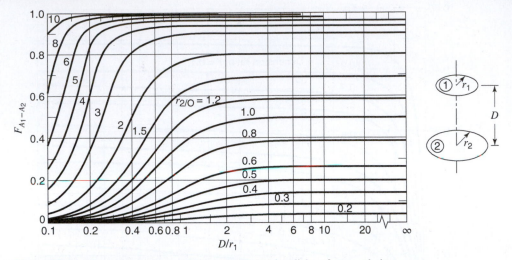

Figure 23.16 View factors for parallel opposed circular disks of unequal size.

that receives energy from i, we may write

$$\sum_{j=1}^{n} F_{ij} = 1 \qquad (23\text{-}16)$$

A general form of the reciprocity relationship may be written as

$$A_i F_{ij} = A_j F_{ji} \qquad (23\text{-}17)$$

These two expressions form the basis of a technique designated *view-factor algebra*. A simplified notation will be introduced, using the symbol G_{ij}, defined as

$$G_{ij} \equiv A_i F_{ij}$$

Equations (23-16) and (23-17) may now be written as

$$\sum G_{ij} = A_i \qquad (23\text{-}18)$$

$$G_{ij} = G_{ji} \qquad (23\text{-}19)$$

The quantity G_{ij} is designated the *geometric flux*. Relations involving geometric fluxes are dictated by energy conservation principles.

Some special symbolism will now be explained. If surface 1 "sees" two surfaces, designated 2 and 3, we may write

$$G_{1-(2+3)} = G_{1-2} + G_{1-3} \qquad (23\text{-}20)$$

This relation says simply that the energy leaving surface 1 and striking both surfaces 2 and 3 is the total of that striking each separately. Equation (23-20) can be reduced further to

$$A_1 F_{1-(2+3)} = A_1 F_{12} + A_1 F_{13}$$

or

$$F_{1-(2+3)} = F_{12} + F_{13}$$

A second expression, involving four surfaces, is reduced to

$$G_{(1+2)-(3+4)} = G_{1-(3+4)} + G_{2-(3+4)}$$

which decomposes further to the form

$$G_{(1+2)-(3+4)} = G_{1-3} + G_{1-4} + G_{2-3} + G_{2-4}$$

Examples of how view-factor algebra can be used follow.

Example 4

Determine the view factors, F_{1-2}, for the finite areas shown.

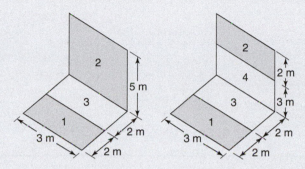

Inspection indicates that, in case (a), view factors F_{2-3} and $F_{2-(1+3)}$ can be read directly from Figure 23.14. The desired view factor, F_{1-2}, can be obtained using view-factor algebra in the following steps.

$$G_{2-(1+3)} = G_{2-1} + G_{2-3}$$

Thus,

$$G_{2-1} = G_{2-(1+3)} - G_{2-3}$$

Finally, by reciprocity, we may solve for F_{1-2} according to

$$G_{1-2} = G_{2-1} = G_{2-(1+3)} - G_{2-3}$$

$$A_1 F_{1-2} = A_2 F_{2-(1+3)} - A_2 F_{2-3}$$

$$F_{1-2} = \frac{A_2}{A_1}[F_{2-(1+3)} - F_{2-3}]$$

From Figure 23.14, we read

$$F_{2-(1+3)} = 0.15 \quad F_{2-3} = 0.10$$

Thus, for configuration (a), we obtain

$$F_{1-2} = \frac{5}{2}(0.15 - 0.10) = 0.125$$

Now, for case (b), the solution steps are

$$G_{1-2} = G_{1-(2+4)} - G_{1-4}$$

which may be written as

$$F_{1-2} = F_{1-(2+4)} - F_{1-4}$$

The result from part (a) can now be utilized to write

$$F_{1-(2+4)} = \frac{A_2 + A_4}{A_1}[F_{(2+4)-(1+3)} - F_{(2+4)-3}]$$

$$F_{1-4} = \frac{A_4}{A_1}[F_{4-(1+3)} - F_{4-3}]$$

Each of the view factors on the right side of these two expressions may be evaluated from Figure 23.14; the appropriate values are

$$F_{(2+4)-(1+3)} = 0.15 \qquad F_{4-(1+3)} = 0.22$$

$$F_{(2+4)-3} = 0.10 \qquad F_{4-3} = 0.165$$

Making these substitutions, we have

$$F_{1-(2+4)} = \frac{5}{2}(0.15 - 0.10) = 0.125$$

$$F_{1-4} = \frac{3}{2}(0.22 - 0.165) = 0.0825$$

The solution to case (b) now becomes

$$F_{1-2} = 0.125 - 0.0825 = 0.0425$$

► 23.8

RADIANT EXCHANGE IN BLACK ENCLOSURES

As pointed out earlier, a surface that views n other surfaces may be described according to

$$F_{11} + F_{12} + \cdots + F_{1i} + \cdots + F_{1n} = 1$$

or

$$\sum_{i=1}^{n} F_{1i} = 1 \qquad (23\text{-}21)$$

Obviously, the inclusion of A_1 with equation (23-12) yields

$$\sum_{i=1}^{n} A_1 F_{1i} = A_1 \qquad (23\text{-}22)$$

Between any two black surfaces the radiant heat exchange rate is given by

$$q_{12} = A_1 F_{12}(E_{b_1} - E_{b_2}) = A_2 F_{21}(E_{b_1} - E_{b_2}) \qquad (23\text{-}23)$$

For surface 1 and any other surface, designated i, in a black enclosure, the radiant exchange is given as

$$q_{1i} = A_1 F_{1i}(E_{b_1} - E_{b_i}) \qquad (23\text{-}24)$$

For an enclosure where surface 1 views n other surfaces, we may write, for the net heat transfer with 1,

$$q_{1-\text{others}} = \sum_{i=1}^{n} q_{1i} = \sum_{i=1}^{n} A_1 F_{1i}(E_{b_1} - E_{b_i}) \qquad (23\text{-}25)$$

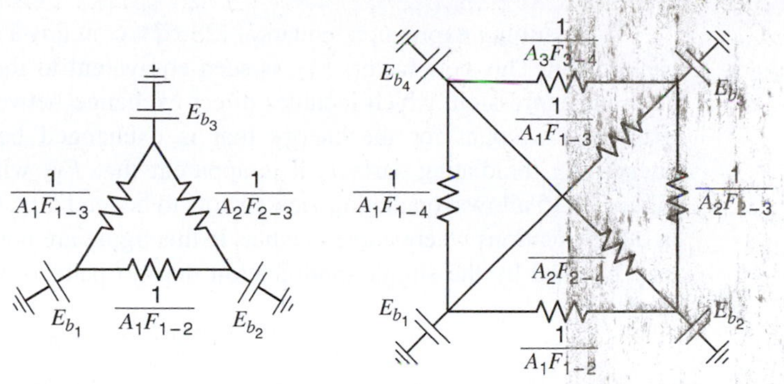

Figure 23.17 Radiation analogs.

Equation (23-25) can be thought of as an analog to Ohm's law where the quantity of transfer, q; the potential driving force, $E_{b_i} - E_{b_j}$; and the thermal resistance, $1/A_1 F_{1j}$; have electrical counterparts I, ΔV, and R, respectively.

Figure 23.17 depicts the analogous electrical circuits for enclosures with three and four surfaces, respectively.

The solution to a three-surface problem—that is, to find q_{12}, q_{13}, q_{23}—although somewhat tedious, can be accomplished in reasonable time. When analyzing enclosures with four or more surfaces, an analytical solution becomes impractical. In such situations one would resort to numerical methods.

▶ **23.9**

RADIANT EXCHANGE WITH RERADIATING SURFACES PRESENT

The circuit diagrams shown in Figure 23.17 show a path to ground at each of the junctions. The thermal analog is a surface that has some external influence whereby its temperature is maintained at a certain level by the addition or rejection of energy. Such a surface is in contact with its surroundings and will conduct heat by virtue of an imposed temperature difference across it.

In radiation applications, we encounter surfaces that effectively are insulated from the surroundings. Such a surface will reemit all radiant energy that is absorbed—usually in a diffuse fashion. These surfaces thus act as reflectors, and their temperatures "float" at some value that is required for the system to be in equilibrium. Figure 23.18 shows a physical situation and the corresponding electric analog for a three-surface enclosure, with one being a nonabsorbing reradiating surface.

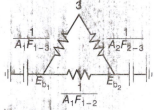

Figure 23.18

Evaluating the net heat transfer between the two black surfaces, q_{1-2}, we have

$$q_{12} = \frac{E_{b_1} - E_{b_2}}{R_{equiv}}$$

$$= \left[A_1 F_{12} + \frac{1}{1/A_1 F_{13} + 1/A_2 F_{23}} \right] (E_{b_1} - E_{b_2})$$

$$= A_1 \left[F_{12} + \frac{1}{1/F_{13} + A_1/A_2 F_{23}} \right] (E_{b_1} - E_{b_2})$$

$$= A_1 \overline{F}_{12} (E_{b_1} - E_{b_2})$$

(23-26)

The resulting expression, equation (23-62), contains a new term, $\overline{F}_{12}$, the *reradiating view factor*. This new factor, $\overline{F}_{12}$, is seen equivalent to the square-bracketed term in the previous expression, which includes direct exchange between surfaces 1 and 2, F_{12}, plus terms that account for the energy that is exchanged between these surfaces via the intervening reradiating surface. It is apparent that $\overline{F}_{12}$ will always be greater than F_{12}. Figure 23.15 allows reradiating view factors to be read directly for some simple geometries. In other situations where curves such as in this figure are not available, the electrical analog may be used by the simple modification that no path to ground exists at the reradiating surface.

▶ 23.10

RADIANT HEAT TRANSFER BETWEEN GRAY SURFACES

In the case of surfaces that are not black, determination of heat transfer becomes more involved. For gray bodies—that is, surfaces for which the absorptivity and emissivity are independent of wavelength—considerable simplifications can be made. The net heat transfer from the surface shown in Figure 23.19 is determined by the difference between the radiation leaving the surface and the radiation incident upon the surface. The *radiosity*, J, is defined as the rate at which radiation leaves a given surface per unit area. The *irradiation*, G, is defined as the rate at which radiation is incident on a surface per unit area. For a gray body, the radiosity, irradiation, and the total emissive power are related by

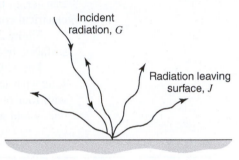

Figure 23.19 Heat transfer at a surface.

$$J = \rho G + \epsilon E_b \tag{23-27}$$

where ρ is the reflectivity and ϵ is the emissivity. The net heat transfer from a surface is

$$\frac{q_{\text{net}}}{A} = J - G = \epsilon E_b + \rho G - G = \epsilon E_b - (1 - \rho)G \tag{23-28}$$

In most cases it is useful to eliminate G from equation (23-28). This yields

$$\frac{q_{\text{net}}}{A} = \epsilon E_b - (1 - \rho)\frac{(J - \epsilon E_b)}{\rho}$$

as $\alpha + \rho = 1$ for an opaque surface:

$$\frac{q_{\text{net}}}{A} = \frac{\epsilon E_b}{\rho} - \frac{\alpha J}{\rho} \tag{23-29}$$

When the emissivity and absorptivity can be considered equal, an important simplification may be made in equation (23-29). Setting $\alpha = \epsilon$, we obtain

$$q_{\text{net}} = \frac{A\epsilon}{\rho}(E_b - J) \tag{23-30}$$

which suggests an analogy with Ohm's law, $V = IR$, where the net heat leaving a surface can be thought of in terms of a current, the difference, $E_b - J$, may be likened to a potential difference, and the quotient $\rho/\epsilon A$ may be termed a resistance. Figure 23.20 illustrates this analogy.

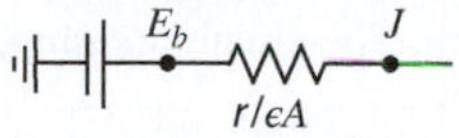

$$q_{net} = \frac{(E_b - J)}{r/\epsilon A}$$

Figure 23.20 Electrical analogy for radiation from a surface.

Now the net exchange of heat via radiation between two surfaces will depend upon their radiosities and their relative "views" of each other. From equation (23-17), we may write

$$q_{1 \rightleftharpoons 2} = A_1 F_{12}(J_1 - J_2) = A_2 F_{21}(J_1 - J_2)$$

We may now write the net heat exchange in terms of the different "resistances" offered by each part of the heat-transfer path as follows:

Rate of heat leaving surface 1: $\qquad q = \dfrac{A_1\epsilon_1}{\rho_1}(E_{b_1} - J_1)$

Rate of heat exchange between surfaces 1 and 2: $\qquad q = A_1 F_{12}(J_1 - J_2)$

Rate of heat received at surface 2: $\qquad q = \dfrac{A_2\epsilon_2}{\rho_2}(J_2 - E_{b_2})$

If surfaces 1 and 2 view each other and no others, then each of the qs in the previous equations is equivalent. In such a case, an additional expression for q can be written in terms of the overall driving force, $E_{b_1} - E_{b_2}$. Such an expression is

$$q = \frac{E_{b_1} - E_{b_2}}{\rho_1/A_1\epsilon_1 + 1/A_1 F_{12} + \rho_2/A_2\epsilon_2} \tag{23-31}$$

where the terms in the denominator are the equivalent resistances due to the characteristics of surface 1, geometry, and the characteristics of surface 2, respectively. The electrical analog to this equation is portrayed in Figure 23.21.

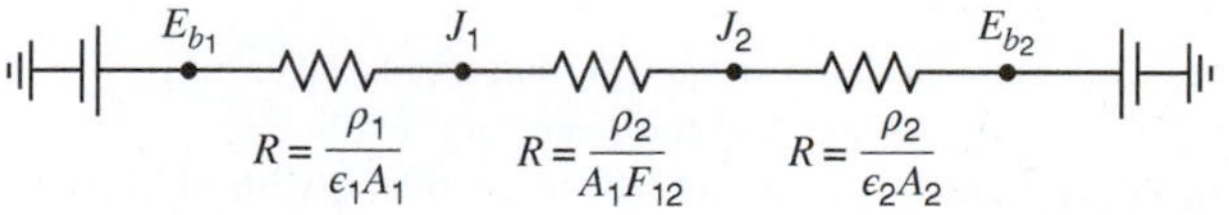

Figure 23.21 Equivalent network for gray-body relations between two surfaces.

The assumptions required to use the electrical analog approach to solve radiation problems are the following:

1. Each surface must be gray.

2. Each surface must be isothermal.

3. Kirchhoff's law must apply—that is, $\alpha = \epsilon$.

4. There is no heat-absorbing medium between the participating surfaces.

Examples 5 and 6, which follow, illustrate features of gray-body problem solutions.

Example 5

Two parallel gray surfaces maintained at temperatures T_1 and T_2 view each other. Each surface is sufficiently large as to be considered infinite. Generate an expression for the net heat transfer between these surfaces.

A simple series electrical circuit is useful in solving this problem. The circuit and important quantities are shown here.

$$R_1 = \rho_1/A_1\epsilon_1 \quad R_2 = 1/A_1F_{12} \quad R_3 = \rho_2/A_2\epsilon_2 \quad E_{b_2} = \sigma T_2^4$$

Utilizing Ohm's law, we obtain the expression

$$q_{12} = \frac{E_{b_1} - E_{b_2}}{\sum R} = \frac{\sigma(T_1^4 - T_2^4)}{\dfrac{\rho 1}{A_1\epsilon_1} + \dfrac{1}{A_1F_{12}} + \dfrac{\rho_2}{A_2\epsilon_2}}$$

Now, noting that for infinite parallel planes $A_1 = A_2 = A$ and $F_{12} = F_{21} = 1$, and writing $\rho_1 = 1 - \epsilon_1$ and $\rho_2 = 1 - \epsilon_2$, we obtain the result

$$q_{12} = \frac{A\sigma(T_1^4 - T_2^4)}{\dfrac{1 - \epsilon_1}{\epsilon_1} + 1 + \dfrac{1 - \epsilon_2}{\epsilon_2}}$$

$$= \frac{A\sigma(T_1^4 - T_2^4)}{\dfrac{1}{\epsilon_1} + \dfrac{1}{\epsilon_2} - 1}$$

Example 6

Two parallel planes measuring 2 m by 2 m are situated 2 m apart. Plate 1 is maintained at a temperature of 1100 K and plate 2 is maintained at 550 K. Determine the net heat transfer from the high-temperature surface under the following conditions:

(a) The plates are black and the surroundings are at 0 K and totally absorbing.
(b) The plates are black and the walls connecting the plates are reradiating.
(c) The plates are gray with emissivities of 0.4 and 0.8, respectively, with black surroundings at 0 K.

Analog electrical circuits for parts (a), (b), and (c) are shown in Figure 23.22.
Heat flux evaluations will require evaluating the quantities F_{12}; F_{1R}; and $\overline{F}_{12}$. The appropriate values are

$$F_{12} = 0.20 \quad \text{from Figure 23.15}$$

$$\overline{F}_{12} = 0.54 \quad \text{from Figure 23.15}$$

and

$$F_{1R} = 1 - F_{12} = 0.80$$

Part (a). The net rate of heat leaving plate 1 is

$$q_{1\,net} = q_{12} + q_{1R}$$
$$= A_1F_{12}(E_{b_1} - E_{b_2}) + A_1F_{1R}E_{b_1}$$
$$= (4\,m^2)(0.2)(5.676 \times 10^{-8}\,W/m^2 \cdot K^4)(1100^4 - 550^4)K^4 + (4\,m^2)(0.8)(5.676 \times 10^{-8}\,W/m^2 \cdot K^4)(1100\,K)^4$$
$$= 62,300\,W + 266,000\,W$$
$$= 328.3\,kW$$

$$F_{1R} = 1 - F_{12} = 0.80$$

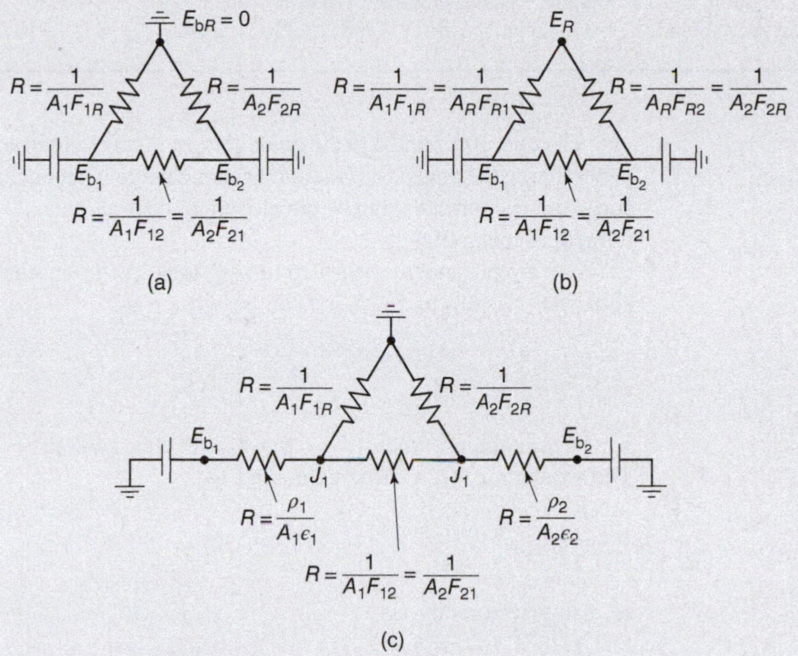

(a)

(b)

(c)

Figure 23.22 Equivalent circuits for Example 6.

Part (b). When reradiating walls are present, the heat flux becomes

$$q_{12} = (E_{b_1} - E_{b_2}) \left[A_1 F_{12} + \cfrac{1}{\cfrac{1}{A_1 F_{1R}} + \cfrac{1}{A_2 F_{2R}}} \right]$$

and, since $A_1 = A_2$ and $F_{1R} = F_{2R}$

$$q_{12} = (E_{b_1} - E_{b_2}) A_1 \left[F_{12} + \frac{F_{1R}}{2} \right]$$

Since $F_{12} + F_{1R} = 1$, the bracketed term is evaluated as

$$F_{12} + \frac{F_{1R}}{2} = 0.2 + \frac{0.8}{2} = 0.6$$

and, finally, the heat flux is

$$q_{12} = (4\,\text{m}^2)(5.678 \times 10^{-8}\ \text{W/m}^2 \cdot \text{K}^4)(1100^4 - 550^4)\text{K}^4(0.6)$$
$$= 187\,\text{kW}$$

We should note that an equivalent expression for the heat flux is

$$q_{12} = A_1 \overline{F}_{12}(E_{b_1} - E_{b_2})$$

and, using the value $\overline{F}_{12} = 0.54$, from Figure 23.15, the result would be

$$q_{12} = 168.3\,\text{kW}$$

This alternate result is the more accurate in that the values of $\overline{F}_{12}$ plotted in Figure 23.15 allow for the temperatures along the reradiating walls to vary from T_1 to T_2. The use of the analog circuit considers the radiating surface to be a constant temperature. Such an assumption, in this example, leads to an error of approximately 11%.

Part (c). An evaluation of the circuit shown in Figure 23.22(c) yields $q_{1,\text{out}} = 131.3\,\text{kW}$.

The concepts related to the quantities, radiosity, and irradiation are particularly useful in generalizing the analysis of radiant heat exchange in an enclosure containing any number of surfaces. The formalism to be developed in this section is directly applicable for solution by numerical methods.

For a representative surface having area, A_i, in an enclosure bounded by n surfaces, equations (23-28) and (23-30) can be written as

$$q_i = \frac{E_{b_i} - J_i}{\rho_i/A_i\epsilon_i} = A_i(J_i - G_i) \tag{23-32}$$

where q_i is the net rate of heat transfer leaving surface i.

The irradiation, G_i, can be expressed as

$$A_i G_i = \sum_{j=1}^{n} J_j A_j F_{ji} \tag{23-33}$$

or, using reciprocity, as

$$A_i G_i = A_i \sum_{j=1}^{n} J_j F_{ij} \tag{23-34}$$

Combining equations (23-32) and (23-34) we obtain

$$q_i = A_i\Big[J_i - \sum_{j=i}^{n} F_{ij} J_j\Big] \tag{23-35}$$

$$= \frac{A_i \epsilon_i}{\rho_i} E_{b_i} - \frac{A_i \epsilon_i}{\rho_i} J_i \tag{23-36}$$

We can now write the two basic expressions for a general surface in an enclosure. If the surface heat flux is known, equation (23-35) can be expressed in the form

$$J_i - \sum_{j=1}^{n} F_{ij} J_j = \frac{q_i}{A_i}$$

or

$$J_i(1 - F_{ii}) - \sum_{\substack{j=1 \\ j \neq i}}^{n} F_{ij} J_j = \frac{q_i}{A_i} \tag{23-37}$$

and, if the temperature at surface i is known, equations (23-35) and (23-36) yield

$$\frac{A_i \epsilon_i}{\rho_i}(E_{b_i} - J_i) = A_i\Big[J_i - \sum_{j=1}^{n} F_{ij} J_j\Big]$$

$$= A_i\Big[J_i(1 - F_{ii}) - \sum_{\substack{j=1 \\ j \neq i}}^{n} F_{ij} J_j\Big]$$

and, finally

$$\left(1 - F_{ii} + \frac{\epsilon_i}{\rho_i}\right)J_i - \sum_{\substack{j=1 \\ j \neq i}}^{n} F_{ij}J_j = \frac{\epsilon_i}{\rho_i}E_{b_i} \tag{23-38}$$

Equations (23-37) and (23-38) comprise the algorithm for evaluating quantities of interest in a many-surface enclosure. The former applies to a surface of known heat flux; the latter is written when the surface temperature is specified.

In these two equations the terms involving the view factor, F_{ii}, have been separated out of the summation. This quantity, F_{ii}, will have a nonzero value in those cases when surface i "sees"; itself—i.e., it is concave. In most cases, F_{ii} will be 0.

When writing equation (23-37) or (23-38) for each surface in an enclosure, a series of n simultaneous equations is generated, involving the unknowns J_i. This set of equations can be represented in matrix form as

$$[A][J] = [B] \tag{23-39}$$

where $[A]$ is the coefficient matrix, $[B]$ is a column matrix involving the right-hand sides of equations (23-37) and (23-38), and $[J]$ is a column matrix of the unknowns, J_i. The solution for the J_i then proceeds according to

$$[J] = [C][B] \tag{23-40}$$

where

$$[C] = [A]^{-1} \tag{23-41}$$

is the inverse of the coefficient matrix.

Example 7 illustrates the application of this approach.

Example 7

Solve the problem posed in Example 5 using the methods developed in this section.

For this case $n = 3$ and the problem formulation will involve three equations—one for each surface. Part (a). Each of the surfaces is at a known temperature in this case, thus equation (23-38) applies. The following conditions are known:

$$T_1 = 1100\,\text{K} \qquad T_2 = 550\,\text{K} \qquad T_3 = 0\,\text{K}$$

$$F_{11} = 0 \qquad F_{21} = 0.2 \qquad F_{31} = 0.2$$

$$F_{12} = 0.2 \qquad F_{23} = 0 \qquad F_{32} = 0.2$$

$$F_{13} = 0.8 \qquad F_{23} = 0.8 \qquad F_{33} = 0.6$$

$$\epsilon_1 = 1 \qquad \epsilon_2 = 1 \qquad \epsilon_3 = 1$$

We can write the following:

$$\left(1 + \frac{\epsilon_1}{\rho_1}\right)J_1 - [F_{12}J_2 + F_{13}J_3] = \frac{\epsilon_1}{\rho_1}E_{b_1}$$

$$\left(1 + \frac{\epsilon_2}{\rho_2}\right)J_2 - [F_{21}J_1 + F_{23}J_3] = \frac{\epsilon_2}{\rho_2}E_{b_2}$$

$$\left(1 - F_{33} + \frac{\epsilon_3}{\rho_3}\right)J_3 - [F_{31}J_1 + F_{32}J_2] = \frac{\epsilon_3}{\rho_3}E_{b_3}$$

which, for the given conditions, reduce to

$$J_1 = E_{b_1} = \sigma T_1^4$$

$$J_2 = E_{b_2} = \sigma T_2^4$$

$$J_3 = 0$$

The net heat leaving plate 1 is thus, according to equation (23-37), equal to

$$q_1 = A_1 [J_1 - F_{12}J_2]$$

$$= A_1 [\sigma T_1^4 - 0.2\sigma T_2^4]$$

$$= 4 \, \text{m}^2 (5.676 \times 10^{-8} \, \text{W/m}^2 \cdot \text{K}^4)[1100^4 - 0.2(550)^4] \text{K}^4$$

$$= 328.3 \, \text{kW}$$

Part (b). Values of T_i and F_{ij} remain the same. The only change from part (a) is that $\epsilon_3 = 0$. The set of equations applying to the three surfaces are again

$$\left(1 + \frac{\epsilon_1}{\rho_1}\right)J_1 - [F_{12}J_2 + F_{13}J_3] = \frac{\epsilon_1}{\rho_1}E_{b_1}$$

$$\left(1 + \frac{\epsilon_2}{\rho_2}\right)J_2 - [F_{21}J_1 + F_{23}J_3] = \frac{\epsilon_2}{\rho_2}E_{b_2}$$

$$\left(1 + F_{33} + \frac{\epsilon_3}{\rho_3}\right)J_3 - [F_{31}J_1 + F_{32}J_2] = \frac{\epsilon_3}{\rho_3}E_{b_3}$$

as before. Substituting values for T_i, F_{ij}, and ϵ_i, we have

$$J_1 = E_{b_1} = \sigma T_1^4$$

$$J_2 = E_{b_2} = \sigma T_2^4$$

$$(1 - F_{33})J_3 - F_{31}J_1 - F_{32}J_2 = 0$$

The expression for q_i is

$$q_1 = A_1 [J_1 - F_{12}J_2 - F_{13}J_3] = A_1 \left[J_1 - F_{12}J_2 - \frac{F_{13}}{1 - F_{33}}(F_{31}J_1 + F_{32}J_2)\right]$$

$$= A_1 \left[J_1 \left(1 - \frac{F_{13}F_{31}}{1 - F_{33}}\right) - J_2 \left(F_{12} + \frac{F_{13}F_{32}}{1 - F_{33}}\right)\right]$$

and, with numerical values inserted, we obtain

$$q_1 = 4(5.676 \times 10^{-8}) \left\{ (1100)^4 \left[1 - \frac{(0.8)(0.2)}{1 - 0.6}\right] - (550)^4 \left[0.2 + \frac{(0.8)(0.2)}{1 - 0.6}\right] \right\} = 187.0 \, \text{kW}$$

Part (c). Values of T_i and F_{ij} remain the same. Emissivities are

$$\epsilon_1 = 0.4 \qquad \epsilon_2 = 0.8 \qquad \epsilon_3 = 1$$

Equations for the three surfaces are, again,

$$\left(1 + \frac{\epsilon_1}{\rho_1}\right)J_1 - [F_{12}J_2 + F_{13}J_3] = \frac{\epsilon_1}{\rho_1}E_{b_1}$$

$$\left(1 + \frac{\epsilon_2}{\rho_2}\right)J_2 - [F_{21}J_1 + F_{23}J_3] = \frac{\epsilon_2}{\rho_2}E_{b_2}$$

$$\left(1 + F_{33} + \frac{\epsilon_3}{\rho_3}\right)J_3 - [F_{31}J_1 + F_{32}J_2] = \frac{\epsilon_3}{\rho_3}E_{b_3}$$

which become

$$\left(1 + \frac{0.4}{0.6}\right)J_1 - (F_{12}J_2 + F_{13}J_3) = \frac{0.4}{0.6}E_{b_1}$$

$$\left(1 + \frac{0.8}{0.2}\right)J_2 - (F_{21}J_1 + F_{23}J_3) = \frac{0.8}{0.2}E_{b_2}$$

$$J_3 = 0$$

We now have

$$1.67J_1 - 0.2J_2 = 0.67E_{b_1}$$

$$5J_2 - 0.2J_1 = 4E_{b_2}$$

Solving these two equations simultaneously for J_1 and J_2, we get

$$J_1 = 33\,900 \text{ W/m}^2$$

$$J_2 = 5510 \text{ W/m}^2$$

and the value for q_i is evaluated as

$$q_1 = \left[33\,900 - \frac{5510}{5}\right]4$$

$$= 131.2 \text{ kW}$$

▶ **23.11**

RADIATION FROM GASES

So far, the interaction of radiation with gases has been neglected. Gases emit and absorb radiation in discrete energy bands dictated by the allowed energy states within the molecule. As the energy associated with, say, the vibrational or rotational motion of a molecule may have only certain values, it follows that the amount of energy emitted or absorbed by a molecule will have a frequency, $\nu = \Delta E/h$, corresponding to the difference in energy ΔE between allowed states. Thus, while the energy emitted by a solid will comprise a continuous spectrum, the radiation emitted and absorbed by a gas will be restricted to bands. Figure 23.23 illustrates the emission bands of carbon dioxide and water vapor relative to black-body radiation at 1500 F.

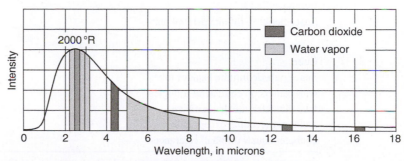

Figure 23.23 Emission bands of CO_2 and H_2O.

The emission of radiation for these gases is seen to occur in the infrared region of the spectrum.

For nonluminous gases, the inert gases and diatomic gases of symmetrical composition such as O_2, N_2, and H_2 may be considered transparent to thermal radiation. Important types of media that absorb and emit radiations are polyatomic gases such as CO_2 and H_2O and unsymmetrical molecules such as CO. These gases are also associated with the products of combustion of hydrocarbons. The determination of the absorption and emission of radiation is very difficult, as it involves the temperature, composition, density, and geometry of the gas. There are several simplifications that allow estimation of radiation in gases to be made in a straightforward manner. These idealizations are as follows:

1. The gas is in thermodynamic equilibrium. The state of the gas may therefore be characterized locally by a single temperature.

2. The gas may be considered gray. This simplification allows the absorption and emission of radiation to be characterized by one parameter as $\alpha = \epsilon$ for a gray body.

In the range of temperatures associated with the products of hydrocarbon combustion, the gray-gas emissivities of H_2O and CO_2 may be obtained from the results of Hottel. A hemispherical mass of gas at 1 atm pressure was used by Hottel to evaluate the emissivity. While the graphs apply strictly only to a hemispherical gas mass of radius L, other shapes can be treated by consideration of a mean beam length L as given in Table 23.5. For geometries not covered in the table, the mean beam length may be approximated by the relation $L = 3.4$ (volume)/(surface area).

Figure 23.24 gives the emissivity of a hemispherical mass of water vapor at 1 atm total pressure and near-zero partial pressure as a function of temperature and the product $p_w L$, where p_w is the partial pressure of the water vapor. For pressures other than atmospheric, Figure 23.25 gives the correction factor, C_w, which is the ratio of the emissivity at total pressure P to the emissivity at a total pressure of 1 atm. Figures 23.26 and 23.27 give the corresponding data for CO_2.

From Figure 23.23, it may be seen that the emission bands of CO_2 and H_2O overlap. When both carbon dioxide and water vapor are present, the total emissivity may be

Table 23.5 Mean beam length, L, for various geometries[†]

Shape	L
Sphere	$\frac{2}{3} \times$ diameter
Infinite cylinder	$1 \times$ diameter
Space between infinite parallel planes	$1.8 \times$ distance between planes
Cube	$\frac{2}{3} \times$ side
Space outside infinite bank of tubes with centers on equilateral triangles; tube diameter equals clearance	$2.8 \times$ clearance
Same as preceding except tube diameter equals one-half clearance	$3.8 \times$ clearance

[†] From H. C. Hottel, "Radiation," Chap. IV in W. H. McAdams (ed.), *Heat Transmission*, 3rd edition, McGraw-Hill Book Company, New York, 1964. By permission of the publishers.

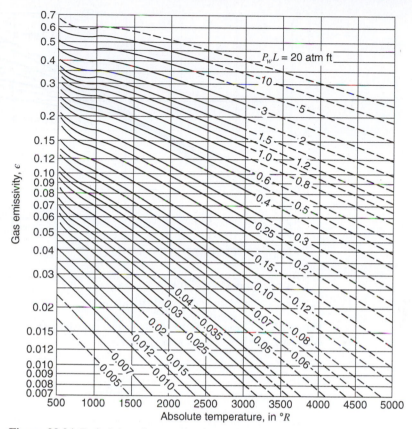

Figure 23.24 Emissivity of water vapor at one atmosphere total pressure and near-zero partial pressure.

determined from the relation

$$\epsilon_{\text{total}} = \epsilon_{H_2O} + \epsilon_{CO_2} - \Delta\epsilon$$

where $\Delta\epsilon$ is given in Figure 23.28.

The results presented here for the gray gas are gross simplifications. For a more complete treatment, textbooks by Siegel and Howell,[2] Modest,[3] and Brewester[4] present the fundamentals of nongray-gas radiation, along with extensive bibliographies.

▶ **23.12**

THE RADIATION HEAT-TRANSFER COEFFICIENT

Frequently in engineering analysis, convection and radiation occur simultaneously rather than as isolated phenomena. An important approximation in such cases is the linearization

[2] R. Siegel and J. R. Howell, *Thermal Radiation Heat Transfer*, 3rd edition, Hemisphere Publishing Corp., Washington, 1992.

[3] M. F. Modest, *Radiative Heat Transfer*, McGraw-Hill, New York, 1993.

[4] M. Q. Brewster, *Thermal Radiative Transfer and Properties*, J. Wiley and Sons, New York, 1992.

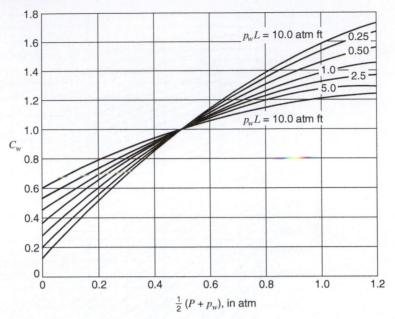

Figure 23.25 Correction factor for converting emissivity of H_2O at one atmosphere total pressure to emissivity at P atmospheres total pressure.

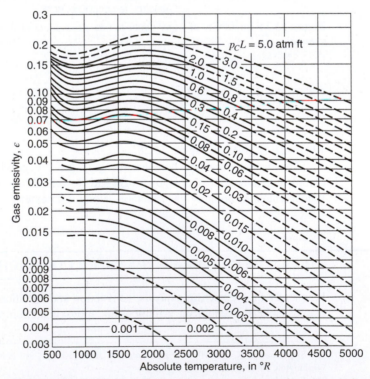

Figure 23.26 Emissivity of CO_2 at one atmosphere total pressure and near-zero partial pressure.

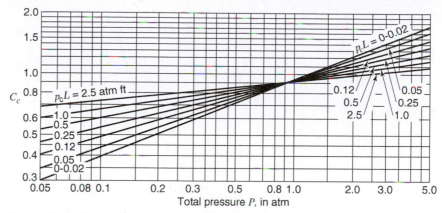

Figure 23.27 Correction factor for converting emissivity of CO_2 at one atmosphere total pressure to emissivity at P atmospheres total pressure.

of the radiation contribution so that

$$h_{\text{total}} = h_{\text{convection}} + h_{\text{radiation}} \tag{23-42}$$

where

$$
\begin{aligned}
h_r &\equiv \frac{q_r/A_1}{(T - T_R)} \\
&= \mathscr{F}_{1-2}\left[\frac{\sigma(T^4 - T_2^4)}{T - T_R}\right]
\end{aligned}
\tag{23-43}
$$

Here T_R is a reference temperature, and T_1 and T_2 are the respective surface temperatures. In effect, equation (23-43) represents a straight-line approximation to the radiant heat transfer as illustrated in Figure 23.29. The factor, $\mathscr{F}$, accounts for geometry and surface condition of the radiating and absorbing surface.

By constructing a tangent to the relation curve at $T = T_1$, the following relations are obtained for h_r and T_R:

$$h_r = 4\sigma T_1^3 \mathscr{F}_{1-2} \tag{23-44}$$

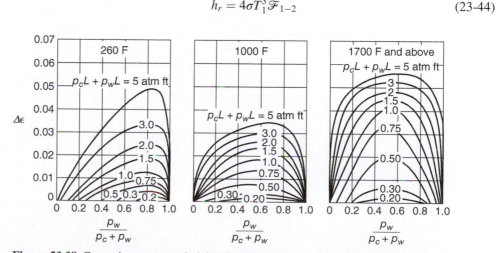

Figure 23.28 Correction to gas emissivity due to spectral overlap of H_2O and CO_2.

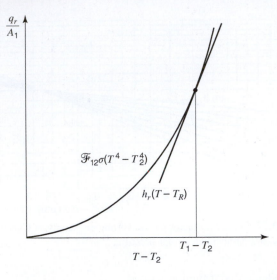

Figure 23.29 Tangent approximation for h_r.

and

$$T_R = T_1 - \frac{T_1^4 - T_2^4}{4T_1^3} \tag{23-45}$$

▶ 23.13

CLOSURE

Radiation heat transfer has been considered in this chapter. Radiant-energy transfer is associated with the portion of the electromagnetic spectrum between 0.1 and 100 μm, which is generally referred to as the thermal band.

The fundamental rate equation for thermal radiation, introduced in Chapter 15, is designated the Stefan–Boltzmann equation; it is expressed as

$$E_b = \sigma T^4 \tag{23-12}$$

where E_b is the black-body emissive power, T is the absolute temperature, and σ is the Stefan–Boltzmann constant, having units of $W/m^2 \cdot K^4$ in the SI system.

Modifications to this relationship were made for nonblack surfaces and for geometric relationships between multiple surfaces in view of each other.

The presence of absorbing and emitting gases between surfaces was also examined. The gases of principal interest in this regard are water vapor and carbon dioxide.

CHAPTER 24

Fundamentals of Mass Transfer

The previous chapters dealing with the transport phenomena of momentum and heat transfer have dealt with one-component phases that possessed a natural tendency to reach equilibrium conditions. When a system contains two or more components whose concentrations vary from point to point, there is a natural tendency for mass to be transferred, minimizing the concentration differences within the system. The transport of one constituent from a region of higher concentration to that of a lower concentration is called *mass transfer*.

Many of our day-to-day experiences involve mass transfer. A lump of sugar added to a cup of coffee eventually dissolves and diffuses uniformly throughout the coffee. Water evaporates from ponds to increase the humidity of the passing air stream. Perfume presents a pleasant fragrance that is imparted throughout the surrounding atmosphere.

Mass transfer is the basis for many biological and chemical processes. Biological processes include the oxygenation of blood and the transport of ions across membranes within the kidney. Chemical processes include the chemical vapor deposition (CVD) of silane (SiH_4) onto a silicon wafer, the doping of a silicon wafer to form a semiconducting thin film, the aeration of wastewater, and the purification of ores and isotopes. Mass transfer underlies the various chemical separation processes where one or more components migrate from one phase to the interface between the two phases in contact. For example, in adsorption or crystallization processes, the components remain at the interface, whereas in gas absorption and liquid–liquid extraction processes, the components penetrate the interface and then transfer into the bulk of the second phase.

If we consider the lump of sugar added to the cup of coffee, experience teaches us that the length of time required to distribute the sugar will depend upon whether the liquid is quiescent or is mechanically agitated by a spoon. The mechanism of mass transfer, as we have also observed in heat transfer, depends upon the dynamics of the system in which it occurs. Mass can be transferred by random molecular motion in quiescent fluids, or it can be transferred from a surface into a moving fluid, aided by the dynamic characteristics of the flow. These two distinct modes of transport, molecular mass transfer and convective mass transfer, are analogous to conduction heat transfer and convective heat transfer. Both of these modes of mass transfer will be described and analyzed. As in the case of heat transfer, we should immediately realize that the two mechanisms often act simultaneously. However, in the confluence of the two modes of mass transfer, one mechanism can dominate quantitatively so that approximate solutions involving only the dominant mode need be used.

MOLECULAR MASS TRANSFER

As early as 1815, Parrot observed qualitatively that whenever a gas mixture contains two or more molecular species whose relative concentrations vary from point to point, an apparently natural process results that tends to diminish any inequalities of composition. This macroscopic transport of mass, independent of any convection within the system, is defined as *molecular diffusion*.

In the specific case of gaseous mixtures, a logical explanation of this transport phenomenon can be deduced from the kinetic theory of gases. At temperatures above absolute zero, individual molecules are in a state of continual, yet random, motion. Within dilute gas mixtures, each solute molecule behaves independently of the other solute molecules, for it seldom encounters them. Collisions between the molecules continually occur. As a result of the collisions, the solute molecules move along a zigzag path, sometimes toward a region of higher concentration, sometimes toward a lower concentration.

As pointed out in Chapters 7 and 15, the transport of momentum and the transport of energy by conduction are also the result of random molecular motion. Accordingly, one should expect the three transport phenomena to depend upon many of the same characteristic properties, such as mean-free path, and that the theoretical analyses of all three phenomena will have much in common.

Let us consider the simple mass-transfer process shown in Figure 24.1. This device, called an Arnold diffusion cell, is used to analyze mass-transfer processes. A volatile liquid (component *A*) rests at the bottom of a tube. At a constant temperature, this liquid exerts a vapor pressure, and molecules of species *A* in the gas phase "diffuse" through air (species *B*) present within tube. The gas space inside the tube is not externally agitated. At the molecular level, we imagine the gas molecules moving randomly as described by the kinetic theory of gases. The net flux of species *A* is the net sum of all molecules of *A* moving through the cross section of tube per unit time along the increasing *z*-direction. Air flows gently over the top of the open tube. As the molecules of species *A* in the gas phase reach the top of the tube, they are swept away by the flowing air stream. Thus, the natural tendency of the molecules of

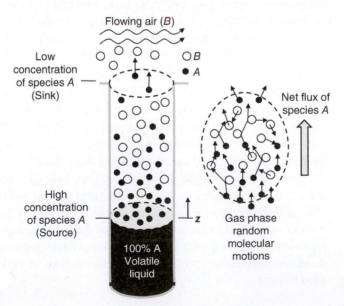

Figure 24.1 Simple mass-transfer process.

species A is to move from the source for mass transfer, the pure liquid interface that generates the vapor species A, to the air stream flowing of the top of the tube that provides a sink for A. We note that both species A and B may have a concentration gradient within the tube. However, since air (species B) has no sink, only species A undergoes mass transfer, although the presence of air (B) may still affect the rate of mass transfer.

In summary, we see that mass transfer is the natural tendency to transfer a given component (species) in a mixture from a region of high concentration, called the source, to a region of low concentration, called the sink, in an effort to bring about a uniform or equilibrium condition. Mass transfer has three requirements: (1) that transfer occur only in a mixture, (2) that at least one substance within the mixture move from its source to its sink, and (3) that the rate of mass transfer—i.e., the "flux" of a given substance—be proportional to the concentration gradient defined by the source and sink for that substance.

The Fick Rate Equation

The laws of mass transfer show the relation between the flux of the diffusing substance and the concentration gradient responsible for this mass transfer. Unfortunately, the quantitative description of molecular diffusion is considerably more complex than the analogous descriptions for the molecular transfer of momentum and energy that occur in a one-component phase. Because molecular mass transfer—or *diffusion*, as it is also called—occurs only in mixtures, its evaluation must involve an examination of the effect of each component. For example, we will often desire to know the diffusion rate of a specific component relative to the velocity of the mixture in which it is moving. Because each component may possess a different mobility, the mixture velocity must be evaluated by averaging the velocities of all of the components present.

To establish a common basis for future discussions, let us first consider definitions and relations that are often used to explain the role of components within a mixture.

Concentrations   Mass transfer occurs only in a mixture. Consider the binary, multi-component, and pseudo-binary mixtures of molecules within a volume element dV, as represented in Figure 24.2. In a binary mixture, only two components are present. In a multi-component mixture, more than two components are present, but no population of any given component dominates. In a pseudo-binary mixture, one component is dominant (C), and the population of the other species is small, relative to this dominant component. For gases, this dominant component is the often called the carrier gas, and for liquids, it is called the solvent.

Mass transfer can occur in gases, liquids, or solids. In the case of gases, the molecules do not touch on another but collide frequently, as described in the kinetic theory of gases. For liquids, the molecules are in a condensed phase but still move freely, albeit at a smaller speed relative to gases. For solids, one or more of the components are connected in a lattice structure or bound together in an amorphous state.

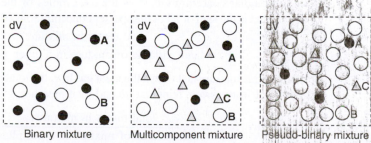

Binary mixture   Multicomponent mixture   Pseudo-binary mixture

Figure 24.2 Elemental volume containing a binary, multicomponent, and pseudo-binary mixture.

In a multicomponent mixture, the concentration of a molecular species can be expressed in many ways. Because each molecule of each species has a mass, a *mass concentration* for each species, as well as for the mixture, can be defined. For species A, the *mass* concentration ρ_A is defined as the mass of A per unit volume of the mixture:

$$\rho_A = \frac{\text{mass of component } A}{\text{unit volume of mixture in a given phase}} \tag{24-1a}$$

The total mass concentration or *density*, ρ, is the total mass of the mixture contained in the unit volume—that is,

$$\rho = \frac{\text{total mass in mixture}}{\text{unit volume of mixture in a given phase}} = \sum_{i=1}^{n} \rho_i \tag{24-1b}$$

where n is the number of species in the mixture. The *mass fraction*, w_A, is the mass concentration of species A divided by the total mass density:

$$w_A = \frac{\rho_A}{\sum\limits_{i}^{n} \rho_i} = \frac{\rho_A}{\rho} \tag{24-2}$$

The sum of the mass fractions, by definition, must be 1:

$$\sum_{i=1}^{n} w_i = 1.0 \tag{24-3}$$

The *molar concentration* of species A, c_A, is defined as the number of moles of A present per unit volume of the mixture:

$$c_A = \frac{\text{moles of component } A}{\text{unit volume of mixture in a given phase}}$$

By definition, one mole of any species contains a mass equivalent to its molecular weight; the mass concentration and molar concentration terms are related by:

$$c_A = \frac{\rho_A}{M_A} \tag{24-4}$$

where M_A is the molecular weight of species A. When dealing with a gas phase, concentrations are often expressed in terms of partial pressures. Under conditions in which the ideal gas law applies, the molar concentration is

$$c_A = \frac{n_A}{V} = \frac{p_A}{RT} \tag{24-5}$$

where p_A is the partial pressure of the species A in the mixture, n_A is the number of moles of species A, V is the gas volume, T is the absolute temperature, and R is the gas constant. The total molar concentration, c, is the total moles of the mixture contained in the unit volume, and is the sum of all of the species concentrations of the components in the mixture:

$$c = \frac{\text{total moles in mixture}}{\text{unit volume of mixture in a given phase}} = \sum_{i=1}^{n} c_i \tag{24-6}$$

For a gaseous mixture that obeys the ideal gas law, the total molar concentration can also be determined from system temperature and total pressure, given by

$$c = \frac{\text{total moles in mixture}}{\text{unit volume of mixture in a gas phase}} = \frac{n}{V} = \frac{P}{RT} \tag{24-7}$$

where P is the system total pressure and T is the absolute temperature. The mole fraction for liquid mixtures, x_A, is defined as

$$x_A = \frac{\text{moles of species } A \text{ in liquid mixture}}{\text{total moles of all species in the liquid mixture}} = \frac{c_A}{c} \qquad (24\text{-}8a)$$

with c_A and c based on the liquid phase. For dilute liquid mixtures of solute A in solvent B, the total molar concentration c approximates the molar solvent concentration c_B. The mole fraction for gas-phase mixtures, y_A, is defined as

$$y_A = \frac{\text{moles of species } A \text{ in gas mixture}}{\text{total moles of all species in the gas mixture}} = \frac{c_A}{c} \qquad (24\text{-}8b)$$

with c_A and c based on the gas phase. For a gaseous mixture that obeys the ideal gas law, the mole fraction, y_A, can be also written in terms of pressures:

$$y_A = p_A/P \qquad (24\text{-}9)$$

The sum of the mole fractions, by definition, must be 1:

$$\sum_{i=1}^{n} x_i = 1.0 \qquad (24\text{-}10a)$$

$$\sum_{i=1}^{n} y_i = 1.0 \qquad (24\text{-}10b)$$

A summary of the various concentration terms and of the interrelations for a binary system containing species A and B is given in Table 24.1. For a binary mixture, the mole fraction is related to the mass fraction by the following relationships:

$$x_A \text{ or } y_A = \frac{w_A/M_A}{w_A/M_A + w_B/M_B} \qquad (24\text{-}11)$$

$$w_A = \frac{x_A M_A}{x_A M_A + x_B M_B} \text{ (liquids)} \qquad (24\text{-}12a)$$

$$w_A = \frac{y_A M_A}{y_A M_A + y_B M_B} \text{ (gases)} \qquad (24\text{-}12b)$$

Table 24.1   Concentrations in a binary mixture of A and B

Molar Liquid Concentrations	Molar Gas Concentrations	Mass Concentrations (Gas or Liquid)
$c = n/V$	$c = n/V = P/RT$	$\rho = m/V$
$c_A = n_A/V$	$c_A = n_A/V = p_A/RT$	$\rho_A = m_A/V$
$c_B = n_B/V$	$c_B = n_B/V = p_B/RT$	$\rho_B = m_B/V$
$x_A = c_A/c$	$y_A = c_A/c = p_A/P$	$w_A = \rho_A/\rho$
$x_B = c_B/c$	$y_B = c_B/c = p_B/P$	$w_B = \rho_B/\rho$
$c = c_A + c_B$	$c = c_A + c_B, P = p_A + p_B$	$\rho = \rho_A + \rho_B$
$1 = x_A + x_B$	$1 = y_A + y_B$	$1 = w_A + w_B$

Example 1

A gas mixture from a hydrocarbon reforming process contains 50% H_2, 40% CO_2, and 10% methane (CH_4) by volume at 400 C (673 K) and 1.5 atm total system pressure. Determine the molar concentration and mass fraction of each species in the mixture, as well as the density of the mixture.

Let $A = H_2$, $B = CO_2$, and $C = CH_4$. Assuming ideal gas behavior, the total molar concentration is

$$c = \frac{P}{RT} = \frac{1.5 \text{ atm}}{(0.08206 \text{ m}^3 \cdot \text{atm/kgmole} \cdot \text{K})(673 \text{ K})} = 2.72 \times 10^{-2} \text{ kgmole/m}^3$$

For an ideal gas, volume percent composition is equivalent to mole percent composition. The molar concentration of species A is

$$c_A = y_A c = (0.50)(2.72 \times 10^{-2} \text{ kgmole/m}^3) = 1.36 \times 10^{-2} \text{ kgmole/m}^3$$

Similarly, $c_B = 1.10 \times 10^{-2}$ kgmole/m^3, and $c_C = 2.72 \times 10^{-3}$ kgmole/m^3. The mass fraction of each species is determined by the mole fraction and molecular weight of each species:

$$w_A = \frac{y_A M_A}{y_A M_A + y_B M_B + y_C M_C} = \frac{(0.50)(2)}{(0.50)(2) + (0.40)(44) + (0.10)(16)} = 0.0495 \frac{\text{g } H_2}{\text{total g}}$$

Likewise, $w_B = 0.871$, and $w_C = 0.0793$. The mass density of the gas mixture is

$$\rho = \rho_A + \rho_B + \rho_C = c_A M_A + c_B M_B + c_C M_C$$

$$= \left(1.36 \times 10^{-2} \frac{\text{kgmole}}{\text{m}^3}\right)\left(\frac{2 \text{ g}}{\text{gmole}}\right) + \left(1.10 \times 10^{-2} \frac{\text{kgmole}}{\text{m}^3}\right)\left(\frac{44 \text{ g}}{\text{gmole}}\right)$$

$$+ \left(2.72 \times 10^{-3} \frac{\text{kgmole}}{\text{m}^3}\right)\left(\frac{16 \text{ g}}{\text{gmole}}\right) = 0.555 \frac{\text{kg}}{\text{m}^3}$$

Example 2

A wastewater stream is contaminated with 200 mg/L of dissolved trichloroethylene (TCE) at 20 C, which is below its solubility limit in water. What are the molar concentration (in SI units) and the mole fraction of TCE in the wastewater, assuming a dilute solution? At 20 C, the mass transfer of liquid water is 998.2 kg/m^3 (Appendix I). The molecular weight of the TCE is 131.4 g/gmole, and the molecular weight of water is 18 g/gmole (18 kg/kgmole).

Let species A represent TCE (solute) and species B represent water (solvent). The molar concentration of TCE in the wastewater (c_A) is determined from the mass concentration (ρ_A):

$$c_A = \frac{\rho_A}{M_A} = \frac{200 \text{ mg } A/\text{L}}{131.4 \text{ g/gmole}} \frac{1 \text{g}}{1000 \text{ mg}} \frac{1 \text{ kgmole}}{1000 \text{ gmole}} \frac{1000 \text{ L}}{1 \text{ m}^3} = 1.52 \times 10^{-3} \frac{\text{kgmole}}{\text{m}^3}$$

For a dilute solution of TCE (solute A) in water (solvent B), the total molar concentration approximates the molar concentration of the solvent:

$$c \cong c_B = \frac{\rho_B}{M_B} = \frac{998.3 \text{ kg/m}^3}{18 \text{ kg/kgmole}} = 55.5 \frac{\text{kgmole}}{\text{m}^3}$$

and so the mole fraction (x_A) is

$$x_A = \frac{c_A}{c} = \frac{1.52 \times 10^{-3} \text{ kgmole/m}^3}{55.5 \text{ kgmole/m}^3} = 2.74 \times 10^{-5}$$

Velocities In a multicomponent system, the various species will normally move at different velocities. Accordingly, an evaluation of a velocity for the gas mixture requires the averaging of the velocities of each species present.

The *mass-average velocity* for a multicomponent mixture is defined in terms of the mass densities and velocities of all components by

$$v = \frac{\sum_{i=1}^{n} \rho_i v_i}{\sum_{i=1}^{n} \rho_i} = \frac{\sum_{i=1}^{n} \rho_i v_i}{\rho} \tag{24-13}$$

where v_i denotes the absolute velocity of species i relative to stationary coordinate axes. This is the velocity that was previously encountered in the equations of momentum transfer. The *molar-average velocity* for a multicomponent mixture is defined in terms of the molar concentrations of all components by

$$V = \frac{\sum_{i=1}^{n} c_i v_i}{\sum_{i=1}^{n} c_i} = \frac{\sum_{i=1}^{n} c_i v_i}{c} \tag{24-14}$$

The velocity of a particular species relative to the mass-average or molar-average velocity is termed a *diffusion velocity*. We can define two different diffusion velocities $v_i - v$, the diffusion velocity of species i relative to the mass-average velocity and $v_i - V$, the diffusion velocity of species i relative to the molar-velocity average. According to Fick's law, a species can have a velocity relative to the mass- or molar-average velocity only if gradients in the concentration exist.

Fluxes The mass (or molar) flux of a given species is a vector quantity denoting the amount of the particular species, in either mass or molar units, that passes per given increment of time through a unit area normal to the vector. The flux may be defined with reference to coordinates that are fixed in space, coordinates that are moving with the mass-average velocity, or coordinates that are moving with the molar-average velocity. The dimensions of molar flux are given below:

$$\begin{pmatrix} \text{molar flux} \\ \text{of } A \text{ in mixture} \end{pmatrix} = \frac{(\text{moles of species } A \text{ transferred})}{(\text{cross-sectional area for flux})(\text{time})} \begin{pmatrix} \dfrac{\text{kgmoles } A}{\text{m}^2\text{-sec}} \end{pmatrix}$$

The basic relation for molecular diffusion defines the molar flux relative to the molar-average velocity, $\mathbf{J}_A$. An empirical relation for this molar flux, first postulated by Fick,[1] and accordingly often referred to as Fick's first law, defines the diffusion of component A in an isothermal, isobaric system as proportional to the concentration gradient:

$$\mathbf{J}_A = -D_{AB} \nabla c_A$$

For diffusion in only the z-direction, the Fick rate equation is

$$J_{A,z} = -D_{AB} \frac{dc_A}{dz} \tag{24-15}$$

where $J_{A,z}$ is the molar flux in the z-direction relative to the molar-average velocity, dc_A/dz is the concentration gradient in the z-direction, and D_{AB}, the proportionality factor, is the *mass diffusivity* or *diffusion coefficient* for component A diffusing through component B. The negative (–) sign sets the direction of flux along positive z, because the concentration gradient of species A is negative as it moves from high concentration to low concentration along increasing z.

---

[1] A. Fick, *Ann. Physik.*, **94**, 59 (1855).

A more general flux relation that is not restricted to isothermal, isobaric systems was proposed by de Groot,[2] who chose to write the flux equation as

$$J_{A,z} = -cD_{AB}\frac{dy_A}{dz} \tag{24-16}$$

As the total concentration c is constant under isothermal, isobaric conditions, equation (24-15) is a special form of the more general relation (24-16). An equivalent expression for $j_{A,z}$, the mass flux in the z-direction relative to the mass-average velocity, is

$$j_{A,z} = -\rho D_{AB}\frac{dw_A}{dz} \tag{24-17}$$

where dw_A/dz is the concentration gradient in terms of the mass fraction. When the density is constant, this relation simplifies to

$$j_{A,z} = -D_{AB}\frac{d\rho_A}{dz}$$

Initial experimental investigations of molecular diffusion were unable to verify Fick's law of diffusion. This was apparently because mass is often transferred simultaneously by two possible means: (1) as a result of the concentration differences as postulated by Fick and (2) by convection differences induced by the density differences that resulted from the concentration variation.

To illustrate the differences between diffusion and bulk flow contributions to flux, consider the following two mass-transfer situations. First, if a balloon, filled with a colored dye, is dropped into a large lake, then the dye will diffuse in all radial directions through the still liquid water from the balloon source. However, if the dye-filled balloon is dropped into a moving stream, the dye would diffuse radially while being carried downstream; thus, both contributions participate simultaneously in the mass transfer.

Stefan (1872) and Maxwell (1877), using the kinetic theory of gases, proved that the mass flux relative to a fixed coordinate was a result of two contributions: the concentration gradient contribution and the bulk motion contribution:

$$\left(\frac{\text{total mass}}{\text{transported}}\right) = \left(\frac{\text{mass transported}}{\text{by diffusion}}\right) + \left(\frac{\text{mass transported}}{\text{bulk motion of fluid}}\right)$$

or

$$N_{A,z} = J_{A,z} + c_A V_z \tag{24-18}$$

For a binary system with a constant average velocity in the z-direction, the molar flux in the z-direction relative to the molar-average velocity may also be expressed by

$$J_{A,z} = c_A(v_{A,z} - V_z) \tag{24-19}$$

Equating expressions (24-16) and (24-19), we obtain

$$J_{A,z} = c_A(v_{A,z} - V_z) = -cD_{AB}\frac{dy_A}{dz}$$

which, upon rearrangement, yields

$$c_A v_{A,z} = -cD_{AB}\frac{dy_A}{dz} + c_A V_z$$

[2] S. R. de Groot, *Thermodynamics of Irreversible Processes*, North-Holland, Amsterdam, 1951.

For this binary system, V_z can be evaluated by equation (24-14) as

$$V_z = \frac{1}{c}\left(c_A v_{A,z} + c_B v_{A,z}\right)$$

or

$$c_A V_z = y_A\left(c_A v_{A,z} + c_B v_{B,z}\right)$$

Therefore,

$$c_A v_{A,z} = -cD_{AB}\frac{dy_B}{dz} + y_A\left(c_A v_{A,z} + c_B v_{B,z}\right) \tag{24-20}$$

The component velocities, $v_{A,z}$ and $v_{B,z}$, are relative to the fixed z axis, and so the quantities $c_A v_{A,z}$ and $c_B v_{B,z}$ are the fluxes of components A and B relative to a fixed z coordinate. Therefore, the flux of species A and B in stationary coordinates are given by

$$N_{A,z} = c_A v_{A,z}$$

and

$$N_{B,z} = c_B v_{B,z}$$

By substitution of these definitions for flux into equation (24-20), we eliminate the component velocities and obtain a relationship for the flux of component A relative to the z axis that reflects the contributions of the component fluxes:

$$N_{A,z} = -cD_{AB}\frac{dy_A}{dz} + y_A\left(N_{A,z} + N_{B,z}\right) \tag{24-21}$$

Equation (24-21), also called the Fick's rate equation, can be simplified based upon the physical system for mass transfer. For example, from Figure 24.1, we see that $N_{B,z} = 0$; this is because since air has no sink, it does not have a flux. Consequently, equation (24-21) reduces to

$$N_{A,z} = -cD_{AB}\frac{dy_A}{dz} + y_A\left(N_{A,z} + 0\right)$$

or

$$N_{A,z} = \frac{-cD_{AB}}{1 - y_A}\frac{dy_A}{dz}$$

which describes the one-dimensional flux equation for diffusion of species A through stagnant species B. We see that the presence of B affects the flux of A relative to Fick's law through the $1-y_A$ term.

Equation (24-21) may be generalized and written in vector form as

$$\mathbf{N}_A = -cD_{AB}\nabla y_A + y_A(\mathbf{N}_A + \mathbf{N}_B) \tag{24-22}$$

It is important to note that the molar flux vector, $\mathbf{N}_A$, is a resultant of the two vector quantities, $-cD_{AB}\nabla y_A$, the molar flux resulting from diffusion along the concentration gradient and $y_A(\mathbf{N}_A + \mathbf{N}_B)$, the molar flux resulting as component A is carried with bulk flow of the species in the mixture. This latter flux term is designated the bulk motion contribution.

Either or both quantities can be a significant part of the total molar flux, $\mathbf{N}_A$. Whenever equation (24-22) is applied to describe molar diffusion, the vector nature of the individual fluxes, $\mathbf{N}_A$ and $\mathbf{N}_B$, must be considered. If species A were diffusing in a multicomponent mixture, the expression equivalent to equation (24-21) is

$$\mathbf{N}_A = -cD_{AM}\nabla y_A + y_A \sum_{i=1}^{n} \mathbf{N}_i$$

where D_{AM} is the diffusion coefficient of A in the mixture.

The mass flux, $\mathbf{n}_A$, relative to a fixed spatial coordinate system, is defined for a binary system in terms of mass density and mass fraction by

$$\mathbf{n}_A = -\rho D_{AB}\nabla w_A + w_A(\mathbf{n}_A + \mathbf{n}_B) \tag{24-23}$$

where

$$\mathbf{n}_A = \rho_A \mathbf{v}_A$$

and

$$\mathbf{n}_B = \rho_B \mathbf{v}_B$$

Under isothermal, isobaric conditions, this relation simplifies to

$$\mathbf{n}_A = -D_{AB}\nabla \rho_A + w_A(\mathbf{n}_A + \mathbf{n}_B)$$

The four equations defining the fluxes, $\mathbf{J}_A$, $\mathbf{j}_A$, $\mathbf{N}_A$, and $\mathbf{n}_A$, are equivalent statements of the Fick rate equation. The diffusion coefficient, D_{AB}, is identical in all four equations. Any one of these equations is adequate to describe molecular diffusion; however, certain fluxes are easier to use for specific cases. The mass fluxes, $\mathbf{n}_A$ and $\mathbf{j}_A$, are used when the Navier–Stokes equations are also required to describe the process. Since chemical reactions are described in terms of moles of the participating reactants, the molar fluxes, $\mathbf{J}_A$ and $\mathbf{N}_A$, are used to describe mass-transfer operations in which chemical reactions are involved. The fluxes relative to coordinates fixed in space, $\mathbf{n}_A$ and $\mathbf{N}_A$, are often used to describe engineering operations within process equipment. The fluxes $\mathbf{J}_A$ and $\mathbf{j}_A$ are used to describe the mass transfer in diffusion cells used for measuring the diffusion coefficient. Table 24.2 summarizes the equivalent forms of the Fick rate equation.

Table 24.2 Equivalent forms of the mass flux equation for binary system A and B

Flux	Gradient	Fick's Rate Equation	Restrictions
$\mathbf{n}_A$	∇w_A	$\mathbf{n}_A = -\rho D_{AB}\nabla w_A + w_A(\mathbf{n}_A + \mathbf{n}_B)$	
	$\nabla \rho_A$	$\mathbf{n}_A = -D_{AB}\nabla \rho_A + w_A(\mathbf{n}_A + \mathbf{n}_B)$	Constant ρ
$\mathbf{N}_A$	∇y_A	$\mathbf{N}_A = -cD_{AB}\nabla y_A + y_A(\mathbf{N}_A + \mathbf{N}_B)$	Constant c
$\mathbf{j}_A$	∇w_A	$\mathbf{j}_A = -\rho D_{AB}\nabla w_A$	
	$\nabla \rho_A$	$\mathbf{j}_A = -D_{AB}\nabla \rho_A$	Constant ρ
$\mathbf{J}_A$	∇y_A	$\mathbf{J}_A = -cD_{AB}\nabla y_A$	
	∇x_A	$\mathbf{J}_A = -cD_{AB}\nabla x_A$	
	∇c_A	$\mathbf{J}_A = -D_{AB}\nabla c_A$	Constant c

Related Types of Molecular Mass Transfer

According to the second law of thermodynamics, systems not in equilibrium will tend to move toward equilibrium with time. A generalized driving force in chemical thermodynamic terms is $-d\mu_c/dz$, where μ_c is the *chemical potential*. The molar diffusion velocity of component A is defined in terms of the chemical potential by

$$v_{A,z} - V_z = u_A \frac{d\mu_c}{dz} = -\frac{D_{AB}}{RT}\frac{d\mu_c}{dz} \tag{24-24}$$

where u_A is the "mobility" of component A, or the resultant velocity of the molecule while under the influence of a unit driving force. Equation (24-24) is known as the Nernst–Einstein relation. The molar flux of A becomes

$$J_{A,z} = c_A\left(v_{A,z} - V_z\right) = c_A \frac{D_{AB}}{RT}\frac{d\mu_c}{dz} \tag{24-25}$$

Equation (24-25) may be used to define all molecular mass-transfer phenomena. As an example, consider the conditions specified for equation (24-15); the chemical potential of a component in a homogeneous ideal solution at constant temperature and pressure is defined by

$$\mu_c = \mu^0 + RT \ln c_A \tag{24-26}$$

where μ^0 is a constant, the chemical potential of the standard state. When we substitute this relation into equation (24-25), the Fick's law for a homogeneous phase is once again obtained:

$$J_{A,z} = -D_{AB}\frac{dc_A}{dz} \tag{24-15}$$

There are a number of other physical conditions, in addition to differences in concentration, which will produce a chemical potential gradient: temperature differences, pressure differences, and differences in the forces created by external fields, such as gravity, magnetic, and electrical fields. We can, for example, obtain mass transfer by applying a temperature gradient to a multicomponent system. This transport phenomenon, the *Soret effect* or *thermal diffusion*, although normally small relative to other diffusion effects, is used successfully in the separation of isotopes. Components in a liquid mixture can be separated with a centrifuge by *pressure diffusion*. There are many well-known examples of mass fluxes' being induced in a mixture subjected to an external force field: separation by sedimentation under the influence of gravity, electrolytic precipitation due to an electrostatic force field, and magnetic separation of mineral mixtures through the action of a magnetic force field. Although these mass-transfer phenomena are important, they are very specific processes.

The molecular mass transfer resulting from concentration differences and described by Fick's law is based on the random molecular motion over small mean-free paths, independent of any containment walls. However, the diffusion of fast neutrons and molecules in extremely small pores or at very low gas density cannot be described by this relationship. We will consider diffusion of molecules in extremely small pores later in this chapter.

▶ **24.2**

THE DIFFUSION COEFFICIENT

In Fick's law, the proportionality constant in a binary mixture of species A and B is known as the diffusion coefficient, D_{AB}. Its fundamental dimensions, which may be obtained from equation (24-15), are

$$D_{AB} = \frac{-J_{A,z}}{dc_A/dz} = \left(\frac{M}{L^2 t}\right)\left(\frac{1}{M/L^3 \cdot 1/L}\right) = \frac{L^2}{t}$$

which are also identical to the fundamental dimensions of the other transport properties: kinematic viscosity, n, and thermal diffusivity, α. The mass diffusivity is commonly reported in units of cm^2/s; the SI units are m^2/s, which gives values that are a factor 10^{-4} smaller.

The diffusion coefficient depends upon the pressure, temperature, and composition of the system. Experimental values for the diffusivities of gases, liquids, and solids are tabulated in Appendix Tables J.1, J.2, and J.3, respectively. As one might expect from the consideration of the mobility of the molecules, diffusion coefficients for gases, which are typically on the order of 10^{-6} to 10^{-5} m^2/s, are much higher than diffusion coefficients for liquids, which are typically on the order of 10^{-10} to 10^{-9} m^2/s. Diffusion coefficients for solids are much lower, and are typically reported in the range of 10^{-14} to 10^{-10} m^2/s. In the absence of experimental data, semi-theoretical expressions have been developed, which give reasonable approximations for the diffusion coefficient, as described below.

Gas Mass Diffusivity

Theoretical expressions for the diffusion coefficient in low-density gaseous mixtures as a function of the system's molecular properties were derived by Sutherland,[3] Jeans,[4] and Chapman and Cowling,[5] based upon the kinetic theory of gases. In the simplest model of gas dynamics, the molecules are regarded as rigid spheres that exert no intermolecular forces. Collisions between these rigid molecules are considered to be completely elastic. With these assumptions, a simplified model for an ideal gas mixture of species A diffusing through its isotope A^* yields an equation for the *self-diffusion coefficient*, defined as

$$D_{AA^*} = \frac{1}{3} \lambda u \tag{24-27}$$

and λ is the mean-free path of length of species A, given by

$$\lambda = \frac{\kappa T}{\sqrt{2}\pi \sigma_A^2 P} \tag{24-28}$$

where u is the mean speed of species A with respect to the molar-average velocity:

$$u = \sqrt{\frac{8 \kappa N T}{\pi M_A}} \tag{24-29}$$

Insertion of equations (24-28) and (24-29) into equation (24-27) results in

$$D_{AA^*} = \frac{2 T^{3/2}}{3 \pi^{3/2} \sigma_A^2 P} \left(\frac{\kappa^3 N}{M_A} \right)^{1/2} \tag{24-30}$$

where M_A is the molecular weight of the diffusing species A, (g/gmole), N is Avogadro's number $(6.022 \times 10^{23}$ molecules/gmole), P is the system pressure, T is the absolute temperature (K), κ is the Boltzmann constant $(1.38 \times 10^{-16}$ ergs/$K)$, and σ_A is the Lennard–Jones molecular diameter of species A.

---

[3] W. Sutherland, *Phil. Mag.*, **36**, 507; **38**, 1 (1894).

[4] J. Jeans, *Dynamical Theory of Gases*, Cambridge University Press, London, 1921.

[5] S. Chapman and T. G. Cowling, *Mathematical Theory of Non-Uniform Gases*, Cambridge University Press, London, 1959.

Using a similar kinetic theory of gases approach for a binary mixture of species A and B composed of rigid spheres of unequal diameters, the gas-phase diffusion coefficient is shown to be

$$D_{AB} = \frac{2}{3} \left(\frac{\kappa}{\pi}\right)^{3/2} N^{1/2} \, T^{3/2} \frac{\left(\frac{1}{2M_A} + \frac{1}{2M_B}\right)^{1/2}}{P\left(\frac{\sigma_A + \sigma_B}{2}\right)^2} \qquad (24\text{-}31)$$

We note from equation (24-31) that the gas-phase diffusion coefficient is dependent on the pressure and the temperature. Specifically, the gas-phase diffusion coefficient is inversely proportional to total system pressure

$$D_{AB} \propto \frac{1}{P} \qquad (24\text{-}32a)$$

and has a power-law dependence on absolute temperature

$$D_{AB} \propto T^{3/2} \qquad (24\text{-}32b)$$

Furthermore, gas-phase diffusion coefficients decrease as the molecular diameter and molecular weight of either A and/or B increases. Finally, equation (24-31) reveals that for gases, $D_{AB} = D_{BA}$. This is not the case for liquid-phase diffusion coefficients.

Correlations for Gas-Phase Binary Diffusion Coefficients Gas-phase diffusion coefficients are fundamentally defined based on a given pair of species in the mixture— e.g., D_{AB} for species A and B. Modern versions of the kinetic theory have been attempted to account for forces of attraction and repulsion between these molecules. Hirschfelder et al.,[6] using the Lennard–Jones potential to evaluate the influence of the molecular forces, presented the following correlation for the diffusion coefficient for gas pairs of nonpolar, nonreacting molecules:

$$D_{AB} = \frac{0.001858 \, T^{3/2} \left(\frac{1}{M_A} + \frac{1}{M_B}\right)^{1/2}}{P\sigma_{AB}^2 \Omega_D} \qquad (24\text{-}33)$$

To use this correlation, the units must be obeyed as follows: D_{AB} is the diffusion coefficient of A through B (cm^2/s); T is the absolute temperature (K); M_A and M_B are the molecular weights of A and B, respectively (g/gmole); P is the absolute pressure (atm); σ_{AB} is the "collision diameter" of the binary pair of species A and B, a Lennard–Jones parameter, in units of Angstroms (Å); and Ω_D is the "collision integral" for molecular diffusion, a dimensionless function of the temperature and of the intermolecular potential field for one molecule of A and one molecule of B. Appendix Table K.1 lists Ω_D as a function of $\kappa T / \epsilon_{AB}$, where κ is the Boltzmann constant, equal to 1.38×10^{-16} ergs/K, and ϵ_{AB} is the energy of molecular interaction for the binary system A and B, a Lennard–Jones parameter, in ergs. Figure 24.3 presents the dependence of Ω_D on the dimensionless temperature, $\kappa T/\epsilon_{AB}$.

The Lennard–Jones parameters, σ and ϵ, are available for many pure gases; some of these are tabulated in Appendix K, Table K.2. In the absence of experimentally determined

---

[6] J. O. Hirschfelder, R. B. Bird, and E. L. Spotz, *Chem. Rev.*, **44**, 205 (1949).

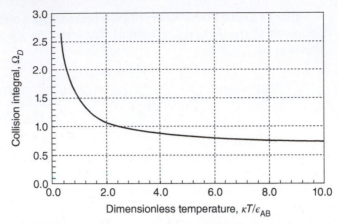

Figure 24.3 Lennard–Jones "collision integral" for diffusion as a function of dimensionless temperature.

values for these parameters, σ and ϵ for pure components may be estimated from the following empirical correlations:

$$\sigma = 1.18 \, V_b^{1/3} \tag{24-34}$$

$$\sigma = 0.841 \, V_c^{1/3} \tag{24-35}$$

$$\sigma = 2.44 \left(\frac{T_c}{P_c} \right)^{1/3} \tag{24-36}$$

$$\epsilon/\kappa = 0.77 \, T_c \tag{24-37}$$

$$\epsilon/\kappa = 1.15 \, T_b \tag{24-38}$$

where V_b is the molecular volume at the normal boiling point (cm^3/gmole), V_c is the critical molecular volume (cm^3/gmole), T_c is the critical temperature (K), T_b is the normal boiling temperature (K), and P_c is the critical pressure (atm). All units must be used as stated in these correlations.

For a binary system composed of nonpolar molecular pairs, the Lennard–Jones parameters of the pure component may be combined empirically by the following relations:

$$\sigma_{AB} = \frac{\sigma_A + \sigma_B}{2} \tag{24-39}$$

and

$$\epsilon_{AB} = \sqrt{\epsilon_A \epsilon_B} \tag{24-40}$$

These relations must be modified for polar-polar and polar-nonpolar molecular pairs; the proposed modifications are discussed by Hirschfelder, Curtiss, and Bird.[7]

The Hirschfelder equation (24-33) is often used to extrapolate experimental data. For moderate ranges of pressure, up to 25 atm, the diffusion coefficient varies inversely with the pressure. Higher pressures apparently require dense gas corrections; unfortunately, no

---

[7] J. O. Hirschfelder, C. F. Curtiss, and R. B. Bird, *Molecular Theory of Gases and Liquids*, John Wiley & Sons, Inc., New York, 1954.

satisfactory correlation is available for high pressures. Equation (24-33) also states that the diffusion coefficient varies with the temperature as $T^{3/2}/\Omega_D$ varies. Simplifying equation (24-33), we can predict the diffusion coefficient at any temperature and at any pressure below 25 atm from a known experimental value by the Hirschfelder extrapolation, given by

$$D_{AB}(T_2, P_2) = D_{AB}(T_1, P_1)\left(\frac{P_1}{P_2}\right)\left(\frac{T_2}{T_1}\right)^{3/2}\frac{\Omega_D(T_1)}{\Omega_D(T_2)} \tag{24-41}$$

In Appendix Table J.1, experimental values of the product $D_{AB}P$ are listed for several gas pairs at a particular temperature. Using equation (24-41), we may extend these values to other temperatures.

Equation (24-33) was developed for dilute gases consisting of nonpolar, spherical monatomic molecules. However, this equation gives good results for most nonpolar, binary gas systems over a wide range of temperatures.[8] Other empirical correlations have been proposed[9] for estimating the diffusion coefficient for nonpolar, binary gas systems at low pressures. The empirical correlation recommended by Fuller, Schettler, and Giddings[10] permits the evaluation of the diffusivity when reliable Lennard–Jones parameters are unavailable. The Fuller–Schettler–Giddings correlation is

$$D_{AB} = \frac{0.001\, T^{1.75}\left(\dfrac{1}{M_A} + \dfrac{1}{M_B}\right)^{1/2}}{P\left[(\Sigma n_i)_A^{1/3} + (\Sigma n_i)_B^{1/3}\right]^2} \tag{24-42}$$

To use this correlation, the units must be obeyed as follows: D_{AB} (cm^2/s), absolute temperature $T(K)$, and total system pressure P (atm). To determine the n_i terms, Fuller et al. recommend the addition of the atomic and structural diffusion-volume increments (n_i) reported in Table 24.3.

Table 24.3   Atomic diffusion volumes for use in estimating D_{AB} by the method of Fuller, Schettler, and Giddings[10]

Atomic and Structure Diffusion-Volume Increments, n_i			
C	16.5	Cl	19.5
H	1.98	S	17.0
O	5.48	Aromatic Ring	−20.2
N	5.69	Heterocyclic Ring	−20.2

Diffusion Volumes for Simple Molecules, n					
H_2	7.07	Ar	16.1	H_2O	12.7
D_2	6.70	Kr	22.8	$C(Cl_2)(F_2)$	114.8
He	2.88	CO	18.9	SF_6	69.7
N_2	17.9	CO_2	26.9	Cl_2	37.7
O_2	16.6	N_2O	35.9	Br_2	67.2
Air	20.1	NH_3	14.9	SO_2	41.1

[8] R. C. Reid, J. M. Prausnitz, and T. K. Sherwood, *The Properties of Gases and Liquids*, 3rd edition, McGraw-Hill Book Company, New York, 1977.

[9] J. H. Arnold, *J. Am. Chem. Soc.*, **52**, 3937 (1930). E. R. Gilliland, *Ind. Eng. Chem.*, **26**, 681 (1934). J. C. Slattery and R. B. Bird, *A.I.Ch.E. J.*, **4**, 137 (1958). D. F. Othmer and H. T. Chen, *Ind. Eng. Chem. Process Des. Dev.*, **1**, 249 (1962).

[10] E. N. Fuller, P. D. Schettler, and J. C. Giddings, *Ind. Eng. Chem.*, **58** (5), 18 (1966).

Example 3

A new gas separation process is being developed to separate ethylene (C_2H_4) from a gas mixture that contains ethylene, small amounts of carbon dioxide (CO_2), and carbon monoxide (CO). As part of the process analysis, the gas-phase diffusion coefficient of CO_2 gas in ethylene and the gas-phase diffusion coefficient of CO in ethylene are needed at 2.0 atm and 77 C (350 K). For the CO_2-C_2H_4 binary pair, estimate the gas-phase binary diffusion coefficient by the Hirschfelder and Fuller–Schettler–Giddings correlations. For the CO-C_2H_4 binary pair, extrapolate data found in Appendix J to estimate the binary gas-phase diffusion coefficient.

For the CO_2-ethylene binary pair, let species A represent CO_2 with molecular weight of 44 g/mole, and species B represent ethylene with molecular weight of 28 g/mole. Let us first use the Hirschfelder equation. From Appendix K, Table K.2, the Lennard–Jones constants needed for the Hirschfelder equation are $\sigma_A = 3.996$ Å, $\sigma_B = 4.232$ Å, $\epsilon_A/\kappa = 190$ K, $\epsilon_B/\kappa = 205$ K. Consequently,

$$\sigma_{AB} = \frac{\sigma_A + \sigma_B}{2} = \frac{3.996 \text{ Å} + 4.232 \text{ Å}}{2} = 4.114 \text{ Å}$$

$$\frac{\kappa T}{\epsilon_{AB}} = \left(\frac{\kappa}{\epsilon_A}\frac{\kappa}{\epsilon_A}\right)^{1/2} T = \left(\frac{1}{190 \text{ K}}\frac{1}{205 \text{ K}}\right)^{1/2}(350 \text{ K}) = 1.77$$

From Appendix K, Table K.1, $\Omega_D = 1.123$. Therefore,

$$D_{AB} = \frac{0.001858\, T^{3/2}\left(\frac{1}{M_A}+\frac{1}{M_B}\right)^{1/2}}{P\sigma_{AB}^2\Omega_D} = \frac{0.001858\,(350)^{3/2}\left(\frac{1}{44}+\frac{1}{28}\right)^{1/2}}{(2.0)(4.114)^2(1.123)} = 0.077 \text{ cm}^2/\text{s}$$

Now, let us compare the Hirschfelder correlation with the Fuller–Schettler–Giddings correlation. From Table 24.3, the atomic diffusion volume for CO_2 is 26.9; the atomic diffusion volume for ethylene (C_2H_4) is estimated the group contribution method using C and H building blocks also given in Table 24.3:

$$(\Sigma n_i)_B = 2 \cdot n_C + 4 \cdot n_H = 2(16.5) + 4(1.98) = 40.92$$

Therefore, the diffusion coefficient estimated by the Fuller–Schettler–Giddings correlation is

$$D_{AB} = \frac{0.001\, T^{1.75}\left(\frac{1}{M_A}+\frac{1}{M_B}\right)^{1/2}}{P\left[(\Sigma n_i)_A^{1/3}+(\Sigma n_i)_B^{1/3}\right]^2} = \frac{0.001\,(350)^{1.75}\left(\frac{1}{44}+\frac{1}{28}\right)^{1/2}}{(2.0)\left[(26.9)^{1/3}+(40.92)^{1/3}\right]^2} = 0.082 \text{ cm}^2/\text{s}$$

The two correlations agree within 7%.

For the CO-C_2H_4 binary pair, with $A = $ CO, and $B = $ ethylene (C_2H_4), from Appendix J, Table J.1, the measured diffusion coefficient is $D_{AB} = 0.151$ cm²/s at 1.0 atm and 273 K. By the Hirschfelder extrapolation, the diffusion coefficient at 2.0 atm and 350 K is

$$D_{AB}(T,P) = D_{AB}(T_o,P_o)\left(\frac{P_o}{P}\right)\left(\frac{T}{T_o}\right)^{3/2}\frac{\Omega_D(T_o)}{\Omega_D(T)}$$

$$= \left(0.151\,\frac{\text{cm}^2}{\text{s}}\right)\left(\frac{1.0 \text{ atm}}{2.0 \text{ atm}}\right)\left(\frac{350 \text{ K}}{273 \text{ K}}\right)^{3/2}\left(\frac{1.112}{1.022}\right) = 0.119 \text{ cm}^2/\text{s}.$$

Additional correlations are available for estimation of Lennard–Jones parameters of polar compounds. Brokaw[11] has suggested a method for estimating diffusion coefficient for binary gas mixtures containing polar compounds. The Hirschfelder equation (24-33) is still

---

[11] R. S. Brokaw, *Ind. Engr. Chem. Process Des. Dev.*, **8**, 240 (1969).

used; however, the collision integral is evaluated by

$$\Omega_D = \Omega_{D,0} + \frac{0.196\delta_{AB}^2}{T^*} \tag{24-43}$$

where

$$\delta_{AB} = \left(\delta_A \delta_B\right)^{1/2}$$

with δ_A and δ_B estimated by

$$\delta = \frac{1.94 \times 10^3 \mu_p^2}{V_b T_b} \tag{24-44}$$

In equations (24-43) and (24-24), μ_p is the dipole moment (Debye), V_b is the specific liquid molar volume of the compound at its normal boiling point (cm^3/gmole), T_b is the normal boiling point temperature (K), and T^* is the reduced temperature, given by

$$T^* = \kappa T / \epsilon_{AB}$$

where

$$\frac{\epsilon_{AB}}{\kappa} = \left(\frac{\epsilon_A}{\kappa} \frac{\epsilon_B}{\kappa}\right)^{1/2} \tag{24-45}$$

and

$$\epsilon/\kappa = 1.18\left(1 + 1.3\,\delta^2\right)T_b$$

Furthermore, in equation (24-43) the reference collision integral, $\Omega_{D,0}$, is estimated by

$$\Omega_{D,0} = \frac{A}{(T^*)^B} + \frac{C}{\exp(DT^*)} + \frac{E}{\exp(FT^*)} + \frac{G}{\exp(HT^*)} \tag{24-46}$$

with $A = 1.06036, B = 0.15610, C = 0.19300, D = 0.47635, E = 1.03587, F = 1.52996,$ $G = 1.76474, H = 3.89411.$

Finally, the collision diameter, σ_{AB}, is estimated by geometric average

$$\sigma_{AB} = \left(\sigma_A \sigma_B\right)^{1/2} \tag{24-47}$$

with each component's characteristic length evaluated by

$$\sigma = \left(\frac{1.585\,V_b}{1 + 1.3\,\delta^2}\right)^{1/3} \tag{24-48}$$

Reid, Prausnitz, and Sherwood[8] noted that the Brokaw equation is fairly reliable, permitting the evaluation of the diffusion coefficients for gases involving polar compounds with errors less than 15%.

Effective Diffusion Coefficients for Multicomponent Gas Mixtures
Diffusion coefficients are fundamentally defined only for two given species—i.e., a binary pair. In a true multicomponent mixture, the "mixture-based diffusion coefficient" for species A in the multicomponent mixture, $D_{A\text{-}M}$, is based on the assembling the individual diffusion coefficients for all binary pairs, using principles derived from the Stefan–Maxwell relationships.

Mass transfer in gas mixtures of several components can be described by theoretical equations involving the diffusion coefficients for the various binary pairs involved in the mixture. Hirschfelder, Curtiss, and Bird[7] present an expression in its most general form for an n-component mixture. Wilke[12] has simplified the theory and has shown that a close approximation to the correct form is given by the relation

$$D_{1-M} = \frac{1}{y_2'/D_{1-2} + y_3'/D_{1-3} + \ldots y_n'/D_{1-n}} \tag{24-49}$$

where D_{1-M} is the mass diffusivity for component 1 in the gas mixture, D_{1-n} is the mass diffusivity for the binary pair, component 1 diffusing through component n, and y_n' is defined as

$$y_n' = \frac{y_n}{y_2 + y_3 + \ldots y_n} = \frac{y_n}{1 - y_1}$$

Liquid-Mass Diffusivity

In contrast to the case for gases, where we have available an advanced kinetic theory for explaining molecular motion, theories of the structure of liquids and their transport characteristics are still inadequate to permit a rigorous treatment. Inspection of published experimental values for liquid diffusion coefficients in Appendix J.2 reveals that they are several orders of magnitude smaller than gas diffusion coefficients and that they depend on concentration due to the changes in viscosity with concentration and changes in the degree of ideality of the solution.

Certain molecules diffuse as molecules, while others that are designated as electrolytes ionize in solutions and diffuse as ions. For example, sodium chloride, NaCl, diffuses in water as the ionic pair Na^+ and Cl^-. Though each ion has a different mobility, the electrical neutrality of the solution indicates that the ions must diffuse at the same rate; accordingly, it is possible to speak of a diffusion coefficient for molecular electrolytes such as NaCl. However, if several ions are present, the diffusion rates of the individual cations and anions must be considered, and molecular diffusion coefficients have no meaning. Needless to say, separate correlations for predicting the relation between the liquid-mass diffusivities and the properties of the liquid solution will be required for electrolytes and nonelectrolytes.

Two theories, the Eyring "hole" theory and the hydrodynamical theory, have been postulated as possible explanations for diffusion of nonelectrolyte solutes in low-concentration solutions. In the Eyring concept, the ideal liquid is treated as a quasi-crystalline lattice model interspersed with holes. The transport phenomenon is then described by a unimolecular rate process involving the jumping of solute molecules into the holes within the lattice model. These jumps are empirically related to Eyring's theory of reaction rate.[13] The hydrodynamical theory states that the liquid diffusion coefficient is related to the solute molecule's mobility—that is, to the net velocity of the molecule while under the influence of a unit driving force. The laws of hydrodynamics provide relations between the force and the velocity. An equation that has been developed from the hydrodynamical theory is the Stokes–Einstein equation:

$$D_{AB} = \frac{\kappa T}{6\,\pi r_A \mu_B} \tag{24-50}$$

where D_{AB} is the diffusivity of solute A in a dilute solution of solvent B, κ is the Boltzmann constant, T is the absolute temperature, r_A is the molecular radius of solute A, and μ_B is the solvent viscosity. This equation has been fairly successful in describing the diffusion of

[12] C. R. Wilke, *Chem. Engr. Prog.*, **46**, 95 (1950).
[13] S. Glasstone, K. J. Laidler, and H. Eyring, *Theory of Rate Processes*, McGraw-Hill Book Company, New York, 1941.

colloidal particles or large round molecules through a solvent that behaves as a continuum relative to the diffusing species. Equation (24-50) also suggests that liquid-phase diffusion coefficients will be nonlinear functions of temperature, as the solvent viscosity is also a strong function of temperature for many liquids, especially water. Furthermore, equation (24-50) suggests that liquid-phase diffusion coefficients will decrease as the molecular size of the solute increases.

Correlations for Liquid-Phase Binary Diffusion Coefficients with Non-Ionic Solutes In contrast to gas-phase diffusion coefficients, correlations for liquid-phase diffusion coefficients must assign the solute and solvent in the liquid-phase mixture, where the solute is treated as being at infinite dilution in the solvent. The theoretical concepts described above suggest that liquid-phase diffusion coefficients are conveniently correlated by

$$\frac{D_{AB}\mu_B}{T} = \kappa f(V) \tag{24-51}$$

where $f(V)$ is a function of the molecular volume of the diffusing solute. Empirical correlations, using the general form of equation (24-51), have been developed, which attempt to predict the liquid diffusion coefficient in terms of the solute and solvent properties. Wilke and Chang[14] have proposed the following correlation for nonelectrolytes in an infinitely dilute solution:

$$\frac{D_{AB}\mu_B}{T} = \frac{7.4 \times 10^{-8}(\Phi_B M_B)^{1/2}}{V_A^{0.6}} \tag{24-52}$$

To use this correlation, the units must be obeyed as follows: D_{AB} is the diffusion coefficient of dissolved solute A in liquid solvent B (cm^2/s); μ_B is the viscosity of the solvent B (cP); T is absolute temperature (K); M_B is the molecular weight of the solvent (g/gmole); V_A is the molal volume of solute A at its normal boiling point (cm^3/gmole); and Φ_B is the "association" parameter for solvent B (dimensionless). Molecular volumes at normal boiling points, V_A, for some commonly encountered compounds, are tabulated in Table 24.4. For other compounds, the atomic volumes of each element present are added together as per the molecular formulas. Table 24.5 lists the contributions for each of the constituent atoms. When certain ring structures are involved, corrections must be made to account for the specific ring configuration.

Table 24.4 Molecular volumes at normal boiling point for some commonly encountered compounds

Compound	Molecular Volume, V_A (cm^3/gmole)	Compound	Molecular Volume, V_A (cm^3/gmole)
Hydrogen, H_2	14.3	Nitric oxide, NO	23.6
Oxygen, O_2	25.6	Nitrous oxide, N_2O	36.4
Nitrogen, N_2	31.2	Ammonia, NH_3	25.8
Air	29.9	Water, H_2O	18.9
Carbon monoxide, CO	30.7	Hydrogen sulfide, H_2S	32.9
Carbon dioxide, CO_2	34.0	Bromine, Br_2	53.2
Carbonyl sulfide, COS	51.5	Chlorine, Cl_2	48.4
Sulfur dioxide, SO_2	44.8	Iodine, I_2	71.5

[14] C. R. Wilke and P. Chang, *A.I.Ch.E.J.*, **1**, 264 (1955).

Table 24.5 Atomic volume increments for estimation of molecular volumes at the normal boiling point for simple substances[15]

Element	Atomic Volume (cm^3/gmole)	Element	Atomic Volume (cm^3/gmole)
Bromine	27.0	Oxygen, except as noted below	7.4
Carbon	14.8	Oxygen, in methyl esters	9.1
Chlorine	21.6	Oxygen, in methyl ethers	9.9
Hydrogen	3.7	Oxygen, in higher ethers	
Iodine	37.0	and other esters	11.0
Nitrogen, double bond	15.6	Oxygen, in acids	12.0
Nitrogen, in primary amines	10.5	Sulfur	25.6
Nitrogen, in secondary amines	12.0		

The following corrections are recommended when using the atomic volumes provided in Table 24.5:

for three-membered ring, as ethylene oxide	deduct 6
for four-membered ring, as cyclobutane	deduct 8.5
for five-membered ring, as furan	deduct 11.5
for pyridine	deduct 15
for benzene ring	deduct 15
for naphthalene ring	deduct 30
for anthracene ring	deduct 47.5

Finally, for equation (24-52), recommended values of the association parameter, Φ_B, are given below for a few common solvents.[8]

Solvent	Φ_B
Water	2.6
Methanol	1.9
Ethanol	1.5
Nonpolar solvents	1.0

If data for V_A are not available, Tyn and Calus[16] recommend the correlation

$$V_A = 0.285 \, V_c^{1.048}$$

where V_c is the critical volume of solute A (cm^3/gmole).

Hayduk and Laudie[17] have proposed a much simpler equation for evaluating infinite dilution liquid-phase diffusion coefficients of nonelectrolytes *in water*:

$$D_{AB} = 13.26 \times 10^{-5} \, \mu_B^{-1.14} \, V_A^{-0.589} \qquad (24\text{-}53)$$

where D_{AB} is the diffusion coefficient of solute A through solvent B (water), in units of cm^2/s, μ_B is the viscosity of water, in units of cP, and V_A is the molal volume of the solute A at normal

[15] G. Le Bas, *The Molecular Volumes of Liquid Chemical Compounds*, Longmans, Green & Company, Ltd., London, 1915.
[16] M. T. Tyn and W. F. Calus, *Processing*, **21** (4), 16 (1975).
[17] W. Hayduk and H. Laudie, *A.I.Ch.E. J.*, **20**, 611 (1974).

boiling point in units of cm^3/gmole. If the solvent is water, this relation is much simpler to use and gives similar results to the Wilke–Chang equation. Scheibel[18] has proposed that the Wilke–Chang relation be modified to eliminate the association factor, Φ_B, yielding

$$\frac{D_{AB}\,\mu_B}{T} = \frac{K}{V_A^{1/3}} \tag{24-54a}$$

where K is determined by

$$K = \left(8.2 \times 10^{-8}\right)\left[1 + \left(\frac{3V_B}{V_A}\right)^{2/3}\right] \tag{24-54b}$$

Equation (24-54b) is used with the exception of the following three cases. First, if benzene is the solvent and $V_A < 2V_B$, use $K = 18.9 \times 10^{-8}$. Second, for all other organic solvents, if $V_A < 2.5V_B$, use $K = 17.5 \times 10^{-8}$. Third, if water is the solvent and $V_A < V_B$, use $K = 25.2 \times 10^{-8}$.

Reid, Prausnitz, and Sherwood[8] recommend the Scheibel equation for solutes diffusing into organic solvents; however, they noted that this equation might evaluate values that have errors up to 20%.

Example 4

Compare estimates the diffusion coefficient of an ethanol-water mixture at 10 C (283 K), under conditions that (1) ethanol is the solute and water is the solvent and (2) water is the solute and ethanol is the solvent. Ethanol has the molecular formula C_2H_5OH and a molecular weight of 46 g/gmole, and water (H_2O) has a molecular weight of 18 g/gmole. At 10 C, the liquid viscosity of water is 1.306×10^{-3} Pa·s (1.306 cP), and the liquid viscosity of ethanol is 1.394×10^{-3} Pa·s (1.394 cP).

The Wilke–Chang correlation will be used for estimation of the diffusion coefficients. This correlation will require the molar volume of each species at its normal boiling point. From Table 24.4, the molecular volume of water is 18.9 cm^3/gmole. From Table 24.5, using the group contribution method, the molar volume of ethanol (EtOH) is estimated as

$$V_{\text{EtOH}} = 2 \cdot V_C + 1 \cdot V_O + 6 \cdot V_H = 2(14.8) + 1(7.4) + 6(3.7) = 59.2 \text{ cm}^2/\text{gmole}$$

For the ethanol-water system, let A = ethanol and B = water. First, if ethanol is the solute, $V_A = 59.2$ cm^3/gmole, and if water is the solvent, $\Phi_B = 2.6$ and $\mu_B = 1.394$ cP. Consequently, by equation (24-52)

$$D_{AB} = \frac{T}{\mu_B}\frac{7.4 \times 10^{-8}(\Phi_B M_B)^{1/2}}{V_A^{0.6}} = \frac{(283)}{(1.394)}\frac{7.4 \times 10^{-8}(2.6 \cdot 18)^{1/2}}{(59.2)^{0.6}} = 8.9 \times 10^{-6} \text{ cm}^2/\text{s}$$

For comparison, from Appendix J, Table J.2, $D_{AB} = 8.3 \times 10^{-6}$ cm^2/s, which is within 10% of the estimated value.

Now consider that water is the solute with $V_B = 18.9$ cm^3/gmole, and ethanol is the solvent with $\Phi_A = 1.5$ and $\mu_A = 1.306$ cP. In context to equation (24-52), D_{BA} is now given by

$$D_{BA} = \frac{T}{\mu_A}\frac{7.4 \times 10^{-8}(\Phi_A M_A)^{1/2}}{V_B^{0.6}} = \frac{(283)}{(1.306)}\frac{7.4 \times 10^{-8}(1.5 \cdot 46)^{1/2}}{(18.9)^{0.6}} = 2.28 \times 10^{-5} \text{ cm}^2/\text{s}$$

This result shows that $D_{AB} \neq D_{BA}$ for liquids.

Most methods for predicting the liquid diffusion coefficients in concentrated solutions have combined the infinite dilution coefficients as functions of composition. For non-associating, ideal liquid mixtures of components A and B, Vignes[19] recommended the following relationship:

$$D_{AB} = \left(D_{AB}\right)^{x_B}\left(D_{BA}\right)^{x_A} \tag{24-55}$$

[18] E. G. Scheibel, *Ind. Eng. Chem.*, **46**, 2007 (1954).
[19] A. Vignes, *Ind. Eng. Chem. Fundam.*, **5**, 189 (1966).

where D_{AB} is liquid-phase diffusion coefficient of A in B at infinite dilution with respect to component A, D_{BA} is the liquid-phase diffusion coefficient of B in A at infinite dilution with respect to B, and x_A and x_B are the mole fraction compositions of A and B, respectively.

Liquid-Phase Binary Diffusion Coefficients for Ionic Solutes The properties of electrically conducting solutions have been studied intensively. Even so, the known relations between electrical conductance and the liquid diffusion coefficient are valid only for dilute solutions of salts in water.

To maintain charge neutrality, both the cation and the anion of the ionic salt must diffuse through water as an ionic pair. The diffusion coefficient of a univalent salt in dilute solution is given by the Nernst–Haskell equation:

$$D_{AB} = \frac{2\,RT}{\left(1/\lambda_+^0 + 1/\lambda_-^0\right)(\mathcal{F})^2} \tag{24-56a}$$

where D_{AB} is the diffusion coefficient of the ion pair in solvent B at infinite dilution, R is the thermodynamic gas constant ($8.316\ J/gmole \cdot K$), T is absolute temperature (K), λ_+^0 and λ_-^0 are the limiting (zero concentration) ionic conductances of the cation and anion, respectively, in the ion pair ($A \cdot cm^2/V \cdot gmole$), and $\mathcal{F}$ is the Faraday's constant ($96,500$ $C/gmole$), taking note that $1\ C = 1\ A \cdot s$ and $1\ J/s = 1\ A \cdot V$. Equation (24-56a) is extended to polyvalent ions by

$$D_{AB} = \frac{(1/n^+ + 1/n^-)RT}{\left(1/\lambda_+^0 + 1/\lambda_-^0\right)(\mathcal{F})^2} \tag{24-56b}$$

where n^+ and n^- are the valences of the cation and anion, respectively. Limiting ionic conductances for selected ionic species[8] are presented in Table 24.6.

Table 24.6 Selected limiting ionic conductances in water at 25 C, reported in units of $A \cdot cm^2/V \cdot gmole$

Cation	λ_+^0	Anion	λ_-^0
H^+	349.8	OH^-	197.6
Li^+	38.7	Cl^-	76.3
Na^+	50.1	Br^-	78.3
K^+	73.5	I^-	76.8
NH_4^+	73.4	NO_3^-	71.4
Ag^+	61.9	HCO_3^-	44.5
Mg^{+2}	106.2	SO_4^{-2}	160
Ca^{+2}	119		
Cu^{+2}	108		
Zn^{+2}	106		

Pore Diffusivity

There are many instances where molecular diffusion occurs inside the pores of porous solids. For example, many catalysts are porous solid pellets containing catalytically active sites on the pore walls. The porous catalyst possesses a high internal surface area to promote chemical reactions at the catalytic surface. The separation of solutes from dilute solution by the process of adsorption is another example. In an adsorption process, the solute sticks to a feature on the solid surface that is attractive to the solute. Many adsorbent materials are porous to provide a high internal surface area for solute adsorption. In both examples, the molecules must diffuse through a gas or liquid phase residing inside the pores. As the pore diameter approaches the diameter of the diffusing molecule, the diffusing molecule can interact with the wall of the pore. Below, we describe two types of pore diffusion: the Knudsen diffusion of gases in cylindrical pores and the hindered diffusion of solutes in solvent-filled cylindrical pores.

Knudsen Diffusion   Consider the diffusion of gas molecules through very small capillary pores. If the pore diameter is smaller than the mean-free path of the diffusing gas molecules and the density of the gas is low, the gas molecules will collide with the pore walls more frequently than with each other. This process is known as Knudsen flow or *Knudsen diffusion*. The gas flux is reduced by the wall collisions.

Pure molecular diffusion, pure Knudsen diffusion, and molecular diffusion in tiny pores influenced by Knudsen diffusion within a straight cylindrical pore are compared in Figure 24.4. The Knudsen number, *Kn*, given by

$$Kn = \frac{\lambda}{d_{\text{pore}}} = \frac{\text{mean-free path length of the diffusing species}}{\text{pore diameter}} \tag{24-57}$$

is a good measure of the relative importance of Knudsen diffusion. If $0.1 < Kn < 1$, then Knudsen diffusion plays a measurable but moderate role in the overall diffusion process. If $Kn > 1$, then Knudsen diffusion becomes important, and if $Kn > 10$, then Knudsen diffusion can dominate. At a given pore diameter, the Kn number goes up as the total system pressure P decreases and absolute temperature T increases. In practice, Knudsen diffusion applies only to gases because the mean-free path for molecules in the liquid state is very small, typically near the molecular diameter of the molecule itself.

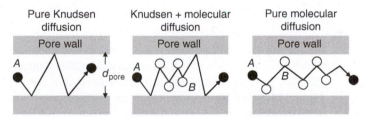

Figure 24.4 Influence of Knudsen and molecular diffusion on the effective diffusion within a straight pore.

The diffusion coefficient for Knudsen diffusion is obtained from the self-diffusion coefficient derived from the kinetic theory of gases:

$$D_{AA^*} = \frac{\lambda u}{3} = \frac{\lambda}{3} \sqrt{\frac{8\kappa NT}{\pi M_A}}$$

For Knudsen diffusion, we replace path length λ with pore diameter d_{pore}, as species A is now more likely to collide with the pore wall as opposed to another molecule. In this instance, the Knudsen diffusivity for diffusing species A, D_{KA}, is

$$D_{KA} = \frac{d_{\text{pore}}}{3} u = \frac{d_{\text{pore}}}{3} \sqrt{\frac{8\kappa NT}{\pi M_A}} \tag{24-58a}$$

Or, since κ and N are physical constants, equation (24-58a) is also given by

$$D_{KA} = 4850 \, d_{\text{pore}} \sqrt{\frac{T}{M_A}} \tag{24-58b}$$

The simplified equation (24-58b) requires that d_{pore} has units of cm, M_A has units of g/gmole, and temperature T has units of K. The Knudsen diffusivity, D_{KA}, is dependent on the pore diameter, species A molecular weight, and temperature.

We can make two comparisons of D_{KA} to the binary gas-phase diffusivity, D_{AB}. First, D_{KA} is not a function of absolute pressure P, or the presence of species B in the binary gas mixture. Second, the temperature dependence for the Knudsen diffusivity is $T^{1/2}$, vs. $T^{3/2}$ for the binary gas-phase diffusivity.

Generally, the Knudsen diffusion process is significant only at low pressure and small pore diameter. However, there are instances where both Knudsen diffusion and molecular diffusion can be important. If we consider that Knudsen diffusion and molecular diffusion compete with one another by a "resistances in series" approach, then it can be shown that the effective diffusion coefficient of species A in a gaseous binary mixture of A and B *within the pore* is determined by

$$\frac{1}{D_{Ae}} = \frac{1 - \alpha y_A}{D_{AB}} + \frac{1}{D_{KA}} \tag{24-59}$$

with

$$\alpha = 1 + \frac{N_B}{N_A}$$

For cases where $\alpha = 0(N_A = -N_B)$, or where y_A is close to zero—e.g., for a dilute mixture of A in carrier gas B—equation (24-59) reduces to

$$\frac{1}{D_{Ae}} = \frac{1}{D_{AB}} + \frac{1}{D_{KA}} \tag{24-60}$$

In general, $D_{Ae} \neq D_{Be}$.

The above relationships for the effective diffusion coefficient are based on diffusion within straight, cylindrical pores aligned in a parallel array. However, in most porous materials, pores of various diameters are twisted and interconnected with one another, and the path for diffusion of the gas molecule within the pores is "tortuous," as shown in Figure 24.5. For these materials, if an average pore diameter is assumed, a reasonable approximation for the effective diffusion coefficient in random pores is

$$D'_{Ae} = \epsilon^2 D_{Ae} \tag{24-61}$$

where ϵ is the volume void fraction of the porous volume within the porous material, which is defined as

$$\epsilon = \frac{\text{the volume occupied by pores within the porous solid}}{\text{total volume of porous solid (solid + pores)}}$$

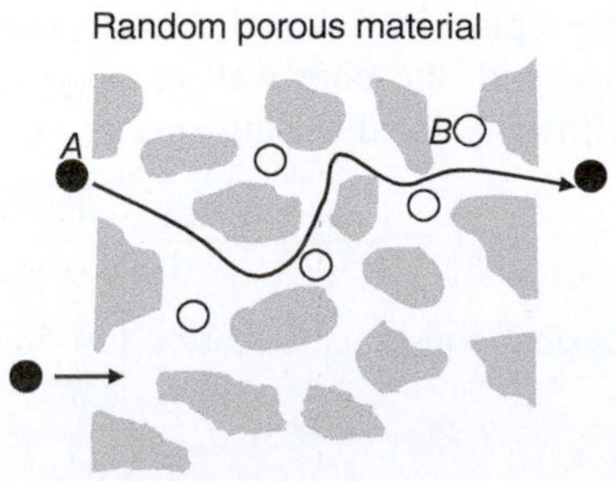

Figure 24.5 Diffusion in random porous materials.

The void fraction (ϵ) is usually experimentally determined for a specific material. In cases of large pore diameters or high pressure where Knudsen diffusion is not important, equation (24-61) reduces to

$$D'_{Ae} = \epsilon^2 D_{AB}$$

The ϵ^2 term accounts for the available cross-sectional area for flux of the fluid mixture through the porous solid, as species A and B can only enter the porous material from the surrounding bulk fluid through the available flux area defined by the pore openings.

Example 5

A dilute mixture of carbon dioxide (CO_2) in ethylene (C_2H_4) gas is diffusing within a random porous material with void fraction of 0.45 at a total system pressure of 2.0 atm and temperature of 350 K. Estimate the effective diffusion coefficient of CO_2 and ethylene within the porous material if the mean pore diameter is 0.20 µm.

Let species A represent CO_2 with molecular weight of 44 g/mole, and species B represent C_2H_4 with a molecular weight of 28 g/mole. From Example 3, the gas-phase binary molecular diffusion coefficient, D_{AB}, is 0.077 cm^2/s at 2.0 atm total system pressure and 350 K absolute temperature. Given that the diffusion process is within a mesoporous material, which has pore diameters less than $1.0\,\mu m$ ($1.0\,\mu m = 1.0 \times 10^{-6}$ m), Knudsen diffusion may contribute to the molecular mass-transfer process. The Knudsen diffusion coefficient for CO_2 is estimated by equation (24-58b), with $d_{pore} = 2.0 \times 10^{-5}$ cm and $M_A = 44$ g/gmole:

$$D_{KA} = 4850\, d_{pore} \sqrt{\frac{T}{M_A}} = 4850\,(2.0 \times 10^{-5}) \sqrt{\frac{350}{44}} = 0.274\ \text{cm}^2/\text{s}$$

Consequently, from equation (24-60), for CO_2 diluted in ethylene, effective diffusion coefficient of species A in a dilute gas mixture within the pore is

$$D_{Ae} = \frac{D_{AB}D_{KA}}{D_{AB} + D_{KA}} = \frac{(0.077\ \text{cm}^2/\text{s})(0.274\ \text{cm}^2/\text{s})}{(0.077\ \text{cm}^2/\text{s}) + (0.274\ \text{cm}^2/\text{s})} = 0.060\ \text{cm}^2/\text{s}$$

We note that $D_{Ae} < D_{AB}$, showing the effect of Knudsen diffusion. To this point, the mean-free path and Knudsen number for CO_2 inside the pore are

$$\lambda = \frac{\kappa T}{\sqrt{2}\pi \sigma_A^2 P} = \frac{\left(1.38 \times 10^{-16} \dfrac{\text{erg}}{K}\dfrac{1\,N\,m}{10^7\,\text{erg}}\right)(350\,K)}{\sqrt{2}\pi \left(0.3996\,\text{nm}\dfrac{1\,m}{10^9\text{nm}}\right)^2 \left(2.0\,\text{atm}\dfrac{101{,}300\,\text{N/m}^2}{\text{atm}}\right)}$$

$$= 3.36 \times 10^{-8}\ \text{m} = 0.0336\ \mu m$$

and

$$Kn = \frac{\lambda}{d_{pore}} = \frac{0.0336\ \mu m}{0.200\ \mu m} = 0.17$$

From this analysis, Knudsen diffusion plays only a moderate role at the conditions of the process even though the pore diameter is only 0.2 µm. Finally, the effective diffusion coefficient for CO_2, as corrected for the void fraction within the random porous material, is

$$D'_{Ae} = \epsilon^2 D_{Ae} = (0.45)^2 (0.060\ \text{cm}^2/\text{s}) = 0.012\ \text{cm}^2/\text{s}$$

The effect of total system pressure on the ratio D_{Ae}/D_{AB} is shown in Figure 24.6. At high total system pressures, the effective diffusion coefficient (D_{Ae}) approaches the molecular diffusion coefficient, D_{AB}. At low total system pressures, Knudsen diffusion becomes important.

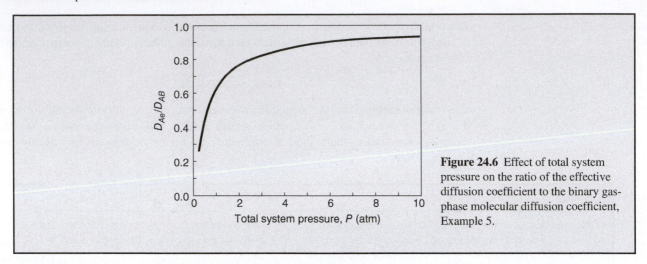

Figure 24.6 Effect of total system pressure on the ratio of the effective diffusion coefficient to the binary gas-phase molecular diffusion coefficient, Example 5.

Hindered Solute Diffusion in Solvent-Filled Pores Consider the diffusion of a solute molecule (species A) through a tiny capillary pore filled with liquid solvent (species B), as shown in Figure 24.7. As the molecular diameter of the solute approaches the diameter of the pore, the diffusive transport of the solute through the solvent is hindered by the presence of the pore and the pore wall. General models for diffusion coefficients describing the "hindered diffusion" of solutes in solvent-filled pores assume the form of

$$\frac{D_{Ae}}{D_{AB}^{o}} = F_1(\varphi)F_2(\varphi) \tag{24-62}$$

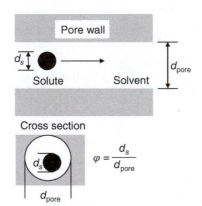

Figure 24.7 Diffusion of solute in solvent-filled pore.

The molecular diffusion coefficient of solute A in the solvent B at infinite dilution, D_{AB}^{o}, is reduced by two correction factors, $F_1(\varphi)$, and $F_2(\varphi)$, both of which are theoretically bounded by 0 and 1. Furthermore, both correction factors are functions of the reduced pore diameter φ

$$\varphi = \frac{d_s}{d_{\text{pore}}} = \frac{\text{solute molecular diameter}}{\text{pore diameter}} \tag{24-63}$$

If $\varphi > 1$, then the solute is too large to enter the pore. This phenomenon is known as solute exclusion, and it is used to separate large biomolecules such as proteins from dilute aqueous mixtures containing solutes of much smaller diameter. As φ approaches 1, both $F_1(\varphi)$ and $F_2(\varphi)$ decrease asymptotically toward zero, so at $\varphi = 1$, the effective diffusion coefficient is zero.

The correction factor $F_1(\varphi)$, the stearic partition coefficient, is based on simple geometric arguments for stearic exclusion:

$$F_1(\varphi) = \frac{\text{flux area available to solute}}{\text{total flux area}} = \frac{\pi(d_{\text{pore}} - d_s)^2}{\pi d_{\text{pore}}^2} = (1 - \varphi)^2 \qquad (24\text{-}64)$$

and holds for $0 \leq F_1(\varphi) \leq 1.0$.

The correction factor $F_2(\varphi)$, the hydrodynamic hindrance factor, is based on the complicated hydrodynamic calculations involving the hindered Brownian motion of the solute within the solvent-filled pore. Equations for $F_2(\varphi)$, assuming diffusion of a rigid spherical solute in a straight cylindrical pore, have been developed. The analytical models are generally asymptotic solutions over a limited range of φ, and ignore electrostatic or other energetic solute-solvent-pore wall interactions, polydispersity of solute diameters, and noncircular pore cross sections. The most common equation, developed by Renkin,[20] is reasonable for the range $0 < \varphi < 0.6$:

$$F_2(\varphi) = 1 - 2.104\varphi + 2.09\varphi^3 - 0.95\varphi^5 \qquad (24\text{-}65)$$

The effect of reduced pore diameter on the effective diffusion coefficient is presented in Figure 24.8.

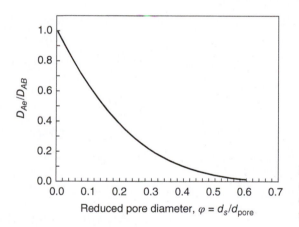

Figure 24.8 Effect of reduced pore diameter on the effective diffusion coefficient, using the Renkin equation for the hydrodynamic hindrance factor.

Example 6

It is desired to separate a mixture of two industrial enzymes, lysozyme and catalase, in a dilute, aqueous solution by a gel filtration membrane. A mesoporous membrane with cylindrical pores of 30 nm diameter is available (Figure 24.9). The following separation factor (α) for the process is proposed:

$$\alpha = \frac{D_{Ae}}{D_{Be}}$$

---

[20] E. M. Renkin, *J. Gen. Physiol.*, **38**, 225 (1954).

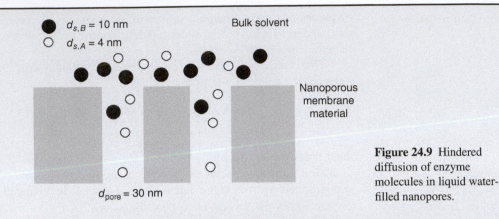

Figure 24.9 Hindered diffusion of enzyme molecules in liquid water-filled nanopores.

Determine the separation factor for this process. The properties of each enzyme as reported by Tanford[21] are given below.

Lysozyme (species A)	Catalase (species B)
$M_A = 14{,}100$ g/gmole	$M_B = 250{,}000$ g/gmole
$d_{s,A} = 4.12$ nm	$d_{s,B} = 10.44$ nm
$D^o_{A-H_2O} = 1.04 \times 10^{-6}$ cm²/s	$D^o_{B-H_2O} = 4.10 \times 10^{-7}$ cm²/s

The transport of large enzyme molecules through pores filled with liquid water represents a hindered diffusion process. The reduced pore diameters for lysozyme and catalase are

$$\varphi_A = \frac{d_{s,A}}{d_{pore}} = \frac{4.12 \text{ nm}}{30.0 \text{ nm}} = 0.137 \text{ and } \varphi_B = \frac{d_{s,B}}{d_{pore}} = \frac{10.44 \text{ nm}}{30.0 \text{ nm}} = 0.348$$

For lysozyme, $F_1(\varphi_A)$, by equation (24-64), and $F_2(\varphi_A)$, by the Renkin equation (24-65), are

$$F_1(\varphi_A) = (1 - \varphi_A)^2 = (1 - 0.137)^2 = 0.744$$

$$F_2(\varphi_A) = 1 - 2.104\varphi_A + 2.09\varphi_A^3 - 0.95\varphi_A^5$$

$$= 1 - 2.104(0.137) + 2.090(0.137)^3 - 0.95(0.137)^5 = 0.716$$

The effective diffusivity of lysozyme in the pore, D_{Ae}, is estimated by equation (24-62)

$$D_{Ae} = D_{A-H_2O}F_1(\varphi_A)F_2(\varphi_A) = 1.04 \times 10^{-6} \frac{\text{cm}^2}{\text{s}} (0.744)(0.716) = 5.54 \times 10^{-7} \frac{\text{cm}^2}{\text{s}}$$

Likewise, for catalase $F_1(\varphi_B) = 0.425, F_2(\varphi_B) = 0.351$, and $D_{Be} = 6.12 \times 10^{-8}$ cm²/s. Finally, the separation factor is

$$\alpha = \frac{D_{Ae}}{D_{Be}} = \frac{5.54 \times 10^{-7} \text{ cm}^2/\text{s}}{6.12 \times 10^{-8} \text{ cm}^2/\text{s}} = 9.06$$

It is interesting to compare the value above with α', the ratio of molecular diffusivities at infinite dilution:

$$\alpha' = \frac{D_{A-H_2O}}{D_{B-H_2O}} = \frac{1.04 \times 10^{-6} \text{ cm}^2/\text{s}}{4.1 \times 10^{-7} \text{ cm}^2/\text{s}} = 1.75$$

The small pore diameter enhances the value for α, because the diffusion of the large catalase molecule is significantly hindered inside the pore relative to the smaller lysozyme molecule.

---

[21] C. Tanford, *Physical Chemistry of Macromolecules*, John Wiley & Sons, New York, 1961.

Solid Mass Diffusivity

The diffusion of atoms within solids underlies the synthesis of many engineering materials. In semiconductor manufacturing processes, "impurity atoms," commonly called *dopants*, are introduced into solid silicon to control the conductivity in a semiconductor device. The hardening of steel results from the diffusion of carbon and other elements through iron. Vacancy diffusion and interstitial diffusion are the two most frequently encountered solid diffusion mechanisms.

In *vacancy diffusion*, also called substitutional diffusion, the transported atom "jumps" from a lattice position of the solid into a neighboring unoccupied lattice site or vacancy, as illustrated in Figure 24.10. The atom continues to diffuse through the solid by a series of

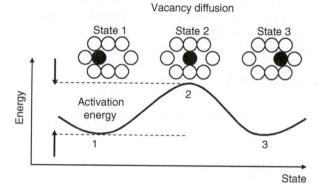

Figure 24.10 Solid-state vacancy diffusion.

jumps into other neighboring vacancies that appear to it from time to time. This normally requires a distortion of the lattice, such as defects and grain boundaries in polycrystalline materials. This mechanism has been mathematically described by assuming a unimolecular rate process and applying Eyring's "activated state" concept, as discussed in the "hole" theory for liquid diffusion. The resulting equation is a complex equation relating the diffusivity in terms of the geometric relations between the lattice positions, the length of the jump path, and the energy of activation associated with the jump.

An atom moves in *interstitial diffusion* by jumping from one interstitial site to a neighboring one, as illustrated in Figure 24.11. This normally involves a dilation or distortion of the lattice. This mechanism is also mathematically described by Eyring's unimolecular rate theory. The energy barrier encountered in interstitial diffusion is typically much smaller than the energy barrier that must be surpassed when an atom jumps between two lattice sites by vacancy diffusion.

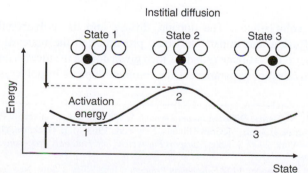

Figure 24.11 Solid-state interstitial diffusion.

Excellent references are available for a more detailed discussion on the diffusion characteristics of atoms in solids.[22]

Appendix Table J.3 lists a few values of binary diffusivities in solids. The diffusion coefficient of solutes into solids has been observed to increase with increasing temperature according to an Arrhenius equation of the form

$$D_{AB} = D_o \, e^{-Q/R\,T} \qquad (24\text{-}66)$$

or in a slope-intercept form of

$$\ln(D_{AB}) = -\left(\frac{Q}{R}\right)\left(\frac{1}{T}\right) + \ln(D_o) \qquad (24\text{-}67)$$

where D_{AB} is solid diffusion coefficient for the diffusing species A within solid B, D_0 is a proportionality constant of units consistent with D_{AB}, Q is the activation energy (J/gmole), R is the thermodynamic constant (8.314 J/gmole $\cdot$ K), and T is the absolute temperature (K).

Figure 24.12 illustrates the dependence of solid-phase diffusion coefficients on temperature, specifically for the diffusion of common elemental impurities, or dopants,

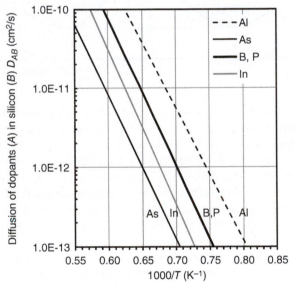

Figure 24.12 Average diffusion coefficients of substitutional dopants in polycrystalline silicon, using data provided by Ghandhi.[23]

in solid silicon. The dopants diffuse into the polycrystalline matrix by substitutional diffusion and impart electronic properties to the material. Data from Figure 24.12 can be used to estimate Q and D_0 for a given dopant in silicon using equation (24-67). Table 24.7 provides these parameters for common dopants in polycrystalline silicon.

[22] R. M. Barrer, *Diffusion In and Through Solids*, Cambridge University Press, London, 1941; P. G. Shewmon, *Diffusion of Solids*, McGraw-Hill Inc., New York, 1963; M. E. Glickman, *Diffusion in Solids*, John Wiley & Sons, New York, 2000; S. Kou, *Transport Phenomena and Materials Processing*, John Wiley & Sons Inc., New York, 1996.
[23] S. K. Ghandhi, *VLSI Fabrication Principles*, John Wiley & Sons, New York, 1983.

Table 24.7   Diffusion parameters common substitutional dopants in polycrystalline silicon, using data obtained from Ghandhi[23]

Dopant	D_0 (cm^2/s)	Q (kJ/gmole)
Al	2.61	319.1
As	0.658	348.1
B, P	11.1	356.2
Ga	0.494	312.6
In	15.7	373.5

Table 24.8 provides experimentally determined parameters needed to evaluate D_{AB} for interstitial solutes in various structures of iron in the solid state, where A is the solute, and B

Table 24.8   Diffusion parameters for interstitial solutes in iron

Structure	Solute	D_0 (mm^2/s)	Q (kJ/gmole)
bcc	C	2.0	84.1
bcc	N	0.3	76.1
bcc	H	0.1	13.4
fcc	C	2.5	144.2

is iron. Values for Q in Tables 24.7 and 24.8 illustrate the significant energy barrier that must be surpassed when an atom jumps between two lattice sites by solid-state diffusion.

Diffusion coefficients and the solubility of solutes in polymers are reported by Rogers[24] and by Crank and Park.[25] Diffusion coefficients of solutes in dilute biological gels are reported by Friedman and Kramer[26] and by Spalding.[27]

► **24.3**

CONVECTIVE MASS TRANSFER

Mass transfer between a moving fluid and a surface or between immiscible moving fluids separated by a mobile interface, for example, in a gas/liquid or liquid/liquid contractor, is often aided by the dynamic characteristics of the moving fluid. This mode of transfer is called *convective mass transfer*, with the transfer always going from a higher to a lower concentration of the species being transferred. Convective transfer depends on both the transport properties and the dynamic characteristics of the flowing fluid.

As in the case of convective heat transfer, a distinction must be made between two types of flow. When an external pump or similar device causes the fluid motion, the process is called *forced convection*. If the fluid motion is due to a density difference, the process is called *free* or *natural convection*.

---

[24] C. E. Rogers, *Engineering Design for Plastics*, Reinhold Press, New York, 1964.
[25] J. Crank and G. S. Park, *Diffusion in Polymers*, Academic Press, New York, 1968.
[26] L. Friedman and E. O. Kramer, *J. Am. Chem. Soc.*, **52**, 1298, 1305, 1311 (1930).
[27] G. E. Spalding, *J. Phys. Chem.*, **73**, 3380 (1969).

The rate equation for convective mass transfer, generalized in a manner analogous to Newton's "law" of cooling, equation (15-11), is

$$N_A = k_c \, \Delta c_A \tag{24-68}$$

Where N_A is the molar mass transfer of species A measured relative to fixed spatial coordinates, Δc_A is the concentration difference between the boundary surface concentration and the average concentration of the fluid stream of the diffusing species A, and k_c is the convective mass-transfer coefficient, which has SI units of m/s.

As in the case of molecular mass transfer, convective mass transfer occurs in the direction of a decreasing concentration. Equation (24-68) defines the mass-transfer coefficient k_c in terms of the mass flux and the concentration difference from the beginning to the end of the mass-transfer path. The reciprocal of the mass-transfer coefficient, $1/k_c$, represents the resistance to the mass transfer through the moving fluid. Chapters 28 and 30 consider the methods of determining the mass coefficient. It is, in general, a function of system geometry, fluid and flow properties, and the concentration difference of transferring species A between the boundary surface and the bulk flowing fluid.

From our experiences in dealing with a fluid flowing past a surface, we can recall that there is always a layer, sometimes extremely thin, close to the surface where the fluid is laminar, and that fluid particles next to the solid boundary are at rest. As this is always true, the mechanism of mass transfer between a surface and a fluid must involve molecular mass transfer through the stagnant and laminar flowing fluid layers. The controlling resistance to convective mass transfer is often the result of this "film" of fluid and the coefficient, k_c, is accordingly referred to as the *film mass-transfer coefficient*.

It is important for the student to recognize the close similarity between the convective mass-transfer coefficient and the convective heat-transfer coefficient. This immediately suggests that the techniques developed for evaluating the convective heat-transfer coefficient may be repeated for convective mass transfer. A complete discussion of convective mass-transfer coefficients and their evaluation is given in Chapters 28 and 30.

▶ **24.4**

CLOSURE

In this chapter, the two modes of mass transport, molecular and convective mass transfer, have been introduced. As diffusion of mass involves a mixture, fundamental relations were presented for concentrations and velocities of the individual species as well as for the mixture. The molecular transport property, D_{AB}, the diffusion coefficient in gas, liquid, and solid systems, has been discussed and correlating equations presented. Concepts underlying the effective diffusion coefficient of gases and liquids within porous materials have also been introduced, and methods to estimate effective diffusion coefficients in dilute systems were described.

Differential Equations of Mass Transfer

In Chapter 9, the general differential equations for momentum transfer are derived by the use of a differential control-volume concept. By an analogous treatment, the general differential equations for heat transfer are generated in Chapter 16. Once again, we shall use this approach to develop the differential equations for mass transfer. By making a mass balance over a differential control volume, we shall establish the equation of continuity for a given species.

Additional differential equations will be obtained when we insert, into the continuity equation, mass flux relationships developed in the previous chapter.

▶ **25.1**

THE DIFFERENTIAL EQUATION FOR MASS TRANSFER

Consider the control volume, $\Delta x \Delta y \Delta z$, through which a mixture including component A is flowing, as shown in Figure 25.1. The control-volume expression for the conservation of mass is

$$\iint_{\text{c.s.}} \rho(\mathbf{v} \cdot \mathbf{n})dA + \frac{\partial}{\partial t} \iiint_{\text{c.v.}} \rho \, dV = 0 \qquad (4\text{-}1)$$

which may be stated in words as

$$\left\{ \begin{array}{c} \text{net rate of mass} \\ \text{efflux from} \\ \text{control volume} \end{array} \right\} + \left\{ \begin{array}{c} \text{net rate of accumulation} \\ \text{of mass within control} \\ \text{volume} \end{array} \right\} = 0$$

If we consider the conservation of a given species A, this relation should also include a term that accounts for the production or disappearance of A by chemical reaction within the volume. The general relation for a mass balance of species A for our control volume may be stated as

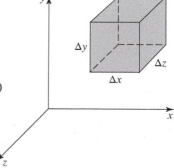

Figure 25.1 A differential control volume.

$$\left\{ \begin{array}{c} \text{net rate of mass} \\ \text{efflux of } A \text{ from} \\ \text{control volume} \end{array} \right\} + \left\{ \begin{array}{c} \text{net rate of} \\ \text{accumulation of } A \text{ within} \\ \text{control volume} \end{array} \right\} - \left\{ \begin{array}{c} \text{rate of chemical} \\ \text{production of } A \\ \text{within the control} \\ \text{volume} \end{array} \right\} = 0 \quad (25\text{-}1)$$

399

The individual terms will be evaluated for constituent A, and a discussion of their meanings will be given below.

The net rate of mass efflux from the control volume may be evaluated by considering the mass transferred across control surfaces. For example, the mass of A transferred across the area $\Delta y \Delta z$ at x will be $\rho_A v_{Ax} \Delta y \Delta z|_x$, or in terms of the flux vector, $\mathbf{n}_A = \rho_A \mathbf{v}_A$, it would be $n_{A,x} \Delta y \Delta z|_x$. The net rate of mass efflux of constituent A will be

$$\text{in the } x\text{-direction:} \quad n_{A,x} \Delta y \Delta z|_{x+\Delta x} - n_{A,x} \Delta y \Delta z|_x$$

$$\text{in the } y\text{-direction:} \quad n_{A,y} \Delta x \Delta z|_{y+\Delta y} - n_{A,y} \Delta x \Delta z|_y$$

and

$$\text{in the } z\text{-direction:} \quad n_{A,z} \Delta x \Delta y|_{z+\Delta z} - n_{A,z} \Delta x \Delta y|_z$$

The rate of accumulation of A in the control volume is

$$\frac{\partial \rho_A}{\partial t} \Delta x \Delta y \Delta z$$

If A is produced within the control volume by a chemical reaction at a rate r_A, where r_A has the units (mass of A produced)/(volume)(time), the rate of production of A is

$$r_A \Delta x \Delta y \Delta z$$

This production term is analogous to the energy generation term that appeared in the differential equation for energy transfer, as discussed in Chapter 16. Substituting each term in equation (25-1), we obtain

$$n_{A,x} \Delta y \Delta z|_{x+\Delta x} - n_{A,x} \Delta y \Delta z|_x + n_{A,y} \Delta x \Delta z|_{y+\Delta y} - n_{A,y} \Delta x \Delta z|_y + n_{A,z} \Delta x \Delta y|_{z+\Delta z}$$
$$- n_{A,z} \Delta x \Delta y|_z + \frac{\partial \rho_A}{\partial t} \Delta x \Delta y \Delta z - r_A \Delta x \Delta y \Delta z = 0 \tag{25-2}$$

Dividing through by the volume, $\Delta x \Delta y \Delta z$, and canceling terms, we have

$$\frac{n_{A,x}|_{x+\Delta x} - n_{A,x}|_x}{\Delta x} + \frac{n_{A,y}|_{y+\Delta y} - n_{A,y}|_y}{\Delta y} + \frac{n_{A,z}|_{z+\Delta z} - n_{A,z}|_z}{\Delta z} + \frac{\partial \rho_A}{\partial t} - r_A = 0 \tag{25-3}$$

Evaluated in the limit as Δx, Δy, and Δz approach zero, this yields

$$\frac{\partial n_{A,x}}{\partial x} + \frac{\partial n_{A,y}}{\partial y} + \frac{\partial n_{A,z}}{\partial z} + \frac{\partial \rho_A}{\partial t} - r_A = 0 \tag{25-4}$$

Equation (25-4) is the *equation of continuity* for component A. As $n_{A,x}, n_{A,y}$, and $n_{A,z}$ are the rectangular components of the mass flux vector, $\mathbf{n}_A$, equation (25-4) may be written

$$\nabla \cdot \mathbf{n}_A + \frac{\partial \rho_A}{\partial t} - r_A = 0 \tag{25-5}$$

A similar equation of continuity may be developed for a second constituent B in the same manner. The differential equations are

$$\frac{\partial n_{B,x}}{\partial x} + \frac{\partial n_{B,y}}{\partial y} + \frac{\partial n_{B,z}}{\partial z} + \frac{\partial \rho_B}{\partial t} - r_B = 0 \tag{25-6}$$

and

$$\nabla \cdot \mathbf{n}_B + \frac{\partial \rho_B}{\partial t} - r_B = 0 \tag{25-7}$$

where r_B is the rate at which B will be produced within the control volume by a chemical reaction. Adding equations (25-5) and (25-7), we obtain

$$\nabla \cdot (\mathbf{n}_A + \mathbf{n}_B) + \frac{\partial(\rho_A + \rho_B)}{\partial t} - (r_A + r_B) = 0 \tag{25-8}$$

For a binary mixture of A and B, we have

$$\mathbf{n}_A + \mathbf{n}_B = \rho_A \mathbf{v}_A + \rho_B \mathbf{v}_B = \rho \mathbf{v}$$
$$\rho_A + \rho_B = \rho$$

and

$$r_A = -r_B$$

by the law of conservation of mass. Substituting these relations into (25-8), we obtain

$$\nabla \cdot \rho \mathbf{v} + \frac{\partial \rho}{\partial t} = 0 \tag{25-9}$$

This is the *equation of continuity for the mixture*. Equation (25-9) is identical to the equation of continuity (9-2) for a homogeneous fluid.

The equation of continuity for the mixture and for a given species can be written in terms of the substantial derivative. As shown in Chapter 9, the continuity equation for the mixture can be rearranged and written

$$\frac{D\rho}{Dt} + \rho \nabla \cdot \mathbf{v} = 0 \tag{9-5}$$

Through similar mathematical manipulations, the equation of continuity for species A in terms of the substantial derivative may be derived. This equation is

$$\frac{\rho D w_A}{Dt} + \nabla \cdot \mathbf{j}_A - r_A = 0 \tag{25-10}$$

We could follow the same development in terms of molar units. If R_A represents the rate of molar production of A per unit volume, and R_B represents the rate of molar production of B per unit volume, the molar-equivalent equations are

for component A

$$\nabla \cdot \mathbf{N}_A + \frac{\partial c_A}{\partial t} - R_A = 0 \tag{25-11}$$

for component B

$$\nabla \cdot \mathbf{N}_B + \frac{\partial c_B}{\partial t} - R_B = 0 \tag{25-12}$$

and for the mixture

$$\nabla \cdot (\mathbf{N}_A + \mathbf{N}_B) + \frac{\partial(c_A + c_B)}{\partial t} - (R_A + R_B) = 0 \tag{25-13}$$

For the binary mixture of A and B, we have

$$\mathbf{N}_A + \mathbf{N}_B = c_A \mathbf{v}_A + c_B \mathbf{v}_B = c \mathbf{V}$$

and

$$c_A + c_B = c$$

However, only when the stoichiometry of the reaction is

$$A \rightleftharpoons B$$

which stipulates that one molecule of B is produced for each mole of A disappearing, can we stipulate that $R_A = -R_B$. In general, the equation of continuity for the mixture in molar units is

$$\nabla \cdot c\mathbf{V} + \frac{\partial c}{\partial t} - (R_A + R_B) = 0 \tag{25-14}$$

▶ **25.2**

SPECIAL FORMS OF THE DIFFERENTIAL MASS-TRANSFER EQUATION

Special forms of the equation of continuity applicable to commonly encountered situations follow. In order to use the equations for evaluating the concentration profiles, we replace the fluxes, $\mathbf{n}_A$ and $\mathbf{N}_A$, by the appropriate expressions developed in Chapter 24. These expressions are

$$\mathbf{N}_A = -cD_{AB}\nabla y_A + y_A(\mathbf{N}_A + \mathbf{N}_B) \tag{24-22}$$

or its equivalent

$$\mathbf{N}_A = -cD_{AB}\nabla y_A + c_A\mathbf{V}$$

and

$$\mathbf{n}_A = -\rho D_{AB}\nabla w_A + w_A(\mathbf{n}_A + \mathbf{n}_B) \tag{24-23}$$

or its equivalent

$$\mathbf{n}_A = -\rho D_{AB}\nabla w_A + \rho_A\mathbf{v}$$

Substituting equation (24-23) into equation (25-5), we obtain

$$-\nabla \cdot \rho D_{AB}\nabla w_A + \nabla \cdot \rho_A\mathbf{v} + \frac{\partial \rho_A}{\partial t} - r_A = 0 \tag{25-15}$$

and substituting equation (24-22) into equation (25-11), we obtain

$$-\nabla \cdot cD_{AB}\nabla y_A + \nabla \cdot c_A\mathbf{v} + \frac{\partial c_A}{\partial t} - R_A = 0 \tag{25-16}$$

Either equation (25-15) or (25-16) may be used to describe concentration profiles within a diffusing system. Both equations are completely general; however, they are relatively unwieldy. These equations can be simplified by making restrictive assumptions. Important forms of the equation of continuity, with their qualifying assumptions, include:

(i) If the density, ρ, and the diffusion coefficient, D_{AB} can be assumed constant, equation (25-15), after dividing each term by the molecular weight of species

A and rearranging becomes

$$\mathbf{v} \cdot \nabla c_A + \frac{\partial c_A}{\partial t} = D_{AB} \nabla^2 c_A + R_A \tag{25-17}$$

(ii) If there is no production term, $R_A = 0$, and if the density and diffusion coefficient are assumed constant, equation (25-17) reduces to

$$\frac{\partial c_A}{\partial t} + \mathbf{v} \cdot \nabla c_A = D_{AB} \nabla^2 c_A \tag{25-18}$$

We recognize that $(\partial c_A / \partial t) + \mathbf{v} \cdot \nabla c_A$ is the substantial derivative of c_A; rewriting the left-hand side of equation (25-18), we obtain

$$\frac{Dc_A}{Dt} = D_{AB} \nabla^2 c_A \tag{25-19}$$

which is analogous to equation (16-14) from heat transfer

$$\frac{DT}{Dt} = \frac{k}{\rho\, c_p} \nabla^2 T \tag{16-14}$$

or

$$\frac{DT}{Dt} = \alpha \nabla^2 T$$

where α is the thermal diffusivity. The similarity between these two equations is the basis for the analogies drawn between heat and mass transfer.

(iii) In a situation in which there is no fluid motion, $\mathbf{v} = 0$, no production term, $R_A = 0$, and no variation in the diffusivity or density, equation (25-18) reduces to

$$\frac{\partial c_A}{\partial t} = D_{AB} \nabla^2 c_A \tag{25-20}$$

Equation (25-20) is commonly referred to as *Fick's second "law" of diffusion*. The assumption of no fluid motion restricts its applicability to diffusion in solids, or stationary liquids, and for binary systems of gases or liquids, where $\mathbf{N}_A$ is equal in magnitude, but acting in the opposite direction to $\mathbf{N}_A$, that is, the case of equimolar counterdiffusion. Equation (25-20) is analogous to Fourier's second "law" of heat conduction:

$$\frac{\partial T}{\partial t} = \alpha \nabla^2 T \tag{16-18}$$

(iv) Equations (25-17), (25-18), and (25-20) may be simplified further when the process to be defined is a steady-state process—that is, $\partial c_A / \partial t = 0$. For constant density and a constant diffusion coefficient, the equation becomes

$$\mathbf{v} \cdot \nabla c_A = D_{AB} \nabla^2 c_A + R_A \tag{25-21}$$

For constant density, constant diffusivity, and no chemical production, $R_A = 0$, we obtain

$$\mathbf{v} \cdot \nabla c_A = D_{AB} \nabla^2 c_A \tag{25-22}$$

If additionally, $\mathbf{v} = 0$, the equation reduces to

$$\nabla^2 c_A = 0 \qquad (25\text{-}23)$$

Equation (25-23) is the *Laplace equation* in terms of molar concentration.

Each of the equations (25-15) through (25-23) has been written in vector form, so each applies to any orthogonal coordinate system. By writing the Laplacian operator, ∇^2, in the appropriate form, the transformation of the equation to the desired coordinate system is accomplished. Fick's second "law" of diffusion written in rectangular coordinates is

$$\frac{\partial c_A}{\partial t} = D_{AB}\left[\frac{\partial^2 c_A}{\partial x^2} + \frac{\partial^2 c_A}{\partial y^2} + \frac{\partial^2 c_A}{\partial z^2}\right] \qquad (25\text{-}24)$$

in cylindrical coordinates is

$$\frac{\partial c_A}{\partial t} = D_{AB}\left[\frac{\partial^2 c_A}{\partial r^2} + \frac{1}{r}\frac{\partial c_A}{\partial r} + \frac{1}{r^2}\frac{\partial^2 c_A}{\partial \theta^2} + \frac{\partial^2 c_A}{\partial z^2}\right] \qquad (25\text{-}25)$$

and in spherical coordinates is

$$\frac{\partial c_A}{\partial t} = D_{AB}\left[\frac{1}{r^2}\frac{\partial}{\partial r}\left(r^2\frac{\partial c_A}{\partial r}\right) + \frac{1}{r^2\sin\theta}\frac{\partial}{\partial \theta}\left(\sin\theta\frac{\partial c_A}{\partial \theta}\right) + \frac{1}{r^2\sin^2\theta}\frac{\partial^2 c_A}{\partial \phi^2}\right] \qquad (25\text{-}26)$$

The general differential equation for mass transfer of component A, or the equation of continuity of A, written in rectangular coordinates is

$$\frac{\partial c_A}{\partial t} = \left[\frac{\partial N_{A,x}}{\partial x} + \frac{\partial N_{A,y}}{\partial y} + \frac{\partial N_{A,z}}{\partial z}\right] = R_A \qquad (25\text{-}27)$$

in cylindrical coordinates is

$$\frac{\partial c_A}{\partial t} = \left[\frac{1}{r}\frac{\partial}{\partial r}\left(rN_{A,r}\right) + \frac{1}{r}\frac{\partial N_{A,\theta}}{\partial \theta} + \frac{\partial N_{A,z}}{\partial z}\right] = R_A \qquad (25\text{-}28)$$

and in spherical coordinates is

$$\frac{\partial c_A}{\partial t} + \left[\frac{1}{r^2}\frac{\partial}{\partial r}\left(r^2 N_{A,r}\right) + \frac{1}{r\sin\theta}\frac{\theta}{\partial \theta}\left(N_{A,\theta}\sin\theta\right) + \frac{1}{r\sin\theta}\frac{\partial N_{A,\phi}}{\partial \phi}\right] = R_A \qquad (25\text{-}29)$$

▶ 25.3

COMMONLY ENCOUNTERED BOUNDARY CONDITIONS

A mass-transfer process is fully described by the differential equations of mass transfer only if the initial boundary and initial conditions are specified. Typically, initial and boundary conditions are used to specify limits of integration or to determine integration constants associated with the mathematical solution of the differential equations for mass transfer. The initial and boundary conditions used for mass transfer are very similar to those used in Section 16.3 for energy transfer. The reader may wish to refer to that section for further discussion of initial and boundary conditions.

The initial condition in mass-transfer processes is the concentration of the diffusing species at the start of the time interval of interest expressed in either mass or molar concentration units. The concentration may be simply equal to a constant value or may be

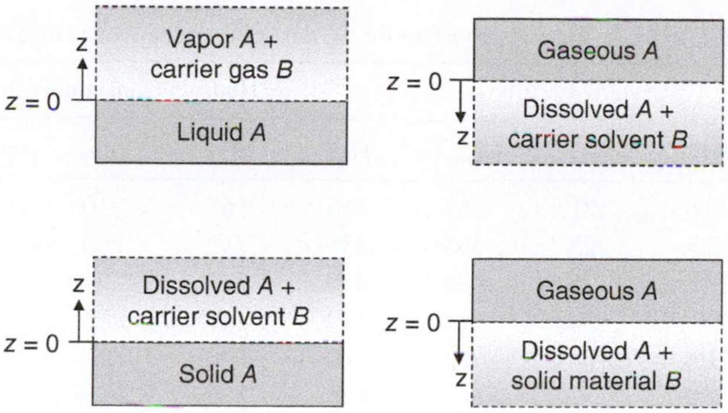

Figure 25.2 Examples of concentration boundary conditions defined by a thermodynamic equilibrium constraint at an interface.

more complex if the initial concentration distribution within the control volume for diffusion is specified. Initial conditions are associated only with unsteady-state or pseudo-steady-state processes.

Four types of boundary conditions are commonly encountered in mass transfer.

1. *The concentration of the transferring species A at a boundary surface is specified.* Surface concentration can assume a variety of units—for example, molar concentration c_{As}, mass concentration ρ_{As}, gas mole fraction y_{As}, liquid mole fraction x_{As}, etc. When the boundary surface is defined by a pure component in one phase and a mixture in the second phase, then the concentration of transferring species A in the mixture at the interface is usually at thermodynamic saturation conditions. Common examples of this type of boundary condition that exists between two phases in contact are provided in Figure 25.2. Specifically, for a gas mixture in contact with a pure volatile liquid A or pure volatile solid A, the partial pressure of species A in the gas at the surface is saturation vapor pressure, P_A, so that $p_{As} = P_A$. For a liquid mixture in contact with a pure solid A, the concentration of species A in the liquid at the surface is the solubility limit of A in the liquid, c_A^*, so that $c_{As} = c_A^*$.

   For a contacting gas and liquid where transferring species A is present in both phases, there are two ways to specify the concentration at the gas–liquid interface. First, if both of the species in the liquid phase are volatile, then the boundary condition at the gas–liquid surface is defined for an ideal liquid mixture by Raoult's law:

   $$p_{As} = x_A P_A$$

   where x_A is the mole fraction in the liquid, P_A is the vapor pressure of species A evaluated at the temperature of the liquid, and p_{As} is the partial pressure of species A in the gas. The partial pressure of species A at the interface is related to surface mole fraction y_{As} by Dalton's law:

   $$y_{As} = \frac{p_{As}}{P}$$

   or to surface concentration c_{As} by the ideal gas law:

   $$c_{As} = \frac{p_{As}}{RT}$$

Table 25.1   Henry's constant for various gases in aqueous solutions (H in bars)

$T(K)$	Henry's Constant, H (bar)							
	NH_3	Cl_2	H_2S	SO_2	CO_2	CH_4	O_2	H_2
273	21	265	260	165	710	22,800	25,500	58,000
280	23	365	335	210	960	27,800	30,500	61,500
290	26	480	450	315	1300	35,200	37,600	66,500
300	30	615	570	440	1730	42,800	45,700	71,600
310		755	700	600	2175	50,000	52,000	76,000
320		860	835	800	2650	56,300	56,800	78,600

Second, for solutions where species A is only weakly soluble in the liquid, Henry's law may be used to relate the mole fraction of A in the liquid to the partial pressure of A in the gas:

$$p_A = H \cdot x_A$$

where coefficient H is known as *Henry's constant*. Values of H in pressure units for selected gaseous solutes dissolved in aqueous solution are listed in Table 25.1. Henry's constants of gaseous solutes in aqueous solution for these and other gaseous solutes are also obtained from solubility data tabulated by Dean.[1] A similar equation may also be used to determine the boundary conditions at a *gas–solid* interface:

$$c_{A,\text{solid}} = S \cdot p_A$$

where $c_{A,\text{solid}}$ is the molar concentration of A within the *solid* at the interface in units of kgmole/m^3 and p_A is the partial pressure of gas-phase species A over the solid in units of Pa. The partition coefficient S, also known as the solubility constant, has units of kgmole/m^3·Pa. Values of S for several gas–solid pairs reported by Barrer[2] are listed in Table 25.2.

2. *A reacting surface boundary is specified.* There are three common situations, all dealing with heterogeneous surface reactions. First, the flux of one species may be related to the flux of another species by chemical reaction stoichiometry. For example, consider the generic chemical reaction at the boundary surface $A + 2B \rightarrow 3C$, where reactants A and B diffuse to the surface, and product C diffuses away from the surface. The fluxes for A and B move in the *opposite direction* to the flux for C. Consequently, the flux N_A is related to the flux of the other species by $N_B = +2N_A$ or $N_C = -3N_A$. Second, a finite rate

Table 25.2   Solubility constants for selected gas–solid combinations

Gas	Solid	$T(K)$	$S = c_{A,\text{solid}}/p_A$ (kg mol/m^3 bar)
O_2	Natural rubber	298	3.12×10^{-3}
N_2	Natural rubber	298	1.56×10^{-3}
CO_2	Natural rubber	298	40.14×10^{-3}
He	Silicon	293	0.45×10^{-3}
H_2	Ni	358	9.01×10^{-3}

[1] J. A. Dean, *Lange's Handbook of Chemistry*, 14th edition, McGraw-Hill, Inc., New York, 1992.
[2] R. M. Barrer, *Diffusion In and Through Solids*, Macmillan Press, New York, 1941.

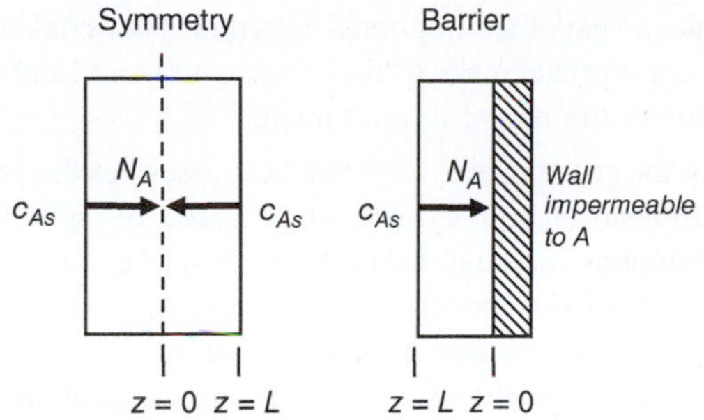

Figure 25.3 Examples of boundary conditions at $z = 0$ where the net flux is zero.

of chemical reaction might exist at the surface, which in turn sets the flux at the surface. For example, if component A is consumed by a first order on a surface at $z = 0$, *and* the positive z-direction is opposite to the direction of flux of A along z, then

$$N_A|_{z=0} = -k_s c_{As}$$

where k_s is a surface reaction rate constant with units of m/s. Third, the reaction may be so rapid that $c_{As} = 0$ if species A is the limiting reagent in the chemical reaction.

3. *The flux of the transferring species is zero at a boundary or at a centerline of symmetry.* This situation can arise at an impermeable boundary, or at the centerline of symmetry of the control volume, where the net flux is equal to zero. In either case, for a one-dimensional flux along z,

$$N_A|_{z=0} = -D_{AB}\frac{\partial c_A}{\partial z}\bigg|_{z=0} = 0 \quad \text{or} \quad \frac{\partial c_A}{\partial z}\bigg|_{z=0} = 0$$

where the impermeable boundary or the centerline of symmetry is located at $z = 0$, as illustrated in Figure 25.3.

4. *The convective mass-transfer flux at the boundary surface is specified.* When a fluid flows over the boundary, the flux can be defined by convection. For example, at some surface located at $z = 0$, the convective mass-transfer flux across the fluid boundary layer is

$$N_A|_{z=0} = k_c\left(c_{As} - c_{A,\infty}\right)$$

where $c_{A,\infty}$ is the bulk concentration of A the flowing fluid, c_{As} is the surface concentration of A at $z = 0$, and k_c is the convection mass-transfer coefficient defined in Section 24.3.

▶ **25.4**

STEPS FOR MODELING PROCESSES INVOLVING MOLECULAR DIFFUSION

Processes involving molecular diffusion can be modeled by the appropriate simplifications to Fick's flux equation and the general differential equation for mass transfer. In general, most molecular diffusion problems involve working through the following five steps:

Step 1: Draw a picture of the physical system. Label the important features, including the system boundaries. Decide where the source and the sink of mass transfer are located.

Step 2: Make a "list of assumptions" based on your consideration of the physical system. As appropriate, make a "list of nomenclature" and update the list as you add more terms to the model development.

Step 3: Pick the coordinate system that best describes the geometry of the physical system: rectilinear (x,y,z), cylindrical (r,z,θ), or spherical (r,θ,ϕ). Then formulate differential material balances to describe the mass transfer within a volume element of the process based on the geometry of the physical system and the assumptions proposed, making use of Fick's flux equation and the general differential equation for mass transfer. Two approaches may be used to simplify the general differential equation for mass transfer. In the first approach, simply reduce or eliminate the terms that do not apply to the physical system. For example,

(a) If the process is steady state, then $\dfrac{\partial c_A}{\partial t} = 0$.

(b) If no chemical reaction occurs uniformly within the control volume for diffusion, then $R_A = 0$.

(c) If the molecular mass-transfer process of species A is one-dimensional in the z-direction, then

$$\nabla \cdot N_A = \frac{\partial N_{A,z}}{\partial z}$$

Likewise, for cylindrical geometry in the r-direction,

$$\nabla \cdot N_A = \frac{1}{r}\frac{\partial (rN_{A,r})}{\partial r}$$

and for radial symmetry in spherical coordinates,

$$\nabla \cdot N_A = \frac{1}{r^2}\frac{\partial (r^2 N_{A,r})}{\partial r}$$

In the second approach, perform a "shell balance" for the component of interest on a differential volume element of the process. Both of these approaches are discussed and illustrated in Chapter 26. Next, the flux equation is simplified by establishing the relationship between the fluxes in the bulk-contribution term. For example, recall the one-dimensional flux of a binary mixture of components A and B:

$$N_{A,z} = -cD_{AB}\frac{dy_A}{dz} + y_A\left(N_{A,z} + N_{B,z}\right)$$

If $N_{A,z} = -N_{B,z}$, then $y_A\left(N_{A,z} + N_{B,z}\right) = 0$. If a differential equation for the concentration profile is desired, then the simplified form of Fick's flux equation must be substituted into the simplified form of the general differential equation for mass transfer. Figure 25.4 illustrates this process.

Step 4: Recognize and specify the boundary conditions and initial conditions. For example,

(a) Known concentration of species A at a surface or interface at $z = 0$, e.g., $c_A = c_{As}$. This concentration can be specified or known by equilibrium relationships such as Henry's law.

Model development pathways for processes
involving diffusion

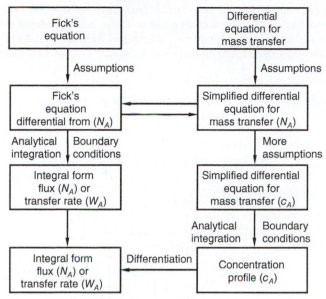

Figure 25.4 Model development pathways for processes involving molecular diffusion.

(b) Symmetry condition at a centerline of the control volume for diffusion, or no net diffusive flux of species A at a surface or interface at $z = 0$, $N_{A,z}|_{z=0} = -D_{AB}\partial c_A/\partial z = 0$.

(c) Convective flux of species A at a surface or interface—e.g., $N_A = k_c(c_{As} - c_{A,\infty})$.

(d) Known flux of species A at a surface or interface—e.g., at $z = 0$, $N_{A,z}|_{z=0} = N_{Ao}$.

(e) Known chemical reaction at a surface or interface. For the rapid disappearance of species A at the surface or interface—e.g., at $z = 0$, $c_{As} = 0$. For a slower chemical reaction at the surface or interface with finite c_{As} at $z = 0$, e.g., $N_A = -k_s c_{As}$, where k_s is a first-order chemical reaction rate constant.

Step 5: Solve the differential equations resulting from the differential material balances and the boundary/initial conditions described to get the concentration profile, the flux, or other parameters of engineering interest. If appropriate, consider asymptotic solutions or limiting cases to more difficult problems first.

The following examples illustrate how physical and chemical processes involving molecular diffusion can be modeled by the appropriate simplifications of Fick's flux equation and the general differential equation for mass transfer. The examples cover many typically encountered boundary conditions in both rectilinear and cylindrical geometry. The examples emphasize the first four steps of model development outlined, and the final model equations are generally left in differential-equation form. Chapters 26 and 27 provide analytical solution techniques for steady-state and unsteady-state diffusion processes. We have taken extra time at the beginning of each example to describe the interesting technology behind the process.

Example 1

Microelectronic devices are fabricated by forming many layers of thin films onto a silicon wafer. Each film has unique chemical and electrical properties. For example, a thin film of solid crystalline silicon (Si), when doped with the appropriate elements—for example, boron or silicon—has semiconductor properties. Silicon thin films are commonly formed by the *chemical vapor deposition*, or CVD, of silane vapor (SiH₄) onto the surface of the wafer. The chemical reaction is

$$SiH_4(g) \rightarrow Si(s) + 2H_2(g)$$

This surface reaction is usually carried out at very low pressure (100 Pa) and high temperature (900 K). In many CVD reactors, the gas phase over the Si film is not mixed. Furthermore, at high temperatures, the surface reaction is very rapid. Consequently, the molecular diffusion of the SiH₄ vapor to the surface often controls the rate of Si film formation. Consider the very simplified CVD reactor shown in Figure 25.5. A mixture of silane and hydrogen gas flows into the reactor. A diffuser provides a quiescent gas space over the growing Si film. Develop a differential model for this process, including statements of assumptions and boundary conditions.

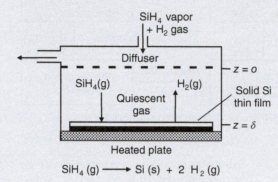

Figure 25.5 Chemical vapor deposition of a silicon thin film from silane gas.

The silane in the feed gas serves as the source for mass transfer, whereas the Si film serves as the sink for silane mass transfer. In contrast, the formation of H₂ at the Si film surface serves as the source for H₂ mass transfer, whereas the feed gas serves as the sink for H₂ mass transfer. The physical system possesses rectilinear geometry, and the major assumptions for model development are listed here.

(1) The reaction occurs only at the surface of growing Si thin film. Consequently, there is no homogeneous reaction of silane within the diffusion zone, so that $R_A = 0$. In this context, the surface reaction is the sink for silane mass transfer. (2) The gas space in the "diffusion zone" is not externally mixed, so that molecular diffusion dominates. (3) The feed gas provides silane in high excess relative to that consumed by reaction, so the silane concentration in the gas space at the diffusion-zone boundary is constant. (4) The flux of silane is one-dimensional along z, as the source and sink for silane mass transfer are aligned at the boundaries along the z-direction. (5) The thickness of the Si film is very thin relative to δ, the diffusion path length along the z-direction. Therefore, δ is essentially constant. (6) The mass-transfer process within the diffusion zone is at steady state.

The assumptions are used to reduce the general forms of the differential equation for mass transfer and Fick's flux equation. The general differential equation for mass transfer in terms of rectilinear coordinates is

$$-\left(\frac{\partial N_{A,x}}{\partial x} + \frac{\partial N_{A,y}}{\partial y} + \frac{\partial N_{A,z}}{\partial z}\right) + R_A = \frac{\partial c_A}{\partial t}$$

For steady-state one-dimensional flux along the z-direction with no homogeneous chemical reaction ($R_A = 0$), the general differential equation for mass transfer reduces to

$$\frac{dN_{A,z}}{dz} = 0$$

which shows that the flux is constant along the z-direction. As the diffusion flux is, with respect to only one dimension, the partial derivative becomes an ordinary derivative. Fick's flux equation for the one-dimensional flux silane through a binary mixture in the gas phase is

$$N_{A,z} = -cD_{AB}\frac{dy_A}{dz} + y_A\left(N_{A,z} + N_{B,z}\right)$$

where species A represents silane vapor (SiH_4) reactant and species B represents the hydrogen gas (H_2) product. The flux of the gaseous reactant is opposite in direction to the flux of the gaseous product. From the reaction stoichiometry and Figure 25.5, $N_{A,z}$ is related to $N_{B,z}$ as follows:

$$\frac{N_{A,z}}{N_{B,z}} = \frac{1 \text{ mol } SiH_4 \text{ reacted (flux in } + z\text{-direction)}}{2 \text{ mol } H_2 \text{ formed (flux in } - z\text{-direction)}} = -\frac{1}{2}$$

Therefore, $N_{B,z} = -2N_{A,z}$ and Fick's flux equation further reduces to

$$N_{A,z} = -cD_{AB}\frac{dy_A}{dz} + y_A\left(N_{A,z} - 2N_{A,z}\right) = -\frac{cD_{AB}}{1 + y_A}\frac{dy_A}{dz}$$

It is interesting to note that increasing y_A decreases the flux. Two boundary conditions must be specified. At the surface of the Si film, the reaction is so rapid that the concentration of silane vapor is zero. Furthermore, the concentration of silane in the feed gas is constant. The two needed boundary conditions are given below:

At the Si film surface, $z = \delta, y_A = y_{As}$
At the diffuser, $z = 0, y_A = y_{Ao}$

The differential model is now specified. Although the analytical solution was not asked for in the problem statement, it is easy to obtain. We first recognize that for this particular system, $N_{A,z}$ is a constant along z. If $N_{A,z}$ is a constant, then Fick's flux equation can be integrated by separation of dependent variable y_A from independent variable z, with integration limits defined by the boundary conditions:

$$N_{A,z}\int_0^\delta dz = \int_{y_{Ao}}^{y_{As}} \frac{-cD_{AB}}{1 + y_A}dy_A$$

If the system temperature T and total system pressure P are constant, then the total molar concentration of the gas, $c = P/RT$, is also constant. Likewise, the binary gas-phase diffusion coefficient of silane vapor in hydrogen gas, D_{AB}, is also constant. The final integrated equation is

$$N_{A,z} = \frac{cD_{AB}}{\delta}\ln\left(\frac{1 + y_{Ao}}{1 + y_{As}}\right)$$

If y_{As} is specified, then $N_{A,z}$ can be determined. With the silane flux $N_{A,z}$ known, parameters of engineering interest, such as the Si film formation rate, can be easily determined. These questions are considered in a problem exercise at the end of Chapter 26. If species A (SiH_4) is dilute with respect to species B (H_2)—i.e., $y_A \ll 1$, then

$$N_{A,z} = -\frac{cD_{AB}}{1 + y_A}\frac{dy_A}{dz} \doteq -\frac{cD_{AB}}{1}\frac{dy_A}{dz} = -D_{AB}\frac{dc_A}{dz}$$

and the integrated form of the flux equation is

$$N_{A,z} = \frac{cD_{AB}}{\delta}\left(y_{Ao} - y_{As}\right)$$

The flux is directly proportional to the concentration difference and inversely proportional to the path length for diffusion.

Example 2

Scalable processes for the chemical vapor deposition (CVD) of thin films of micrometer thickness on silicon wafers used in the manufacture of microelectronic devices are often carried out within a diffusion furnace. Within the diffusion furnace, the silicon wafers, typically thin disks of crystalline silicon 15–20 cm diameter, are vertically stacked on a support tray, as shown in

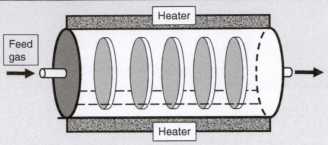

Figure 25.6 Diffusion furnace containing vertically stacked wafers.

Figure 25.6. The reactant gas is introduced into the enclosed furnace, which is maintained at high temperature to promote rates of diffusion and reaction, and often at low pressure, depending upon the process chemistry. The flowing feed gas concentration is considered constant in the gas region surrounding the stacked wafers. However, in the quiescent gas space between the wafers (interwafer region of distance L), the reactant and product gases diffuse to and from the wafer surface, creating a concentration profile in two dimensions. Consequently, there as a constant source and sink for the gaseous reactants.

Examples of CVD processes carried out in a diffusion furnace are silicon thin form formation described in Example 1 in this chapter, and silicon dioxide (SiO_2) thin film formation, which serves as an electronic insulator layer. One particular surface reaction process for deposition of a solid SiO_2 thin film is the decomposition of the gas-phase reactant TEOS, $Si(OC_2H_5)_4$.

$$Si(OC_2H_5)_4(g)(A) \rightarrow 4C_2H_4(g)(C) + 2H_2O(g)(D) + SiO_2(s)(E)$$

The gas phase is often diluted in an inert gas such as helium (He, species B). By principles shown in Example 1, Fick's flux equation is

$$N_{A,z} = -cD_{A-\text{mixture}} \frac{\partial y_A}{\partial z} + y_A \left(N_{A,z} + N_{B,z} + N_{C,z} + N_{D,z} + N_{E,z} \right)$$

with flux $N_{A,z}$ along the positive z-direction. Relative to $N_{A,z}$, $N_{B,z} = 0$, $N_{C,z} = -4N_{A,z}$, $N_{D,z} = -2N_{A,z}$, and $N_{E,z} = 0$, so that

$$N_{A,z} = -\frac{cD_{A-\text{mix}}}{1 + 5y_A} \frac{\partial y_A}{\partial z}$$

If $y_A \ll 1$, then

$$N_{A,z} \doteq -\frac{cD_{AB}}{1} \frac{\partial y_A}{\partial z} = -D_{AB} \frac{\partial c_A}{\partial z}$$

If this process is carried out where the helium diluent is in large excess with respect to reactant A so that species B is the dominant species in the mixture, then the mixture is considered to be pseudo-binary even though four gas-phase species are present. In this case where $y_A \ll 1$, Fick's flux equation simplifies to Fick's first law.

We now turn our attention to the differential material balance on the process. The system control volume for mass transfer is the interwafer region. Since the disk is circular, the gas space between the wafer assumes the shape of a cylinder, and so cylindrical geometry will be used. Due to symmetry considerations, only one quarter of this two-dimensional region needs to be considered, as shown in Figure 25.7. The source for mass transfer of species A is the feed gas, and the sink for species A is the reaction of A to a solid at the wafer surface. The assumptions for analysis are (1) constant source and sink for reactant species A, leading to a steady-state process with $\partial c_A/\partial t = 0$, (2) two-dimensional flux for species A, as the source and sink are oriented perpendicular to each other, (3) no homogeneous reaction of species A within the control volume ($R_A = 0$), and (4) dilute process with Fick's flux equation approximated by Fick's first law.

Based upon these assumptions for analysis, the general differential equation for mass transfer in cylindrical coordinates (25-28), stated as

$$\frac{\partial c_A}{\partial t} + \left[\frac{1}{r} \frac{\partial (rN_{A,r})}{\partial r} + \frac{1}{r} \frac{\partial N_{A,\theta}}{\partial \theta} + \frac{\partial N_{A,z}}{\partial z} \right] = R_A$$

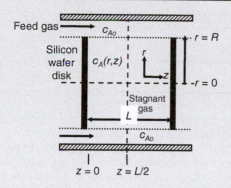

Figure 25.7 Interwafer region and control volume for Example 2.

reduces to

$$\frac{1}{r}\frac{\partial (r N_{A,r})}{\partial r} + \frac{\partial N_{A,z}}{\partial z} = 0$$

The component flux equations for reactant A in the r- and z-directions are, respectively,

$$N_{A,r} = -D_{AB}\frac{\partial c_A}{\partial r} \text{ and } N_{A,z} = -D_{AB}\frac{\partial c_A}{\partial z}$$

Inserting the flux equation into the general differential equation for mass transfer yields

$$-\frac{1}{r}\frac{\partial}{\partial r}\left(r D_{AB}\frac{\partial c_A}{\partial r}\right) - \frac{\partial}{\partial z}\left(D_{AB}\frac{\partial c_A}{\partial z}\right) = 0$$

and so the differential equation for mass transfer in terms of the concentration profile is

$$D_{AB}\left[\frac{\partial^2 c_A(r,z)}{\partial r^2} + \frac{1}{r}\frac{\partial c_A(r,z)}{\partial r} + \frac{\partial^2 c_A(r,z)}{\partial z^2}\right] = 0$$

The boundary conditions must now be developed to specify the differential equation for mass transfer. By symmetry arguments, only the shaded quadrant shown in Figure 25.7 must be analyzed. Since the highest derivative order on c_A is two for the r-direction and two for the z-direction, four boundary conditions, two for c_A in the r-direction, and two for c_A in the z-direction, must be specified:

$$r = R, 0 < z < L/2, c_A(R,z) = c_{Ao}$$

(feed gas concentration)

$$r = 0, 0 < z < L/2, N_A|_{r=0,z} = -D_{AB}\frac{\partial c_A(0,z)}{\partial r} = 0 \quad \therefore \quad \frac{\partial c_A(0,z)}{\partial r} = 0$$

(no net flux at symmetry point at the center of the wafer)

$$z = L/2, 0 \leq r \leq R, N_A|_{z=L/2,r} = -D_{AB}\frac{\partial c_A(r,L/2)}{\partial r} = 0 \quad \therefore \quad \frac{\partial c_A(r,L/2)}{\partial r} = 0$$

(no net flux at symmetry point midway between two wafers)

$$z = 0, 0 \leq r \leq R, c_A(r,0) = c_{As}$$

(at surface of wafer)

For a diffusion-limited surface reaction, $c_{As} \approx 0$. If the reaction is not limited by the diffusion of the reactant to the surface, then c_{As} not known. However, if the surface reaction can be approximated as first-order with respect to c_{As}, then the diffusion flux of species A to the surface the z-direction is balanced by the reaction rate at the surface:

$$z = 0, 0 \leq r \leq R, N_A|_{z=0,r} = -k_s c_A(r,0) = -D_{AB}\frac{\partial c_A(r,0)}{\partial z}$$

which eliminates need for c_{As} to be known explicitly. Although the solution for $c_A(r,z)$ will not be provided here, it can be obtained in closed analytical form by the separation of variables method for partial differential equations, or by numerical methods for partial differential equations.

Example 3

An emerging area of biotechnology called "tissue engineering" develops new processes to grow organized living tissues of human or animal origin. A typical configuration is the engineered tissue bundle. Engineered tissue bundles have several potential biomedical applications, including the production of replacement body tissue (skin, bone marrow, etc.) for transplantation into the human body, or in the future may serve as artificial organs for direct implantation into the human body.

Living tissues require oxygen to stay alive. The mass transport of oxygen (O_2) to the tissue is an important design consideration. One potential system is schematically illustrated in Figure 25.8. Thin tubes arranged on a triangular pitch pass longitudinally through the tissue bundle. The tubes serve as a "scaffold" for supporting the living tissue matrix and supply oxygen and nutrients to the tissue at the same time. Let us focus on a single O_2 delivery tube with tissue surrounding it, as illustrated in Figure 25.8. Pure oxygen (O_2) gas flows through the tube. The tube wall is extremely permeable to O_2, and the O_2 partial pressure through the porous tube wall can be taken as the O_2 partial pressure inside the tube. Oxygen is only sparingly soluble in the tissue, which is mostly water. The concentration of dissolved O_2 at $r = R_1$, is

$$c_{As} = \frac{p_A}{H}$$

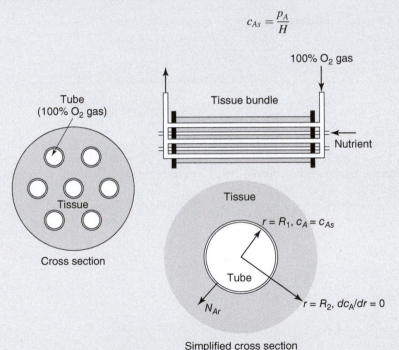

Figure 25.8 Oxygen transport within an engineered tissue bundle.

where H is the Henry's law constant for the dissolution of O_2 in living tissue at the process temperature and p_A is the partial pressure of O_2 in the tube. The dissolved O_2 diffuses through the tissue and is metabolically consumed. The metabolic consumption of dissolved O_2 is described by a kinetic rate equation of the form

$$R_A = -\frac{R_{A,\max} c_A}{K_A + c_A}$$

This important rate equation for biological systems, called the Michaelis–Menten equation, arises from complex interactions of the reactant substrate (species A) with enzymes that catalyze the conversion of A to products. However, this simple equation adequately describes the reaction kinetics of many biological processes. In this relationship, K_A is the binding constant of the substrate to the enzyme(s) of interest with units of concentration, and $R_{A,max}$ is the maximum possible reaction rate once the biologically mediated reaction has become saturated by the substrate. There are two limiting cases. First, if $K_A \ll c_A$, which occurs at high substrate concentrations or for substrates with a high affinity to the enzyme(s) of interest, the reaction does not depend on c_A, and so $R_A \doteq R_{A,max}$, i.e., is zero order with respect to c_A. In the second limiting case, $K_A \gg c_A$, so that

$$R_A = -\frac{R_{A,max}c_A}{K_A + c_A} = -\frac{R_{A,max}}{K_A}c_A = -k_1 c_A$$

and so the reaction is first order with respect to c_A.

A key parameter in the design of the engineered tissue bundle is the spacing between the tubes. If the tube spacing is too wide, the dissolved O_2 concentration will go to near zero and starve the tissue. Therefore, it is important to know the radial concentration profile, $c_A(r)$ of dissolved O_2. Develop a differential model to predict $c_A(r)$.

The physical system possesses cylindrical geometry, and the following assumptions for model development are listed here. (1) The source for O_2 mass transfer is the pure O_2 gas inside the tube, and the sink for mass transfer is the metabolic consumption of dissolved oxygen by the tissue. If the O_2 partial pressure p_A is maintained constant inside the tube along longitudinal coordinate z, then the flux of oxygen through the tissue is one-dimensional along the radial (r) direction. (2) Tissue remains viable and maintains constant physical properties. (3) The O_2 transfer process is at steady state. (4) The tissue is stationary, and the dissolved O_2 concentration is dilute. (5) At $r = R_1$, the tube material is thin and highly permeable to O_2 so that the dissolved O_2 concentration in the tissue is in equilibrium with the O_2 partial pressure in the tube. (6) At $r = R_2$, there is no net flux of O_2.

The general differential equation for mass transfer in cylindrical coordinates is

$$-\left(\frac{1}{r}\frac{\partial}{\partial r}(rN_{A,r}) + \frac{1}{r}\frac{\partial N_{A,\theta}}{\partial \theta} + \frac{\partial N_{A,z}}{\partial z}\right) + R_A = \frac{\partial c_A}{\partial t}$$

For steady-state one-dimensional flux along the r-direction, the general equation for mass transfer reduces to

$$-\frac{1}{r}\frac{\partial}{\partial r}(rN_{A,r}) + R_A = 0$$

For a one-dimensional system, the partial derivatives can be replaced with ordinary derivatives.

Alternatively, we can perform a material balance for dissolved O_2 on the differential element of volume $2\pi Lr\Delta r$ shown in Figure 25.8 and get the same result. Specifically, for steady-state one-dimensional flux along the r-direction with a homogeneous reaction R_A within the differential volume element, we have

$$2\pi LrN_{Ar}|_{r=r} - 2\pi LrN_{Ar}|_{r=r+\Delta r} + R_A 2\pi Lr\Delta r = 0$$

Dividing through by $2\pi L\Delta r$, and rearranging, we get

$$-\left(\frac{rN_{A,r}|_{r=r+\Delta r} - rN_{A,r}|_{r=r}}{\Delta r}\right) + R_A r = 0$$

Finally, taking the limits as $\Delta r \to 0$ yields

$$-\frac{1}{r}\frac{d}{dr}(rN_{Ar}) + R_A = 0$$

For one-dimensional flux of dissolved O_2 through the stagnant tissue in cylindrical coordinates along the r-direction, Fick's flux equation reduces to

$$N_{A,r} = -D_{AB}\frac{dc_A}{dr} + \frac{c_A}{c}(N_{A,r}) = -D_{AB}\frac{dc_A}{dr}$$

because O_2 is only sparingly soluble in the tissue so that $c_A \ll c$, where c is the total molar concentration of the tissue, which approximates the molar concentration of water. In cylindrical geometry, $N_{A,r}$ is not constant along diffusion path r, because (a) cross-sectional area for flux is increasing along r and (b) the R_A term is present. As a result, the flux equation cannot be integrated, as was the case in Example 1. It is now necessary to combine Fick's flux equation and the differential equation for mass transfer in order to get the concentration profile

$$-\frac{1}{r}\frac{d}{dr}\left(-rD_{AB}\frac{dc_A}{dr}\right) + R_A = 0$$

or

$$D_{AB}\left[\frac{d^2 c_A}{dr^2} + \frac{1}{r}\frac{dc_A}{dr}\right] - \frac{R_{A,max}c_A}{K_A + c_A} = 0$$

The concentration profile $c_A(r)$ is now expressed as a second-order differential equation. Therefore, two boundary conditions on $c_A(r)$ must be specified:

at

$$r = R_1, N_{A,r}\big|_{r=R_1} = 0 = -D_{AB}\frac{dc_A}{dr}\bigg|_{r=R_1} \quad \therefore \frac{dc_A}{dr}\bigg|_{r=R_1} = 0$$

at

$$r = R_2, c_A = c_{As} \doteq c_A^* = \frac{p_A}{H}$$

The analytical solution for $c_A(r)$ and its extension to predicting the overall rate of oxygen consumption in the tissue bundle has been left as a problem exercise in Chapter 26.

▶ **25.5**

CLOSURE

The general differential equation for mass transfer was developed to describe the mass balances associated with a diffusing component in a mixture. Special forms of the general differential equation for mass transfer that apply to specific situations were presented. Commonly encountered boundary conditions for molecular diffusion processes were also listed. From this theoretical framework, a five-step method for mathematically modeling processes involving molecular diffusion was proposed. Three examples illustrating how the differential form of Fick's flux equation presented in Chapter 24 and the general differential equation for mass transfer presented in this chapter are reduced to simple differential equations that describe the molecular diffusion aspects of a specific process. The approaches presented in this chapter serve as the basis for problem solving in Chapters 26 and 27.

CHAPTER 26

Steady-State Molecular Diffusion

In this chapter, we will direct our attention to describing the steady-state transfer of mass from a differential point of view. To accomplish this task, the differential equation and the boundary conditions that describe the physical situation must be established. The approach will parallel those previously used in Chapter 8 for the analysis of a differential fluid element in laminar flow and in Chapter 17 for the analysis of a differential volume element of a quiescent material for steady-state heat conduction.

During our discussion of steady-state diffusion, two approaches will be used to simplify the differential equations of mass transfer as recommended in Section 24.4. First, Fick's flux equation and the general differential equation for mass transfer can be simplified by eliminating the terms that do not apply to the physical situation. Second, a material balance can be performed on a differential volume element of the control volume for mass transfer. In using both approaches, the student will become more familiar with the various terms in the general differential equation for mass transfer:

$$\nabla \cdot \mathbf{N}_A + \frac{\partial c_A}{\partial t} - R_A = 0 \tag{25-11}$$

To gain confidence in treating mass-transfer processes, we will initially treat the simplest case, steady-state diffusion in only one direction, which is free of any chemical production occurring uniformly throughout the process—i.e., $R_A = 0$. We will then obtain solutions for increasingly more complex mass-transfer operations.

▶ 26.1

ONE-DIMENSIONAL MASS TRANSFER INDEPENDENT OF CHEMICAL REACTION

In this section, steady-state molecular mass transfer through simple systems in which the concentration and the mass flux are functions of a single space coordinate will be considered. Although all four fluxes, $\mathbf{N}_A$, $\mathbf{n}_A$, $\mathbf{J}_A$, and $\mathbf{j}_A$, may be used to describe mass-transfer operations, only the molar flux relative to a set of axes fixed in space, N_A will be used in the following discussions. In a binary system, the z component of this flux is expressed by equation (24-21):

$$N_{A,z} = -cD_{AB}\frac{dy_A}{dz} + y_A\left(N_{A,z} + N_{B,z}\right) \tag{24-21}$$

Unimolecular Diffusion

The diffusion coefficient or mass diffusivity for a gas may be experimentally measured in an Arnold diffusion cell. This cell is illustrated schematically in Figures 24.1 and 26.1.

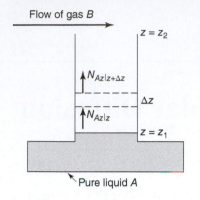

Flow of gas B

$z = z_2$

$N_{Az}|_{z+\Delta z}$

Δz

$N_{Az}|_z$

$z = z_1$

Pure liquid A

Figure 26.1 Arnold diffusion cell.

The narrow tube, which is partially filled with pure liquid A, is maintained at a constant temperature and pressure. Gas B, which flows across the open end of the tube, has a negligible solubility in liquid A and is also chemically inert to A. Component A vaporizes and diffuses into the gas phase; the rate of vaporization may be physically measured and may also be mathematically expressed in terms of the molar mass flux.

Recall that the general differential equation for mass transfer is given by

$$\nabla \cdot \mathbf{N}_A + \frac{\partial c_A}{\partial t} - R_A = 0 \tag{25-11}$$

In rectilinear coordinates, this equation is

$$\frac{\partial N_{A,x}}{\partial x} + \frac{\partial N_{A,y}}{\partial y} + \frac{\partial N_{A,z}}{\partial z} + \frac{\partial c_A}{\partial t} - R_A = 0 \tag{25-27}$$

Assume that (1) the mass-transfer process is at steady state with $\partial c_A / \partial t = 0$; (2) there is no chemical production of A in the diffusion path so that $R_A = 0$; and (3) the diffusion is only in the z-direction, so that we are only concerned with the z component of the mass flux vector, $N_{A,z}$. For this physical situation, equation (25-11) reduces to

$$\frac{dN_{A,z}}{dz} = 0 \tag{26-1}$$

We can also generate this governing differential equation by considering the mass transfer occurring in the differential control volume of $S\Delta z$, where S is the uniform cross-sectional area of the control volume and Δz is the depth of the control volume. A mass balance over this control volume for a steady-state operation, free of any chemical production of A, yields

$$SN_{A,z}|_{z+\Delta z} - SN_{A,z}|_z = 0$$

Dividing through by the control volume, $S\Delta z$, and taking limit as Δz approaches zero, we once again obtain equation (26-1).

A similar differential equation could also be generated for component B:

$$\frac{dN_{B,z}}{dz} = 0 \tag{26-2}$$

and, accordingly, the molar flux of B is also constant over the entire diffusion path from z_1 to z_2. Considering only the plane at z_1 and the restriction that gas B is insoluble in liquid A, we realize that $N_{B,z}$ at plane z_1 is zero and conclude that $N_{B,z}$, the net flux of B, is zero throughout the diffusion path; accordingly, component B is a *stagnant* gas.

For a binary mixture of A and B, the constant molar flux of A in the z-direction was described in Chapter 24 by the equation

$$N_{A,z} = -cD_{AB}\frac{dy_A}{dz} + y_A\left(N_{A,z} + N_{B,z}\right) \tag{24-21}$$

this equation reduces, when $N_{B,z} = 0$, to

$$N_{A,z} = -\frac{cD_{AB}}{1 - y_A}\frac{dy_A}{dz} \tag{26-3}$$

This equation may be integrated between the two boundary conditions:

$$\text{at } z = z_1, \quad y_A = y_{A_1}$$

and

$$\text{at } z = z_2, \quad y_A = y_{A_2}$$

Assuming the diffusion coefficient to be independent of concentration, and realizing from equation (26-1) that $N_{A,z}$ is constant along the diffusion path z, we obtain, by integrating,

$$N_{A,z} \int_{z_1}^{z_2} dz = cD_{AB} \int_{y_{A_1}}^{y_{A_2}} \frac{-dy_A}{1 - y_A} \tag{26-4}$$

Upon integration, and solving for $N_{A,z}$, we obtain

$$N_{A,z} = \frac{cD_{AB}}{(z_2 - z_1)} \ln \frac{(1 - y_{A_2})}{(1 - y_{A_1})} \tag{26-5}$$

The log-mean average concentration of component B is defined as

$$y_{B,lm} = \frac{y_{B_2} - y_{B_1}}{\ln(y_{B_2}/y_{B_1})}$$

or, in the case of a binary mixture, this equation may be expressed in terms of component A as follows:

$$y_{B,lm} = \frac{(1 - y_{A_2}) - (1 - y_{A_1})}{\ln[(1 - y_{A_2})/(1 - y_{A_1})]} = \frac{y_{A_1} - y_{A_2}}{\ln[(1 - y_{A_2})/(1 - y_{A_1})]} \tag{26-6}$$

Inserting equation (26-6) into equation (26-5), we obtain

$$N_{A,z} = \frac{cD_{AB}}{z_2 - z_1} \frac{(y_{A_1} - y_{A_2})}{y_{B,lm}} \tag{26-7}$$

Equation (26-7) may also be written in terms of pressures. For an ideal gas

$$c = \frac{n}{V} = \frac{P}{RT}$$

and

$$y_A = \frac{p_A}{P}$$

The equation equivalent to equation (26-7) is

$$N_{A,z} = \frac{D_{AB}P}{RT(z_2 - z_1)} \frac{(p_{A_1} - p_{A_2})}{p_{B,lm}} \tag{26-8}$$

Equations (26-7) and (26-8) are commonly referred to as equations for *steady-state diffusion of one gas through a second stagnant gas*. Many mass-transfer operations involve the diffusion of one gas component through another nondiffusing component; *absorption* and *humidification* are typical operations defined by these two equations.

Equation (26-8) has also been used to describe the convective mass-transfer coefficients by the *"film concept"* or film theory. In Figure 26.2, the flow of gas over a liquid surface is

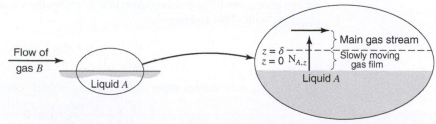

Figure 26.2 Film model for mass transfer of component A into a moving gas stream.

illustrated. The "film concept" is based upon a model in which the entire resistance to diffusion from the liquid surface to the main gas stream is assumed to occur in a stagnant or laminar film of constant thickness δ. In other words, for this model, δ is a fictitious length that represents the thickness of a fluid layer offering the same resistance to molecular diffusion as is encountered in the combined process of molecular diffusion and diffusion due to the mixing by the moving fluid. If this model is accurate, the convective mass-transfer coefficient may be expressed in terms of the gas diffusion coefficient. If $z_2 - z_1$ is set equal to δ, equation (26-8) becomes

$$N_{A,z} = \frac{D_{AB}P}{RT\, p_{B,lm}\, \delta}\left(p_{A_1} - p_{A_2}\right)$$

and from equation (24-68), we have

$$N_{A,z} = k_c(c_{A_1} - c_{A_2})$$

and

$$N_{A,z} = \frac{k_c}{RT}\left(p_{A_1} - p_{A_2}\right)$$

Comparison reveals that the film coefficient is expressed as

$$k_c = \frac{D_{AB}P}{p_{B,lm}\delta} \tag{26-9}$$

when the diffusing component A is transported through a nondiffusing gas B. Although this model is physically unrealistic, the "film concept" has had educational value in supplying a simple picture of a complicated process. The film concept has proved frequently misleading in suggesting that the convective mass-transfer coefficient is always directly proportional to the mass diffusivity. Other models for the convective coefficient will be discussed in this chapter and in Chapter 28. At that time we will find that k_c is a function of the diffusion coefficient raised to an exponent varying from 0.5 to 1.0.

Frequently, to complete the description of the physical operation in which mass is being transported, it is necessary to express the concentration profile. Recalling equation (26-1):

$$\frac{dN_{A,z}}{dz} = 0 \tag{26-1}$$

and equation (26-3):

$$N_{A,z} = -\frac{cD_{AB}}{1 - y_A}\frac{dy_A}{dz} \tag{26-3}$$

we can obtain the differential equation that describes the variation in concentration along the diffusing path. This equation is

$$\frac{d}{dz}\left(-\frac{cD_{AB}}{1 - y_A}\frac{dy_A}{dz}\right) = 0 \tag{26-10}$$

As c and D_{AB} are constant under isothermal and isobaric conditions, the equation reduces to

$$\frac{d}{dz}\left(\frac{1}{1 - y_A}\frac{dy_A}{dz}\right) = 0 \tag{26-11}$$

This second-order equation may be integrated twice with respect to z to yield

$$-\ln(1 - y_A) = C_1 z + C_2 \tag{26-12}$$

The two constants of integration (C_1 and C_2) are evaluated, using the boundary conditions

$$\text{at } z = z_1, \quad y_A = y_{A_1}$$

and

$$\text{at } z = z_2, \quad y_A = y_{A_2}$$

Substituting the resulting constants into equation (26-12), we obtain the following expression for the concentration profile of component A:

$$\left(\frac{1 - y_A}{1 - y_{A_1}}\right) = \left(\frac{1 - y_{A_2}}{1 - y_{A_1}}\right)^{(z - z_1)/(z_2 - z_1)} \tag{26-13}$$

or, as $y_A + y_B = 1$

$$\left(\frac{y_B}{y_{B_1}}\right) = \left(\frac{y_{B_2}}{y_{B_1}}\right)^{(z - z_1)/(z_2 - z_1)} \tag{26-14}$$

Representative plots of the concentration profile for unimolecular diffusion, in terms of the mole fractions of A and B, are provided in Figure 26.3 at concentrated and dilute concentrations of species A in the mixture.

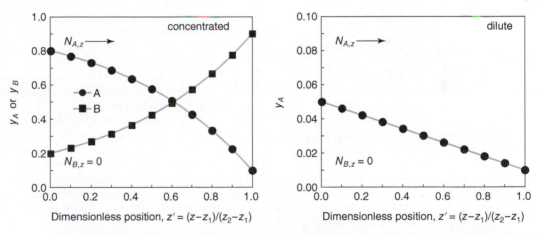

Figure 26.3 Concentration profiles for unimolecular diffusion of component A in concentrated and dilute systems.

Equations (26-13) and (26-14) describe logarithmic concentration profiles for both species. The average concentration of one of the species along the diffusion path may be evaluated, as an example for species B, by

$$\bar{y}_B = \frac{\displaystyle\int_{z_1}^{z_2} y_B(z)\,dz}{\displaystyle\int_{z_1}^{z_2} dz} \tag{26-15}$$

Upon substitution of equation (26-14) into equation (26-15), we obtain

$$\bar{y}_B = y_{B_1} \frac{\int_{z_1}^{z_2} \left(\frac{y_{B_2}}{y_{B_1}}\right)^{(z-z_1)/(z_2-z_1)} dz}{z_2 - z_1}$$

$$= \frac{(y_{B_2} - y_{B_1})(z_2 - z_1)}{\ln(y_{B_2}/y_{B_1})(z_2 - z_1)} = \frac{y_{B_2} - y_{B_1}}{\ln(y_{B_2}/y_{B_1})}$$

$$= y_{B,lm}$$

(26-6)

The following example problem illustrates the application of the foregoing analysis to a mass-transfer situation.

Example 1

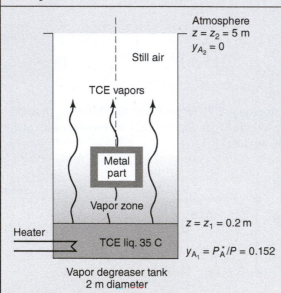

Atmosphere
$z = z_2 = 5$ m
$y_{A_2} = 0$

Still air

TCE vapors

Metal part

Vapor zone

Heater

TCE liq. 35 C

$z = z_1 = 0.2$ m

$y_{A_1} = P_A^*/P = 0.152$

Vapor degreaser tank
2 m diameter

Figure 26.4 TCE emissions from a vapor degreaser.

Vapor degreasers like the one shown in Figure 26.4 are widely used for cleaning metal parts. Liquid solvent rests at the bottom of the degreaser tank. A heating coil immersed in the solvent vaporizes a small portion of the solvent and maintains a constant temperature, so that the solvent exerts a constant vapor pressure. The cold parts to be cleaned are suspended in the solvent vapor zone where the concentration of solvent vapors is highest. The solvent condenses on the part, dissolves the grease, and then drips back down into the tank, thereby cleaning the part. Vapor degreasers are often left open to the atmosphere for ease of dipping and removing parts and because covering them might release an explosive mixture. When the degreaser is not in use, molecular diffusion of the solvent vapor through the stagnant air inside the headspace can result in significant solvent emissions, because the surrounding atmosphere serves as an infinite sink for the mass-transfer process. As the amount solvent in the degreaser tank is large relative to the amount of vapor emitted, a steady-state diffusion process with a constant diffusion path length takes place.

At present, a cylindrical degreaser tank with a diameter of 2.0 m and total height of 5.0 m is in operation, and the solvent level height is kept constant at 0.2 m. The temperatures of the solvent and headspace of the degreaser are both constant at 35 C. The solvent used for vapor degreasing is trichloroethylene (TCE). Current regulations require that the degreaser not emit more than 1.0 kg TCE per day. Does the estimated emission rate of the degreaser exceed this limit? TCE has a molecular weight of 131.4 g/mol and a vapor pressure of 115.5 mm Hg at 35 C. The binary-diffusion coefficient TCE in air is 0.088 cm^2/s at 35 C, as determined by the Fuller–Schettler–Giddings correlation.

The source of TCE mass transfer is the liquid solvent at the bottom of the tank, and the sink for mass transfer is the surrounding atmosphere outside the tank. The steady-state molecular diffusion flux of TCE vapor through the stagnant gas headspace of the degreaser in the z-direction is described by

$$N_{A,z} = \frac{cD_{AB}}{z_2 - z_1} \ln\left(\frac{1 - y_{A_2}}{1 - y_{A_1}}\right)$$

with species A representing TCE vapor and species B representing air. The total molar concentration of the gas, c, is determined from the ideal gas law.

$$c = \frac{P}{RT} = \frac{1.0 \text{ atm}}{\dfrac{0.08206 \text{ m}^3 \cdot \text{atm}}{\text{kgmole} \cdot \text{K}}(273 + 35)\text{K}} = 0.0396 \frac{\text{kgmole}}{\text{m}^3}$$

The mole fraction of TCE vapor at the solvent surface (y_{A_1}) is determined from the vapor pressure of the solvent at 35 C.

$$y_{A_1} = \frac{P_A}{P} = \frac{115.1 \text{ mm Hg}}{1.0 \text{ atm}} \frac{1.0 \text{ atm}}{760 \text{ mm Hg}} = 0.152$$

The mole fraction of TCE vapor at the exit of the degreaser tank is taken as zero ($y_{A_2} = 0$), as the surrounding atmosphere serves as an infinite sink for mass transfer. The path length for diffusion is simply the difference between the solvent level height and the top of the degreaser tank:

$$z_2 - z_1 = 5.0 \text{ m} - 0.2 \text{ m} = 4.8 \text{ m}$$

From these input values, the flux of TCE vapor from the degreaser is

$$N_{A,z} = \frac{\left(0.0396 \dfrac{\text{kgmole}}{\text{m}^3}\right)\left(0.088 \dfrac{\text{cm}^2}{\text{s}} \dfrac{1 \text{ m}^2}{(100 \text{ cm})^2}\right)}{4.8 \text{ m}} \ln\left(\frac{1 - 0}{1 - 0.152}\right)$$

$$= 1.197 \times 10^{-8} \frac{\text{kgmole TCE}}{\text{m}^2 \cdot \text{s}}$$

The TCE emissions rate (W_A) is the product of the flux and the cross-sectional area of the degreaser tank of diameter D:

$$W_A = N_{A,z} \frac{\pi D^2}{4} = 1.197 \times 10^{-8} \frac{\text{kgmole TCE}}{\text{m}^2 \text{s}} \times \frac{\pi (2.0 \text{ m})^2}{4} \times \left(\frac{131.4 \text{ kg TCE}}{\text{kgmole TCE}}\right)\left(\frac{3600 \text{ s}}{1 \text{ h}}\right)\left(\frac{24 \text{ h}}{\text{day}}\right)$$

$$= 0.423 \frac{\text{kg TCE}}{\text{day}}$$

The estimated TCE vapor emission rate is below the current regulatory limit of 1.0 kg TCE per day. In a real degreaser, it may be difficult to ensure a completely still gas space, as local air currents induced from a variety of sources may occur. The air currents would increase the mass-transfer flux by convection. Consequently, this analysis considers only the limiting case for the minimum vapor emissions from a diffusion-limited process.

Pseudo-Steady-State Diffusion

In many mass-transfer operations, one of the boundaries may move with time. If the length of the diffusion path changes a small amount over a long period of time, a pseudo-steady-state diffusion model may be used. When this condition exists, equation (26-7) describes the mass flux in the stagnant gas film. Reconsider Figure 26.1, with a moving liquid surface as illustrated in Figure 26.5. Two surface levels are shown, one at time t_0 and the other at time t_1. If the difference in the level of liquid A over the time interval considered is only a small fraction of the total diffusion path, and $t_1 - t_0$ is a relatively long period of time, at any instant in that period the molar flux in the gas phase may be evaluated by

$$N_{A,z} = \frac{cD_{AB}(y_{A_1} - y_{A_2})}{Z\, y_{B,lm}} \qquad (26\text{-}7)$$

Where Z is the length of the diffusion path at time t, given by $Z = z_2 - z$.

From an unsteady-state material balance on species A (i.e., In − Out = Accumulation) the molar flux $N_{A,z}$ is

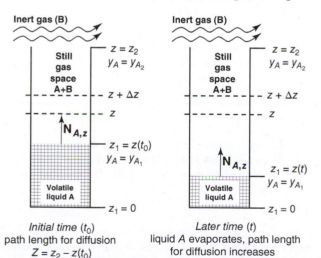

Figure 26.5 Arnold diffusion cell with moving liquid surface.

related to the amount of A leaving the liquid by

$$0 - N_{A,z}S = \frac{\rho_{A,L}}{M_A}\frac{S\,dz}{dt} \tag{26-16}$$

where $\rho_{A,L}/M_A$ is the molar density of A in the liquid phase, and S is the cross-sectional area for flux. Recognizing that $dz/dt = -dZ/dt$, equation (26-16) can also be written as

$$N_{A,z} = \frac{\rho_{A,L}}{M_A}\frac{dZ}{dt}$$

Under pseudo-steady-state conditions, equations (26-7) and (26-16) may be combined to give

$$\frac{\rho_{A,L}}{M_A}\frac{dZ}{dt} = \frac{cD_{AB}\left(y_{A_1} - y_{A_2}\right)}{Z\,y_{B,lm}} \tag{26-17}$$

Following separation of independent variable t from dependent variable Z, equation (26-17) may be integrated from $t = t_o = 0$ to $t = t$, and correspondingly from $Z_o = z_2 - z(t_o)$ to $Z = z_2 - z(t_o)$ as follows:

$$\int_{t_o}^{t} dt = \frac{\rho_{A,L}y_{B,lm}/M_A}{cD_{A,B}\left(y_{A_1} - y_{A_2}\right)}\int_{Z_o}^{Z} Z\,dZ$$

This yields

$$t - t_o = \frac{\rho_{A,L}\,y_{B,lm}/M_A}{cD_{A,B}\left(y_{A_1} - y_{A_2}\right)}\left(\frac{Z^2 - Z_o^2}{2}\right) \tag{26-18}$$

Therefore, for the pseudo-steady-state diffusion process, a plot of $(Z^2 - Z_o^2)$ vs. time t should be linear, with slope given by

$$slope = \frac{\rho_{A,L}y_{B,lm}/M_A}{cD_{A,B}\left(y_{A_1} - y_{A_2}\right)} \tag{26-19}$$

In the laboratory, the diffusion coefficient D_{AB} can be backed out from the least-squares slope of the linearly regressed data, plotted as $(Z^2 - Z_o^2)$ vs. t.

As illustrated by the Arnold diffusion cell above, pseudo-steady-state diffusion processes usually involve the slow depletion of the source or sink for the mass-transfer process with time. Below, we consider another process that is modeled by pseudo-steady-state diffusion, the thermal oxidation of a silicon wafer.

Example 2

The formation of a silicon oxide (SiO_2) thin film on a silicon (Si) wafer surface is an important step in the fabrication of solid-state microelectronic devices. A thin film of SiO_2 serves as a barrier to dopant diffusion or as a dielectric insulator to isolate various devices being formed on the wafer. In one common process, silicon is oxidized by exposure to oxygen (O_2) gas at temperatures above 700 C:

$$Si(s) + O_2(g) \rightarrow SiO_2(s)$$

Molecular O_2 dissolves into the SiO_2 solid, diffuses through the SiO_2 film, and then reacts with Si at the Si/SiO_2 interface, as shown in Figure 26.6. Assuming that the diffusion of O_2 through the SiO_2 film limits the oxidation process, develop a model to predict the thickness of the SiO_2 layer (δ) as a function of time at 1000 C. The density of solid $SiO_2(\rho_B)$ is 2.27 g/cm³, and the molecular weight of $SiO_2(M_B)$ is 60 g/mole. The molecular diffusion coefficient of O_2 in SiO_2 (D_{AB}) is 2.7×10^{-9} cm²/s at

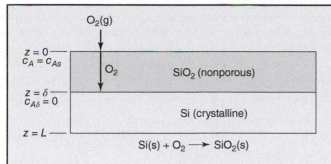

$$Si(s) + O_2 \longrightarrow SiO_2(s)$$

Figure 26.6 Thermal oxidation of a silicon wafer.

$1000°C$, and the maximum solubility of O_2 in $SiO_2(c_{As})$ is 9.6×10^{-8} gmole O_2/cm^3 solid at 1000 C and 1.0 atm O_2 partial pressure, using data provided by Norton.[1]

The physical system is represented in the rectilinear coordinate system. The model development follows the approach outlined earlier in Section 25.4. The assumptions for model development are listed here. (1) The oxidation of Si to SiO_2 occurs only at the Si/SiO_2 interface. The unreacted Si at the interface serves as the sink for molecular mass transfer of O_2 through the film. (2) The O_2 in the gas phase above the wafer represents an infinite source for O_2 transfer. The O_2 molecules "dissolve" into the nonporous SiO_2 solid at the gas/solid interface. (3) The rate of SiO_2 formation is controlled by the rate of molecular diffusion of O_2 (species A) through the solid SiO_2 layer (species B) to the unreacted Si layer. The reaction is very rapid, so that the concentration of molecular O_2 at the interface is zero, that is, $C_{A,\delta} = 0$. Furthermore, there are no mass-transfer resistances in the gas film above the wafer surface, as O_2 is a pure component in the gas phase. (4) The flux of O_2 through the SiO_2 layer is one-dimensional along coordinate z. (5) The rate of SiO_2 film formation is slow enough that at a given film thickness δ there is no accumulation of reactants or products within the SiO_2 film. However, the thickness of the film will still increase with time. Consequently, this is a "pseudo-steady-state" process. (6) The overall thickness of the wafer does not change as the result of the formation of the SiO_2 layer. (7) The process is isothermal.

Based on the previous assumptions, the general differential equation for mass transfer reduces to

$$\frac{dN_{A,z}}{dz} = 0$$

and Fick's flux equation for one-dimensional diffusion of O_2 (species A) through crystalline solid SiO_2 (species B) is

$$N_{A,z} = -D_{AB}\frac{dc_A}{dz} + \frac{c_A}{c}\left(N_{A,z} + N_{B,z}\right) = -D_{AB}\frac{dc_A}{dz} + \frac{c_A}{c}N_{A,z}$$

Usually, the concentration of molecular O_2 in the SiO_2 layer is dilute enough that c_A/c term is very small in magnitude relative to the other terms. Therefore, Fick's flux equation reduces to

$$N_{A,z} = -D_{AB}\frac{dc_A}{dz}$$

It is interesting to note here that unimolecular diffusion (UMD) flux mathematically simplifies to the equimolar counterdiffusion (EMCD) flux at dilute concentration of the diffusing species. As N_A is constant along z, the differential flux equation can be integrated directly following separation of dependent variable c_A from independent variable z:

$$\int_0^\delta N_{A,z}dz = -D_{AB}\int_{c_{As}}^0 dc_A$$

or simply

$$N_{A,z} = \frac{D_{AB}\, c_{As}}{\delta}$$

which describes the flux of O_2 through the SiO_2 layer of thickness δ. The surface concentration C_{As} refers to the concentration of O_2 dissolved in solid phase SiO_2 (gmole O_2/cm^3 solid).

[1] F. J. Norton, *Nature,* **191**, 701 (1961).

We know that δ increases slowly with time, even though there is no accumulation term for O_2 in the SiO_2 layer. In other words, the process operates under the pseudo-steady-state assumption. In order to discover how δ increases with time, consider an unsteady-state material balance for SiO_2 within the wafer (In − Out + Production = Accumulation), given by

$$0 - 0 + N_{A,z} \cdot S \cdot \frac{1.0 \text{ mole } SiO_2(B)}{1.0 \text{ mole } O_2(A)} = \frac{dm_B}{dt} \text{ with } \frac{dm_B}{dt} = \frac{d\left(\frac{\rho_B S \delta}{M_B}\right)}{dt}$$

where ρ_B is the density of solid SiO_2 (2.27 g/cm^3), M_B is the molecular weight of the SiO_2 layer (60 g/mole), and S is the surface area of the wafer. Given the stoichiometry of the reaction, one mole of SiO_2 is formed for every mole of O_2 consumed. Combining this material balance with the flux equation gives

$$\frac{\rho_B}{M_B} \frac{d\delta}{dt} = \frac{D_{AB} c_{As}}{\delta}$$

Separation of dependent variable δ from the independent variable t, followed by integration from $t = 0, \delta = 0$, to $t = t, \delta = \delta$, gives

$$\int_0^\delta \delta d\delta = \frac{M_B D_{AB} c_{As}}{\rho_B} \int_0^t dt$$

or

$$\delta = \sqrt{\frac{2M_B D_{AB} c_{As}}{\rho_B} t}$$

The above equation predicts that the thickness of the SiO_2 thin film is proportional to the square root of time. Recall that the molecular diffusion coefficient of O_2 in $SiO_2 (D_{AB})$ is 2.7×10^{-9} cm^2/s at 1000 C, and the solubility of O_2 in $SiO_2 (c_{As})$ is 9.6×10^{-8} gmole O_2/cm^3 solid at 1000 C. Figure 26.7 compares the predicted film thickness δ vs. time to process data provided by Hess[2] for 1.0 atm O_2 at 1000 C. As one can see, the model adequately predicts the data trend. The film is very thin, less than 0.5 µm, in part because the value for the term $D_{AB} \cdot c_{As}$ is so small.

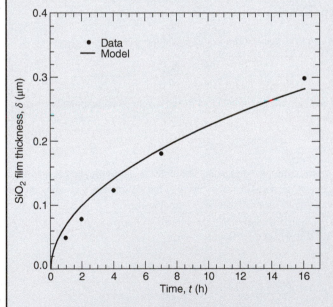

Figure 26.7 Silicon dioxide (SiO_2) film thickness vs. time at 1000 C.

This example illustrates how a chemical reaction at a boundary surface can serve as the driving force for molecular diffusion. This concept is formally presented in Section 26.2.

---

[2] D. W. Hess, *Chem. Eng. Educ.*, **24**, 34 (1990).

Equimolar Counterdiffusion

A common example of equimolar counterdiffusion is encountered in the *distillation* of two constituents whose molar latent heats of vaporization are essentially equal. This physical situation stipulates that the flux of one gaseous component is equal to but acting in the opposite direction from the other gaseous component—that is, $N_{A,z} = -N_{B,z}$. Equation (25-11)

$$\nabla \cdot \mathbf{N}_A + \frac{\partial c_A}{\partial t} - R_A = 0 \tag{25-11}$$

for the case of steady-state mass transfer without chemical reaction may be reduced to

$$\nabla \cdot \mathbf{N}_A = 0$$

For the transfer in the z-direction, this equation reduces to

$$\frac{dN_{A,z}}{dz} = 0 \tag{26-1}$$

This relation stipulates that $N_{A,z}$ is constant along the path of transfer. The molar flux, $N_{A,z}$, for a binary system at constant temperature and pressure is described by

$$N_{A,z} = -D_{AB}\frac{dc_A}{dz} + y_A\left(N_{A,z} + N_{B,z}\right) \tag{24-21}$$

The substitution of the restriction, $N_{A,z} = -N_{B,z}$, into the above equation gives an equation describing the flux of A when *equimolar-counterdiffusion* conditions exist:

$$N_{A,z} = -D_{AB}\frac{dc_A}{dz} \tag{26-20}$$

Equation (26-20) may be integrated, using the boundary conditions

$$\text{at } z = z_1, \quad c_A = c_{A_1}$$

and

$$\text{at } z = z_2, \quad c_A = c_{A_2}$$

giving

$$N_{A,z}\int_{z_1}^{z_2} dz = -D_{AB}\int_{c_{A_1}}^{c_{A_2}} dc_A$$

from which we obtain

$$N_{A,z} = \frac{D_{AB}}{(z_2 - z_1)}\left(c_{A_1} - c_{A_2}\right) \tag{26-21}$$

When the ideal gas law is obeyed, the molar concentration of A is related to the partial pressure of A by

$$c_A = \frac{n_A}{V} = \frac{p_A}{RT}$$

Substituting this expression for c_A into equation (26-21), we obtain

$$N_{A,z} = \frac{D_{AB}}{RT(z_2 - z_1)}\left(p_{A_1} - p_{A_2}\right) \tag{26-22}$$

Equations (26-21) and (26-22) are commonly referred to as the *equations for steady-state equimolar counterdiffusion*.

The concentration profile for equimolar-counterdiffusion processes may be obtained by substituting equation (26-20) into the differential equation that describes transfer in the z-direction:

$$\frac{dN_{A,z}}{dz} = 0$$

or

$$\frac{d^2 c_A}{dz^2} = 0$$

This second-order equation may be integrated twice with respect to z to yield

$$c_A = C_1 z + C_2$$

The two constants of integration (C_1 and C_2) are evaluated, using the boundary conditions

$$\text{at } z = z_1, \quad c_A = c_{A_1}$$
$$\text{at } z = z_2, \quad c_A = c_{A_2}$$

to obtain the linear concentration profile

$$\frac{c_A - c_{A_1}}{c_{A_1} - c_{A_2}} = \frac{z - z_1}{z_1 - z_2} \tag{26-23}$$

Equations (26-21) and (26-23) may be used to describe any process where the bulk-contribution term is zero. Besides the equimolar-counterdiffusion phenomenon, a negligible bulk-contribution term is also encountered when transferring species A is dilute in the mixture. Note that dilute unimolecular diffusion given by equation (26-3) also is represented by equation (26-20) if transferring species A is dilute with respect to species B:

$$N_{A,z} = -\left(\frac{c D_{AB}}{1 - y_A} \frac{dy_A}{dz} \right) = -c D_{AB} \frac{dy_A}{dz} = -D_{AB} \frac{dc_A}{dz} \quad \text{if } y_A \ll 1$$

This situation is particularly true for the diffusion of a solute into a liquid or solid, which is usually modeled as a dilute unimolecular mass-transfer process.

It is interesting to note that when we consider the "film concept" for mass transfer with equimolar counterdiffusion, the definition of the convective mass-transfer coefficient is different from that for diffusion in a stagnant gas film. In the case of equimolar counterdiffusion,

$$k^0 = \frac{D_{AB}}{\delta} \tag{26-24}$$

The superscript on the mass-transfer coefficient is used to designate that there is no net molar transfer into the film due to the equimolar counterdiffusion. Comparing equation (26-24) with equation (26-9), we realize that these two defining equations yield the same results only when the concentration of A is very small and $p_{B,lm}$ is essentially equal to P.

► **26.2**

ONE-DIMENSIONAL SYSTEMS ASSOCIATED WITH CHEMICAL REACTION

Many diffusional operations involve the simultaneous diffusion of a molecular species and the disappearance or appearance of the species through a chemical reaction either within or at the boundary of the phase of interest. We distinguish between the two types of chemical reactions, defining the reaction that occurs uniformly throughout a given phase as a *homogeneous reaction* and the reaction that takes place in a restricted region within or at a boundary of the phase as a *heterogeneous reaction*.

The rate of appearance of species A by a homogeneous reaction appears in the general differential equation of mass transfer as the source term, R_A:

$$\nabla \cdot \mathbf{N}_A + \frac{\partial c_A}{\partial t} - R_A = 0 \tag{25-11}$$

Examples of the source term, R_A, include the first-order conversion of reactant A to product D, so $R_A = -k_1 c_A$, where k_1 is the first-order rate constant in typical units of 1/s, and the second-order reaction of reactants A and B to form the product P with $R_A = -k_2 c_A c_B$, where k_2 is the second-order rate constant in typical units of $cm^3/mole \cdot s$.

The rate of disappearance of A by a heterogeneous reaction on a surface or at an interface does not appear in the general differential equation, as R_A involves only reactions within the control volume. Heterogeneous reactions typically involve a species in a fluid phase that reacts with a solid-phase species at the surface, or the reaction of a species on a catalytic surface to form a product. Homogeneous and heterogeneous reactions are compared in Figure 26.8. A heterogeneous reaction enters the analysis as a boundary condition and provides information on the fluxes of the species involved in the reaction; for example, if the surface reaction is $O_2(g) + C(s) \rightarrow CO_2(g)$, the flux of $CO_2(g)$ will be the same as the flux of $O_2(g)$, but leaving in the opposite direction.

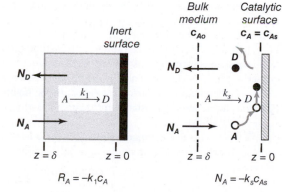

Figure 26.8 Homogeneous reaction within the control volume for diffusion vs. heterogeneous reaction at a catalytic surface.

Simultaneous Diffusion and Heterogeneous, First-Order Chemical Reaction: Diffusion with Varying Area

Many industrial processes involve the diffusion of a reactant to a surface where a chemical reaction occurs. As both diffusion and reaction steps are involved in the overall process, the relative rates of each step are important. When the reaction rate is instantaneous relative to the rate of diffusion, then the process is *diffusion-controlled*. In contrast, when the reaction rate of the transferring species at the surface limits the mass-transfer rate, then the process is *reaction-controlled*.

Many systems for mass transfer also involve steady-state radial diffusion within a cylinder or sphere, where the cross-sectional area for flux increases along the increasing radial direction. Example 3 below illustrates the steady-state molecular diffusion and surface reaction in the radial direction within a cylindrical and spherical particle.

Example 3

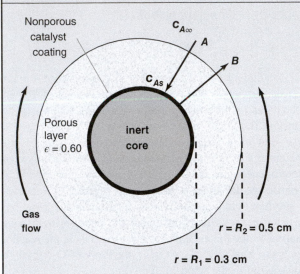

Figure 26.9 Cross section of catalyst particles for Example 3.

A catalyst particle is suspended in a gas stream of mixture A and B that is maintained at constant concentration $c_{A,\infty}$. A radial cross section of the particle is provided in Figure 26.9. The core of the particle is nonporous and is coated with a catalyst. A highly porous layer with void fraction (ϵ) of 0.60 surrounds the catalyst coating. The catalyst coating catalyzes the *heterogeneous* reaction $A \xrightarrow{k_s} B$, where k_s is the surface reaction rate constant. Consider both a spherical and cylindrical particle, where $r = R_1$ represents the radial position of the catalyst coating and $r = R_2$ represents the outer radius of the porous layer. The length of the cylindrical particle is L. Reactant A in the surrounding fluid diffuses through the porous layer to the catalyst surface. At the catalyst surface of the particle, $r = R_1$, the concentration of species A is c_{As}, and the reaction rate of species A at the catalyst surface is described by $N_A|_{r=R_1} = -k_s c_{As}$, with the negative sign indicating that the flux of A is opposite to the direction of increasing r.

Develop equations to predict the flux of species A at the catalyst surface ($r = R_1$), and then compare the flux for a reaction carried out at 300 C and 1.0 atm with $k_s = 2.0$ cm/s, $D_{AB} = 0.30$ cm^2/s, and 50 mole% A maintained in the gas phase.

The system control volume for mass transfer is the porous layer of the particle from $r = R_1$ to $r = R_2$. The source for mass transfer of A is the surrounding fluid, and the sink for A is the reaction of A to product B at the catalyst surface. The assumptions common to both the spherical particle and the cylindrical particle are (1) constant source and sink for species A, leading to a steady-state process with $\partial c_A/\partial t = 0$, (2) one-dimensional flux of species A and B along coordinate r, (3) no *homogeneous* reaction of species A or B within the noncatalytic porous layer ($R_A = 0$), (4) the effective diffusion coefficient in the highly porous layer approximates $D_{Ae} \doteq \epsilon^2 D_{AB}$, where ϵ is the porosity of the porous layer, and (5) constant temperature (T) and total system pressure (P).

In order to simplify the differential equation for mass transfer, we will use a spherical coordinate system for the spherical particle and a cylindrical coordinate system for the cylindrical particle. Based on assumptions 1–3 above, appropriate relationships, based on simplification of equations (25-26) and (25-28) are

$$\text{Sphere:} \quad \frac{1}{r^2}\frac{\partial}{\partial r}\left(r^2 N_{A,r}\right) = 0 \quad (r^2 N_{A,r} \text{ constant along } r) \tag{26-25}$$

$$\text{Cylinder:} \quad \frac{1}{r}\frac{\partial}{\partial r}\left(r N_{A,r}\right) = 0 \quad (r N_{A,r} \text{ constant along } r) \tag{26-26}$$

Equations (26-25) and (26-26) can also be obtained by a material balance on a differential volume element of the control volume, a "shell" bounded by radial position r and $r + \Delta r$. For a sphere,

$$4\pi r^2 N_{A,r}\big|_r (IN) - 4\pi r^2 N_{A,r}\big|_{r+\Delta r} (OUT) + 0 (GEN) = 0 (ACCUMULATION)$$

dividing by $4\pi^2 \Delta r$, and rearranging yields

$$-\left(\frac{r^2 N_{A,r}\big|_{r+\Delta r} - r^2 N_{A,r}\big|_r}{\Delta r}\right)\Bigg|_{\lim \Delta r \to 0} = 0$$

Finally, taking the limit as $\Delta r \rightarrow 0$ yields and rearrangement yields the differential equation for mass transfer of the spherical particle. Likewise, for a cylindrical partilcle,

$$2\pi rLN_{A,r}\big|_r\,(IN) - 2\pi rLN_{A,r}\big|_{r+\Delta r}\,(OUT) + 0\,(GEN) = 0\,(ACCUMULATION)$$

dividing by $2\pi L\Delta r$ and taking the limit as $\Delta r \rightarrow 0$ yields its differential equation for mass transfer.

The above equations show that the flux $N_{A,r}$ is *not* constant along radial position r. However, the total mass-transfer rate W_A, which is the product of flux and cross-sectional area for flux, is constant along r by continuity. Specifically, for a sphere,

$$W_A = 4\pi r^2 N_{A,r} = 4\pi R_1^2 N_{A,r}\big|_{r=R_1} = 4\pi R_2^2 N_{A,r}\big|_{r=R_2}$$

Likewise, for a cylinder,

$$W_A = 2\pi rL N_{A,r} = 2\pi R_1 L N_{A,r}\big|_{r=R_1} = 2\pi R_2 L N_{A,r}\big|_{r=R_2}$$

In comparison, for equation (26-1), in rectilinear coordinates, $N_{A,z}$ is constant along z, since the cross-sectional area for flux does not vary with position.

Due to the reaction stoichiometry, note that the flux of B in the positive r-direction is equal to the flux of A in the negative r-direction, so that $N_{B,r} = -N_{A,r}$. Consequently, the flux equation in the r-direction is

$$N_{A,r} = -D_{Ae}\frac{\partial c_A}{\partial r} + \frac{c_A}{c}(N_{A,r} + N_{B,r}) = -D_{Ae}\frac{\partial c_A}{\partial r} = -D_{Ae}\frac{dc_A}{dr}$$

The partial derivative $\partial/\partial r$ is replaced within an ordinary derivative d/dr, since there is only one independent variable, r.

Finally, two boundary conditions must be given for species A to completely specify the differential process model:

$$\text{at } r = R_1(0.3\,\text{cm}), \quad c_A = c_{As}$$

$$\text{at } r = R_2(0.5\,\text{cm}), \quad c_A = c_{A,\infty}$$

Given the forms of the general differential equation for mass transfer, the differential flux equation can be integrated directly without the need to get an integral model for the concentration profile first. For the spherical catalyst particle,

$$W_A = 4\pi R_1^2 N_{A,r}\big|_{r=R_1} = 4\pi r^2 N_{A,r} = -4\pi r^2 D_{Ae}\frac{dc_A}{dr}$$

Separation of dependent variable c_A from independent variable r, and insertion of the boundary conditions as the limits of integration, yield

$$W_A \int_{R_1}^{R_2} \frac{1}{r^2}\,dr = -4\pi D_{Ae}\int_{c_{AS}}^{c_{A,\infty}} dc_A$$

The final integral is

$$W_A = \frac{4\pi D_{Ae}}{1/R_2 - 1/R_1}(c_{A,\infty} - c_{As}) \tag{26-27}$$

or

$$N_{A,r}\big|_{r=R_1} = \frac{D_{Ae}(c_{A,\infty} - c_{As})}{R_1^2(1/R_2 - 1/R_1)} \tag{26-28}$$

Likewise, for the cylindrical catalyst particle,

$$W_A = 2\pi R_2 L N_{A,r}\big|_{r=R_1} = 2\pi rL N_{A,r} = -2\pi rL D_{Ae}\frac{dc_A}{dr}$$

$$W_A \int_{R_1}^{R_2} \frac{1}{r} dr = -2\pi L D_{Ae} \int_{c_{AS}}^{c_{A,\infty}} dc_A$$

$$W_A = \frac{2\pi L D_{Ae}}{\ln(R_1/R_2)} (c_{A,\infty} - c_{As}) \tag{26-29}$$

or

$$N_A|_{r=R_1} = \frac{D_{Ae}}{R_1 \ln(R_1/R_2)} (c_{A,\infty} - c_{As}) \tag{26-30}$$

At this point, the flux cannot be calculated, since c_{As} is not known. However, since

$$c_{As} = \frac{-N_A|_{r=R_1}}{k_s} \tag{26-31}$$

it can be readily shown that can be c_{As} eliminated. For the spherical particle,

$$N_{A,r}|_{r=R_1} = \frac{-c_{A,\infty}}{\left(\dfrac{R_1^2 (1/R_1 - 1/R_2)}{D_{Ae}} + \dfrac{1}{k_s} \right)} \tag{26-32}$$

And for the cylindrical particle,

$$N_A|_{r=R_1} = \frac{-c_{A\infty}}{\left(\dfrac{R_1 \ln(R_2/R_1)}{D_{Ae}} + \dfrac{1}{k_s} \right)} \tag{26-33}$$

At the conditions of the process,

$$c_{A,\infty} = y_{A,\infty} c = \frac{y_{A,\infty} P}{RT} = \frac{(0.50)(1.0 \text{ atm})}{\left(\dfrac{82.06 \text{ cm}^3 \cdot \text{atm}}{\text{gmole} \cdot \text{K}} \right)(300 + 273)\text{K}} = 1.06 \times 10^{-5} \text{ gmole/cm}^3$$

From Figure 24.5, for a random porous material where Knudsen diffusion effects are not important,

$$D_{Ae} \doteq \epsilon^2 D_{AB} = (0.60)^2 (0.30 \text{ cm}^2/\text{s}) = 0.108 \text{ cm}^2/\text{s}$$

The flux of A to the spherical particle at $r = R_1$ is

$$N_{A,r}|_{r=R_1} = \frac{-(1.06 \times 10^{-5} \text{ gmole/cm}^3)}{\left(\dfrac{(0.3 \text{ cm})^2 (1/0.3 \text{ cm} - 1/0.5 \text{ cm})}{(0.108 \text{ cm}^2/\text{s})} + \dfrac{1}{2.0 \text{ cm/s}} \right)} = -6.60 \times 10^{-6} \text{ gmole/cm}^2 \cdot \text{s}$$

with mole fraction of A at the catalytic surface, $y_{As} = \dfrac{-N_{A,r}|_{r=R_1}}{k_s c} = 0.155$.

The (−) indicates the direction of flux is opposite increasing radial position r. For comparison, the flux of A to the cylindrical particle at $r = R_1$ is

$$N_A|_{r=R_1} = \frac{-(1.06 \times 10^{-5} \text{ gmole/cm}^3)}{\left(\dfrac{(0.3 \text{ cm})\ln(0.5 \text{ cm}/0.3 \text{ cm})}{0.108 \text{ cm}^2/\text{s}} + \dfrac{1}{2.0 \text{ cm/s}} \right)} = -6.78 \times 10^{-6} \text{ gmole/cm}^2 \cdot \text{s}$$

with $y_{As} = 0.159$.

It is interesting to note that as k_s increases, c_{As} moves toward zero, which is the diffusion-limited condition.

Diffusion with a Homogeneous, First-Order Chemical Reaction

In the unit operation of absorption, one of the constituents of a gas mixture is preferentially dissolved in a contacting liquid. Depending upon the chemical nature of the involved molecules, the absorption may or may not involve chemical reactions. When there is a production or disappearance of the diffusing component, equation (25-11) may be used to analyze the mass transfer within the liquid phase. The following analysis illustrates mass transfer that is accompanied by a homogeneous chemical reaction.

Consider a layer of the absorbing medium as illustrated in Figure 26.10. At the liquid surface, the composition of A is c_{Ao}. The thickness of the film, δ, is defined so that beyond this film the concentration of A is always zero. If there is very little fluid motion within the film, and if the concentration of A in the film is assumed dilute, then the molar flux within the film is described by

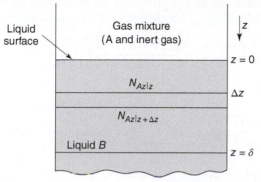

Figure 26.10 Absorption with homogeneous chemical reaction.

$$N_{A,z} = -D_{AB}\frac{dc_A}{dz}$$

For one-directional steady-state mass transfer, the general differential equation of mass transfer reduces to

$$-\frac{dN_{A,z}}{dz} + R_A = 0 \tag{26-34}$$

The disappearance of component A by a first-order reaction is defined by

$$R_A = -k_1 c_A \tag{26-35}$$

where k_1 is the chemical reaction rate constant. Substitution of equations (26-33) and (26-35) into equation (26-34) gives a second-order differential equation that describes simultaneous mass transfer accompanied by a first-order chemical reaction:

$$-\frac{d}{dz}\left(D_{AB}\frac{dc_A}{dz}\right) + k_1 c_A = 0 \tag{26-36}$$

If the diffusion coefficient is constant, equation (26-36) reduces to

$$D_{AB}\frac{d^2 c_A}{dz^2} - k_1 c_A = 0 \tag{26-37}$$

The general solution to equation (26-37) is

$$c_A(z) = C_1 \cosh\left(z\sqrt{k_1/D_{AB}}\right) + C_2 \sinh\left(z\sqrt{k_1/D_{AB}}\right) \tag{26-38}$$

The boundary conditions

$$\text{at } z = 0, \quad c_A = c_{Ao}$$

and

$$\text{at } z = \delta, \quad c_A = 0$$

permit the evaluation of the two constants of integration, C_1 and C_2. The constant C_1 is equal to c_{Ao}, and C_2 is equal to $-c_{Ao}/\tanh\left(\delta\sqrt{k_1/D_{AB}}\right)$, where δ is the thickness of the liquid film. Substituting these constants into equation (26-38), we obtain an equation for the concentration profile:

$$c_A(z) = c_{Ao}\cosh\left(z\sqrt{k_1/D_{AB}}\right) - \frac{c_{Ao}\sinh\left(z\sqrt{k_1/D_{AB}}\right)}{\tanh\left(\delta\sqrt{k_1/D_{AB}}\right)} \qquad (26\text{-}39)$$

The molar mass flux at the liquid surface can be determined by differentiating equation (26-39) and evaluating the derivative, $(dc_A/dz)|_{z=0}$. The derivative of c_A with respect to z is

$$\frac{dc_A}{dz} = +c_{Ao}\sqrt{k_1/D_{AB}}\,\sinh\left(z\sqrt{k_1/D_{AB}}\right) - \frac{c_{Ao}\sqrt{k_1/D_{AB}}\,\cosh\left(z\sqrt{k_1/D_{AB}}\right)}{\tanh\left(\delta\sqrt{k_1/D_{AB}}\right)}$$

which, when z equals zero, becomes

$$\left.\frac{dc_A}{dz}\right|_{z=0} = 0 - \frac{c_{Ao}\sqrt{k_1/D_{AB}}}{\tanh\left(\delta\sqrt{k_1/D_{AB}}\right)} = -\frac{c_{Ao}\sqrt{k_1/D_{AB}}}{\tanh\left(\delta\sqrt{k_1/D_{AB}}\right)} \qquad (26\text{-}40)$$

Substituting equation (26-40) into equation (26-33), we obtain

$$N_{A,z}\big|_{z=0} = \frac{D_{AB}c_{Ao}}{\delta}\left[\frac{\delta\sqrt{k_1/D_{AB}}}{\tanh\left(\delta\sqrt{k_1/D_{AB}}\right)}\right] \qquad (26\text{-}41)$$

If the boundary condition at $z = \delta$ is changed so that there is no next flux at this boundary

$$N_{A,z}\big|_{z=\delta} = -D_{AB}\left.\frac{dc_A}{dz}\right|_{z=\delta} = 0$$

then equations (26-39) and (26-41) will also change. This type of boundary condition commonly arises in many physical systems involving diffusion and homogeneous reaction, as illustrated in Example 4.

It is interesting to consider the simpler mass-transfer operation involving the absorption of A into liquid B *without* an accompanying chemical reaction. If no homogeneous chemical reaction is present $(R_A = 0)$, the molar flux of A is easily determined by integrating equation (26-20) between the two boundary conditions, giving

$$N_{A,z} = \frac{D_{AB}c_{Ao}}{\delta}$$

It is apparent by comparing the two equations that the term

$$\left[\left(\delta\sqrt{k_1/D_{AB}}\right)/\tanh\left(\delta\sqrt{k_1/D_{AB}}\right)\right]$$

shows the influence of the chemical reaction. This term is a dimensionless quantity, often called the Hatta number.[3] As the rate of the chemical reaction increases, the reaction rate constant, k_1, increases and the hyperbolic tangent term, $\tanh\left(\delta\sqrt{k_1/D_{AB}}\right)$, approaches the value of 1.0. Accordingly, equation (26-41) reduces to

$$N_{A,z}\big|_{z=0} = \sqrt{D_{AB}\,k_1}\,(c_{Ao} - 0)$$

---

[3] S. Hatta, *Technol. Rep. Tohoku Imp. Univ.*, **10**, 199 (1932).

A comparison of this equation with equation (24-68)

$$N_A = k_c(c_{A_1} - c_{A_2})$$

(24-68)

reveals that the film coefficient, k_c, is proportional to the diffusion coefficient raised to the 0.5 power. With a relatively rapid chemical reaction, component A will disappear after penetrating only a short distance into the absorbing medium; thus, a second model for convective mass transfer has been proposed, the *penetration theory model*, in which k_c is considered a function of D_{AB} raised to the 0.5 power. In our earlier discussion of another model for convective mass transfer, the film theory model, the mass-transfer coefficient was a function of the diffusion coefficient raised to the first power. We shall reconsider the penetration model in Section 26.4 and also in Chapter 28, when we discuss convective mass-transfer coefficients.

The following example considers diffusion with a homogeneous first-order chemical reaction under a different set of boundary conditions.

Example 4

Dilute concentrations of toxic organic solutes can often be degraded by a "biofilm" attached to an inert, nonporous solid surface. A biofilm consists of living cells immobilized in a gelatinous matrix. Biofilms are not very thick, usually less than a few millimeters. A toxic organic solute (species A) diffuses into the biofilm and is degraded to harmless products, hopefully CO_2 and water, by the cells within the biofilm. For engineering applications, the biofilm can be approximated as a homogeneous substance (i.e., species B). The rate of degradation of the toxic solute per unit volume of the biofilm is described by a kinetic rate equation of the form

$$R_A = -\frac{R_{A,\max} c_A}{K_A + c_A}$$

where $R_{A,\max}$ is the maximum possible degradation rate of species A in the biofilm and K_A (mole/cm^3) is the half-saturation constant for the degradation of species A within the biofilm at hand.

Consider the simple "rotating disk" process unit shown in Figure 26.11 for the treatment of phenol (species A) in wastewater. The biofilm contains a microorganism rich in the enzyme peroxidase that oxidatively degrades phenol. The concentration of species A in the bulk-fluid phase over the biofilm is constant if the fluid phase is well mixed. However, the concentration of A within the biofilm will decrease along the depth of the biofilm z as species A is degraded. There are no

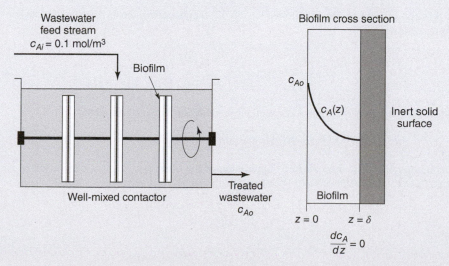

Figure 26.11 Biofilm for wastewater treatment.

resistances to convective mass transfer across the fluid boundary layer between the bulk fluid and the biofilm surface. Furthermore, phenol is equally soluble in both water and the biofilm, and the density difference between the biofilm and water can be neglected, so that the surface concentration of phenol in the aqueous phase equals the surface concentration of phenol in the gel phase just inside the biofilm—that is, at $z = 0$, $c_{As} = c_{Ao}$.

It is desired to treat 0.10 m^3 per hour of wastewater containing 0.10 gmole/m^3 of the toxic substance phenol. If the biofilm thickness is 2.0 mm (0.002 m), what is the required surface area of the biofilm necessary to achieve the desired outlet concentration of 0.02 gmole/m^3? The kinetic and mass-transport properties for the biofilm are $K_A = 0.3 \text{ gmole/m}^3$, $R_{A,\text{max}} = 5.7 \times 10^{-3} \text{ gmole/m}^3 \cdot \text{s}$, and $D_{AB} = 2.0 \times 10^{-10} \text{ m}^2/\text{s}$ at the process temperature of 25 C.

The source for phenol mass transfer is the inlet wastewater stream, whereas the sink for mass transfer is the steady-state consumption of phenol within the biofilm. First, a steady-state process material balance is performed on the contactor to determine the required rate of phenol degradation. A material balance on species A on the process unit is

$$(\text{rate of phenol degraded}) = (\text{rate of phenol added to process}) - (\text{rate of phenol exiting process})$$

or

$$W_A = v_i c_{Ai} - v_o c_{Ao} = v_o(c_{Ai} - c_{Ao}) = \frac{0.1 \text{ m}^3}{\text{h}}(0.1 - 0.02)\frac{\text{gmole}}{\text{m}^3} = 8.0 \times 10^{-3}\frac{\text{gmole}}{\text{h}}$$

where the subscript "i" represents the inlet stream and subscript "o" represents the outlet stream. Note that c_{Ao} is the bulk concentration of phenol inside the contactor. The biofilm possesses slab geometry best described by rectilinear coordinates. The rate of degradation W_A is proportional to the flux of species A into the biofilm at $z = 0$:

$$W_A = S N_{A,z} = S\left(-D_{AB}\frac{dc_A}{dz}\bigg|_{z=0}\right)$$

where S is the required surface area of the biofilm. At low concentrations where $K_A \gg c_A$, the above rate equation approximates a first-order process with respect to c_A:

$$R_A = -\frac{R_{A,\text{max}} c_A}{K_A + c_A} \cong -\frac{R_{A,\text{max}}}{K_A} c_A = -k_1 c_A$$

with k_1 equal to

$$k_1 = \frac{R_{A,\text{max}}}{K_A} = \frac{5.7 \times 10^{-3}\dfrac{\text{gmole}}{\text{m}^3 \cdot \text{s}}}{0.3\dfrac{\text{gmole}}{\text{m}^3}} = 1.9 \times 10^{-2}\text{ s}^{-1}$$

The flux can be obtained from the concentration profile. Recall equation (26-37) for one-dimensional, steady-state diffusion with a homogenous first-order chemical reaction:

$$D_{AB}\frac{d^2 c_A}{dz^2} - k_1 c_A = 0 \tag{26-37}$$

Recall also that this homogeneous second-order differential equation has a general solution of the form

$$c_A(z) = C_1 \cosh\left(z\sqrt{k_1/D_{AB}}\right) + C_2 \sinh\left(z\sqrt{k_1/D_{AB}}\right) \tag{26-38}$$

where C_1 and C_2 are integration constants to be determined by application of the boundary conditions as described earlier. The biofilm is immobilized onto a nonporous solid surface. Therefore, the flux at $z = \delta$ is zero. Consequently, the boundary conditions are

$$\text{at } z = \delta, \quad \frac{dc_A}{dz} = 0$$
$$\text{at } z = 0, \quad c_A = c_{As} = c_{Ao}$$

Note that the boundary conditions previously discussed are different from the ones described earlier to develop equations (26-39) and (26-41). The concentration profile based on this set of boundary conditions is

$$c_A(z) = \frac{c_{Ao}\cosh(\delta - z)\sqrt{k_1/D_{AB}}}{\cosh\left(\delta\sqrt{k_1/D_{AB}}\right)} \tag{26-42}$$

The mathematical development of equation (26-42) is left to the reader. Differentiation of equation (26-42), and evaluating the derivative at $z = 0$ yields

$$\left.\frac{dc_A}{dz}\right|_{z=0} = -c_{Ao}\sqrt{k_1/D_{AB}}\,\tanh\left(\delta\sqrt{k_1/D_{AB}}\right)$$

From this, the flux of phenol into the biofilm is

$$N_A|_{z=0} = \frac{D_{AB}c_{Ao}}{\delta}\left(\delta\sqrt{k_1/D_{AB}}\right)\tanh\left(\delta\sqrt{k_1/D_{AB}}\right) \tag{26-43}$$

It is useful first to calculate the dimensionless parameter, ϕ, the Thiele modulus, given by

$$\phi = \delta\sqrt{\frac{k_1}{D_{AB}}} = 0.002\text{ m}\sqrt{\frac{1.9 \times 10^{-2}\dfrac{1}{\text{s}}}{2 \times 10^{-10}\dfrac{\text{m}^2}{\text{s}}}} = 19.49$$

Equation (26-43) can be also expressed in terms of the Thiele modulus as

$$N_A|_{z=0} = \frac{D_{AB}c_{Ao}}{\delta}(\phi)\tanh(\phi)$$

The Thiele modulus represents the ratio of reaction rate to diffusion rate. When ϕ is less than 0.1, $\tanh(\phi)$ approaches ϕ and

$$N_A|_{z=0} = \frac{D_{AB}c_{Ao}}{\delta}\left(\delta\sqrt{k_1 D_{AB}}\right)^2 = \frac{D_{AB}c_{Ao}}{\delta}\phi^2$$

This situation arises when the rate of reaction is low relative to the rate of diffusion, so that the process is reaction-rate controlling. When ϕ is greater than 5, $\tanh(\phi) \approx 1$ and

$$N_{A,z}|_{z=0} = \frac{D_{AB}c_{Ao}}{\delta}\left(\delta\sqrt{k_1 D_{AB}}\right) = \frac{D_{AB}c_{Ao}}{\delta}\phi$$

This situation arises with the diffusion rate is low relative to the rate of reaction, so that the process is diffusion-rate controlling. In our example, the molecular diffusion flux of phenol through the biofilm very strongly influences the overall phenol degeneration rate. The flux of phenol into the biofilm is

$$N_{A,z} = \frac{\left(2.0 \times 10^{-10}\dfrac{\text{m}^2}{\text{s}}\right)\left(0.020\dfrac{\text{gmole}}{\text{m}^3}\right)}{0.002\text{ m}}(19.49)\tanh(19.49) = 3.9 \times 10^{-8}\frac{\text{gmole}}{\text{m}^2 \cdot \text{s}}$$

Finally, the required surface area of the biofilm is backed out from the required degradation rate and the flux:

$$S = \frac{W_A}{N_{A,z}} = \frac{8.0 \times 10^{-3}\dfrac{\text{gmole}}{\text{h}}\dfrac{1\text{ h}}{3600\text{ s}}}{3.9 \times 10^{-8}\dfrac{\text{gmole}}{\text{m}^2 \cdot \text{s}}} = 57.0\text{ m}^2$$

The steady-state concentration profile $c_A(z)$ within the biofilm is shown in Figure 26.12. It is interesting to note that concentration profile rapidly goes to zero within the first millimeter of the biofilm, again illustrating a strong diffusional resistance to the phenol degradation reaction.

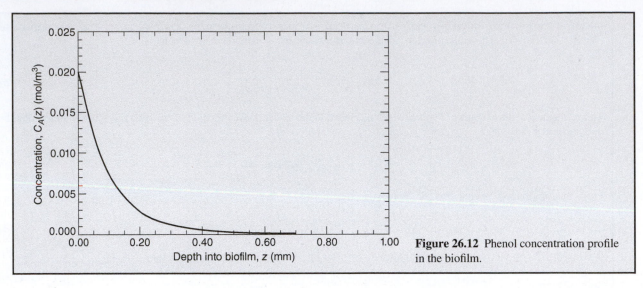

Figure 26.12 Phenol concentration profile in the biofilm.

▶ **26.3**

TWO- AND THREE-DIMENSIONAL SYSTEMS

In Sections 26.1 and 26.2, we have discussed problems in which the concentration and the mass transfer were functions of a single-space variable. Although many problems fall into this category, there are systems involving irregular boundaries or nonuniform concentrations along the boundary for which the one-dimensional treatment may not apply. In such cases, the concentration profile may be a function of two or even three spatial coordinates.

In this section, we shall review some of the methods for analyzing molecular mass transfer in two- and three-dimensional systems. As the transfer of heat by conduction is analogous to molecular mass transfer, we shall find the analytical, analogical, and numerical techniques described in Chapter 17 to be directly applicable.

An analytical solution to any transfer problem must satisfy the general differential equation describing the transfer as well as the boundary conditions specified by the physical situation. A complete treatment of the analytical solutions for two- and three-dimensional systems requires a prior knowledge of partial differential equation and complex variable theory. As most of this material is too advanced for an introductory course, we shall limit our discussions to a relatively simple two-dimensional example described below. For treatment of more complicated problems, the student is referred to the two excellent treatises by Crank[4] and Jost.[5]

Recall the general differential equation for mass transfer in rectilinear coordinates, given by equation (25-27):

$$\frac{\partial c_A}{\partial t} + \frac{\partial N_{A,x}}{\partial x} + \frac{\partial N_{A,y}}{\partial y} + \frac{\partial N_{A,z}}{\partial z} = R_A$$

The simplest type of multidimensional diffusion occurs under conditions where the process is assumed to be at steady state, with no homogeneous reaction or production of the diffusing species within the control volume, and no bulk motion of fluid through

[4] J. Crank, *The Mathematics of Diffusion,* Oxford University Press, London, 1957.
[5] W. Jost, *Diffusion in Solids, Liquids and Gases,* Academic Press, New York, 1952.

the control volume. Furthermore, if there is no z-component flux of A, the general differential equation of mass transfer for species A in the x and y directions reduces to

$$\frac{\partial N_{A,x}}{\partial x} + \frac{\partial N_{A,y}}{\partial y} = 0 \qquad (26\text{-}44)$$

Consider that the control volume is a slab shown in Figure 26.13. A constant source and sink for diffusing species A will determine the directionality of the component fluxes, which will be specified through the boundary conditions. The z-direction of the flux is not considered, which could be realized by extending z-direction infinitely out of the x-y plane in the $+/-$ directions, or by sealing off the x-y plane to allow flux only through the y-z and x-z planes.

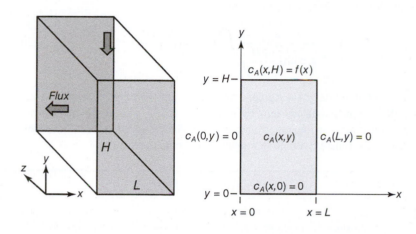

Figure 26.13 Two-dimensional diffusion.

For the additional assumption where Fick's flux equation reduces to Fick's first law—e.g., for a dilute mixture of species A in medium B, the component fluxes for diffusing species A in stagnant medium B are

$$N_{A,x} = -D_{AB}\frac{\partial c_A}{\partial x} \quad \text{and} \quad N_{A,y} = -D_{AB}\frac{\partial c_A}{\partial y}$$

In terms of concentration $c_A(x,y)$, the differential equation for mass transfer is

$$\frac{\partial^2 c_A}{\partial x^2} + \frac{\partial^2 c_A}{\partial y^2} = 0 \qquad (26\text{-}45)$$

which is a form of the familiar Laplace equation (25-23). The particular solution to the Laplace equation assumes a unique form once the required four boundary conditions are specified. These boundary conditions can include a known concentration of species A at a given boundary, or a known flux of species A at a given boundary. For example, consider the following boundary conditions where the concentration is known at each boundary

$$x = 0, \quad 0 \le y \le H, \quad c_A(0,y) = c_{As}$$

$$x = L, \quad 0 \le y \le H, \quad c_A(L,y) = c_{As}$$

$$y = 0, \quad 0 < x < L, \quad c_A(x,0) = c_{As}$$

$$y = H, \quad 0 < x < L, \quad c_A(x,H) = f(x)$$

where $f(x)$ is a function that describes the concentration profile at $y = H$ along x, not including $x = 0$ and $x = L$.

The analytical solution to the partial differential equation with the associated boundary conditions is obtained by the separation of variables technique. The solution will not be detailed here, but details of the method for the problem above are provided by Kreyszig.[6] The solution method is facilitated by the use of homogeneous boundary conditions at $x = 0$, $x = L$, and $y = 0$, which can be easily obtained by setting $c_{As} = 0$, or by a variable transformation—e.g., $u(x, y) = c_A(x, y) - c_{As}$. The analytical solution under these boundary conditions is

$$c_A(x, y) = \sum_{n=1}^{\infty} A_n \sin\left(\frac{n\pi x}{L}\right) \sinh\left(\frac{n\pi y}{L}\right) \qquad (26\text{-}46)$$

where A_n are the Fourier coefficients, given by

$$A_n = \frac{2}{L \sinh\left(\dfrac{n\pi H}{L}\right)} \int_0^L f(x) \sin\left(\frac{n\pi x}{L}\right) dx \qquad (26\text{-}47)$$

The above differential model development and analytical solution formulation can be extended to a three-dimensional system by also considering flux along the z-direction. Example 4 below illustrates the implementation of the analytical solution for two-dimensional mass steady-state mass transfer with no homogeneous chemical reaction.

Example 5

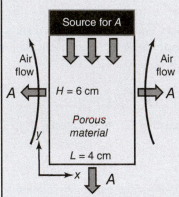

Figure 26.14 Mass-transfer system for Example 5.

The device shown in Figure 26.14 is used to continuously release volatile component A into a flowing air stream by a diffusion-limited process. A reservoir that serves as a source for solid A rests on top of a porous slab of dimensions $L = 4.0$ cm in the x-direction and $H = 6.0$ cm in the y-direction. The solid sublimes provide a vapor at saturation vapor pressure of 0.10 atm, to provide a concentration c_A^* at the top of the porous slab. The vapor diffuses through the porous slab, and then out through the exposed faces of the slab. Assuming that volatile species A of vapor pressure of 0.10 atm is immediately dispersed into the air stream maintained at 27 C and 1.0 atm so that $c_{As} \approx 0$, estimate the concentration profile $c_A(x, y)$ at 2.0 cm, 4.0 cm, and 5.5 cm from the base ($y = 0$).

The physical system is described equations (26-46) to (26-47) and the underlying assumption that $c_{As} \approx 0$ for the exposed surfaces. The control volume for mass transfer is the porous slab. Since the source for A (reservoir) and the sink for A (flowing gas stream) are oriented perpendicular to one another, and the x-y plane is sealed, the flux will have x and y components with two-dimensional concentration profile $c_A(x, y)$. Simplification of equation (26-47) for $f(x) = c_A^*$ gives

$$A_n = \frac{2}{L \sinh\left(\dfrac{n\pi H}{L}\right)} \int_0^L c_A^* \sin\left(\frac{n\pi x}{L}\right) dx = -\frac{4 c_A^*}{\sinh\left(\dfrac{n\pi H}{L}\right)} \frac{(-1)^n}{n\pi} \quad n = 1, 3, 5 \ldots$$

and so equation (26-46) becomes

$$c_A(x, y) = -4 c_A^* \sum_{n=1}^{\infty} \frac{(-1)^n}{n\pi \sinh\left(\dfrac{n\pi H}{L}\right)} \sin\left(\frac{n\pi x}{L}\right) \sinh\left(\frac{n\pi y}{L}\right) \quad n = 1, 3, 5 \ldots \qquad (26\text{-}48)$$

---

[6] E. Kreyszig, *Advanced Engineering Mathematics*, 6th edition, John Wiley & Sons, New York, 1988.

The concentration profile $c_A(x, y)$ was calculated by computer on a spreadsheet program to allow the infinite series solution to converge to a value for $c_A(x, y)$ at a given value of x and y. The results of the calculations are plotted in Figure 26.15.

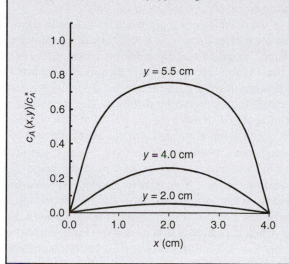

Figure 26.15 Normalized two-dimensional concentration profiles for Example 5.

▶ **26.4**

SIMULTANEOUS MOMENTUM, HEAT, AND MASS TRANSFER

In previous sections, we have considered steady-state mass transfer independent of the other transport phenomena. Many physical situations involve the simultaneous transfer of mass and either energy or momentum, and in a few cases, the simultaneous transfer of mass, energy, and momentum. The drying of a wet surface by a hot, dry gas is an excellent example in which all three transport phenomena are involved. Energy is transferred to the cooler surface by convection and radiation; mass and its associated enthalpy are transferred back into the moving gas stream. The simultaneous transport processes are more complex, requiring the simultaneous treatment of each transport phenomenon involved.

In this section, we consider two examples involving the simultaneous transfer of mass and a second transport phenomenon.

Simultaneous Heat and Mass Transfer

Generally, a diffusion process is accompanied by the transport of energy, even within an isothermal system. As each diffusing species carries its own individual enthalpy, a heat flux at a given plane is described by

$$\frac{\mathbf{q}_D}{A} = \sum_{i=1}^{n} \mathbf{N}_i \bar{H}_i \tag{26-49}$$

where $\mathbf{q}_D/A$ is the heat flux due to the diffusion of mass past the given plane and $\bar{H}_i$ is the partial molar enthalpy of species i in the mixture. When a temperature difference exists, energy will also be transported by one of the three heat-transfer mechanisms. For example, the equation for total energy transport by conduction and molecular diffusion becomes

$$\frac{\mathbf{q}}{A} = -k\nabla T + \sum_{i=1}^{n} \mathbf{N}_i H_i \tag{26-50}$$

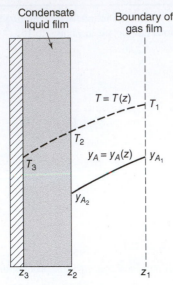

Condensate liquid film

Boundary of gas film

$T = T(z)$ | T_1

T_2

$y_A = y_A(z)$ | y_{A_1}

T_3

y_{A_2}

z_3 z_2 z_1

Figure 26.16 Vapor condensation on a cold surface.

If the heat transfer is by convection, the first energy-transport term in equation (26-50) would be replaced by the product of the convective heat-transfer coefficient and a ΔT driving force.

A process important in many engineering processes as well as in day-to-day events involves the condensation of a vapor upon a cold surface. Examples of this process include the "sweating" on cold water pipes and the condensation of moist vapor on a cold window pane. Figure 26.16 illustrates the process that involves a film of condensed liquid flowing down a cold surface and a film of gas through which the condensate is transferred by molecular diffusion. This process involves the simultaneous transfer of heat and mass.

The following conditions will be stipulated for this particular steady-state physical situation. Pure component A will condense from a binary gas mixture. The composition, y_A, and the temperature, T_1, are known at the plane z_1. The temperature of the condensing surface, T_3, is also known. By heat-transfer considerations, the convective heat-transfer coefficients for the condensate liquid film and the gas film can be calculated from equations given in Chapter 20. For example, in the gas phase, when the carrier gas is air and the vapor content of the diffusing species is relatively low, the heat-transfer coefficient for natural convection can be estimated by equation (20-5):

$$\text{Nu}_L = 0.68 + \frac{0.670\,\text{Ra}_L^{1/4}}{\left[1 + (0.492/\text{Pr})^{9/16}\right]^{4/9}} \tag{20-5}$$

Using the general differential equation for mass transfer, equation (25-11), we see that the differential equation that describes the mass transfer in the gas phase is

$$\frac{dN_{A,z}}{dz} = 0 \tag{26-51}$$

Equation (26-51) stipulates that the mass flux in the z-direction is constant over the diffusion path. To complete the description of the process, the proper form of Fick's flux equation must be chosen. If component A is diffusing through a stagnant gas, the flux is defined by equation (26-3):

$$N_{A,z} = \frac{-cD_{AB}}{1 - y_A}\frac{dy_A}{dz} \tag{26-3}$$

As a temperature profile exists within the film, and the diffusion coefficient and total gas concentration vary with temperature, this variation with z must often be considered. Needless to say, this complicates the problem and requires additional information before equation (26-3) can be integrated.

When the temperature profile is known or can be approximated, the variation in the diffusion coefficient can be treated. For example, if the temperature profile is of the form

$$\frac{T}{T_1} = \left(\frac{z}{z_1}\right)^n \tag{26-52}$$

the relation between the diffusion coefficient and the length parameter may be determined by using equation (24-41) as follows:

$$D_{AB} = D_{AB}\big|_{T_1}\left(\frac{T}{T_1}\right)^{3/2} = D_{AB}\big|_{T_1}\left(\frac{z}{z_1}\right)^{3n/2} \tag{26-53}$$

The variation in the total concentration due to the temperature variation can be evaluated by

$$c = \frac{P}{RT} = \frac{P}{RT_1 (z/z_1)^n}$$

The flux equation now becomes

$$N_{A,z} = \frac{-P\,D_{AB}|_{T_1}}{RT_1(1-y_A)} \left(\frac{z}{z_1}\right)^{n/2} \frac{dy_A}{dz} \tag{26-54}$$

This is the same approach used in Chapter 15, which discussed heat transfer by conduction when the thermal conductivity was a variable.

Over a small temperature range, an average diffusion coefficient and the total molar concentration may be used. With this assumption, equation (26-3) simplifies to

$$N_{A,z} = -\frac{(cD_{AB})_{avg}}{(1-y_A)} \frac{dy_A}{dz} \tag{26-55}$$

Integrating this equation between the boundary conditions

$$\text{at } z = z_1, \quad y_A = y_{A_1}$$

and

$$\text{at } z = z_2, \quad y_A = y_{A_2}$$

we obtain the relation

$$N_{A,z} = \frac{(cD_{AB})_{avg}\left(y_{A_1} - y_{A_2}\right)}{(z_2 - z_1)y_{B,lm}} \tag{26-56}$$

The temperature, T_2, is needed for evaluating $(cD_{AB})_{avg}$, the temperature difference between the liquid surface and the adjacent vapor, and the vapor pressure of species A at the liquid surface. This temperature may be evaluated from heat-transfer considerations. The total energy flux through the liquid surface also passes through the liquid film. This can be expressed by

$$\frac{q_z}{A} = h_{\text{liquid}}(T_2 - T_3) = h_c(T_1 - T_2) + N_{A,z}M_A(H_1 - H_2) \tag{26-57}$$

where h_{liquid} is the convective heat-transfer coefficient in the liquid film, h_c, is the natural convective heat-transfer coefficient in the gas film, M_A is the molecular weight of A, and H_1 and H_2 are the enthalpies of the vapor at plane 1 and the liquid at plane 2, respectively, for species A per unit mass. It is important to realize that there are two contributions to the energy flux entering the liquid surface from the gas film, convective heat transfer, and the energy carried by the condensing species.

To solve equation (26-57), a trial-and-error solution is required. If a value for the temperature of the liquid surface is assumed, T_2, h_c, and $(cD_{AB})_{avg}$ may be calculated. The equilibrium composition, y_{A_2}, can be determined from thermodynamic relations. For example, if Raoult's law holds,

$$p_{A_2} = x_A P_A$$

where x_A for a pure liquid is 1.0, and the partial pressure of A above the liquid surface is equal to the vapor pressure P_A. By Dalton's law, the mole fraction of A in the gas immediately above the liquid is

$$y_{A_2} = \frac{p_{A_2}}{P} = \frac{P_A}{P}$$

where P is the total pressure of the system and P_A is the vapor pressure of A at the assumed temperature T_2. Knowing $(cD_{AB})_{avg}$ and y_{A_2}, we can evaluate $N_{A,z}$ by equation (26-56). The liquid-film heat-transfer coefficients can be evaluated, using equations presented in Chapter 20. A value is now known for each term in equation (26-57). When the left- and right-hand sides of the equation are equal, the correct temperature of the liquid surface has been assumed. If the initially assumed temperature does not yield an equality, additional values must be assumed until equation (26-57) is satisfied.

There are several industrial-unit operations in which heat and mass transfer between gas and liquid phases occur simultaneously. Distillation, humidification or dehumidification of air, and water cooling are such operations. In early space exploration, the cooling of the reentry vehicles by sublimation of ablative material is another example where simultaneous transfer played an important engineering role.

In the following example, consideration of the simultaneous transfer of mass and heat is required to predict the flux relations described by Fick's law.

Simultaneous Momentum and Mass Transfer

In several mass-transfer operations, mass is exchanged between two phases. An important example that we have previously encountered is *absorption*, the selective dissolution of one of the components of a gas mixture by a liquid. A wetted-wall column, as illustrated in Figure 26.17, is commonly used to study the mechanism of this mass-transfer operation, as it provides a well-defined area of contact between the two phases. In this operation, a thin liquid film flows along the wall of the column while in contact with a gas mixture. The time of contact between the two phases is relatively short during normal operation. As only a small quantity of mass is absorbed, the properties of the liquid are assumed to be unaltered. The velocity of the falling film will thus be virtually unaffected by the diffusion process.

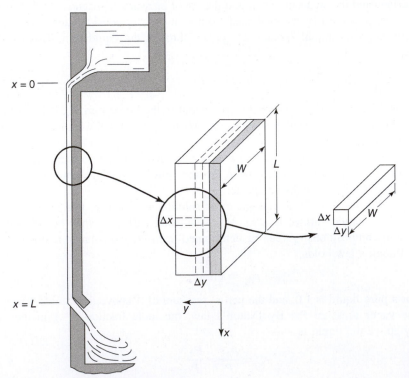

Figure 26.17 Absorption into a falling-liquid film.

The process involves both momentum and mass transfer. In Chapter 8, the laminar flow of a fluid down an inclined plane was discussed. When the angle of inclination is 90 degrees, the results obtained in Section 8.2 can be used to describe the falling-film velocity profile. With this substitution, the differential equation for momentum transfer becomes

$$\frac{d\tau_{yx}}{dy} + \rho g = 0$$

and the boundary conditions that must be satisfied are

$$\text{at } y = 0, \quad v_x = 0$$

and

$$\text{at } y = \delta, \quad \frac{\partial v_x}{\partial y} = 0$$

The final expression for the velocity profile is given by

$$v_x = \frac{\rho g \delta^2}{\mu}\left[\frac{y}{\delta} - \frac{1}{2}\left(\frac{y}{\delta}\right)^2\right]$$

The maximum velocity will be at the edge of the film where $y = \delta$

$$v_{max} = \frac{\rho g \delta^2}{2\mu}$$

Substituting this result into the velocity profile, we obtain another form of the expression for v_x

$$v_x = 2v_{max}\left[\frac{y}{\delta} - \frac{1}{2}\left(\frac{y}{\delta}\right)^2\right] \tag{26-58}$$

The differential equation for mass transfer can be obtained by using the general differential equation of mass transfer and eliminating the irrelevant terms or by making a balance over the control volume, $\Delta x \, \Delta y W$, as shown in Figure 26.17. It is important to note that the y component of the mass flux, $N_{A,y}$, is associated with the negative y-direction, according to the axes previously established in our fluid-flow considerations. The mass balance over the control volume is

$$N_{A,x}|_{x+\Delta x} \, W\Delta y - N_{A,x}|_x \, W\Delta y + N_{A,y}|_{y+\Delta y} \, W\Delta x - N_{A,y}|_y \, W\Delta x = 0$$

Dividing by $W\Delta x\Delta y$ and letting Δx and Δy approach zero, we obtain the differential equation

$$\frac{\partial N_{A,x}}{\partial x} + \frac{\partial N_{A,y}}{\partial y} = 0 \tag{26-59}$$

The one-directional molar fluxes in the x- and y-directions are respectively defined by

$$N_{A,x} = -D_{AB}\frac{\partial c_A}{\partial x} + x_A\left(N_{A,x} + N_{B,x}\right)$$

and

$$N_{A,y} = -D_{AB}\frac{\partial c_A}{\partial y} + x_A\left(N_{A,y} + N_{B,y}\right)$$

As previously mentioned, the time of contact between the vapor and liquid is relatively short; thus, a negligible concentration gradient will develop in the x-direction, and so the flux equation in the x-direction will reduce to

$$N_{A,x} = x_A(N_{A,x} + N_{B,x}) = c_A v_x \tag{26-60}$$

The convective transport term in the negative y-direction, $x_A(N_{A,y} + N_{B,y})$, involves multiplying two extremely small values and is negligible; thus, the flux equation in the y-direction becomes

$$N_{A,y} = -D_{AB}\frac{\partial c_A}{\partial y} \tag{26-61}$$

Substituting flux equations (26-60) and (26-61) into equation (26-59), we obtain

$$\frac{\partial(c_A v_x)}{\partial x} - D_{AB}\frac{\partial^2 c_A}{\partial y^2} = 0$$

or, as v_x is dependent upon y only,

$$v_x\frac{\partial c_A}{\partial x} - D_{AB}\frac{\partial^2 c_A}{\partial y^2} = 0 \tag{26-62}$$

The velocity profile, as defined by equation (26-58), may be substituted into equation (26-62), yielding

$$2 v_{\max}\left[\frac{y}{\delta} - \frac{1}{2}\left(\frac{y}{\delta}\right)^2\right]\frac{\partial c_A}{\partial x} = D_{AB}\frac{\partial^2 c_A}{\partial y^2} \tag{26-63}$$

The boundary conditions for mass transfer into the falling film are

$$\text{at } x = 0, \quad c_A = 0$$

$$\text{at } y = 0, \quad \frac{\partial c_A}{\partial y} = 0$$

and

$$\text{at } y = \delta, \quad c_A = c_{Ao}$$

The specific case in which solute A penetrates only a short distance into the liquid film because of a slow rate of diffusion or a short time of exposure can be treated by the *penetration theory* model developed by Higbie.[7] As solute A is transferred into the film at $y = \delta$, the effect of the falling film on the diffusing species is such that the fluid may be considered to be flowing at the uniform velocity, $v_{\max}$. Figure 26.18 illustrates the penetration depth. Solute A will not be affected by the presence of the wall; thus, the fluid may be considered to be of infinite depth in the y-direction. With these simplifications, equation (26-63) reduces to

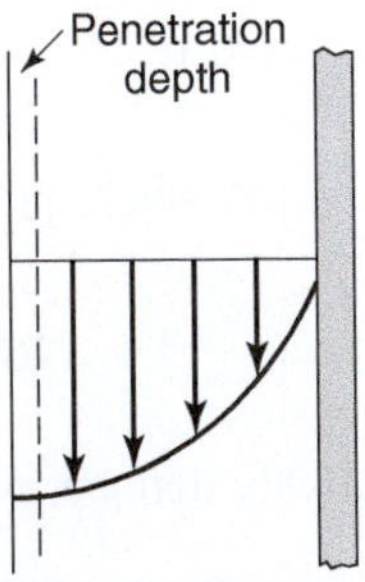

$$v_{\max}\frac{\partial c_A}{\partial x} = D_{AB}\frac{\partial^2 c_A}{\partial y^2} \tag{26-64}$$

Figure 26.18 Penetration depth in a falling film.

with the boundary conditions at $x = 0$, $\quad c_A = 0$:

$$\text{at } y = \delta, \quad c_A = c_{Ao}$$

$$\text{at } y = -\infty, \quad c_A = 0$$

[7] R. Higbie, *Trans. A.I.Ch.E.*, **31**, 368–389 (1935).

Equation (26-64) can be transformed into a form commonly encountered in unsteady-state mass transfer. If ξ is set equal to $\delta - y$, the transformed equation and boundary conditions become

$$v_{max} \frac{\partial c_A}{\partial x} = D_{AB} \frac{\partial^2 c_A}{\partial \xi^2} \tag{26-65}$$

with

$$\text{at } x = 0, \quad c_A = 0$$
$$\text{at } \xi = 0, \quad c_A = c_{Ao}$$

and

$$\text{at } \xi = \infty, \quad c_A = 0$$

This partial differential equation can be solved by using Laplace transforms. On applying the transforms in the x-direction, we obtain an ordinary differential equation in the s domain:

$$v_{max} \, s \, \bar{c}_A - 0 = D_{AB} \frac{d^2 \bar{c}_A(\xi, s)}{d\xi^2}$$

or

$$\frac{d^2 \bar{c}_A}{d\xi^2} - \frac{v_{max} s \, \bar{c}_A}{D_{AB}} = 0 \tag{26-66}$$

This ordinary equation is readily solved to give

$$\bar{c}_A = A_1 \exp\left(\xi \sqrt{\frac{v_{max} s}{D_{AB}}} \right) + B_1 \exp\left(-\xi \sqrt{\frac{v_{max} s}{D_{AB}}} \right) \tag{26-67}$$

The constants A_1 and B_1 are evaluated, using the two transformed boundary conditions:

$$\text{at } \xi = 0, \quad \bar{c}_A(0, s) = \frac{c_{Ao}}{s}$$
$$\text{at } \xi = \infty, \quad \bar{c}_A(\infty, s) = 0$$

yielding the solution

$$\bar{c}_A = \frac{c_{Ao}}{s} \exp\left(-\xi \sqrt{\frac{v_{max} s}{D_{AB}}} \right) \tag{26-68}$$

Equation (26-68) can be transformed back to the x domain by taking the inverse Laplacian, yielding

$$c_A(x, \xi) = c_{Ao} \left[1 - \text{erf}\left(\frac{\xi}{\sqrt{\dfrac{4D_{AB}x}{v_{max}}}} \right) \right] \tag{26-69}$$

The error function, a mathematical form that is commonly encountered in transient problems, was discussed in Chapter 18. Similar to other mathematical functions, tables have been prepared of the error function and one of these tables is presented in Appendix L.

The local mass flux at the surface, where $\xi = 0$, or $y = \delta$, is obtained by differentiating equation (26-69) with respect to ξ and then inserting the derivative into the flux equation (26-61):

$$N_{A,y}\big|_{\xi=0} = N_{A,y}\big|_{y=\delta} = -D_{AB}\frac{\partial c_A}{\partial y}\bigg|_{y=\delta}$$

The unidirectional flux becomes

$$N_{A,y}\big|_{y=\delta} = c_{Ao}\sqrt{\frac{D_{AB}v_{\max}}{\pi x}} \tag{26-70}$$

A comparison of equation (26-70) with the convective mass-transfer equation

$$N_{A,y} = k_c(c_{A_1} - c_{A_2}) \tag{24-68}$$

reveals that

$$k_c = \sqrt{\frac{D_{AB}v_{\max}}{\pi x}} \tag{26-71}$$

In equation (26-71), we see that the convective mass-transfer coefficient, k_c, is proportional to the diffusion coefficient (D_{AB}) raised to the 0.5 power. This dependency was also shown earlier in Section 26.2 for the diffusion of a solute into a liquid accompanied by a rapid chemical reaction. *Penetration theory* considers that the solute only penetrates a short distance into the liquid phase due to a short residence time of exposure of the solute with the liquid, or because the solute rapidly disappears by a chemical reaction within the liquid. Consequently, penetration theory proposes that liquid-phase mass-transfer coefficient for transferring species A will assume the form of equation (26-71).

▶ **26.5**

CLOSURE

In this chapter, we have considered solutions to steady-state molecular mass-transfer problems. The defining differential equations were established by simplifying the general differential equation for mass transfer or through the use of a control-volume expression for the conservation of mass. It is hoped that this two-pronged attack will provide the student with an insight into the various terms contained in the general differential equation, and thus enable the reader to decide whether the terms are relevant or irrelevant to any specific situation.

One-directional systems both with and without chemical production were considered. Two models of convective mass transfer, film theory and penetration theory, were introduced. These models will be used in Chapter 28 to evaluate and explain convective mass-transfer coefficients.

CHAPTER 27

Unsteady-State Molecular Diffusion

In Chapter 26, we restricted our attention to describing the steady-state molecular diffusion where the concentration at a given point was constant with time. In this chapter, we shall consider problems and their solutions that involve the concentration varying with time, thus resulting in *unsteady-state molecular diffusion* or *transient diffusion*. Many common examples of unsteady-state transfer will be cited. These generally fall into two categories: a process that is in an unsteady state only during its initial startup, and a process in which the concentration is continually changing throughout its duration.

The time-dependent differential equations are simple to derive from the general differential equation of mass transfer. The equation of continuity for component A in terms of mass,

$$\nabla \cdot \mathbf{n}_A + \frac{\partial \rho_A}{\partial t} - r_A = 0 \qquad (25\text{-}5)$$

or in terms of moles,

$$\nabla \cdot \mathbf{N}_A + \frac{\partial c_A}{\partial t} - R_A = 0 \qquad (25\text{-}11)$$

contains the concentration time-dependent of the unsteady-state accumulation term. It also contains the net rate of mass efflux of species A, which accounts for the variation of the concentration with spatial directions. The solution to the resulting partial differential equations is generally difficult, involving relatively advanced mathematical techniques. We will consider the solutions to some of the less-complex mass-transfer processes. A detailed discussion of the mathematics of diffusion is beyond the scope of this book. An excellent reference on the subject is a treatise by Crank.[1]

▶ **27.1**

UNSTEADY-STATE DIFFUSION AND FICK'S SECOND LAW

Although the differential equations for unsteady-state diffusion are easy to establish, most solutions to these equations have been limited to situations involving simple geometries and boundary conditions, and a constant diffusion coefficient. Many solutions are for

---

[1] J. Crank, *The Mathematics of Diffusion*, 2nd edition, Oxford University Press, 1975.

one-directional mass transfer as defined by Fick's second "law" of diffusion:

$$\frac{\partial c_A}{\partial t} = D_{AB} \frac{\partial^2 c_A}{\partial z^2} \tag{27-1}$$

This partial differential equation describes a physical situation in which there is no bulk motion contribution—that is, $v = 0$, and no chemical reaction, that is, $R_A = 0$. This situation is encountered when the diffusion takes place in solids, in stationary liquids, or in systems involving equimolar counterdiffusion. Due to the extremely slow rate of diffusion within liquids, the bulk motion contribution of Fick's first law (i.e., $y_A \Sigma N_i$) approaches the value of zero for dilute solutions; accordingly, this system also satisfies Fick's second law of diffusion.

It may be advantageous to express equation (27-1) in terms of other concentration units. For example, the mass density of species A, ρ_A, is equal to $c_A M_A$; by multiplying both sides of equation (27-1) by the constant molecular weight of A, we obtain

$$\frac{\partial \rho_A}{\partial t} = D_{AB} \frac{\partial^2 \rho_A}{\partial z^2} \tag{27-2}$$

If the density of the given phase remains essentially constant during the mass-transfer period, the density of species A can be divided by the total density, ρ_A/ρ; this ratio is the weight fraction of A, w_A, and our equation becomes

$$\frac{\partial w_A}{\partial t} = D_{AB} \frac{\partial^2 w_A}{\partial z^2} \tag{27-3}$$

Equations (27-1) to (27-3) are similar in form to Fourier's second "law" of heat conduction, given by

$$\frac{\partial T}{\partial t} = \alpha \frac{\partial^2 T}{\partial z^2}$$

thereby establishing an analogy between transient molecular diffusion and heat conduction.

The solution to Fick's second law usually has one of two standard forms. It may involve error functions or related special integrals that are suitable for small values of time or small penetration depths, or it may appear in the form of a trigonometric series that converge at large values of time. Analytical solutions are commonly obtained by Laplace transform techniques or by the separation of variables technique. The analytical solution to Fick's second law is described for transient diffusion into a semi-infinite medium and for transient diffusion into a finite-dimensional medium.

▶ **27.2**

TRANSIENT DIFFUSION IN A SEMI-INFINITE MEDIUM

An important case of transient mass diffusion amenable to analytical solution is the one-dimensional mass transfer of a solute into a semi-infinite, stationary liquid or solid medium where surface concentration of the solute is fixed, as illustrated in Figure 27.1. For example, we might like to describe the absorption of oxygen gas into a deep tank of still water, the "doping" of phosphorous into a silicon wafer, or the solid-phase diffusion process involved in the case-hardening of mild steel within a carburizing atmosphere. In these systems, the transferring solute does not penetrate very far into the diffusion medium,

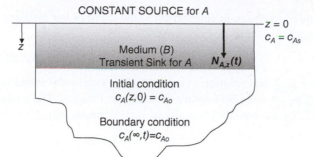

Figure 27.1 Transient diffusion in a semi-infinite medium.

and so the opposing boundary of the medium does not exert any influence on the diffusion process, which enables the semi-infinite medium assumption. This most commonly occurs at relatively short diffusion times, or in systems where the diffusion coefficient is relatively small, particularly in liquid- or solid-phase diffusion media.

Figure 27.2 depicts the concentration profiles as the time increases for a semi-infinite medium that has a uniform initial concentration of c_{Ao} and a constant surface concentration of c_{As}. The differential equation to be solved is

$$\frac{\partial c_A}{\partial t} = D_{AB} \frac{\partial^2 c_A}{\partial z^2} \tag{27-1}$$

Equation (27-1) is subject to the initial condition

$$t = 0, c_A(z,0) = c_{Ao} \text{ for all } z$$

Two boundary conditions are required. The first boundary condition at the surface is

$$\text{at } z = 0, c_A(0,t) = c_{As} \text{ for } t > 0$$

A second boundary condition in the z-direction must be specified. It is obtained by assuming that the diffusing solute penetrates only a very small distance during the finite

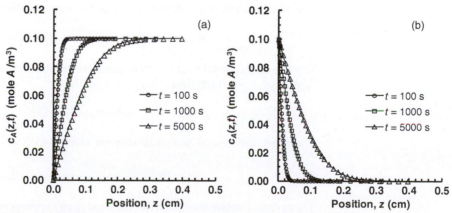

Figure 27.2 Representative concentration profiles for transient diffusion of species (A) in semi-infinite medium (B) with diffusion coefficient $D_{AB} = 1.0 \times 10^{-6}$ cm²/s. (a) Medium is a source for transfer of A with $c_{Ao} = 0.10$ mole/m³ and $c_{As} = 0$. (b) Medium is a sink for transfer of A, with $c_{Ao} = 0$, and $c_{As} = 0.10$ mole/m³.

time of exposure in comparison to the depth of the medium; this assumption provides the boundary condition

$$\text{at } z = \infty, \, c_A(\infty, t) = c_{Ao} \text{ for all } t$$

The analytical solution of equation (27-1) under the stated boundary and initial conditions can be obtained by Laplace transform techniques. The analytical solution is facilitated by making the boundary conditions homogeneous using the simple transformation

$$\theta = c_A - c_{Ao} \tag{27-4}$$

The partial differential equation and its initial and boundary conditions become

$$\frac{\partial \theta}{\partial t} = D_{AB} \frac{\partial^2 \theta}{\partial z^2} \tag{27-5}$$

with

$$\theta(z, 0) = 0$$
$$\theta(0, t) = c_{As} - c_{Ao}$$
$$\theta(\infty, t) = 0$$

The Laplace transformation of equation (27-5) with respect to time yields

$$s\bar{\theta} - 0 = D_{AB} \frac{d^2 \bar{\theta}}{dz^2}$$

which readily transforms into the ordinary differential equation

$$\frac{d^2 \bar{\theta}}{dz^2} - \frac{s}{D_{AB}} \bar{\theta} = 0$$

with the transformed boundary conditions of

$$\bar{\theta}(z = 0) = \frac{c_{As} - c_{Ao}}{s}$$

and

$$\bar{\theta}(z = \infty) = 0$$

The general analytical solution of this differential equation is

$$\bar{\theta} = A_1 \exp\left(+z\sqrt{s/D_{AB}}\right) + B_1 \exp\left(-z\sqrt{s/D_{AB}}\right)$$

The boundary condition at $z = \infty$ requires integration constant A_1 to be zero. The boundary condition at $z = 0$ requires

$$B_1 = \frac{(c_{As} - c_{Ao})}{s}$$

Therefore, the general analytical solution reduces to

$$\bar{\theta} = \frac{(c_{As} - c_{Ao})}{s} \exp\left(-z\sqrt{s/D_{AB}}\right)$$

The inverse Laplace transform can be found in any appropriate Laplace transform table. The result is

$$\theta = (c_{As} - c_{Ao}) \text{erfc}\left(\frac{z}{2\sqrt{D_{AB}t}}\right)$$

which can be expressed as the dimensionless concentration change with respect to the initial concentration of species A, c_{Ao}, as

$$\frac{c_A - c_{Ao}}{c_{As} - c_{Ao}} = \text{erfc}\left(\frac{z}{2\sqrt{D_{AB}\,t}}\right) = 1 - \text{erf}\left(\frac{z}{2\sqrt{D_{AB}\,t}}\right) \tag{27-6}$$

or with respect to the surface concentration of species A, c_{As}, as

$$\frac{c_{As} - c_A}{c_{As} - c_{Ao}} = \text{erf}\left(\frac{z}{2\sqrt{D_{AB}\,t}}\right) = \text{erf}(\phi) \tag{27-7}$$

Equation (27-7) is analogous to heat conduction in a semi-infinite wall given by equation (18-20). The argument of the error function, given by the dimensionless quantity

$$\phi = \frac{z}{2\sqrt{D_{AB}\,t}} \tag{27-8}$$

contains the independent variables of position (z) and time (t). Transient diffusion into a semi-infinite medium described by equation (27-7) works best under the condition that

$$L > \sqrt{D_{AB} \cdot t}$$

where L is the actual thickness of the medium, which is mathematically assumed to be semi-infinite. The error function is defined by

$$\text{erf}(\phi) = \frac{2}{\sqrt{\pi}} \int_0^\phi e^{-\xi^2}\, d\xi \tag{27-9}$$

where ϕ is the argument of the error function, and ξ is the dummy variable for ϕ. The error function is bounded by $\text{erf}(0) = 0$ and $\text{erf}(\infty) = 1.0$, and a plot of the error function is provided in Figure 27.3. The error function is approximated by

$$\text{erf}(\phi) = \frac{2}{\sqrt{\pi}}\left(\phi - \frac{\phi^3}{3}\right) \quad \text{if } \phi \leq 0.5$$

and

$$\text{erf}(\phi) = 1 - \frac{1}{\phi\sqrt{\pi}}e^{-\phi^2} \quad \text{if } \phi > 1.0$$

A short table of the values of erf (ϕ) is also presented in Appendix L.

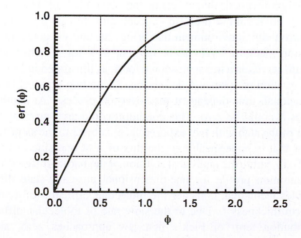

Figure 27.3 Error function.

The one-dimensional diffusion flux of species A into the semi-infinite medium at the surface of the medium $(z = 0)$ is

$$N_A|_{z=0} = -D_{AB}\frac{dc_A}{dz}\bigg|_{z=0}$$

To obtain the diffusive flux of species A into the semi-infinite medium, the derivative of c_A with respect to z is needed. By applying the chain rule of differentiation of the error function to equation (27-7), we obtain

$$\frac{dc_A}{dz}\bigg|_{z=0} = -\frac{(c_{As} - c_{Ao})}{\sqrt{\pi D_{AB}\,t}}$$

which is inserted into the flux equation to obtain

$$N_{A,z}|_{z=0} = \sqrt{\frac{D_{AB}}{\pi t}}(c_{As} - c_{Ao}) \qquad (27\text{-}10)$$

The total amount of species A transferred with time t can be determined by integrating the flux with time

$$m_A(t) - m_{Ao} = S\int_0^t N_{A,z}|_{z=0}\,dt = S\int_0^t \sqrt{\frac{D_{AB}}{\pi t}}(c_{As} - c_{Ao})dt = S\sqrt{\frac{4D_{AB}t}{\pi}}(c_{As} - c_{Ao}) \qquad (27\text{-}11)$$

where m_{Ao} is the initial amount of species A in the semi-infinite medium. If m_{Ao} is not equal to zero, then the total volume of the medium, or its depth, must be known even if the system is modeled as semi-infinite.

▶ **27.3**

TRANSIENT DIFFUSION IN A FINITE-DIMENSIONAL MEDIUM UNDER CONDITIONS OF NEGLIGIBLE SURFACE RESISTANCE

Analytical solutions for time-dependent mass-transfer processes have been obtained for simple geometrical shapes using the separation of variables technique. These bodies, initially possessing a uniform concentration of c_{Ao}, are subjected to a sudden change in the surrounding environment that brings the surface concentration to c_{As}. If $c_{As} > c_{Ao}$, the diffusion medium serves as a transient sink for A, whereas if $c_{As} < c_{Ao}$, the diffusion medium serves as a transient source for A, as illustrated in Figure 27.4. The constant surface concentration of species A defines one of the two types of boundary conditions for a one-dimensional, time-dependent mass-transfer process. At another boundary, the net flux of species A could be zero. This condition could arise from symmetry considerations at the center of the slab with two exposed faces subject to the same surface concentration, or by a barrier that is impermeable to the flux of A at one side of the slab.

To illustrate the analytical solution of the partial differential equation that describes the concentration profile for one-dimensional unsteady-state diffusion by the separation of variables technique, consider the molecular diffusion of a solute through a solid slab of uniform thickness L. Due to the slow rate of molecular diffusion within solids, the bulk contribution term of Fick's first law approaches zero, and so our solution for the

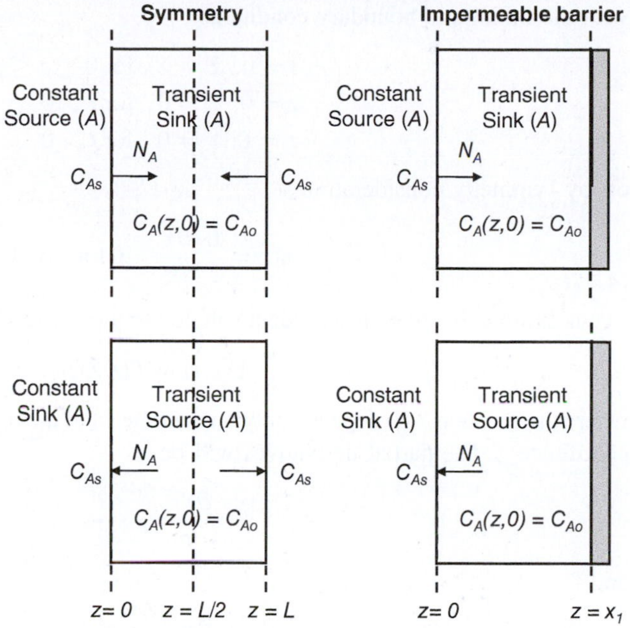

Figure 27.4 Transient diffusion in finite-dimensional medium, comparing transient source for A vs. transient sink for A.

concentration profile will satisfy the partial differential equation

$$\frac{\partial c_A}{\partial t} = D_{AB}\frac{\partial^2 c_A}{\partial z^2}$$

(27-1)

with initial and boundary conditions

$$t = 0, \ c_A = c_{Ao} \text{ for } 0 \leq z \leq L$$
$$z = 0, \ c_A = c_{As} \text{ for } t > 0$$
$$z = L, \ c_A = c_{As} \text{ for } t > 0$$

By symmetry, the following boundary condition at $z = L/2$ also holds:

$$\text{at } z = \frac{L}{2}, \frac{\partial c_A}{\partial z} = 0 \text{ for } t > 0$$

This transport process is analogous to the heating of a body under conditions of negligible surface resistance as discussed in Chapter 18. The boundary conditions above are simplified by expressing the concentrations in terms of the dimensionless concentration change Y, given by

$$Y = \frac{c_A - c_{As}}{c_{Ao} - c_{As}}$$

The partial differential equation becomes

$$\frac{\partial Y}{\partial t} = D_{AB}\frac{\partial^2 Y}{\partial z^2}$$

(27-12)

with the initial and boundary conditions

$$t = 0, \ Y = Y_o \ \text{for} \ 0 \le z \le L$$
$$z = 0, \ Y = 0 \ \ \text{for} \ t > 0$$
$$z = L, \ Y = 0 \ \ \text{for} \ t > 0$$

or, by symmetry considerations

$$\text{at} \ z = \frac{L}{2}, \frac{\partial Y}{\partial z} = 0 \ \text{for} \ t > 0$$

Let us assume that there is a product solution to the partial differential equation of the form

$$Y(z,t) = T(t)Z(z)$$

where the function $T(t)$ depends only on the time t and the function $Z(z)$ depends only on the coordinate z. The partial derivatives will be

$$\frac{\partial Y}{\partial t} = Z\frac{\partial T}{\partial t}$$

and

$$\frac{\partial^2 Y}{\partial z^2} = T\frac{\partial^2 Z}{\partial z^2}$$

Substitution into equation (27-12) yields

$$Z\frac{\partial T}{\partial t} = D_{AB}T\frac{\partial^2 Z}{\partial z^2}$$

which may be divided by $D_{AB}\,T\,Z$ to give

$$\frac{1}{D_{AB}T}\frac{\partial T}{\partial t} = \frac{1}{Z}\frac{\partial^2 Z}{\partial z^2} \tag{27-13}$$

The left-hand side of this equation depends only on time t and the right-hand side depends only on position z. If t varies, the right-hand side of the equation remains constant, and if z varies, the left-hand side remains constant. Accordingly, both sides must be independent of z and t and are equal to an arbitrary constant, $-\lambda^2$. This produces two separate ordinary differential equations, one for time t,

$$\frac{1}{D_{AB}T}\frac{dT}{dt} = -\lambda^2$$

with a general solution of

$$T(t) = C_1 e^{-D_{AB}\lambda^2 t}$$

and one for position z,

$$-\frac{1}{Z}\frac{d^2 Z}{dz^2} = -\lambda^2$$

with the general solution of

$$Z(z) = C_2\cos(\lambda z) + C_3\sin(\lambda z)$$

Substituting these two solutions into our product solution, we obtain

$$Y = T(t)Z(z) = \left[C_1'\cos(\lambda z) + C_2'\sin(\lambda z)\right]e^{-D_{AB}\lambda^2 t} \tag{27-14}$$

The constants C_1' and C_2' and the parameter λ are obtained by applying the boundary and initial conditions to the general solution. The first boundary condition, $Y = 0$ at $z = 0$, requires C_1' to be zero. The second boundary condition, $Y = 0$ at $z = L$, stipulates that $(\lambda L) = 0$, because C_2' cannot be equal to zero. Otherwise, if C_2' were equal to zero, the entire equation would be zero, yielding a trivial solution. In this context, $\sin(\lambda L)$ will be equal to zero only when

$$\lambda = \frac{n\pi}{L} \text{ for } n = 1, 2, 3, \ldots$$

To obtain C_2', the property of orthogonality must be applied, yielding the complete solution

$$Y = \frac{c_A - c_{As}}{c_{Ao} - c_{As}} = \frac{2}{L}\sum_{n=1}^{\infty}\sin\left(\frac{n\pi z}{L}\right)e^{-(n\pi/2)^2 X_D}\int_0^L Y_o\sin\left(\frac{n\pi z}{L}\right)dz \tag{27-15}$$

where L is the thickness of the sheet and X_D is the relative time ratio, given by

$$X_D = \frac{D_{AB}t}{x_1^2}$$

with x_1 being the characteristic length of $L/2$. If the sheet has a uniform initial concentration along z, the final solution is

$$\frac{c_A - c_{As}}{c_{Ao} - c_{As}} = \frac{4}{\pi}\sum_{n=1}^{\infty}\frac{1}{n}\sin\left(\frac{n\pi z}{L}\right)e^{-(n\pi/2)^2 X_D}, \qquad n = 1, 3, 5, \ldots \tag{27-16}$$

which is analogous to the heat conduction equation (18-13) obtained for the heating of a body under conditions of negligible surface resistance.

The flux at any position z is

$$N_{A,z} = -D_{AB}\frac{\partial c_A}{\partial z}$$

For the concentration profile given by equation (27-16), the flux at any position z and time t within the slab is

$$N_{A,z} = \frac{4D_{AB}}{L}(c_{As} - c_{Ao})\sum_{n=1}^{\infty}\cos\left(\frac{n\pi z}{L}\right)e^{-(n\pi/2)^2 X_D}, \qquad n = 1, 3, 5, \ldots \tag{27-17}$$

The cumulative amount of species A transferred over time can be obtained by integration of the flux at the surface ($z = 0$) with time by setting $z = 0$ in equation (27-17). At the center of the slab $z = L/2$, N_A is equal to zero. Mathematically, it is equal to zero because the cosine term in equation (27-17) vanishes at intervals of $\pi/2$; physically, it is equal to zero because the *net flux* is equal to zero at the centerline. Consequently, the following boundary condition at $z = L/2$ also holds:

$$\left.\frac{\partial c_A}{\partial z}\right|_{z=L/2} = 0$$

Recall from Section 25.3 that the mathematical boundary condition also arises when the flux at a boundary is equal to zero because of the presence of barrier that is impermeable to the transfer of diffusing species A. Consequently, equations (27-16) and (27-17) can also be used for the physical situation where a slab of thickness x_1 has an impermeable barrier at the boundary $z = x_1$ shown in Figure 27.4.

The following examples illustrate processes that are governed by one-dimensional unsteady-state diffusion of a dilute solute into semi-infinite or finite-dimensional media. The phosphorous doping of silicon wafers illustrates molecular diffusion into a semi-infinite medium, whereas the timed drug release from a spherical capsule illustrates molecular diffusion from a finite-dimensional medium. We take a little extra time at the beginning of each example to describe the interesting technology behind the process.

Example 1

In the fabrication of solid-state microelectronic devices and solar photovoltaic cells, semiconducting thin films can be made by impregnating either phosphorous or boron into a silicon wafer. This process is called *doping*. The doping of phosphorous atoms into crystalline silicon makes an *n-type semiconductor*, whereas the doping of boron atoms into crystalline silicon makes a *p-type semiconductor*. The formation of the semiconducting thin film is controlled by the molecular diffusion of the dopant atoms through crystalline-silicon matrix.

Methods to deliver phosphorous atoms to the silicon wafer surface include chemical vapor deposition and ion implantation. In one typical process, phosphorous oxychloride, $POCl_3$, which has a normal boiling point of 105 C is vaporized. The $POCl_3$ vapors are fed into a chemical vapor deposition (CVD) reactor at elevated temperature and reduced system pressure (e.g., 0.10 atm), where $POCl_3$ decomposes on the silicon surface according to the reaction

$$Si(s) + 2POCl_3(g) \rightarrow SiO_2(s) + 3Cl_2 + 2P(s)$$

A SiO_2 coating rich in molecular phosphorous (P) is formed over the crystalline-silicon surface. The molecular phosphorous then diffuses through the crystalline silicon to form the Si–P thin film. So the coating is the source for mass transfer of phosphorous, and the silicon wafer is the sink for mass transfer of phosphorous.

As one can see in Figure 27.5, the process for making Si–P thin films can be quite complex with many species diffusing and reacting simultaneously. But consider a simplified case where the P-atom concentration is constant at the interface. As the diffusion coefficient of P atoms in crystalline silicon is very low, and only a thin film of Si–P is desired, phosphorous atoms do not penetrate very far into the silicon. Therefore, the phosphorous atoms cannot "see" through the entire thickness of the wafer, and the Si solid serves as a semi-infinite sink for the diffusion process. It is desirable to predict the properties of the Si–P thin film as a function of doping conditions. The concentration profile of the doped phosphorous atoms is particularly important for controlling the electrical conductivity of the semiconducting thin film.

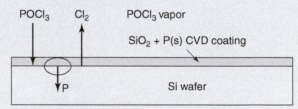

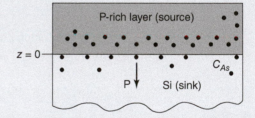

Figure 27.5 Phosphorous doping of a silicon wafer.

Consider the phosphorous doping of crystalline silicon at 1100 C, a temperature high enough to promote phosphorous diffusion. The surface concentration of phosphorous (c_{As}) silicon is 2.5×10^{20} atoms P/cm³ solid Si, which is relatively dilute, as pure solid silicon is 5.0×10^{22} atoms Si/cm³ solid. Furthermore, the phosphorous-rich coating is considered as an infinite source relative to the amount of P atoms transferred, so that c_{As} is constant. Predict the depth of the Si–P thin film after 1 h, if the target concentration is 1% of the surface value $(2.5 \times 10^{18}$ atoms P/cm³ solid Si$)$, and the concentration profile of P atoms after 1.0 h.

Based on the assumptions given in the problem statement, the partial differential equation describing the one-dimensional, unsteady-state concentration profile $c_A(z, t)$ of phosphorous (species A) in solid silicon (species B) is

$$\frac{\partial c_A}{\partial t} = D_{AB} \frac{\partial^2 c_A}{\partial z^2} \tag{27-1}$$

For a semi-infinite medium, the initial and boundary conditions are

$$t = 0, \quad c_A(z, 0) = c_{Ao} = 0 \text{ for all } z$$
$$z = 0, \quad c_A(0, t) = c_{As} = 2.5 \times 10^{20} \text{ atoms P/cm}^3 \text{ solid Si, for } t > 0$$
$$z = \infty, \quad c_A(\infty, t) = c_{Ao} = 0 \text{ for all } t$$

If the diffusion coefficient D_{AB} is a constant, then the analytical solution is

$$\frac{c_{As} - c_A}{c_{As} - c_{Ao}} = \text{erf}\left(\frac{z}{2\sqrt{D_{AB}t}}\right) = \text{erf}(\phi) \tag{27-7}$$

Note that the depth z is imbedded in ϕ, the argument of the error function. The value for $\text{erf}(\phi)$ is calculated from the dimensionless concentration change

$$\frac{c_{As} - c_A}{c_{As} - c_{Ao}} = \frac{2.5 \times 10^{20} \text{ atoms P/cm}^3 - 2.5 \times 10^{18} \text{ atoms P/cm}^3}{2.5 \times 10^{20} \text{ atoms P/cm}^3 - 0} = 0.990 = \text{erf}(\phi)$$

From the table in Appendix L, the argument of the error function at $\text{erf}(\phi) = 0.990$ is $\phi = 1.82$. From Figure 24.12, the solid-diffusion coefficient of P atoms (species A) in crystalline silicon (species B) is 6.5×10^{-13} cm^2/s at 1100 C (1373 K). The depth z can be backed out from ϕ by

$$z = 2\phi\sqrt{D_{AB}\,t} = 2 \cdot 1.822\sqrt{\left(6.5 \times 10^{-13} \frac{\text{cm}^2}{\text{s}} \frac{10^8 \,\mu\text{m}^2}{1 \text{ cm}^2}\right)\left(1\text{ h} \frac{3600 \text{ s}}{1 \text{ h}}\right)} = 1.76 \,\mu\text{m}$$

Prediction of the concentration profile after 1.0 h requires the calculation of ϕ at different values of z, followed by the calculation of $\text{erf}(\phi)$ and finally $c_A(z, t)$. Repetitive calculation of $\text{erf}(\phi)$ is best carried out with the help of a mathematics software package or a spreadsheet program. The predicted phosphorous concentration profile is compared with the data of Errana and Kakati[2] obtained under similar process conditions, as shown in Figure 27.6. It is known that the molecular-diffusion coefficient of phosphorous in crystalline silicon is a function of phosphorous concentration. The concentration-dependent diffusion coefficient creates a "dip" in the observed phosphorous concentration profile. A more detailed model of this phenomenon is provided by Middleman and Hochberg.[3]

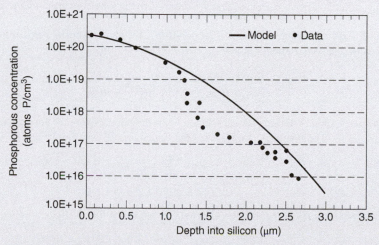

Figure 27.6 Phosphorous doping profile in silicon wafer after 1 h at 1100 C.

[2] G. Errana and D. Kakati, *Solid State Technol.*, **27**(12), 17 (1984).
[3] S. Middleman and A. K. Hochberg, *Process Engineering Analysis in Semiconductor Device Fabrication*, McGraw-Hill, New York, 1993.

Example 2

One way to deliver a timed dosage of a drug within the human body is to ingest a capsule and allow it to settle in the gastrointestinal system. Once inside the body, the capsule slowly releases the drug by a diffusion-limited process. A suitable drug carrier is a spherical bead of a nontoxic gelatinous material that can pass through the gastrointestinal system without disintegrating. A water-soluble drug (solute A) is uniformly dissolved within the gel and has an initial concentration c_{Ao}. The drug loaded within the bead is the source for mass transfer, whereas the fluid surrounding the bead is the sink for mass transfer. This is an unsteady-state process, as the source for mass transfer is contained within the diffusion control volume itself.

Consider a limiting case where the resistance to film mass transfer of the drug through the liquid boundary layer surrounding the capsule surface to the bulk surrounding the fluid is negligible. Furthermore, assume that the drug is immediately consumed or swept away once it reaches the bulk solution so that in essence the surrounding fluid is an infinite sink. In this particular limiting case, c_{As} is equal to zero, so after a long time the entire amount of drug initially loaded into the bead will be depleted. If radial symmetry is assumed, then the concentration profile is only a function of the r-direction (Figure 27.7).

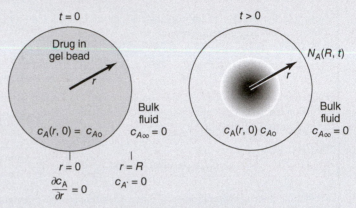

Figure 27.7 Drug release from a spherical gel bead.

It is desired to design a spherical capsule for the timed release of the drug dimenhydrinate, commonly called Dramamine, which is used to treat motion sickness. A conservative total dosage for one capsule is 10 mg, where 50% of the drug (5.0 mg) must be released to the body within 3 h. Determine the size of the bead and the initial concentration of Dramamine in the bead necessary to achieve this dosage. The diffusion coefficient of Dramamine (species A) in the gel matrix (species B) is 3.0×10^{-7} cm^2/s at a body temperature of 37 C. The solubility limit of Dramamine in the gel is 100 mg/cm^3, whereas the solubility of Dramamine in water is only 3.0 mg/cm^3.

The model must predict the amount of drug released vs. time, bead diameter, initial concentration of the drug within the bead, and the diffusion coefficient of the drug within the gel matrix. The physical system possesses spherical geometry. The development of the differential material balance model and the assumptions associated with it follow the approach presented in Section 25.4.

The general differential equation for mass transfer reduces to the following partial differential equation for the one-dimensional unsteady-state concentration profile $c_A(r,t)$:

$$\frac{\partial c_A}{\partial t} = D_{AB}\left(\frac{\partial^2 c_A}{\partial r^2} + \frac{2}{r}\frac{\partial c_A}{\partial r}\right) \tag{27-18}$$

Key assumptions include radial symmetry, dilute solution of the drug dissolved in the gel matrix, and no degradation of the drug inside the bead ($R_A = 0$). The boundary conditions at the center ($r = 0$) and the surface ($r = R$) of the bead are

$$r = 0, \frac{\partial c_A}{\partial r} = 0, \qquad t \geq 0$$

$$r = R, c_A = c_{As} = 0, t > 0$$

At the center of the bead, we note the condition of symmetry where the flux $N_A(0,t)$ is equal to zero. The initial condition is

$$t = 0, c_A = c_{Ao}, 0 \leq r \leq R$$

The analytical solution for the unsteady-state concentration profile $c_A(r,t)$ is obtained by separation-of-variables technique described earlier. The details of the analytical solution in spherical coordinates are provided by Crank.[1] The result is

$$Y = \frac{c_A - c_{Ao}}{c_{As} - c_{Ao}} = 1 + \frac{2R}{\pi r}\sum_{n=1}^{\infty}\frac{(-1)^n}{n}\sin\left(\frac{n\pi r}{R}\right)e^{-D_{AB}n^2\pi^2 t/R^2}, \quad r \neq 0, n = 1, 2, 3, \ldots \tag{27-19}$$

At the center of the spherical bead ($r = 0$), the concentration is

$$Y = \frac{c_A - c_{Ao}}{c_{As} - c_{Ao}} = 1 + 2 \sum_{n=1}^{\infty} (-1)^n e^{-D_{AB}n^2\pi^2 t/R^2}, \quad r = 0, n = 1, 2, 3, \ldots \quad (27\text{-}20)$$

Once the analytical solution for the concentration profile is known, calculations of engineering interest can be performed, including the rate of drug release and the cumulative amount of drug release over time. The rate of drug release, W_A is the product of the flux at the surface of the bead ($r = R$) and the surface area of the spherical bead:

$$W_A(t) = 4\pi R^2 N_{A,r}\big|_{r=R} = 4\pi R^2 \left(-D_{AB}\frac{\partial c_A(R,t)}{\partial r}\right) \quad (27\text{-}21)$$

It is not so difficult to differentiate the concentration profile, $c_A(r, t)$, with respect to radial coordinate r, set $r = R$, and then insert back into the above expression for $W_A(t)$ to ultimately obtain

$$W_A(t) = 8\pi R c_{Ao} D_{AB} \sum_{n=1}^{\infty} e^{-D_{AB}n^2\pi^2 t/R^2} \quad (27\text{-}22)$$

The above equation shows that the rate of drug release will decrease as time increases until all of the drug initially loaded into the bead is depleted, at which point W_A will go to zero. Initially, the drug is uniformly loaded into the bead. The initial amount of drug loaded in the bead is the product of the initial concentration and the volume of the spherical bead:

$$m_{Ao} = c_{Ao}V = c_{Ao}\frac{4}{3}\pi R^3$$

The cumulative amount of drug release from the bead over time is the integral of the drug release rate over time:

$$m_{Ao} - m_A(t) = \int_0^t W_A(t)dt$$

After some effort, the result is

$$\frac{m_A(t)}{m_{Ao}} = \frac{6}{\pi^2} \sum_{n=1}^{\infty} \frac{1}{n^2} e^{-D_{AB}n^2\pi^2 t/R^2} \quad (27\text{-}23)$$

The analytical solution is expressed as an infinite series summation that converges as n goes to infinity. In practice, convergence to a single numerical value can be attained by carrying the series summation out to only a few terms, especially if the dimensionless parameter $D_{AB}t/R^2$ is relatively large. It is a straightforward task to implement the infinite series summation on computer using a spreadsheet program.

The cumulative drug release vs. time profile is shown in Figure 27.8. The drug-release profile is affected by the dimensionless parameter $D_{AB}t/R^2$. If the diffusion coefficient D_{AB} is fixed for a given drug and gel matrix, then the critical engineering-design parameter we can manipulate is the bead radius R. As R increases, the rate of drug release decreases; if it is desired to release 50% of Dramamine from a gel bead within 3 h, a bead radius of 0.326 cm (3.26 mm) is required, as shown in Figure 27.8. Once the bead radius R is specified, the initial concentration of Dramamine required in the bead can be backed out:

$$c_{Ao} = \frac{m_{Ao}}{V} = \frac{3m_{Ao}}{4\pi R^3} = \frac{3(10\text{ mg})}{4\pi(0.326\text{ cm})^3} = \frac{68.9\text{ mg}}{\text{cm}^3}$$

In summary, a 6.52 mm diameter bead with an initial concentration of 68.9 mg A/cm^3 will dose out the required 5.0 mg of Dramamine within 3 h. The concentration profile along the r-direction at different points in time is provided in Figure 27.9. The concentration profile was calculated on a computer using a spreadsheet. The concentration profile decreases as time increases and then flattens out to zero after the drug is completely released from the bead.

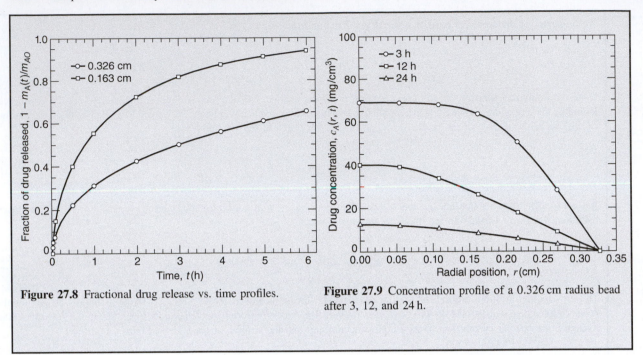

Figure 27.8 Fractional drug release vs. time profiles.

Figure 27.9 Concentration profile of a 0.326 cm radius bead after 3, 12, and 24 h.

▶ **27.4**

CONCENTRATION–TIME CHARTS FOR SIMPLE GEOMETRIC SHAPES

In our analytical solutions, the unaccomplished change, Y, was found to be a function of the relative time, X_D. The mathematical solutions, for the unsteady-state mass transfer in several simple shapes with certain restrictive boundary conditions, have been presented in a wide variety of charts to facilitate their use. Two forms of these charts are available in Appendix F.

The concentration–time charts present solutions for the flat plate, sphere, and long cylinder. As the defining partial differential equations for heat conduction and molecular diffusion are analogous, these charts may be used to solve either transport phenomenon. For molecular diffusion, the charts are in terms of four dimensionless ratios:

$$Y = \text{unaccomplished concentration change} = \frac{c_{As} - c_A}{c_{As} - c_{Ao}}$$

$$X_D = \text{relative time} = \frac{D_{AB}t}{x_1^2}$$

$$n = \text{relative position} = \frac{x}{x_1}$$

$$m = \text{relative resistance} = \frac{D_{AB}}{k_c x_1}$$

The characteristic length, x_1, is the distance from the point of symmetry. For shapes where the transport takes place from two opposite faces, x_1 is the distance from the midpoint to the surfaces from which the transfer occurs. For shapes where the transport takes place from only one of the faces, the x_1 value in the dimensionless ratios is calculated as if the thickness were twice the true value, that is, for a slab of thickness $2a$, the relative

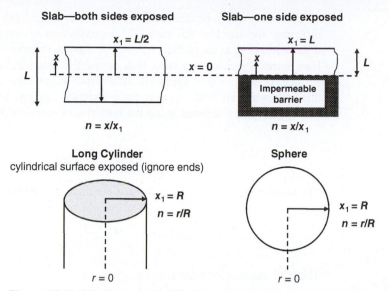

Figure 27.10 One-dimensional diffusion geometries for concentration–time charts.

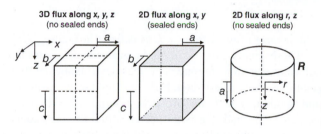

Figure 27.11 Two- and three-dimensional diffusion geometries for concentration–time charts.

time, X_D, is $D_{AB}t/4a^2$. The definitions of the geometries for use of the concentration–time charts for one-dimensional diffusion are shown in Figure 27.10.

The extension of these geometries for two and three dimensions is shown in Figure 27.11.

Transient diffusion calculations using the concentration–time charts can accommodate convective mass-transfer processes at the boundary. The physical situation is illustrated in Figure 27.12. It is important to note the diffusion coefficient, D_{AB}, refers to the diffusion of species A in diffusion medium B and not the diffusion coefficient of species A in the carrier fluid flowing over the surface (species D in Figure 27.12). The relative resistance, m, is the ratio of the convective mass-transfer resistance to the internal molecular mass-transfer resistance. For a process with negligible convective mass-transfer resistance, the convective mass-transfer coefficient, k_c, will be very large relative to D_{AB}, so

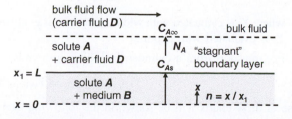

Figure 27.12 Physical situation for convection mass-transfer resistance at the boundary.

m will be assumed to be zero in processes where molecular diffusion controls the flux of the diffusing species. For this case of no convection resistance, the concentration of the diffusing species at the surface, c_{As}, will be equal to the concentration in the bulk fluid, $c_{A,\infty}$. If m is much greater than zero, then convective mass-transfer resistances exist, and the surface concentration of species A differs from its concentration in the bulk fluid phase. In this case where $m > 0$, c_{As} may not be known, and the unaccomplished concentration change, Y, is now defined using the bulk fluid concentration, $c_{A,\infty}$.

$$Y = \frac{c_{A,\infty} - c_A}{c_{A,\infty} - c_{Ao}}$$

The relative resistance m arises from a consideration of the flux at the boundary, given by

$$N_A(L, t) = -D_{AB} \left.\frac{\partial c_A}{\partial x}\right|_{x=L} = k_c(c_{As} - c_{A,\infty})$$

The Biot number for mass transfer, given by

$$Bi = \frac{k_c x_1}{D_{AB}} = \frac{1}{m}$$

is the inverse of m, with $x_1 = L$ as defined in Figure 27.12.

These charts may be used to evaluate concentration profiles for cases involving molecular mass transfer into as well as out of bodies of the specific shapes if the following conditions are satisfied. First, Fick's second law of diffusion is assumed—that is, there is no fluid motion, $v = 0$, no production term, $R_A = 0$, and constant molecular diffusion coefficient. Second, the body has an initial uniform concentration, c_{Ao}. And finally, the boundary is subjected to a new condition that remains constant with time t. Although the charts were drawn for one-dimensional transport, they may be combined to yield solutions for two- and three-dimensional transfer. In two dimensions, Y_a evaluated with the width, $x_1 = a$ and Y_b evaluated with the depth, $x_1 = b$, are combined to give

$$Y = Y_a Y_b \tag{27-24}$$

A summary of these combined solutions follows:

1. For transport from a rectangular bar with sealed ends,

$$Y_{\text{bar}} = Y_a Y_b \tag{27-25}$$

where Y_a is evaluated with width $x_1 = a$ and Y_b is evaluated with thickness $x_1 = b$.

2. For transport from a rectangular parallelepiped,

$$Y_{\text{parallelepiped}} = Y_a Y_b Y_c \tag{27-26}$$

where Y_a is evaluated with width $x_1 = a$, Y_b is evaluated with thickness $x_1 = b$, and Y_c is evaluated with depth $x_1 = c$.

3. For transport from a cylinder, including both ends,

$$Y_{\text{cylinder plus ends}} = Y_{\text{cylinder}} \, Y_a \tag{27-27}$$

The use of these charts will be shown in the following example.

Example 3

Recall the drug capsule described in Example 2. The present drug capsule consists of a 0.652-cm-diameter spherical bead (radius of 0.326 cm) containing a uniform initial concentration of 68.9 mg/cm³ Dramamine. (a) What is the residual concentration of Dramamine at the center of the spherical bead after 48 h? (b) Consider now that the capsule is now a cube 0.652 cm on a side. Recalculate part (a) above. (c) Consider that the capsule is now a cylindrical tablet of diameter 0.652 cm and thickness 0.3 cm. Recalculate part (a) above. The diffusion coefficient of Dramamine (species A) in the gel matrix (species B) is 3×10^{-7} cm²/s at a body temperature of 37 C. The three capsule configurations are presented in Figure 27.13.

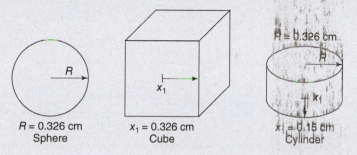

| R = 0.326 cm | x_1 = 0.326 cm | x_1 = 0.15 cm | **Figure 27.13** Three |
| Sphere | Cube | Cylinder | capsule configurations. |

This problem is readily solved using the charts given in Appendix F.

(a) For the spherical capsule, calculate the relative time (X_D), relative position (n), and relative resistance (m) based on the spherical coordinate system:

$$X_D = \frac{D_{AB}t}{R^2} = \frac{\left(3 \times 10^{-7} \, \frac{cm^2}{s}\right)\left(48 \, h \, \frac{3600 \, s}{1 \, h}\right)}{(0.326 \, cm)^2} = 0.488$$

$$n = \frac{r}{R} = \frac{0 \, cm}{0.326 \, cm} = 0 \, (\text{center of sphere})$$

$$m = \frac{D_{AB}}{k_c R} \cong 0$$

From Appendix F, Figure F.3, the value for Y, which in this case is the unaccomplished concentration change at the center of the spherical bead, is about 0.018. We can now calculate c_A:

$$Y = 0.018 = \frac{c_{As} - c_A}{c_{As} - c_{Ao}} = \frac{0 - c_A}{0 - 68.9 \, mg/cm^3}$$

The residual Dramamine concentration at the center of the bead after 48 h (c_A) is 1.24 mg/cm³.

(b) For the cube-shaped capsule, the distance from the midpoint of the cube to any of the six faces is 0.652 cm/2, or 0.326 cm. The relative time X_D is now defined as

$$X_D = \frac{D_{AB} \, t}{x_1^2} = \frac{\left(3 \times 10^{-7} \, \frac{cm^2}{s}\right)\left(48 \, h \, \frac{3600 \, s}{1 \, h}\right)}{(0.326 \, cm)^2} = 0.488$$

Values for n and m are unchanged, with $n = 0$ and $m = 0$. As all of the faces of the cube are of equal dimension, let

$$Y = Y_a Y_b Y_c = Y_a^3$$

From the Appendix F, Figure F.1, given $X_D = 0.488$, $m = 0$, and $n = 0$, Y_a is 0.4 for a flat plate of semi-thickness $x_1 = a = 0.326$ cm. Extending this value to a three-dimensional cube using the above relationship, we have

$$Y = Y_a^3 = (0.4)^3 = 0.064$$

Finally,

$$Y = 0.064 = \frac{c_{As} - c_A}{c_{As} - c_{Ao}} = \frac{0 - c_A}{0 - 68.9 \text{ mg/cm}^3}$$

with $C_A = 4.41$ mg/cm^3 after 48 h.

(c) For a cylindrical capsule with exposed ends, $R = 0.326$ cm for the radial coordinate, and $x_1 = a = 0.15$ cm for the axial coordinate. The relative times are

$$X_D = \frac{D_{AB}t}{R^2} = \frac{\left(3.0 \times 10^{-7} \dfrac{\text{cm}^2}{\text{s}}\right)\left(48 \text{ h} \dfrac{3600 \text{ s}}{1 \text{ h}}\right)}{(0.326 \text{ cm})^2} = 0.488$$

for the cylindrical dimension and

$$X_D = \frac{D_{AB}t}{x_1^2} = \frac{\left(3.0 \times 10^{-7} \dfrac{\text{cm}^2}{\text{s}}\right)\left(48 \text{ h} \dfrac{3600 \text{ s}}{1 \text{ h}}\right)}{(0.15 \text{ cm})^2} = 2.30$$

for the axial dimension. Values for n and m are unchanged, with $n = 0$ and $m = 0$. From Appendix F, Figure F.1 and Figure F.2, respectively, $Y_{\text{cylinder}} = 0.1$ for the cylindrical dimension and $Y_a = 0.006$ for the axial dimension. Therefore,

$$Y = Y_{\text{cylinder}} Y_a = (0.1)(0.006) = 0.0006$$

and finally

$$Y = 0.006 = \frac{c_{As} - c_A}{c_{As} - c_{Ao}} = \frac{0 - c_A}{0 - 68.9 \text{ mg/cm}^3}$$

with $c_A = 0.413$ mg/cm^3 after 48 h. As $Y_a \ll Y_{\text{cylinder}}$, the flux directed out of the exposed ends of the cylindrical tablet along the axial dimension dominates.

The above calculations assume that convective mass-transfer resistances associated with external fluid flow over the surface of the capsule are negligible. Problems in Chapter 30 will reconsider the drug release for unsteady-state diffusion and convection in series.

▶ **27.5**

CLOSURE

In this chapter, we have considered unsteady-state molecular diffusion. The partial differential equations that described the transient processes were obtained from the combination of Fick's equation with the general differential equation for mass transfer. Two approaches for the analytical solution to Fick's second law of diffusion were presented. Charts for solving unsteady-state mass-transfer problems were also introduced.

Convective Mass Transfer

Convective mass transfer, initially introduced in Section 24.3, involves the transport of material between a boundary surface and a moving fluid or between two immiscible moving fluids separated by a mobile interface. In this chapter, we will discuss the transfer within a single phase where the mass is exchanged between a boundary surface and a moving fluid, and the flux is related to an *individual mass-transfer convective coefficient*. In Chapter 29, we will consider the mass transfer between two contacting phases where the flux is related to an *overall mass-transfer convective coefficient*.

The rate equation for convective mass transfer, generalized in a manner analogous to Newton's law of cooling, is

$$N_A = k_c \Delta c_A \tag{24-68}$$

where the mass flux, N_A, is the molar-mass flux of species A, measured relative to fixed spatial coordinates, Δc_A is the concentration difference between the boundary surface concentration and the average concentration of the diffusing species in the moving fluid stream, and k_c is the convective mass-transfer coefficient. Recalling the discussions of the convective heat-transfer coefficient, as defined by Newton's law of cooling, we should realize that the determination of the analogous convective mass-transfer coefficient is not a simple undertaking. Both the heat- and the mass-transfer coefficients are related to the properties of the fluid, the dynamic characteristics of the flowing fluid, and the geometry of the specific system of interest.

In light of the close similarity between the convective heat- and mass-transfer equations used to define these two convective coefficients, we may expect that the analytical treatment of the heat-transfer coefficient in Chapter 19 might be applied to analyze the mass-transfer coefficient. Considerable use will also be made of developments and concepts from Chapters 9 to 13.

▶ **28.1**

FUNDAMENTAL CONSIDERATIONS IN CONVECTIVE MASS TRANSFER

The are many physical situations in which convection involving flow of a fluid over a surface drives the mass-transfer process, including evaporation of a volatile liquid surface into a flowing gas stream, sublimation of a volatile solid into a flowing gas stream, dissolution of a solid into a flowing liquid, and the transfer of gaseous reactants and products to and from a highly reactive surface.

When the mass transfer involves a solute being transferred into a moving fluid, the convective mass-transfer coefficient is defined by

$$N_A = k_c \left(c_{As} - c_{A,\infty} \right) \tag{28-1}$$

In this equation, the flux N_A represents the moles of solute A leaving the interface per unit time and unit interfacial area. The concentration of the solute A in the fluid at the interface is c_{As}, and c_A represents the concentration at some position within the bulk fluid phase. The interface or boundary surface is the source for solute A, and the moving fluid is the sink for transfer of A.

When the moving fluid encounters an immobile surface, the velocity of the fluid will decrease, creating a boundary layer. Film theory suggests that this boundary layer is conceptually localized in the slow-moving film of fluid existing near the surface. Boundary-layer theory considers that if the fluid flows parallel to the surface, then a velocity distribution is created, where the fluid velocity is zero at the surface, but approaches the bulk velocity (v_∞) at the edge of the hydrodynamic boundary layer. The hydrodynamic boundary layer develops along the direction of flow. If mass transfer is involved, a concentration profile also exists, defined by a concentration boundary layer, with concentrations c_{As} and $c_{A,\infty}$ at the surface and the bulk fluid, respectively. Film and boundary-layer theory for flow over a flat plate are contrasted in Figure 28.1.

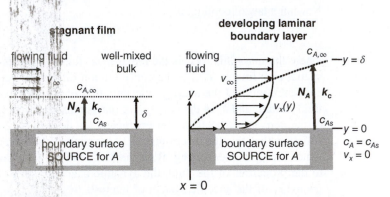

Figure 28.1 Film theory and boundary-layer theory representation of convective mass transfer for flow over a flat surface.

From our early discussions dealing with a fluid flowing past a surface, we may recall that the fluid particles immediately adjacent to the solid boundary are stationary, and a thin layer of fluid close to the surface will be in laminar flow regardless of the nature of the free stream. The mass transfer through this film will involve molecular transport, and it will play a role in any convection process. If the fluid flow is laminar, all of the transport between the surface and the moving fluid will be by molecular means. If the flow is turbulent, the mass will be transported by eddies present within the turbulent core of the stream. As in the case of heat transfer, higher mass-transfer rates are associated with turbulent conditions. The distinction between laminar and turbulent flow will be an important consideration in any convective situation.

The hydrodynamic boundary layer, analyzed in Chapter 12, plays a major role in convective mass transfer. Additionally, we shall define and analyze the concentration boundary layer, which will be vital to the analysis of the convective mass-transfer process. This layer is similar, but not necessarily equal in thickness to the thermal boundary layer that was discussed in Chapter 19.

There are four methods of evaluating convective mass-transfer coefficients that will be discussed in this chapter:

1. Dimensional analysis coupled with experiment
2. Exact laminar boundary-layer analysis
3. Approximate boundary-layer analysis
4. Analogy between momentum, energy, and mass transfer

The following example illustrates the basic features of a convective mass-transfer process associated with flow over a flat surface, where the material forming the flat surface is the source for the transferred component (species A), which must diffuse through the boundary layer to the bulk flowing fluid.

Example 1

A thin coating of a photoresist polymer dissolved in the volatile solvent methyl ethyl ketone (MEK) covers of 15 cm by 15 cm square silicon wafer to thickness of approximately 0.25 mm. The wet coating contains 30 wt% polymer solid and 70 wt% solvent, and the total mass loading of the coating is 0.50 g. The solvent can be considered to be in a liquid state. The solvent will be evaporated from the polymer surface by flowing air parallel to the surface of the wafer, as shown in Figure 28.2. The solvent vapor must transfer from the coating surface through the hydrodynamic boundary layer (dotted line in Figure 28.2) formed by

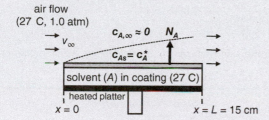

Figure 28.2 Evaporation of MEK solvent vapor from wet photoresist coating into a flowing air stream, Example 1.

the air flow over the surface. During the solvent evaporation process, a heating plate beneath the wafer maintains the coating at a uniform temperature of 27 C, and the vapor pressure exerted by the solvent at this temperature is constant at 99.4 mm Hg. The flowing air is also at 27 C and 1.0 atm total system pressure. The rate of solvent evaporation is limited by the convective mass-transfer process. The flow of air is high enough that relative to the rate of solvent transfer, the concentration of solvent vapor in the bulk gas ($c_{A,\infty}$) is very small and can be assumed to be near zero. Furthermore, any resistances associated with the molecular diffusion of the solvent through the very thin polymer coating can be neglected.

Experimental tests demonstrated that the solvent in the polymer coating evaporated after 45 s following exposure to the flowing air stream. Based on the assumptions above, what is an estimate of convective mass-transfer coefficient needed for this process?

First, the molar flux of solvent (MEK, species A, molecular weight 72 g/gmole) from the coating will be estimated from the test result. The source of mass transfer is the solvent in the polymer coating, and the sink is the surrounding air stream. The convective mass-transfer process will be assumed to be steady state so the solvent evaporation rate is constant up until all the solvent in the wet polymer coating is evaporated. A process material balance on the solvent (A) as moles A/time is

$$\begin{pmatrix} \text{rate of solvent transfer} \\ \text{from surface to flowing} \\ \text{airsteam} \end{pmatrix} = \begin{pmatrix} \text{rate of change of solvent} \\ \text{loading within polymer} \\ \text{coating} \end{pmatrix}$$

$$-N_A S = \frac{dm_A}{dt}$$

where m_A is the molar amount of solvent remaining in the coating, and S is the cross-sectional area for mass-transfer flux. Since the convective flux N_A is constant with time, integration from the initial time ($t = 0$, $m_A = m_{Ao}$) to the final time when all the solvent has evaporated ($t = t_f$, $m_A = 0$) gives

$$-N_A S \int_0^{t_f} dt = \int_{m_{Ao}}^0 dm_A$$

or

$$N_A = \frac{m_{Ao}}{t_f S} = \frac{4.86 \times 10^{-3} \text{ gmole MEK}}{(45 \text{ s})(15 \text{ cm})^2} = 4.80 \times 10^{-7} \frac{\text{gmole}}{\text{cm}^2 \cdot \text{s}}$$

where m_{Ao} is determined by

$$m_{Ao} = 0.50 \text{ g coating} \frac{0.70 \text{ g MEK}}{\text{g coating}} \cdot \frac{1 \text{ gmole MEK}}{72 \text{ g MEK}} = 4.86 \times 10^{-3} \text{ gmole MEK}$$

The convective mass-transfer flux N_A is defined by

$$N_A = k_c(c_{As} - c_{A,\infty})$$

where c_{As} is determined from the vapor pressure (P_A) exerted by the solvent, assuming the ideal gas law:

$$c_{As} = c_A^* = \frac{P_A}{RT} = \frac{\left(99.4 \text{ mm Hg} \frac{1.0 \text{ atm}}{760 \text{ mm Hg}}\right)}{\left(82.06 \frac{\text{cm}^3 \cdot \text{atm}}{\text{gmole} \cdot \text{K}}\right)(300 \text{ K})} = 5.31 \times 10^{-6} \frac{\text{gmole}}{\text{cm}^3}$$

Since the flux N_A is known, the convective mass-transfer coefficient required to achieve N_A is

$$k_c = \frac{N_A}{(c_{As} - c_{A,\infty})} = \frac{4.86 \times 10^{-7} \frac{\text{gmole}}{\text{cm}^2\text{s}}}{5.31 \times 10^{-6} \frac{\text{gmole}}{\text{cm}^3} - 0} = 0.090 \text{ cm/s}$$

We will see in this chapter that k_c will be a function of the molecular diffusion coefficient in the fluid phase, the properties of the flowing fluid, and the fluid velocity.

► **28.2**

SIGNIFICANT PARAMETERS IN CONVECTIVE MASS TRANSFER

Dimensionless parameters are often used to correlate convective transport data. In momentum transfer, we encountered the Reynolds and the Euler numbers. In the correlation of convective heat-transfer data, the Prandtl and the Nusselt numbers were important. Some of the same parameters, along with some newly defined dimensionless ratios, will be useful in the correlation of convective mass-transfer data. In this section, we shall consider the physical interpretation of three such ratios.

The molecular diffusivities of the three transport phenomena have been defined as momentum diffusivity, given by

$$\nu = \frac{\mu}{\rho}$$

thermal diffusivity, given by

$$\alpha = \frac{k}{\rho C_p}$$

and mass diffusivity, given by D_{AB}.

As we have noted earlier, each of the diffusivities has the dimensions of L^2/t; thus, a ratio of any two of these must be dimensionless. The ratio of the molecular diffusivity of momentum to the molecular diffusivity of mass is designated the *Schmidt number*:

$$Sc = \frac{\text{momentum diffusivity}}{\text{mass diffusivity}} = \frac{\nu}{D_{AB}} = \frac{\mu}{\rho D_{AB}} \qquad (28\text{-}2)$$

The Schmidt number plays a role in convective mass transfer analogous to that of the Prandtl number in convective heat transfer, which is defined as

$$Pr = \frac{\text{momentum diffusivity}}{\text{thermal diffusivity}} = \frac{\nu}{\alpha} = \frac{\mu C_p}{k}$$

The ratio of the thermal diffusivity to the molecular diffusivity of mass is designated the *Lewis number*:

$$Le = \frac{\text{thermal diffusivity}}{\text{mass diffusivity}} = \frac{\alpha}{D_{AB}} = \frac{k}{\rho C_p D_{AB}} \qquad (28\text{-}3)$$

The Lewis number is encountered when a process involves the simultaneous convective transfer of mass and energy. The Schmidt and the Lewis numbers are observed to be combinations of fluid properties; thus, each number may be treated as a property of the diffusing system.

Consider the mass transfer of solute A from a solid to a fluid flowing past the surface of the solid. The concentration profile is shown in Figure 28.3. For such a case, the mass transfer between the surface and the fluid may be written as

$$N_A = k_c\left(c_{A,s} - c_{A,\infty}\right) \qquad (28\text{-}1)$$

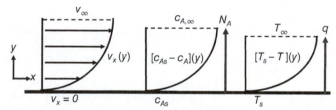

Figure 28.3 Velocity, concentration, and temperature profiles for a fluid flowing past a solid surface.

Since the mass transfer at the surface is by molecular diffusion, the mass transfer may also be described by

$$N_A = -D_{AB}\frac{dc_A}{dy}\bigg|_{y=0}$$

When the boundary concentration, c_{As}, is constant, this equation can be written as

$$N_A = -D_{AB}\frac{d(c_A - c_{As})}{dy}\bigg|_{y=0} \qquad (28\text{-}4)$$

Equations (28-4) and (28-1) may be equated, as they define the same flux of component A leaving the surface and entering the fluid. This gives the relation

$$k_c\left(c_{As} - c_{A,\infty}\right) = -D_{AB}\frac{d(c_A - c_{As})}{dy}\bigg|_{y=0} \qquad (28\text{-}5)$$

which may be rearranged into the following form:

$$\frac{k_c}{D_{AB}} = \frac{-d(c_A - c_{As})/dy|_{y=0}}{\left(c_{As} - c_{A,\infty}\right)}$$

Multiplying both sides of this equation by a characteristic length, L, we obtain the following dimensionless expression:

$$\frac{k_c L}{D_{AB}} = \frac{-d(c_A - c_{As})/dy|_{y=0}}{\left(c_{As} - c_{A,\infty}\right)/L} \tag{28-6}$$

The right-hand side of equation (28-6) is the ratio of the concentration gradient at the surface to the overall concentration gradient. Accordingly, it may be considered a ratio of the molecular mass-transport resistance to the convective mass-transfer resistance of the fluid. This dimensionless ratio is referred to as the *Sherwood number*, Sh, and is defined as

$$\mathrm{Sh} = \frac{k_c L}{D_{AB}} \tag{28-7}$$

Since the development of equation (28-7) parallels the development of equation (19-5) for the Nusselt number encountered in convective heat transfer, the Sherwood number has also been referred to as the *mass-transfer Nusselt number*, Nu_{AB}.

These three parameters—Sc, Sh, and Le—are encountered in the analyses of convective mass transfer in the following sections. The Schmidt number is evaluated in the following example to reveal the relative magnitude of its value in the gas and liquid phases.

Example 2

Compare the Schmidt number (Sc) for methanol vapor in air at 308 K (35 C) and 1.0 atm, and methanol dissolved in liquid water at 308 K.

Methanol (MeOH) is species A. From Appendix J, for MeOH in air at 1.0 atm, $D_{A-air} = 0.162$ cm^2/s at 298 K. At 308 K

$$D_{A-air}(P, T) = D_{A-air}(P, T_{ref})\left(\frac{T}{T_{ref}}\right)^{1.5} = 0.162 \frac{\mathrm{cm}^2}{\mathrm{s}}\left(\frac{308\ \mathrm{K}}{298\ \mathrm{K}}\right)^{1.5} = 0.170 \frac{\mathrm{cm}^2}{\mathrm{s}}$$

From Appendix I, for air at 1.0 atm and 308 K, the kinematic viscosity (ν) is 0.164 cm^2/s. The Schmidt number for MeOH vapor in air is

$$\mathrm{Sc} = \frac{\nu_{air}}{D_{A-air}} = \frac{0.164\ \mathrm{cm}^2/\mathrm{s}}{0.170\ \mathrm{cm}^2/\mathrm{s}} = 0.966$$

Now compare this result for the Schmidt number of MeOH in liquid water. From Appendix I, for liquid water at 308 K, the kinetic viscosity (ν) is 9.12×10^{-3} cm^2/s, and the viscosity (μ) is 9.09×10^{-3} g/cm · s (0.909 cP). For MeOH, the molar volume at boiling point is $V_A = 59.2$ cm^3/gmole. The liquid-phase diffusion coefficient is estimated by the Hayduk–Laudie correlation (Chapter 24), given by

$$D_{A-water} = 13.26 \times 10^{-5}\mu_{water}^{-1.14}V_A^{-0.589} = 13.26 \times 10^{-5}(0.909)^{-1.14}(59.2)^{-0.589} = 1.34 \times 10^{-5}\ \mathrm{cm/s}$$

The Schmidt number is

$$\mathrm{Sc} = \frac{\nu_{water}}{D_{A-water}} = \frac{9.12 \times 10^{-3}\ \mathrm{cm}^2/\mathrm{s}}{1.34 \times 10^{-5}\ \mathrm{cm/s}} = 683$$

Values of the Schmidt number for liquids are typically much higher than those for gases.

▶ **28.3**

DIMENSIONAL ANALYSIS OF CONVECTIVE MASS TRANSFER

Dimensionless analysis predicts the various dimensionless parameters that are helpful in correlating experimental data. There are two important mass-transfer processes that we will consider, mass transfer into a stream flowing under forced convection, and mass transfer into a phase that is moving under natural-convection conditions.

Transfer into a Stream Flowing Under Forced Convection

Consider the transfer of mass from the walls of a circular conduit to a fluid flowing through the conduit. The transfer is a result of the concentration driving force, $c_{As} - c_A$. The important variables, their symbols, and their dimensional representations are listed here.

Variable	Symbol	Dimension
Tube diameter	D	L
Fluid density	ρ	M/L^3
Fluid viscosity	μ	M/Lt
Fluid velocity	v	L/t
Fluid diffusivity	D_{AB}	L^2/t
Mass-transfer coefficient	k_c	L/t

The above variables include terms descriptive of the system of geometry, the fluid velocity, the fluid properties, and the quantity that is of primary interest, k_c.

By the Buckingham method of grouping the variables as presented in Chapter 11, we can determine that there will be three dimensionless groups. With D_{AB}, ρ, and D as the core variables, the three π groups to be formed are

$$\pi_1 = D_{AB}^a \, \rho^b \, D^c \, k_c$$

$$\pi_2 = D_{AB}^d \, \rho^e \, D^f \, v$$

and

$$\pi_3 = D_{AB}^g \, \rho^h \, D^i \, \mu$$

Writing π_1 in dimensional form, we have

$$1 = \left(\frac{L^2}{t}\right)^a \left(\frac{M}{L^3}\right)^b (L)^c \left(\frac{L}{t}\right)$$

By equating the exponents of the fundamental dimensions on both sides of the equation, we have, for length L,

$$0 = 2a - 3b + c + 1$$

time t:

$$0 = -a - 1$$

and mass M:

$$0 = b$$

The solution of these equations for the three unknown exponents yields $a = -1, b = 0$, and $c = 1$.

Thus,

$$\pi_1 = k_c \, L/D_{AB}$$

which is the Sherwood number, Sh. The two other π groups could be determined in the same manner, yielding

$$\pi_2 = \frac{Dv}{D_{AB}}$$

and

$$\pi_3 = \frac{\mu}{\rho D_{AB}} = Sc$$

the Schmidt number. Dividing π_2 by π_3, we obtain

$$\frac{\pi_2}{\pi_3} = \left(\frac{Dv}{D_{AB}}\right)\left(\frac{D_{AB}\rho}{\mu}\right) = \frac{Dv\,\rho}{\mu} = Re$$

which is the familiar Reynolds number. The result of the dimensional analysis of forced-convection mass transfer in a circular conduit indicates that a correlating relation could be of the form

$$Sh = f(Re, Sc) \tag{28-8}$$

which is analogous to the heat-transfer correlation

$$Nu = f(Re, Pr) \tag{19-7}$$

Transfer into a Phase Whose Motion Is Due to Natural Convection

Natural convection currents will develop if any variation in density within a liquid or gas phase exists. The density variation may be due to temperature differences or relatively large concentration differences.

In the case of natural convection involving mass transfer from a vertical plane wall to an adjacent fluid, the variables will differ from those used in the forced-convection analysis. The important variables, their symbols, and dimensional representations are listed below.

Variable	Symbol	Dimension
Characteristic length	L	L
Fluid diffusivity	D_{AB}	L^2/t
Fluid density	ρ	M/L^3
Fluid viscosity	μ	M/LT
Buoyant force	$\Delta\rho\,g$	M/L^2t^2
Mass-transfer coefficient	k_c	L/t

By the Buckingham theorem, there will be three dimensionless groups. With D_{AB}, L, and μ as the core variables, the three π groups to be formed are

$$\pi_1 = D_{AB}^a \, L^b \, \mu^c \, k_c$$
$$\pi_2 = D_{AB}^d \, L^e \, \mu^f \, \rho$$
$$\pi_3 = D_{AB}^g \, L^h \, \mu^i \, \Delta\rho$$

Solving for the three π groups, we obtain

$$\pi_1 = \frac{k_c L}{D_{AB}} = \text{Sh}$$

which is the Sherwood number,

$$\pi_2 = \frac{\rho D_{AB}}{\mu} = \frac{1}{\text{Sc}}$$

which is the reciprocal of the Schmidt number, and

$$\pi_3 = \frac{L^3 \Delta \rho g}{\mu D_{AB}}$$

Multiplying π_2 and π_3, we obtain the Grashof number for natural convection

$$\pi_2 \pi_3 = \left(\frac{\rho D_{AB}}{\mu} \right) \frac{L^3 g \Delta \rho}{\mu D_{AB}} = \frac{L^3 \rho g \Delta \rho}{\mu^2} = \text{Gr}$$

The result of the dimensional analysis of natural-convection mass transfer suggests a correlating relation of the form

$$\text{Sh} = f(\text{Gr}, \text{Sc}) \tag{28-9}$$

For both forced and natural convection, relations have been obtained by dimensional analysis suggest that a correlation of experimental data may be in terms of three variables instead of the original six. This reduction in variables has aided investigators who have suggested correlations of these forms to provide many of the empirical equations reported in Chapter 30.

► 28.4

EXACT ANALYSIS OF THE LAMINAR CONCENTRATION BOUNDARY LAYER

Exact analysis of the laminar boundary layer for flow over a flat surface is used to establish a fundamental basis for the convective mass-transfer coefficient, k_c, and to specify the dimensionless groups and associated variables that k_c depends upon. This analysis begins with hydrodynamic considerations, and then couples momentum transfer with mass transfer. Specifically, Blasius developed an exact solution for the hydrodynamic boundary layer for laminar flow parallel to a flat surface. This solution was discussed in Section 12.5. An extension of the Blasius solution was made in Section 19.4 to explain convective heat transfer. In an exactly analogous manner, we shall also extend the Blasius solution to include convective mass transfer for the same geometry and laminar flow.

The boundary-layer equations considered in the steady-state momentum transfer included the two-dimensional, incompressible continuity equation is given by

$$\frac{\partial v_x}{\partial x} + \frac{\partial v_y}{\partial y} = 0 \tag{12-11a}$$

The equation of motion in the x direction for constant v and pressure is given by

$$v_x \frac{\partial v_x}{\partial x} + v_y \frac{\partial v_x}{\partial y} = \nu \frac{\partial^2 v_x}{\partial y^2} \tag{12-11b}$$

For the thermal boundary layer, the equation describing the energy transfer in a steady, incompressible, two-dimensional, isobaric flow with constant thermal diffusivity was

$$v_x \frac{\partial T}{\partial x} + v_y \frac{\partial T}{\partial y} = \alpha \frac{\partial^2 T}{\partial y^2} \tag{19-15}$$

An analogous differential equation applies to mass transfer within a concentration boundary layer if no production of the diffusing component occurs and if the $\partial^2 c_A/\partial x^2$ is much smaller in magnitude than $\partial^2 c_A/\partial y^2$. This equation, written for steady, incompressible two-dimensional flow with constant mass diffusivity, is

$$v_x \frac{\partial c_A}{\partial x} + v_y \frac{\partial c_A}{\partial y} = D_{AB} \frac{\partial^2 c_A}{\partial y^2} \tag{28-10}$$

The developing boundary layers are shown schematically in Figure 28.1 and Figure 28.3. For momentum transfer, the boundary conditions on the fluid velocity are

$$y = 0, \ \frac{v_x}{v_\infty} = 0 \quad \text{and} \quad y = \infty, \ \frac{v_x}{v_\infty} = 1$$

Since the velocity in the x direction at the wall ($v_{x,s}$) is zero, these boundary conditions are rewritten as

$$y = 0, \ \frac{v_x - v_{x,s}}{v_\infty - v_{x,s}} = 0 \quad \text{and} \quad y = \infty, \ \frac{v_x - v_{x,s}}{v_\infty - v_{x,s}} = 1$$

Recall that for heat transfer, the boundary conditions on the fluid temperature are

$$y = 0, \ \frac{T - T_s}{T_\infty - T_s} = 0 \quad \text{and} \quad y = \infty, \ \frac{T - T_s}{T_\infty - T_s} = 1$$

Similarly, for mass transfer, the boundary conditions on the concentration of solute A in the fluid mixture are

$$y = 0, \ \frac{c_A - c_{As}}{c_{A,\infty} - c_{As}} = 0 \quad \text{and} \quad y = \infty, \ \frac{c_A - c_{As}}{c_{A,\infty} - c_{As}} = 1$$

Equations (12-11a), (19-15), and (28-10) can be written in terms of dimensionless velocity, temperature, and concentration ratios. For dimensionless velocity,

$$v_x \frac{\partial \left(\frac{v_x - v_{x,s}}{v_\infty - v_{x,s}} \right)}{\partial x} + v_y \frac{\partial \left(\frac{v_x - v_{x,s}}{v_\infty - v_{x,s}} \right)}{\partial y} = \nu \frac{\partial^2 \left(\frac{v_x - v_{x,s}}{v_\infty - v_{x,s}} \right)}{\partial y^2}$$

or if

$$V = \left(\frac{v_x - v_{x,s}}{v_\infty - v_{x,s}} \right)$$

then

$$v_x \frac{\partial V}{\partial x} + v_y \frac{\partial V}{\partial y} = \nu \frac{\partial^2 V}{\partial y^2} \tag{28-11}$$

with the boundary conditions given by $y = 0$, $V = 0$, and $y = \infty$, $V = 1$. Similarly, for dimensionless temperature,

$$\theta = \frac{T - T_s}{T_\infty - T_s}$$

$$v_x \frac{\partial \theta}{\partial x} + v_y \frac{\partial \theta}{\partial y} = \alpha \frac{\partial^2 \theta}{\partial y^2} \tag{28-12}$$

with the boundary conditions given by $y = 0$, $\theta = 0$, and $y = \infty$, $\theta = 1$. Finally, for dimensionless concentration

$$\hat{C} = \frac{c_A - c_{As}}{c_{A,\infty} - c_{As}}$$

$$v_x \frac{\partial \hat{C}}{\partial x} + v_y \frac{\partial \hat{C}}{\partial y} = D_{AB} \frac{\partial^2 \hat{C}}{\partial y^2} \qquad (28\text{-}13)$$

with the boundary conditions given by $y = 0$, $\hat{C} = 0$, and $y = \infty$, $\hat{C} = 1$.

The similarity in the three differential equations (28-11), (28-12), and (28-13) and their associated boundary conditions suggests that similar solutions should be obtained for the three transfer phenomena. In Chapter 19, the Blasius solution was successfully applied to explain convective heat transfer when the Prandtl number, defined as the ratio of momentum to thermal diffusivity, was equal to 1. The same type of solution should also describe convective mass transfer when the Schmidt number, defined as ratio of the momentum to mass diffusivity, is also equal to 1. Using the nomenclature defined in Chapter 12, let

$$f' = 2\frac{v_x}{v_\infty} = 2\frac{v_x - v_{x,s}}{v_\infty - v_{x,s}} = 2\frac{c_A - c_{As}}{c_{A,\infty} - c_{As}} \qquad (28\text{-}14)$$

and

$$\eta = \frac{y}{2}\sqrt{\frac{v_\infty}{\nu x}} = \frac{y}{2x}\sqrt{\frac{x\,v_\infty}{\nu}} = \frac{y}{2x}\sqrt{Re_x} \qquad (28\text{-}15)$$

Recall that the Blasius solution to the momentum boundary layer at the surface ($y = 0$), given by

$$\frac{df'}{d\eta} = f''(0) = \frac{d[2(v_x/v_\infty)]|_{y=0}}{d\left[(y/2x)\sqrt{Re_x}\right]\big|_{y=0}} = 1.328 \qquad (19\text{-}18)$$

suggests an analogous solution for the concentration boundary layer:

$$\frac{df'}{d\eta} = f''(0) = 1.328 = \frac{\dfrac{d}{dy}\left(2\dfrac{c_A - c_{As}}{c_{A,\infty} - c_{As}}\right)\Big|_{y=0}}{\dfrac{d}{dy}\left(\dfrac{y}{2x}\sqrt{Re_x}\right)\Big|_{y=0}} = \frac{\left(\dfrac{2}{c_{A,\infty} - c_{As}}\right)\left(\dfrac{dc_A}{dy}\right)\Big|_{y=0}}{\left(\dfrac{1}{2x}\sqrt{Re_x}\right)\Big|_{y=0}} \qquad (28\text{-}16)$$

Solving for the concentration gradient at $y = 0$, we obtain

$$\frac{dc_A}{dy}\bigg|_{y=0} = (c_{A,\infty} - c_{As})\left(\frac{0.332}{x}Re_x^{1/2}\right) \qquad (28\text{-}17)$$

It is important to recall that the Blasius solution does not involve a velocity term in the y direction at the surface. Accordingly, equation (28-17) involves the important assumption that the rate at which mass enters or leaves the boundary layer at the surface is so small, i.e., $v_y \approx 0$, that it does not alter the velocity profile predicted by the Blasius solution.

When the velocity in the y-direction at the surface ($v_{y,s}$) is essentially zero, the bulk contribution term in Fick's equation for the mass flux in the y-direction is also zero. Consequently, the mass transfer from the flat surface into the laminar boundary layer is described by

$$N_{A,y} = -D_{AB}\frac{\partial c_A}{\partial y}\bigg|_{y=0} \qquad (28\text{-}18)$$

Substitution of equation (28-17) into equation (28-18) gives

$$N_{A,y} = D_{AB}\left(\frac{0.332\,\text{Re}_x^{1/2}}{x}\right)(c_{As} - c_{A,\infty}) \tag{28-19}$$

The mass-transfer flux of the diffusing component was also defined in terms of the mass-transfer coefficient by

$$N_{A,y} = k_c(c_{As} - c_{A,\infty}) \tag{28-1}$$

Equations (28-19) and (28-1) are equated to give

$$k_{c,x} = \frac{D_{AB}}{x}\left[0.332\,\text{Re}_x^{1/2}\right] \tag{28-20}$$

or

$$\frac{k_{c,x}\,x}{D_{AB}} = \text{Sh}_x = 0.332\,\text{Re}_x^{1/2} \tag{28-21}$$

In equation (28-20), the mass-transfer coefficient is a function of position x, and so is designated $k_{c,x}$, the *local* mass-transfer coefficient. Equation (28-21) is restricted to systems having a Schmidt number equal to 1, and for dilute systems where the mass-transfer rate between the flat surface and boundary layer is small relative to the rate of fluid flow.

In most physical operations involving mass transfer, the low mass-transfer Blasius type of solution is used to define the transfer into the laminar boundary layer. For a fluid with a Schmidt number other than 1, the concentration boundary layer is related to the hydrodynamic boundary layer by

$$\frac{\delta}{\delta_c} = \text{Sc}^{1/3} \tag{28-22}$$

where δ is the thickness of the hydrodynamic boundary layer and δ_c is the thickness of the concentration boundary layer. Thus, the Blasius η term in equation (28-15) must be multiplied by the term $\text{Sc}^{1/3}$. A plot of the dimensionless concentration vs. $\eta\text{Sc}^{1/3}$, for no velocity contribution in the y-direction ($v_{y,s} = 0$) is shown in Figure 28.4. Consequently, at $y = 0$, the concentration gradient is

$$\left.\frac{\partial c_A}{\partial y}\right|_{y=0} = (c_{A,\infty} - c_{As})\left(\frac{0.332}{x}\text{Re}_x^{1/2}\text{Sc}^{1/3}\right) \tag{28-23}$$

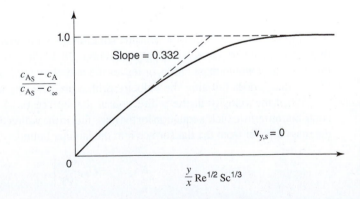

Figure 28.4 Concentration variation for laminar flow over a flat plate.

Equation (28-23), when combined with equations (28-1) and (28-18), yields

$$\frac{k_{c,x} x}{D_{AB}} = \text{Sh}_x = 0.332\,\text{Re}_x^{1/2}\text{Sc}^{1/3} \tag{28-24}$$

Equation (28-24) is the dimensionless mass-transfer relationship for the local Sherwood number (Sh_x) at position x for laminar flow over a flat plate where $\text{Re}_x < 2.0 \times 10^5$. Hartnett and Eckert[1] consider the physical situations and mass-transfer analysis where there is a finite velocity imposed in the y-direction emanating from the surface—i.e., $v_{y,s} > 0$—but this analysis will not be considered here.

For laminar flow over a flat plate, recall that from Chapter 12, the Blasius solution for the thickness of the laminar hydrodynamic boundary layer is

$$\frac{\delta}{x} = \frac{5}{\sqrt{\text{Re}_x}} \tag{12-28}$$

Consequently, by combining equations (28-22) and (12-28), the concentration boundary-layer thickness for laminar flow over a flat plate can be estimated by

$$\delta_c = 5\,\text{Sc}^{-1/3}\sqrt{\frac{\nu x}{v_\infty}} \tag{28-25}$$

The Schmidt number affects the mass-transfer process by setting the thickness of the concentration boundary layer relative to the hydrodynamic boundary layer, as shown in Figure 28.5. If $\text{Sc} < 1$, then the concentration profile continues to develop beyond the

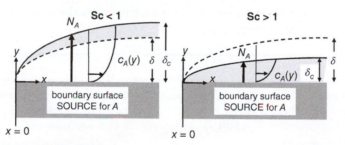

Figure 28.5 Location of concentration boundary layer (δ_c) relative to hydrodynamic boundary layer (δ) for $\text{Sc} < 1$, and $\text{Sc} > 1$.

hydrodynamic boundary layer. If $\text{Sc} = 1$, then the concentration and boundary layers are of the same thickness. If $\text{Sc} > 1$, then the concentration boundary layer resides within the hydrodynamic boundary layer. This latter case is common for convective mass transfer into flowing liquids, where the Schmidt number is typically much higher than 1.

Equations (28-24) and (28-25) show that both the local mass-transfer coefficient and concentration boundary layer are a function of position x down the length of the flat surface, with $k_{c,x} \propto x^{-1/2}$ and $\delta_c \propto x^{1/2}$. Often, it is necessary to know the average mass-transfer coefficient for fluid flow over a flat surface of length L. The integral average mass-transfer coefficient is defined by

$$k_c = \frac{\int_0^L k_{c,x}(x)\,dx}{\int_0^L dx} = \frac{\int_0^L \left(\frac{D_{AB}}{x}0.332\,\text{Re}_x^{1/2}\text{Sc}^{1/3}\right)dx}{L} = \frac{\int_0^L \left(\frac{D_{AB}}{x}0.332\left(\frac{v_\infty x}{\nu}\right)^{1/2}\text{Sc}^{1/3}\right)dx}{L} \tag{28-26}$$

[1] J. P. Hartnett and E. R. G. Eckert, *Trans. ASME*, **13**, 247 (1957).

Further simplification to collect terms, followed by integration with respect to x gives

$$k_c = \frac{0.332\, D_{AB} Sc^{1/3} \left(\dfrac{v_\infty}{\nu}\right)^{1/2} \displaystyle\int_0^L x^{-1/2}\,dx}{L} = 0.664 \frac{D_{AB}}{L} \left(\frac{v_\infty L}{\nu}\right)^{1/2} Sc^{1/3} \qquad (28\text{-}27)$$

The Reynolds number for flow over a flat plate of length L is

$$Re_L = \frac{v_\infty L}{\nu} \qquad (28\text{-}28)$$

Consequently, rearrangement of equation (28-27) gives

$$\frac{k_c L}{D_{AB}} = Sh_L = 0.664\, Re_L^{1/2} Sc^{1/3} \qquad (28\text{-}29)$$

Equation (28-29) is the dimensionless mass-transfer relationship for the average Sherwood number (Sh_L) for a flat plate of length L where $Re_L < 2.0 \times 10^5$.

In comparing equation (28-24) to equation (28-29), it is noted that at $x = L$, the local Sherwood number is twice the average Sherwood number:

$$Sh_L = 2 Sh_x|_{x=L} \qquad (28\text{-}30)$$

Equations (28-24) and (28-29) have been experimentally verified.[2] It is interesting to note that this analysis produced results of the same form predicted in Section 28.3 by dimensional analysis for forced-convection mass transfer:

$$Sh = f(Re, Sc) \qquad (28\text{-}8)$$

The following example illustrates how convective mass transfer can be integrated into other chemical processes.

Example 3

A chemical vapor deposition (CVD) process is used to deposit pure silicon thin films on wafers for electronic device applications. One way to deposit pure silicon onto a surface is by the heterogeneous surface reaction:

$$SiHCl_3(g) + H_2(g) \rightarrow Si(s) + 3\,HCl(g)$$

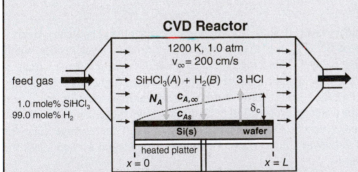

CVD Reactor

Figure 28.6 Chemical vapor deposition of solid silicon onto a wafer surface by a flowing gas stream mixture of trichlorosilane and H_2, Example 3.

In the present process, 1.0 mole% trichlorosilane ($SiHCl_3$) vapor (species A) diluted in H_2 gas (species B) is fed to the CVD reactor shown in Figure 28.6 to establish a bulk velocity of 200 cm/s. The trichlorosilane is reduced by the H_2 gas to elemental silicon solid (Si), which is deposited as a thin film on a 15 cm by 15 cm wafer. Although the reaction produces HCl gas, it is significantly diluted by the H_2 gas, and so the H_2 gas can also be considered as the carrier gas. The flow over the silicon wafer is approximated as flow over a flat plate. The reaction process is maintained at 1.0 atm and 1200 K. The reactor contains elaborate safety systems to handle the toxic and flammable gases. At 1200 K, the surface reaction constant for the first-order decomposition reaction with respect to $SiHCl_3$ vapor concentration is $k_s = 0.83$ cm/s, with the surface reaction rate law defined as $R'_{A,s} = -k_s c_{As}$, based on the studies of Stein.[3]

---

[2] W. J. Christian and S. P. Kezios, *A.I.Ch.E.J.* **5**, 61 (1959).
[3] A. M. Stein, J. *Electrochem.* Soc., **111**, 483 (1964).

Based on the above information, determine the average rate of solid silicon deposition onto the wafer surface.

The trichlorosilane must diffuse through the boundary layer created by flow of the feed gas over the wafer surface. At the wafer surface, the trichlorosilane will decompose to solid Si. The feed gas serves as the source for mass transfer for of $SiHCl_3$, and the reaction at the Si surface serves as the sink. At steady state, the rate of $SiHCl_3$ flux to the Si surface will be balanced by the rate of reaction at the surface:

$$N_A = k_c(c_{A,\infty} - c_{As})$$

and

$$N_A = -R'_{A,s}$$

or

$$k_c(c_{A,\infty} - c_{As}) = k_s c_{As}$$

The concentration of trichlorosilane at the wafer surface, c_{As}, is unknown. Solving for c_{As}, we obtain

$$c_{As} = c_{A,\infty}\frac{k_c}{k_c + k_s}$$

By substitution of this relationship into the flux equation, c_{As} is eliminated, and the flux equation is now

$$N_A = c_{A,\infty}\frac{k_c k_s}{k_c + k_s} \tag{28-31}$$

It is noted that as $k_s \gg k_c$, $c_{As} \rightarrow 0$, and the process becomes convection mass-transfer rate limited.

The average convective mass-transfer coefficient for flux of $SiHCl_3$ across the boundary layer, k_c, will now be estimated. At 1200 K and 1.0 atm, the gas-phase diffusion coefficient of trichlorosilane in H_2 (D_{AB}) is 4.20 cm^2/s, as determined by the Fuller–Shettler–Giddings correlation. The kinematic viscosity of the H_2 carrier gas (ν_B) is 11.098 cm^2/s (Appendix I). Relevant dimensionless groups—Re, Sc, and Sh—must now be calculated. First, the Reynolds number (Re) is estimated for flow over a flat plate to determine the regime for flow:

$$\mathrm{Re}_L = \frac{v_\infty L}{\nu_B} = \frac{(200\ \mathrm{cm/s})(15.0\ \mathrm{cm})}{11.098\ \mathrm{cm^2/s}} = 270.3$$

The flow is laminar across the Si wafer, since $\mathrm{Re}_L < 2.0 \times 10^5$. The Schmidt (Sc) is

$$\mathrm{Sc} = \frac{\nu}{D_{AB}} \doteq \frac{\nu_B}{D_{AB}} = \frac{11.098\ \mathrm{cm^2/s}}{4.20\ \mathrm{cm^2/s}} = 2.64$$

For laminar flow over a flat plate, the average Sherwood number (Sh) is

$$\mathrm{Sh} = \frac{k_c L}{D_{AB}} = 0.664\,\mathrm{Re}^{1/2}\,\mathrm{Sc}^{1/3} = 0.664(270.3)^{1/2}(2.64)^{1/3} = 15.1$$

The value for k_c is obtained from the Sherwood number:

$$k_c = \frac{\mathrm{Sh}\,D_{AB}}{L} = \frac{(15.1)(4.20\ \mathrm{cm^2/s})}{(15.0\ \mathrm{cm})} = 4.23\ \mathrm{cm/s}$$

With k_c known, the molar flux of $SiHCl_3$ is determined. The consumption of $SiHCl_3$ to Si deposition relative to the molar flow rate of $SiHCl_3$ in the feed gas is assumed to be small. Therefore, the bulk concentration of $SiHCl_3$ is considered to be constant, and is estimated by

$$c_{A,\infty} = y_{A,\infty}c = y_{A,\infty}\frac{P}{RT} = 0.01\frac{(1.0\ \mathrm{atm})}{\left(\dfrac{82.06\ \mathrm{cm^3 \cdot atm}}{\mathrm{gmole \cdot K}}\right)(1200\ \mathrm{K})} = 1.016 \times 10^{-7}\ \mathrm{gmole/cm^3}$$

The gas-phase concentration of SiHCl$_3$ at the surface is

$$c_{As} = c_{A,\infty} \frac{k_c}{k_c + k_s} = 1.016 \times 10^{-7} \frac{\text{gmole}}{\text{cm}^3} \frac{4.23 \text{ cm/s}}{4.23 \text{ cm/s} + 0.83 \text{ cm/s}} = 8.49 \times 10^{-8} \frac{\text{gmole}}{\text{cm}^3}$$

The average flux of SiHCl$_3$ is

$$N_A = 1.016 \times 10^{-7} \text{ gmole/cm}^3 \frac{(4.23 \text{ cm/s})(0.83 \text{ cm/s})}{(4.23 \text{ cm/s}) + (0.83 \text{ cm/s})} = 7.05 \times 10^{-8} \frac{\text{gmole}}{\text{cm}^2 \cdot \text{s}}$$

The silicon deposition rate, expressed as the rate of Si film thickness vs. time, is readily calculated from the SiHCl$_3$ flux:

$$r_{Si} = \frac{N_A v_{Si} M_{Si}}{\rho_{Si}} = \frac{\left(7.05 \times 10^{-8} \dfrac{\text{gmole}}{\text{cm}^2 \cdot \text{s}}\right)\left(\dfrac{1 \text{ gmole Si}}{1 \text{ gmole SiHCl}_3}\right)\left(\dfrac{28 \text{ g Si}}{\text{gmole Si}}\right)\left(\dfrac{3600 \text{ s}}{1 \text{ h}}\right)\left(\dfrac{10^4 \text{ μm}}{1.0 \text{ cm}}\right)}{\left(2.33 \dfrac{\text{g Si}}{\text{cm}^3}\right)}$$

$$= 30.5 \text{ μm/h}$$

where ρ_{Si} is the density of solid silicon, M_{Si} is the molecular weight of Si, and v_{Si} is the stoichiometric coefficient for conversion of SiHCl$_3$ to Si. Note that the rate of Si thin-film thickness formation is measured at the micron scale (μm).

Since the flux is based on a mean k_c averaged over the whole length of the wafer, the thin-film deposition rate also represents an average rate from $x = 0$ to $x = L$. However, for flow over a flat plate, the local mass-transfer coefficient $k_{c,x}$ will decrease proportional to $x^{1/2}$ as x increases from $x > 0$ to $x = L = 15$ cm. In contrast, the concentration boundary-layer thickness (δ_c) will increase as x increases. For laminar flow, the working equations are

$$\text{Sh}_x = \frac{k_{c,x} x}{D_{AB}} = 0.332 \, \text{Re}^{1/2} \, \text{Sc}^{1/3} \tag{28-24}$$

$$k_{c,x} = \frac{0.332 \, D_{AB}}{x} \left(\frac{v_\infty x}{\nu_B}\right)^{1/2} \text{Sc}^{1/3}$$

$$\delta_c = \delta\left(\text{Sc}^{-1/3}\right) = \frac{5 \, x}{\text{Sc}^{1/3} \sqrt{\text{Re}_x}} \tag{28-25}$$

Plots of $k_{c,x}$, δ_c, and δ down position x of the wafer are provided in Figure 28.7. Given these insights, what would the Si film thickness profile look like down position x of the wafer? Furthermore, note that at $x = 0$, the local mass-transfer coefficient is not defined; what process would control the solid Si deposition rate at this location?

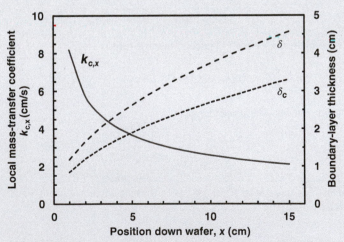

Figure 28.7 Local mass-transfer coefficient vs. position along the wafer in the direction of the gas stream, Example 3.

APPROXIMATE ANALYSIS OF THE CONCENTRATION BOUNDARY LAYER

When the flow is other than laminar or the configuration is other than a flat plate, few exact solutions presently exist for mass transport in a boundary layer. Approximate analysis of the concentration boundary layer is an alternative approach for developing a fundamental basis for convective mass transport. The approximate method developed by von Kármán to describe the hydrodynamic boundary layer can be used for analyzing the concentration boundary layer. The use of this approach was discussed in Chapters 12 and 19.

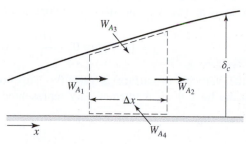

Figure 28.8 The concentration boundary-layer control volume.

Consider a control volume that is located in the concentration boundary layer as illustrated in Figure 28.8. This volume, designated by the dashed line, has a width of Δx, a height equal to the thickness of the concentration boundary layer, δ_c, and a unit depth. A steady-state molar mass balance over the control volume produces the relation

$$W_{A,1} + W_{A,3} + W_{A,4} = W_{A,2} \qquad (28\text{-}32)$$

where W_A is the molar rate of mass transfer of component A. At each surface, the molar rate is expressed as

$$W_{A,1} = \int_0^{\delta_c} c_A v_x dy \bigg|_x$$

$$W_{A,2} = \int_0^{\delta_c} c_A v_x dy \bigg|_{x+\Delta x}$$

$$W_{A,3} = c_{A,\infty} \left[\frac{\partial}{\partial x} \int_0^{\delta_c} v_x dy \right] \Delta x$$

and

$$W_{A,4} = k_c \left(c_{As} - c_{A,\infty} \right) \Delta x$$

In terms of these molar rates, equation (28-32) may be rewritten as

$$\int_0^{\delta_c} c_A v_x dy \bigg|_x + c_{A,\infty} \left[\frac{\partial}{\partial x} \int_0^{\delta_c} v_x dy \right] \Delta x + k_c \left(c_{As} - c_{A,\infty} \right) \Delta x = \int_0^{\delta_c} c_A v_x dy \bigg|_{x+\Delta x}$$

Rearranging, dividing each term by Δx, and evaluating the results in the limit as approaches zero, we obtain

$$\frac{d}{dx} \int_0^{\delta_c} c_A v_x dy = \frac{d}{dx} \int_0^{\delta_c} c_{A,\infty} v_x dy + k_c \left(c_{As} - c_{A,\infty} \right) \qquad (28\text{-}33)$$

Recall that

$$k_c\left(c_{As} - c_{A,\infty}\right) = -D_{AB}\frac{d(c_A - c_{As})}{dy}\bigg|_{y=0} \tag{28-5}$$

Therefore, equation (28-33) also assumes the form

$$\frac{d}{dx}\int_0^{\delta_c} \left(c_A - c_{A,\infty}\right)v_x\,dy = -D_{AB}\frac{d(c_A - c_{As})}{dy}\bigg|_{y=0} \tag{28-34}$$

Equation (28-34), the von Kármán integral for convective mass transfer, is analogous to equation (19-30). At this point in the analysis, the velocity and concentration profiles are unknown. However, there are four boundary conditions needed to establish the velocity profile that are known. The two boundary conditions at the surface, given by

$$y = 0, \quad v_x = 0, \quad \frac{\partial^2 v_x}{\partial y^2} = 0$$

specify that the velocity at surface is zero (no slip condition). The two boundary conditions for the bulk fluid, mathematically represented at $y = \infty$, and approximated as valid at $y = \delta$

$$y = \delta, \quad v_x = v_\infty, \quad \frac{\partial v_x}{\partial y} = 0$$

suggest the localization of the velocity gradient within $0 < y < \delta$, where it assumed that the bulk fluid properties dominate for $y > \delta$. Similarly, needed boundary conditions needed to establish the concentration profile are from $y = 0$ to $y = \delta_c$ are

$$y = 0, \quad c_A - c_{As} = 0, \quad \frac{\partial^2(c_A - c_{As})}{\partial y^2} = 0$$

$$y = \delta_c, \quad c_A - c_{As} = c_{A,\infty} - c_{As}, \quad \frac{\partial(c_A - c_{As})}{\partial y} = 0$$

If we reconsider the laminar flow parallel to a flat surface, we can use the von Kármán integral for convective mass transfer, given by equation (28-34) to obtain an approximate solution. The results can be compared to the exact solution, equation (28-24), to indicate how well we have approximated the velocity and the concentration profiles. As our first approximation, let us consider a power-series expression for the concentration variation with y;

$$c_A - c_{As} = a + by + cy^2 + dy^3$$

within laminar flow. Application of the boundary conditions to determine the constants $a, b, c,$ and d results in the following expression for the concentration profile:

$$\frac{c_A - c_{As}}{c_{A,\infty} - c_{As}} = \frac{3}{2}\left(\frac{y}{\delta_c}\right) - \frac{1}{2}\left(\frac{y}{\delta_c}\right)^3 \tag{28-35a}$$

or

$$\frac{c_A - c_{A,\infty}}{c_{As} - c_{A,\infty}} = 1 - \frac{3}{2}\left(\frac{y}{\delta_c}\right) + \frac{1}{2}\left(\frac{y}{\delta_c}\right)^3 \tag{28-35b}$$

Furthermore, differentiation of equation (28-35a) and evaluation at $y = 0$ gives

$$-D_{AB}\frac{d(c_A - c_{As})}{dy}\bigg|_{y=0} = -\frac{3D_{AB}(c_{A,\infty} - c_{As})}{2\delta_c} = k_c(c_{As} - c_{A,\infty}) \qquad (28\text{-}36a)$$

or

$$k_c = \frac{3D_{AB}}{2\delta_c} \qquad (28\text{-}36b)$$

If the velocity profile is assumed in the same power-series form, then the resulting expression, as obtained in Chapter 12, is

$$\frac{v_x}{v_\infty} = \frac{3}{2}\left(\frac{y}{\delta}\right) - \frac{1}{2}\left(\frac{y}{\delta}\right)^3 \qquad (12\text{-}40)$$

From Chapter 12, we also recall that from this approximate velocity profile $v_x(y)$ is used to approximate the hydrodynamic boundary-layer thickness as

$$\frac{\delta}{x} = \frac{4.64}{\sqrt{Re_x}} \qquad (12\text{-}41)$$

or, in equivalent form,

$$\delta^2 = \frac{280}{13}\frac{\nu x}{v_\infty} \qquad (12\text{-}41a)$$

If it is assumed that δ_c/δ is not a function of x, after considerable effort it can be shown that insertion of equations (28-35b), (28-36a), and (12-40) into equation (28-34), and simplification, gives

$$\left(\frac{\delta_c}{\delta}\right)^3\frac{d\delta}{dx} = \frac{140}{13}\frac{D_{AB}}{v_\infty\delta}$$

Finally, combination of this relationship with equation (12-41a) and subsequent integration results in the relationship

$$\left(\frac{\delta_c}{\delta}\right)^3 = \frac{D_{AB}}{\nu} \qquad (28\text{-}37)$$

which we saw earlier in equivalent form as equation (28-12). Finally, combination of equations (28-36b), (28-37), and (12-41) yields

$$\frac{k_c x}{D_{AB}} = 0.323\left(\frac{v_\infty x}{\nu}\right)^{1/2}\left(\frac{\nu}{D_{AB}}\right)^{1/3} \qquad (28\text{-}38)$$

In terms of dimensionless groups Sh_x, Re_x, and Sc, equation (28-38) is

$$Sh_x = 0.323\,Re_x^{1/2}\,Sc^{1/3} \qquad (28\text{-}39)$$

which is close to the exact solution expressed in equation (28-24).

Although this result is not exact, it is sufficiently close to the exact solution to indicate that the integral method may be used with some degree of confidence in other situations in which an exact solution is unknown. The accuracy of the method depends entirely on the ability to assume good velocity and concentration profiles.

The von Kármán integral equation (28-29) has been used to obtain an approximate solution for the turbulent boundary layer over a flat plate. With the velocity profile approximated by

$$v_x = \alpha + \beta y^{1/7}$$

and the concentration profile approximated by

$$c_A - c_{A,\infty} = \eta + \xi y^{1/7}$$

it can be shown that local Sherwood number in the turbulent layer by approximate boundary-layer analysis is

$$Sh_x = 0.0289 \, Re_x^{4/5} Sc^{1/3} \tag{28-40}$$

Development of equation (28-40) is left as a homework problem at the end of this chapter. For fully developed turbulent flow, $Re_x > 3.0 \times 10^6$.

Convective mass transfer in the transition from laminar to turbulent flow is described in Example 4.

Example 4

For flow parallel to a flat plate, the local convective mass-transfer coefficient ($k_{c,x}$) is a function of position x. For laminar flow, where $Re_x < 2.0 \times 10^5$

$$Sh_x = \frac{k_c x}{D_{AB}} = 0.332 \, Re_x^{1/2} \, Sc^{1/3} \tag{28-24}$$

and for fully developed turbulent flow, where $Re_x > 3.0 \times 10^6$, the experimentally validated correlation is

$$Sh_x = \frac{k_c x}{D_{AB}} = 0.0292 \, Re_x^{4/5} Sc^{1/3} \tag{30-1}$$

where

$$Re_x = \frac{v_\infty x}{\nu}$$

Develop an expression to estimate the average mass-transfer coefficient (k_c) in the flow regime between laminar and fully developed turbulent flow, where $2.0 \times 10^5 < Re_x < 3.0 \times 10^6$, and then plot out the average Sh vs. Re for Sc = 1.0.

The transition from laminar to turbulent flow over a flat plate is visualized in Figure 12.5. To begin, we note that the location on the flat plate where the flow is no longer fully laminar ($x = L_t$) is bounded by Re:

$$Re_t = 2.0 \times 10^5 = \frac{v_\infty L_t}{\nu}$$

or

$$L_t = 2.0 \times 10^5 \frac{\nu}{v_\infty}$$

The average mass-transfer coefficient is obtained by integration of the local mass-transfer coefficient ($k_{c,x}$) over position x. The integral is broken up into a laminar component from $x = 0$ to $x = L_t$, and a turbulent component, from $x = L_t$ to $x = L$:

$$k_c = \frac{\int_0^L k_{c,x}(x)dx}{\int_0^L dx} = \frac{\int_0^L k_{c,\text{lam}}(x)\,dx + \int_{L_t}^L k_{c,\text{turb}}(x)\,dx}{L} \tag{28-41}$$

with

$$k_{c,\text{lam}}(x) = 0.332\frac{D_{AB}}{x}(Re_x)^{1/2}Sc^{1/3}$$

$$k_{c,\text{turb}}(x) = 0.0292\frac{D_{AB}}{x}(Re_x)^{4/5}Sc^{1/3}$$

Substitution of these two equations into equation (28-41) gives

$$k_c = \frac{\int_0^{L_t} \frac{0.332\, D_{AB}(\mathrm{Re}_x)^{1/2}}{x}\mathrm{Sc}^{1/3}dx + \int_{L_t}^{L} \frac{0.0292\, D_{AB}(\mathrm{Re}_x)^{4/5}}{x}\mathrm{Sc}^{1/3}dx}{L}$$

$$k_c = \frac{0.332\, D_{AB}\left(\frac{\mathrm{v}_\infty}{\nu}\right)^{1/2}\mathrm{Sc}^{1/3}\int_0^{L_t} x^{-1/2}dx + 0.0292\, D_{AB}\left(\frac{\mathrm{v}_\infty}{\nu}\right)^{4/5}\mathrm{Sc}^{1/3}\int_{L_t}^{L} x^{-1/5}dx}{L}$$

$$k_c = \frac{0.664\, D_{AB}\left(\frac{\mathrm{v}_\infty}{\nu}\right)^{1/2}\mathrm{Sc}^{1/3}L_t^{1/2} + 0.0365\, D_{AB}\left(\frac{\mathrm{v}_\infty}{\nu}\right)^{4/5}\mathrm{Sc}^{1/3}\left[(L)^{4/5} - (L_t)^{4/5}\right]}{L} \tag{28-42}$$

$$k_c = \frac{0.664\, D_{AB}(\mathrm{Re}_t)^{1/2}\mathrm{Sc}^{1/3} + 0.0365\, D_{AB}\,\mathrm{Sc}^{1/3}\left[(\mathrm{Re}_L)^{4/5} - (\mathrm{Re}_t)^{4/5}\right]}{L}$$

or

$$\frac{k_c L}{D_{AB}} = \mathrm{Sh}_L = 0.664(\mathrm{Re}_t)^{1/2}\mathrm{Sc}^{1/3} + 0.0365\,\mathrm{Sc}^{1/3}\left[(\mathrm{Re}_L)^{4/5} - (\mathrm{Re}_t)^{4/5}\right] \tag{28-43}$$

with

$$\mathrm{Re}_L = \frac{\mathrm{v}_\infty L}{\nu}$$

which is valid for $\mathrm{Re}_L > 2.0 \times 10^5$. For laminar flow where $\mathrm{Re}_L \leq 2.0 \times 10^5$, the average mass-transfer coefficient is

$$k_c = \frac{\int_0^L k_{c,\mathrm{lam}}(x)\,dx}{\int_0^L dx} = 0.664\frac{D_{AB}}{L}\mathrm{Re}_L^{1/2}\,\mathrm{Sc}^{1/3}$$

or

$$\mathrm{Sh}_L = 0.664\,\mathrm{Re}_L^{1/2}\,\mathrm{Sc}^{1/3} \tag{28-29}$$

For fully developed turbulent flow where $\mathrm{Re}_L \geq 3.0 \times 10^6$, the average mass-transfer coefficient is approximated as

$$k_c = \frac{\int_0^L k_{c,\mathrm{turb}}(x)\,dx}{\int_0^L dx} = 0.0365\frac{D_{AB}}{L}(\mathrm{Re}_L)^{4/5}\mathrm{Sc}^{1/3}$$

or

$$\mathrm{Sh}_L = 0.0365\,\mathrm{Re}_L^{4/5}\,\mathrm{Sc}^{1/3} \tag{28-44}$$

In fully developed turbulent flow, values of Sh_L from equation (28-44) approach equation (28-43), and the laminar contribution to the convective mass-transfer process can be neglected. Representative plots of Sh_L vs. Re_L for $\mathrm{Sc} = 1.0$ are provided in Figure 28.9.

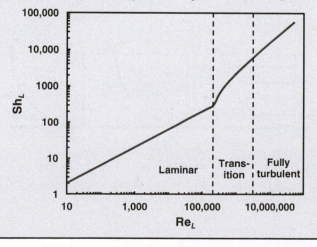

Figure 28.9 Average Sherwood number (Sh_L) at $\mathrm{Sc} = 1.0$ for flow over a flat plate, showing transition from laminar to fully turbulent regimes, Example 4.

MASS-, ENERGY-, AND MOMENTUM-TRANSFER ANALOGIES

In the previous analyses of convective mass transfer, we have recognized the similarities in the differential equations for momentum, energy, and mass transfer and in the boundary conditions when the transport gradients were expressed in terms of dimensionless variables. These similarities have permitted us to predict solutions for the similar transfer processes. In this section, we shall consider several analogies among transfer phenomena that have been proposed because of the similarity in their mechanisms. The analogies are useful in understanding the transfer phenomena, and as a satisfactory means for predicting behavior of systems for which limited quantitative data are available.

The similarity among the transfer phenomena and, accordingly, the existence of the analogies require that the following five conditions exist within the system:

1. There is no energy or mass produced within the system. This, of course, implies that no homogeneous reaction occurs.

2. There is no emission or absorption of radiant energy.

3. There is no viscous dissipation.

4. The velocity profile is not affected by the mass transfer; thus, there is only a low rate of mass transfer.

5. The physical properties are constant. As there may be slight changes in the physical properties due to variations in temperature or concentration, this condition can be approximated by using average concentration and film temperature properties.

Reynolds Analogy

The first recognition of the analogous behavior of momentum and energy transfer was reported by Reynolds.[4] Although this analogy is limited in application, it has served as the catalyst for seeking better analogies, and it has been useful in analyzing complex boundary-layer phenomena of aerodynamics.

A simplified representation of the hydrodynamic, concentration, and thermal boundary layers for the respective momentum-, mass-, and heat-transfer processes associated with fluid flow over a flat surface is presented in Figure 28.10. Reynolds postulated that the

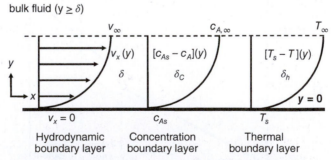

Figure 28.10 Representation of hydrodynamic, concentration, and thermal boundary layers for $Sc = Pr = 1.0$ associated with the Reynolds analogy.

[4] O. Reynolds, *Proc. Manchester Lit. Phil. Soc.*, **8** (1874).

mechanisms for transfer of momentum and energy were identical. We have observed in our earlier discussions on laminar boundary layers that this is true if the Prandtl number, Pr, is unity. From our previous consideration in Section 28.4, we can extend the Reynolds postulation to include the mechanism for the transfer of mass if the Schmidt number, Sc, is also equal to 1. For example, if we consider the laminar flow over a flat plate, the concentration and velocity profiles within the boundary layers are related by

$$\frac{\partial}{\partial y}\left(\frac{c_A - c_{As}}{c_{A,\infty} - c_{As}}\right)\bigg|_{y=0} = \frac{\partial}{\partial y}\left(\frac{v_x}{v_\infty}\right)\bigg|_{y=0} \tag{28-45}$$

Recalling that at the boundary next to the plate, where $y = 0$, we may express the mass flux in terms of either the mass-diffusivity or the mass-transfer coefficient by

$$N_{A,y} = -D_{AB}\frac{\partial(c_A - c_{As})}{\partial y}\bigg|_{y=0} = k_c\left(c_{As} - c_{A,\infty}\right) \tag{28-5}$$

When Sc = 1, we note that $D_{AB} = \mu/\rho$. We can then combine equations (28-45) and (28-5) to obtain an expression that relates the mass-transfer coefficient to the velocity gradient at the surface:

$$k_c = \frac{\mu}{\rho v_\infty}\frac{\partial v_x}{\partial y}\bigg|_{y=0} \tag{28-46}$$

The coefficient of skin friction was related to this same velocity gradient by

$$C_f = \frac{\tau_0}{\rho v_\infty^2/2} = \frac{2\mu}{\rho v_\infty^2}\frac{\partial v_x}{\partial y}\bigg|_{y=0} \tag{12-2}$$

Using this definition, we can rearrange equation (28-46) to obtain the mass-transfer Reynolds analogy for systems with a Schmidt number of 1:

$$\frac{k_c}{v_\infty} = \frac{C_f}{2} \tag{28-47}$$

Equation (28-47) is analogous to the energy-transfer Reynolds's analogy for systems with a Prandtl number of 1. This analogy was discussed in Chapter 19 and may be expressed by

$$\frac{h}{\rho\, v_\infty c_p} = \frac{C_f}{2} \tag{19-37}$$

The Reynolds analogy for mass transfer is only valid if the Schmidt number equals 1 and the resistance to flow is that due to skin friction—i.e., no form drag is involved. This was experimentally verified by von Kármán for a fully turbulent flow with Sc = Pr = 1.

Turbulent-Flow Considerations

In many practical applications, the flow is turbulent rather than laminar. Although many investigators have contributed considerably to the understanding of turbulent flow, so far no one has succeeded in predicting convective transfer coefficients or friction factors by direct analysis. This is not too surprising when we recall from our earlier discussions on turbulent flow, in Section 13.1, that the flow at any point is subject to irregular fluctuations in direction and velocity. Accordingly, any particle of the fluid undergoes a series of random movements, superimposed on the main flow. These eddy movements bring about mixing throughout the turbulent core. This process is often referred to as "eddy diffusion."

The value of the eddy-mass diffusivity will be very much larger than the molecular diffusivity in the turbulent core.

In an effort to characterize this type of motion, Prandtl proposed the mixing-length hypothesis as discussed in Chapter 12. In this hypothesis, any velocity fluctuation v'_x is due to the y-directional motion of an eddy through a distance equal to the mixing length L. The fluid eddy, possessing a mean velocity, $\bar{v}_x|_y$, is displaced into a stream where the adjacent fluid has a mean velocity, $\bar{v}_x|_{y+L}$. The velocity fluctuation is related to the mean-velocity gradient by

$$v'_x = \bar{v}_x|_{y+L} - \bar{v}_x|_y = \pm L\frac{d\bar{v}_x}{dy} \tag{12-52}$$

The total shear stress in a fluid was defined by the expression

$$\tau = \mu\frac{d\bar{v}_x}{dy} - \rho\overline{v'_x v'_y} \tag{12-51}$$

The substitution gives

$$\tau = \rho\left(\nu + Lv'_y\right)\frac{d\bar{v}_x}{dy}$$

or

$$\tau = \rho(\nu + \epsilon_M)\frac{d\bar{v}_x}{dy} \tag{28-48}$$

where $\epsilon_M = Lv'_y$ is designated at the eddy momentum diffusivity. It is analogous to the molecular momentum diffusivity, ν.

Similarly, we may now similarly analyze mass transfer in turbulent flow, as this transport mechanism is also due to the presence of the fluctuations or eddies. In Figure 28.11, the curve represents a portion of the turbulent concentration profile with mean flow in the x-direction. The instantaneous rate of transfer of component A in the y-direction is

$$N_{A,y} = c'_A v'_y \tag{28-49}$$

where $c_A = \bar{c}_A + c'_A$, the temporal average plus the instantaneous fluctuation in the concentration of component A. We can again use the concept of the mixing length to define the concentration fluctuation by the following relation:

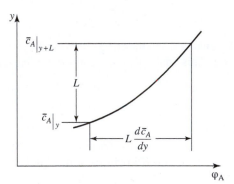

Figure 28.11 Portion of turbulent concentration profile curve, showing the Prandtl mixing length.

$$c'_A = \bar{c}_A|_{y+L} - \bar{c}_A|_y = L\frac{d\bar{c}_A}{dy} \tag{28-50}$$

By insertion of equation (28-49) into equation (28-50), we obtain an expression for the turbulent transfer of mass by eddy transport. The total mass transfer normal to the direction of flow is

$$N_{A,y} = -D_{AB}\frac{d\bar{c}_A}{dy} - \bar{v}'_y L\frac{d\bar{c}_A}{dy}$$

or

$$N_{A,y} = -(D_{AB} + \epsilon_D)\frac{d\bar{c}_A}{dy} \tag{28-51}$$

where $\epsilon_D = \overline{Lv'_y}$ is designated as the eddy mass diffusivity.

By similar reasoning, an expression was derived in Chapter 19 for convective heat transfer:

$$\frac{q_y}{A} = -\rho\, c_p(\alpha + \epsilon_H)\frac{d\bar{T}}{dy} \tag{19-50}$$

where α is the molecular thermal diffusivity and ϵ_H is the eddy thermal diffusivity.

Eddy diffusion plays an important role in a number of mass-transfer processes. For instance, there is mass transfer between a fluid flowing past solids in heterogeneous catalytic reactors, blast furnaces, driers, and so on. As a result of eddy diffusion, transport in the turbulent region is rapid, reducing any gradient in composition. As the wall is approached, the turbulence is progressively damped until it essentially disappears in the immediate neighborhood of the solid surface, and the transport is almost entirely by molecular diffusion. The majority of the resistance to transfer occurs in the boundary layer near the surface where the magnitude of the concentration gradient is greatest.

The Prandtl and von Kármán Analogies

In Chapter 19, the Prandtl analogy for heat and momentum transfer was developed when consideration was given to the effect of both the turbulent core and the laminar sublayer. The same reasoning with regard to mass and momentum transfer can be used to develop a similar analogy. The result of the Prandtl analysis for mass transfer is

$$\frac{k_c}{v_\infty} = \frac{C_f/2}{1 + 5\sqrt{C_f/2}(\mathrm{Sc} - 1)} \tag{28-52}$$

which is analogous to equation (19-57) for heat transfer. It can also be shown that equation (28-52) can also be written in terms of Sh, Re, and Sc, given by

$$\mathrm{Sh} = \frac{(C_f/2)\mathrm{Re\,Sc}}{1 + 5\sqrt{C_f/2}(\mathrm{Sc} - 1)} \tag{28-53}$$

von Kármán extended the Prandtl analogy by considering the so-called "buffer layer" in addition to the laminar sublayer and the turbulent core. This led to the development of the *von Kármán analogy*, given by

$$\mathrm{Nu} = \frac{(C_f/2)\mathrm{Re\,Pr}}{1 + 5\sqrt{C_f/2}\{\mathrm{Pr} - 1 + \ln[(1 + 5\mathrm{Pr})/6]\}}$$

for momentum and energy transfer, which can be obtained through equation (19-59). A similar von Kármán analysis for mass transfer yields

$$\mathrm{Sh} = \frac{(C_f/2)\mathrm{Re\,Sc}}{1 + 5\sqrt{C_f/2}\{\mathrm{Sc} - 1 + \ln[(1 + 5\,\mathrm{Sc})/6]\}} \tag{28-54}$$

Chilton–Colburn Analogy

Chilton and Colburn,[5] using experimental data, sought modifications to Reynolds's analogy that would not have the restrictions that Pr and Sc numbers must be equal to 1. They defined the *j-factor for mass transfer* as

$$j_D = \frac{k_c}{v_\infty} (Sc)^{2/3} \qquad (28\text{-}55)$$

This factor is analogous to the j_H factor for heat transfer defined by equation (19-40). Based on the data collected in both laminar and turbulent flow regimes, Chilton and Colburn found that

$$j_D = \frac{k_c}{v_\infty} (Sc)^{2/3} = \frac{C_f}{2} \qquad (28\text{-}56)$$

The analogy is valid for gases and liquids within the range $0.6 < Sc < 2500$. Equation (28-56) can be shown to satisfy the "exact solution" for laminar flow over a flat plate

$$Sh_x = 0.332 \, Re_x^{1/2} \, Sc^{1/3} \qquad (28\text{-}24)$$

If both sides of this equation are divided by $Re_x Sc^{1/3}$, we obtain

$$\frac{Sh_x}{Re_x Sc^{1/3}} = \frac{0.332}{Re_x^{1/2}} \qquad (28\text{-}57)$$

This equation reduces to the *Chilton–Colburn analogy* when we substitute into the above expression the Blasius solution for the laminar boundary layer

$$\frac{Sh_x}{Re_x Sc^{1/3}} = \frac{Sh_x}{Re_x Sc} Sc^{2/3} = \frac{k_c}{v_\infty} = \frac{C_f}{2}$$

or

$$\left(\frac{k_c x}{D_{AB}}\right) \left(\frac{\mu}{x v_\infty \rho}\right) \left(\frac{\rho D_{AB}}{\mu}\right) (Sc)^{2/3} = \frac{k_c Sc^{2/3}}{v_\infty} = \frac{C_f}{2} \qquad (28\text{-}58)$$

The complete Chilton–Colburn analogy is

$$j_H = j_D = \frac{C_f}{2} \qquad (28\text{-}59)$$

which relates all three types of transport in one expression. Equation (28-59) is exact for flat plates and is satisfactory for systems of other geometry provided no form drag is present. For systems where form drag is present, it has been found that neither j_H or j_D is equal to $C_f/2$; however, when form drag is present

$$j_H = j_D \qquad (28\text{-}60)$$

or

$$\frac{h}{\rho v_\infty c_p} (Pr)^{2/3} = \frac{k_c}{v_\infty} (Sc)^{2/3} \qquad (28\text{-}61)$$

Equation (28-61) relates convective heat and mass transfer. It permits the evaluation of one unknown transfer coefficient through information obtained for another transfer phenomenon.

---

[5] A. P. Colburn, *Trans. A.I.Ch.E.*, **29**, 174–210 (1933); T. H. Chilton and A. P. Colburn, *Ind. Eng. Chem.*, **26**, 1183 (1934).

It is experimentally validated for gases and liquids within the ranges $0.60 < \text{Sc} < 2500$ and $0.6 < \text{Pr} < 100$.

As previously stated, one of the five conditions that should exist if the analogies are to be used requires the physical properties of the fluid stream to be constant. If there are only slight variations in the properties due to the variations in the overall film temperature, one may minimize this restrictive condition by evaluating the physical properties at the mean film temperature.

The Chilton–Colburn analogy for heat and mass transfer has been observed to hold for many different geometries; for example, flow over flat plate, flow in circular pipe and annulus, and flow around cylinders.

The following example illustrates the coupling of convective heat- and mass-transfer processes that utilizes the Chilton–Colburn analogy to link the heat-transfer coefficient to the mass-transfer coefficient.

Example 5

The "transpiration cooling" process shown in Figure 28.12 is used to cool volatile liquids in hot, arid environments. The well-mixed liquid reservoir, insulated as shown, and flowing gas stream are separated by a thin, semipermeable barrier that allows vapor, but not the liquid, to pass through it. At liquid temperature (T), the volatile liquid exerts a vapor pressure P_A, and a small amount of liquid vaporizes, using heat transferred from the hot flowing gas stream to the cooler liquid reservoir to drive the vaporization process.

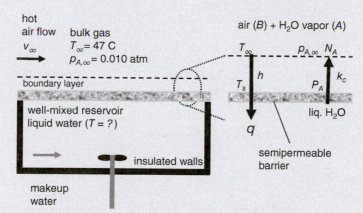

Figure 28.12 Transpiration cooling process for chilling water, Example 5.

In the present process, the reservoir is loaded with liquid water, and a flowing air stream at 47 C (320 K) and 1.0 atm is contacted with the device. The gas stream has a heat-transfer coefficient of $h = 20 \text{ W/m}^2 \cdot \text{K}$, and a water vapor partial pressure of 0.010 atm. It is assumed that the semipermeable barrier offers no resistance to heat or mass transfer, so that $T = T_s$.

Determine the lowest steady-state water temperature allowed by these process conditions, and the flux of water vapor to the gas stream.

An energy balance must be performed on the process. Let species A represent water vapor and species B represent air. At steady state, the heat transferred by convection across the gas stream boundary layer to the liquid is balanced by energy needed to drive the mass transfer of vapor across the boundary layer. The energy balance is described by

$$k_c\left(c_{As} - c_{A,\infty}\right) S\, \Delta H_{v,A} M_A = \text{Sh}(T_\infty - T_s)$$

where S is the surface area of the semipermeable barrier surface, and $\Delta H_{v,A}$ is the latent heat of vaporization of the liquid at temperature T_s. To determine the water temperature T_s, it is first noted that the Chilton–Colburn analogy for heat and mass transfer, equation (28-61), can also be expressed as

$$h = k_c\, C_{p,B}\, \rho_B \left(\frac{\text{Sc}}{\text{Pr}}\right)^{2/3}$$

This form of the Chilton–Colburn analogy is combined with the energy balance to yield

$$(c_{As} - c_{A,\infty})\Delta H_{v,A} M_A = C_{p,B}\rho_B \left(\frac{Sc}{Pr}\right)^{2/3}(T_\infty - T_s)$$

Since the mass-transfer coefficient k_c and heat-transfer coefficient h share the same flow field in the gas stream, the bulk gas velocity (v_∞) is not present. It is also noted that the water vapor composition at the surface (c_{As}) is determined by the vapor pressure of the water, which is a function of T_s. Since the vapor pressure of water can be correlated to temperature through the Antoine equation:

$$c_{As}(T_s) = \frac{P_A(T_s)}{RT} \quad \text{with} \quad P_A(T_s) = 10^{\left(a - \frac{b}{T_s+c}\right)}$$

with literature values of the Antoine constants for water given as $a = 8.10765$, $b = 1750.286$, $c = 235.00$, with P_A in units of mm Hg and T_s in units of °C. Finally, if it is assumed that the gas stream properties can be taken at the bulk gas stream temperature of T_∞, then the solution is implicit in T_s, and can be determined by the root of the equation once all of the other thermophysical properties are known, as detailed below.

At 320 K, from Appendix I, for air, $\rho_B = 1.1032$ kg/m^3, $C_{p,B} = 1007.3$ J/kg · K, $v_B = 1.7577 \times 10^{-5}$ m^2/s, and Pr = 0.703. At 320 K, for H$_2$O, $\Delta H_{v,A} = 2390$ kJ/kg. From Appendix J, the diffusion coefficient of H$_2$O vapor in air is $D_{AB} = 0.26$ cm^2/s at 298 K and 1.0 atm, which scales to 0.29 cm^2/s (2.9×10^{-5} m^2/s) at 320 K.

Consequently,

$$Sc = \frac{v_B}{D_{AB}} = \frac{1.7577 \times 10^{-5} \text{ m}^2/\text{s}}{2.9 \times 10^{-5} \text{ m}^2/\text{s}} = 0.606$$

and

$$c_{A,\infty} = \frac{P_A}{RT} = \frac{(0.010 \text{ atm})}{\left(8.206 \times 10^{-5} \dfrac{\text{m}^3 \cdot \text{atm}}{\text{gmole} \cdot \text{K}}\right)(320 \text{ K})} = 0.381 \frac{\text{gmole}}{\text{m}^3}$$

Finally, insertion of all known parameters into the energy balance yields

$$\left(c_{As}(T_s) - \frac{3.81 \times 10^{-4} \text{ kgmole}}{\text{m}^3}\right)\left(\frac{2390 \times 10^3 \text{ J}}{\text{kg}}\frac{18 \text{ kg}}{\text{kgmole}}\right)$$

$$= \left(\frac{1007.3 \text{ J}}{\text{kg} \cdot \text{K}}\right)\left(\frac{1.1032 \text{ kg}}{\text{m}^3}\right)\left(\frac{0.606}{0.703}\right)^{2/3}(320 \text{ K} - T_s)$$

Solving this equation gives $T_s = 20.7$ C, which is obtained from the root of the nonlinear equation.

With the steady-state liquid temperature T_s now known, the flux of water vapor can be estimated. At $T_s = 20.7$ C, the water vapor pressure (P_A) is 0.024 atm, and so the concentration of water vapor at the surface (c_{As}) is 1.00 gmole/m^3. The mass-transfer coefficient (k_c) is scaled the heat-transfer coefficient (h) using the Chilton–Colburn analogy:

$$k_c = \frac{h}{C_{p,B}\rho_B}\left(\frac{Pr}{Sc}\right)^{2/3} = \frac{(20 \text{ J/s} \cdot \text{m}^2 \cdot \text{K})}{\left(\dfrac{1007.3 \text{ J}}{\text{kg} \cdot \text{K}}\right)\left(\dfrac{1.1032 \text{ kg}}{\text{m}^3}\right)}\left(\frac{0.703}{0.606}\right)^{2/3} = 0.020 \text{ m/s}$$

Finally, the flux of water vapor is determined by

$$N_A = k_c(c_{As} - c_{A,\infty}) = \left(0.020 \frac{\text{m}}{\text{s}}\right)(1.00 - 0.381)\left(\frac{\text{gmole}}{\text{m}^3}\right) = 0.0124 \frac{\text{gmole}}{\text{m}^2 \cdot \text{s}}$$

In Figure 28.12, a small flow of makeup water is added to the process to account for the evaporative loss.

MODELS FOR CONVECTIVE MASS-TRANSFER COEFFICIENTS

We have seen in this chapter that mass-transfer coefficients can be predicted from first principles for fluids in laminar flow. However, empirical mass-transfer coefficients determined from experimental investigations are commonly used in the design of mass-transfer processes involving turbulent flow. A theoretical basis for mass-transfer coefficients requires a better understanding of the mechanism of turbulence, as they are directly related to the dynamic characteristics of the flow.

Alternative models exist for explaining convective mass-transfer processes. *Film theory* is based upon the presence of a conceptual film of fluid that offers the same resistance to mass transfer as actually exists in the entire flowing fluid, as shown in Figure 28.1. In other words, all resistance to mass transfer is assumed to exist in a fictitious stagnant fluid film in which the transport is entirely by molecular diffusion. This conceptual film thickness, δ, must extend beyond the laminar sublayer to include an equivalent resistance encountered as the concentration changes within the buffer layer and the turbulent core. For diffusion through a nondiffusing layer or stagnant gas, this theory predicts the mass-transfer coefficient to be

$$k_c = \frac{D_{AB}}{\delta}\frac{P}{p_{B,1m}} \tag{26-9}$$

as developed in Chapter 26. For equimolar counterdiffusion, the mass-transfer coefficient is expressed as

$$k_c^0 = \frac{D_{AB}}{\delta} \tag{26-24}$$

In both cases, the convective mass-transfer coefficient is directly related to the molecular diffusion coefficient, D_{AB}. However, the overly simplistic nature of film theory does not physically explain convective mass transfer in a flowing fluid, and so other theories and models have been postulated to describe this phenomenon.

Penetration theory was originally proposed by Higbie[6] to explain the mass transfer in the liquid phase during gas absorption. It has been applied to turbulent flow by Danckwerts[7] and many other investigators when the diffusing component penetrates only a short distance into the phase of interest because of its rapid disappearance through chemical reaction or its relatively short time of contact. Higbie considered mass to be transferred into the liquid phase by unsteady-state molecular transport. Danckwerts subsequently applied this unsteady-state concept to the absorption of component A in a turbulent liquid stream. The Danckwerts model assumed that the motion of the liquid is constantly bringing fresh liquid eddies from the interior up to the surface, where they displace the liquid elements previously on the surface, as shown in Figure 28.13. While on the surface, each element of the liquid becomes exposed to the second phase and mass is transferred into the liquid as though it were stagnant and infinitely deep, with the rate of transfer dependent upon the exposure time, t_{exp}. With this concept

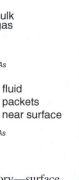

Figure 28.13 Penetration theory—surface renewal of fluid elements.

[6] R. Higbie, *Trans. A.I.Ch.E.*, **31**, 368–389 (1935).
[7] P. V. Danckwerts, *Ind. Eng. Chem.*, **43**, 1460–1467 (1951).

in mind, the average mass-transfer flux at the interface between the liquid and the gas phases was expressed as

$$N_A = \sqrt{\frac{4D_{AB}}{\pi\, t_{exp}}}(c_{As} - c_{A,\infty}) = k_c(c_{As} - c_{A,\infty}) \qquad (28\text{-}62)$$

with

$$k_c = \sqrt{\frac{4D_{AB}}{\pi t_{exp}}}$$

where t_{exp} is now the average exposure time of a fluid element from the bulk fluid that is brought up to the gas–liquid interface. A major weakness of penetration theory is that the exposure time cannot be theoretically predicted and must be determined experimentally.

Equation (28-63) predicts that the mass-transfer coefficient is proportional to $D_{AB}^{0.5}$. The surface-renewal concept has been very successful in the explanation and analysis of convective mass transfer, particularly when the mass transport is accompanied by chemical reactions in the liquid phase.[8]

Penetration theory can also be compared to mass transfer into a falling liquid film, as described in Chapter 26. Recall that for mass transfer into a falling liquid film, solute A is transferred from the gas phase into the falling liquid film, with c_{As} defined as the concentration of solute A dissolved in the liquid that is in equilibrium with the partial pressure of solute A in the bulk gas phase. Figure 28.14 shows this mass-transfer process, which compares the parabolic velocity profile with a simplified case where the velocity profile is flat and the solute penetrates only a short distance into the falling liquid film.

At the gas–liquid interface ($y = \delta$), the local flux of solute A from the gas phase into the liquid phase at position x down the length of the falling liquid film is given by

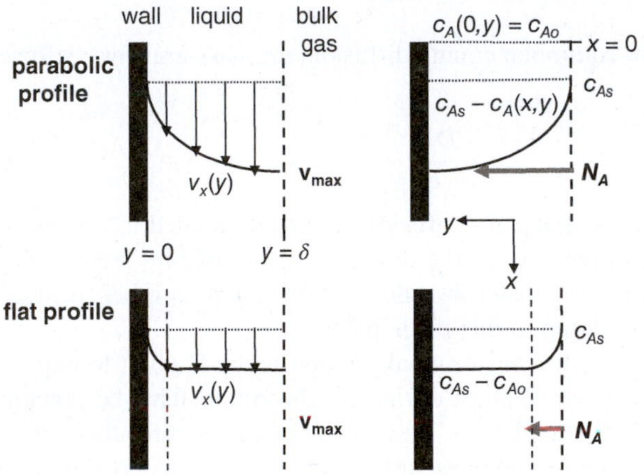

Figure 28.14 Mass transfer into falling liquid film.

$$N_A = (c_{As} - c_{Ao})\sqrt{\frac{D_{AB}v_{max}}{\pi x}} \qquad (28\text{-}63)$$

or

$$N_A = k_{c,x}(c_{As} - c_{Ao})$$

with

$$k_{c,x} = \sqrt{\frac{D_{AB}v_{max}}{\pi x}} \qquad (28\text{-}64)$$

The maximum velocity (v_{max}) is given by

$$v_{max} = \frac{\rho g \delta^2}{2\mu}$$

The average mass-transfer coefficient is obtained by integration of $k_{c,x}$ down the length L of the falling liquid film

$$k_c = \frac{1}{L}\int_0^L k_{c,x}\,dx = \sqrt{\frac{4D_{AB}v_{max}}{\pi L}} \qquad (28\text{-}65)$$

---

[8] P. V. Danckwerts, *Gas–Liquid Reactions*, McGraw-Hill, New York, 1970.

Equation (28-65) suggests that k_c is proportional to $D_{AB}^{0.5}$, similar to penetration theory. Note also that L/v_{max} is equivalent to t_{exp}.

In both the film and penetration models, the mass transfer involves an interface between two moving fluids. When one of the phases is a solid, the fluid velocity parallel to the surface at the interface must be zero; accordingly, we should expect the need of a third model, the *boundary-layer model*, for correlating data involving a solid subliming into a gas or a solid dissolving into a liquid. For diffusion through a laminar boundary layer, from Section 28.4 the average mass-transfer coefficient was found to be

$$k_c = 0.664 \frac{D_{AB}}{L} \text{Re}_L^{1/2} \text{Sc}^{1/3} \tag{28-27}$$

This shows the mass-transfer coefficient is proportional $D_{AB}^{2/3}$, which is typical of convective mass-transfer correlations.

Table 28.1 provides a summary of the three models proposed for mass-transfer coefficients. Each model has its own specific diffusion coefficient dependency; this dependency is sometimes used to scale mass-transfer coefficients from one solute to another that are exposed to the same hydrodynamic flows.

Table 28.1 Models for convective mass-transfer coefficients (dilute systems)

Model	Basic Form	$f(D_{AB})$	Notes
Film theory	$k_c = \dfrac{D_{AB}}{\delta}$	$k_c \propto D_{AB}$	δ unknown
Falling liquid film	$k_c = \sqrt{\dfrac{4D_{AB}v_\infty}{\pi L}}$	$k_c \propto D_{AB}^{1/2}$	Solute does not penetrate very far into the liquid film
Penetration theory	$k_c = \sqrt{\dfrac{4D_{AB}}{\pi t_{exp}}}$	$k_c \propto D_{AB}^{1/2}$	t_{exp} unknown
Boundary-layer theory	$k_c = 0.664\dfrac{D_{AB}}{L}\text{Re}_L^{1/2}\text{Sc}^{1/3}$	$k_c \propto D_{AB}^{2/3}$	Best way to scale k_c from one solute to another exposed to same hydrodynamic flow

► **28.8**

CLOSURE

In this chapter, we have discussed the principles of mass transfer by forced convection, the significant parameters that help describe convective mass transfer, and the models proposed to explain the mechanism of convective transport. We have seen that the transfer of mass by convection is intimately related to the dynamic characteristics of the flowing fluid, particularly to the fluid in the vicinity of the boundary. Because of the close similarities in the mechanisms of momentum, energy, and mass transfer, we were able to use the same four methods for evaluating convective mass-transfer coefficients that were originally developed to analyze convective heat-transfer coefficients. In all four analyses, the mass-transfer coefficient was correlated by the general equation

$$\text{Sh} = f(\text{Re}, \text{Sc})$$

Mass transfer into turbulent streams was discussed, and the eddy mass diffusivity was defined. Analogies were presented for convective mass transfer into turbulent streams.

CHAPTER 29

Convective Mass Transfer
Between Phases

In Chapter 28, convective mass transfer within a single phase was considered; in this case, mass is exchanged between a boundary surface and a moving fluid, and the flux is related to an *individual film mass-transfer convective coefficient*. Many mass-transfer operations, however, involve the transfer of material between two contacting phases where the flux may be related to an *overall mass-transfer convective coefficient*. These phases may be a gas stream contacting a liquid stream or two liquid streams if they are immiscible. In this chapter, we shall consider the mechanism of steady-state mass transfer between the phases and the interrelations between the individual convective coefficients for each phase and the overall convective coefficient.

Chapter 30 presents empirical equations for the individual mass-transfer convective coefficients involved in the interphase transfer. These equations have been established from experimental investigations. Chapter 31 presents methods of applying these interphase concepts to the design of mass-transfer equipment.

▶ 29.1

EQUILIBRIUM

To begin our discussion of convective mass transfer between two phases in contact, consider the wetted-wall tower presented in Figure 29.1. A liquid solvent stream containing a dissolved solute *A* is sent to the top of the tower and fed into weir containing a narrow slit aligned around the inner circumference of the tube. The gravity-driven flow of liquid through the slit creates a thin liquid film that uniformly wets the inner surface of the tube and flows downward. A gas steam containing gaseous solute *A* is introduced into the bottom of the tower and flows upward, contacting the falling liquid film. The flow of gas and liquid creates a hydrodynamic boundary layer, or "film," in both the gas and liquid phases, formed at each side of the gas–liquid interface. Solute *A*, which is present in both the gas phase and liquid phases, can be exchanged between the contacting phases if there is a departure from their bulk concentrations and their concentrations at thermodynamic equilibrium.

The transport of mass within a phase, by either molecular or convective transport mechanisms, has been shown to be directly dependent upon the concentration gradient responsible for the mass transfer. When equilibrium within the system is established, the concentration gradient and, in turn, the net diffusion rate of the diffusing species becomes zero. Transfer between two phases also requires a departure from equilibrium that might exist between the average or bulk concentrations within each phase. As the deviations from

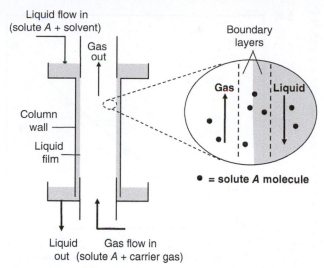

Figure 29.1 Wetted-wall tower for transfer of solute A between contacting gas and liquid phases.

equilibrium provides the concentration driving force within a phase, it is necessary to consider interphase equilibrium in order to describe mass transfer between the phases.

Initially, let us consider the equilibrium characteristics of a particular system and then we will generalize the results for other systems. For example, let the initial system composition include air and ammonia in the gas phase and only water in the liquid phase. Ammonia is soluble in water, but is also volatile, and so can exist in both the gas and liquid phases. When first brought into contact, some of the ammonia will be transferred into the water phase in which it is soluble and some of the water will be vaporized into the gas phase. If the gas–liquid mixture is contained within an isothermal, isobaric container, a dynamic equilibrium between the two phases will eventually be established. A portion of the molecules entering the liquid phase returns to the gas phase at a rate dependent upon the concentration of the ammonia in the liquid phase and the vapor pressure exerted by the ammonia in the aqueous solution. Similarly, a portion of the water vaporizing into the gas phase condenses into the solution. Dynamic equilibrium is indicated by a constant concentration of ammonia in the liquid phase and a constant concentration or partial pressure of ammonia in the gas phase. This equilibrium condition can be altered by adding more ammonia to the isothermal, isobaric container. After a period of time, a new dynamic equilibrium will be established with a different concentration of ammonia in the liquid and a different partial pressure of ammonia in the gas. Obviously, one could continue to add more ammonia to the system; each time a new equilibrium will be reached. Since ammonia is more much volatile than water, and the components of air are not appreciably soluble in water relative to ammonia, it is often assumed that the solvent is nonvolatile, and that the carrier gas (air) and the solvent (water) are immiscible. Consequently, it is assumed that only ammonia (solute A) can be exchanged between phases.

Figure 29.2 presents the equilibrium distribution of ammonia (NH_3) between the gas phase and liquid water, with compositions in terms of the partial pressure of ammonia in the gas phase (p_A), and the molar concentration of the dissolved ammonia in the liquid phase (c_{AL}). At liquid-phase concentrations below 1.5 kgmole NH_3/m^3, the equilibrium distribution of ammonia dissolved in the liquid phase is a linear function of the partial pressure of ammonia in the gas phase over the liquid. The dashed line is the extrapolation of this linear relationship to the nonlinear portion of the equilibrium line.

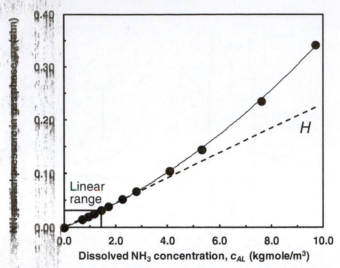

Figure 29.2 Ammonia solubility in water vs. partial pressure of ammonia at 30 C.

There are many graphical forms of equilibrium data due to the many ways of expressing concentrations in each of the phases. We will find use for many types of equilibrium plots in Chapter 31. Representative equilibrium data for several gaseous solutes dissolved in water are provided in Problem 29.1.

Equations relating the equilibrium concentrations in the two phases have been developed, and are presented in physical chemistry and thermodynamic textbooks. For the case of nonideal gas and liquid phases, the relations are generally complex. However, in cases involving ideal gas and liquid phases, some fairly simple yet useful relations are known. For example, when the liquid phase is ideal, Raoult's law applies

$$p_A = x_A P_A \tag{29-1}$$

where p_A is the *equilibrium* partial pressure of component A in the vapor phase above the liquid phase, x_A is the equilibrium mole fraction of A in the liquid phase, and P_A is the vapor pressure of pure A at the equilibrium temperature. When the gas phase is ideal, Dalton's law is obeyed:

$$p_A = y_A P \tag{29-2}$$

where y_A is the mole fraction of A in the gas phase and P is the total pressure of the system. When both phases are ideal, the two equations may be combined to obtain a equilibrium relation between the concentration terms, x_A and y_A, at constant pressure and temperature, the combined Raoult–Dalton equilibrium law stipulates

$$y_A P = x_A P_A \tag{29-3}$$

Another equilibrium relation for gas and liquid phases where dilute solutions are involved is Henry's law. This law is expressed by

$$p_A = H\, c_{AL} \tag{29-4}$$

where H is the Henry's law constant, and c_{AL} is the concentration of solute A in the liquid phase at equilibrium. In Figure 29.2, the Henry's law region is nominally below an equilibrium liquid-phase concentration of 1.5 kgmole NH_3/m^3. The slope of the dashed line, obtained by linear regression of the equilibrium distribution data in this linear range, is H. Therefore in equation (29-4), H will have units of pressure/concentration. Henry's law can assume different units depending on the units of the gas and liquid phases. For example, recognizing that

$$c_{AL} = x_A C_L$$

where C_L is the total molar concentration of the liquid. Henry's law can also be expressed as

$$p_A = (H \cdot C_L) x_A \qquad (29\text{-}5)$$

The term $H \cdot C_L$ is often also called H, but is defined by $p_A = H x_A$, with H having units of pressure. Finally, combining equations (29-2) and (29-5) results in

$$y_A P = (H \cdot C_L) x_A$$

or

$$y_A = \left(\frac{H \cdot C_L}{P}\right) x_A = m\, x_A \qquad (29\text{-}6)$$

where m is the equilibrium distribution coefficient, which is dimensionless. For dilute solutions, the total molar liquid concentration C_L approximates that of the solvent liquid B. Consequently,

$$m = \frac{H \cdot C_L}{P} \cong \frac{H}{P} \frac{\rho_{B,\text{liq}}}{M_B}$$

Note that H is a fundamental physical property, whereas m is dependent upon total system pressure, P.

A complete discussion of equilibrium relations will be left to physical chemistry and thermodynamics textbooks. However, the following basic concepts common to all systems involving the distribution of a component between two phases are descriptive of interphase mass transfer:

1. At a fixed set of conditions, such as temperature and pressure, Gibbs's phase rule stipulates that a set of equilibrium relations exists, which may be shown in the form of an equilibrium distribution curve.

2. When the system is in equilibrium, there is no net mass transfer between the phases.

3. When a system is not in equilibrium, components or a component of the system will be transported in such a manner as to cause the system composition to shift toward equilibrium. If sufficient time is permitted, the system will eventually reach equilibrium.

▶ **29.2**

TWO-RESISTANCE THEORY

Many mass-transfer operations involve the transfer of material between two contacting phases. For example, in gas absorption, a solute is transferred from the gas phase into a liquid phase, where the gas serves as the source for solute, and the liquid phase serves as the sink for mass transfer. In contrast, for liquid stripping, the solute is transferred from the liquid to the gas, where the liquid is the source for solute, and the gas phase serves as the sink. These processes are contrasted in Figure 29.3. The interphase transfer involves three

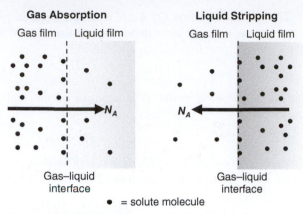

Gas Absorption

Liquid Stripping

Gas–liquid interface

Gas–liquid interface

● = solute molecule

Figure 29.3 Gas absorption vs. liquid stripping for transfer of solute *A* between contacting gas and liquid phases.

transfer steps: (1) the transfer of mass from the bulk conditions of one phase to the interfacial surface, (2) the transfer across the interface into the second phase, and (3) the transfer to the bulk conditions of the second phase.

A two-resistance theory, initially suggested by Whitman,[1,2] is often used to explain this process. The theory has two principal assumptions: (1) the rate of mass transfer between the two phases is controlled by the rates of diffusion through the phases on each side of the interface and (2) no resistance is offered to the transfer of the diffusing component across the interface. The concentration gradient driving force required to produce the mass transfer of component *A* from the gas phase to the bulk liquid phase, as illustrated in Figure 29.4, includes a partial pressure gradient from the bulk gas composition p_A, to the

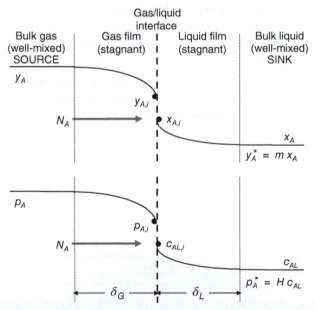

Figure 29.4 Concentration gradients between contacting gas and liquid phases where solute *A* is transferred from the bulk gas to the bulk liquid.

[1] W. G. Whitman, *Chem. Met. Engr.*, **29**(4), 197 (1923).
[2] W. K. Lewis and W. G. Whitman, *Ind. Eng. Chem.*, **16**, 1215 (1924).

interfacial gas composition, $p_{A,i}$, and a concentration gradient in the liquid from $c_{AL,i}$, at the interface to the bulk liquid concentration, c_{AL}. On the basis of Whitman's second assumption of no resistance to mass transfer at the interfacial surface, the local compositions of solute A right at the gas and liquid sides of the gas–liquid interface, represented by $p_{A,i}$ and $c_{AL,i}$, are at the equilibrium condition and are described by the thermodynamic relationships discussed in Section 29.1.

When the transfer is from the gas phase to the liquid phase, then $p_{A,i}$ will be greater than $p_{A,i}$, and $c_{AL,i}$ will be greater than c_{AL}. In contrast, when the transfer is from the liquid phase to the gas phase, as shown in Figure 29.5, then c_{AL} will be greater than $c_{AL,i}$ and $p_{A,i}$ will be greater than p_A.

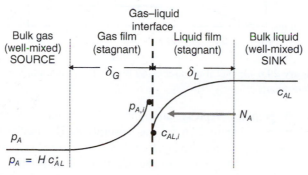

Figure 29.5 Concentration gradients between contacting gas and liquid phases where solute A is transferred from the bulk liquid to the bulk gas.

Individual Film Mass-Transfer Coefficients

Consider a gas phase that contains a mixture carrier gas and solute A, and a liquid phase that contains solute A dissolved in a solvent. Restricting our discussion to the steady-state transfer of component A from the gas phase to the liquid phase, we can describe the mass-transfer flux across the gas and liquid films by the following equations:

$$N_A = k_G(p_A - p_{A,i}) \tag{29-7}$$

and

$$N_A = k_L(c_{AL,i} - c_{AL}) \tag{29-8}$$

where k_G is the *convective mass-transfer coefficient in the gas film*, in units of moles of A transferred/(time)(interfacial area)(pressure)(and k_L is the *convective mass-transfer coefficient in the liquid film*, in units of moles of A transferred)/(time)(interfacial area) (concentration).

In this gas–liquid interphase mass-transfer process, it is assumed that the carrier gas and solvent are immiscible, and the solvent is nonvolatile. Therefore, the carrier gas and the solvent do not participate in mass transfer across phases. Under these conditions, the transfer of solute A through the carrier gas in the gas film, and the transfer of solute A through the solvent in the liquid phase, are both considered to be unimolecular mass-transfer processes, as described in Chapter 26. If air is the carrier gas and water is the solvent, we acknowledge that oxygen gas in air is sparingly soluble in water, and water has a finite vapor pressure. However, the presence of dissolved oxygen in water and water vapor in air are not assumed to affect the transfer of solute A.

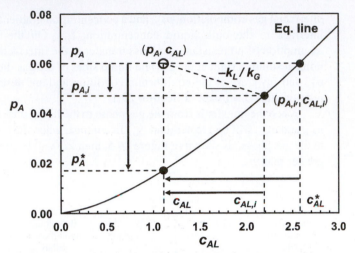

Figure 29.6 Graphical representation of gas–liquid interphase mass-transfer process for gas absorption.

A graphical representation of the gas–liquid interphase mass-transfer process is presented in Figure 29.6 for gas absorption. The partial pressure difference in the gas film, $p_A - p_{A,i}$, is the driving force necessary to transfer component A from the bulk gas conditions to the gas side of the interface separating the two phases. The concentration difference, $c_{AL,i} - c_{AL}$, is the driving force necessary to continue the transfer of A from the liquid side of the interface into the bulk liquid phase. These film mass-transfer driving forces are indicated by the short arrows in Figure 29.6.

Under steady-state conditions, the flux of mass in one phase must equal the flux of mass in the second phase. Combining equations (29-7) and (29-8) obtains

$$N_A = k_G(p_A - p_{A,i}) = -k_L(c_{AL} - c_{AL,i}) \tag{29-9}$$

The ratio of the two convective mass-transfer coefficients may be obtained from equation (29-9) by rearrangement, giving

$$-\frac{k_L}{k_G} = \frac{p_A - p_{A,i}}{c_{AL} - c_{AL,i}} \tag{29-10}$$

In Figure 29.6, the application of equation (29-10) for the evaluation of the interfacial compositions for a specific set of bulk compositions is illustrated. The operating point (c_{AL}, p_A), located *above* the equilibrium line, represents conditions where the transfer is from the gas phase to the bulk liquid phase. Equation (29-10) can be thought of as the point-slope form of a line extending from operating point (c_{AL}, p_A) to interface composition point $(c_{AL,i}, p_{A,i})$ with slope $-k_L/k_G$, as shown in Figure 29.6. Hence, if the individual film mass-transfer coefficients are known, the interface compositions can be graphically estimated.

We note here that an operating point representing the bulk conditions found in a liquid stripping operation, where transfer of the solute is from the liquid phase to the gas phase, would be located *below* the equilibrium line.

Based on the definitions provided in Table 29.1, equations (29-7) and (29-8) can also be expressed in terms of mole fraction driving forces for the gas and liquid phases:

$$N_A = k_y(y_A - y_{A,i}) \tag{29-11}$$

$$N_A = k_x(x_{A,i} - x_A) \tag{29-12}$$

Table 29.1 Individual mass-transfer coefficients

Gas film		
Driving force	Flux equation	Units of k
Partial pressure (p_A)	$N_A = k_G(p_A - p_{A,i})$	kgmole/m$^2 \cdot$ s $\cdot$ atm
Concentration (c_A)	$N_A = k_c(c_{AG} - c_{AG,i})$	kgmole/(m$^2 \cdot$ s $\cdot$ (kgmole/m^3)) or m/s
Mole fraction (y_A)	$N_A = k_y(y_A - y_{A,i})$	kgmole/m$^2 \cdot$ s

Liquid film		
Concentration (c_{AL})	$N_A = k_L(c_{AL,i} - c_{AL})$	kgmole/(m$^2 \cdot$ s $\cdot$ (kgmole/m^3)) or m/s
Mole fraction (x_A)	$N_A = k_x(x_{A,i} - x_A)$	kgmole/m$^2 \cdot$ s

with

$$N_A = k_y\left(y_A - y_{A,i}\right) = -k_x\left(x_A - x_{A,i}\right)$$

and

$$-\frac{k_x}{k_y} = \frac{y_A - y_{A,i}}{x_A - x_{A,i}} \tag{29-13}$$

Table 29.1 provides the definitions of the most often encountered individual-phase mass-transfer coefficients, based on the phase and the dependent variable used to describe the mass-transfer driving force.

In Table 29.1, the flux N_A is the same through the gas film and the liquid film. Consequently, by equating the flux for a given definition, interrelationships between mass-transfer coefficients can be obtained. For the gas phase,

$$N_A = k_y\left(y_A - y_{A,i}\right) = k_G\left(y_A - y_{A,i}\right) = k_G P\left(y_A - y_{A,i}\right)$$

and

$$N_A = k_c\left(c_A - c_{A,i}\right) = k_y C\left(y_A - y_{A,i}\right) = k_y \frac{P}{RT}\left(y_A - y_{A,i}\right)$$

Therefore,

$$k_G = \frac{k_y}{P} = \frac{k_c}{RT} \tag{29-14a}$$

Similarly, for the liquid phase, one can show that

$$k_L = \frac{k_x}{C_L} \cong k_x \frac{M_B}{\rho_{B,\text{liq}}} \tag{29-14b}$$

For dilute solutions, the total molar concentration C_L is approximated by the molar concentration of the solvent.

Overall Mass-Transfer Coefficients

It is quite difficult to physically measure the partial pressure and the liquid concentration of transferring solute A at the gas–liquid interface. It is therefore convenient to employ overall coefficients based on an overall driving force between the bulk gas- and liquid-phase compositions for solute A. Obviously, one cannot express the overall driving force as $p_A - c_{AL}$ due to the difference in concentration units. In Figure 29.6, one observes that on the equilibrium line, the bulk liquid composition c_{AL} would be in equilibrium with the partial pressure term p_A^*. Therefore, at the total pressure and temperature of the system, p_A^* represents c_{AL} in units consistent with p_A. Accordingly, a flux equation that includes the resistance to diffusion in both phases, and that is in terms of partial pressure driving force, is defined by

$$N_A = K_G(p_A - p_A^*) \tag{29-15}$$

where p_A^* represents the partial pressure of solute A that would be in equilibrium with c_{AL}, the bulk concentration of solute A in the liquid phase, and K_G is the overall mass-transfer coefficient based on the overall partial pressure driving force, in units of moles of A transferred/(time)(interfacial area)(pressure).

Similarly, the overall bulk gas partial pressure of solute A, p_A, would be equilibrium with the liquid concentration term c_{AL}^*. Therefore, c_{AL}^* represents p_A in units consistent with c_{AL}. A flux equation that includes the resistance to diffusion in both phases, and that is in terms of the liquid-phase concentration driving force, is defined by

$$N_A = K_L(c_{AL}^* - c_{AL}) \tag{29-16}$$

where c_{AL}^* represents the liquid concentration of solute A that would be in equilibrium with partial pressure of solute A in the bulk gas phase (p_A), and K_L is the overall mass-transfer coefficient based on a liquid driving force, in units of moles of A transferred/(time)(interfacial area)(mole A/volume). The "overall" mass-transfer driving forces, $(p_A - p_A^*)$ based on the gas phase, and $(c_{AL}^* - c_{AL})$ based on the liquid phase, are represented by the long arrows in Figure 29.6.

Linear Equilibrium A relation between the overall mass-transfer coefficients and the individual phase mass-transfer coefficients can be obtained when the equilibrium relation is linear, as expressed by

$$p_{A,i} = H\, c_{AL,i} \tag{29-17}$$

at the gas–liquid interface. This condition is always encountered at low concentrations where Henry's law is obeyed; the proportionality constant is then the Henry's law constant, H. For a linear equilibrium relationship, it may also be stated that

$$p_A^* = H\, c_{AL} \tag{29-18}$$

and

$$p_A = H\, c_{AL}^* \tag{29-19}$$

To obtain a relationship for K_G in terms of k_G, k_L, and H, rearrangement of equation (29-15) gives

$$\frac{1}{K_G} = \frac{p_A - p_A^*}{N_A} = \frac{p_A - p_{A,i}}{N_A} + \frac{p_{A,i} - p_A^*}{N_A}$$

and combining with equations (29-17) and (29-18) gives

$$\frac{1}{K_G} = \frac{(p_A - p_{A,i})}{N_A} + \frac{H(c_{AL,i} - c_{AL})}{N_A} \tag{29-20}$$

The substitution of equations (29-7) and (29-8) into equation (29-20) relates K_G to the individual film coefficients and local equilibrium at the gas–liquid interface by the relationship

$$\frac{1}{K_G} = \frac{1}{k_G} + \frac{H}{k_L}$$

(29-21)

A similar expression for K_L may be derived as follows:

$$\frac{1}{K_L} = \frac{c_{AL}^* - c_{AL}}{N_A} = \frac{(c_{AL} - c_{AL,i})}{N_A} + \frac{(c_{AL,i} - c_{A,L})}{N_A} = \frac{(p_A - p_{A,i})}{H \cdot N_A} + \frac{(c_{AL,i} - c_{A,L})}{N_A}$$

or

$$\frac{1}{K_L} = \frac{1}{H \cdot k_G} + \frac{1}{k_L}$$

(29-22)

Equations (29-15) and (29-16) can also be expressed in terms of overall mole fraction driving forces for the gas and liquid phases:

$$N_A = K_y\left(y_A - y_A^*\right)$$

(29-23)

$$N_A = K_x\left(x_A^* - x_A\right)$$

(29-24)

with

$$y_A^* = m \cdot x_A$$

(29-25)

and

$$y_A = m \cdot x_A^*$$

(29-26)

The overall mass-transfer coefficient based on gas-phase mole fraction driving force (K_y) is

$$\frac{1}{K_y} = \frac{1}{k_y} + \frac{m}{k_x}$$

(29-27)

and the overall mass-transfer coefficient based on the liquid-phase mole fraction driving force (K_x) is

$$\frac{1}{K_x} = \frac{1}{m \cdot k_y} + \frac{1}{k_x}$$

(29-28)

Nonlinear Equilibrium For a nonlinear equilibrium line, if the operating point is changed, then the overall mass-transfer coefficient will be a function of gas and liquid composition even if the gas and liquid film mass-transfer coefficients have no concentration dependence. Consider the curved equilibrium line shown in Figure 29.7 with operating point (x_A, y_A).

In the case of a gas absorption process where the operating point lies above the equilibrium line, the equilibrium line is approximated as linear from x_A to $x_{A,i}$ to with slope m', and from $x_{A,i}$ to x_A^* to with slope m'':

$$m' = \frac{y_{A,i} - y_A^*}{x_{A,i} - x_A}$$

(29-29)

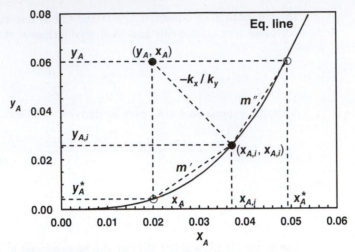

Figure 29.7 Graphical representation of gas–liquid interphase mass-transfer process with nonlinear equilibrium line.

and

$$m'' = \frac{y_A - y_{A,i}}{x_A^* - x_{A,i}}$$

(29-30)

as shown in Figure 29.7. Consequently, the overall mass-transfer coefficients K_y and K_x are developed by considering

$$\frac{1}{K_y} = \frac{y_A - y_A^*}{N_A} = \frac{(y_A - y_{A,i})}{N_A} + \frac{(y_{A,i} - y_A^*)}{N_A} = \frac{(y_A - y_{A,i})}{N_A} + \frac{m'(x_{A,i} - x_A)}{N_A}$$

which simplifies to

$$\frac{1}{K_y} = \frac{1}{k_y} + \frac{m'}{k_x}$$

(29-31)

and

$$\frac{1}{K_x} = \frac{x_A^* - x_A}{N_A} = \frac{(x_A^* - x_{A,i})}{N_A} + \frac{(x_{A,i} - x_A)}{N_A} = \frac{(y_A - y_{A,i})}{m'' \cdot N_A} + \frac{(x_{A,i} - x_A)}{N_A}$$

which simplifies to

$$\frac{1}{K_x} = \frac{1}{m'' \cdot k_y} + \frac{1}{k_x}$$

(29-32)

Note that in the linear equilibrium range, $m = m' = m''$.

Mass-Transfer Driving Forces and Relative Resistances   Figure 29.6 illustrates the driving forces associated with each phase and the overall driving forces. The ratio of the resistances in an individual phase to the total resistance may be determined by

$$\frac{\text{resistance in the gas phase}}{\text{total resistance in both phases}} = \frac{\Delta p_{A,\text{gas film}}}{\Delta p_{A,\text{total}}} = \frac{p_A - p_{A,i}}{p_A - p_A^*} = \frac{1/k_G}{1/K_G}$$

(29-33)

and

$$\frac{\text{resistance in the liquid phase}}{\text{total resistance in both phases}} = \frac{\Delta c_{AL,\text{liquid film}}}{\Delta c_{AL,\text{total}}} = \frac{c_{AL,i} - c_{AL}}{c_{AL}^* - c_{AL}} = \frac{1/k_L}{1/K_L} \tag{29-34}$$

for the gas and liquid phases, respectively. Since the mass-transfer resistances are in series, the overall mass-transfer coefficient referenced to a specific phase will always be smaller than its individual film mass-transfer coefficient. Equations (29-21) and (29-22) stipulate that the relative magnitudes of the individual phase resistances depend on the solubility of the gas, as indicated by the magnitude of the Henry's law constant H. For a system involving a highly soluble solute gas, such as ammonia in water, H is small, whereas for a system involving a sparingly soluble gas, such as carbon dioxide in water, H is large. From equation (29-21), for very small values for H, it is noted that $K_G \rightarrow k_G$, and we may conclude that the gas-phase resistance is essentially equal to the overall resistance in such a system. When this is true, the major resistance to mass transfer lies in the gas phase, and such a system is said to be *gas-phase controlled*. On Figure 29.6, this will occur when the driving force $(c_{AL,i} - c_{AL}) \rightarrow 0$ and $(p_A - p_{A,i}) \rightarrow (p_A - p_A^*)$, with slope $-k_L/k_G \rightarrow \infty$ (vertical line).

Systems that involve gases of low solubility, such as carbon dioxide in water, may have such a large value of H that from equation (29-22), $K_L \rightarrow k_L$, and so the system is said to be *liquid-phase controlled*. On Figure 29.6, this will occur when the driving force $(p_A - p_{A,i}) \rightarrow 0$ and $(c_{AL,i} - c_{AL}) \rightarrow (c_{AL}^* - c_{AL})$, with slope $-k_L/k_G \rightarrow 0$. If 100% solute A exists within the gas phase as a pure component, and if the solvent is nonvolatile, then the system is by definition liquid-phase controlling, since no mixture exists in the gas phase.

Of course, in many systems, both phase resistances are important and must be considered when evaluating the total resistance.

In Chapter 28, the individual phase convective coefficients were shown to be dependent on the nature of the diffusing component, on the phase through which the component is diffusing, and also on the flow conditions of the phase. Even when the individual film coefficients are essentially independent of the concentration, the overall coefficients may vary with the concentration unless the equilibrium line is straight. In this context, if the equilibrium line is nonlinear, K_G and K_L are referred to as *local* overall mass-transfer coefficients.

Gas Absorption vs. Liquid Stripping

In a liquid stripping process, solute A dissolved in the bulk liquid is the source for mass transfer, and the bulk gas phase is the sink for mass transfer. The operating point—e.g., (p_A, c_{AL})—or (y_A, x_A) is below the equilibrium line. In this context, the flux relationships are

$$N_A = k_G(p_{A,i} - p_A) = K_G(p_A^* - p_A)$$
$$N_A = k_y(y_{A,i} - y_A) = K_y(y_A^* - y_A) \tag{29-35}$$

$$N_A = k_L(c_{AL} - c_{AL,i}) = K_L(c_{AL} - c_{AL}^*)$$
$$N_A = k_x(x_A - x_{A,i}) = K_x(x_A - x_A^*) \tag{29-36}$$

For linear equilibrium, equations for the overall mass-transfer coefficients $(K_G, K_L, K_y,$ and $K_x)$ are unchanged. However, for a nonlinear equilibrium line, the definitions for m' and m'' change, and it is left to the reader to show that

$$\frac{1}{K_y} = \frac{1}{k_y} + \frac{m''}{k_x} \tag{29-37}$$

$$\frac{1}{K_x} = \frac{1}{k_x} + \frac{1}{m'k_y} \tag{29-38}$$

with

$$m' = \frac{y_{A,i} - y_A}{x_{A,i} - x_A^*} \quad \text{and} \quad m'' = \frac{y_A^* - y_{A,i}}{x_A - x_{A,i}}$$

for the liquid stripping process.

The application of the two-resistance theory for both absorption and stripping of a component will be illustrated in the following two examples.

Example 1

A liquid stripping process is used to transfer hydrogen sulfide (H_2S) dissolved in water into a contacting air stream. The solvent (water) and carrier gas (air) are assumed to be immiscible, and the solvent is assumed to be nonvolatile. At the present conditions of operation, the composition of H_2S in the bulk gas phase is 1.0 mole%, and in the liquid phase is 0.0006 mole%, or 112 mg H_2S/L. The individual mass-transfer coefficients are $k_x = 0.30$ kgmole/m$^2 \cdot$s for the liquid film, and $k_y = 4.5 \times 10^{-3}$ kgmole/m$^2 \cdot$s for the gas film. The temperature is 20 C, and the total system pressure is 1.5 atm. The density of the liquid approaches the density of liquid water at 20 C, 992.3 kg/m^3. The equilibrium line for the H_2S-water-air system at 20 C is provided in Figure 29.8. The molecular weight of H_2S is 34 g/gmole. The characteristics of this gas–liquid mass-transfer process are developed in parts (a) to (d) below.

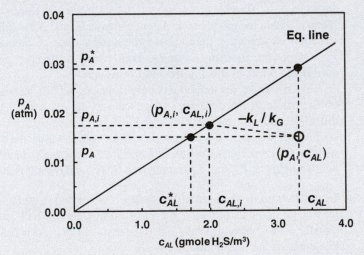

Figure 29.8 Graphical representation of gas–liquid interphase mass-transfer process for transfer of hydrogen sulfide (H_2S) dissolved in water into air, Example 1.

(a) Plot the operating point, in composition terms of p_A for the gas phase, and c_{AL} for the liquid phase, directly on Figure 29.8. The following changes on the composition units on the composition units are required. For the bulk liquid phase,

$$c_{AL} = x_A C_L \cong x_A \frac{\rho_{B,liq}}{M_B} = (6.0 \times 10^{-5}) \left(\frac{992.3 \text{ kg/m}^3}{18 \text{ kg/gmole}}\right) \left(\frac{1000 \text{ gmole}}{\text{kgmole}}\right) = 3.31 \frac{\text{gmole}}{\text{m}^3}$$

For the bulk gas phase,

$$p_A = y_A P = 0.01(1.5 \text{ atm}) = 0.015 \text{ atm}$$

The operating point (p_A, c_{AL}) is plotted on Figure 29.8 as the open circle. Note that since this is a liquid stripping process, the operating point is below the equilibrium line.

(b) Determine m, the equilibrium distribution coefficient.

The equilibrium line in p_A-c_{AL} coordinates is linear, and it is determined that the slope $H = 8.8$ m³ · atm/gmole. Therefore,

$$m = \frac{H \cdot C_L}{P} \simeq \frac{H \rho_{B,\text{liq}}}{P \, M_B} = \left(\frac{8.8 \text{ m}^3 \cdot \text{atm/kgmole}}{1.5 \text{ atm}}\right)\left(\frac{992.3 \text{ kg/m}^3}{18 \text{ kg/kgmole}}\right) = 323.4$$

with $C_L = 55.1$ kgmole/m³.

(c) Estimate the overall mass-transfer coefficients K_G and K_L.

There are many pathways to obtain K_G and K_L. One option is to convert k_y and k_x to k_G and k_L:

$$k_G = \frac{k_y}{P} = \frac{4.5 \times 10^{-3} \text{ kgmole/m}^2 \cdot \text{s}}{1.5 \text{ atm}} = 3.0 \times 10^{-3} \frac{\text{kgmole}}{\text{m}^2 \cdot \text{s} \cdot \text{atm}}$$

$$k_L = \frac{k_x}{C_L} = \frac{0.30 \text{ kgmole/m}^2 \cdot \text{s}}{55.1 \text{ kgmole/m}^3} = 5.44 \times 10^{-3} \text{ m/s}$$

and then use equation (29-21):

$$\frac{1}{K_G} = \frac{1}{k_G} + \frac{H}{k_L} = \frac{1}{3.0 \times 10^{-3} \text{ kgmole/m}^2 \cdot \text{s} \cdot \text{atm}} + \frac{8.8 \text{ m}^3 \cdot \text{atm/kgmole}}{5.44 \times 10^{-3} \text{ m/s}}$$

to obtain $K_G = 5.13 \times 10^{-4}$ kgmole/m² · s · atm, and equation (29-22):

$$\frac{1}{K_L} = \frac{1}{H \cdot k_G} + \frac{1}{k_L} = \frac{1}{(8.8 \text{ m}^3 \cdot \text{atm/kgmole})(3.0 \times 10^{-3} \text{ kgmole/m}^2 \cdot \text{s} \cdot \text{atm})} + \frac{1}{5.44 \times 10^{-3} \text{ m/s}}$$

to obtain $K_L = 4.51 \times 10^{-3}$ m/s. Another approach would be to estimate K_y and K_x, and then convert via $K_G = K_y/P$ and $K_L = K_x/C_L$, which will return the same result.

(d) Determine the flux from the liquid phase to the gas phase, and the interface compositions, $p_{A,i}$ and $c_{AL,i}$.

Using values for K_G and K_L, the flux is

$$N_A = K_G(p_A - p_A^*) = \left(5.13 \times 10^{-4} \frac{\text{kgmole}}{\text{m}^2 \cdot \text{s} \cdot \text{atm}}\right)(0.015 - 0.029) \text{ atm} = -7.2 \times 10^{-6} \frac{\text{kgmole}}{\text{m}^2 \cdot \text{s} \cdot \text{atm}}$$

$$N_A = K_L(c_{AL}^* - c_{AL}) = (4.51 \times 10^{-3} \text{ m/s})(1.71 - 3.31)\frac{\text{gmole}}{\text{m}^3}\frac{1 \text{ kgmole}}{1000 \text{ gmole}} = -7.2 \times 10^{-6} \frac{\text{kgmole}}{\text{m}^2 \cdot \text{s}}$$

with

$$p_A^* = H \cdot c_{AL} = \left(8.8 \frac{\text{m}^3 \cdot \text{atm}}{\text{kgmole}}\right)\left(3.31 \frac{\text{gmole}}{\text{m}^3}\right)\left(\frac{1 \text{ kgmole}}{1000 \text{ gmole}}\right) = 0.029 \text{ atm}$$

and

$$c_{AL}^* = \frac{p_A}{H} = \left(\frac{0.015 \text{ atm}}{8.8 \text{ m}^3 \cdot \text{atm/kgmole}}\right)\left(\frac{1000 \text{ gmole}}{1 \text{ kgmole}}\right) = 1.71 \frac{\text{gmole}}{\text{m}^3}$$

Note that the flux calculated based on either the gas phase or the liquid phase returns the same result. Note also that the flux is negative, which indicates that the flux of H_2S is moving from the liquid to the gas phase. Finally, the interface compositions are backed out from the flux using equations (29-7) and (29-8):

$$p_{A,i} = p_A - \frac{N_A}{k_G} = 0.015 \text{ atm} - \frac{-7.2 \times 10^{-6} \text{ kgmole/m}^2 \cdot \text{s}}{3.0 \times 10^{-3} \text{ kgmole/m}^2 \cdot \text{s} \cdot \text{atm}} = 0.017 \text{ atm}$$

$$c_{AL,i} = c_{AL} + \frac{N_A}{k_L} = 3.31 \frac{\text{gmole}}{\text{m}^3} + \frac{-7.2 \times 10^{-6} \text{ kgmole/m}^2 \cdot \text{s}}{5.44 \times 10^{-3} \text{ m/s}} = 1.99 \frac{\text{gmole}}{\text{m}^3}$$

All composition points are plotted out on Figure 29.8. Since $c_{AL,i}$ is approaching c_{AL}^*, the process is moving toward the liquid-phase controlling condition. In comparing this diagram with Figure 29.6, we see the differences in mass-transfer driving forces between liquid stripping and gas absorption.

Example 2

A gas absorption process is used to remove ammonia (NH_3) from a gaseous mixture of ammonia and air, using liquid water as the solvent. In the present process, the bulk gas stream contains 30 mole% NH_3, and the bulk liquid stream contains 5.0 mole% dissolved NH_3. The equilibrium distribution of NH_3 between gas and liquid phases at 30 C and 1.0 atm total system pressure, in mole fraction coordinates, is presented in Figure 29.9. At the conditions of the process, the liquid film individual mass-transfer coefficient in mole fraction coordinates is $k_x = 0.030$ kgmole/m$^2 \cdot$ s, and the gas film individual mass-transfer coefficient in mole fraction coordinates is $k_y = 0.010$ kgmole/m$^2 \cdot$ s.

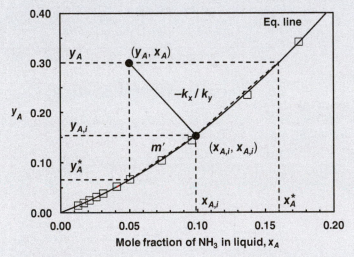

Figure 29.9 Graphical representation of gas–liquid interphase mass-transfer process for transfer of ammonia (NH_3) from a gas mixture to water, Example 2.

Determine the interface compositions in the gas and liquid phase, $x_{A,i}$ and $y_{A,i}$, the overall mass-transfer coefficient based on the gas-phase mole fraction driving force, K_y, and the flux of NH_3 from the bulk gas to liquid phase.

Since the equilibrium line is nonlinear, a graphical solution method will be used. We begin by plotting the operating point, $x_A = 0.05$, $y_A = 0.30$ on the distribution plot provided in Figure 29.9. By equation (29-13), a line from the operating point extends down to the interface composition point $(x_{A,i}, y_{A,i})$ with slope

$$-\frac{k_x}{k_y} = \frac{y_A - y_{A,i}}{x_A - x_{A,i}} = -\frac{0.030 \text{ kgmole/m}^2 \cdot \text{s}}{0.01 \text{ kgmole/m}^2 \cdot \text{s}} = -3.0$$

By graphical solution at the intersection of this line and the equilibrium line, $x_{A,i} = 0.10$, and $y_{A,i} = 0.15$. Also, from the graph, $y_A^* = 0.065$ at $x_A = 0.05$, and $x_A^* = 0.16$ at $y_A = 0.30$. The equilibrium line is nonlinear, so the overall mass-transfer coefficient K_y will depend upon composition according to equation (29-31). The local equilibrium relationship, m', is estimated by equation (29-29):

$$m' = \frac{y_{A,i} - y_A^*}{x_{A,i} - x_A} = \frac{0.15 - 0.065}{0.10 - 0.05} = 1.70$$

Subsequently, K_y is estimated by equation (29-31):

$$\frac{1}{K_y} = \frac{1}{k_y} + \frac{m'}{k_x} = \frac{1}{0.010 \text{ kgmole/m}^2 \cdot \text{s}} + \frac{1.70}{0.030 \text{ kgmole/m}^2 \cdot \text{s}}$$

with $K_y = 6.4 \times 10^{-3}$ kgmole/m² · s. Note that $K_y < k_y$, and the relative resistance, based on equation (29-33), for the gas phase is

$$\frac{1/k_y}{1/K_y} = \frac{K_y}{k_y} = \frac{6.4 \times 10^{-3} \text{ kgmole/m}^2 \cdot \text{s}}{1.0 \times 10^{-2} \text{ kgmole/m}^2 \cdot \text{s}} \times 100\% = 64\% \text{ gas-phase controlling}$$

Finally, the flux from the bulk gas to liquid phase is

$$N_A = K_y(y_A - y_A^*) = (6.4 \times 10^{-3} \text{ kgmole/m}^2 \cdot \text{s})(0.30 - 0.065) = 1.5 \times 10^{-3} \text{ kgmole/m}^2 \cdot \text{s}$$

which checks with the flux through the gas film:

$$N_A = k_y(y_A - y_{A,i}) = (0.010 \text{ kgmole/m}^2 \cdot \text{s})(0.30 - 0.15) = 1.5 \times 10^{-3} \text{ kgmole/m}^2 \cdot \text{s}$$

Interphase Mass Transfer in Process Material Balances Gas–liquid interphase mass-transfer processes are realized in a variety of process configurations, as described in Chapters 30 and 31. These include bubble columns, gas-sparged stirred tanks, surface-aerated stirred tanks, and gas–liquid packed towers. To start, we consider a steady-state, well-mixed process with a defined surface area for gas–liquid mass-transfer, as illustrated in Figure 29.10. In an enclosed tank, both gas and liquid are fed into the tank, and the gas

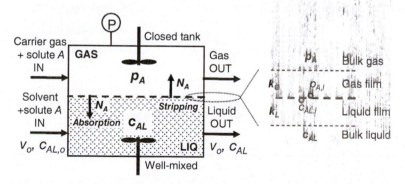

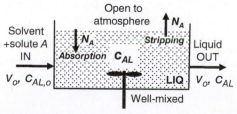

Figure 29.10 Steady-state, well-mixed, gas–liquid mass-transfer process across a defined interface.

headspace and liquid inside the tank are considered well mixed. Depending on the concentrations of the transferring solute in either the liquid or the gas, the process can be gas absorption or liquid stripping. In an open tank, the liquid phase is considered well mixed, and the solute transfers from the liquid to the gas as a liquid stripping process. If the tank is open to the atmosphere, usually the transferring solute concentration in the bulk gas is considered to be near zero—i.e., the surrounding gas phase is an infinite sink for interphase mass transfer.

Example 3 illustrates how the gas–liquid interface mass-transfer concepts presented in this chapter can be integrated into process material balances for design and analysis of gas–liquid mass-transfer equipment.

Example 3

A process similar to that shown in Figure 29.10 is used to transfer CO_2 from a compressed flue gas mixture into a tank of liquid water. The gas space over the liquid inside the tank is considered well mixed, and the CO_2 composition exiting the headspace contains 10 mole% CO_2 and 90% N_2 at 2.0 atm total system pressure and 25 C. Water containing no dissolved CO_2 enters the tank at a flow rate (ν_o) of 0.20 m³/h. The liquid is well mixed, but the gas is not bubbled into the liquid, so the liquid surface defines the gas–liquid interphase mass-transfer area, S. The diameter of the cylindrical tank (d) is 3.0 m, and the liquid depth is 1.0 m. It can be assumed that the N_2 gas does not dissolve appreciably in the liquid, and the solvent is nonvolatile.

What is the concentration of dissolved CO_2 in the liquid stream exiting the tank, c_{AL}, and what is the maximum dissolved CO_2 concentration, c_{AL}^*?

At the conditions of agitation, the gas film mass-transfer coefficient is $k_c = 0.05$ m/s, and the liquid film mass-transfer coefficient is $k_L = 1.5 \times 10^{-5}$ m/s. The Henry's law constant for dissolution of CO_2 gas into liquid water[3] at 25 C is 1630 atm, based on the definition $p_{A,i} = H\,x_{A,i}$. The density of the liquid water at 25°C is 995.2 kg/m³.

For convenience, the calculations will be based on the liquid phase in the tank, with $A = CO_2$, and $B = H_2O$. The process analysis and solution strategy will require determination of the overall mass-transfer coefficient K_L, and the development of a material balance model to predict the dissolved CO_2 concentration in the liquid stream exiting the tank.

For a linear equilibrium relationship defined by Henry's law, K_L is given by equation (29-22)

$$\frac{1}{K_L} = \frac{1}{H \cdot k_G} + \frac{1}{k_L}$$

with H defined in terms of $p_{A,i} = H\,c_{AL,i}$. The needed conversions for obtaining H and k_G are

$$H = \frac{1630\,atm}{c_L} = 1630\,atm\,\frac{M_B}{\rho_{B,liq}} = 1630\,atm\,\frac{18\text{ kg/kgmole}}{995.2\text{ kg/m}^3} = 29.5\,\frac{\text{m}^3 \cdot atm}{\text{kgmole}}$$

$$k_G = \frac{k_c}{RT} = \frac{\left(0.05\,\dfrac{\text{m}}{\text{s}}\right)}{\left(0.08206\,\dfrac{\text{m}^3 \cdot atm}{\text{kgmole} \cdot \text{K}}\right)(298\text{ K})} = 2.05 \times 10^{-3}\,\frac{\text{kgmole}}{\text{m}^2 \cdot \text{s} \cdot atm}$$

---

[3] R. H. Perry and C. H. Chilton, *Chemical Engineer's Handbook*, 5th edition, McGraw-Hill Book Company, New York, 1973.

and so

$$\frac{1}{K_L} = \frac{1}{(29.5 \text{ m}^3 \cdot \text{atm/kgmole})(2.05 \times 10^{-3} \text{ kgmole/m}^2 \cdot \text{s} \cdot \text{atm})} + \frac{1}{(1.5 \times 10^{-5} \text{m/s})}$$

or $K_L = 1.5 \times 10^{-5}$ m/s. Note that $K_L \approx k_L$, so that the liquid film mass transfer coefficient k_L approaches the overall liquid mass transfer coefficient.

A material balance on CO_2 in the liquid phase of the tank makes the following assumptions: (1) steady-state process, (2) well-mixed liquid in the tank, so that the dissolved CO_2 concentration in the tank equals the dissolved CO_2 concentration in the outlet liquid stream, (3) constant liquid volume, (4) dilute process, and (5) no reaction of CO_2. Under these assumptions, the material balance for CO_2 (moles CO_2/time) is

$$\begin{pmatrix} \text{dissolved } CO_2 \text{ in inlet} \\ \text{liquid flow (IN)} \end{pmatrix} - \begin{pmatrix} \text{interphase mass transfer of} \\ CO_2 \text{ in gas to } CO_2 \text{ in liquid (IN)} \end{pmatrix} + \begin{pmatrix} \text{dissolved } CO_2 \text{ in outlet} \\ \text{liquid flow (OUT)} \end{pmatrix}$$
$$+ \begin{pmatrix} \text{reaction of } CO_2 \\ \text{in liquid (GEN)} \end{pmatrix} = \begin{pmatrix} \text{accmulation of } CO_2 \\ \text{in liquid (ACC)} \end{pmatrix}$$

or

$$v_o \, c_{AL,o} + N_A \, S \, - v_o \, c_{AL} + 0 = 0$$

In an absorption process, the CO_2 in the gas serves as the source, and CO_2 dissolved in the liquid serves as the sink. The flux N_A from bulk gas phase to bulk liquid phase based on the overall liquid phase driving force is defined by

$$N_A = K_L (c_{AL}^* - c_{AL}) \text{ with } c_{AL}^* = p_A/H$$

Combining these relationships gives

$$v_o \, (c_{AL,o} - c_{AL}) + K_L \left(\frac{p_A}{H} - c_{AL} \right) S \; = 0$$

And solving for c_{AL}, one obtains

$$c_{AL} = \frac{v_o \, c_{AL,o} + \dfrac{K_L \, S \, p_A}{H}}{v_o + K_L \, S} = \frac{v_o \, c_{AL,o} + K_L \, S \, c_{AL}^*}{v_o + K_L \, S}$$

The material balance equation requires the gas–liquid surface area, S, and bulk gas CO_2 partial pressure, p_A:

$$S = \pi d^2/4 = \pi (3.0 \text{ m})^2/4 = 7.07 \text{ m}^2$$

$$p_A = y_A P = (0.10)(2.0 \text{ atm}) = 0.20 \text{ atm}$$

The inlet concentration of CO_2 dissolved in the liquid, $c_{AL,o}$, is 0. Finally, the outlet concentration is

$$c_{AL} = \frac{(0.2 \text{ m}^3/\text{h})(0) + \dfrac{(1.5 \times 10^{-5}\text{m/s})(7.07 \text{ m}^2)(0.20 \text{ atm})}{29.5 \text{ m}^3 \cdot \text{atm/kgmole}} \dfrac{1000 \text{ gmole}}{1 \text{ kgmole}}}{(0.2 \text{ m}^3/\text{h})(1 \text{ h }/3600 \text{ s}) + (1.5 \times 10^{-5}\text{m/s})(7.07 \text{ m}^2)} = 4.45 \frac{\text{gmole A}}{\text{m}^3}$$

The maximum possible outlet concentration is $c_{AL}*$, determined by Henry's law:

$$c_{AL}^* = \frac{p_A}{H} = \frac{0.20 \text{ atm}}{(29.5 \text{ m}^3 \cdot \text{atm/kgmole})(1 \text{ kgmole} / 1000 \text{ gmole})} = 6.78 \frac{\text{gmole A}}{\text{m}^3}$$

In this analysis, c_{AL} will approach $c_{AL}*$ as the liquid flow rate v_o decreases or overall mass transfer coefficient K_L increases.

CLOSURE

In this chapter, we have considered the mechanism of steady-state mass transfer of solute A between a contacting gas and liquid phase. The two-resistance theory was presented. This theory defines the mass transfer in each phase as a function of the concentration driving force and the individual mass-transfer coefficient according to the equations

$$N_A = k_G(p_A - p_{A,i})$$

and

$$N_A = k_L(c_{AL,i} - c_{AL})$$

Local equilibrium was assumed at the gas–liquid interface, with linear equilibrium relationships described by Henry's law of the form

$$p_{A,i} = H\, c_{AL,i}$$

The overall mass-transfer coefficients were defined by

$$N_A = K_G(p_A - p_A^*)$$

and

$$N_A = K_L(c_{AL}^* - c_{AL})$$

where $p_A^* = H\, c_{AL}$ and $p_A = H\, c_{AL}^*$. The overall mass-transfer coefficients were related to the individual film mass-transfer coefficients by the relations

$$\frac{1}{K_G} = \frac{1}{k_G} + \frac{H}{k_L}$$

and

$$\frac{1}{K_L} = \frac{1}{H\, k_G} + \frac{1}{k_L}$$

Convective Mass-Transfer Correlations

$\mathbf{T}$hus far, we have considered convective mass transfer from an analytical viewpoint, and from relations developed for the analogous transport of momentum or convective heat. Although these considerations have given an insight into the mechanisms of convective mass transport, the validity of the analysis must be proven by comparison with experimental data. In this chapter, we will present dimensionless correlations for mass-transfer coefficients based upon experimental results. There will be no attempt to review all of the mass-transfer investigations, as reviews are available in several excellent references.[1–3] However, correlations will be presented to show that the forms of many of these equations are indeed predicted by the analytical expressions derived in Chapter 28. Additional correlations will be given for those situations that have not been successfully treated analytically.

In Chapter 28, several dimensionless numbers are introduced as important parameters used in correlating convective-transport data. Before presenting the convective mass-transfer correlations, let us summarize in Table 30.1 the dimensionless variables that have been used frequently in reported correlations, where L is given as the characteristic length for fluid flow around the surface associated with the convective mass-transfer process. The Sherwood and Stanton numbers are dimensionless parameters involving the mass-transfer coefficient. The Schmidt, Lewis, and Prandtl numbers are required when two separate convective transport processes are simultaneously involved, and the Reynolds, Peclet, and Grashof numbers are used to describe flow. The two j-factors are included in Table 30.1 because they are often used to develop a new correlation for mass transfer based on a previously established heat-transfer coefficient, as discussed in Chapter 28.

[1] W. S. Norman, *Absorption, Distillation and Cooling Towers,* John Wiley & Sons, New York, 1961; C. J. Geankoplis, *Mass Transfer Phenomena,* Holt, Rinehart, and Winston, New York, 1972; A. H. P. Skelland, *Diffusional Mass Transfer,* John Wiley & Sons, New York, 1974; T. K. Sherwood, R. L. Pigford, and C. R. Wilke, *Mass Transfer,* McGraw-Hill Inc., New York, 1975; L. J. Thibodeaux, *Chemodynamics—Environmental Movement of Chemicals in Air, Water and Soil,* John Wiley & Sons, New York, 1979; R. E. Treybal, *Mass Transfer Operations,* McGraw-Hill Book Company, New York, 1980.

[2] E. L. Cussler, *Diffusion-Mass Transfer in Fluid Systems,* 2nd edition, Cambridge University Press, 1997.

[3] S. Middleman, *An Introduction to Heat and Mass Transfer,* John Wiley & Sons, New York, 1998.

Table 30.1 Dimensionless numbers used in correlating mass-transfer data

Name	Symbol	Dimensionless Group
Reynolds number	Re	$\dfrac{v_\infty \rho L}{\mu} = \dfrac{v_\infty L}{\nu}$
Sherwood number	Sh	$\dfrac{k_c L}{D_{AB}}$
Schmidt number	Sc	$\dfrac{\mu}{\rho D_{AB}} = \dfrac{\nu}{D_{AB}}$
Lewis number	Le	$\dfrac{\alpha}{D_{AB}} = \dfrac{k}{\rho c_p D_{AB}}$
Prandtl number	Pr	$\dfrac{\nu}{\alpha} = \dfrac{\mu c_p}{k}$
Peclet number	Pe_{AB}	$\dfrac{v_\infty L}{D_{AB}} = ReSc$
Stanton number	St_{AB}	$\dfrac{k_c}{v_\infty}$
Grashof number	Gr	$\dfrac{L^3 \rho\, g\, \Delta\rho}{\mu^2}$
Mass-transfer j-factor	j_D	$\dfrac{k_c}{v_\infty}(Sc)^{2/3}$
Heat-transfer j-factor	j_H	$\dfrac{h}{\rho c_p v_\infty}(Pr)^{2/3}$

▶ **30.1**

MASS TRANSFER TO PLATES, SPHERES, AND CYLINDERS

Extensive data have been obtained for the transfer of mass between a moving fluid and certain shapes, such as flat plates, spheres, and cylinders. The techniques employed include sublimation of a solid, vaporization of a liquid into air, and the dissolution of a solid into water. By correlating the data in terms of dimensionless parameters, these empirical equations can be extended to other moving fluids and geometrically similar surfaces.

Flat Plate

Several investigators have measured the evaporation from a free liquid surface or sublimation from a flat, volatile solid surface into a controlled air stream. The mass-transfer coefficients obtained from these experiments compare favorably with the mass-transfer coefficients theoretically predicted for laminar and turbulent boundary layers. As described in Chapter 28, for laminar flow, the appropriate correlation for the mean mass-transfer coefficient over characteristic length L is

$$Sh_L = \frac{k_c L}{D_{AB}} = 0.664\, Re_L^{1/2} Sc^{1/3} \quad (Re_L < 2 \times 10^5) \tag{28-29}$$

and for fully-developed turbulent flow, the appropriate correlation for the mean mass-transfer coefficient is

$$Sh_L = \frac{k_c L}{D_{AB}} = 0.0365\, Re_L^{0.8} Sc^{1/3} \quad Re_L > 3.0 \times 10^6 \tag{30-1}$$

with Re_L defined as

$$\text{Re}_L = \frac{\rho \, \text{v}_\infty L}{\mu}$$

where L is the length of the flat plate in the direction of flow. At a local distance x from the leading edge of the flat plate, the exact solution to the laminar boundary-layer problem resulting in the theoretical prediction for the local Sherwood number is given by

$$\text{Sh}_x = \frac{k_c x}{D_{AB}} = 0.332 \, \text{Re}_x^{1/2} \text{Sc}^{1/3} \tag{28-24}$$

and for turbulent flow, the local Sherwood number from approximate boundary-layer analysis is given by

$$\text{Sh}_x = \frac{k_c x}{D_{AB}} = 0.0292 \, \text{Re}_x^{0.8} \text{Sc}^{1/3} \tag{30-2}$$

where the local Reynolds number, Re_x, is defined as

$$\text{Re}_x = \frac{\rho \, \text{v}_\infty x}{\mu}$$

The above equations may also be expressed in terms of the j-factor by recalling

$$j_D = \frac{k_c}{\text{v}_\infty} \text{Sc}^{2/3} = \frac{k_c L}{D_{AB}} \frac{\mu}{L \text{v}_\infty \rho} \frac{D_{AB}\rho}{\mu} \left(\frac{\mu}{\rho D_{AB}}\right)^{2/3} = \frac{\text{Sh}_L}{\text{Re}_L \text{Sc}^{1/3}} \tag{28-56}$$

Upon rearrangement of equations (28-29) and (30-1), we obtain

$$j_D = 0.664 \, \text{Re}_L^{-1/2} \, (\text{laminar}, \text{Re}_L < 2 \times 10^5) \tag{30-3}$$

and

$$j_D = 0.365 \, \text{Re}_L^{-0.2} \, (\text{turbulent}, \text{Re}_L > 3 \times 10^6) \tag{30-4}$$

These equations may be used if the Schmidt number is in the range $0.6 < \text{Sc} < 2500$. The j-factor for mass transfer is also equal to the j-factor for heat transfer in the Prandtl number range of $0.6 < \text{Pr} < 100$ and is equal to $C_f/2$.

Chapter 28 illustrates the use of the boundary-layer equations for evaluating the point value and the average value mass-transfer convective coefficients for a flow over a flat plate. In most situations, the hydrodynamic and concentration boundary layers both start at the same position along x, the direction of fluid flow. However, there are some situations in which the hydrodynamic and the concentration boundary layers have different starting points, as illustrated in Figure 30.1. The fluid flows over a portion of an inert surface *before* flowing over a surface that can also serve as a source or sink for mass transfer. Consequently, the

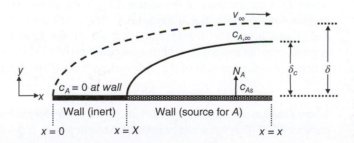

Figure 30.1 Laminar flow over a flat plate with the initiation of the hydrodynamic boundary layer before the concentration boundary layer.

hydrodynamic boundary layer begins to develop before the concentration boundary layer, so that the boundary condition for the concentration of transferring species A becomes

$$0 \leq x < X, c_A = c_{A\infty}$$

$$X \leq x < \infty, c_A = c_{As}$$

Also, consider an analogous situation for flow over a flat plate where there is an unheated starting length prior to the heated zone. In this case, the temperatures at the wall surface are

$$0 \leq x < X, T_{\text{wall}} = T_\infty$$

$$X \leq x < \infty, T_{\text{wall}} = T_s$$

and the hydrodynamic and thermal boundary layers have different starting points. The local Nusselt number for heat transfer is

$$\text{Nu}_x = 0.332 \, \text{Re}_x^{1/2} \left(\frac{\text{Pr}}{1 - \left(\frac{X}{x} \right)^{3/4}} \right)^{1/3} \tag{30-5}$$

using the analytical approach outlined in Chapter 19. From this result, the local Sherwood number for the analogous mass-transfer phenomenon is

$$\text{Sh}_x = 0.332 \, \text{Re}_x^{1/2} \left(\frac{\text{Sc}}{1 - \left(\frac{X}{x} \right)^{3/4}} \right)^{1/3} \tag{30-6}$$

Single Sphere

The Sherwood number (Sh) and Reynolds number (Re) for external flow around a sphere are defined as

$$\text{Sh} = \frac{k_c D}{D_{AB}}$$

and

$$\text{Re} = \frac{\rho \, v_\infty D}{\mu}$$

where D is the diameter of the sphere, D_{AB} is the diffusion coefficient of the transferring species A in gaseous or liquid species B, v_∞ is the bulk fluid velocity flowing over the sphere, and ρ and μ are the density and viscosity of the fluid mixture, respectively, usually approximated as species B at dilute concentration of A. Mass-transfer correlations for single spheres consider the sum of the molecular diffusion and forced convection contributions

$$\text{Sh} = \text{Sh}_o + C \, \text{Re}^m \text{Sc}^{1/3}$$

where C and m are correlating constants. If there is no forced convection, then the Sherwood number is 2.0. This value can be derived theoretically by considering the molecular diffusion flux of species A from a sphere into an infinite sink of stagnant fluid B. Accordingly, the generalized equation becomes

$$Sh = 2.0 + C\,Re^m\,Sc^{1/3}$$

For mass transfer into liquid streams, the equation of Brian and Hales[4]

$$Sh = \frac{k_L D}{D_{AB}} = \left(4.0 + 1.21\,Pe_{AB}^{2/3}\right)^{1/2} \tag{30-7}$$

correlates data where the mass-transfer Peclet number, Pe_{AB}, is less than 10,000. From Table 30.1, recall Pe_{AB} is the product of the Reynolds and Schmidt numbers, ReSc. For Peclet numbers greater than 10,000, Levich[5] recommends the simpler relationship

$$Sh = \frac{k_L D}{D_{AB}} = 1.01\,Pe_{AB}^{1/3} \tag{30-8}$$

For mass transfer into gas streams, the Froessling equation[6]

$$Sh = \frac{k_c D}{D_{AB}} = 2.0 + 0.552\,Re^{1/2}\,Sc^{1/3} \tag{30-9}$$

correlates the data at Reynolds numbers ranging from 2 to 800 and Schmidt numbers ranging from 0.6 to 2.7. Data of Evnochides and Thodos[7] have extended the Froessling equation to a Reynolds number range of 1500 to 12,000 under a Schmidt number range of 0.6 to 1.85.

Equations (30-7) to (30-9) can be used to describe forced convection mass-transfer coefficients only when the effects of free or natural convection are negligible—that is,

$$Re \geq 0.4\,Gr^{1/2}\,Sc^{-1/6} \tag{30-10}$$

The following correlation of Steinberger and Treybal[8] is recommended when the transfer occurs in the presence of natural convection:

$$Sh = Sh_o + 0.347\left(Re\,Sc^{1/2}\right)^{0.62} \tag{30-11}$$

where Sh_o is dependent on GrSc, given by

$$Sh_o = 2.0 + 0.569(GrSc)^{0.25}, \quad GrSc \leq 10^8 \tag{30-12}$$

$$Sh_o = 2.0 + 0.0254(GrSc)^{1/3}(Sc)^{0.244}, \quad GrSc \geq 10^8 \tag{30-13}$$

From Table 30.1, the Grashof number is defined as

$$Gr = \frac{D^3 \rho_L\,g\,\Delta\rho}{\mu_L^2} = \frac{D^3 \rho_L\,g\,(\rho_L - \rho_G)}{\mu_L^2}$$

where density, ρ_L, and viscosity, μ_L, are taken at the bulk conditions of the flowing fluid, and ρ_G is the density of gas, taken at the temperature and pressure of gas bubble. The prediction for Sh is valid when $2 < Re < 3 \times 10^4$ and $0.6 < Sc < 3200$.

[4] P. L. T. Brian and H. B. Hales, *A.I.Ch.E. J.*, **15**, 419 (1969).
[5] V. G. Levich, *Physicochemical Hydrodynamics*, Prentice-Hall, Englewood Cliffs, NJ, 1962.
[6] N. Froessling, *Gerlands Beitr. Geophys.*, **52**, 170 (1939).
[7] S. Evnochides and G. Thodos, *A.I.Ch.E. J.*, **5**, 178 (1960).
[8] R. L. Steinberger and R. E. Treybal, *A.I.Ch.E. J.*, **6**, 227 (1960).

Example 1

Estimate the distance a spherical drop of liquid water, originally 1.0 mm in diameter, must fall in quiet, dry air at 323 K in order to reduce its volume by 50%. Assume that the velocity of the drop is its terminal velocity evaluated at its mean diameter and that the water temperature remains at 293 K. Evaluate all gas properties at the average gas–film temperature of 308 K.

The physical system requires a combined analysis of momentum and mass transport. The liquid water droplet is the source for mass transfer, the surrounding air serves as an infinite sink, and water vapor (species A) is the transferring species. The rate of evaporation is sufficiently small so that the water droplet is considered isothermal at 293 K. The terminal velocity of the droplet is given by

$$v_o = \sqrt{\frac{4\, d_p(\rho_w - \rho_{air})g}{3\, C_D\, \rho_{air}}}$$

where d_p is the diameter of the particle, ρ_w is the density of the water droplet, ρ_{air} is the density of the surrounding fluid (air), g is the acceleration due to gravity, and C_D is the drag coefficient, which is a function of the Reynolds number of the spherical particle, as illustrated in Figure 12.4. The particle diameter is taken at the arithmetic mean of the evaporating droplet diameter:

$$\bar{d}_p = \frac{d_{p,o} + d_p}{2} = \frac{d_{p,o} + (0.5)^{1/3} d_{p,o}}{2} = 8.97 \times 10^{-4}\ \text{m}$$

At 293 K, the density of the water droplet (ρ_w) is 995 kg/m^3. At 308 K, the density of the air is 1.14 kg/m^3, and the viscosity of air is 1.91×10^{-5} kg/m·s. Substitution of these values into the terminal velocity equation yields

$$v_o = \sqrt{\frac{(4)\left(8.97 \times 10^{-4}\ \text{m}\right)\left(9.95 \times 10^2\ \text{kg/m}^3 - 1.14\ \text{kg/m}^3\right)\left(9.8\ \text{m/s}^2\right)}{(3)(1.14\ \text{kg/m}^3)C_D}} = \sqrt{\frac{10.22\ \text{m}^2/\text{s}^2}{C_D}}$$

To obtain v_o, trial and error is required. First, a value for v_o is guessed, the Reynolds number is calculated, and C_D is read from Figure 12.4. To start, guess $v_o = 3.62$ m/s. The Reynolds number is

$$\text{Re} = \frac{d_p v_o \rho_{air}}{\nu_{air}} = \frac{(8.97 \times 10^{-4}\ \text{m})(3.62\ \text{m/s})(1.14\ \text{kg/m}^3)}{1.19 \times 10^{-5}\ \text{kg/m·s}}$$

and from Figure 12.4, $C_D = 0.78$. Now recalculate v_o:

$$v_o = \sqrt{\frac{10.22\ \text{m}^2/\text{s}^2}{C_D}} = \sqrt{\frac{10.22\ \text{m}^2/\text{s}^2}{0.78}} = 3.62\ \text{m/s}$$

Therefore, the guessed value for v_o is correct. The Schmidt number must now be calculated. From Appendix J.1, the gas diffusivity (D_{AB}) for water vapor in air at 298 K is 2.60×10^{-5} m^2/s, which is corrected to the desired temperature by

$$D_{AB} = \left(2.60 \times 10^{-5}\ \text{m}^2/\text{s}\right)\left(\frac{308\ \text{K}}{298\ \text{K}}\right)^{3/2} = 2.73 \times 10^{-5}\ \text{m}^2/\text{s}$$

The Schmidt number is

$$\text{Sc} = \frac{\mu_{air}}{\rho_{air} D_{AB}} = \frac{\left(1.91 \times 10^{-5}\ \text{kg/m·s}\right)}{\left(1.14\ \text{kg/m}^3\right)\left(0.273 \times 10^{-4}\ \text{m}^2/\text{s}\right)} = 0.61$$

The Fröessling equation (30-9) can now be used to evaluate the mass-transfer coefficient for transfer of water vapor from the surface of the droplet to the surrounding air:

$$\frac{k_c d_p}{D_{AB}} = 2 + 0.552\, \text{Re}^{1/2}\text{Sc}^{1/3}$$

or

$$k_c = \frac{D_{AB}}{d_p}\left(2 + 0.552\,Re^{1/2}Sc^{1/3}\right)$$

$$= \frac{(0.273 \times 10^{-4}\ m^2/s)}{8.97 \times 10^{-4}\ m}\left(2.0 + 0.552(194)^{1/2}(0.61)^{1/3}\right) = 0.276\ m/s$$

The average rate of water evaporation from the droplet is

$$W_A = 4\pi r_p^2 N_A = 4\pi r_p^2 k_c(c_{As} - c_{A\infty})$$

The dry air concentration, $c_{A,\infty}$, is zero, and the surrounding is assumed to be an infinite sink for mass transfer. The gas-phase concentration of water vapor at the liquid droplet surface is evaluated from the vapor pressure of water at 293 K:

$$c_{As} = \frac{P_A}{RT} = \frac{2.33 \times 10^3\ Pa}{\left(8.314\,\dfrac{Pa \cdot m^3}{gmole \cdot K}\right)(293\ K)} = 0.956\,\frac{gmole}{m^3}$$

When we substitute the known values into the rate of evaporation equation, we obtain

$$W_A = 4\pi\left(4.48 \times 10^{-4}\ m\right)^2(0.276\ m/s)(0.956\ gmole/m^3 - 0) = 6.65 \times 10^{-7}\ gmole/s$$

or 1.2×10^{-8} kg/s on a mass basis. The amount of water evaporated is

$$m_A = \rho_w \Delta V = \rho_w(V(t_1) - V(t_2)) = \frac{\rho_w V(t_1)}{2}$$

$$= \frac{\rho_w}{2}\frac{4\pi}{3}r_p^3 = \frac{4\pi}{6}\left(9.95 \times 10^2\ kg/m^3\right)\left(4.48 \times 10^{-4}\ m\right)^3 = 1.87 \times 10^{-7}\ kg$$

The time necessary to reduce the volume by 50% is

$$t = \frac{m_A}{W_A} = \frac{1.87 \times 10^{-7}\ kg}{1.20 \times 10^{-8}\ kg/s} = 15.6\ s$$

Finally, the distance of the fall is equal to $v_o t$ or 56.5 m.

Spherical Bubbles

Consider a process where a gas is bubbled into a column of liquid, as shown in Figure 30.2. Usually the spherical gas bubbles are produced in swarms or clusters by the orifice that introduces the gas into the liquid. Gas bubbles rise up in the liquid by natural convection process where a hydrodynamic boundary layer is formed in the liquid surrounding the outer surface of the gas bubble.

Unlike the single, rigid sphere previously described, viscous circulation also occurs within the deformable bubble as it rises through the liquid. Consequently, a single-sphere correlation fails to describe the mass transport accurately in the vicinity of the gas–liquid interface of a rising bubble. Calderbank and Moo-Young[9] recommend the following two-point correlation for the mass-transfer coefficient associated with the transfer of a sparingly soluble gaseous solute A into solvent B by a swarm of gas bubbles in a natural convection process.

For gas bubble diameters less than 2.5 mm, use

$$Sh = \frac{k_L d_b}{D_{AB}} = 0.31\,Gr^{1/3}Sc^{1/3} \tag{30-14}$$

---

[9] P. H. Calderbank and M. Moo-Young, *Chem. Eng. Sci.*, **16**, 39 (1961).

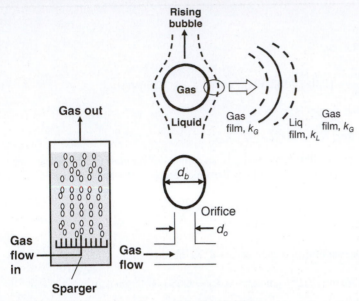

Figure 30.2 Gas bubbles dispersed in liquid column.

For bubble diameters greater or equal to 2.5 mm, use

$$\text{Sh} = \frac{k_L d_b}{D_{AB}} = 0.42\,\text{Gr}^{1/3}\text{Sc}^{1/2} \qquad (30\text{-}15)$$

In the above correlations, the Grashof number is defined as

$$\text{Gr} = \frac{d_b^3 \rho_L g \Delta\rho}{\mu_L^2} = \frac{d_b^3 \rho_L g(\rho_L - \rho_G)}{\mu_L^2}$$

where $\Delta\rho$ is the difference of the density of the bulk liquid and the density of the gas inside the bubble, with density (ρ_L) and viscosity (μ_L) determined at the bulk average properties of the liquid mixture. For dilute solutions, the fluid properties of the solvent approximate the fluid properties of the liquid mixture. The diffusion coefficient D_{AB} is with respect to dissolved gaseous solute A in solvent B.

Single Cylinder

Several investigators have studied the sublimation from a solid cylinder into air flowing normal to its axis. Additional results on the dissolution of solid cylinders into a turbulent water stream have been reported. Bedingfield and Drew[10] correlated the available data to obtain

$$\frac{k_G P(\text{Sc})^{0.56}}{G_M} = \frac{k_c(\text{Sc})^{0.56}}{v_\infty} = 0.281(\text{Re}_D)^{-0.4} \qquad (30\text{-}16)$$

which is valid for $400 < \text{Re}_D < 25{,}000$ and $0.6 < \text{Sc} < 2.6$. In this correlation, P is the system total pressure and G_M is the superficial molar velocity of the gas flowing normal to the cylinder in units of kgmole/m$^2 \cdot$s. The Reynolds number for flow normal to a solid cylinder, Re_D, is defined as

$$\text{Re}_D = \frac{\rho v_\infty D}{\mu}$$

where D is cylinder diameter, v_∞ is the fluid velocity normal to the solid cylinder, and ρ and μ for the gas stream evaluated at the film average temperature.

The full analogy among momentum, heat, and mass transfer breaks down when the flow is around bluff bodies, such as spheres and cylinders. The total drag force includes the form drag in addition to the skin friction and so the j-factor will not equal $C_f/2$. However, the analogy between heat and mass transfer, $j_H = j_D$, still holds. Accordingly, the mass-transfer coefficient for a single cylinder that does not satisfy the specified ranges for equation (30-16) can be evaluated by using the Chilton–Colburn analogy and the appropriate heat-transfer relations described in Chapter 20.

[10] C. H. Bedingfield and T. B. Drew, *Ind. Eng. Chem.*, **42**, 1164 (1950).

Example 2

In a humidification apparatus, liquid water flows in a thin film down the outside of a vertical, circular cylinder. Dry air at 310 K and 1.013×10^5 Pa (1.0 atm) flows at right angles to the 0.076-m-diameter, 1.22-m-long vertically aligned cylinder at a velocity of 4.6 m/s. The liquid film temperature is 290 K. Calculate the rate at which liquid must be supplied to the top of the cylinder if the entire surface of the cylinder is to be used for the evaporating process and no water may drip off from the bottom of the cylinder.

The liquid film on the outside of the cylinder represents the source for mass transfer, and the air stream flowing normal to the cylinder represents an infinite sink. The properties of the air stream are evaluated at the film-average temperature of 300 K. The properties of air may be obtained from Appendix I, with $\rho = 1.1769$ kg/m^3 and $\nu = 1.5689 \times 10^{-5}$ m^2/s at 300 K and 1 atm. The Reynolds number is

$$\mathrm{Re}_D = \frac{v_\infty D}{\nu_{air}} = \frac{(4.6\ \text{m/s})(0.076\ \text{m})}{1.5689 \times 10^{-5}\ \text{m}^2/\text{s}} = 22{,}283$$

From Appendix J, Table J.1, the diffusion coefficient of water vapor in air at 298 K and 1.0 atm is 2.6×10^{-5} m^2/s, which, corrected for temperature, becomes

$$D_{AB} = \left(2.60 \times 10^{-5}\ \text{m}^2/\text{s}\right)\left(\frac{300\ \text{K}}{298\ \text{K}}\right)^{3/2} = 2.63 \times 10^{-5}\ \text{m}^2/\text{s}$$

The Schmidt number is

$$\mathrm{Sc} = \frac{\nu_{air}}{D_{AB}} = \frac{1.5689 \times 10^{-5}\ \text{m}^2/\text{s}}{2.63 \times 10^{-5}\ \text{m}^2/\text{s}} = 0.60$$

The superficial molar velocity of the air normal to the cylinder is

$$G_M = \frac{v_\infty \rho_{air}}{M_{air}} = \frac{(4.6\ \text{m/s})(1.1769\ \text{kg/m}^3)}{(29\ \text{kg/kg mole})} = 0.187\ \frac{\text{kgmole}}{\text{m}^2\text{s}}$$

Upon substitution of the known values into equation (30-16), we can solve for the gas-phase film mass-transfer coefficient:

$$k_G = \frac{G_M}{P\ \mathrm{Sc}^{0.56}} 0.281(\mathrm{Re}_D)^{-0.4}$$

or

$$k_G = \frac{0.281(0.187\ \text{kgmole/m}^2 \cdot \text{s})(22{,}283)^{-0.4}}{(1.013 \times 10^5\ \text{Pa})(0.60)^{0.56}} = 1.26 \times 10^{-8}\ \frac{\text{kgmole}}{\text{m}^2 \cdot \text{s} \cdot \text{Pa}}$$

The flux of water can be evaluated by

$$N_A = k_G\left(p_{As} - p_{A,\infty}\right)$$

The vapor pressure of water at 290 K is 1.73×10^3 Pa $(p_{As} = P_A)$, and the partial pressure of the dry air $(p_{A\infty})$ is zero, as the surrounding air stream is assumed to be an infinite sink for mass transfer. Consequently,

$$N_A = \left(1.26 \times 10^{-8}\ \frac{\text{kgmole}}{\text{m}^2 \cdot \text{s} \cdot \text{Pa}}\right)\left(1.73 \times 10^3\ \text{Pa} - 0\right) = 2.18 \times 10^{-5}\ \frac{\text{kgmole}}{\text{m}^2 \cdot \text{s}}$$

Finally, the mass-feed rate of water for a single cylinder is the product of the flux, and the external surface area of the cylinder is

$$W_A = N_A M_A(\pi D L) = \left(2.18 \times 10^{-5}\ \frac{\text{kgmole}}{\text{m}^2 \cdot \text{s}}\right)\left(\frac{18\ \text{kg}}{\text{kgmole}}\right)(\pi)(0.076\ \text{m})(1.22\ \text{m})$$

$$= 1.14 \times 10^{-4}\ \text{kg/s}$$

MASS TRANSFER INVOLVING FLOW THROUGH PIPES

Mass transfer from the inner walls of a tube to a moving fluid has been studied extensively. Most of the data have been obtained for vaporization of liquids into air, although some data have also been obtained for mass transfer of a soluble solid into a moving liquid, where the solid coats the inner surface of the tube. Under *turbulent* gas flow conditions, Gilliland and Sherwood[11] studied the vaporization of nine different liquids into air flowing through the inside of a tube and obtained the correlation

$$\frac{k_c D}{D_{AB}}\frac{p_{B,lm}}{P} = 0.023\,\text{Re}^{0.83}\text{Sc}^{0.44} \tag{30-17}$$

with

$$\text{Re} = \frac{\rho\, v_\infty D}{\mu}$$

and

$$\text{Sh} = \frac{k_c D}{D_{AB}}$$

where D is the inner diameter of the pipe, v_∞ is the bulk velocity of fluid flowing though the pipe, $p_{B,lm}$ is the log mean composition of the carrier gas B, P is the total system pressure, and D_{AB} is the mass diffusivity of the diffusing component A in the flowing carrier gas B. The Reynolds and Schmidt numbers are evaluated at the bulk conditions of the gas inside the pipe; for dilute solutions, the density and viscosity of the carrier gas can be assumed, and $p_{B,lm}$ for the carrier gas approximates total system pressure P. The correlation is valid for gases where $2000 < \text{Re} < 35{,}000$ and $0.6 < \text{Sc} < 2.5$.

In a subsequent study, Linton and Sherwood[12] extended the range of the Schmidt number when they investigated the dissolution of benzoic acid, cinnamic acid, and β-naphthol in various solvents flowing through a tube. The combined experimental results of Gilliland and Sherwood, and of Linton and Sherwood, were correlated by the relation

$$\text{Sh} = \frac{k_L D}{D_{AB}} = 0.023\,\text{Re}^{0.83}\text{Sc}^{1/3} \tag{30-18}$$

which is valid for liquids where $2000 < \text{Re} < 35{,}000$ and $1000 < \text{Sc} < 2260$. The Reynolds and Schmidt numbers are evaluated at the bulk conditions of the liquid inside the pipe. Again, for dilute solutions, the density and viscosity of the fluid approximates those properties of the solvent carrier B.

We note here that the Sherwood number for mass transfer (Sh) is analogous to the Nusselt number for heat transfer (Nu). The similarity between equation (30-18) and the Dittus–Boelter equation for energy transfer (20-28) illustrate the analogous behavior of these two transport phenomena.

For fully developed *laminar* fluid flow inside a pipe, the convective mass coefficient k_c can be determined from first principles. In Figure 30.3, the velocity distribution in laminar flow along the radial direction is given by

---

[11] E. R. Gilliland and T. K. Sherwood, *Ind. Eng. Chem.*, **26**, 516 (1934).
[12] W. H. Linton and T. K. Sherwood, *Chem. Eng. Prog.*, **46**, 258 (1950).

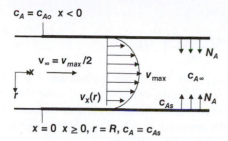

$c_A = c_{Ao}$ $x < 0$

$v_\infty = v_{max}/2$

v_{max} $c_{A\infty}$

$v_x(r)$ c_{As}

N_A

$x = 0$ $x \geq 0$, $r = R$, $c_A = c_{As}$

Figure 30.3 Laminar flow mass transfer within a pipe.

$$v_x(r) = v_{max}\left[1 - \left(\frac{r}{R}\right)^2\right] = 2 \cdot v_\infty \cdot \left[1 - \left(\frac{r}{R}\right)^2\right] \tag{8-7}$$

Graetz–Nusselt analysis considers the flux of the transferring species from the tube wall to the laminar flow field. The result of this theoretical analysis shows that the mass-transfer coefficient depends on axial position x down the length of the pipe:

$$k_{c,x} = \frac{1}{\Gamma(4/3)}\left(\frac{8}{9}\frac{v_\infty D_{AB}^2}{D \cdot x}\right)^{1/3} \tag{30-19}$$

where $\Gamma(4/3) = 0.89$. Integration of equation (30-19) over axial position x gives average mass-transfer coefficient of the length of pipe, L:

$$k_c = \frac{1}{L}\int_0^L k_c(x)dx = 1.62\left(\frac{v_\infty D_{AB}^2}{L D}\right)^{1/3} \tag{30-20}$$

In dimensionless form, equation (30-20) compares closely with the experimentally validated Sieder–Tate equation for laminar mass transfer inside a pipe, given by

$$\mathrm{Sh} = 1.86\left(\frac{v_\infty D^2}{L D_{AB}}\right)^{1/3} = 1.86\left(\frac{D v_\infty}{L}\frac{D\nu}{\nu D_{AB}}\right)^{1/3} = 1.86\left(\frac{D}{L}\mathrm{Re}\,\mathrm{Sc}\right)^{1/3} \tag{30-21}$$

Equation (30-21) is valid for Re < 2000 and ReSc$(D/L) > 10$. Equation (30-21) is analogous to the Sieder–Tate equation (20-27) for laminar flow heat transfer inside a pipe. Equation (30-21) can also be expressed in terms of the Graetz number (Gz), given by $D/L \cdot$ReSc or $D/L \cdot$Pe$_{AB}$. Examples 3 and 6 provide more details on estimation of convective mass-transfer coefficients for laminar and turbulent flow within a pipe.

▶ **30.3**

MASS TRANSFER IN WETTED-WALL COLUMNS

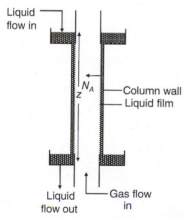

Liquid flow in

N_A

z

Column wall
Liquid film

Liquid flow out

Gas flow in

Figure 30.4 Wetted-wall column for interphase mass transfer of gas and liquid.

Much of the data for interphase mass transfer of a solute between gas and liquid carrier streams have been obtained using wetted-wall columns. In a wetted-wall column, gas flows into the bottom of the tube and moves upward, as shown in Figure 30.4. Liquid is loaded into the top of the column, and a weir evenly distributes the flow of liquid around the inner perimeter of the tube, forming a falling liquid film that evenly wets the inner surface of the tube down its length. The liquid film is often somewhat thin, often less than a few millimeters, and the liquid velocity is relatively high owing to gravitational acceleration. There are two principal reasons for using wetted-wall columns in interphase mass-transfer investigations. First, the contacting area between the two phases can be accurately measured. Second, the experiments can be readily set up for steady-state operation.

The convective mass-transfer coefficient for the gas film based on either turbulent or laminar flow is defined by the correlating equations (30-17) and (30-21), respectively. A suitable correlation proposed by Vivian and Peaceman[13] for convective mass transfer of a gaseous solute into a falling liquid film evenly

---

[13] L. E. Scriven and R. L. Pigford, *A.I.Ch.E. J.*, **4**, 439 (1958).

wetting the inner surface of a tube is given by

$$\text{Sh} = \frac{k_L z}{D_{AB}} = 0.433 (\text{Sc})^{1/2} \left(\frac{\rho_L^2 \, g \, z^3}{\mu_L^2} \right)^{1/6} (\text{Re}_L)^{0.4} \qquad (30\text{-}22)$$

where z is the length of contact down the falling film, D_{AB} is the mass diffusivity of the diffusing component A into liquid solvent B, ρ_L is the density of the liquid, μ_L is the viscosity of liquid, g is the acceleration due to gravity, and Sc is the Schmidt number for the solute dissolved in the liquid, evaluated at the liquid film temperature. The Reynolds number of the liquid flowing down the inner surface of the wetted tube is defined as

$$\text{Re}_L = \frac{4\Gamma}{\mu_L} = \frac{4w}{\pi D \mu_L}$$

where w is the mass flow rate of liquid (kg/s), D is the inner diameter of the cylindrical column, and Γ is the mass flow rate of liquid per unit wetted perimeter of the column (kg/m · s).

The liquid–film mass-transfer coefficients predicted by equation (30-22) were found to be 10–20% lower than the theoretical equation for absorption of a gaseous solute into a laminar falling liquid film, as discussed in Section 26.4. This may have been due to ripples along the liquid surface or to disturbances in the liquid flow at the two ends of the wetted-wall column. These discrepancies between the theoretical and measured rates of mass transfer have often led to the suggestion that a resistance to the mass transfer exists at the gas–liquid interface. However, investigations by Scriven and Pigford[13] and others have substantiated that the interfacial resistance is negligible in normal interphase mass-transfer operations.

Example 3

Trichloroethylene (TCE), a common industrial solvent, is often found at low concentrations in industrial waste waters. Stripping is a common process for removing sparingly soluble, volatile organic solutes such as TCE from aqueous solution. A wetted-wall column is used to study the stripping of TCE from water to air at a constant temperature of 293 K and total system pressure of 1.0 atm. The column inner diameter is 4.0 cm, and the height is 2.0 m. In the present process, the volumetric air flow rate into the column is 500 cm³/s (5.0×10^{-4} m³/s), and the volumetric flow rate of water is 50 cm³/s (5.0×10^{-5} m³/s). At 293 K, the density of liquid water is 998.2 kg/m³, and so the mass flow rate of water wetting the column is $w = 0.05$ kg/s. Estimate K_L, the overall liquid-phase mass-transfer coefficient for TCE across the liquid and gas film. Assume that water loss by evaporation is negligible.

Relevant physical property data are provided below. The process is very dilute so that the bulk gas has the properties of air and the bulk liquid has the properties of water. The equilibrium solubility of TCE in water is described by Henry's law of the form

$$p_{A,i} = H \, x_{A,i}$$

where H is 550 atm at 293 K. The binary gas-phase diffusivity of TCE in air is 8.0×10^{-6} m²/s at 1.0 atm and 293 K, as determined by the Fuller–Shettler–Giddings correlation. The binary liquid-phase diffusivity of TCE in water at 293 K is 8.9×10^{-10} m²/s, as determined by the Hayduk–Laudie correlation.

With this physical property information in hand, our strategy is to estimate the gas–film coefficient k_G, the liquid film coefficient k_L, and then the overall mass-transfer coefficient K_L. First, the bulk velocity of the gas is

$$v_\infty = \frac{4Q_g}{\pi D^2} = \frac{4 \left(5.0 \times 10^{-4} \dfrac{\text{m}^3}{\text{s}} \right)}{\pi (0.04 \, \text{m})^2} = 0.40 \, \text{m/s}$$

The Reynolds number for air flow through the inside of the wetted-wall column is

$$\text{Re} = \frac{\rho_{\text{air}} v_\infty D}{\mu_{\text{air}}} = \frac{\left(1.19 \dfrac{\text{kg}}{\text{m}^3} \right) \left(0.40 \dfrac{\text{m}}{\text{s}} \right) (0.04 \, \text{m})}{\left(1.84 \times 10^{-5} \dfrac{\text{kg}}{\text{m} \cdot \text{s}} \right)} = 1035$$

and the Schmidt number for TCE in air is

$$Sc = \frac{\mu_{air}}{\rho_{air} D_{TCE-air}} = \frac{1.84 \times 10^{-5} \frac{kg}{m \cdot s}}{\left(1.19 \frac{kg}{m^3}\right)\left(8.08 \times 10^{-6} \frac{m^2}{s}\right)} = 1.91$$

where the properties of air are found from Appendix I. As the gas flow is laminar (Re < 2000), equation (30-21) for laminar flow inside a pipe is appropriate for estimation of k_c. Therefore,

$$k_c = \frac{D_{AB}}{D} 1.86 \left(\frac{D}{L} Re\, Sc\right)^{1/3} = \frac{8.0 \times 10^{-6}\, m^2/s}{0.04\, m} 1.86 \left(\frac{0.04\, m}{2.0\, m}(1035)(1.91)\right)^{1/3} = 1.27 \times 10^{-3}\, th/s$$

The conversion to k_G is

$$k_G = \frac{k_c}{RT} = \frac{1.27 \times 10^{-3} \frac{m}{s}}{\left(0.08206 \frac{m^3 \cdot atm}{kgmole \cdot K}\right)(293\, K)} = 5.28 \times 10^{-5} \frac{kgmole}{m^2 \cdot s \cdot atm}$$

The liquid–film mass-transfer coefficient is now estimated. The Reynolds number for the falling liquid film is

$$Re_L = \frac{4w}{\pi D \mu_L} = \frac{4(0.05\, kg/s)}{\pi(0.04\, m)(9.93 \times 10^{-4}\, kg/m \cdot s)} = 1600$$

and the Schmidt number is

$$Sc = \frac{\mu_L}{\rho_L D_{TCE-H_2O}} = \frac{\left(9.93 \times 10^{-4} \frac{kg}{m \times s}\right)}{\left(998.2 \frac{kg}{m^3}\right)\left(8.90 \times 10^{-10} \frac{m^2}{s}\right)} = 1118$$

where the properties of liquid water at 293 K are found in Appendix I. Equation (30-22) is appropriate for estimation of k_L for the falling liquid film inside the wetted-wall column:

$$k_L = \frac{D_{AB}}{z} 0.433 (Sc)^{1/2} \left(\frac{\rho_L^2 g\, z^3}{\mu_L^2}\right)^{1/6} (Re_L)^{0.4}$$

$$= \frac{8.9 \times 10^{-10} \frac{m^2}{s}}{2.0\, m} 0.433(1118)^{1/2} \left(\frac{\left(998.2 \frac{kg}{m^3}\right)^2 \left(9.8 \frac{m}{s^2}\right)(2.0\, m)^3}{\left(9.93 \times 10^{-4} \frac{kg}{m \cdot s}\right)^2}\right)^{1/6} (1600)^{0.4}$$

$$= 2.55 \times 10^{-5}\, m/s$$

Since the process is dilute, the Henry's law constant in units consistent with k_L and k_G is

$$H = 550\, atm \frac{M_{H_2O}}{\rho_{L,H_2O}} = (550\, atm)\left(\frac{18\, kg/gmole}{998.2\, kg/m^3}\right) = 9.92 \frac{m^3 \cdot atm}{kgmole}$$

The overall liquid-phase mass-transfer coefficient, K_L, is estimated by equation (29-22):

$$\frac{1}{K_L} = \frac{1}{k_L} + \frac{1}{Hk_G} = \frac{1}{2.55 \times 10^{-5} \frac{m}{s}} + \frac{1}{\left(9.92 \frac{atm \cdot m^3}{kgmole}\right)\left(5.28 \times 10^{-5} \frac{kgmole}{m^2 \cdot s \cdot atm}\right)}$$

or $K_L = 2.43 \times 10^{-5}$ m/s. Since $K_L \rightarrow k_L$, the process is liquid-phase mass-transfer controlling, which is characteristic of interphase mass-transfer processes involving a large value for Henry's law constant.

▶ **30.4**

MASS TRANSFER IN PACKED AND FLUIDIZED BEDS

Packed and fluidized beds are commonly used in industrial mass-transfer operations, including adsorption, ion exchange, chromatography, and gaseous reactions that are catalyzed by solid surfaces. The basic configurations of packed and fluidized beds are compared in Figure 30.5.

Numerous investigations have been conducted for measuring mass-transfer coefficients in packed beds and correlating the results. In general, the agreement among the investigators is often poor, which is to be expected when one realizes the experimental difficulties. Sherwood, Pigford, and Wilke[14] presented a graphical representation of most of the data for mass transfer in packed beds with single-phase fluid and gas flows. They found that a single straight line through the experimental points did a fair job of representing all the data. This line is represented by a fairly simple equation:

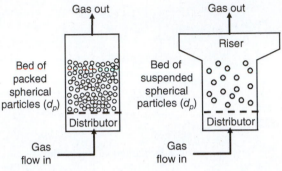

Figure 30.5 Packed vs. fluidized bed.

$$j_D = 1.17 \, \text{Re}^{-0.415} \quad (10 < \text{Re} < 2500) \quad (30\text{-}23)$$

with

$$\text{Re} = \frac{\rho \, d_p \, u_{ave}}{\mu}$$

where u_{ave} is the average superficial gas velocity of the empty bed (m/s), and d_p is the average diameter of the particle equivalent to a sphere. If it is assumed that the particles are spherical, then an approximation of the surface area for mass transfer from particle to fluid is given by

$$A_i = V \frac{A_{sphere}}{V_e} = V \frac{\pi}{d_p}$$

where V is the volume of the empty bed, A_{sphere} is the surface area of a sphere, and V_e is the volume of the box element surrounding one sphere. Most of the earlier correlations for packed beds failed to account for variations in the void fraction of the bed. The void fraction in the packed bed is designated as ϵ, the volume of void space between the solid particles divided by the total volume of void space plus the solid particles. Values for ϵ range from about 0.3 to 0.5 in most packed beds.

Mass transfer between liquids and beds of spheres was investigated by Wilson and Geankoplis[15] who correlated their data by

$$\epsilon \, j_D = \frac{1.09}{\text{Re}} \quad (30\text{-}24a)$$

for $165 < \text{Sc} < 70{,}600$ and $0.35 < \epsilon < 0.75$, and by

$$\epsilon \, j_D = \frac{0.25}{(\text{Re})^{0.31}} \quad (30\text{-}24b)$$

[14] T. K. Sherwood, R. L. Pigford, and C. R. Wilke, *Mass Transfer*, McGraw-Hill Book Company, New York, 1975.
[15] E. J. Wilson and C. J. Geankoplis, *Ind. Eng. Chem. Fund.*, **5**, 9 (1966).

for $55 < \text{Re} < 1500$ and $165 < \text{Sc} < 10{,}690$. The correlation of Gupta and Thodos[16]

$$\epsilon\, j_D = \frac{2.06}{(\text{Re})^{0.575}} \tag{30-24c}$$

is recommended for mass transfer between gases and beds of spheres in the Reynolds number range $90 < \text{Re} < 4000$. Data above this range indicate a transitional behavior and are reported in a graphical form by Gupta and Thodos.[17]

Mass transfer in both gas and liquid fluidized beds of spheres has been correlated by Gupta and Thodos[18] with the equation

$$\epsilon\, j_D = 0.010 + \frac{0.863}{(\text{Re})^{0.58} - 0.483} \tag{30-25}$$

In a fluidized bed, the upward flow of fluid suspends the particles if the fluid velocity is above the minimum fluid velocity. A riser section increases the cross-sectional area so that the velocity is lowered below to minimum fluid velocity to retain the particles. For fluidized beds where there are significant particle interactions, Kunii and Levenspiel[19] provide a more detailed discussion of heat- and mass-transfer processes.

▶ **30.5**

GAS–LIQUID MASS TRANSFER IN BUBBLE COLUMNS AND STIRRED TANKS

Gas–liquid interphase mass-transfer processes described in Chapter 29 are often carried out by bubbling the gas into a liquid to increase the surface area for mass transfer, as shown previously in Figure 30.2. For example, an important industrial process is the aeration of water, which is used in wastewater treatment and aerobic fermentation operations. Air is bubbled into the bottom of a vessel containing liquid water. Oxygen gas inside the air bubble absorbs into the water, where it is sparingly soluble. Usually, the air bubbles are produced in swarms or clusters by a gas sparger. A gas sparger is typically a tube or ring with orifices. The balance of surface tension and buoyant forces determine bubble formation.

There are two types of gas–liquid mass-transfer operations where a gas is bubbled into the liquid. The first is the bubble column, where the gas is sparged into the bottom of a tube filled with liquid (Figure 30.2). The bubbles rise up the liquid-filled tube by natural convection and agitate the liquid, and the liquid phase is considered well mixed. The second is the aerated stirred tank, where gas is sparged into a tank filled with liquid that is mixed with a rotating, submerged impeller. The stirred tank promotes gas/liquid contact by breaking up the rising gas bubbles released at the bottom of the tank and dispersing them throughout the liquid volume.

In both of these gas–liquid mass-transfer operations, the interphase mass-transfer area is defined by

$$a = \frac{A_i}{V} = \frac{\text{area available for interphase mass transfer } (\text{m}^2)}{\text{liquid volume } (\text{m}^3)} \tag{30-26}$$

[16] A. S. Gupta and G. Thodos, *A.I.Ch.E. J.*, **9**, 751 (1963).

[17] A. S. Gupta and G. Thodos, *Ind. Eng. Chem. Fund.*, **3**, 218 (1964).

[18] A. S. Gupta and G. Thodos, *A.I.Ch.E. J.*, **8**, 608 (1962).

[19] D. Kunii and O. Levenspiel, *Fluidization Engineering*, Wiley, New York (1969).

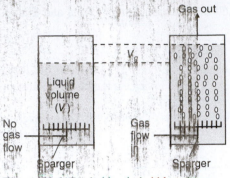

Figure 30.6 Gas holdup in bubble column.

Capacity coefficients based on a "concentration driving force" mass-transfer coefficient (e.g., k_L) have units of reciprocal time, with a typical units conversion of the following:

$$k_L a = k_L \frac{A_i}{V} = \left(\frac{m}{s}\right)\left(\frac{m^2}{m^3}\right) = s^{-1}$$

Capacity coefficients cannot be used to calculate flux N_A directly. Instead, they are used to compute the total rate of mass transfer, W_A, by

$$W_A = N_A \frac{A_i}{V} V = k_L a\, V\left(c_{AL}^* - c_{AL}\right) \qquad (30\text{-}27)$$

In bubble columns, the liquid film mass-transfer coefficient k_L is estimated by equations (30-14) and (30-15), and the parameter "a" is estimated from the gas holdup in the aerated liquid. When the gas is bubbled into the liquid, the aerated liquid level rises, as shown in Figure 30.6. The gas holdup ratio, ϕ_g, is defined as the volume of gas bubbles per unit volume of liquid:

$$\phi_g = \frac{V_g}{V}$$

The interphase mass-transfer area per unit volume for bubbles of average diameter d_b is

$$\frac{A_i}{V} = \frac{V_g}{V} \cdot \frac{\text{bubble area}}{\text{bubble volume}} = \phi_g \frac{\pi d_b^2}{\pi d_b^3/6} = \frac{6\phi_g}{d_b} \qquad (30\text{-}28)$$

For a bubble column using water-like liquids without external agitation, ϕ_g is estimated by

$$\frac{\phi_g}{(1-\phi_g)^4} = 0.20\left(\frac{g\,D^2\,\rho_L}{\sigma_L}\right)^{1/8}\left(\frac{g\,D^3}{\nu_L^2}\right)^{1/12}\left(\frac{u_{gs}}{\sqrt{g\,D}}\right) \qquad (30\text{-}29)$$

where σ_L is the surface tension of the liquid, D is the diameter of the column, and u_{gs} is the superficial gas velocity through the empty column. The gas holdup ratio is typically less than 0.20 for most sparging operations. The gas bubble diameter d_b can also be estimated through correlation. For low gas flow rates, where the rising gas bubbles are separated, by equating the buoyant force with the force to due surface tension at the orifice, the bubble diameter is

$$d_b = \left(\frac{6\,d_o\sigma_L}{g(\rho_L - \rho_G)}\right)^{1/3} \qquad (30\text{-}30)$$

where d_o is the diameter of the gas sparger orifice that creates the bubble. The stable gas bubble diameter in a bubble column without external agitation is also approximated by the Sauter mean diameter, given by

$$\frac{d_b}{D} = 26\left(\frac{g\,D^2\,\rho_L}{\sigma_L}\right)^{-0.50}\left(\frac{g\,D^3}{\nu_L^2}\right)^{-0.12}\left(\frac{u_{gs}}{\sqrt{g\,D}}\right)^{-0.12} \qquad (30\text{-}31)$$

The reader is encouraged to consult Treybal[20] for further information.

In aerated vessels where gas–liquid mass transfer is augmented with mechanical stirring, due to the continual collision of the gas bubbles resulting from gas sparging and mechanical agitation of the submerged impeller, the interfacial area for mass transfer is impossible to estimate or measure. Consequently, measured mass-transfer coefficients for

---

[20] R. E. Treybal, *Mass Transfer Operations*, McGraw-Hill Book Company, New York, 1980.

aerated stirred tanks are reported as capacity coefficients—for example, $k_L a$—where the mass-transfer coefficient (k_L) is lumped together with the parameter a.

Van't Riet[21] reviewed many studies of gas–liquid mass-transfer processes associated with oxygen transfer to low-viscosity liquids in agitated vessels. The following correlations for the liquid–film capacity coefficients are valid for the interphase mass transfer of oxygen into liquid water. For a stirred vessel of coalescing air bubbles, a suitable correlation is

$$(k_L a)_{O_2} = 2.6 \times 10^{-2} \left(\frac{P_g}{V}\right)^{0.4} \left(u_{gs}\right)^{0.5} \tag{30-32}$$

which is valid for $V < 2.6 \, \text{m}^3$ of liquid and $500 < P_g/V < 10{,}000 \, \text{W/m}^3$. For a stirred vessel of noncoalescing air bubbles, a suitable correlation is

$$(k_L a)_{O_2} = 2 \times 10^{-3} \left(\frac{P_g}{V}\right)^{0.7} \left(u_{gs}\right)^{0.2} \tag{30-33}$$

which in turn is valid for $V < 4.4 \, \text{m}^3$ of liquid and $500 < P_g/V < 10{,}000 \, \text{W/m}^3$. In both correlations, the following units must be strictly followed: $(k_L a)_{O_2}$ is the liquid-phase film-capacity coefficient for O_2 in water in units of s^{-1}, P_g/V is the power consumption of the aerated vessel per unit *liquid* volume in units of W/m^3, and u_{gs} is the superficial velocity of the gas flowing through the *empty* vessel in units of m/s, which can be obtained by dividing the volumetric flow rate of the gas by the cross-sectional area of the vessel. These correlations agree with experimental data at 20–40% accuracy for equations (30-32) and (30-33), respectively, regardless of the type of impeller used (e.g., paddle, marine, or flat-blade disk turbine impeller).

The power input per unit liquid volume (P/V) is a complex function of impeller diameter (d_i), impeller rotation rate (N), impeller geometry, liquid viscosity, liquid density, and aeration rate. A correlation shown in Figure 30.7 for the non-aerated, nonvortexing agitation of a Newtonian fluid provides a reasonable approximation for estimating the non-aerated power input P. In Figure 30.7, the impeller Reynolds number is defined as

$$\text{Re}_i = \frac{\rho_L d_i^2 N}{\mu_L} \tag{30-34}$$

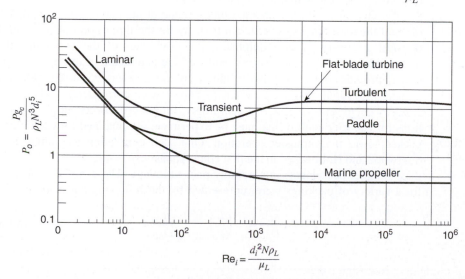

Figure 30.7 Power number vs. Reynolds number for impellers immersed in single-phase liquids.[22]

[21] K. Van't Riet, *Ind. Eng. Chem. Proc. Des. Dev.*, **18**, 357 (1979).
[22] J. H. Ruston, E. W. Costich, and H. J. Everett, *Chem. Eng. Prog.*, **46**, 467 (1950).

and the Power number P_o is defined as

$$P_o = \frac{P_L g_c}{\rho_L N^3 d_i^5} \tag{30-35}$$

For SI units, g_c is equal to 1. Aerating a stirred tank of liquid lowers the impeller power input. Nagata[23] suggests the following correlation of dimensionless groups for estimation of gased power input (P_g) as a function of gas volumetric flow rate (Q_g, e.g., units m³/s) for a flat-blade disk turbine impeller:

$$\log_{10}\left(\frac{P_g}{P}\right) = -192\left(\frac{d_i}{d_T}\right)^{4.38}\left(\frac{d_i^2 N \rho_L}{\mu_L}\right)^{0.115}\left(\frac{d_i N^2}{g}\right)^{1.96\left(\frac{d_i}{d_T}\right)}\left(\frac{Q_g}{d_i^3 N}\right) \tag{30-36}$$

where d_T is the diameter of the vessel. An alternative means to obtain P_g/V is to measure the power input into the stirred tank, but this is often not practical for equipment-design purposes. Chapter 31, Section 31.2, provides more details on design and analysis of gas–liquid mass-transfer operations in well-mixed tanks. The interested reader is also referred to Treybal[20] and Ruston et al.[22] for discussion of the effects of impeller type on the hydrodynamics and power requirements in mechanically agitated vessels.

▶ 30.6

CAPACITY COEFFICIENTS FOR PACKED TOWERS

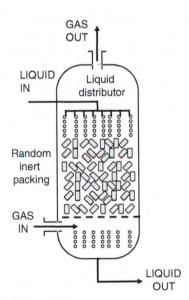

Figure 30.8 Countercurrent flow gas–liquid packed tower.

Although the wetted-wall column has a readily defined interfacial surface area for gas–liquid mass transfer, the corresponding area in other types of gas–liquid contacting equipment is difficult to estimate or measure. For example, a packed tower can provide significant surface area for gas–liquid interphase mass transfer, but the interface area for this process cannot be readily determined because of the complex, multiphase fluid motions involved. In a countercurrent flow packed tower, the gas flows upward and the liquid flows downward through a bed of randomly packed material, as shown in Figure 30.8.

The interfacial surface area per unit volume of the mass-transfer equipment (a), and the mass-transfer coefficient (e.g., k_L for the liquid) both depend on the physical geometry of the equipment, the flow rates of the two contacting, immiscible gas and liquid streams, and the size and shape of the inert packing used to provide contact area. In the bubble column and aerated stirred-tank mass-transfer operations described earlier, k_L and a, were correlated together as the *capacity coefficient*—e.g., $k_L a$. Accordingly, for packed towers, empirical equations for capacity coefficients must be experimentally obtained for each type of packed tower mass-transfer operation. One such correlation was obtained by Sherwood and Holloway[24] in the first comprehensive investigation of liquid–film mass-transfer coefficients in randomly packed towers. The experimental results for a variety of packings were represented by the following correlation:

$$\frac{k_L a}{D_{AB}} = \alpha\left(\frac{L}{\mu}\right)^{1-n}\left(\frac{\mu}{\rho D_{AB}}\right)^{0.5} \tag{30-37}$$

[23] S. Nagata, *Mixing: Principles and Application,* John Wiley & Sons, 1975.
[24] T. K. Sherwood and F. A. Holloway, *Trans. A.I.Ch.E.,* **36**, 21, 39 (1940).

Table 30.2 Packing coefficients for equation (30-37)

Packing	α	n
2.0 in rings	80	0.22
1.5 in rings	90	0.22
1.0 in rings	100	0.22
0.5 in rings	280	0.35
0.375 in rings	550	0.46
1.5 in saddles	160	0.28
1.0 in saddles	170	0.28
0.375 in saddles	150	0.28
3.0 in spiral tiles	110	0.28

The units in the correlation given by equation (30-37) are all in English units and must be strictly followed: $k_L a$ is the mass-transfer capacity coefficient, in h^{-1}, L is the liquid mass flow rate per cross-sectional area of the empty tower, in $lb_m/ft^2 \cdot h$; μ is the viscosity of the liquid, in $lb_m/ft \cdot h$; ρ is the density of the liquid, in lb_m/ft^3; and D_{AB} is the liquid mass diffusivity of transferring component A into inert liquid B, in ft^2/h. The values of the constant α and the exponent n for various packings are given in Table 30.2.

Further correlations for capacity coefficients can be found in treatises on mass-transfer operations in the discussion of each specific operation and each specific type of tower.[25] A more detailed discussion of the design and analysis of randomly packed towers for gas–liquid mass transfer is provided in Chapter 31.

▶ **30.7**

STEPS FOR MODELING MASS-TRANSFER PROCESSES INVOLVING CONVECTION

In many real processes, the flux is coupled to a material balance on the control volume of the physical system. Processes of this type are modeled similarly to the five-step procedure described earlier in Section 25.4.

Step 1: Draw a picture of the physical system. Label the important features, including the boundary surface where convective mass transfer occurs. Decide where the source and the sink for mass transfer are located.

Step 2: Make a "list of assumptions" based on your consideration of the physical system. Assumptions can be added as the model develops.

Step 3: Formulate material balances on the species undergoing mass transfer, and then incorporate the appropriate mass-transfer correlation(s) into the material balance. Processes dominated by convective mass transfer generally fall into two types: (1) the well-mixed control volume of uniform concentration of the transferring species—i.e., a stirred tank or (2) the differential control volume with a

---

[25] T. K. Sherwood, R. L. Pigford, and C. R. Wilke, *Mass Transfer,* McGraw-Hill Book Company, New York, 1975; R. E. Treybal, *Mass Transfer Operations,* McGraw-Hill Book Company, New York, 1980; C. J. King, *Separation Processes,* McGraw-Hill Book Company, New York, 1971; W. S. Norman, *Absorption, Distillation and Cooling Towers,* Wiley, 1961; A. H. P. Skelland, *Diffusional Mass Transfer,* Wiley, New York, 1974.

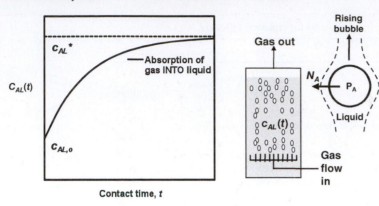

Figure 30.9 Example of well-mixed, unsteady-state mass-transfer process: absorption of bubbled gas into a tank of liquid.

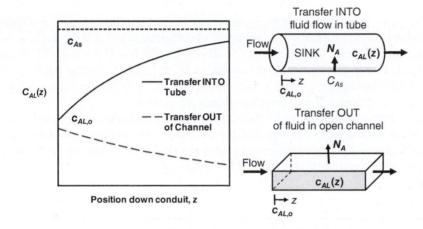

Figure 30.10 Examples of steady-state convective mass transfer where the concentration changes with position.

one-dimensional variation in concentration of the transferring species—i.e., flow through a conduit. Examples of these two process types are shown in Figures 30.9 and 30.10, respectively. Once the material balance is set up, substitute the convective mass-transfer relationship, $N_A = k_c \Delta c_A$, into the material-balance model, and carefully define Δc_A, taking into consideration the concentrations of the transferring species in the fluid at the boundary surface and in the bulk phase. Finally, specify the appropriate correlation for k_c, keeping in mind restrictions on Re, Sc, geometry, and the phase of the transferring species.

Step 4: Recognize and specify the process boundary and initial conditions. These are distinct from the concentration values at the boundary surface for convective mass transfer, which should be specified in Step 3.

Step 5: Solve the algebraic or differential equation(s) resulting from the material balance(s) to obtain the concentration profile, flux, or other parameters of engineering interest. In many cases, the mass-transfer coefficient k_c can be estimated before model development.

Examples 4 to 6 illustrate how convective mass-transfer relationships are integrated into process-material balances.

Example 4

One step in the manufacture of microelectronic devices is microlithography, which traces a microscopic-circuit pattern on the silicon wafer. In one typical process, a thin polymer film, typically less than a thickness 2 μm, is coated over the surface of the silicon wafer. A microscopic template, called a mask, is placed over the surface and irradiated. Radiation that passes through the very tiny holes in the mask hits the photoresist. For a negative photoresist, the radiation initiates reactions that greatly increase the molecular weight of polymer, rendering the photoresist insoluble in an organic solvent. The unreacted photoresist is then dissolved away from the silicon wafer with an organic solvent, and the circuit pattern is revealed by the reacted, insoluble photoresist.

We are interested in using a "spinning disk" mass-transfer device shown in Figure 30.11 to study the photoresist-dissolution process within a closed tank of organic solvent. Consider a limiting case where all of the photoresist on surface of the wafer is soluble in the organic solvent. The negative photoresist is polystyrene (species A), and the organic solvent is methyl ethyl ketone (MEK, species B). The initial thickness of the photoresist coating (ℓ_o) is 2.0 μm, the diameter of the wafer (d) is 10 cm, and the volume of the solvent in the tank (V) is 500 cm^3. If the dissolution process is controlled by the convective mass-transfer rate at the polymer–solvent interface, determine the time required to completely dissolve the photoresist if the disk rotates at 0.5 rev/s (30 rpm). The solubility limit of the developed photoresist in the solvent (c_A^*) is 0.04 g/cm^3, and diffusivity of the photoresist in the solvent at infinite dilution (D_{AB}) is 2.93×10^{-7} cm^2/s at a molecular weight of 7.0×10^5 g/gmole, as reported by Tu and Ouano.[26] The viscosity of the solvent (μ) is 5.0×10^{-3} g/cm·s, the density of the solvent (ρ) is 0.805 g/cm^3, and the density of the solid polymer $\rho_{A,\text{solid}}$ is 1.05 g/cm^3. All physical properties are valid at the process temperature of 298 K. The mass-transfer correlation for a spinning disk is given by Levich:[5]

$$\text{Sh} = \frac{k_c d}{D_{AB}} = 0.62\,\text{Re}^{1/2}\text{Sc}^{1/3} \tag{30-38a}$$

with

$$\text{Re} = \frac{d^2 \omega}{\nu} \tag{30-38b}$$

where ω is the angular rotation rate (radians/time) of the disk.

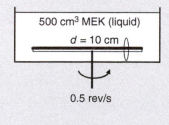

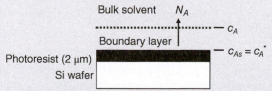

Figure 30.11 Dissolution of photoresist coating on spinning silicon wafer disk into methyl ethyl ketone (MEK) solvent.

The strategy for solving this problem is to develop a material-balance model for the process, and then incorporate the appropriate mass-transfer correlation(s) into the material-balance calculations.

The physical system represents a closed process where the polymer coating on the surface of the disk is the source for mass transfer, and the surrounding well-mixed organic solvent of constant volume is the sink for mass transfer. Under these

---

[26] Y. O. Tu and A. C. Ouano, *IBM J. Res. Dev.*, **21**, 131 (1977).

assumptions, the unsteady-state material balance on the dissolved photoresist in the solvent phase of the well-mixed tank is

$$\begin{pmatrix} \text{rate of photoresist} \\ \text{added to solvent} \end{pmatrix} + \begin{pmatrix} \text{rate of photoresist} \\ \text{exiting solvent tank} \end{pmatrix} = \begin{pmatrix} \text{rate of accumulation of} \\ \text{photoresist within solvent} \end{pmatrix}$$

or

$$N_A \frac{\pi d^2}{4} - 0 = \frac{d(c_A V)}{dt}$$

where c_A represents the concentration of dissolved photoresist in the solvent at time t. The only input term is convective mass-transfer flux from the surface of the spinning disk to the solvent:

$$N_A = k_c(c_{As} - c_A)$$

At the polymer–solvent interface, the dissolved concentration of the photoresist is at its solubility limit, $c_A^* = c_{As}$. Furthermore, as the source is a pure component, the surface concentration remains constant. The material balance reduces to

$$k_c(c_{As} - c_A) \frac{\pi d^2}{4} = V \frac{dc_A}{dt}$$

Separation of the dependent variable c_A from the independent variable t, followed by integration from the initial condition, $t = 0$, $c_A = c_{Ao}$, to the final condition where all the photoresist is dissolved, $t = t_f$, $c_A = c_{Af}$, yields

$$-\int_{c_{Ao}}^{c_{Af}} \frac{-dc_A}{c_{As} - c_A} = \frac{k_c \pi d^2}{4V} \int_{t_o}^{t_f} dt$$

and, finally,

$$t_f - t_o = \frac{4V}{\pi d^2 k_c} \ln\left(\frac{c_{As} - c_{Ao}}{c_{As} - c_{Af}}\right)$$

The final concentration c_{Af} and convective mass-transfer coefficient, k_c, must now be determined. First, c_{Af} is estimated by an overall material balance for the photoresist on the solid wafer and dissolved in the solution:

$$c_{Af} V - c_{Ao} V = m_{Ao} - m_{Af}$$

where m_A is the remaining mass of solid photoresist on the wafer and m_{Ao} is the initial mass, given by

$$m_{Ao} = \rho_{A,\text{solid}} \frac{\pi d^2}{4} l_o$$

where l_o is the original thickness of the photoresist. When all the photoresist is dissolved, $m_{Af} = 0$. If there is initially no photoresist dissolved in the solvent, then $c_{Ao} = 0$ and c_{Af} is

$$c_{Af} = \frac{m_{Ao}}{V} = \frac{\rho_{A,\text{solid}} \pi d^2 l_o}{4V} = \frac{\left(1.05 \frac{\text{g}}{\text{cm}^2}\right) \pi (10\,\text{cm})^2 \left(2.0\,\mu\text{m} \frac{1\,\text{cm}}{10^4\,\mu\text{m}}\right)}{4(500\,\text{cm}^3)} = 3.3 \times 10^{-5} \frac{\text{g}}{\text{cm}^3}$$

The final concentration is well below the solubility limit of 0.04 g/cm³, and so all of the photoresist will dissolve. In order to calculate k_c, Sc and Re are needed. For a dilute system, the fluid properties are essentially the properties of the solvent, so Sc and Re are

$$\text{Sc} = \frac{\mu}{\rho D_{AB}} = \frac{\left(5.0 \times 10^{-3} \frac{\text{g}}{\text{cm}\cdot\text{s}}\right)}{\left(0.805 \frac{\text{g}}{\text{cm}^3}\right)\left(2.93 \times 10^{-7} \frac{\text{cm}^2}{\text{s}}\right)} = 21,199$$

$$\text{Re} = \frac{d^2 \omega \rho}{\mu} = \frac{(10\,\text{cm})^2 \left(\frac{0.5\,\text{rev}}{\text{s}} \frac{2\pi\,\text{rad}}{\text{rev}}\right)\left(0.805 \frac{\text{g}}{\text{cm}^3}\right)}{\left(5.0 \times 10^{-3} \frac{\text{g}}{\text{cm}\cdot\text{s}}\right)} = 50,580$$

Consequently,

$$Sh = \frac{k_c d}{D_{AB}} = 0.62\,Re^{1/2}Sc^{1/3} = 0.62(50{,}580)^{1/2}(21{,}199)^{1/3} = 3859$$

or

$$k_c = \frac{Sh\,D_{AB}}{d} = \frac{3859\left(2.93 \times 10^{-7}\,\dfrac{cm^2}{s}\right)}{10\ cm} = 1.13 \times 10^{-4}\ cm/s$$

Finally, the time required to completely dissolve the photoresist is

$$t_f = \frac{(4)(500\ cm^3)}{\pi(10\ cm)^2\left(1.13 \times 10^{-4}\,\dfrac{cm}{s}\right)} \ln\left(\frac{(0.04 - 0)g/cm^3}{(0.04 - 3.3 \times 10^{-5})g/cm^3}\right) = 46\ s$$

Notice that the concentration difference $c_{As} - c_A$ is relatively constant, because c_A is very small. It is left to the reader to show that

$$W_A = k_c c_{As} \frac{\pi d^2}{4}$$

and

$$t_f = \frac{4m_{Ao}}{k_c c_{As} \pi d^2}$$

for the limiting case where $c_{As} \gg c_A$, i.e., the surrounding solvent represents an *infinite sink* for mass transfer.

Example 5

A holding pond containing a suspension of aerobic microorganisms is used to biologically degrade dissolved organic materials in wastewater, as shown in Figure 30.12. The pond is 15.0 m in diameter and 3.0 m deep. At present, there is no inflow or outflow of wastewater. The process uses an array of air spargers uniformly distributed on the base of the pond to deliver 3.0-mm (0.003 m) air bubbles into the wastewater. The rising bubbles mix the liquid phase. Only a very small portion of the O_2 gas within the air bubbles (0.21 atm O_2 partial pressure) actually dissolves into the liquid. The dissolved oxygen is then consumed by the microorganisms. The gas holdup of the aerated pond (ϕ_g) is 0.005 m^3 gas/m^3 liquid at the present aeration rate. The gas–liquid interphase mass-transfer process is 100% liquid film controlling.

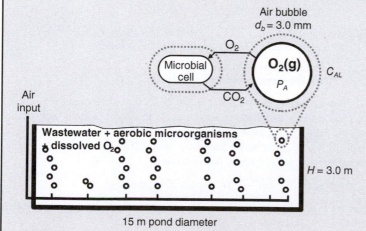

Figure 30.12 Oxygen transfer in an aerated pond containing microorganisms.

We are concerned with satisfying the biological oxygen demand, or "BOD," associated with the microbial degradation process. The BOD is expressed as the molar consumption rate of dissolved O_2 per unit volume of liquid, and has units of gmole $O_2/m^3 \cdot s$. If the dissolved oxygen concentration must be maintained at no less than 0.05 gmole O_2/m^3, what BOD can the aerated pond support, and what is the volumetric flow rate of air into the pond to maintain the gas holdup? At the process temperature of 25 C, the Henry's law constant for O_2 gas dissolved in water is $H = 0.78$ atm $\cdot m^3$/gmole O_2, the molecular diffusion coefficient of dissolved O_2 in water is $D_{AB} = 2.1 \times 10^{-9}$ m^2/s (A = dissolved O_2, B = water), the density of liquid water is $\rho_{B,liq} = 1000$ kg/m^3, and the viscosity of liquid water is $\mu_{B,liq} = 825 \times 10^{-6}$ kg/m $\cdot$ s. At 25 C and 1.0 atm, the density of air is $\rho_{air} = 1.2$ kg/m^3.

The process analysis and solution strategy will require determination of the overall volumetric mass-transfer coefficient $K_L a$, and the development of a model to predict the dissolved O_2 concentration within the pond water. From Chapter 28, if the gas–liquid interphase mass-transfer process is liquid film controlling, then $K_L \approx k_L$, the liquid film mass-transfer coefficient. Furthermore, for a dilute process, the physical properties of the liquid mixture approximate the physical properties of liquid water. For 3.0-mm diameter (d_b) rising air bubbles, equation (30-15) is used to estimate k_L through determination of the Sherwood number (Sh) via the Grashof number (Gr) and Schmidt number (Sc):

$$\text{Gr} = \frac{d_b^3 \rho_L g (\rho_L - \rho_G)}{\mu_L^2} = \frac{(3.0 \times 10^{-3} \text{ m})^3 (1000 \text{ kg/m}^3)(9.8 \text{ m/s}^2)((1000 - 1.2)\text{kg/m}^3)}{(825 \times 10^{-6} \text{ kg/m} \cdot \text{s})^2}$$

$$= 3.88 \times 10^5$$

$$\text{Sc} = \frac{\mu}{\rho D_{AB}} = \frac{\mu_{B,liq}}{\rho_{B,liq} D_{AB}} = \frac{925 \times 10^{-6} \text{ kg/m} \cdot \text{s}}{(1000 \text{ kg/m}^3)(2.1 \times 10^{-9} \text{ m}^2/\text{s})} = 441$$

$$\text{Sh} = \frac{k_L d_b}{D_{AB}} = 0.42 \, \text{Gr}^{1/3} \, \text{Sc}^{1/2} = 0.42(3.88 \times 10^5)^{1/3}(441)^{1/2} = 643$$

$$k_L = \frac{\text{Sh} \, D_{AB}}{d_b} = \frac{(643)(2.1 \times 10^{-9} \text{ m/s})}{3.0 \times 10^{-3} \text{ m}} = 4.5 \times 10^{-4} \text{ m/s}$$

The definition for Sc is based on the liquid phase, since Sh is based on the liquid film surrounding the gas bubble. The interphase mass-transfer area per unit volume of liquid is estimated from the gas holdup (ϕ_g) by equation (30-28)

$$a = \frac{6\phi_g}{d_b} = \frac{6(0.005 \text{ m}^3/\text{m}^3)}{3.0 \times 10^{-3} \text{ m}} = 10 \text{ m}^2/\text{m}^3$$

Consequently,

$$k_L a = (4.5 \times 10^{-4} \text{ m/s})(10 \text{ m}^2/\text{m}^3) = 4.5 \times 10^{-3} \text{ s}^{-1}$$

A model to predict the dissolved oxygen concentration in the pond water requires a material balance on O_2, based on the well-mixed liquid phase of the pond:

$$\left(\begin{array}{c} \text{rate of } O_2 \text{ carried by water} \\ \text{into control volume (IN)} \end{array} \right) + \left(\begin{array}{c} \text{mass transfer of } O_2 \\ \text{into control volume (IN)} \end{array} \right) - \left(\begin{array}{c} \text{rate of } O_2 \text{ carried by water} \\ \text{into control volume (OUT)} \end{array} \right)$$

$$+ \left(\begin{array}{c} \text{rate of } O_2 \text{ generation within} \\ \text{volume element (GEN)} \end{array} \right) = \left(\begin{array}{c} \text{rate of } O_2 \text{ accumulation within} \\ \text{volume element (ACC)} \end{array} \right)$$

All terms are in units of moles O_2/time. The source for O_2 mass transfer is the O_2 in air bubbles, and the sink is the microorganisms that consume O_2. In this process, four primary assumptions are made: (1) the liquid phase is well mixed, so dissolved oxygen concentration is uniform within the pond, (2) the total liquid volume is constant, (3) the biological oxygen demand is constant, (4) the process is at a nominal steady state, because the rate of O_2 input will be balanced by the rate of O_2 consumption. Under these assumptions, the material balance on well-mixed liquid phase is

$$0 + N_A A_i - 0 + R_A V - = 0$$

where the R_A term is the BOD. If we recall that $N_A = K_L(c_{AL}^* - c_{AL})$, $a = A_i/V$, and $c_{AL}^* = P_A/H$, then insertion of these relationships into the material balance, and rearrangement, yields

$$R_A = -k_L a \left(\frac{P_A}{H} - c_{AL} \right) \qquad (30\text{-}39)$$

Equation (30-39) enables estimation of the required R_A

$$R_A = \left(4.5 \times 10^{-3}\ \text{s}^{-1}\right)\left(\frac{0.21\ \text{atm}}{0.78\ \text{atm} \cdot \text{m}^3/\text{gmole}} - 0.05\ \text{gmole/m}^3\right) = -9.87 \times 10^{-4}\ \text{gmole/m}^3\ \text{s}$$

The negative sign on R_A indicates that O_2 is being consumed. The total O_2 transfer rate is

$$W_A = k_L a \left(\frac{P_A}{H} - c_{AL}\right) V = -R_A \frac{\pi D^2 H}{4} = -\left(-9.87 \times 10^{-4}\ \text{gmole } O_2/\text{m}^3\text{s}\right)\left(\frac{\pi (15\ \text{m})^2 (3.0\ \text{m})}{4}\right)$$

$$= 0.523\ \text{gmole } O_2/\text{s}$$

The superficial air velocity (u_{gs}) into the pond can be determined by the gas holdup through equation (30-29). In equation (30-29), the surface tension of water at 25 C, which can be found in many published handbooks, such as the International Critical Tables, is $\sigma_L = 72 \times 10^{-3}$ N/m. To back out u_{gs}, rearrangement of equation (30-29) yields

$$u_{gs} = \frac{\phi_g}{(1 - \phi_g)^4} = 5.0\ \sqrt{g\,D}\left(\frac{g\,D^2\,\rho_L}{\sigma_L}\right)^{-1/8}\left(\frac{g\,D^3}{\nu_L^2}\right)^{-1/12} = \frac{0.005}{(1 - 0.005)^4}\,5.0\,\sqrt{(9.8\ \text{m/s}^2)(15\ \text{m})}$$

$$\times \left(\frac{(9.8\ \text{m/s}^2)(15\ \text{m})^2(1000\ \text{kg/m}^3)}{72\ \text{kg} \cdot \text{m/s}}\right)^{-1/8}\left(\frac{(9.8\ \text{m/s}^2)(15\ \text{m})^3}{8.25 \times 10^{-7}\ \text{m}^2/\text{s}}\right)^{-1/12} = 0.011\ \text{m/s}$$

Finally, with u_{gs} known, the volumetric flow rate of air is

$$Q_g = u_{gs} \frac{\pi D^2}{4} = 0.011\ \text{m/s} \frac{\pi (15\ \text{m})^2}{4} = 1.96\ \text{m}^3\ \text{air/s}$$

It is left to the reader to verify that the molar flow of O_2 delivered in volumetric inlet flow rate of air is in large excess to the interphase O_2 transfer rate into the liquid. Consequently, the O_2 partial pressure inside the bubble does not measurably decrease.

Example 6

The "bubbleless" shell-and-tube membrane aeration system shown in Figure 30.13 is used to transfer oxygen gas (O_2) to liquid water. Water containing no dissolved oxygen is added to the tube side at the entrance at a bulk velocity of 50 cm/s. The inner diameter of the tube is 1.0 cm, and the tube wall thickness is 1.0 mm (0.10 cm). The tube wall is made of silicone, a polymer that is highly permeable to O_2 gas but not to water vapor. Pure oxygen gas (100% O_2) is maintained at a constant pressure of 1.50 atm in the annular space surrounding the tube. The O_2 gas partitions into silicone polymer, and then diffuses through the tube wall to reach the flowing water, as shown in Figure 30.13.

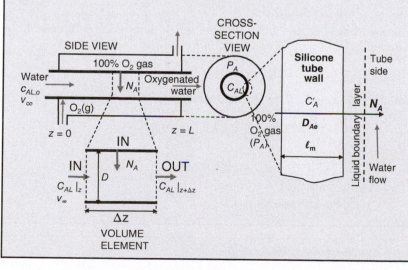

Figure 30.13 Bubbleless shell-and-tube membrane mass-transfer system for aeration of water.

As the fluid flows down the length of the tube, the absorption of oxygen will increase the concentration of the dissolved oxygen. Determine the length of tubing necessary for the dissolved oxygen concentration to reach 30% of saturation with respect to the O_2 gas. The process is maintained at 25 C. At 25 C, the Henry's law constant for O_2 gas dissolved in water is $H = 0.78$ atm·m³/gmole, the molecular diffusion coefficient for dissolved O_2 (A) in water (B) is $D_{AB} = 2.1 \times 10^{-5}$ cm²/s, and the kinematic viscosity of water is $\nu_B = 9.12 \times 10^{-3}$ cm²/s. The solubility constant of O_2 gas in the silicone polymer is $S_m = 0.029$ atm·m³-silicone/gmole, which is defined by the equilibrium relationship $P_A = S_m/C_A'$. The effective diffusion coefficient of O_2 in the silicone polymer is $D_{Ae} = 5.0 \times 10^{-6}$ cm²/s.[27]

The process analysis and solution strategy will require determination of the overall mass-transfer coefficient K_L, and the development of a model to predict the dissolved O_2 concentration in the flowing water down the length of the tube. The flux of O_2 from gas to the bulk flowing water is determined by two mass-transfer resistances in series, the diffusion of O_2 through the silicone tube wall, and the convective mass transfer of O_2 to the flowing water inside the tube. If the overall mass-transfer flux is defined based on the liquid phase, then

$$N_A = K_L \left(c_{AL}^* - c_{AL} \right)$$

where $c_{AL}^* = P_A/H$, and K_L is the overall mass-transfer coefficient based on the liquid-phase driving force, defined by

$$\frac{1}{K_L} = \frac{1}{k_L} + \frac{1}{k_m} \tag{30-40}$$

with

$$k_m = \frac{H D_{Ae}}{S_m l_m} \tag{30-41}$$

In equations (30-40) and (30-41), k_L is the convective mass-transfer coefficient for O_2 mass transfer into the water flowing through the tube, and k_m is the membrane permeation coefficient, which describes the diffusion of O_2 through the silicone tube wall of thickness ℓ_m, and includes solubility constant for partitioning of O_2 gas into the silicone polymer (S_m), the effective diffusion coefficient of O_2 in the silicone polymer (D_{Ae}), and solubility constant for partitioning of O_2 gas into the fluid (H). The derivation of equations (30-40) and (30-41), which ignore the radius of curvature of the tube, is left as an exercise for the reader.

To determine k_L, the Reynolds number (Re) and Schmidt number (Sc) for flow inside the tube must first be estimated. The Reynolds number is

$$Re = \frac{v_\infty D}{\nu_L} = \frac{(50 \text{ cm/s})(1.0 \text{ cm})}{(9.12 \times 10^{-3} \text{ cm}^2/\text{s})} = 5482$$

and the Schmidt number (Sc) is

$$Sc = \frac{\nu_L}{D_{AB}} = \frac{9.12 \times 10^{-3} \text{ cm}^2/\text{s}}{2.1 \times 10^{-5} \text{ cm}^2/\text{s}} = 434$$

The process is considered to be in turbulent flow, since Re > 2100. Therefore, it is necessary to use the Linton–Sherwood correlation for turbulent flow of liquid flowing inside of a tube, given by equation (30-18):

$$Sh = 0.023 \, Re^{0.83} \, Sc^{1/3} = 0.023(5482)^{0.83}(434)^{1/3} = 221$$

The convective mass transfer coefficient k_L is backed out from the Sherwood number

$$k_L = \frac{Sh}{D} D_{AB} = \frac{221}{1.0 \text{ cm}} 2.1 \times 10^{-5} \text{ cm}^2/\text{s} = 4.64 \times 10^{-3} \text{ cm/s}$$

The membrane permeation constant is

$$k_m = \frac{H D_{Ae}}{S_m l_m} = \frac{(0.78 \text{ atm} \cdot \text{m}^3/\text{gmole})(5.0 \times 10^{-6} \text{ cm}^2/\text{s})}{(0.029 \text{ atm} \cdot \text{m}^3/\text{gmole})(0.1 \text{ cm})} = 1.35 \times 10^{-3} \text{ cm/s}$$

[27] C. K. Yeom et al., *J. Membrane Sci.*, **166**, 71 (2000).

And so, finally, K_L is

$$K_L = \frac{k_L k_m}{k_L + k_m} = \frac{(4.64 \times 10^{-3} \text{ cm/s})(1.35 \times 10^{-3} \text{ cm/s})}{(4.64 \times 10^{-3} \text{ cm/s}) + (1.35 \times 10^{-3} \text{ cm/s})} = 1.04 \times 10^{-3} \text{ cm/s}$$

As an aside, note that as k_m increases, K_L approaches k_L and the mass-transfer resistance offered by the membrane becomes small. Increasing k_m can be accomplished by decreasing the membrane thickness ℓ_m or by using a membrane material with a lower S_m—i.e., a material that is more soluble for O_2.

A model to predict dissolved oxygen profile down the length of the tube requires a material balance for O_2 on a differential volume element of the tube, as shown in Figure 30.13 and as stated below:

$$\begin{pmatrix} \text{rate of } O_2 \text{ carried by fluid} \\ \text{into volume element (IN)} \end{pmatrix} + \begin{pmatrix} \text{mass transfer of } O_2 \\ \text{into volume element (IN)} \end{pmatrix} - \begin{pmatrix} \text{rate of } O_2 \text{ carried by fluid} \\ \text{into volume element (OUT)} \end{pmatrix}$$

$$+ \begin{pmatrix} \text{rate of } O_2 \text{ generation within} \\ \text{volume element (GEN)} \end{pmatrix} = \begin{pmatrix} \text{rate of } O_2 \text{ accumulation within} \\ \text{volume element (ACC)} \end{pmatrix}$$

All terms are in units of moles O_2/time. The source for O_2 mass transfer is the O_2 gas, and the sink is the flowing water. The model development requires three primary assumptions: (1) the process is at steady state, (2) the concentration profile of interest is along the z-direction only and represents the local bulk concentration and fluid properties, (3) there is no reaction of O_2 in the water, and (4) the process is dilute with respect to dissolved O_2. Under these assumptions, the differential material balance is

$$\frac{\pi D^2}{4} v_\infty c_{AL} \Big|_z + N_A \pi D \, \Delta z - \frac{\pi D^2}{4} v_\infty c_{AL} \Big|_{z+\Delta z} + 0 = 0$$

Rearrangement yields

$$-\frac{\pi D^2}{4} v_\infty \left(\frac{c_{AL}|_{z+\Delta z} - c_{AL}|_z}{\Delta z} \right) + K_L \pi D \left(c_{AL}^* - c_{AL} \right) = 0$$

When $\Delta z \to 0$,

$$-\frac{dc_{AL}}{dz} + \frac{4K_L}{D v_\infty} \left(c_{AL}^* - c_{AL} \right) = 0$$

At constant p_A, c_{AL}^* is also constant, and so separation of dependent variable c_{AL} from independent variable z, followed by integration gives

$$\int_{c_{AL,o}}^{c_{AL}} \frac{-dc_{AL}}{c_{AL}^* - c_{AL}} = -\frac{4K_L}{v_\infty D} \int_0^L dz$$

or

$$\ln \left(\frac{c_{AL}^* - c_{AL,o}}{c_{AL}^* - c_{AL}} \right) = -\frac{4K_L L}{v_\infty D} \tag{30-42}$$

For $p_A = 1.5$ atm,

$$c_{AL}^* = \frac{p_A}{H} = \frac{1.5 \text{ atm}}{0.78 \text{ atm} \cdot \text{m}^3/\text{gmole}} = 1.92 \text{ gmole } O_2/\text{m}^3$$

and so at 30% of saturation c_{AL} at $z = L$ is $0.3 c_{AL}^*$, or 0.577 gmole O_2/m^3. Therefore, the required length L is

$$L = \frac{v_\infty D}{4 K_L} \ln \left(\frac{c_{AL}^* - c_{AL,o}}{c_{AL}^* - c_{AL}} \right) = \frac{(50 \text{ cm/s})(1.0 \text{ cm})}{4 \cdot (1.04 \times 10^{-3} \text{ cm/s})} \ln \left(\frac{(1.92 - 0) \text{ gmole/m}^3}{(1.92 - 0.577) \text{ gmole/m}^3} \right) = 4296 \text{ cm}$$

or 43 m. Rearrangement of equation (30-42) for the concentration profile gives

$$c_{AL}(z) = c_{AL}^* - \left(c_{AL}^* - c_{AL,o}\right)\exp\left(-\frac{4K_L z}{v_\infty D}\right)$$

Two plots of $c_{AL}(z)$ are provided in Figure 30.14, one ending at 50 m, and the second ending at 500 m, to reveal the approach of $c_{AL}(z)$ to c_{AL}^*.

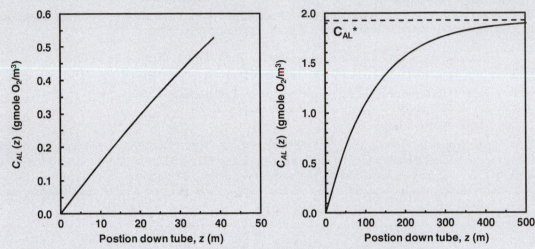

Figure 30.14 Dissolved oxygen profile in bubbleless membrane for aeration of water.

► **30.8**

CLOSURE

In this chapter, we have presented correlating equations for convective mass-transfer coefficients obtained from experimental investigations. The correlations have verified the validity of the analysis of convective transport as presented in Chapter 20. In Chapter 31, methods will be presented for applying the capacity coefficient correlations to the design of mass-transfer equipment.

Mass-Transfer Equipment

In earlier chapters, the theory currently used to explain the mechanism of convective mass transfer between phases was introduced, and correlations for the interphase convective mass-transfer coefficients were provided. In this chapter, we will develop methods for applying the transport equations to the design of gas–liquid mass-transfer equipment. It is important to realize that design procedures are not restricted to the design of new equipment, for they may also be applied in analyzing existing equipment for possible improvement in performance.

The presentation or development of mass transfer from the defining equations to the final design equations, which is presented in this chapter, is completely analogous to our earlier treatment of energy transfer. Convective mass-transfer coefficients are defined in Chapter 28. These definitions and the methods of analysis are similar to those presented in Chapter 19 for convective heat-transfer coefficients. An overall driving force and an overall transfer coefficient expressed in terms of the individual coefficients were developed to explain the transfer mechanisms of both mass- and energy-transport processes. By integrating the appropriate energy-transfer relation in Chapter 23, we were able to evaluate the area of a heat exchanger. Similarly, in this chapter we should expect to find similar mass-transfer relations that can be integrated to yield the required length of a mass exchanger.

▶ **31.1**

TYPES OF MASS-TRANSFER EQUIPMENT

A substantial number of industrial operations in which the compositions of solutions and/or mixtures are changed involve interphase mass transfer. Typical examples of such operations could include (1) the transfer of a solute from the gas phase into a liquid phase, as encountered in absorption, dehumidification, and distillation; (2) the transfer of a solute from the liquid phase into a gas phase, as encountered in desorption or stripping and humidification; (3) the transfer of a solute from one liquid phase into a second immiscible liquid phase (such as the transfer from an aqueous phase to a hydrocarbon phase), as encountered in liquid–liquid extraction; (4) the transfer of a solute from a solid into a fluid phase, as encountered in drying and leaching; and (5) the transfer of a solute from a fluid onto the surface of a solid, as encountered in adsorption and ion exchange.

Mass-transfer operations are commonly encountered in towers or tanks that are designed to provide intimate contact of the two phases. This equipment may be classified into one of the four general types according to the method used to produce the interphase contact. Many varieties and combinations of these types exist or are possible; we will restrict our discussion to the major classifications.

Bubble towers consist of large open chambers through which the liquid phase flows and into which the gas is dispersed into the liquid phase in the form of fine bubbles. The small gas bubbles provide the desired contact area. Mass transfer takes place both during the bubble formation and as the bubbles rise up through the liquid. The rising bubbles create mixing action within the liquid phase, thus reducing the liquid-phase resistance to mass transfer. Bubble towers are used with systems in which the liquid phase normally controls the rate of mass transfer; for example, they are used for the absorption of relatively insoluble gases, such as in the oxygenation of water. Figure 31.1 illustrates the contact time and the

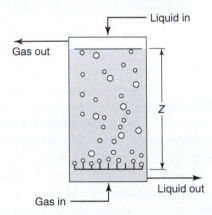

Figure 31.1 Bubble tower.

direction of phase flow in a bubble tower. As one would expect, the contact time, as well as the contact area, plays an important role in determining the amount of mass transferred between the two phases. The basic mass-transfer mechanism involved in bubble towers is also encountered in *batch bubble tanks or ponds* where the gas is dispersed at the bottom of the tanks. Such equipment is commonly encountered in aeration and in wastewater treatment operations.

In a *spray tower*, the gas phase flows up through a large open chamber and the liquid phase is introduced by spray nozzles or other atomizing devices. The liquid, introduced as fine droplets, falls countercurrently to the rising gas stream. The spray nozzle is designed to subdivide the liquid into a large number of small drops; for a given liquid flow rate, smaller drops provide a greater interphase contact area across which mass is transferred. However, as also encountered in bubble towers, care in design must be exercised to avoid producing drops so fine that they become entrained in the exiting, countercurrent stream. Figure 31.2 illustrates the contact time and the direction of phase flow in the spray tower. Resistance to transfer within the gas phase is reduced by the swirling motion of the falling liquid droplets. Spray towers are used for the mass transfer of highly soluble gases where the gas-phase resistance normally controls the rate of mass transfer.

Packed towers are the third general type of mass-transfer equipment, which involves a continuous, countercurrent contact of two immiscible phases. These towers are vertical columns that have been filled with packing as illustrated in Figure 31.3. A variety of packing materials is used, ranging from specially designed ceramic and plastic packing, as illustrated in Figure 31.4, to crushed rock. The chief purpose of the packing is to provide a large contact area between the two immiscible phases. The liquid is distributed over the packing and flows down the packing surface as thin films or subdivided streams. The gas generally flows upward, countercurrent to the falling liquid. Thus, this type of equipment may be used for gas–liquid systems in which either of the phase resistances controls or in which both resistances are important.

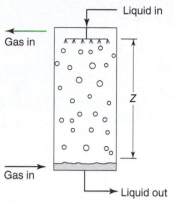

Figure 31.2 Spray tower.

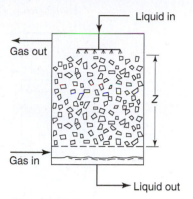

Figure 31.3 Countercurrent packed tower.

Raschig ring   Berl saddle   Lessing ring   Pall ring

Figure 31.4 Common industrial tower packing.

Bubble-plate and *sieve-plate towers* are commonly used in industry for gas–liquid mass-transfer operations. They represent the combined transfer mechanisms observed in the spray and the bubble towers. At each plate, bubbles of gas are formed at the bottom of a liquid pool by forcing the gas through small holes drilled in the plate or under slotted caps immersed in the liquid. Interphase mass-transfer occurs during the bubble formation, and as the bubbles rise up through the agitated liquid pool. Additional mass-transfer takes place above the liquid pool because of spray carryover produced by the active mixing of the liquid and gas on the plate. Such plates are arranged one above the other in a cylindrical shell as schematically illustrated in Figure 31.5. The liquid flows downward, crossing first the upper plate and then the plate below. The vapor rises through each plate. As Figure 31.5 illustrates, the contact of the two phases is stepwise. Such towers cannot be designed by equations that are obtained by integrating over a continuous area of interphase contact. Instead, they are designed by stagewise calculations that are developed and used in design courses of stagewise operations. We shall not consider the design of plate towers in this book; our discussions will be limited to continuous-contact equipment.

► **31.2**

GAS–LIQUID MASS-TRANSFER OPERATIONS IN WELL-MIXED TANKS

Aeration is a common gas–liquid contacting operation where compressed air is introduced to the bottom of a tank of liquid water through small-orifice dispersers, such as perforated pipes, porous sparger tubes, or porous plates. These dispersers produce small bubbles of gas that rise through the liquid. Often, rotating impellers break up the bubble swarms and

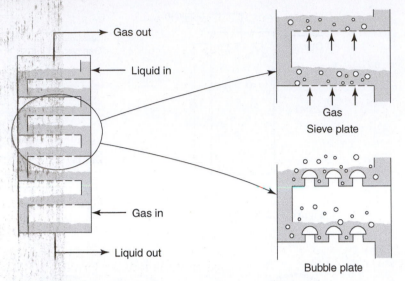

Gas out

Liquid in

Gas

Sieve plate

Bubble plate

Gas in

Liquid out

Figure 31.5 Plate towers.

disperse the bubbles throughout the liquid volume. Gas–liquid mass-transfer processes induced by aeration include *absorption* and *stripping*. In gas absorption, a solute in the aeration gas is transferred from the gas to the liquid. Often, the solute is the oxygen gas in air, which is sparingly soluble in water. The absorption of oxygen into water underlies many processes important to biochemical engineering. In liquid stripping, the volatile dissolved solute is transferred from the liquid to the aeration gas. Stripping underlies many wastewater treatment processes important to environmental engineering.

A well-mixed gas–liquid process is shown in Figure 31.6. The gas–liquid contacting pattern is gas-dispersed, meaning that the gas is dispersed within a continuous liquid phase.

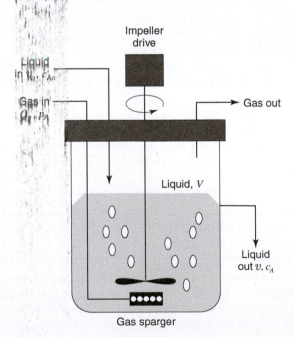

Impeller drive

Liquid in v_o, c_{Ao}

Gas in Q, p_A

Gas out

Liquid, V

Liquid out v, c_A

Gas sparger

Figure 31.6 Aerated stirred tank.

Consequently, the material balances for solute mass-transfer are based on the liquid phase. Recall from Section 29.2 that the interphase mass-transfer flux for solute A across the gas- and liquid-phase films, based on the overall liquid-phase driving force, is

$$N_A = K_L\left(c_A^* - c_A\right) \tag{29-16}$$

Where c_A refers to the molar concentration of solute A in the liquid. The transfer rate for solute A is

$$W_A = K_L \frac{A_i}{V} V\left(c_A^* - c_A\right) = K_L a \cdot V\left(c_A^* - c_A\right) \tag{30-27}$$

with

$$c_A^* = \frac{p_A}{H}$$

where p_A is the partial pressure of solute A in the bulk gas phase. Recall from Section 30.5 that the interphase mass-transfer area per unit volume is hard to measure, and so capacity coefficients—e.g., $K_L a$—are used.

Well-mixed gas–liquid contacting processes can be either continuous or batch with respect to the liquid phase. For a batch process, the liquid flow is turned off, but the gas flow remains on. In this case, the unsteady-state material balance on solute A in the liquid phase is

$$N_A A_i + R_A V = \frac{d(c_A V)}{dt}$$

which is subject to the initial condition $t = 0, c_A = c_{A,o}$. If the liquid volume V is constant, then

$$K_L a \cdot V\left(c_A^* - c_A\right) + R_A V = V \frac{dc_A}{dt}$$

Furthermore, if the partial pressure of solute A is constant so that c_A^* is constant, and there is no homogeneous reaction of dissolved solute A in the liquid phase so that $R_A = 0$, then

$$\int_{c_{A,o}}^{c_A} \frac{-dc_A}{c_A^* - c_A} = -K_L a \int_0^t dt$$

which upon integration yields

$$\ln\left(\frac{c_A^* - c_{A,o}}{c_A^* - c_A}\right) = K_L a \cdot t$$

or

$$c_A = c_A^* - \left(c_A^* - c_{A,o}\right)e^{-K_L a \cdot t} \tag{31-1}$$

In equation (31-1), the concentration of A in the liquid (c_A) exponentially approaches c_A^* as time t goes to infinity.

For a continuous process with one liquid inflow stream and one liquid outflow stream, the steady-state material balance is

$$c_{Ao} \dot{V}_o + N_A A_i - c_A \dot{V} + R_A V = 0$$

where c_{Ao} is now the inlet concentration of solute A. For a dilute process, the inlet liquid volumetric flow rate ($\dot{V}_o$) approaches outlet volumetric flow rate. Consequently,

$$\dot{V}_o(c_{Ao} - c_A) + K_La \cdot V(c_A^* - c_A) + R_A V = 0 \tag{31-2}$$

If $R_A = 0$, then the predicted outlet concentration is

$$c_A = \frac{\dfrac{\dot{V}_o}{V} c_{Ao} + K_La \cdot c_A^*}{\dfrac{\dot{V}_o}{V} + K_La} \tag{31-3}$$

An application of well-mixed gas–liquid contacting operations to the design of an aerobic fermenter is provided in Example 1.

Example 1

The design of aeration systems for aerobic-fermentation processes is based on gas–liquid mass-transfer. Microorganisms grow in a liquid suspension and feed on dissolved nutrients such as glucose and mineral salts. *Aerobic* microorganisms in liquid suspension also require dissolved oxygen for growth. If oxygen is not supplied at a rate sufficient to support cell growth, the cells will die.

In the present process, *Aerobacter aerogenes* is being cultivated within a continuous flow fermenter of 3.0 m³ liquid volume (V) and tank diameter (d_T) of 1.5 m. Fresh nutrient medium containing a trace amount of dissolved O_2 at concentration 0.010 gmole O_2/m³ enters the fermenter at a flow rate of 1.8 m³/h. At steady-state conditions, the aerobic fermenter operates at a cell concentration (c_X) of 5.0 kg/m³ of liquid culture. The cell concentration is determined by the specific growth rate of the organism and the nutrient composition of the liquid medium, details of which will not be presented here. The liquid cell suspension consumes oxygen proportional to the cell concentration according to the rate equation

$$R_A = -q_o c_X$$

where q_o is the specific oxygen consumption rate of the cells, equal to 20 gmole O_2/kg cells · h, which is assumed to be constant. Determine the K_La value necessary to ensure that the dissolved oxygen concentration in the liquid culture (c_A) is maintained at 0.050 gmole/m³. Also, determine the power input into a 3.0 m³ fermenter if the gas flow rate into the fermenter is 1.0 m³ air/min at the process conditions of 298 K and 1.0 atm. Assume that the bubbles are noncoalescing. At 298 K, Henry's law constant for dissolution of O_2 in the liquid nutrient medium is 0.826 m³atm/gmole.

The required K_La is backed out from a material balance on dissolved oxygen (species A) within the well-mixed liquid phase of the fermenter. Recall equation (31-2):

$$\dot{V}_o(c_{Ao} - c_A) + K_LaV(c_A^* - c_A) + R_A V = 0$$

Inserting $R_A = -q_o c_X$ and solving for the required K_La yields

$$K_La = \frac{q_o c_X - \dfrac{\dot{V}_o}{V}(c_{Ao} - c_A)}{c_A^* - c_A} \tag{31-4}$$

The saturation concentration of dissolved oxygen is determined by Henry's law:

$$c_A^* = \frac{p_A}{H} = \frac{0.21 \text{ atm}}{0.826 \dfrac{\text{m}^3 \cdot \text{atm}}{\text{gmole}}} = 0.254 \frac{\text{gmole } O_2}{\text{m}^3}$$

The partial pressure of oxygen (p_A) is presumed constant, as the rate of O_2 transferred to the sparingly soluble liquid is very small in comparison to the molar flow rate of O_2 in the aeration gas. Finally,

$$K_La = \frac{\left(\dfrac{20 \text{ gmole } O_2}{\text{kg cells} \cdot \text{h}} \dfrac{5.0 \text{ kg cells}}{\text{m}^3} - \dfrac{1.8 \text{ m}^3/\text{h}}{3.0 \text{ m}^3}(0.010 - 0.050)\dfrac{\text{gmole } O_2}{\text{m}^3}\right) \dfrac{1 \text{ h}}{3600 \text{ s}}}{(0.254 - 0.050)\dfrac{\text{gmole } O_2}{\text{m}^3}} = 0.136 \text{ s}^{-1}$$

The power input to the aerated tank is backed out from the correlation

$$(k_La)_{O_2} = 2 \times 10^{-3} \left(\frac{P_g}{V}\right)^{0.7} (u_{gs})^{0.2} \tag{30-33}$$

where k_La has units of s^{-1}, P_g/V has units of W/m^3, and u_{gs} has units of m/s. The superficial velocity of the gas through the empty tank is

$$u_{gs} = \frac{4Q_g}{\pi d_T^2} = \frac{(4)\left(\dfrac{1.0\,\text{m}^3}{\text{min}}\dfrac{1\,\text{min}}{60\,\text{s}}\right)}{\pi(1.5\,\text{m})^2} = 0.0094\,\frac{\text{m}}{\text{s}}$$

If the gas is sparingly soluble in the liquid, the interphase mass-transfer process is liquid-phase controlling so that $K_La \cong k_La$. Therefore,

$$0.136 = 2.0 x 10^{-3} \left(\frac{P_g}{V}\right)^{0.7} (0.0094)^{0.2}$$

or

$$\frac{P_g}{V} = 1572\,\frac{\text{W}}{\text{m}^3}$$

The total required power input (P_g) for the 3.0 m^3 aerated fermenter is 4716 W.

Eckenfelder[1] developed a general correlation for the transfer of oxygen from air bubbles rising in a column of still water:

$$K_L \frac{A}{V} = \frac{\theta_g\, Q_g^{1+n}\, h^{0.78}}{V} \tag{31-5}$$

where θ_g is a correlating constant dependent on the type of disperser, Q_g is the gas flow rate in standard ft^3/min, n is a correlating constant that is dependent on the size of the small orifices in the disperser, and h is the depth below the liquid surface at which the air is introduced to the aeration tank. Typical data for a sparger-aeration unit, correlated according to equation (31-5), are presented in Figure 31.7. Equation (31-5) and Figure 31.7 are useful in the mass-transfer design of aeration basins, as described in Example 2 below.

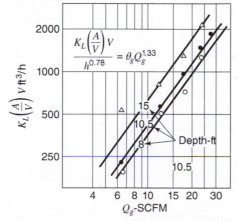

Figure 31.7 Oxygen transfer factor of a single sparger unit in an aeration tank.

[1] W. W. Eckenfelder, Jr., *J. Sanit. Engr. Div.,* Amer. Soc. Civ. Engr., **85**, SA4, 89 (1959).

Example 2

An aeration pond of 20,000 ft^3 (566 m^3) liquid volume is aerated with 15 spargers, each using compressed air at a rate of 15 standard ft^3 air/min (7.08×10^{-3} m^3/s). The spargers will be located 15.0 ft (4.57 m) below the surface of the pond. Estimate the time required to raise the dissolved oxygen concentration from 2.0 to 5.0 mg O$_2$/L if the water temperature is 293 K. At 298 K, Henry's law constant for dissolution of O$_2$ water is 0.826 m^{3}atm/gmole.

It is first necessary to estimate the mass-transfer coefficient for O$_2$ gas–liquid transfer the aeration pond. Using Figure 31.7, the transfer factor, $K_L(A/V)V$, for a single sparger is 1200 ft^3/h at $Q_g = 15$ ft^3/min and h = 15 ft. For the entire aeration system,

$$K_L\left(\frac{A}{V}\right) = \frac{K_L(A/V)V}{V}n = \frac{(1200\ \text{ft}^3/\text{h})(1\ \text{h}/3600\ \text{s})(15\ \text{spargers})}{(20,000\ \text{ft}^3)} = 2.50 \times 10^{-4}\ \text{s}^{-1}$$

To determine the time required for the aeration process, the dissolved O$_2$ concentration at the operating and equilibrium (saturation) conditions must now be specified in consistent units. First, the average hydrostatic pressure of the rising air bubble in the pond is equal to the arithmetic mean of the pressure at the top (liquid surface) and the bottom (submerged discharge point for the sparger):

$$P_{\text{ave}} = \frac{P_{\text{top}} + P_{\text{bottom}}}{2} = P_{\text{top}} + \frac{1}{2}\rho_L g h$$

$$P_{\text{ave}} = 1.0123 \times 10^5\ \text{Pa} + \frac{1}{2}\left(\frac{998.2\ \text{kg}}{\text{m}^3}\right)\left(\frac{9.81\ \text{m}}{\text{s}^2}\right)(4.55\ \text{m}) = 1.236 \times 10^5\ \text{Pa}$$

The O$_2$ composition in air bubble is 21 mole%. The concentration of dissolved oxygen in equilibrium with O$_2$ in the air bubble is determined by Henry's law

$$c_A^* = \frac{p_A}{H} = \frac{y_A P_{\text{ave}}}{H} = \frac{(0.21)\left(1.236 \times 10^5\ \text{Pa}\right)\left(1.0\ \text{atm}/1.013 \times 10^5\ \text{Pa}\right)}{(0.826\ \text{m}^3 \cdot \text{atm/gmole})} = 0.310\frac{\text{gmole O}_2}{\text{m}^3}$$

The molar concentration of O$_2$ dissolved in water at the final condition is

$$c_A = \rho_{A,o}/M_A = \left(\frac{5.0\ \text{mg O}_2}{\text{L}}\frac{1000\ \text{L}}{\text{m}^3}\frac{1\ \text{g}}{1000\ \text{mg}}\right)\left(\frac{1\ \text{gmole}}{32\ \text{g O}_2}\right) = 0.156\frac{\text{gmole}}{\text{m}^3}$$

Similarly, the molar concentration of O$_2$ at the initial condition is $c_{A,o} = 0.0625$ gmole O$_2$/m^3.

Finally, using equation (31-1), the required time is

$$t = \ln\left(\frac{c_A^* - c_{A,o}}{c_A^* - c_A}\right)\left(\frac{1}{K_L\left(\frac{A}{V}\right)}\right) = \ln\left(\frac{(0.310 - 0.0625)\text{gmole/m}^3}{(0.310 - 0.156)\text{gmole/m}^3}\right)\left(\frac{1}{2.50 \times 10^{-4}\ \text{s}^{-1}}\right) = 1900\ \text{s}$$

► **31.3**

MASS BALANCES FOR CONTINUOUS-CONTACT TOWERS: OPERATING-LINE EQUATIONS

There are four important fundamentals that constitute the basis for continuous-contact equipment design:

1. Material and enthalpy balances, involving the equations of conservation of mass and energy

2. Interphase equilibrium

3. Mass-transfer equations

4. Momentum-transfer equations

Interphase equilibrium relations are defined by laws of thermodynamics, as discussed in Section 29.1. Momentum-transfer equations, as presented in Section 9.3, are used to define the pressure drop within the equipment. We shall not treat this subject in this chapter, as it was previously discussed. The material and enthalpy balances are important, as they provide expressions for evaluating the bulk compositions of the two contacting phases at any plane in the tower as well as the change in bulk compositions between two planes in the tower. The mass-transfer equations will be developed in differential form, combined with a differential material balance, and then integrated over the area of interfacial contact to provide the length of contact required in the mass exchanger.

Countercurrent Flow

Consider any steady-state mass-transfer operation that involves the countercurrent contact of two insoluble phases as schematically shown in Figure 31.8. The two insoluble phases will be identified as phase G and phase L.

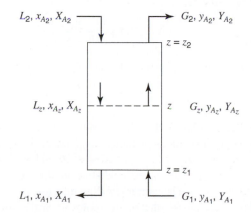

Figure 31.8 Steady-state countercurrent process.

At the bottom of the mass-transfer tower, the flow rates and concentrations are defined as follows:

G_1 is the total moles of gas-phase G entering the tower per hour per cross-sectional area of the tower, i.e., the superficial molar velocity of phase G;

L_1 is the total moles of liquid-phase L leaving the tower per hour per cross-sectional area of the tower, i.e., the superficial molar velocity of phase L;

y_{A_1} is the mole fraction of component A in G_1, expressed as moles of A per total moles in phase G;

x_{A_1} is the mole fraction of component A in L_1, expressed as moles of A per total moles in phase L.

Similarly at the top of the tower, the total superficial molar velocity of each phase will be G_2 and L_2, and the mole fraction compositions of each stream will be y_{A_2} and x_{A_2}. An overall macroscopic mass balance per cross-sectional area of tower for component A around the steady-state mass exchanger, in which there is no chemical production or disappearance of A, requires

$$\begin{pmatrix} \text{molar flow rate of A} \\ \text{entering the tower} \end{pmatrix} = \begin{pmatrix} \text{molar flow rate of A} \\ \text{leaving the tower} \end{pmatrix}$$

or

$$y_{A_1}G_1 + x_{A_2}L_2 = y_{A_2}G_2 + x_{A_1}L_1 \tag{31-6}$$

A mass balance for component A around plane $z = z_1$ and the arbitrary plane z stipulates

$$y_{A_1}G_1 + x_{A,z}L_z = y_{A,z}G_z + x_{A_1}L_1 \tag{31-7}$$

or

$$y_A = \frac{L}{G}x_A + \frac{y_{A_1}G_1 - x_{A_1}L_1}{G}$$

Simpler relations, and certainly easier equations to use, may be expressed in terms of *solute-free concentration units*, also known as mole ratios. The concentration of each phase will be defined as follows. For the gas, Y_A is the moles of A in G per mole of A-free G—that is,

$$Y_A = \frac{y_A}{1 - y_A} \tag{31-8}$$

For the liquid, X_A is the moles of A in L per mole of A-free L—that is,

$$X_A = \frac{x_A}{1 - x_A} \tag{31-9}$$

The molar flow rates to be used with the solute-free concentration units are L_S and G_S, where L_S is the superficial molar velocity of phase L on a solute-free basis—that is, the moles of solvent in phase L per hour per cross-sectional area of the tower—and is equal to $L(1 - x_A)$, where both L and x_A are evaluated at the same plane in the tower: that is, $L_1(1 - x_{A_1})$ or $L_2(1 - x_{A_2})$. The carrier gas flow rate, G_S, is the superficial molar velocity of phase G on a solute-free basis—that is, the moles of carrier solvent in phase G per hour per cross-sectional area of the tower—and is equal to $G(1 - x_A)$, where both G and y_A are evaluated at the same plane in the tower. The overall balance on component A may be written using the solute-free terms as

$$Y_{A_1}G_S + X_{A_2}L_S = Y_{A_2}G_S + X_{A_1}L_S \tag{31-10}$$

or

$$G_S(Y_{A_1} - Y_{A_2}) = L_S(X_{A_1} - X_{A_2})$$

or

$$\frac{L_S}{G_S} = \frac{Y_{A_1} - Y_{A_2}}{X_{A_1} - X_{A_2}}$$

Equation (31-10) is an equation of a straight line that passes through points (X_{A_1}, Y_{A_1}) and (X_{A_2}, Y_{A_2}) with a slope of L_S/G_S. A mass balance of component A around plane z_1 and an arbitrary plane $z = z$ in solute-free terms is

$$Y_{A_1}G_S + X_{A,z}L_S = Y_{A,z}G_S + X_{A_1}L_S \tag{31-11}$$

which may be rewritten as

$$G_S\left(Y_{A_1} - Y_{A,z}\right) = L_S\left(X_{A_1} - X_{A,z}\right)$$

or

$$Y_A = \frac{L_S}{G_S} X_A + \frac{Y_{A_1} G_S - X_{A_1} L_S}{G_S}$$

or

$$\frac{L_S}{G_S} = \frac{Y_{A_1} - Y_{A,z}}{X_{A_1} - X_{A,z}}$$

Equation (31-11) describes an equation of a straight line, one that passes through the points (X_{A_1}, Y_{A_1}) and $(X_{A,z}, Y_{A,z})$ with a slope L_S/G_S. As it defines operating conditions within the equipment, it is designated the *operating line for countercurrent operations*. In our earlier discussions on interphase transfer in Section 29.2, the operating point is one of many points that lie on the operating line.

It is important to recognize the difference between equations (31-7) and (31-11). Although both equations describe the mass balance for component A, only equation (31-11) is an equation of a straight line. When written in the solute-free units, X_A and Y_A, the operating line is straight because the mole ratios are both referenced to constant quantities, L_S and G_S. When written in mole-fraction units, x_A and y_A, the total moles in a phase, L or G, change as the solute is transferred into or out of the phase; this produces a curved operating line in x-y coordinates.

Figure 31.9 illustrates the location of the operating line relative to the equilibrium line when the transfer of the solute is from phase G to phase L, as in the case of gas absorption. The bulk composition, located on the operating line, must be greater than the equilibrium concentration in order to provide the mass-transfer driving force.

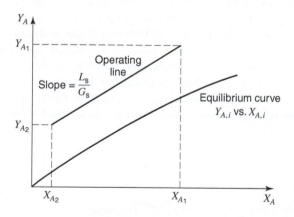

Figure 31.9 Steady-state countercurrent process, transfer from phase G to L.

Figure 31.10 illustrates the location of the operating line relative to the equilibrium line when the transfer of the solute is from phase L to phase G, as in desorption or liquid stripping. The location of the operating line below the equilibrium line assures the correct driving forces.

A mass balance for component A over the differential length, dz, is easily obtained by differentiating equation (31-11). This differential equation

$$L_S\, dX_A = G_S\, dY_A \tag{31-11a}$$

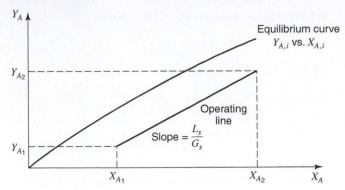

Figure 31.10 Steady-state countercurrent process, transfer from phase L to phase G.

relates the moles transferred in countercurrent operations per time per cross-sectional area available in the length dz.

In the design of mass-transfer equipment, the flow rate of at least one terminal stream and three of the four entering and exiting compositions must be fixed by the process requirements. The necessary flow rate of the second phase is often a design variable. For example, consider the case in which phase G, with a known G_S, changes in composition from Y_{A_1} to Y_{A_2} by transferring solute to a second phase that enters the tower with composition X_{A_2}. According to equation (31-10), the operating line must pass through point (X_{A_2}, Y_{A_2}) and must end at the ordinate Y_{A_1}. Three possible operating lines are shown in Figure 31.11. Each line has a different slope, L_S/G_S, and since G_S is fixed by the process requirement, each line represents a different quantity, L_S, of the second phase. In fact, as the slope decreases, L_S decreases. The *minimum* L_S that may be used corresponds to the operating line ending at point P_3. This quantity of the second phase corresponds to an operating line that "touches" the equilibrium line, and so X_{A_1} is represented as $X_{A_1,\min}$ at this "pinch point." If we recall from Chapter 29 the definition of driving forces, we should immediately recognize that the closer the operating line is to the equilibrium curve, the smaller will be the mass-transfer driving force. At the point of tangency, the diffusional driving force is zero; thus, mass-transfer between the two phases cannot occur. This represents a limiting condition, the *minimum L_S/G_S ratio for mass-transfer*. For gas absorption, this minimum condition refers to the inert solvent—i.e., $L_{S,\min}$—but for liquid

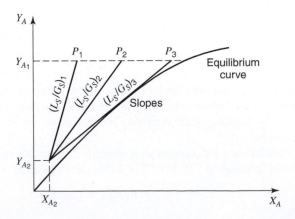

Figure 31.11 Operating-line locations.

stripping this minimum conditions refers to the inert carrier gas: i.e., $G_{S,min}$. Consequently, for gas absorption, at fixed G_S, the slope L_S/G_S will be at its smallest value at the pinch point, with minimum solvent flow rate $L_{S,min}$ at $X_{A_1,min}$. For liquid stripping at fixed L_S, the slope L_S/G_S will be at its largest value at the pinch point, with minimum carrier gas flow rate $G_{S,min}$ at $Y_{A_2,min}$.

Example 3

Ammonia (NH_3) will be absorbed from an air mixture at 293 K and 1.0 atm pressure in a countercurrent packed tower, using water as the absorption solvent. The total inlet gas molar flow rate is 5.03×10^{-4} kgmole/s, and the NH_3 composition in the inlet gas is 3.52% by volume. Ammonia-free water at a mass flow rate of 9.46×10^{-3} kg/s will be used as the absorption solvent. If the ammonia concentration in the outlet gas is reduced to 1.29% by volume, determine the ratio of the operating solvent molar flow rate to the minimum molar solvent flow rate, Equilibrium distribution data for the NH_3-water-air system at 293 K and 1.0 atm are given in Figure 31.12.

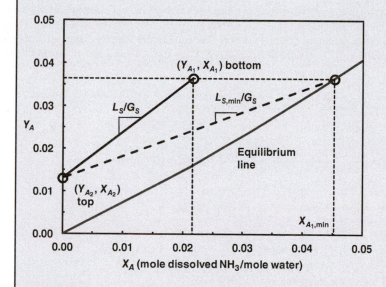

Figure 31.12 Operating and equilibrium lines for Example 3. Equilibrium distribution data from Perry and Chilton.[2]

The analysis begins with a process material balance. First, the carrier gas flow molar rate is determined by

$$AG_S = AG_1\left(1 - y_{A_1}\right) = 5.03 \times 10^{-4} \text{ kgmole/s}(1 - 0.0352) = 4.85 \times 10^{-4} \text{ kgmole/s}$$

where A refers to the cross-sectional area of the tower, and the subscript A refers to the transferring solute. The total molar flow rate of the solvent entering the top of the tower is

$$AL_S = AL_2(1 - x_{A_2}) = AL_2(1 - 0) = AL_2 = \left(9.46 \times 10^{-3} \frac{\text{kg}}{\text{s}}\right)\left(\frac{1 \text{ kgmole}}{18 \text{ kg}}\right) = 5.26 \times 10^{-4} \text{ kgmole/s}$$

The mole ratio compositions of the streams of known mole fraction are

$$Y_{A_1} = \frac{y_{A_1}}{1 - y_{A_1}} = \frac{0.0352}{1 - 0.0352} = 0.0365$$

---

[2] R. H. Perry and C. H. Chilton, *Chemical Engineer's Handbook*, 5th edition, McGraw-Hill Book Company, New York, 1973, Table 3-124.

$$Y_{A_2} = \frac{y_{A_2}}{1 - y_{A_2}} = \frac{0.0129}{1 - 0.0129} = 0.0131$$

$$X_{A_2} = \frac{x_{A_2}}{1 - x_{A_2}} = \frac{0}{1 - 0} = 0$$

The mole ratio for solute A in the liquid exiting the bottom of the tower, X_{A_1}, is determined by a material balance on solute A around the terminal streams, given by

$$AG_S Y_{A_1} + AL_S X_{A_2} = AG_S Y_{A_2} + AL_S X_{A_1}$$

or

$$X_{A_1} = \frac{AG_S(Y_{A_1} - Y_{A_2}) + AL_S X_{A_2}}{AL_S} = \frac{4.85 \times 10^{-4}(0.0365 - 0.0131) + 0}{5.26 \times 10^{-4}} = 0.0216$$

With the terminal stream compositions now known, the equilibrium and operating lines in mole ratio coordinates are shown in Figure 31.12. Given that the shape equilibrium line is nearly linear in the composition range of interest, the minimum solvent flow rate will occur when the composition of the solute in the liquid exiting the tower is in equilibrium with the composition of the solute A in the gas entering the tower. Consequently, $X_{A_1,\min} = 0.0453$ at $Y_{A_1} = 0.0365$. With $X_{A_1,\min}$ known, the minimum solvent flow rate is estimated by material balance

$$AG_S Y_{A_1} + AL_{S,\min} X_{A_2} = AG_S Y_{A_2} + AL_{S,\min} X_{A_1,\min}$$

or

$$AL_{S,\min} = \frac{AG_S(Y_{A_1} - Y_{A_2})}{X_{A_1,\min} - X_{A_2}} = \frac{4.85 \times 10^{-4}(0.0365 - 0.0131)}{(0.0453 - 0)} = 2.51 \times 10^{-4} \text{ kgmole/s}$$

Finally, the ratio of the operating solvent flow rate to the minimum solvent flow rate is

$$\frac{AL_S}{AL_{S,\min}} = \frac{5.26 \times 10^{-4} \text{ kgmole/s}}{2.51 \times 10^{-4} \text{ kgmole/s}} = 2.09$$

Thus, the operating solvent flow rate is 109% above the minimum solvent flow rate.

Cocurrent Flow

For steady-state mass-transfer operations involving cocurrent contact of two immiscible phases as shown in Figure 31.13, the overall mass balance for component A is

$$L_S X_{A_2} + G_S Y_{A_2} = L_S X_{A_1} + G_S Y_{A_1} \tag{31-12}$$

which may be rewritten as

$$L_S(X_{A_2} - X_{A_1}) = G_S(Y_{A_1} - Y_{A_2})$$

or

$$-\frac{L_S}{G_S} = \frac{(Y_{A_1} - Y_{A_2})}{(X_{A_1} - X_{A_2})}$$

$L_2, L_s, x_{A_2}, X_{A_2}$ ◄   ► $G_2, G_s, y_{A_2}, Y_{A_2}$

$z = z_2$

$L_z, L_s, x_{A,z}, X_{A,z}$ $G_z, G_s, y_{A,z}, Y_{A,z}$

$z = z_1$

$L_1, L_s, x_{A_1}, X_{A_1}$ $G_1, G_s, y_{A_1}, Y_{A_1}$ **Figure 31.13** Steady-state cocurrent process.

The mass balance on component A around planes 1 and an arbitrary plane z stipulates that

$$L_S X_{A,z} + G_S Y_{A,z} = L_S X_{A_1} + G_S Y_{A_1} \tag{31-13}$$

or

$$L_S\left(X_{A,z} - X_{A_1}\right) = G_S\left(Y_{A_1} - Y_{A,z}\right)$$

or

$$-\frac{L_S}{G_S} = \frac{\left(Y_{A_1} - Y_{A,z}\right)}{\left(X_{A_1} - X_{A,z}\right)}$$

Equations (31-12) and (31-13) both represent straight lines that pass through a common point (X_{A_1}, Y_{A_1}) and have the same slope, $-L_S/G_S$; equation (31-13) is the general expression that relates to the composition of the two contacting phases at any plane in the equipment. It is designated as the *operating-line equation for cocurrent operations*. Figures 31.14 and 31.15 illustrate the location of the operating line relative to the equilibrium line. A mass balance for component A over the differential length, dz, for

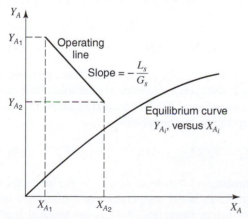

Figure 31.14 Steady-state cocurrent process, transfer from phase G to phase L.

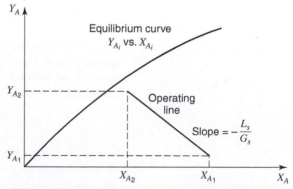

Figure 31.15 Steady-state cocurrent process transfer from phase L to phase G.

cocurrent flow, given by

$$L_S \, dX_A = -G_S \, dY_A \qquad (31\text{-}13a)$$

verifies that the slope of the operating line for cocurrent operation is $-L_S/G_S$.

As in the case of countercurrent flow, there is a minimum L_S/G_S ratio for cocurrent mass-transfer operations established from the fixed-process variables: G_S, Y_{A_1}, Y_{A_2}, and X_{A_1}. Its evaluation involves the same procedure as discussed for countercurrent flow.

▶ **31.4**

ENTHALPY BALANCES FOR CONTINUOUS-CONTACTS TOWERS

Many mass-transfer operations are isothermal. This is especially true when we are dealing with dilute mixtures. However, when large quantities of solute are transferred, the heat of mixing can produce a temperature rise in the receiving phase. If the temperature of the phase changes, the equilibrium solubility of the solute will be altered, and in turn the diffusion driving forces will be altered.

Consider the steady-state countercurrent process illustrated in Figure 31.16. An enthalpy balance around the planes $z = z_2$ and z is

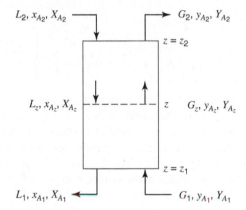

Figure 31.16 Steady-state countercurrent process.

$$L_2 H_{L_2} + G H_G = G_2 H_{G_2} + L H_L \qquad (31\text{-}14)$$

where H is the molal enthalpy of the stream at its particular phase, temperature, pressure, and concentration. The enthalpies are normally based upon a reference of pure solute-free carrier solvent and pure solute at a chosen base temperature, T_o. The normal enthalpy of a liquid mixture is evaluated above this base temperature by the relation

$$H_L = C_{p,L}(T_L - T_o)M_{\text{avg}} + \Delta H_S \qquad (31\text{-}15)$$

where H_L is the enthalpy of the liquid stream, in kJ/mole of L; $C_{P,L}$ is the heat capacity of the mixture on a mass basis, in kJ/kg·K; T_L is the temperature of the mixture in K; M_{avg} is the average molecular weight of the mixture; and ΔH_S is the integral heat of solution, evaluated at the base temperature, T_o, and at the concentration of the mixture, in kJ/mole.

The enthalpy of a gas mixture with the same base temperature and standard state of the solute is expressed as

$$H_G = \left[y_A C_{p,G,A} M_A + (1 - y_A)(C_{p,G,B})(M_B) \right](T_G - T_0) + y_A h_{f,G,A} M_A \qquad (31\text{-}16)$$

where H_G is the enthalpy of the gas stream, in kJ/mole of G; $C_{p,G,A}$ is the heat capacity in the gas phase in kJ/kg · K; T_G is the temperature of the gas mixture in K; M_A is the molecular weight of solute A; and $h_{f,G,A}$ is the heat of vaporization of the solute in kJ/kg. The integral heat of solution, ΔH_S, is zero for ideal solutions and essentially zero for gas mixtures. For nonideal solutions, it is a negative quantity if heat is evolved on mixing and a positive quantity if heat is absorbed on mixing.

Equations (31-14) to (31-16) may be used to compute the temperature of a given phase at any plane within the mass-transfer equipment. The calculations involve the simultaneous application of the mass balance in order to know the flow rate of the stream associated with the particular enthalpy term.

► 31.5

MASS-TRANSFER CAPACITY COEFFICIENTS

Recall that the individual mass-transfer coefficient, k_G, was defined by the expression

$$N_A = k_G (p_A - p_{A,i}) \qquad (29\text{-}7)$$

and the overall mass-transfer coefficient was defined by a similar equation in terms of the overall driving force in partial pressure units

$$N_A = K_G (p_A - p_A^*) \qquad (29\text{-}15)$$

In both expressions, the interphase mass transfer was expressed as moles of A transferred per unit time per unit area per unit driving force in terms of partial pressure. To use these equations in the design of mass exchangers, the interphase contact area must be known. Although the wetted-wall column described in Chapters 29 and 30 has a defined interfacial surface area, the corresponding area in other types of equipment is virtually impossible to measure. For this reason the factor a must be introduced to represent the interfacial surface area per unit volume of the mass-transfer equipment. In this physical situation, the mass-transfer rate within a differential height, dz, per unit cross-sectional area of the mass exchanger is $N_A \cdot a \cdot dz$, which has the following dimensional terms:

$$N_A \left(\frac{\text{moles of A transferred}}{(\text{time})(\text{interfacial area})} \right) a \left(\frac{\text{interfacial area}}{\text{volume}} \right) dz(\text{length}) = \frac{\text{moles of A transferred}}{(\text{time})(\text{cross-sectional area})}$$

or, in terms of the mass-transfer coefficients,

$$N_A a \cdot dz = k_G a (p_A - p_{A,i}) \, dz \qquad (31\text{-}17)$$

and

$$N_A a \cdot dz = K_G a (p_A - p_A^*) \, dz \qquad (31\text{-}18)$$

As both the factor a and the mass-transfer coefficient depend on the geometry of the mass-transfer equipment and on the flow rates of the two contacting, immiscible streams, they are commonly combined as a product. The *individual capacity coefficient*, $k_G a$, and the *overall capacity coefficient*, $K_G a$, are each experimentally evaluated as a combined process variable. The units of the gas-phase capacity coefficient are

$$k_G a \left(\frac{\text{moles of A transferred}}{(\text{time})(\text{interfacial area})(\text{pressure})} \right) \left(\frac{\text{interfacial area}}{\text{volume}} \right) = \frac{\text{moles of A transferred}}{(\text{time})(\text{volume})(\text{pressure})}$$

The most often encountered units are moles $A/\text{m}^3 \cdot \text{s} \cdot \text{atm}$. The capacity coefficients in terms of liquid concentration driving forces are similarly defined by

$$N_A \, a \cdot dz = k_L a \left(c_{AL,i} - c_{AL} \right) dz \tag{31-19}$$

and

$$N_A \, a \cdot dz = K_L a \left(c_{AL}^* - c_{AL} \right) dz \tag{31-20}$$

The most common units for the liquid-phase capacity coefficients $k_L a$ and $K_L a$ are s^{-1}. Mass-transfer capacity coefficients in terms of mole fraction driving force ($k_y a$, $k_x a$, $K_y a$, $K_x a$) commonly have units of moles/$\text{m}^3 \cdot \text{s}$, those based on a mole ratio driving force ($k_Y a$, $k_X a$, $K_Y a$, $K_X a$) are commonly expressed in units of moles/$\text{m}^3 \cdot \text{s}$ or moles/$\text{m}^3 \cdot \text{s} \cdot$ (mole A/mole solvent or carrier gas).

► 31.6

CONTINUOUS-CONTACT EQUIPMENT ANALYSIS

The moles of the diffusing component A transferred per time per cross-sectional area have been defined by both material balance and the mass-transfer equations. For equipment involving the continuous contact between the two immiscible phases, these two equations may be combined and the resulting expression integrated to provide a defining relation for the unknown height of the mass exchanger.

Constant Overall Capacity Coefficient

Consider an isothermal, countercurrent mass exchanger used to achieve a separation in a system that has a constant overall mass-transfer coefficient $K_Y a$ through the concentration range involved in the mass-transfer operations. The mass balance for component A over the differential length dz is described by

$$\frac{\text{moles of } A \text{ transferred}}{(\text{time})(\text{cross-sectional area})} = L_S \, dX_A = G_S \, dY_A$$

The mass-transfer of component A in the differential length dz is defined by moles of A transferred:

$$\frac{\text{moles of A transferred}}{(\text{time})(\text{cross-sectional area})} = N_A a \cdot dz = K_Y a \left(Y_A - Y_A^* \right) dz \tag{31-21}$$

As equation (31-21) involves the mass flux of component A, N_A, it not only defines the quantity of A transferred per time and cross-sectional area but it also indicates the direction of mass transfer. If the driving force, $Y_A - Y_A^*$, is positive, the transfer of A must be from the bulk composition in the G phase to the bulk composition in the L phase. The two differential quantities, $L_S\,dX_A$ and $G_s\,dY_A$, stipulate only the quantity of A transferred per time and cross-sectional area; each term must have either a positive or a negative sign assigned to it to indicate the direction of the A transfer.

Let us consider the transfer of A from the G phase to the L phase; $-G_s dY_A$ indicates that the G phase is losing A. Therefore, the mass transfer of component A in the differential length dz may be defined by

$$-G_S\,dY_A = K_Y a\left(Y_A - Y_A^*\right)dz$$

or

$$dz = \frac{-G_S}{K_Y a}\frac{dY_A}{Y_A - Y_A^*} \tag{31-22}$$

This equation may be integrated over the length of the mass exchanger, with the assumption of a constant overall capacity coefficient:

$$\int_{z_1}^{z_2} dz = \frac{-G_S}{K_Y a}\int_{Y_{A_1}}^{Y_{A_2}}\frac{dY_A}{Y_A - Y_A^*}$$

or

$$z = (z_2 - z_1) = \frac{G_S}{K_Y a}\int_{Y_{A_2}}^{Y_{A_1}}\frac{dY_A}{Y_A - Y_A^*} \tag{31-23}$$

The evaluation of the right-hand side of this equation often requires a numerical or a graphical integration. As discussed in Section 31.3, we may evaluate $Y_A - Y_A^*$ from the plot of Y_A vs. X_A as illustrated in Figure 31.17.

The vertical distance between the operating line and the equilibrium line represents the overall driving force in Y-units. Each value at a given bulk concentration, Y_A, and its reciprocal, $1/\left(Y_A - Y_A^*\right)$, may then be plotted vs. Y_A, as illustrated in Figure 31.18.

After obtaining the area under the curve in Figure 31.18, we may evaluate the depth of packing (z) in the mass exchanger by equation (31-23).

When the transfer is from the L phase to the G phase, as in the case of desorption or stripping, G, dy_A and dY_A will be positive; the mass transfer of component A in the differential length dz will be

$$G_S\,dY_A = K_Y a\left(Y_A^* - Y_A\right)dz$$

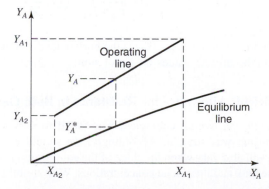

Figure 31.17 Evaluation of $Y_A - Y_A^*$ the overall driving force.

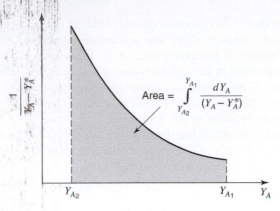

Figure 31.18 Evaluation of the integral.

or

$$dz = \frac{G_S}{K_Y a}\frac{dY_A}{Y_A^* - Y_A} = \frac{-G_S}{K_Y a}\frac{dY_A}{Y_A - Y_A^*}$$

and so the same design equation (31-23) will be obtained.

The length of the mass exchanger can also be determined by an equation written in terms of the overall liquid capacity coefficient, $K_X a$. For the transfer of A from the G phase into the L phase,

$$-L_S\, dX_A = K_X a\left(X_A^* - X_A\right)dz$$

or

$$dz = \frac{-L_S}{K_X a}\frac{dX_A}{X_A^* - X_A} \tag{31-24}$$

If the capacity coefficient is constant over the concentration range involved in the mass-transfer operation

$$\int_{z_1}^{z_2} dz = \frac{-L_S}{K_X a}\int_{X_{A_1}}^{X_{A_2}}\frac{dX_A}{X_A^* - X_A}$$

or

$$z = \frac{-L_S}{K_X a}\int_{X_{A_1}}^{X_{A_2}}\frac{dX_A}{X_A^* - X_A} \tag{31-25}$$

The overall driving force, $X_A^* - X_A$, is the horizontal difference between the operating line and the equilibrium line values on a plot similar to Figure 31.17.

Variable Overall Capacity Coefficient: Allowance for Resistance in Both Gas and Liquid Phase

In Chapter 29, the overall coefficient was found to vary with concentration unless the equilibrium line was straight. Accordingly, we should expect that the overall capacity coefficient will also vary when the slope of the equilibrium line varies within the region that includes the bulk and interfacial concentrations. With slightly curved equilibrium lines, one may safely use the design equations (31-23) and (31-25). However, in the case of equilibrium lines with more pronounced curvature, the exact calculations should be based on one of the individual capacity coefficients.

The mass balance for component A over the differential length dz is

$$L_S\, dX_A = G_S\, dY_A \tag{31-11a}$$

Differentiating equation (31-8), we obtain

$$dY_A = \frac{dy_A}{(1 - y_A)^2}$$

This relation may be substituted into equation (31-11a) to give

$$L_S\, dX_A = G_S \frac{dy_A}{(1 - y_A)^2} \tag{31-26}$$

The mass transfer of component A in the differential length, dz, was defined in terms of the individual gas-phase capacity coefficient by

$$N_A a \cdot dz = k_G a (p_A - p_{A,i}) dz \tag{31-17}$$

Combining equation (31-17) and equation (31-26), we obtain, upon rearranging,

$$dz = \frac{-G_S\, dy_A}{k_G a (p_A - p_A)(1 - y_A)^2}$$

or

$$dz = \frac{-G_S\, dy_A}{k_G a \cdot P (y_A - y_{A,i})(1 - y_A)^2} \tag{31-27}$$

Equation (31-27) in the integral form is

$$z = z_2 - z_1 = \int_{y_{A_2}}^{y_{A_1}} \frac{G_S\, dy_A}{k_G a \cdot P (y_A - y_{A,i})(1 - y_A)^2} \tag{31-28}$$

As discussed in Chapter 29, the interfacial compositions, $y_{A,i}$ and $x_{A,i}$ may be found for each point on the operating line by drawing a line from the point toward the equilibrium line. The slope of this line is $-k_L/k_G$ on a p_A vs. c_A plot, or is $-k_x/k_y$ (given by $-c_L k_L/k_G P$) on a y_A vs. x_A plot. In Figure 31.19, the location of the interfacial composition on each phase plot is illustrated. It is important to recall from the discussion in Section 31.3

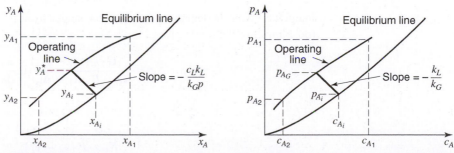

Figure 31.19 Determination of interfacial composition for transfer from phase G to phase L.

that the operating line is not straight on plots of y_A vs. x_A and p_A vs. c_A, except when we are dealing with relatively dilute gas and liquid mixtures. Knowing the interfacial composition, $y_{A,i}$, for every bulk composition y_A in the gas stream, we may numerically or graphically integrate equation (31-28) to obtain the length of the mass exchanger.

Logarithmic-Mean Driving Force

Although the graphical integration procedure must be employed in most practical design calculations involving streams that have high mole fraction compositions of solute A, it is sometimes possible to use a much simpler equation based upon a logarithmic-mean driving force. When the two contacting streams are relatively dilute, the equilibrium curve and the operating line may both be linear in terms of the mole fractions over the range of concentration involved in the mass-transfer operation. Under these conditions, $G_1 \approx G_2 \approx G$ and $L_1 \approx L_2 \approx L$. The mass balance for component A may be approximated by

$$L(x_{A_1} - x_A) = G(y_{A_1} - y_A) \tag{31-29}$$

or

$$L\,dx_A = G\,dy_A \tag{31-30}$$

The rate of interphase transfer may be expressed in terms of the overall gas-phase capacity coefficient by

$$N_A a \cdot dz = K_G a (p_A - p_A^*)dz$$

or

$$N_A a \cdot dz = K_G a \cdot P(y_A - y_A^*)dz \tag{31-31}$$

As the operating and equilibrium lines are straight, the difference in the ordinates of the two lines will vary linearly in composition. Designating the difference $y_A - y_A^*$ by Δ, we see that this linearity stipulates

$$\frac{d\Delta}{dy_A} = \frac{(y_A - y_A^*)_1 - (y_A - y_A^*)_2}{y_{A_1} - y_{A_2}} = \frac{\Delta_1 - \Delta_2}{y_{A_1} - y_{A_2}} \tag{31-32}$$

For the transfer of A from the G phase into the L phase, we can combine the rate of mass transfer from equation (31-31) with equation (31-32) and the differential mass balance for A to obtain

$$dz = \frac{-G}{K_G a \cdot P} \frac{dy_A}{y_A - y_A^*} = \frac{-G}{K_G a \cdot P} \frac{(y_{A_1} - y_{A_1})}{\Delta_1 - \Delta_2} \cdot \frac{d\Delta}{\Delta} \tag{31-33}$$

Integration over the length of the mass exchanger from the bottom (position 1) to the top (position 2) of the tower gives

$$z = \frac{G}{K_G a \cdot P} \frac{(y_{A_1} - y_{A_2})}{\Delta_1 - \Delta_2} \ln \frac{\Delta_1}{\Delta_2}$$

or

$$z = \frac{G}{K_G a \cdot P} \frac{(y_{A_1} - y_{A_2})}{(y_A - y_A^*)_{lm}} \tag{31-34}$$

where

$$\Delta_{lm} = \frac{\Delta_1 - \Delta_2}{\ln \Delta_1/\Delta_2} = \left(y_A - y_A^* \right)_{lm}$$

or

$$\left(y_A - y_A^* \right)_{lm} = \frac{\left(y_A - y_A^* \right)_1 - \left(y_A - y_A^* \right)_2}{\ln\left[\left(y_A - y_A^* \right)_1 / \left(y_A - y_A^* \right)_2 \right]} \qquad (31\text{-}35)$$

A similar expression in terms of the overall liquid-phase capacity coefficient is

$$z = \frac{L}{K_L a \cdot c} \frac{\left(x_{A_1} - x_{A_2} \right)}{\left(x_A^* - x_A \right)_{lm}} \qquad (31\text{-}36)$$

where

$$\left(x_A^* - x_A \right)_{lm} = \frac{\left(x_A^* - x_A \right)_1 - \left(x_A^* - x_A \right)_2}{\ln\left[\left(x_A^* - x_A \right)_1 / \left(x_A^* - x_A \right)_2 \right]} \qquad (31\text{-}37)$$

Equations (31-36) and (31-37) are convenient to use for design and analysis of liquid stripping operations in packed towers.

Example 4

A gas mixture containing 8.25 mole% NH_3 in air will be treated in a countercurrent packed absorption tower to reduce the NH_3 composition in the outlet gas to 0.30 mole%, using water as the absorption solvent. The tower is 0.50 m in diameter, and will be operated at 293 K and 1.0 atm total system pressure. The total molar flow rate of the feed gas added to the bottom of the tower is 8.325×10^{-3} kgmole/s, and the mass flow rate of NH_3-free water added to the top of the tower is 0.203 kg/s, which is above the minimum solvent flow rate. At these flow conditions, the mass-transfer capacity coefficient based on the overall gas-phase driving force in mole-ratio units is $K_Y a = 0.080$ kgmole/m$^3 \cdot$ s. Determine the height of the packed tower needed to accomplish the separation. Equilibrium distribution data for the solute NH_3 between water and air at 293 K and 1.0 atm in mole ratio coordinates were provided in Example 3.

It is first necessary to characterize the composition of each terminal stream. The mole ratio compositions of the known streams are

$$Y_{A_1} = \frac{y_{A_1}}{1 - y_{A_1}} = \frac{0.0825}{1 - 0.0825} = 0.090$$

$$Y_{A_2} = \frac{y_{A_2}}{1 - y_{A_2}} = \frac{0.0030}{1 - 0.0030} = 0.0030$$

$$X_{A_2} = \frac{x_{A_2}}{1 - x_{A_2}} = \frac{0}{1 - 0} = 0$$

The cross-sectional area of the 0.5-m-diameter tower is

$$A = \frac{\pi D^2}{4} = \frac{\pi (0.50 \text{ m})^2}{4} = 0.196 \text{ m}^2$$

The superficial molar velocities of the carrier gas and the solvent at this tower diameter are

$$G_S = G_1(1 - y_{A_1}) = \frac{A G_1 (1 - y_{A_1})}{A} = \frac{\left(8.325 \times 10^{-3} \text{ kgmole/s} \right) (1 - 0.0825)}{0.196 \text{ m}^2} = 0.0389 \text{ kgmole/m}^2 \cdot \text{s}$$

$$L_S = \frac{AL_2(1 - x_{A_2})}{A} = \frac{AL_2}{A} = \frac{(0.203 \text{ kg/s})(1 \text{ kgmole/18 kg})}{(0.196 \text{ m}^2)} = 0.0575 \text{ kgmole/m}^2 \cdot \text{s}$$

The mole ratio of solute A in the outlet liquid stream is determined by material balance on the terminal streams:

$$X_{A_1} = \frac{AG_S(Y_{A_1} - Y_{A_2}) + AL_S X_{A_2}}{AL_S} = \frac{(0.0389)(0.090 - 0.00301) + 0}{(0.0574)} = 0.059$$

Based on the terminal stream compositions, the equilibrium and operating lines are shown in Figure 31.20.

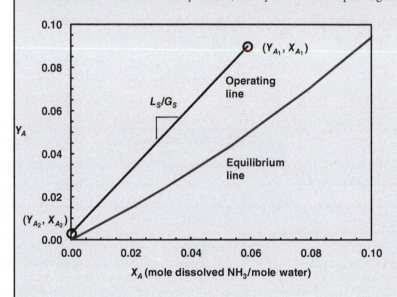

Figure 31.20 The operating line for Example 4.

Since the inlet gas composition of NH_3 is relatively high at 8.25 mole%, the total gas flow rate will decrease appreciably from the bottom to the top of the tower. Furthermore, the equilibrium line is nonlinear over the NH_3 inlet and outlet concentration range. Based on these two considerations, equation (31-23) is used to estimate the packed-tower height, z. The integral in equation (31-23) is the number of transfer units for the separation based on the overall gas-phase mole ratio driving force, and is given by

$$N_{YG} = \int_{Y_{A_1}}^{Y_{A_2}} \frac{dY_A}{Y_A - Y_A^*} \tag{31-38}$$

The number of transfer units, N_{YG}, can be considered as the integral of the reciprocal of the mass-transfer driving force $Y_A - Y_A^*$ from the top to the bottom of the tower. Accordingly, a plot of $1/(Y_A - Y_A^*)$, vs. Y_A from Y_{A_2} to Y_{A_1} is provided in Figure 31.21, using information provided in Table 31.1, where $A = NH_3$. The integral is the area under the curve shown in Figure 31.21, which can be obtained by graphical methods or by numerical integration. In Table 31.1, X_A and Y_A represent the operating line extending from the top to the bottom of the tower, and Y_A^* is the composition of solute A in the gas phase in equilibrium with operating liquid composition X_A. By trapezoid rule numerical integration using the base points provided in Table 31.1, the number of transfer units based on the mole ratio driving force for solute A is

$$N_{YG} = \int_{0.0030}^{0.090} \frac{dY_A}{Y_A - Y_A^*} = 5.9$$

With the number of transfer units known, the packed-tower height required to accomplish the separation is

$$z = \frac{G_S}{K_Y a} \int_{Y_{A_2}}^{Y_{A_1}} \frac{dY_A}{Y_A - Y_A^*} = \frac{G_S}{K_Y a} N_{YG} = \frac{(0.0389 \text{ kgmole/s})}{(0.080 \text{ kgmole/m}^3 \cdot \text{s})}(5.9) = 2.8 \text{ m}$$

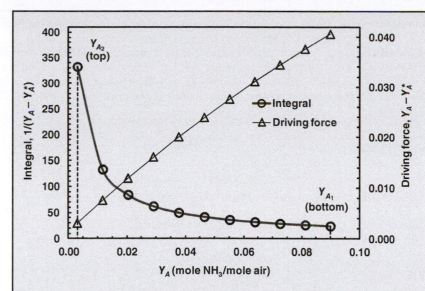

Figure 31.21 Evaluation of the integral for Example 4.

Table 31.1 Example 4 gas compositions, as mole ratios

X_A	Y_A	Y_A^*	$Y_A - Y_A^*$	$1/(Y_A - Y_A^*)$
0.0000	0.0030	0.0000	0.0030	332.3
0.0059	0.0117	0.0042	0.0075	133.1
0.0118	0.0204	0.0085	0.0119	84.3
0.0177	0.0291	0.0131	0.0160	62.4
0.0235	0.0378	0.0177	0.0200	49.9
0.0294	0.0465	0.0226	0.0239	41.9
0.0353	0.0552	0.0276	0.0275	36.3
0.0412	0.0638	0.0328	0.0311	32.2
0.0471	0.0725	0.0381	0.0344	29.1
0.0530	0.0812	0.0437	0.0376	26.6
0.0589	0.0899	0.0493	0.0406	24.6

In Figure 31.21, we see that if the driving force for mass transfer is small, then the contribution to the integral is large. We also note that the overall gas-phase driving force $\left(Y_A - Y_A^*\right)$ is approximately linear with gas-phase mole ratio Y_A, with the highest driving force at the bottom of the tower.

Example 5

An existing packed tower with a packing height of 12.0 m was used to test a gas absorption process for removal of solute A from a gas stream. In the present test, the mole fraction of solute A in the gas phase was reduced from 2.0 mole% to 0.5 mole%. Gas was fed to the bottom of the tower at superficial molar velocity of 0.0136 kgmole/m$^2 \cdot$ s, and pure solvent containing no solute A was fed to the top of the tower at a superficial molar velocity of 0.0272 kgmole/m$^2 \cdot$ s. The total system pressure of the process was maintained at 1.2 atm. At the temperature and total system pressure of the process, the equilibrium distribution of the solute between the solvent and the carrier gas is described by Henry's law of the form

$$y_A^* = 1.5\, x_A$$

Using the test data provided above, estimate the overall mass-transfer capacity coefficient based on the gas-phase driving force, $K_G a$ at the test conditions.

To use the design equations for a packed tower, it is first necessary to characterize the compositions and flow rates of all terminal streams. From the initial specifications of the process, the following flow rates and compositions are known: $G_1 = 0.0136 \, \text{kgmole/m}^2 \cdot \text{s}$, $L_2 = 0.0272 \, \text{kgmole/m}^2 \cdot \text{s}$, $y_{A_1} = 0.020$, $y_{A_2} = 0.0050$, $x_{A_2} = 0$. To characterize the remaining unknown streams, we first estimate the carrier gas and solvent flow rates, which are inert components:

$$G_S = G_1(1 - y_{A_1}) = (0.0136 \, \text{kgmole/m}^2 \cdot \text{s})(1 - 0.020) = 0.0133 \, \text{kgmole/m}^2 \cdot \text{s}$$

$$L_S = L_2(1 - x_{A_2}) = (0.0272 \, \text{kgmole/m}^2 \cdot \text{s})(1 - 0.0) = 0.0272 \, \text{kgmole/m}^2 \cdot \text{s}$$

Consequently,

$$G_2 = G_S/(1 - y_{A_2}) = (0.0133 \, \text{kgmole/m}^2 \cdot \text{s})/(1 - 0.0050) = 0.0134 \, \text{kgmole/m}^2 \cdot \text{s}$$

With G_2 known, an overall material balance around the terminal streams is used to determine L_1:

$$G_1 + L_2 = G_2 + L_1$$

$$L_1 = G_1 - G_2 + L_2 = 0.0136 - 0.0134 + 0.0272 = 0.0274 \, \text{kgmole/m}^2 \cdot \text{s}$$

Finally, the outlet mole fraction of solute A in the liquid (x_{A_1}) is determined by

$$L_S = L_1(1 - x_{A_1})$$

or

$$x_{A_1} = 1 - L_S/L_1 = 1 - (0.0272)/(0.0274) = 0.0073$$

Since the equilibrium line is linear, the pinch point at the minimum solvent flow rate will occur at the gas feed, so that

$$x_{A_1,\text{min}} = y_{A_1}/1.75 = 0.020/1.75 = 0.0114$$

Therefore, since $x_{A_1,\text{min}} > x_{A_1}$, we are assured that operating solvent flow rate is above the minimum solvent flow rate.

All mole fraction compositions are at or below 2.0 mole%, which indicates a dilute system. Furthermore, the equilibrium line is linear. Based on these two considerations, the appropriate design equation for the countercurrent gas absorption tower is equation (31-34). In order to estimate the log-mean mole fraction driving force $(y_A - y_A^*)_{lm}$, values for y_A^* at the top and bottom ends of the tower are needed. Therefore,

$$y_{A_1}^* = 1.75 x_{A_1} = 1.75(0.0073) = 0.0130$$

$$y_{A_2}^* = 1.75 x_{A_2} = 1.75(0.0) = 0.0$$

Consequently, from equation (31-35), the overall log-mean mole fraction driving force for mass-transfer of solute A based on the gas phase is

$$\left(y_A - y_A^*\right)_{lm} = \frac{\left(y_A - y_A^*\right)_1 - \left(y_A - y_A^*\right)_2}{\ln\left[\left(y_A - y_A^*\right)_1/\left(y_A - y_A^*\right)_2\right]} = \frac{(0.020 - 0.0130) - (0.020 - 0)}{\ln[(0.020 - 0.0130)/(0.020 - 0.0)]} = 0.0123$$

Since the gas flow rate does not change significantly from the bottom to the top of the tower, an average value is used

$$G = \frac{G_1 + G_2}{2} = \frac{0.0136 + 0.0134}{2} = 0.0135 \, \text{kgmole/m}^2 \cdot \text{s}$$

Finally, $K_G a$ is backed out from by a rearranged form of equation (31-34):

$$K_G a = \frac{G}{z \cdot P} \frac{\left(y_{A_1} - y_{A_2}\right)}{\left(y_A - y_A^*\right)_{lm}} = \frac{(0.0135 \, \text{kgmole/m}^2 \cdot \text{s})}{(12.0 \, \text{m})(1.2 \, \text{atm})} \frac{(0.020 - 0.0050)}{(0.0123)} = 1.14 \times 10^{-3} \frac{\text{kgmole}}{\text{m}^3 \cdot \text{s} \cdot \text{atm}}$$

As an aside, for dilute systems, the packed-tower height is commonly described as the product

$$z = H_{OG}N_{OG} \tag{31-39}$$

where N_{OG} is the number of transfer units, given by

$$N_{OG} = \frac{(y_{A_1} - y_{A_2})}{(y_A - y_A^*)_{lm}} = \frac{(0.020 - 0.0050)}{(0.0123)} = 1.22 \tag{31-40}$$

and H_{OG} is the packing height of the transfer unit, given by

$$H_{OG} = \frac{G}{K_G a \cdot P} = \frac{(0.0135 \text{ kgmole/m}^2 \text{s})}{(1.14 \times 10^{-3} \text{ kgmole/m}^3 \cdot \text{s} \cdot \text{atm})(1.2 \text{ atm})} = 9.86 \text{ m} \tag{31-41}$$

Packed-Tower Diameter

The packed tower is the most commonly encountered continuous-contacting equipment in gas–liquid operations. A variety of packing material is used, ranging from specially designed ceramic or plastic packing, as illustrated in Figure 31.4, to crushed rock. The packing is chosen to promote a large area of contact between the phases, with a minimum resistance to the flow of the two phases. Table 31.2 lists some of the properties of packing frequently used in the industry.

We have previously established that the height of a continuous-contact tower is determined by the rate of mass transfer. The diameter of the tower is established to handle the flow rates of the two phases to be treated.

As illustrated in Figure 31.22, the pressure drop encountered by the gas phase as it flows through the packing is influenced by the flow rates of both phases. This is to be expected, because both phases will be competing for the free cross section that is available for the streams to flow through. Let us consider a tower operating with a fixed liquid flow rate, L'; below the region marked A, the quantity of liquid retained in the packed bed will remain reasonably constant with changing gas velocities. As the gas flow rate increases, the interphase friction increases and a greater quantity of liquid is held up in the packing. This is known as *loading*. Finally, at a certain value of the gas flow rate, G', the holdup is so high that the tower starts to fill with liquid. The tower cannot be operated above this *flooding velocity* (G_f'), which is a function of the liquid velocity, the fluid properties, and the characteristics of the packing.

In Figure 31.23, a correlation is given for the flooding velocity in a random-packed tower. Figure 31.23 is commonly called the Flooding correlation. Gas absorbers and liquid strippers are designed to operate well below the pressure drop that is associated with flooding. Typically, they are designed for gas pressure drops of 200–400 N/m^2 per meter of packed depth. The abscissa on Figure 21.23 involves a ratio of the superficial liquid and gas-mass flow rates and the densities of the gas and liquid phases. The ordinate involves the superficial gas-mass flow rate, the liquid-phase viscosity, the liquid and gas densities, a packing characteristic, c_f, which can be obtained from Table 31.2, and two constants. For SI units, g_c equals 1.0, and J equals 1. For U.S. English units, μ_L is in centipoises (cP), densities are in lb_m/ft^3, mass flow rates are in $lb_m/ft^2 \cdot h$, g_c equals 4.18×10^8, and J equals 1.502. When using the Flooding correlation, these unit conventions must be followed strictly, as the units will not cancel.

Table 31.2 Tower packing characteristics[y]

Packing	Nominal size, in (mm)					
	0.25 (6)	0.50 (13)	0.75 (19)	1.00 (25)	1.50 (38)	2.00 (50)
Raschig rings						
Ceramic						
ϵ	0.73	0.63	0.73	0.73	0.71	0.74
c_f	1600	909	255	155	95	65
a_p ft^2/ft^3	240	111	80	58	38	28
Metal						
ϵ	0.69	0.84	0.88	0.92		
c_f	700	300	155	115		
a_p ft^2/ft^3	236	128	83.5	62.7		
Berl saddles						
Ceramic						
ϵ	0.60	0.63	0.66	0.69	0.75	0.72
c_f	900	240	170	110	65	45
a_p ft^2/ft^3	274	142	82	76	44	32
Intalox saddles						
Ceramic						
ϵ	0.75	0.78	0.77	0.775	0.81	0.79
c_f	725	200	145	98	52	40
a_p ft^2/ft^3	300	190	102	78	59.5	36
Plastic						
ϵ				0.91		0.93
c_f				33		56.5
a_p ft^2/ft^3				63		33
Pall rings						
Plastic						
ϵ				0.90	0.91	0.92
ϵ				52	40	25
c_f				63	39	31
Metal						
ϵ				0.94	0.95	0.96
c_f				48	28	20
a_p ft^2/ft^3				63	39	31

[y]R. E. Treybal, *Mass-Transfer Operations*, McGraw-Hill Book Company, New York, 1980.

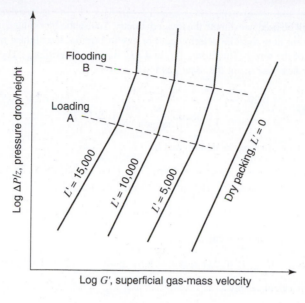

Figure 31.22 Typical gas pressure drop for countercurrent, packed tower.

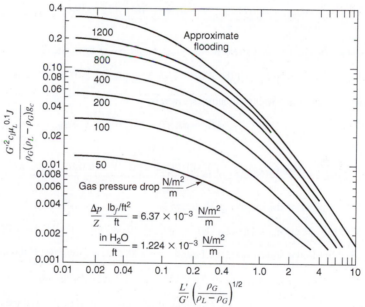

Figure 31.23 Flooding correlation for random-packed towers.

Example 6 will illustrate how to estimate the diameter of a packed tower using the Flooding correlation.

Example 6

A packed tower is used to reduce the ammonia (NH_3) concentration in a gas stream from 4.0 to 0.30 mole%. Pure liquid water is fed to the top of the tower at a rate of 0.231 kg/s, and the gas is fed countercurrently to the bottom of the tower at a volumetric rate of 0.20 m^3/s. The tower is packed with 1.0-inch Raschig rings, and is operated at 293 K and 1.0 atm. The bulk properties of the liquid can be taken as the properties of liquid water. At 298 K, for liquid water, the density is 998.2 kg/m^3, and the viscosity is 993×10^{-6} kg/m·s.

Calculate the diameter of the packed tower if the gas pressure drop is limited to 200 N/m^2 per meter of packing.

The Flooding correlation will be used to estimate the tower diameter. To use the Flooding correlation, the mass flow rates of the gas and liquid in countercurrent flow must be determined. For gas absorption, the highest mass flow rates of both gas and liquid are located at the bottom of the tower. Therefore, AG'_1 and AL'_1 must be determined. The inlet molar flow rate of gas is determined from the volumetric flow rate using the ideal gas law:

$$AG_1 = \dot{V}_1 \frac{P}{RT} = \left(0.20 \frac{m^3}{s}\right) \frac{(1.0\ atm)}{\left(0.08206 \dfrac{m^3 \cdot atm}{kgmole \cdot K}\right)(293\ K)} = 8.32 \times 10^{-3}\ kgmole/s$$

The average molecular weight of the inlet gas is determined by

$$M_{w,G_1} = y_{A_1} M_A + (1 - y_{A_1}) M_B = (0.040)(17) + (1 - 0.040)(29) = 28.5\ kg/kgmole$$

where component B refers to the carrier gas (air). The mass flow rate of entering gas is

$$AG'_1 = AG_1 \cdot M_{w,G_1} = \left(8.32 \times 10^{-3}\ kgmole/s\right)(28.5\ kg/kgmole) = 0.237\ kg/s$$

The inlet molar flow rate of the pure water solvent ($x_{A_2} = 0$) is

$$AL_2 = AL'_2 / M_{w,L_2} = (0.231\ kg/s)/(18\ kg/kgmole) = 1.28 \times 10^{-2}\ kgmole/s$$

The outlet gas molar flow rate is calculated by

$$AG_2 = \frac{AG_S}{1 - y_{A_2}} = \frac{AG_1(1 - y_{A_1})}{1 - y_{A_2}} = \frac{\left(8.32 \times 10^{-3}\right)(1 - 0.040)}{(1 - 0.003)} = 8.01 \times 10^{-3}\ kgmole/s$$

From an overall balance on the terminal streams, the outlet liquid molar flow rate is

$$AL_1 = AL_2 + (AG_1 - AG_2) = 1.28 \times 10^{-2} + \left(8.32 \times 10^{-3} - 8.01 \times 10^{-3}\right) = 1.31 \times 10^{-2}\ kgmole/s$$

A balance of solute A around the terminal streams is given by

$$y_{A_1} AG_1 + x_{A_2} AL_2 = y_{A_2} AG_2 + x_{A_1} AL_1$$

Or, since $x_{A_2} = 0$,

$$x_{A_1} = \frac{y_{A_1} AG_1 - y_{A_2} AG_2}{AL_1} = \frac{(0.040)\left(8.32 \times 10^{-3}\right) - (0.0030)\left(8.01 \times 10^{-3}\right)}{1.31 \times 10^{-2}} = 0.024$$

The average molecular weight of the outlet liquid stream is

$$M_{w,L_1} = x_{A_1} M_A + (1 - x_{A_1}) M_B = (0.024)(17) + (1 - 0.024)(18) = 18\ kg/kgmole$$

where component B refers to the solvent (water). Finally, the outlet mass flow rate of liquid is

$$AL'_1 = AL_1 \cdot M_{w,L_1} = \left(1.31 \times 10^{-2}\ kgmole/s\right)(18\ kg/kgmole) = 0.237\ kg/s$$

With the mass flow rates known, the x-axis on the Flooding correlation will now be determined. First, the ratio of the liquid to gas-mass flow rate at the bottom of the tower is

$$\frac{L'}{G'} = \frac{AL'}{AG'} = \frac{0.237\ kg/s}{0.237\ kg/s} = 1.00$$

Note that this ratio can be evaluated without knowing the diameter or cross-sectional area of the empty tower. Next, the density of the gas stream entering the tower is

$$\rho_G = \frac{P}{RT} M_{w,G_1} = \frac{(1.0\ atm)}{\left(0.08206 \dfrac{m^3 \cdot atm}{kgmole \cdot K}\right)(293\ K)} (28.5\ kg/kgmole) = 1.19\ kg/m^3$$

Therefore, the x-axis on the Flooding correlation (Figure 31.23) is

$$\frac{L'}{G'} \left(\frac{\rho_G}{\rho_L - \rho_G}\right)^{1/2} = 1.00 \left(\frac{1.19}{998.2 - 1.19}\right)^{1/2} = 0.034$$

which is a dimensionless quantity. At a pressure drop of $200 \, \text{N/m}^2$ per meter of packing depth, the y-axis value of the Flooding correlation is 0.049 if the x-axis value is 0.034. Consequently,

$$0.049 = \frac{(G')^2 c_f (\mu_L)^{0.1} J}{\rho_G (\rho_L - \rho_G) g_c}$$

This expression for the y-axis is rearranged to determine the required superficial molar velocity of gas, G':

$$G' = \sqrt{\frac{0.049 \rho_G (\rho_L - \rho_G) g_c}{c_f (\mu_L)^{0.1} J}}$$

From Table 31.2, $c_f = 155$ for 1.0-inch Raschig rings. Therefore, the required superficial mass velocity of the gas is

$$G' = \sqrt{\frac{0.049 \left(1.19 \, \frac{\text{kg}}{\text{m}^3}\right) \left(998.2 - 1.19 \, \frac{\text{kg}}{\text{m}^3}\right) (1.00)}{(155) \left(993 \times 10^{-6} \, \text{kg/m} \cdot \text{s}\right)^{0.1} (1.0)}} = 0.865 \, \text{kg/m}^2 \text{s}$$

The cross-sectional area of the tower is backed out from G' by

$$A = \frac{AG'}{G'} = \frac{0.237 \, \text{kg/s}}{0.865 \, \text{kg/m}^2 \cdot \text{s}} = 0.274 \, \text{m}^2$$

and so the tower diameter is

$$D = \sqrt{\frac{4A}{\pi}} = \sqrt{\frac{4(0.274 \, \text{m}^2)}{\pi}} = 0.59 \, \text{m}$$

Example 7

Examples 3 to 5 have focused on gas absorption processes. In the final example of this chapter, we will consider a liquid stripping process, focusing on determination of the minimum and operating carrier gas flow rate.

A liquid stream containing dissolved ammonia (NH_3, solute A) in water must be reduced in composition from 0.080 moles A/mole solvent to 0.020 moles A/mole solvent, using air as the stripping gas at 293 K and 1.0 atm to transfer the volatile NH_3 from the liquid to the gas phase. The process will be carried out in a packed tower with countercurrent flow of gas and liquid. If inlet liquid flow rate fed to the top of the tower is fixed at $0.010 \, \text{kgmole/m}^2 \cdot \text{s}$, determine the minimum inlet gas flow rate of air fed to the bottom of the tower, and the mole ratio of NH_3 in the exiting stripping gas at an operating rate of 2.0 times the minimum gas flow rate. Equilibrium distribution data were provided in Example 3.

The equilibrium distribution curve in mole ratio coordinates for the ammonia-water-air system at 293 K and 1.0 atm is provided in Figure 31.24. At the bottom of the tower, $Y_{A_1} = 0$ and $X_{A_1} = 0.020$. Based on the shape of the operating curve, the pinch point will occur at the top of the tower where the gas exits and the feed liquid enters. Consequently, at $X_{A_2} = 0.080$, we see

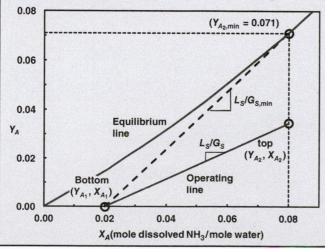

Figure 31.24 Operating and equilibrium lines for Example 7. Equilibrium distribution data from Perry and Chilton.[2]

that $Y_{A_2,min} = 0.071$, which is on the equilibrium line, as shown in Figure 31.24. The minimum carrier gas flow rate at this $Y_{A_2,min}$ is determined by material balance. First, the solvent flow rate is

$$L_S = L_2(1 - x_{A_2}) = \frac{L_2}{1 + X_{A_2}} = \frac{0.010 \text{ kgmole/m}^2 \cdot \text{s}}{1 + 0.080} = 0.00926 \text{ kgmole/m}^2 \cdot \text{s}$$

The minimum carrier gas flow rate is

$$G_{S,min} = L_S \frac{(X_{A_2} - X_{A_1})}{(Y_{A_2,min} - Y_{A_1})} = (0.00926) \frac{(0.080 - 0.020)}{(0.071 - 0.0)} = 0.0078 \text{ kgmole/m}^2 \cdot \text{s}$$

and the operating carrier gas flow rate is

$$G_S = 2 \cdot G_{S,min} = 2(0.0078 \text{ kgmole/m}^2 \cdot \text{s}) = 0.0156 \text{ kgmole/m}^2 \cdot \text{s}$$

Finally, the mole ratio composition of NH_3 in the outlet gas at the operating carrier gas flow rate is

$$Y_{A_2} = Y_{A_1} + \frac{L_S}{G_S}(X_{A_2} - X_{A_1}) = 0.0 + \frac{(0.00926)}{(0.0156)}(0.080 - 0.020) = 0.0356 \frac{\text{mole } NH_3}{\text{mole air}}$$

The operating line at Y_{A_2} is also shown in Figure 31.24.

▶ **31.7**

CLOSURE

Continuous-contact mass exchangers are designed by integrating an equation that relates the mass balance and the mass-transfer relations for a differential area of interfacial contact. In this chapter, we have described the four major types of mass-transfer equipment. The fundamental equations for the design of continuous gas–liquid contacting processes in packed towers have been developed.

Nomenclature

a	interphase mass-transfer area per unit volume; ft^2/ft^3, m^2/m^3.
$\mathbf{a}$	acceleration; ft/s^2, m/s^2.
a_p	packing characteristic; ft^2/ft^3, m^2/m^3.
A	area; ft^2, m^2.
A_i	interphase mass-transfer area; ft^2, m^2.
A_p	projected area of surface; ft^2, m^2; equation (12-3).
c	total molar concentration; lb mol/ft^3; mol/m^3.
c_A	concentration of A in equilibrium with the bulk composition of gas phase, $P_{A.G}$; lb mol/ft^3; mol/m^3.
c_{A_o}	concentration of A at time $t = 0$; lb mol/ft^3; mol/m^3.
$c_{A,i}$	liquid molar concentration of A at the interface; lb mol/ft^3, mol/m^3; Section 29.2.
$c_{A,L}$	liquid molar concentration of A in the bulk stream; lb mol/ft^3, mol/m^3; Section 29.2.
$c_{A,s}$	concentration of A at the surface; lb mol/ft^3, mol/m^3.
$c_{A,\infty}$	concentration of A in the bulk stream; lb mol/ft^3, mol/m^3.
c_i	molar concentration of species i; lb mol/ft^3, mol/m^3; equation (24-4).
c_p	heat capacity; Btu/lb F, J/kg K.
$\underline{C}$	dimensionless concentration; dimensionless.
C	average random molecular velocity; m/s; Sections 7.3 and 15.2.
C_C	capacity rate of cold fluid stream; Btu/h F, kW/K; equation (22-1).
C_D	drag coefficient; dimensionless; equation (12-3).
C_f	coefficient of skin friction; dimensionless; equation (12-2).
C_f	packing characteristic; dimensionless.
C_H	capacity rate of hot fluid stream; Btu/h F, kW/K; equation (22-1).
C_{sf}	correlating coefficient for nucleate boiling; dimensionless; Table 21.1.
d_c	diameter of cylinder; ft, m.
d_p	diameter of spherical particle; ft, m.
D	tube diameter; ft, m.
D_{AB}	mass diffusivity or diffusion coefficient for component A diffusing through component B; ft^2/h, m^2/s; equation (24-15).
D_{Ae}	effective diffusion coefficient of species A within straight pores; ft^2/s, m^2/s.
D'_{Ae}	effective diffusion coefficient of species A within random pores; ft^2/s, m^2/s.
$D_{A,mix}$	diffusion coefficient of species A in a multicomponent mixture; ft^2/s, m^2/s.
D_{eq},	equivalent diameter; ft, m; equation (13-18).
D_{KA}	Knudsen diffusion coefficient of species A; ft^2/s, m^2/s.
D^o_{AB}	diffusion coefficient of solute A in solvent B at infinite dilution; ft^2/s, m^2/s.
d_{pore}	pore diameter; Å, nm.
d_s	molecular diameter; Å, nm.
e	pipe roughness; in., mm; Section 13.1.

e	specific energy or energy per unit mass; Btu/lb$_m$, J/kg; Section 6.1.
E	total energy of system; Btu, J; Section 6.1.
E	total emissive power; Btu/h ft^2, W/m^2; equation (23-2).
E	electrical potential; V; Section 15.5.
E_b	black-body emissive power; Btu/h ft^2, W/m^2; equation (23-12).
f	dependent variable used in the Blasius solution of boundary layer; dimensionless; equation (12-13).
f'	similarity parameter for convective analysis of boundary layer, prime denotes derivative with respect to η; dimensionless; equation (19-16).
f_D	Darcy friction factor; dimensionless; equation (13-4).
f_f	Fanning friction factor; dimensionless; equation (13-3).
F	force; lb$_f$, N; Section 1.2.
F	correction factor for compact heat-exchanger configurations; dimensionless; equation (22-14).
F_{ii}	view factor for radiant heat transfer; dimensionless; Section 23.7.
$\bar{F}_{ij}$	reradiating view factor; dimensionless; Section 23.9.
$\mathbf{g}$	acceleration due to gravity; ft/s^2, m/s^2.
g_c	dimensional conversion factor; 32.2 ft lb$_m$/lb$_f$ s^2, 1 kg $\cdot$ m/s^2 $\cdot$ N.
G	irradiation; Btu/h ft^2, W/m^2; Section 23.7.
G	mass velocity; lb$_m$/ft^2 h, g/m^2 $\cdot$ s.
G	total moles of the gas phase per time per cross-sectional area; lb mol/ft^2 h, g mol/m^2 $\cdot$ s.
G'	superficial gas-mass flow rate; lb$_m$/h ft^2; Section 31.6.
G_b	mass velocity of bubbles; lb$_m$/ft^2 s, kg/m^2 $\cdot$ s; equation (21-4).
G_M	molar velocity; lb mol/ft^2 h, g mol/m^2 $\cdot$ s.
G_s	moles of gas stream on a solute-free basis per time per cross-sectional area; lb mol/ft^2 h, g mol/m^2 $\cdot$ s.
h	convective heat-transfer coefficient; Btu/h ft^2 F, W/m^2 $\cdot$ K; equation (15-11).
$h_{fg,\,\text{solute}}$	heat of vaporization of solute; Btu/lb$_m$, kJ/kg.
h_L	head loss, $\Delta P/\rho$; ft lb$_f$/lb$_m$, Pa/(kg/m$_3$); Section 13.1.
h_r	radiation heat-transfer coefficient; Btu/h ft^2 F, W/m^2 $\cdot$ K; Section 23.12.
H	Henry's law constant; concentration of gas phase/concentration of liquid phase.
$\mathbf{H}$	moment of momentum; lb$_m$ ft^2/s, kg $\cdot$ m^2/s; equation (5-7).
H_i	enthalpy of species i; Btu, J.
ΔH_s	integral heat of solution; Btu/lb mol of solute; J/g mol of solute.
$\Delta H_{v,\,A}$	enthalpy of vaporization for species A; Btu/lb mol, J/g mol.
H_i	partial molar enthalpy of species i; Btu/lb mol, J/mol.
I	intensity of radiation; Btu/h ft^2, W/m^2; Section 23.3.
j'	j-factor for heat transfer with tube bundles; dimensionless; Figures 20.12 and 20.13.
j_D	j-factor for mass transfer, Chilton-Colburn analogy; dimensionless.
j_H	j-factor for heat transfer, Colburn analogy; dimensionless; equation (19-38).
$\mathbf{j}_i$	mass flux relative to the mass-average velocity; lb$_m$/ft^2 h, kg/m^2 $\cdot$ s; equation (24-17).
J	radiosity; Btu/h ft^2, W/m^2; Section 23.10.
$\mathbf{J}_i$	molar flux relative to the molar-average velocity; lb mol/h ft^2, mol/m^2 $\cdot$ s; equation (24-15).
k	thermal conductivity; Btu/h ft F, W/m $\cdot$ K; equation (15-1).
k	rate constant for chemical reaction, used to define r_A and R_A; Section 25.1.

k^0	mass-transfer coefficient with no net mass transfer into film; lb mol/ft^2 s Δ_{c_A}; mol/m$^2 \cdot$ s $\cdot \Delta\mathbf{c_A}$.
k_c	convective mass-transfer coefficient; lb mol/ft^2 h Δc_A, mol/m$^2 \cdot$ s $\cdot$ mol/m^3.
$\bar{k}_c$	mean convective mass-transfer coefficient; lb mol/ft^2 h Δ_{c_A}, mol/m$^2 \cdot$ s $\cdot$ mol/m^3.
k_G	convective mass-transfer coefficient in the gas phase; lb mol/ft^2 h atm, mol/m$^2 \cdot$ s $\cdot$ Pa.
k_L	convective mass-transfer coefficient in the liquid phase; lb mol/ft^2 h lb mol/ft^3, mol/m$^2 \cdot$ s $\cdot$ mol/m^3.
$k_G a$	individual gas-capacity coefficient; lb mol/h ft^3 atm, mol/s $\cdot$ m$^3 \cdot$ Pa.
$k_L a$	individual liquid-capacity coefficient; lb mol/h ft$^3 \Delta c_A$, mol/s $\cdot$ m$^3 \cdot$ mol/m^3.
K_G	overall mass-transfer coefficient in the gas phase; lb mol/h ft^2 atm, mol/s $\cdot$ m$^2 \cdot$ Pa.
K_L	overall mass-transfer coefficient in the liquid phase; lb mol/h ft$^2 \Delta c_A$, mol/s $\cdot$ m$^2 \cdot$ mol/m^3.
$K_G a$	overall gas-capacity coefficient; lb mol/h ft^3 atm, mol/s $\cdot$ m$^3 \cdot$ Pa.
$K_L a$	overall liquid-capacity coefficient; lb mol/h ft$^3 \Delta c_A$, mol/s $\cdot$ m$^3 \cdot$ mol/m^3.
$K_X a$	overall liquid-capacity coefficient based on ΔX_A driving force; lb mol/h ft$^3 \Delta X_A$, mol/s $\cdot$ m$^3 \cdot \Delta X_A$.
$K_Y a$	overall gas-capacity coefficient based on ΔY_A driving force; lb mol/h ft$^3 \Delta Y_A$, mol/s $\cdot$ m$^3 \cdot \Delta Y_A$.
L	mixing length; equations (12-52), (19-41), and (28-43).
L	characteristic length; ft, m.
L	total moles of liquid phase per time per cross-sectional area; lb mol/h ft^2, mol/s $\cdot$ m^2.
L_{eq}	equivalent length; ft, m; equation (13-17).
L_m	molar liquid-mass velocity; lb mol/h ft^2, mol/s $\cdot$ m^2.
L_s	moles of liquid phase on a solute-free basis per time per cross-sectional area; lb mol/h ft^2, mol/s $\cdot$ m^2.
m	mass of molecule; Section 7.3.
m	relative resistance $= D_{AB}/k_c x_1$; dimensionless; Section 27.4.
m	slope of the equilibrium line; units of gas concentration per units of liquid concentration.
$\mathbf{M}$	moment; lb$_m$ ft^2/s^2, kg $\cdot$ m^2/s^2.
M_i	molecular weight of species i; lb/lb mol, kg/kg mol.
n	packed-bed constant; dimensionless; equation (30-33).
n	number of species in a mixture; equations (24-1), (24-3), and (24-6).
n	relative position $= x/x_1$; dimensionless; Section 27.4.
N	molecules per unit volume; Section 7.3.
$\mathbf{n}$	outward directed unit normal vector; Sections 4.1, 5.1, and 6.1.
n_i	number moles of species i.
$\mathbf{n}_i$	mass flux relative to a set of stationary axes; lb$_m$/h ft^2, kg/s $\cdot$ m^2.
$\mathbf{N}_i$	molar flux relative to a set of stationary axes; lb mol/h ft^2, mol/s $\cdot$ m^2.
NTU	number of transfer units; dimensionless; Section 22.4.
p_A^*	partial pressure of A in equilibrium with bulk composition in liquid phase, $c_{A,L}$; atm, Pa.
$p_{A,G}$	partial pressure of component A in the bulk gas stream; atm, Pa; Section 29.2.
$p_{A,i}$	partial pressure of component A at the interface; atm, Pa; Section 29.2.
p_i	partial pressure of species i; atm, Pa.
$p_{B,lm}$	log mean of partial pressure of the nondiffusing gas; atm, Pa.

P	total pressure; atm, Pa.
P	power input for stirred tank of liquid; N · m/s.W.
P_A	vapor pressure of pure volatile liquid species A; lb$_f$/in.^{2}, Pa.
P	total linear momentum of system; lb$_m$ ft/s, kg · m/s; equation (5-1).
P_c	critical pressure; atm, Pa.
P_i	vapor pressure of species i; atm, Pa.
P_g	power input for aerated stirred tank of liquid; N · m/s., W.
q	heat flow rate; Btu/h, W; equation (15-1).
$\dot{q}$	volumetric energy generation rate; Btu/h ft^3, W/m^3; equation (16-1).
Q	heat transfer; Btu, J; Section 6.1.
Q	activation energy for solid diffusion coefficient; J/g mol.
r	radial distance in both cylindrical and spherical coordinates; ft, m.
r	radius; ft, m.
r_{crit}	critical radius of insulation; ft, m; equation (17-13).
R	radius of sphere, ft, m.
R	gas constant; 0.73 atm ft^3/lb mol F, 8.314 Pa · m^3/mol · K.
R_t	thermal resistance; h F/Btu, K/W; equation (15-16).
r_A	rate of the production of mass A within the control volume; lb$_m$/ft^3 h, kg/m^3 · s.
R_A	rate of production of moles A within control volume; lb mol/ft^3 h, mol/m^3 · s.
s	surface renewal factor.
S	shape factor; ft or m; equation (15-19).
S	partition coefficient for dissolution of a gas into a solid; kg mol/m^3 · Pa.
t	time; h, s.
t_{exp}	time of exposure; s.
T	absolute temperature; °R, K.
T	dimensionless temperature; dimensionless.
T_b	normal boiling temperature; K.
T_c	critical temperature; K.
T_f	film temperature; F, K; equation (19-28).
T_{sat}	temperature of saturated liquid-vapor mixtures; F, K; Figure 21.1.
u	mean molecular speed; ft/s, m/s.
U	overall heat-transfer coefficient; Btu/h ft^2 F, W/m^2 · K; equation (15-17).
V	volumetric flow rate of fluid; ft^3/s, m^3/s.
v_x	x component of velocity, $\mathbf{v}$; ft/s, m/s.
v_y	y component of velocity, $\mathbf{v}$; ft/s, m/s.
v_z	z component of velocity, $\mathbf{v}$; ft/s, m/s.
v_∞	free stream velocity of flowing fluid; ft/s, m/s.
u^+	dimensionless velocity.
v	velocity, ft/s, m/s.
$\mathbf{v}$	mass-average velocity for multicomponent mixture; ft/s, m/s equation (24-13).
$\mathbf{v}_i$	velocity of species i; ft/s, m/s.
$\mathbf{v}_i - \mathbf{v}$	diffusion velocity of species i relative to mass-average velocity; ft/s, m/s; Section 24.1.
$\mathbf{v}_i - \mathbf{V}$	diffusion velocity of species i relative to the molar-average velocity; ft/s, m/s; Section 24.1.
V	volume; ft^3, m^3.
V_b	molecular volume at the normal boiling point, cm^3/g mol.
V_c	critical molecular volume; cm^3/g mol.
$\mathbf{V}$	molar-average velocity; ft/s, m/s equation (24-14).
W	work done; Btu, J; Section 6.1.

w_A	mass rate of flow of species A; lbm/h, g/s.
W_s	shaft work; Btu, J; Section 6.1.
W_δ	normal stress work; Btu, J; Section 6.1.
W_τ	shear work; Btu, J; Section 6.1.
x	rectangular coordinate.
x_A	mole fraction in either liquid or solid phase; dimensionless; equation (24-7).
X_A	mole of A/mole of A-free liquid.
X_D	relative time, $D_{AB}t/x_1^2$; dimensionless; Section 27.3.
y	rectangular coordinate.
y^+	dimensionless distance; equation (12-60).
y_A	mole fraction in the gas phase.
$y_{B,lm}$	log-mean mole fraction of the carrier gas.
y_n'	log fraction of component n in a gas mixture on species i-free basis; equation (24-49).
Y	parameter in heat exchanger analysis; dimensionless; equation (22-12).
Y	unaccomplished change; dimensionless; Section 27.4.
Y_A	mole of A/mole of A-free gas.
z	distance in the z-direction; ft, m.
z	rectangular coordinate.
Z	wall collision frequency; equation (7-8).
Z	parameter in heat-exchanger analysis; dimensionless; equation (22-13).
α	absorptivity; dimensionless; Section 23.2.
α	thermal diffusivity; ft^2/h, m^2/s; equation (16-17).
α	ratio of fluxes, N_B/N_A; dimensionless.
α	packed-bed constant.
β	bulk modulus of elasticity; lb$_f$/ft, N/m equation (1-11a).
β	coefficient of thermal expansion; 1/F, 1/K; equation (19-10).
δ	boundary-layer thickness; ft, m; equation (12-28).
δ	thickness of stagnant or laminar layer; ft, m.
δ_c	concentration boundary-layer thickness; ft, m.
δ_i	thermal boundary-layer thickness; ft, m; equation (19-22).
Δ_{lm}	log-mean concentration difference, $(y_A - y_A^*)_{lm}$; dimensionless; equation (31-35).
ϵ	emissivity; dimensionless; equation (23-2).
ϵ	volume void.
ϵ	packing characteristic; dimensionless.
ϵ_{AB}	a Lennard–Jones parameter; erg.
ϵ_D	eddy mass diffusivity, ft^2/h, m^2/s.
ϵ_H	eddy thermal diffusivity; ft^2/h, m^2/s; Section 19.7.
ϵ_i	a Lennard–Jones parameter; erg.
ϵ_M	eddy momentum diffusivity or eddy viscosity; ft^2/h, m^2/s; equation (12-52).
η	dependent variable used by Blasius in solution of boundary layer; dimensionless; equation (12-12).
η	similarity parameter for convection analysis; dimensionless; equation (19-17).
η_F	fin efficiency; dimensionless; Section 17.3, Figure 17.11.
θ	temperature parameter $= T - T_\infty$; F, K; Section 17.3.
θ	fractional void space of a catalyst.
θ	angle in cylindrical or spherical coordinates; rad.
θ_g	correlating constant.
φ	reduced pore diameter; dimensionless.

κ	Boltzmann constant; 1.38×10^{-16} erg/K.
λ	molecular mean-free path; Sections 7.3, 15.2, and 24.2.
λ	wave length of thermal radiation; μm; Section 23.4.
λ	ionic conductance; $(A/cm^2)(V/cm)(g \text{ equivalent}/cm^3)$.
λ_{Ts}	latent heat of vaporization; Btu/lb mol, J/g mol.
μ	viscosity; lb_m/ft s, Pa · s; equation (7-4).
μ_B	viscosity of solvent B; cp.
μ_c	chemical potential of given species; Btu/mol, J/mol.
ν	frequency; Hz; Section 23.1.
ν	kinematic viscosity, μ/ρ; ft^2/s, m^2/s.
π	pi groups in dimensional analysis; Sections 11.3, 13.1, 19.3, and 28.3.
ρ	density of a fluid; lb_m/ft^3, kg/m^3; Section 1.2.
ρ	mass density of mixture; lb_m/ft^3, kg/m^3.
ρ	reflectivity; dimensionless; Section 23.2.
ρ_i	mass concentration of species i, lb_m/ft^3, kg/m^3.
σ	surface tension; lb_f/ft, N/m.
σ	Stefan–Boltzmann constant; 0.1714×10^{-8} Btu/h ft^2 F^4, 5.672×10^{-8} $W/m^2 \cdot K^4$; equation (15-13).
σ_A	Lennard–Jones molecular diameter of species A; Å, nm.
σ_{AB}	Lennard–Jones parameter; Å.
σ_i	a Lennard–Jones parameter; Å.
σ_{ii}	normal stress; $lb_f/in.^2$, N/m^2; Section 1.2.
τ	transmissivity; dimensionless; Section 23.2.
τ_{ij}	shear stress; $lb_f/in.^2$, N/m^2; Section 1.2.
τ_0	shear stress at the surface; $lb_f/in.^2$, N/m^2; equation (12-30).
ϕ	velocity potential; Section 10.4.
ϕ	angle in spherical coordinates; rad.
ϕ	argument of error function; dimensionless.
ω	angular velocity; 1/s.
ω_i	mass fraction of species i; dimensionless.
$\omega/2$	vorticity; equation (10-4).
Γ	flow rate of condensate film per width; lb_m/ft s, kg/m · s; equation (21-13).
Δ	$y_A - y_A^*$ dimensionless; equation (31-31).
ΔT_{lm}	logarithmic-mean temperature difference; F, K; equation (22-9).
$\mathscr{E}$	heat-exchanger effectiveness; dimensionless; equation (22-17).
Φ_B	association parameter.
Ψ	stream function; Section 10.2.
Ω	solid angle; rad; Section 23.3.
Ω_D	collision integral: Appendix K.
Ω_k	Lennard–Jones collision integral; equation (15-7) and Appendix K.
Ω_μ	Lennard–Jones collision integral; equation (7-10) and Appendix K.

DIMENSIONLESS PARAMETERS

Bi	Biot number, $(hV/A)/k$; equation (18-7).
Eu	Euler number, $P/\rho v^2$; equation (11-5).
Fo	Fourier number, $\alpha t/(V/A)^2$; equation (18-8).
Fr	Froude number, v^2/gL; equation (11-4).
Gr	Grashof number, $\beta g \rho^2 L^3 \Delta T/\mu^2$; equation (19-12).
Gr_{AB}	mass-transfer Grashof number, $L^3 g \Delta \rho_A/\rho \nu^3$; equation (28-9).

Gz   Graetz number, $(\pi/4)(D/x)$ Re Pr; Section 20.2.

Le   Lewis number, $k/\rho c_p D_{AB}$; equation (28-3).

Nu   Nusselt number, hL/k; equation (19-6).

Nu_{AB} mass-transfer Nusselt number, $k_c L/D_{AB}$; equation (28-7).

Pe   Peclet number, $Dv\rho c_p/k = $ Re Pr; Section 20.2.

Pe_{AB} mass-transfer Peclet number, $Dv/D_{AB} = $ Re Sc; Table 30.1.

Pr   Prandtl number, $\nu/\alpha = \mu c_p/k$; equation (19-1).

Re   Reynolds number, $Lv\rho/\mu = Lv/\nu$; equation (11-7).

Sc   Schmidt number, $\mu/\rho D_{AB}$; equation (28-2).

Sh   Sherwood number, $k_c L/D_{AB}$; Section 28.3.

St   Stanton number, $h/\rho v c_p$; equation (19-8).

St_{AB} mass-transfer Stanton number, k_c/v_∞.

MATHEMATICAL OPERATIONS

D/Dt substantial derivative; equation (9-4).

div **A**   or $\nabla \cdot \mathbf{A}$, divergence of a vector.

erf ϕ the error function of ϕ; Appendix L.

exp x or e^x, exponential function of x.

ln x logarithm of x to the base e.

$\log_{10} x$ logarithm of x to base 10.

$$\nabla = \frac{\partial}{\partial x}\mathbf{e}_x + \frac{\partial}{\partial y}\mathbf{e}_y + \frac{\partial}{\partial z}\mathbf{e}_z.$$

SS **Student solution available in interactive e-text.**

1.1 The number of molecules crossing a unit area per unit time in one direction is given by

$$N = \frac{1}{4} n \bar{v}$$

where n is the number of molecules per unit volume and v is the average molecular velocity. As the average molecular velocity is approximately equal to the speed of sound in a perfect gas, estimate the number of molecules crossing a circular hole 10^{-3} in in diameter. Assume that the gas is at standard conditions. At standard conditions, there are 4×10^{20} molecules per in^3.

SS **1.2** Which of the quantities listed below are flow properties and which are fluid properties?

pressure	temperature	velocity
density	stress	speed of sound
specific heat	pressure gradient	

1.3 For a fluid of density ρ in which solid particles of density ρ_s are uniformly dispersed, show that if x is the mass fraction of solid in the mixture, the density is given by

$$\rho_{\text{mixture}} = \frac{\rho_s \rho}{\rho x + \rho_s (1 - x)}$$

1.4 An equation linking water density and pressure is

$$\frac{P + B}{P_1 + B} = \left(\frac{\rho}{\rho_1} \right)^7$$

where the pressure is in atmospheres and $B = 3000 \text{ atm}$. Determine the pressure in psi required to increase water density by 1% above its nominal value.

1.5 What pressure change is required to change the density of air by 10% under standard conditions?

1.6 Using the information given in Problem 1.1 and the properties of the standard atmosphere given in Appendix G, estimate the number of molecules per cubic inch at an altitude of 250,000 ft.

1.7 Show that the unit vectors $\mathbf{e}_r$ and $\mathbf{e}_\theta$ in a cylindrical coordinate system are related to the unit vectors $\mathbf{e}_x$ and $\mathbf{e}_y$ by

$$\mathbf{e}_r = \mathbf{e}_x \cos \theta + \mathbf{e}_y \sin \theta$$

and

$$\mathbf{e}_\theta = -\mathbf{e}_x \sin \theta + \mathbf{e}_y \cos \theta$$

1.8 Using the results of Problem 1.7, show that $d\mathbf{e}_r / d\theta = \mathbf{e}_\theta$ and $d\mathbf{e}_\theta / d\theta = -\mathbf{e}_r$.

1.9 Using the geometric relations given below and the chain rule for differentiation, show that

$$\frac{\partial}{\partial x} = -\frac{\sin \theta}{r} \frac{\partial}{\partial \theta} + \cos \theta \frac{\partial}{\partial r}$$

and

$$\frac{\partial}{\partial y} = \frac{\cos \theta}{r} \frac{\partial}{\partial \theta} + \sin \theta \frac{\partial}{\partial r}$$

when $r^2 = x^2 + y^2$ and $\tan \theta = y/x$.

1.10 Transform the operator ∇ to cylindrical coordinates (r, θ, z), using the results of Problems 1.7 and 1.9.

1.11 Find the pressure gradient at point (a, b) when the pressure field is given by

$$P = \rho_\infty v_\infty^2 \left(\sin \frac{x}{a} \sin \frac{y}{b} + 2 \frac{x}{a} \right)$$

where $\rho_\infty, v_\infty, a,$ and b are constants.

1.12 Find the temperature gradient at point (a, b) at time $t = (L^2/\alpha) \ln e$ when the temperature field is given by

$$T = T_0 e^{-\alpha t / 4L^2} \sin \frac{x}{a} \cos \text{h} \frac{y}{b}$$

where $T_0, \alpha, a,$ and b are constants.

1.13 Are the fields described in Problems 1.11 and 1.12 dimensionally homogeneous? What must the units of ρ_∞ be in order that the pressure be in pounds per square foot when v_∞ is given in feet per second (Problem 1.11)?

1.14 A scalar field is given by the function $\phi = 3x^2 y + 4y^2$.

a. Find $\nabla \phi$ at the point $(3, 5)$.

b. Find the component of $\nabla \phi$ that makes a $-60°$ angle with the x axis at the point $(3, 5)$.

1.15 If the fluid of density ρ in Problem 1.3 obeys the perfect gas law, obtain the equation of state of the mixture—that is, $P = f(\rho_s, (RT/M), \rho_m, x)$. Will this result be valid if a liquid is present instead of a solid?

1.16 Using the expression for the gradient in polar coordinates (Appendix A), find the gradient of $\psi (r, \theta)$ when

$$\psi = A r \sin \theta \left(1 - \frac{a^2}{r^2} \right).$$

where is the gradient maximum? The terms A and a are constant.

1.17 Given the following expression for the pressure field where x, y, and z are space coordinates, t is time, and P_0, ρ, V_∞, and L are constants. Find the pressure gradient

$$P = P_0 + \frac{1}{2} \rho V_\infty^2 \left[2 \frac{xyz}{L^3} + 3 \left(\frac{x}{L} \right)^2 + \frac{V_\infty t}{L} \right]$$

1.18 A vertical cylindrical tank having a base diameter of 10 m **SS** and a height of 5 m is filled to the top with water at 20 C. How much water will overflow if the water is heated to 80 C?

1.19 A liquid in a cylinder has a volume of 1200 cm^3 at 1.25 MPa and a volume of 1188 cm^3 at 2.50 MPa. Determine its bulk modulus of elasticity.

1.20   A pressure of 10 MPa is applied to 0.25 m^3 of a liquid, causing a volume reduction of 0.005 cm^3. Determine the bulk modulus of elasticity.

1.21   The bulk modulus of elasticity for water is 2.205 GPa. Determine the change in pressure required to reduce a given volume by 0.75%.

1.22   Water in a container is originally at 100 kPa. The water is subjected to a pressure of 120 MPa. Determine the percentage decrease in its volume.

1.23   Determine the height to which water at 68 C will rise in a clean capillary tube having a diameter of 0.2875 cm.

1.24   Two clean and parallel glass plates, separated by a gap of 1.625 mm, are dipped in water. If $\sigma = 0.0735$ N/m, determine how high the water will rise.

1.25   Determine the capillary rise for a water–air–glass interface at 40 C in a clean glass tube having a radius of 1 mm.

1.26   Determine the difference in pressure between the inside and outside of a soap film bubble at 20 C if the diameter of the bubble is 4 mm.

1.27   At 60 C the surface tension of water is 0.0662 N/m and that of mercury is 0.47 N/m. Determine the capillary height changes in these two fluids when they are in contact with air in a glass tube of diameter 0.55 mm. Contact angles are 0° for water and 130° for mercury.

1.28   Determine the diameter of the glass tube necessary to keep the capillary-height change of water at 30 C less than 1 mm.

1.29   An experimental fluid is used to create a spherical bubble with a diameter of 0.25 cm. When in contact with a surface made of plastic, it has a contact angle of 30 degrees. The pressure inside the bubble is 101,453 Pa and outside the bubble the pressure is atmospheric. In a particular experiment, you are asked to calculate how high this experimental fluid will rise in a capillary tube made of the same plastic as used in the surface described above. The diameter of the capillary tube is 0.2 cm. The density of the experimental fluid used in this experiment is 750 kg/m^3.

1.30   A colleague is trying to measure the diameter of a capillary tube, something that is very difficult to physically accomplish. Since you are a Fluid Dynamics student, you know that the diameter can be easily calculated after doing a simple experiment. You take a clean glass capillary tube and place it in a container of pure water and observe that the water rises in the tube to a height of 17.5 mm. You take a sample of the water and measure the mass of 100 mls to be 97.18 grams and you measure the temperature of the water to be 80 C. Please calculate the diameter of your colleague's capillary tube.

2.1   On a certain day, the barometric pressure at sea level is 30.1 in Hg and the temperature is 70 F. The pressure gage in an airplane in flight indicates a pressure of 10.6 psia, and the temperature gage shows the air temperature to be 46 F. Estimate as accurately as possible the altitude of the airplane above sea level.

2.2   The open end of a cylindrical tank 2 ft in diameter and 3 ft high is submerged in water as shown. If the tank weighs 250 lb, to what depth h will the tank submerge? The air barometric pressure is 14.7 psia. The thickness of the tank wall may be neglected. What additional force is required to bring the top of the tank flush with the water surface?

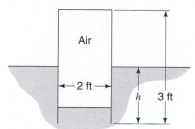

2.3   In Problem 2.2, find the depth at which the net force on the tank is zero.

2.4   If the density of sea water is approximated by the equation of state $\rho = \rho_0 \exp[(p - p_{atm})/\beta]$, where β is the compressibility, determine the pressure and density at a point 32,000 ft below the surface of the sea. Assume $\beta = 30,000$ psi.

2.5   The change in density due to temperature causes the take-off and landing speeds of a heavier-than-air craft to increase as the square root of the temperature. What effect do temperature-induced density changes have on the lifting power of a rigid lighter-than-air craft?

SS **2.6**   The practical depth limit for a suited diver is about 185 m. What is the gage pressure in sea water at that depth? The specific gravity of sea water is 1.025.

2.7   Matter is attracted to the center of Earth with a force proportional to the radial distance from the center. Using the known value of g at the surface where the radius is 6330 km, compute the pressure at Earth's center, assuming the material behaves like a liquid, and that the mean specific gravity is 5.67.

2.8   The deepest known point in the ocean is 11,034 m in the Mariana Trench in the Pacific. Assuming sea water to have a constant density of 1050 kg/m^3, determine the pressure at this point in atmospheres.

SS **2.9**   Determine the depth change to cause a pressure increase of 1 atm for (a) water, (b) sea water (SG = 1.0250), and (c) mercury (SG = 13.6).

2.10   Find the pressure at point A.

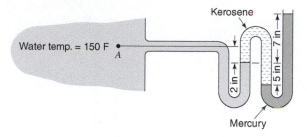

2.11   Using a liquid having a specific gravity of 1.2 and inverting a tube full of this material as shown in the figure, what is the value of h if the vapor pressure of the liquid is 3 psia?

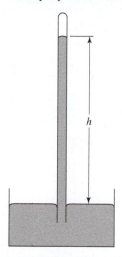

2.12   What is the pressure p_A in the figure? The specific gravity **SS** of the oil is 0.8.

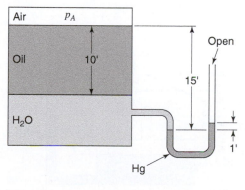

2.13   Find the difference in pressure between tanks A and B, if **SS** $d_1 = 2$ ft, $d_2 = 6$ in, $d_3 = 2.4$ in, and $d_4 = 4$ in.

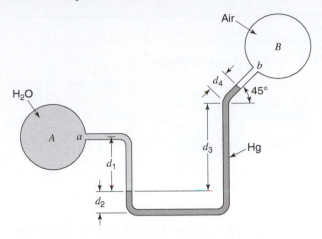

2.14 What is the pressure difference between points A and B if $d_1 = 1.7$ ft, $d_2 = 1$ in, and $d_3 = 6.3$ in?

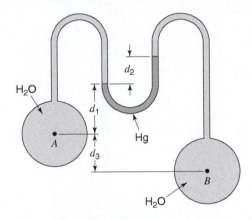

SS **2.15** A differential manometer is used to measure the pressure change caused by a flow constriction in a piping system as shown. Determine the pressure difference between points A and B in psi. Which section has the higher pressure?

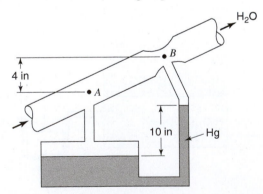

2.16 The car shown in the figure is accelerated to the right at a uniform rate. What way will the balloon move relative to the car?

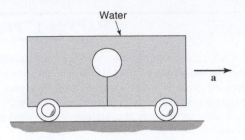

2.17 The tank is accelerated upward at a uniform rate. Does the manometer level go up or down?

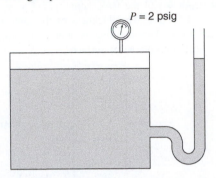

2.18 Glass viewing windows are to be installed in an aquarium. Each window is to be 0.6 m in diameter and to be centered 2 m below the surface level. Find the force and location of the force acting on the window.

2.19 Find the minimum value of h for which the gate shown will rotate counterclockwise if the gate cross section is (a) rectangular, 4 ft × 4 ft; (b) triangular, 4 ft at the base × 4 ft high. Neglect bearing friction.

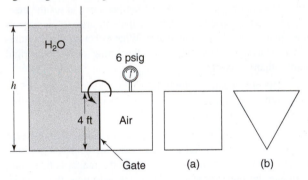

2.20 A circular log of radius r is to be used as a barrier as shown in the figure below. If the point of contact is at O, determine the required density of the log.

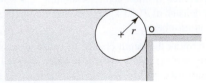

2.21 A rectangular block of concrete 3 ft × 3 ft × 6 in has its 6 in side half buried at the bottom of a lake 23 ft deep. What force is required to lift the block free of the bottom? What force is required to maintain the block in this position? (Concrete weighs $150 \, lb/ft^3$.)

2.22 A dam spillway gate holds back water of depth h. The gate weights 500 lb/ft and is hinged at A. At what depth of water, will the gate rise and permit the water to flow under it?

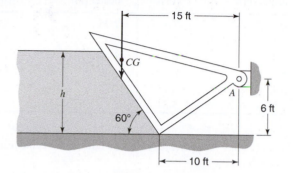

2.23 It is desired to use a 0.75-m-diameter beach ball to stop a drain in a swimming pool. Obtain an expression that relates the drain diameter D and the minimum water depth h for which the ball will remain in place.

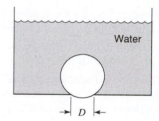

2.24 A watertight bulkhead 22 ft high forms a temporary dam for some construction work. The top 12 ft behind the bulkhead consists of sea water with a density of 2 slugs/ft³, but the bottom 10 ft begin a mixture of mud and water can be considered a fluid of density 4 slugs/ft³. Calculate the total horizontal load per unit width and the location of the center of pressure measured from the bottom.

2.25 The circular gate ABC has a 1 m radius and is hinged at B. Neglecting atmospheric pressure, determine the force P just sufficient to keep the gate from opening when $h = 12 \, m$.

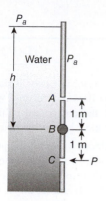

2.26 The figure below shows an open triangular channel in which the two sides, AB and AC, are held together by cables, spaced 1 m apart, between B and C. Determine the cable tension.

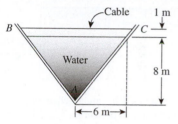

2.27 The dam shown below is 100 m wide. Determine the magnitude and location of the force on the inclined surface.

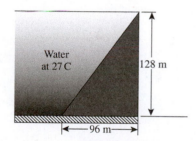

2.28 The float in a toilet tank is a sphere of radius R and is made of a material with density ρ. An upward buoyant force F is required to shut the ballcock valve. The density of water is designated ρ_w. Develop an expression for x, the fraction of the float submerged, in terms of R, ρ, F, g, and ρ_w.

2.29 A cubical piece of wood with an edge L in length floats in the water. The specific gravity of the wood is 0.90. What moment M is required to hold the cube in the position shown? The right-hand edge of the block is flush with the water.

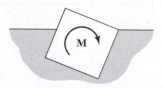

4.1   The velocity vector in a two-dimensional flow is given by the expression $\mathbf{v} = 10\mathbf{e}_x + 7x\mathbf{e}_y$ m/s when x is measured in meters. Determine the component of the velocity that makes a $-30°$ angle with the x axis at the point (2, 2).

4.2   Using the velocity vector of the previous problem, determine (a) the equation of the streamline passing through point (2, 1); (b) the volume of flow that crosses a plane surface connecting points (1, 0) and (2, 2).

SS **4.3**   Water is flowing through a large circular conduit with a velocity profile given by the equation $v = 9(1 - r^2/16)$ fps. What is the average water velocity in the 1.5-ft pipe?

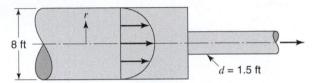

4.4   Water enters a 4 in. square channel as shown at a velocity of 10 fps. The channel converges to a 2 in. square configuration as shown at the discharge end. The outlet section is cut at 30° to the vertical as shown, but the mean velocity of the discharging water remains horizontal. Find the exiting water's average velocity and total rate of flow.

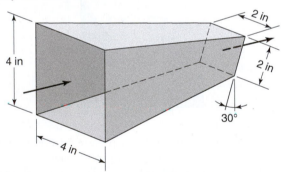

4.5   Water enters one end of a perforated pipe 0.2 m in diameter with a velocity of 6 m/s. The discharge through the pipe wall is approximated by a linear profile. If the flow is steady, find the discharge velocity.

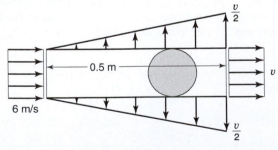

4.6   The velocities in a circular duct of 20 in diameter were measured as follows:

Distance from center (in)	Velocity (fps)	Distance from center (in)	Velocity (fps)
0	7.5	7.75	5.47
3.16	7.10	8.37	5.10
4.45	6.75	8.94	4.50
5.48	6.42	9.49	3.82
6.33	6.15	10.00	2.40
7.07	5.81	. . .	. . .

Find (a) the average velocity; (b) the flow rate in cubic feet per second.

4.7   Salt water containing 1.92 lb/gal of salt flows at a fixed rate of 2 gal/min into a 100 gal tank, initially filled with fresh water. The density of the incoming solution is 71.8 lb/ft³. The solution, kept uniform by stirring, flows out at a fixed rate of 19.2 lb/min.

a.   How many pounds of salt will there be in the tank at the end of 1 h and 40 min?

b.   What is the upper limit for the number of pounds of salt in the tank if the process continues indefinitely?

c.   How much time will elapse while the quantity of salt in the tank changes from 100 to 150 lb?

4.8   In the piston and cylinder arrangement shown below, the large piston has a velocity of 2 fps and an acceleration of 5 fps². Determine the velocity and acceleration of the smaller piston.

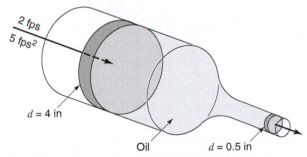

4.9   Show that in a one-dimensional steady flow the following equation is valid:

$$\frac{dA}{A} + \frac{dv}{v} + \frac{d\rho}{\rho} = 0$$

4.10   Using the symbol M for the mass in the control volume, **SS** show that equation (4-1) may be written as

$$\frac{\partial M}{\partial t} + \iint_{c.s.} d\dot{m} = 0$$

4.11 A shock wave moves down a pipe as shown below. The fluid properties change across the shock, but they are not functions of time. The velocity of the shock is v_w. Write the continuity equation and obtain the relation between ρ_2, ρ_1, v_2, and v_w. The mass in the control volume at any time is $M = \rho_2 Ax + \rho_1 Ay$. Hint: Use a control volume that is moving to the right at velocity v_w.

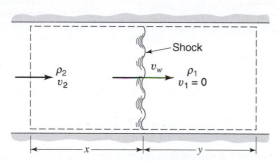

4.12 The velocity profile in circular pipe is given by $v = v_{max}(1 - r/R)^{1/7}$, where R is the radius of the pipe. Find the average velocity in the pipe in terms of v_{max}.

4.13 In the figure below, the x-direction velocity profiles are shown for a control volume surrounding a cylinder. If the flow is incompressible, what must the rate of flow be across the horizontal control-volume surface?

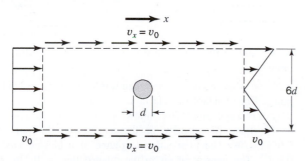

4.14 Two very long parallel plates of length $2L$ are separated a distance b. The upper plate moves downward at a constant rate v. A fluid fills the space between the plates. Fluid is squeezed out between the plates. Determine the mass flow rate and maximum velocity under the following conditions:

a. The exit velocity is uniform.

b. The exit velocity is parabolic.

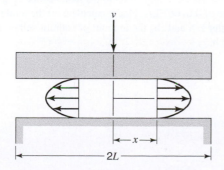

4.15 A thin layer of liquid, draining from an inclined plane, has a velocity profile $v_x \approx v_0(2y/h - y^2/h^2)$, where v_0 is the surface velocity. If the plane has width 10 cm into the paper, determine the volume rate of flow in the film. Suppose that $h = 2$ cm and the flow rate is 2 L/min. Estimate v_0.

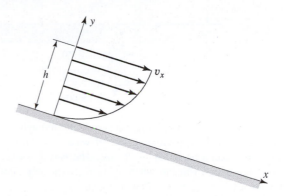

4.16 The V-shaped tank has width w into the paper and is filled from the inlet pipe at volume flow Q. Derive expressions for (a) the rate of change dh/dt and (b) the time required for the surface to rise from h_1 to h_2.

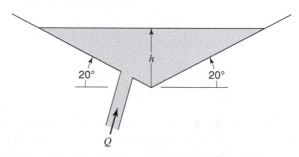

4.17 A bellows may be modeled as a deforming wedge-shaped volume. The check valve on the left (pleated) end is closed during the stroke. If w is the bellows width into the paper, derive an expression for outlet mass flow $\dot{m}_0$ as a function of stroke $\theta(t)$.

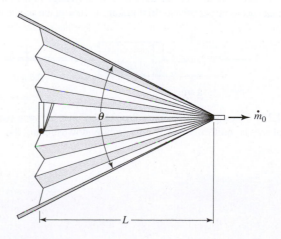

SS **4.18** Water flows steadily through the piping junction, entering section 1 at 0.0013 m³/s. The average velocity at section 2 is 2.1 m/s. A portion of the flow is diverted through the showerhead, which contains 100 holes of 1 mm diameter. Assuming uniform shower flow, estimate the exit velocity from the showerhead jets.

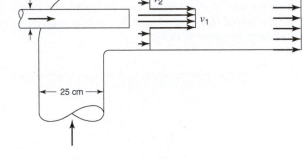

SS **4.19** The jet pump injects water at $v_1 = 40$ m/s through a 7.6 cm pipe and entrains a secondary flow of water $v_2 = 3$ m/s in the annular region around the small pipe. The two flows become fully mixed downstream, where v_3 is approximately constant. For steady incompressible flow, compute v_3.

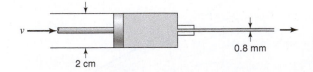

4.20 A vertical, cylindrical tank closed at the bottom is partially filled with an incompressible liquid. A cylindrical rod of diameter d_i (less than tank diameter, d_0) is lowered into the liquid at a velocity v Determine the average velocity of the fluid escaping between the rod and the tank walls (a) relative to the bottom of the tank and (b) relative to the advancing rod.

SS **4.21** The hypodermic needle shown below contains a liquid serum ($\rho = 1$ g/cm³). If the serum is to be injected steadily at 6 cm³/s, how fast should the plunger be advanced (a) if leakage in the plunger clearance is neglected and (b) if leakage is 10% of the needle flow?

4.22 Incompressible steady flow in the inlet between parallel plates is uniform, $v_0 = 8$ cm/s, while downstream, the flow develops into the parabolic profile $v_x = az(z_0 - z)$, where a is a constant. What is the maximum value of v_x?

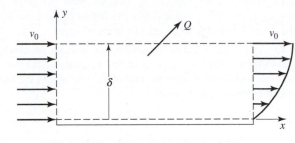

4.23 An incompressible fluid flows past a flat plate, as in the figure below, with a uniform inlet profile and a polynomial exit profile

$$v_x = v_0 \left(\frac{3\eta - \eta^3}{2} \right) \text{ where } \eta = \frac{y}{\delta}$$

Compute the volume flow Q across the top surface of the control volume. The plate has width b into the paper.

4.24 Rework Problem 4.14 if the plates are circular and have radius L.

4.25 You are going to pressurize a tube using a mechanical pump. The standard bicycle tube is wrapped around a 24 in diameter metal rim. The inflow to the pump is constant at 1 ft³/min. The density of the air leaving the pump is 0.075 lb$_m$/ft³. The final inflated volume of a bicycle tube is 0.6 ft³. The density of air in the inflated tube is 0.4 lb$_m$/ft³. The atmospheric pressure surrounding the system is 14.7 lb$_f$/in². Assuming that there are no leaks anywhere in the system, please calculate the time necessary to pressurize the tube if there initially is no air in the tube.

4.26 A large tank of unknown total volume is initially filled with 6000 g of a 10% by mass sodium sulfate solution. Into this tank a 50% sodium sulfate solution is added at a rate of 40 g/min. At the single outlet to the tank flows a 20 g/L solution at a rate of 0.01667 L/s. Please calculate (a) the total mass in the tank after 2 h and (b) the amount of sodium sulfate in the tank after 2 h.

CHAPTER 5   PROBLEMS

SS **5.1** A two-dimensional object is placed in a 4-ft-wide water tunnel as shown. The upstream velocity, v_1, is uniform across the cross section. For the downstream velocity profile as shown, find the value of v_2.

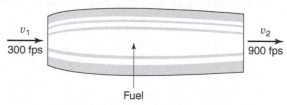

$v_1 = 20$ fps · 2 ft · 4 ft · v_2

5.2 If, in the system for Problem 5.1, the total drag on the object is measured to be 800 N/m of length normal to the direction of flow, and frictional forces at the walls are neglected, find the pressure difference between inlet and outlet sections.

5.3 Rework Problem 5.1 if the exit velocity profile is given by

$$v = v_2\left(1 - \cos\frac{\pi y}{4}\right)$$

when y is measure vertically from the center line of the water tunnel.

5.4 A stationary jet engine is shown. Air with a density of 0.0805 lb/ft^3 enters as shown. The inlet and outlet cross-sectional areas are both 10.8 ft^2. The mass of fuel consumed is 2% of the mass of air entering the test section. For these conditions, calculate the thrust developed by the engine tested.

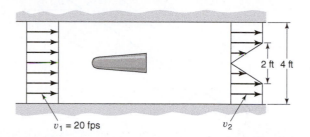

v_1 300 fps · v_2 900 fps · Fuel

5.5 The pump in the boat shown pumps 6 ft^3/s of water through a submerged water passage, which has an area of 0.25 ft^2 at the bow of the boat and 0.15 ft^2 at the stern. Determine the tension in the restraining rope, assuming that the inlet and exit pressures are equal.

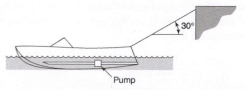

30° · Pump

5.6 Oil (sp. gr. $= 0.8$) flows smoothly through the circular reducing section shown at 3 ft^3/s. If the entering and leaving velocity profiles are uniform, estimate the force that must be applied to the reducer to hold it in place.

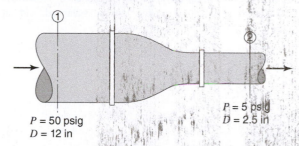

① ② $P = 50$ psig $D = 12$ in · $P = 5$ psig $D = 2.5$ in

SS **5.7** At the end of a water pipe of 3-in diameter is a nozzle that discharges a jet having a diameter of 1½ in into the open atmosphere. The pressure in the pipe is 60 psig (pounds per square inch gage), and the rate of discharge is 400 gal/min. What are the magnitude and direction of the force necessary to hold the nozzle to the pipe?

5.8 A water jet pump has an area $A_j = 0.06$ ft^2 and a jet velocity $v_j = 90$ fps, which entrains a secondary stream of water having a velocity $v_s = 10$ fps in a constant-area pipe of total area $A = 0.6$ ft^2. At section 2, the water is thoroughly mixed. Assuming one-dimensional flow and neglecting wall shear

a. find the average velocity of mixed flow at section 2;

b. find the pressure rise $(P_2 - P_1)$, assuming the pressure of the jet and secondary stream to be the same at section 1.

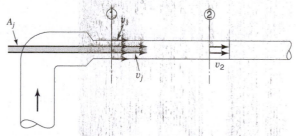

① A_j v_s v_j ② v_2

5.9 If the plate shown is inclined at an angle as shown, what are the forces F_x and F_y necessary to maintain its position? The flow is frictionless.

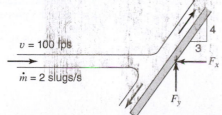

$v = 100$ fps · $\dot{m} = 2$ slugs/s · 4 · 3 · F_x · F_y

5.10   A steady, incompressible, frictionless, two-dimensional jet of fluid with breadth h, velocity v, and unit width impinges on a flat plate held at an angle α to its axis. Gravitational forces are to be neglected.

a.   Determine the total force on the plate, and the breadths a, b, of the two branches.

b.   Determine the distance 1 to the center of pressure (c.p.) along the plate from the point 0. (The center of pressure is the point at which the plate can be balanced without requiring an additional moment.)

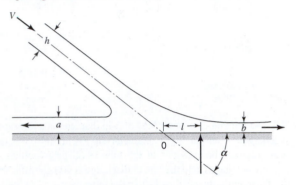

5.11   A plate moves perpendicularly toward a discharging jet at the rate of 5 fps. The jet discharges water at the rate of 3 ft^3/s and a speed of 30 fps. Find the force of the fluid on the plate and compare it with what it would be if the plate were stationary. Assume frictionless flow.

5.12   The shock wave illustrated below is moving to the right at v_w fps. The properties in front and in back of the shock are not a function of time. By using the illustrated control volume, show that the pressure difference across the shock is

$$P_2 - P_1 = \rho_1 v_w v_2$$

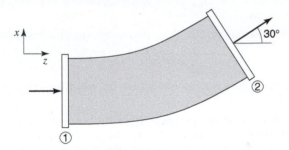

5.13   If the shock-wave velocity in Problem 5.13 is approximated by the speed of sound, determine the pressure change causing a velocity change of 10 fps in

a.   air at standard conditions

b.   water

5.14   Consider the differential control volume shown on the next page. By applying the conservation of mass and the momentum theorem, show that

$$dP + \rho v\, dv + g\, dy = 0$$

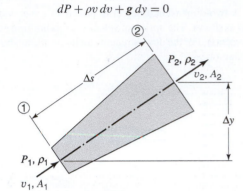

5.15   Water flows steadily through the horizontal 30° pipe bend shown below. At station 1, the diameter is 0.3 m, the velocity is 12 m/s, and the pressure is 128 kPa gage. At station 2, the diameter is 0.38 m and the pressure is 145 kPa gage. Determine the forces F_x and F_z necessary to hold the pipe bend stationary.

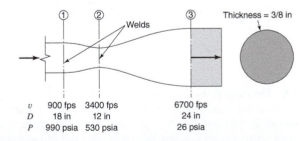

5.16   The rocket nozzle shown below consists of three welded sections. Determine the axial stress at junctions 1 and 2 when the rocket is operating at sea level. The mass flow rate is 770 lb$_m$/s. **SS**

	①	②	③	Thickness = 3/8 in
v	900 fps	3400 fps	6700 fps	
D	18 in	12 in	24 in	
P	990 psia	530 psia	26 psia	

5.17   The pressure on the control volume illustrated below is constant. The x components of velocity are as illustrated. Determine the force exerted on the cylinder by the fluid. Assume incompressible flow.

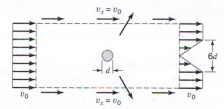

5.18 Water flows in a pipe at 3 m/s. A valve at the end of the pipe is suddenly closed. Determine the pressure rise in the pipe.

5.19 A dam discharge into a channel of constant width as shown. It is observed that a region of still water backs up behind the jet to a height H. The velocity and height of the flow in the channel are given as v and h, respectively, and the density of the water is ρ. Using the momentum theorem and the control surface indicated, determine H. Neglect the horizontal momentum of the flow that is entering the control volume from above, and assume friction to be negligible. The air pressure in the cavity below the crest of falling water is to be taken as atmospheric.

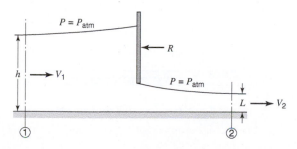

5.20 A liquid of density ρ flows through a sluice gate as shown. The upstream and downstream flows are uniform and parallel, so that the pressure variations at stations 1 and 2 may be considered hydrostatic.

a. Determine the velocity at station 2.

b. Determine the force per unit width, R, necessary to hold the sluice gate in place.

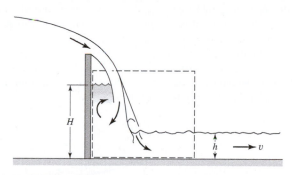

5.21 As can often be seen in a kitchen sink when the faucet is running, a high-speed channel flow (v_1, h_1) may "jump" to a

low-speed, low-energy condition (v_2, h_2). The pressure at sections 1 and 2 is approximately hydrostatic, and wall friction is negligible. Use the continuity and momentum relations to find h_2 and v_2 in terms of (h_1, v_1).

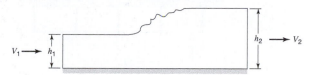

5.22 For the pipe-flow-reducing section $D_1 = 8$ cm, $D_2 = 5$ cm, and $p_2 = 1$ atm. If $v_1 = 5$ m/s and the manometer reading is $h = 58$ cm, estimate the total force resisted by the flange bolts.

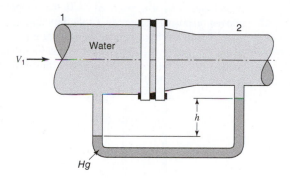

5.23 An open tank car as shown below travels to the right at a uniform velocity of 4.5 m/s. At the instant shown the car passes under a jet of water issuing from a stationary 0.1-m-diameter pipe with a velocity of 20 m/s. What force is exerted on the tank by the water jet? SS

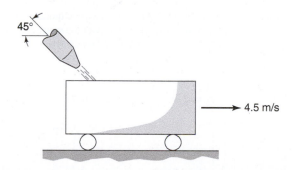

5.24 An open tank L ft long as shown below travels to the right at a velocity v_c fps. A jet of area A_j exhausts fluid of density ρ at a velocity v_j fps relative to the car. The tank car, at the same time, collects fluids from an overhead sprinkler that directs fluid downward with velocity v_s. Assuming that the sprinkler flow is uniform over the car area, A_c, determine the net force of the fluid on the tank car.

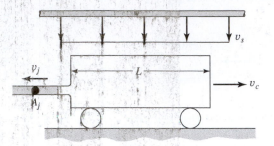

5.25 A liquid column of height h is confined in a vertical tube of cross-sectional area A by a stopper. At $t=0$, the stopper is suddenly removed, exposing the bottom of the liquid to atmospheric pressure. Using a control-volume analysis of mass and vertical momentum, derive the differential equation for the downward motion $v(t)$ of the liquid. Assume one-dimensional, incompressible, frictionless flow.

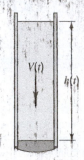

5.26 Sea water, $\rho = 64 \, \text{lb}_m/\text{ft}^3$, flows through the impeller of a centrifugal pump at the rate of 800 gal/min. Determine the torque exerted on the impeller by the fluid and the power required to drive the pump. Assume that the absolute velocity of the water entering the impeller is radial. The dimensions are as follows:

$$\omega = 1180 \, \text{rpm} \quad t_3 = 0.6 \, \text{in}$$
$$r_1 = 2 \, \text{in.} \quad \theta_2 = 135°$$
$$r_2 = 8 \, \text{in} \quad t_1 = 0.8 \, \text{in}$$

5.27 In Problem 5.27 determine

a. the angle θ_1 such that the entering flow is parallel to the vanes

b. the axial load on the shaft if the shaft diameter is 1 in and the pressure at the pump inlet is atmospheric

5.28 A water sprinkler consists of two 1/2-in diameter jets at the ends of a rotating hollow rod as shown below. If the water leaves at 20 fps, what torque would be necessary to hold the sprinkler in place?

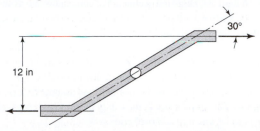

5.29 A lawn sprinkler consists of two sections of curved pipe **SS** rotating about a vertical axis as shown below. The sprinkler rotates with an angular velocity ω, and the effective discharge area is A, so the water is discharged at a rate $Q = 2v_r A$, where v_r is the velocity of the water relative to the rotating pipe. A constant friction torque M_f resists the motion of the sprinkler. Find an expression for the speed of the sprinkler in terms of the significant variables.

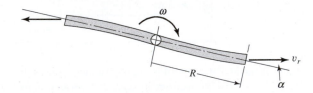

5.30 The pipe shown below has a slit of thickness 1/4 in so shaped that a sheet of water of uniform thickness 1/4 in issues out radially from the pipe. The velocity is constant along the pipe as shown below and a flow rate of 8 ft³/s enters at the top. Find the moment on the tube about the axis BB from the flow of water inside the pipe system.

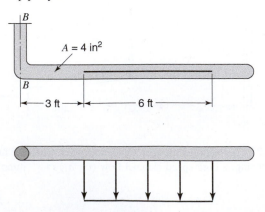

5.31 Water flows at 30 gal/min through the 0.75 in diameter double-pipe bend. The pressures are $p_1 = 30\,\text{lb}_f/\text{in}^2$ and $p_2 = 24\,\text{lb}_f/\text{in}^2$. Compute the torque T at point B necessary to keep the pipe from rotating.

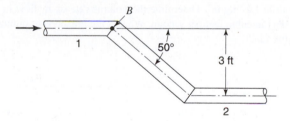

5.32 The illustration in the next column shows a vane with a turning angle θ that moves with a steady speed v_c. The vane receives a jet that leaves a fixed nozzle with speed v.

a. Assume that the vane is mounted on rails as shown in the sketch. Show that the power transmitted to the cart is maximum when $v_c/v = 1/3$.

b. Assuming that there are a large number of such vanes attached to a rotating wheel with peripheral speed, v_c, show that the power transmitted is maximum when $v_c/v = 1/2$.

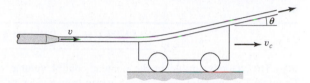

5.33 A water jet at 60 F with a flow rate of 250 ft³/s and a velocity of 75 ft/s hits a stationary V-shaped splitter such that half of the fluid is directed upward and the other half is directed downward as shown in the figure. Both of these streams have a final velocity of 75 ft/s. Assume steady, incompressible flow, that gravitational effects are negligible, and that the entire system is open to the atmosphere where the pressure is 2116.8 lb_f/ft^2. Calculate the x and y components of force required to hold the V-shaped splitter in place.

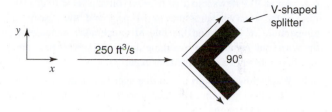

6.1 The velocity profile in the annular control volume of Example 3 is approximately linear, varying from a velocity of zero at the outer boundary to a value of $\omega d/2$ at the inner boundary. Develop an expression for the fluid velocity, $v(r)$, where r is the distance from the center of the shaft.

SS **6.2** Sea water, $\rho = 1025$ kg/m³, flows through a pump at 0:21 m³/s. The pump inlet is 0.25 m in diameter. At the inlet the pressure is -0.15 m of mercury. The pump outlet, 0.152 m in diameter, is 1.8 m above the inlet. The outlet pressure is 175 kPa. If the inlet and exit temperatures are equal, how much power does the pump add to the fluid?

SS **6.3** Air at 70 F flows into a 10-ft³ reservoir at a velocity of 110 fps. If the reservoir pressure is 14 psig and the reservoir temperature of 70 F, find the rate of temperature increase in the reservoir. Assume the incoming air is at reservoir pressure and flows through a 8-in-diameter pipe.

SS **6.4** Water flows through a 2-in-diameter horizontal pipe at a flow rate of 35 gal/min. The heat transfer to the pipe can be neglected, and frictional forces result in a pressure drop of 10 psi. What is the temperature change of the water?

6.5 During the flow of 200 ft³/s of water through the hydraulic turbine shown on the next page, the pressure indicated by gage A is 12 psig. What should gage B read if the turbine is delivering 600 hp at 82% efficiency? Gage B is designed to measure the total pressure—that is, $P + \rho v^2/2$ for an incompressible fluid.

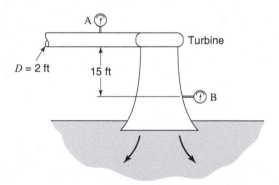

6.6 During the test of a centrifugal pump, a Bourdon pressure gage just outside the casing of the 12-in-diameter suction pipe reads -6 psig (i.e., vacuum). On the 10-in-diameter discharge pipe, another gage reads 40 psig. The discharge pipe is 5 ft above the suction pipe. The discharge of water through the pump is measured to be 4 ft³/s . Compute the horsepower input of the test pump.

SS **6.7** A fan draws air from the atmosphere through a 0.30-m-diameter round duct that has a smoothly rounded entrance. A differential manometer connected to an opening in the wall of the duct shows a vacuum pressure of 2.5 cm of water. The density of air is 1.22 kg/m³. Determine the volume rate of air flow in the duct in cubic feet per second. What is the horsepower output of the fan?

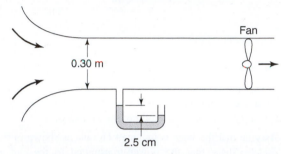

6.8 Find the change in temperature between stations (1) and (2) in terms of the quantities A_1, A_3, v_1, v_3, c_v, and θ. The internal energy is given by $c_v T$. The fluid is water and $T_1 = T_3, P_1 = P_3$.

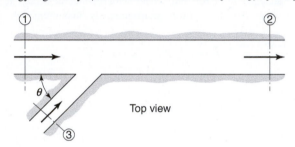

Top view

6.9 A liquid flows from A to B in a horizontal pipe line shown **SS** at a rate of 3 ft³/s with a friction loss of 0.45 ft of flowing fluid. For a pressure head at B of 24 in, what will be the pressure head at A?

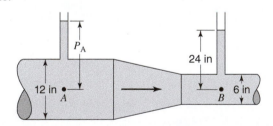

6.10 In Problem 6.26, compute the upward force on the device from water and air. Use the results of Problem 6.26 as well as any other data given in that problem that you may need. Explain why you cannot profitably use Bernoulli's equation here for a force calculation.

6.11 A Venturi meter with an inlet diameter of 0.6 m is designed **SS** to handle 6 m³/s of standard air. What is the required throat diameter if this flow is to give a reading of 0.10 m of alcohol in a differential manometer connected to the inlet and the throat? The specific gravity of alcohol may be taken as 0.8.

6.12 The pressurized tank shown has a circular cross section of 6 ft in diameter. Oil is drained through a nozzle of 2 in diameter in the side of the tank. Assuming that the air pressure is maintained constant, how long does it take to lower the oil surface in the tank by 2 ft? The specific gravity of the oil in the tank is 0.85, and that of mercury is 13.6.

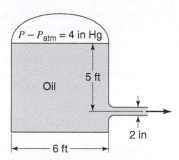

6.13 An automobile is driving into a 45-mph headwind at 40 mph. If the barometer reads 29 in Hg and the temperature is 40 F, what is the pressure at a point on the auto where the wind velocity is 120 fps with respect to the auto?

6.14 Water is discharged from a 1.0-cm-diameter nozzle that is inclined at a 30° angle above the horizontal. If the jet strikes the ground at a horizontal distance of 3.6 m and a vertical distance of 0.6 m from the nozzle as shown on the next page, what is the rate of flow in cubic meters per second? What is the total head of the jet? (See equation (6-11b).)

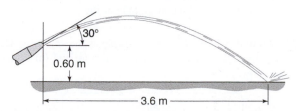

6.15 The pump shown in the figure delivers water at 59 F at a rate of 550 gal/min. The inlet pipe has an inside diameter of 5.95 in and it is 10 ft long. The inlet pipe is submerged 6 ft into the water and is vertical. Estimate the pressure inside the pipe at the pump inlet.

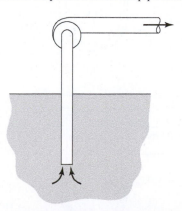

6.16 In the previous problem, determine the flow rate at which the pump inlet pressure is equal to the vapor pressure of the water. Assume that friction causes a head loss of 4 ft. The vapor pressure of water at 59 F is 0.247 psi.

6.17 Using the data of Problem 6.27, determine the velocity head of the fluid leaving the impeller. What pressure rise would result from such a velocity head?

6.18 In order to maneuver a large ship while docking, pumps are used to issue a jet of water perpendicular to the bow of the ship as shown in the figure. The pump inlet is located far enough away from the outlet that the inlet and outlet do not interact. The inlet is also vertical so that the net thrust of the jets on the ship is independent of the inlet velocity and pressure. Determine the pump horsepower required per pound of thrust. Assume that the inlet and outlet are at the same depth. Which will produce more thrust per horsepower: a low-volume, high-pressure pump, or a high-volume, low-pressure pump?

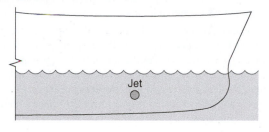

Side

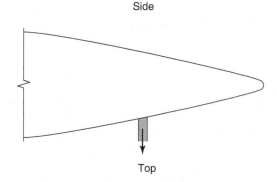

Top

6.19 Determine the head loss between stations (1) and (2) in Problem 5.7.

6.20 Multnomah Falls in Oregon has a sheer drop of 165 m. Estimate the change in water temperature caused by this drop.

6.21 An "air cushion" vehicle is designed to traverse terrain while floating on a cushion of air. Air, supplied by a compressor, escapes through the clearing between the ground and the skirt of the vehicle. If the skirt has a rectangular shape 3×9 m, the vehicle mass is 8100 kg and the ground clearance is 3 cm, determine the airflow rate needed to maintain the cushion and the power given by the compressor to the air. Assume that the air speeds within the cushion are very low.

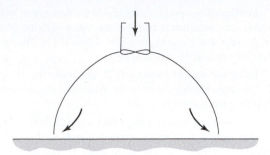

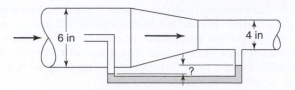

6.22 The solution to Problem 5.22 is

$$h_2 = \frac{h_1}{2}\left[\left(1 + \frac{8v_1^2}{gh_1}\right)^{1/2} - 1\right]$$

Show that Bernoulli's equation applied between sections 1 and 2 does not give this result. Derive all expressions for the change in total head across a hydraulic jump.

6.23 Residential water use, exclusive of fire protection, runs about 80 gallons per person per day. If the water is delivered to a residence at 60 psig, estimate the monthly energy required to pump the water from atmospheric pressure to the delivery pressure. Neglect line losses and elevation changes. Assume the pumps are 75% efficient and are driven by electric motors with 90% efficiency.

6.24 A 1968 Volkswagen sedan is driving over a 7300-ft-high mountain pass at a speed of v m/s into a headwind of W m/s. Compute the gage pressure in mPa at a point on the auto where the velocity relative to the auto is $v - W$ m/s. The local air density is 0.984 kg/m³.

6.25 A liquid is heated in a vertical tube of constant diameter, 15 m long. The flow is upward. At the entrance the average velocity is 1 m/s, the pressure 340,000 Pa, and the density is 1001 kg/m³. If the increase in internal energy is 200,000 J/kg, find the heat added to the fluid.

6.26 Water flows steadily up the vertical pipe and is then deflected to flow outward with a uniform radial velocity. If friction is neglected, what is the flow rate of water through the pipe if the pressure at A is 10 psig?

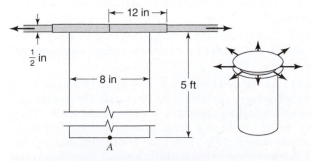

6.27 Water flows through the pipe contraction shown at a rate of 1 ft³/s. Calculate the differential manometer reading in inches of mercury, assuming no energy loss in the flow. Be sure to give the correct direction of the manometer reading.

6.28 The figure illustrates the operation of an air lift pump. Compressed air is forced into a perforated chamber to mix with the water so that the specific gravity of the air–water mixture above the air inlet is 0.5. Neglecting any pressure drop across section (1), compute the discharge velocity v of the air–water mixture. Can Bernoulli's equation be used across section (1)?

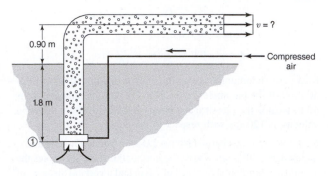

6.29 Rework Problem 6.28 with the assumption that the momentum of the incoming air at section (1) is zero. Determine the exit velocity, v, and the magnitude of the pressure drop at section (1).

6.30 Air of density 1.21 kg/m³ is flowing as shown. If $v = 15$ m/s, determine the readings on manometers (a) and (b) in the figures below.

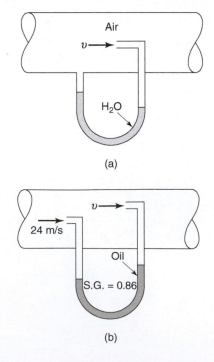

6.31 Referring to the figure below, assume the flow to be frictionless in the siphon. Find the rate of discharge in cubic feet per second, and the pressure head at B if the pipe has a uniform diameter of 1 in. How long will it take for the water level to decrease by 3 ft? The tank diameter is 10 ft.

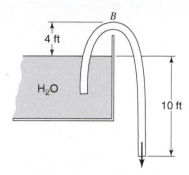

6.32 In Problem 6.31, find the rate of discharge if the frictional head loss in the pipe is 3.2 v^2/g where v is the flow velocity in the pipe.

6.33 Assume that the level of water in the tank remains the same and that there is no friction loss in the pipe, entrance, or nozzle. Determine

a. The volumetric discharge rate from the nozzle

b. The pressure and velocity at points A, B, C, and D

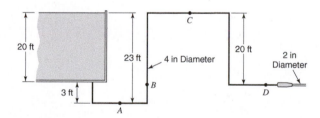

6.34 Water in an open cylindrical tank 15 ft in diameter discharges into the atmosphere through a nozzle 2 in diameter. Neglecting friction and the unsteadiness of the flow, find the time required for the water in the tank to drop from a level of 28 ft above the nozzle to the 4-ft level.

6.35 A fluid of density ρ_1 enters a chamber where the fluid is heated so that the density decreases to ρ_2. The fluid then escapes through a vertical chimney that has a height L. Neglecting friction and treating the flow processes as incompressible except for the heating, determine the velocity, v, in the stack. The fluid velocity entering the heating chamber may be neglected, and the chimney is immersed in fluid of density ρ_1.

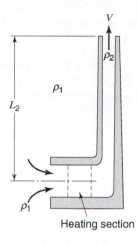

Heating section

6.36 Repeat the previous problem without the assumption that the velocity in the heating section is negligible. The ratio of the flow area of the heating section to the chimney flow area is R.

6.37 Consider a 4-cm pipe that runs between a tank open to the atmosphere and a station open to the atmosphere 10 m below the water surface in the tank. Assuming frictionless flow, what will be the mass flow rate? If a nozzle with a 1-cm diameter is placed at the pipe exit, what will be the mass flow rate? Repeat the problem if a head loss of 3 v^2/g occurs in the pipe where v is the flow velocity in the pipe.

6.38 The tank in the previous problem feeds two lines, a 4-cm pipe that exits 10 m below the water level in the tank and a second line, also 4 cm in diameter runs from the tank to a station 20 m below the water level in the tank. The exits of both lines are open to the atmosphere. Assuming frictionless flow, what is the mass flow rate in each line?

6.39 A client has asked you to find the pressure change in a SS pumping station. The outlet from the pump is 20 ft above the inlet. A Newtonian fluid is being pumped at steady state. At the inlet to the pump where the diameter is 6 in, the temperature of the fluid is 80 F, the viscosity is 1.80×10^{-3} lb_m/ft s, the density is 50 lb_m/ft^3, the heat capacity is 0.580 Btu/lb_m F, and the kinematic viscosity is 3.60×10^{-5} ft^2/s. At the outlet to the pump, where the diameter is 4 in, the temperature of the fluid is 100 F, the viscosity is 1.30×10^{-3} lb_m/ft sec, the density is 49.6 lb_m/ft^3, the heat capacity is 0.610 Btu/lb_m F, and the kinematic viscosity is 2.62×10^{-5} ft^2/s. The flow rate through the system is constant at 20 ft^3/sec. The pump provides work to the fluid at 3.85×10^8 lb_m ft^2/s^3, and the heat transferred is 2.32×10^6 Btu/h. You may neglect viscous work in your analysis. Under these circumstances, calculate the pressure change between the inlet and the outlet of the pumping station.

6.40 Butyl alcohol, a Newtonian fluid, is being pumped at steady state with the density of 50 lb_m/ft^3, viscosity of 1.80×10^{-3} lb_m/ft s, heat capacity of 0.58 Btu/lb_m F, and kinematic viscosity of 3.60×10^{-5} ft^2/s. The inlet to the pump is a pipe with a diameter of 6 in, and the outlet is a pipe with a diameter of 2 in. The outlet is 10 ft above the inlet. The pump provides work to the fluid at 7.1×10^8 lb_m ft^2/s^3. The flow rate through the system is constant at 20 ft^3/s. During the pumping process, the fluid undergoes a 20 F increase in temperature. Under these circumstances, please calculate the pressure change between the inlet and the outlet of the pumping station. You may neglect any heat transfer to the control volume from the surroundings and viscous work, as these are very small.

7.1 Sketch the deformation of a fluid element for the following cases:

a. $\partial v_x/\partial y$ is much larger than $\partial v_y/\partial x$

b. $\partial v_y/\partial x$ is much larger than $\partial v_x/\partial y$

7.2 For a two-dimensional, incompressible flow with velocity $v_x = v_x(y)$, sketch a three-dimensional fluid element and illustrate the magnitude, direction, and surface of action of each stress component.

7.3 Show that the axial strain rate in a one-dimensional flow, $v_x = v_x(x)$, is given by $\partial v_x/\partial x$. What is the rate of volume change? Generalize for a three-dimensional element, and determine the rate of volume change.

7.4 Using a cylindrical element, show that Stokes's viscosity relation yields the following shear-stress components:

$$\tau_{r\theta} = \tau_{\theta r} = \mu\left[r\frac{\partial}{\partial r}\left(\frac{v_\theta}{r}\right) + \frac{1}{r}\frac{\partial v_r}{\partial\theta}\right]$$

$$\tau_{\theta z} = \tau_{z\theta} = \mu\left[\frac{\partial v_\theta}{\partial z} + \frac{1}{r}\frac{\partial v_z}{\partial\theta}\right]$$

$$\tau_{zr} = \tau_{rz} = \mu\left[\frac{\partial v_z}{\partial r} + \frac{\partial v_r}{\partial z}\right]$$

7.5 Estimate the viscosity of nitrogen at 175 K using equation (7-10).

SS **7.6** Calculate the viscosity of oxygen at 350 K and compare with the value given in Appendix I.

7.7 What is the percentage change in the viscosity of water when the water temperature rises from 60 to 120 F?

7.8 At what temperature is the kinematic viscosity of glycerin the same as the kinematic viscosity of helium?

7.9 According to the Hagen–Poiseuille laminar flow model, the volumetric flow rate is inversely proportional to the viscosity. What percentage change in volumetric flow rate occurs in a laminar flow as the water temperature changes from near freezing to 140 F?

SS **7.10** Repeat the preceding problem for air.

7.11 An automobile crankshaft is 3.175 cm in diameter. A bearing on the shaft is 3.183 cm in diameter and 2.8 cm long. The bearing is lubricated with SAE 30 oil at a temperature of 365 K. Assuming that the shaft is centrally located in the bearing, determine how much heat must be removed to maintain the bearing at constant temperature. The shaft is rotating at 1700 rpm, and the viscosity of the oil is 0.01 Pa · s.

7.12 If the speed of the shaft is doubled in Problem 7.11, what will be the percentage increase in the heat transferred from the bearing? Assume that the bearing remains at constant temperature.

7.13 Two ships are traveling parallel to each other and are connected by flexible hoses. Fluid is transferred from one ship to the other for processing and then returned. If the fluid is flowing at 100 kg/s, and at a given instant the first ship is making 4 m/s whereas the second ship is making 3.1 m/s, what is the net force on ship one when the above velocities exist?

7.14 An auto lift consists of 36.02-cm-diameter ram that slides in a 36.04-cm-diameter cylinder. The annular region is filled with oil having a kinematic viscosity of 0.00037 m²/s and a specific gravity of 0.85. If the rate of travel of the ram is 0.15 m/s, estimate the frictional resistance when 3.14 m of the ram is engaged in the cylinder.

7.15 If the ram and auto rack in the previous problem together have a mass of 680 kg, estimate the maximum sinking speed of the ram and rack when gravity and viscous friction are the only forces acting. Assume 2.44 m of the ram is engaged.

7.16 The conical pivot shown in the figure has angular velocity ω and rests on an oil film of uniform thickness h. Determine the frictional moment as a function of the angle α, the viscosity, the angular velocity, the gap distance, and the shaft diameter.

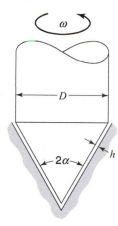

7.17 For water flowing in a 0.1-in-diameter tube, the velocity distribution is parabolic Example 2, Chapter 4. If the average velocity is 2 fps, determine the magnitude of the shear stress at the tube wall.

7.18 What pressure drop per foot of tube is caused by the shear stress in Problem 7.17?

7.19 The rate of shear work per unit volume is given by the product τv. For a parabolic velocity profile in a circular tube Example 2, Chapter 4, determine the distance from the wall at which the shear work is maximum.

7.20 A Newtonian oil with a density of 60 lb$_m$/ft³, viscosity **SS** of 0.206×10^{-3} lb$_m$/ft s and kinematic viscosity of 0.342×10^{-5} ft²/s undergoes steady shear between a horizontal fixed lower plate and a moving horizontal upper plate. The upper plate is moving with a velocity of 3 ft/s. The distance between the plates is 0.03 in, and the area of the upper plate in contact with the fluid is 0.1 ft². Assume incompressible, isothermal, inviscid, frictionless flow.

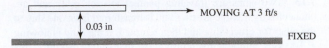

a. What is shear stress exerted on the fluid under these conditions?

b. What is the force of the upper plate on the fluid?

CHAPTER 8 PROBLEMS

SS **8.1** Express equation (8-9) in terms of the flow rate and the pipe diameter. If the pipe diameter is doubled at constant pressure drop, what percentage change will occur in the flow rate?

8.2 A 40-km-long pipeline delivers petroleum at a rate of 4000 barrels per day. The resulting pressure drop is 3.45×10^6 Pa. If a parallel line of the same size is laid along the last 18 km of the line, what will be the new capacity of this network? Flow in both cases is laminar and the pressure drop remains 3.45×10^6 Pa.

8.3 A 0.635 cm hydraulic line suddenly ruptures 8 m from a reservoir with a gage pressure of 207 kPa. Compare the laminar and inviscid flow rates from the ruptured line in cubic meters per second.

8.4 A common type of viscosimeter for liquids consists of a relatively large reservoir with a very slender outlet tube, the rate of outflow being determined by timing the fall in the surface level. If oil of constant density flows out of the viscosimeter shown at the rate of 0.273 cm³/s, what is the kinematic viscosity of the fluid? The tube diameter is 0.18 cm.

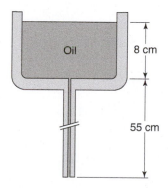

8.5 Derive the expressions for the velocity distribution and for the pressure drop for a Newtonian fluid in fully developed laminar flow in the annular space between two horizontal, concentric pipes. Apply the momentum theorem to an annular fluid shell of thickness Δr and show that the analysis of such a control volume leads to

$$\frac{d}{dr}(r\tau) = r\frac{\Delta P}{L}$$

The desired expressions may then be obtained by the substitution of Newton's viscosity law and two integrations.

8.6 A thin rod of diameter d is pulled at constant velocity through a pipe of diameter D. If the wire is at the center of the pipe, find the drag per unit length of wire. The fluid filling the space between the rod and the inner pipe wall has density ρ and viscosity μ.

8.7 The viscosity of heavy liquids, such as oils, is frequently measured with a device that consists of a rotating cylinder inside a large cylinder. The annular region between these cylinders is filled with liquid and the torque required to rotate the inner cylinder at constant speed is computed, a linear velocity profile being assumed. For what ratio of cylinder diameters is the assumption of a linear profile accurate within 1% of the true profile?

8.8 Two immiscible fluids of different density and viscosity are flowing between two parallel plates. Express the boundary conditions at the interface between the two fluids.

8.9 Determine the velocity profile for fluid flowing between **SS** two parallel plates separated by a distance $2h$. The pressure drop is constant.

8.10 Fluid flows between two parallel plates, a distance h apart. The upper plate moves at velocity, v_0; the lower plate is stationary. For what value of pressure gradient will the shear stress at the lower wall be zero?

8.11 Derive the equation of motion for a one-dimensional, inviscid, unsteady compressible flow in a pipe of constant cross-sectional area neglect gravity.

8.12 A continuous belt passes upward through a chemical bath at velocity v_0 and picks up a film of liquid of thickness h, density ρ, and viscosity μ. Gravity tends to make the liquid drain down, but the movement of the belt keeps the fluid from running off completely. Assume that the flow is a well-developed laminar flow with zero pressure gradient, and that the atmosphere produces no shear at the outer surface of the film.

a. State clearly the boundary conditions at $y=0$ and $y=h$ to be satisfied by the velocity.

b. Calculate the velocity profile.

c. Determine the rate at which fluid is being dragged up with the belt in terms of μ, ρ, h, v_0.

8.13 The device in the schematic diagram on the next page is a viscosity pump. It consists of a rotating drum inside of a stationary case. The case and the drum are concentric. Fluid enters at A, flows through the annulus between the case and the drum, and leaves at B. The pressure at B is higher than that at A, the difference being ΔP. The length of the annulus is L. The width of the annulus h is very small compared to the diameter of the drum, so that the flow in the annulus is equivalent to the flow between two flat plates.

Assume the flow to be laminar. Find the pressure rise and efficiency as a function of the flow rate per unit depth.

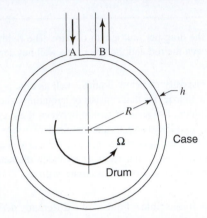

8.14 Oil is supplied at the center of two long plates. The volumetric flow rate per unit length is Q and the plates remain a constant distance, b, apart. Determine the vertical force per unit length as a function of the Q, μ, L, and b.

8.15 A viscous film drains uniformly down the side of a vertical rod of radius R. At some distance down the rod, the film approaches a terminal or fully developed flow such that the film thickness, h, is constant and $v_z = f(r)$. Neglecting the shear resistance due to the atmosphere, determine the velocity distribution in the film.

8.16 Determine the maximum film velocity in Problem 8.15.

8.17 Benzene flows steadily and continuously at 100 F through a 3000-ft horizontal pipe with a constant diameter of 4 in. The pressure drop across the pipe under these conditions is 300 lb_f/ft^2. Assuming fully developed, laminar, incompressible flow, calculate the volumetric flow rate and average velocity of the fluid in the pipe.

8.18 A Newtonian fluid in continuous, incompressible laminar flow is moving steadily through a very long 700 m, horizontal pipe. The inside radius is 0.25 m for the entire length, and the pressure drop across the pipe is 1000 Pa. The average velocity of the fluid is 0.5 m/s. What is the viscosity of this fluid? **SS**

8.19 Benzene, which is an incompressible Newtonian fluid, flows steadily and continuously at 150 F through a 3000 ft pipe with a constant diameter of 4 in with a volumetric flow rate of 3.5 ft^3/s. Assuming fully developed laminar flow, and that the no-slip boundary condition applies, calculate the change in pressure across this pipe system. **SS**

9.1 Apply the law of conservation of mass to an element in a polar coordinate system and obtain the continuity equation for a steady, two-dimensional, incompressible flow.

SS **9.2** In Cartesian coordinates, show that

$$v_x \frac{\partial}{\partial x} + v_y \frac{\partial}{\partial y} + v_z \frac{\partial}{\partial z}$$

may be written $(\mathbf{v} \cdot \nabla)$. What is the physical meaning of the term $(\mathbf{v} \cdot \nabla)$?

9.3 In an incompressible flow, the volume of the fluid is constant. Using the continuity equation, $\nabla \cdot \mathbf{v} = 0$, show that the fluid volume change is zero.

SS **9.4** Find $D\mathbf{v}/Dt$ in polar coordinates by taking the derivative of the velocity. (Hint: $\mathbf{v} = v_r (r, \theta, t)\,\mathbf{e}_r + v_\theta (r, \theta, t)\mathbf{e}_\theta$. Remember that the unit vectors have derivatives.)

SS **9.5** For flow at very low speeds and with large viscosity (the so-called creeping flows) such as occur in lubrication, it is possible to delete the inertia terms, $D\mathbf{v}/Dt$ from the Navier–Stokes equation. For flows at high velocity and small viscosity, it is not proper to delete the viscous term $v\nabla^2\mathbf{v}$. Explain this.

SS **9.6** Using the Navier–Stokes equations and the continuity equation, obtain an expression for the velocity profile between two flat, parallel plates.

9.7 Does the velocity distribution in Example 2 satisfy continuity?

9.8 The atmospheric density may be approximated by the relation $\rho = \rho_0 \exp(-y/\beta)$, where $\beta = 22{,}000$ ft. Determine the rate at which the density changes with respect to body falling at v fps. If $v = 20{,}000$ fps at $100{,}000$ ft, evaluate the rate of density change.

9.9 In a velocity field where $\mathbf{v} = 400[(y/L)^2\mathbf{e}_x + (x/L)^2\mathbf{e}_y]$ fps, determine the pressure gradient at the point $(L, 2L)$. The y axis is vertical, the density is $64.4\,\text{lb}_m/\text{ft}^3$ and the flow may be considered inviscid.

9.10 Write equations (9-17) in component form for Cartesian coordinates.

9.11 Derive equation (2-3) from equation (9-27).

9.12 In polar coordinates, the continuity equation is

$$\frac{1}{r}\frac{\partial}{\partial r}(rv_r) + \frac{1}{r}\frac{\partial v_\theta}{\partial \theta} = 0$$

Show that

a. if $v_\theta = 0$, then $v_r = F(\theta)/r$

b. if $v_r = 0$, then $v_\theta = f(r)$

9.13 Using the laws for the addition of vectors and equation (9-19), show that in the absence of gravity,

a. The fluid acceleration, pressure force, and viscous force all lie in the same planc

b. In the absence of viscous forces the fluid accelerates in the direction of decreasing pressure

c. A static fluid will always start to move in the direction of decreasing pressure

9.14 Obtain the equations for a one-dimensional steady, viscous, compressible flow in the x-direction from the Navier–Stokes equations. (These equations, together with an equation of state and the energy equation, may be solved for the case of weak shock waves.)

9.15 Obtain the equations for one-dimensional inviscid, unsteady, compressible flow.

9.16 Using the Navier–Stokes equations as given in Appendix E, work Problems 8.17 and 8.18.

9.17 Using the Navier–Stokes equations, find the differential equation for a radial flow in which $v_z = v_\theta = 0$, and $v_r = f(r)$. Using continuity, show that the solution to the equation does not involve viscosity.

9.18 Using the Navier–Stokes equations in Appendix E, solve Problem 8.13.

9.19 For the flow described in Problem 8.13, obtain the differential equation of motion if $v_\theta = f(r, t)$.

9.20 Determine the velocity profile in a fluid situated between two coaxial rotating cylinders. Let the inner cylinder have radius R_1 and angular velocity Ω_1; let the outer cylinder have radius R_2 and angular velocity Ω_2.

9.21 Beginning with the appropriate form of the Navier– **SS** Stokes equations, develop an equation in the appropriate coordinate system to describe the velocity of a fluid that is flowing in the annular space as shown in the figure. The fluid is Newtonian, and is flowing in steady, incompressible, fully developed, laminar flow through an infinitely long vertical round pipe annulus of inner radius R_I and outer radius R_O. The inner cylinder (shown in the figure as a gray solid) is solid, and the fluid flows between the inner and outer walls as shown in the figure. The center cylinder moves downward in the same direction as the fluid with a velocity v_0. The outside wall of the annulus is stationary. In developing your equation, please state the reason for eliminating any terms in the original equation.

Fluid flow around center section

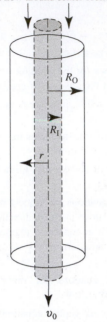

9.22 Two immiscible fluids are flowing down between two flat, infinitely long flat parallel plates as shown in the figure below. The plate on the left is moving down with a velocity of v_A, and the plate on the right is moving up with a velocity v_B. The dotted line is the interface between the two fluids. The fluids maintain constant widths as they flow downward and L_1 and L_2 are not equal. The flow is incompressible, parallel, fully developed, and laminar. You may ignore surface tension, and assume that the fluids are not open to the atmosphere and that the interface between the fluids is vertical at all times. Derive equations for the velocity profiles of both fluids when the flow is at steady state.

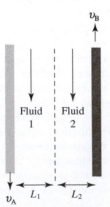

9.23 A wide moving belt passes through a container of a viscous liquid. The belt moves vertically upward with constant velocity v_w, as illustrated in the figure. Because of viscous forces, the belt picks up a thin film of fluid having a thickness h. Use the appropriate form of the Navier–Stokes equations to derive an expression for the velocity of the fluid film as it is dragged up the belt. Assume the flow is laminar, steady, continuous, incompressible, and fully developed.

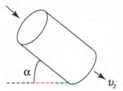

9.24 Consider steady, continuous, incompressible, fully developed laminar flow of a Newtonian fluid in an infinitely long round pipe of diameter D inclined at an angle α. The fluid is not open to the atmosphere and flows down the pipe due to an applied pressure gradient and from gravity. Derive an expression for the shear stress using the appropriate form of the Navier–Stokes equations.

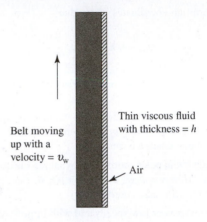

10.1 In polar coordinates, show that

$$\nabla \cdot \mathbf{v} = \frac{1}{r}\left[\frac{\partial(r v_\theta)}{\partial r} - \frac{\partial v_r}{\partial \theta}\right]\mathbf{e}_z$$

10.2 Determine the fluid rotation at a point in polar coordinates, using the method illustrated in Figure 10.1.

SS **10.3** Find the stream function for a flow with a uniform free-stream velocity v_∞. The free-stream velocity intersects the x-axis at an angle α.

10.4 In polar coordinates, the continuity equation for steady, incompressible flow becomes

$$\frac{1}{r}\frac{\partial}{\partial r}(r v_r) + \frac{1}{r}\frac{\partial v_\theta}{\partial \theta} = 0$$

Derive equations (10-10), using this relation.

SS **10.5** The velocity potential for a given two-dimensional flow field is

$$\phi = \left(\frac{5}{3}\right)x^3 - 5xy^2$$

Show that the continuity equation is satisfied and determine the corresponding stream function.

10.6 Make an analytical model of a tornado using an irrotational vortex (with velocity inversely proportional to distance from the center) outside a central core (with velocity directly proportional to distance). Assume that the core diameter is 200 ft and the static pressure at the center of the core is 38 psf below ambient pressure. Find

a. The maximum wind velocity

b. The time it would take a tornado moving at 60 mph to lower the static pressure from -10 to -38 psfg

c. The variation in stagnation pressure across the tornado; Euler's equation may be used to relate the pressure gradient in the core to the fluid acceleration

10.7 For the flow about a cylinder, find the velocity variation along the streamline leading to the stagnation point. What is the velocity derivative $\partial v_r/\partial r$ at the stagnation point?

10.8 In Problem 10.7, explain how one could obtain $\partial v_\theta/\partial\theta$ at the stagnation point, using only r and $\partial v_r/\partial r$.

10.9 At what point on the surface of the circular cylinder in a potential flow does the pressure equal the free-stream pressure?

10.10 For the velocity potentials given below, find the stream function and sketch the streamlines:

a. $\phi = v_\infty L\left[\left(\dfrac{x}{L}\right)^3 - \dfrac{3xy^2}{L^3}\right]$

b. $\phi = v_\infty \dfrac{xy}{L}$

c. $\phi = \dfrac{v_\infty L}{2}\ln(x^2 + y^2)$

10.11 The stream function for an incompressible, two-dimensional flow field is

$$\psi = 2r^3\sin 3\theta$$

For this flow field, plot several streamlines for $0 \le \theta \le \pi/3$.

10.12 For the case of a source at the origin with a uniform free-stream plot the streamline $\psi = 0$.

10.13 In Problem 10.12, how far upstream does the flow from the source reach?

10.14 Determine the pressure gradient at the stagnation point of Problem 10.10(a).

10.15 Calculate the total lift force on the Arctic hut shown below as a function of the location of the opening. The lift force results from the difference between the inside pressure and the outside pressure. Assume potential flow, and that the hut is in the shape of a half-cylinder.

10.16 Consider three equally spaced sources of strength m placed at $(x, y) = (-a, 0)$, $(0, 0)$, and $(a, 0)$. Sketch the resulting streamline pattern. Are there any stagnation points?

10.17 Sketch the streamlines and potential lines of the flow due to a line source at $(a, 0)$ plus an equivalent sink at $(-a, 0)$. **SS**

10.18 The stream function for an incompressible, two-dimensional flow field is

$$\psi = 3x^2 y + y$$

For this flow field, sketch several streamlines.

10.19 A line vortex of strength K at $(x, y) = (0, a)$ is combined with opposite strength vortex at $(0, -a)$. Plot the streamline pattern and find the velocity that each vortex induces on the other vortex.

10.20 A source of strength 1.5 m²/s at the origin is combined with a uniform stream moving at 9 m/s in the x-direction. For the half-body that results, find

a. The stagnation point

b. The body height as it crosses the y-axis

c. The body height at large x

d. The maximum surface velocity and its position (x, y)

10.21 When a doublet is added to a uniform stream so that the source part of the doublet faces the stream, a cylinder flow results. Plot the streamlines when the doublet is reversed so that the sink faces the stream.

10.22 A 2-m-diameter horizontal cylinder is formed by bolting two semicylindrical channels together on the inside. There are 12 bolts per meter of width holding the top and bottom together. The inside pressure is 60 kPa (gage). Using potential theory for the outside pressure, compute the tension force in each bolt if the free-stream fluid is sea-level air and the free-stream wind speed is 25 m/s.

10.23 For the stream function given by

$$\psi = 6x^2 - 6y^2$$

determine whether this flow is rotational or irrotational.

10.24 In Example 3 we began finding the equation for the stream function by integrating equation (3). Repeat this example, but instead begin by integrating equation (4) and show that no matter which equation you begin with, the results are identical.

SS **11.1** The power output of a hydraulic turbine depends on the diameter D of the turbine, the density ρ of water, the height H of water surface above the turbine, the gravitational acceleration g, the angular velocity ω of the turbine wheel, the discharge Q of water through the turbine, and the efficiency η of the turbine. By dimensional analysis, generate a set of appropriate dimensionless groups.

11.2 Through a series of tests on pipe flow, H, Darcy derived an equation for the friction loss in pipe flow as

$$h_L = f \frac{L}{D} \frac{v^2}{2g}$$

in which f is a dimensionless coefficient that depends on (a) the average velocity v of the pipe flow; (b) the pipe diameter D; (c) the fluid density ρ; (d) the fluid viscosity μ; and (e) the average pipe wall uneveness e (length). Using the Buckingham π theorem, find a dimensionless function for the coefficient f.

SS **11.3** The pressure rise across a pump P (this term is proportional to the head developed by the pump) may be considered to be affected by the fluid density ρ, the angular velocity ω, the impeller diameter D, the volumetric rate of flow Q, and the fluid viscosity μ. Find the pertinent dimensionless groups, choosing them so that P, Q, and μ each appear in one group only. Find similar expressions, replacing the pressure rise first by the power input to the pump, then by the efficiency of the pump.

11.4 The rate at which metallic ions are electroplated from a dilute electrolytic solution onto a rotating disk electrode is usually governed by the mass diffusion rate of ions to the disk. This process is believed to be controlled by the following variables:

	Dimension
k = mass-transfer coefficient	L/t
D = diffusion coefficient	L^2/t
d = disk diameter	L
a = angular velocity	$1/t$
ρ = density	M/L^3
μ = viscosity	M/Lt

Obtain the set of dimensionless groups for these variables where k, μ, and D are kept in separate groups. How would you accumulate and present the experimental data for this system?

11.5 The performance of a journal bearing around a rotating shaft is a function of the following variables: Q, the rate of flow lubricating oil to the bearing in volume per unit time; D, the bearing diameter; N, the shaft speed in revolutions per minute; μ, the lubricant viscosity; ρ, the lubricant density; and σ, the surface tension of the lubricating oil. Suggest appropriate parameters to be used in correlating experimental data for such a system.

11.6 The mass M of drops formed by liquid discharging by gravity from a vertical tube is a function of the tube diameter D, liquid density, surface tension, and the acceleration of gravity. Determine the independent dimensionless groups that would allow the surface-tension effect to be analyzed. Neglect any effects of viscosity.

11.7 The functional frequency n of a stretched string is a **SS** function of the string length L, its diameter D, the mass density ρ, and the applied tensile force T. Suggest a set of dimensionless parameters relating these variables.

11.8 The power P required to run a compressor varies with **SS** compressor diameter D, angular velocity ω, volume flow rate Q, fluid density ρ, and fluid viscosity μ. Develop a relation between these variables by dimensional analysis, where fluid viscosity and angular velocity appear in only one dimensionless parameter.

11.9 A large amount of energy E is suddenly released in the air as in a point of explosion. Experimental evidence suggests that the radius r of the high-pressure blast wave depends on time t as well as the energy E and ρ the density of the ambient air.

a. Using the Buckingham method, find the equation for r as a function of t, ρ, and E.

b. Show that the speed of the wave front decreases as r increases.

11.10 The size d of droplets produced by a liquid spray nozzle is thought to depend upon the nozzle diameter D, jet velocity V, and the properties of the liquid ρ, μ, and σ. Rewrite this relation in dimensionless form. Take D, ρ, and V as repeating variables.

11.11 A car is traveling along a road at 22.2 m/s. Calculate the Reynolds number

a. based on the length of the car.

b. based on the diameter of the radio antenna.

The car length is 5.8 m and the antenna diameter is 6.4 mm.

11.12 In natural-convection problems, the variation of density due to the temperature difference ΔT creates an important buoyancy term in the momentum equation. If a warm gas at T_H moves through a gas at temperature T_0 and if the density change is only due to temperature changes, the equation of motion becomes

$$\rho \frac{Dv}{Dt} = -\nabla P + \mu \nabla^2 v + \rho g \left(\frac{T_H}{T_0} - 1 \right)$$

Show that the ratio of gravity (buoyancy) to inertial forces acting on a fluid element is

$$\frac{Lg}{V_0^2} \left(\frac{T_H}{T_0} - 1 \right)$$

where L and V_0 are reference lengths and velocity, respectively.

11.13 A 1/6-scale model of a torpedo is tested in a water tunnel to determine drag characteristics. What model velocity corresponds to a torpedo velocity of 20 knots? If the model resistance is 10 lb, what is the prototype resistance?

11.14 During the development of a 300-ft ship, it is desired to test a 10% scale model in a towing tank to determine the drag characteristics of the hull. Determine how the model is to be tested if the Froude number is to be duplicated.

11.15 A 25% scale model of an undersea vehicle that has a maximum speed of 16 m/s is to be tested in a wind tunnel with a pressure of 6 atm to determine the drag characteristics of the full-scale vehicle. The model is 3 m long. Find the air speed required to test the model and find the ratio of the model drag to the full-scale drag.

11.16 An estimate is needed on the lift provided by a hydrofoil wing section when it moves through water at 60 mph. Test data are available for this purpose from experiments in a pressurized wind tunnel with an airfoil section model geometrically similar to but twice the size of the hydrofoil. If the lift F_1 is a function of the density ρ of the fluid, the velocity v of the flow, the angle of attack θ, the chord length D, and the viscosity μ, what velocity of flow in the wind tunnel would correspond to the hydrofoil velocity for which the estimate is desired? Assume the same angle of attack in both cases, that the density of the air in the pressurized tunnel is 5.0×10^{-3} slugs/ft³, that its kinematic viscosity is 8.0×10^{-5} ft²/s, and that the kinematic viscosity of the water is approximately 1.0×10^{-5} ft²/s. Take the density of water to be 1.94 slugs/ft³.

11.17 A model of a harbor is made on the length ratio of 360:1. Storm waves of 2 m amplitude and 8 m/s velocity occur on the breakwater of the prototype harbor. Significant variables are the length scale, velocity, and g, the acceleration of gravity. The scaling of time can be made with the aid of the length scale and velocity scaling factors.

a. Neglecting friction, what should be the size and speed of the waves in the model?

b. If the time between tides in the prototype is 12 h, what should be the tidal period in the model?

11.18 A 40% scale model of an airplane is to be tested in a flow regime where unsteady flow effects are important. If the full-scale vehicle experiences the unsteady effects at a Mach number of 1 at an altitude of 40,000 ft, what pressure must the model be tested at to produce an equal Reynolds number? The model is to be tested in air at 70 F. What will the timescale of the flow about the model be relative to the full-scale vehicle?

11.19 A model ship propeller is to be tested in water at the same temperature that would be encountered by a full-scale propeller. Over the speed range considered, it is assumed that there is no dependence on the Reynolds or Euler numbers, but only on the Froude number (based on forward velocity V and propeller diameter d). In addition, it is thought that the ratio of forward to rotational speed of the propeller must be constant (the ratio V/Nd, where N is propeller rpm).

a. With a model 041 m in diameter, a forward speed of 2.58 m/s and a rotational speed of 450 rpm is recorded. What are the forward and rotational speeds corresponding to a 2.45-m diameter prototype?

b. A torque of 20 N·m is required to turn the model, and the model thrust is measured to be 245 N. What are the torque and thrust for the prototype?

11.20 A coating operation is creating materials for the electronics industry. The coating requires a specific volumetric flow rate Q, solution density ρ, solution viscosity μ, substrate coating velocity v, solution surface tension σ, and the length of the coating channel L. Determine the dimensionless groups formed from the variables involved using the Buckingham method. Choose the groups so that Q, σ, and μ appear in one group only.

SS **12.1** If Reynolds's experiment was performed with a 38-mm-ID pipe, what flow velocity would occur at transition?

SS **12.2** Modern subsonic aircraft have been refined to such an extent that 75% of the parasite drag (portion of total aircraft drag not directly associated with producing lift) can be attributed to friction along the external surfaces. For atypical subsonic jet, the parasite drag coefficient based on wing area is 0.011. Determine the friction drag on such an aircraft

a. at 500 mph at 35,000 ft

b. at 200 mph at sea level

The wing area is 2400 ft^2.

SS **12.3** Consider the flow of air at 30 m/s along a flat plate. At what distance from the leading edge will transition occur?

12.4 Find a velocity profile for the laminar boundary layer of the form

$$\frac{v_x}{v_x\delta} = c_1 + c_2y + c_3y^2 + c_4y^3$$

when the pressure gradient is not zero.

12.5 Evaluate and compare with the exact solution δ, C_{fx}, and C_{fL} for the laminar boundary layer over a flat plate, using the velocity profile

$$v_x = \alpha \sin by.$$

12.6 There is fluid evaporating from a surface at which $v_x|_{y=0} = 0$, but $v_x|_{y=0} \neq 0$. Derive the von Kármán momentum relation.

12.7 The drag coefficient for a smooth sphere is shown below. Determine the speed at the critical Reynolds number for a 42-mm-diameter sphere in air.

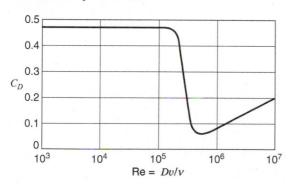

12.8 Plot a curve of drag vs. velocity for a 1.65-in.-diameter sphere in air between velocities of 50 fps and 400 fps.

SS **12.9** For what wind velocities will a 12.7-mm-diameter cable be in the unsteady wake region of Figure 12.2?

12.10 Estimate the drag force on a 3-ft radio antenna with an **SS** average diameter of 0.2 in. at a speed of 60 mph.

12.11 A 2007 Toyota Prius has a drag coefficient of 0.26 at **SS** road speeds, using a reference area of 2.33 m^2. Determine the horsepower required to overcome drag at a velocity of 30 m/s. Compare this figure with the case of head and tail winds of 6 m/s.

12.12 The lift coefficient is defined as $C_L = $ (lift force)$/\frac{1}{2}\rho v_x^2 A_r$. If the lift coefficient for the auto in the previous problem is 0.21, determine the lift force at a road speed of 100 mph.

12.13 The auto in Problem 12.11 has shown a sensitivity to yaw angle. At a yaw angle of 20°, the lift coefficient increased to 1.0. What is the lift force at 100 mph for this case?

12.14 What diameter circular plate would have the same drag as the auto of Problem 12.11?

12.15 Estimate the normal force on a circular sign 8 ft in diameter during a hurricane wind (120 mph).

12.16 A 1998 Lexus LS400 has a drag coefficient of 0.28 and a reference area of 2.4 m^2. Determine the horsepower required to overcome drag when driving at 70 mph at sea level.

a. On a hot summer day $T \cong 100°F$

b. On a cold winter day $T \cong 0°F$

12.17 A baseball has a circumference of 9¼ in. and a weight of 5¼ ounces. At 95 mph determine

a. the Reynolds number

b. the drag force

c. the type of flow (see the illustration for Problem 12.7)

12.18 Golfball "dimples" cause the drag drop (see Figure 12.4 and the illustration for Problem 12.7) to occur at a lower Reynolds number. The table below gives the drag coefficient for a rough sphere as a function of the Reynolds number. Plot the drag for a 1.65-in.-diameter sphere as a function of velocity. Show several comparison points for a smooth sphere.

Re · 10^{-4}	7.5	10	15	20	25
C_D	0.48	0.38	0.22	0.12	0.10

12.19 The lift coefficient on a rotating sphere is very approximately given by

$$C_L \cong 0.24\left(\frac{R\Omega}{V}\right) - 0.05 \quad \text{over the range of } 1.5 > \left(\frac{R\Omega}{V}\right) > 0.2$$

Here R is the sphere radius and Ω is rotation rate of the sphere. For the baseball in Problem 12.17, determine the rotation rate for a baseball thrown at 110 mph to have the lift equal the weight. How many rotations would such a ball make in 60 ft 6 in.?

12.20 If the vertical velocity at the wall is not zero such as would be the case with suction or blowing, what modifications occur to equation (12-33)?

12.21 If the turbulence intensity is 10%, what fraction of the total kinetic energy of the flow is due to the turbulence?

12.22 In a house, water flows through a copper tube with a 0.75-in. ID, at a flow rate of 2 gpm. Determine the Reynolds number for

a. hot water ($T \cong 120°F$)

b. cold water ($T \cong 45°F$)

12.23 Plot the boundary-layer thickness along a flat plate for the flow of air at 30 m/s assuming

a. laminar flow

b. turbulent flow

Indicate the probable transition point.

12.24 For the fully developed flow of water in a smooth 0.15-m pipe at a rate of $0.006 \, \text{m}^3/\text{s}$, determine the thickness of

a. the laminar sublayer

b. the buffer layer

c. the turbulent core

12.25 Using Blasius's correlation for shear stress (equation (12-68)), develop an expression for the local skin-friction coefficient in pipes. In pipes, the average velocity is used for the friction coefficient and the Reynolds number. Use the one-seventh-power law.

12.26 For a thin a plate 6 in. wide and 3 ft long, estimate the friction force in air at a velocity of 40 fps, assuming

a. turbulent flow

b. laminar flow

The flow is parallel to the 6-in. dimension.

12.27 Using a sine profile the laminar flow and a one-seventh-power law for turbulent flow, make a dimensionless plot of the momentum and kinetic energy profiles in the boundary layer a Reynolds number of 10^5.

12.28 Estimate the friction drag on a wing by considering the following idealization. Consider the wing to be a rectangular flat plate, 7 ft by 40 ft, with a smooth surface. The wing is flying at 140 mph at 5000 ft. Determine the drag, assuming

a. a laminar boundary layer

b. a turbulent boundary layer

12.29 Compare the boundary-layer thicknesses and local skin-friction coefficients of a laminar boundary layer and a turbulent boundary layer on a smooth flat plate at a Reynolds number of 10^6. Assume both boundary layers to originate at the leading edge of the flat plate.

12.30 Use the one-seventh-power-law profile and compute the drag force and boundary layer thickness on a plate 20 ft long and 10 ft wide (for one side) if it is immersed in a flow of water of 20 ft/s velocity. Assume turbulent flow to exist over the entire length of the plate. What would the drag be if laminar flow could be maintained over the entire surface?

12.31 The turbulent shear stress in a two-dimensional flow is given by

$$(\tau_{yx})_{\text{turb}} = \rho \epsilon_M \frac{\partial \bar{v}_x}{\partial y} = -\rho \overline{v_x' v_y'}$$

Expanding v_x' and v_y' in a Taylor series in x and y near the wall and with the aid of the continuity equation

$$\frac{\partial v_x'}{\partial x} + \frac{\partial v_y'}{\partial y} = 0$$

show that, near the wall, $\epsilon_M \sim y^3 +$ higher-order terms in y. How does this compare with the mixing-length theory?

12.32 Evaluate the velocity derivative, $\partial \bar{v}_x / \partial y$, for the power-law velocity profile at $y=0$ and $y=R$.

12.33 Using the Blasius shear-stress relation (12–68) and the power-law velocity profile, determine the boundary-layer thickness on a flat plate as a function of the Reynolds number and the exponent n.

SS **13.1**   An oil with kinematic viscosity of 0.08×10^{-3} ft^2/s and a density of 57 lb$_m$/ft^3 flows through a horizontal tube 0.24 in in diameter at the rate of 10 gal/h. Determine the pressure drop in 50 ft of tube.

SS **13.2**   A lubricating line has an inside diameter of 0.1 in and is 30 in long. If the pressure drop is 15 psi, determine the flow rate of the oil. Use the properties given in Problem 13.1.

SS **13.3**   The pressure drop in a section of a pipe is determined from tests with water. A pressure drop of 13 psi is obtained at a flow rate of 28.3 lb$_m$/s. If the flow is fully turbulent, what will be the pressure drop when liquid oxygen ($\rho = 70$ lb$_m$/ft^3) flows through the pipe at the rate of 35 lb$_m$/s?

13.4   A 280-km-long pipeline connects two pumping stations. If 0.56 m^3/s are to be pumped through a 0.62-m-diameter line, the discharge station is 250 m lower in elevation than the upstream station, and the discharge pressure is to be maintained at 300,000 Pa, determine the power required to pump the oil. The oil has a kinematic viscosity of 4.5×10^{-6} m^2/s and a density of 810 kg/m^3. The pipe is constructed of commercial steel. The inlet pressure may be taken as atmospheric.

13.5   In the previous problem, a 10-km-long section of the pipeline is replaced during a repair process with a pipe with internal diameter of 0.42 m. Determine the total pumping power required when using the modified pipeline. The total pipeline length remains 280 km.

13.6   Oil having a kinematic viscosity of 6.7×10^{-6} m^2/s and density of 801 kg/m^3 is pumped through a pipe of 0.71 m diameter at an average velocity of 1.1 m/s. The roughness of the pipe is equivalent to that of a commercial steel pipe. If pumping stations are 320 km apart, find the head loss (in meters of oil) between the pumping stations and the power required.

13.7   The cold-water faucet in a house is fed from a water main through the following simplified piping system:

a.   A 160 ft length of 3/4-in-ID copper pipe leading from the main line to the base of the faucet.

b.   Six 90° standard elbows.

c.   One wide-open angle valve (with no obstruction).

d.   The faucet. Consider the faucet to be made up of two parts: (1) a conventional globe valve and (2) a nozzle having a cross-sectional area of 0.10 in^2.

The pressure in the main line is 60 psig (virtually independent of flow), and the velocity there is negligible. Find the maximum rate of discharge from the faucet. As a first try, assume for the pipe $f_f = 0.007$. Neglect changes in elevation throughout the system.

13.8   Water at the rate of 118 ft^3/min flows through a smooth SS horizontal tube 250 ft long. The pressure drop is 4.55 psi. Determine the tube diameter.

13.9   Calculate the inlet pressure to a pump 3 ft above the level of a sump. The pipe is 6 in in diameter, 6 ft long, and made of commercial steel. The flow rate through the pump is 500 gal/min. Use the (incorrect) assumption that the flow is fully developed.

13.10   The pipe in Problem 6.33 is 35 m long and made of commercial steel. Determine the flow rate.

13.11   The siphon of Problem 6.31 is made of smooth rubber hose and is 23 ft long. Determine the flow rate and the pressure at point B.

13.12   A galvanized rectangular duct of 8 in^2 is 25 ft long and SS carries 600 ft^3/min of standard air. Determine the pressure drop in inches of water.

13.13   A cast-iron pipeline 2 m long is required to carry 3 million gal of water per day. The outlet is 175 ft higher than the inlet. The costs of three sizes of pipe when in place are as follows:

10-in diameter	$11.40 per ft
12-in diameter	$14.70 per ft
14-in diameter	$16.80 per ft

Power costs are estimated at $0.07 per kilowatt hour over the 20-year life of the pipeline. If the line can be bonded with 6.0% annual interest, what is the most economical pipe diameter? The pump efficiency is 80%, and the water inlet temperature is expected to be constant at 42 F.

13.14   Estimate the flow rate of water through 50 ft of garden hose from a 40-psig source for

a.   A 1/2-in-ID hose

b.   A 3/4-in-ID hose

13.15   Two water reservoirs of height $h_1 = 60$ m and $h_2 = 30$ m are connected by a pipe that is 0.35 m in diameter. The exit of the pipe is submerged at distance $h_3 = 8$ m from the reservoir surface.

a.   Determine the flow rate through the pipe if the pipe is 80 m long and the friction factor $f_f = 0.004$. The pipe inlet is set flush with the wall.

b.   If the relative roughness $e/D = 0.004$, determine the friction factor and flow rate.

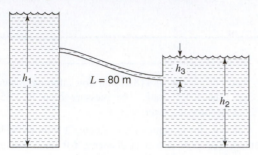

13.16 An 8-km-long, 5-m-diameter headrace tunnel at the Paute River hydroelectric project in Ecuador supplies a power station 668 m below the entrance of the tunnel. If the tunnel surface is concrete, find the pressure at the end of the tunnel if the flow rate is 90 m³/s.

13.17 Determine the flow rate through a 0.2-m gate valve with upstream pressure of 236 kPa when the valve is

a. open

b. 1/4 closed

c. 1/2 closed

d. 3/4 closed

13.18 Water at 20 C flows through a cast-iron pipe at a velocity of 34 m/s. The pipe is 400 m long and has a diameter of 0.18 m. Determine the head loss due to friction.

13.19 Water at 20 C is being drained from an open tank through a cast-iron pipe 0.6 m diameter and 30 m long. The surface of the water in the pipe is at atmospheric pressure and at an elevation of 46.9 m, and the pipe discharges to the atmosphere at an elevation 30 m. Neglecting minor losses due to configuration, bends, valves, and fittings, determine the volumetric flow rate of the water leaving the pipe.

13.20 A 15-cm diameter wrought-iron pipe is to carry water at 20 C. Assuming a level pipe, determine the volumetric flow rate at the discharge if the pressure loss is not permitted to exceed 30.0 kPa per 100 m.

13.21 A level 10-m-long water pipe has a manometer at both the inlet and the outlet. The manometers indicate pressure head of 1.5 and 0.2 m, respectively. The pipe diameter is 0.2 m and the pipe roughness is 0.0004 m. Determine the mass flow rate in the pipe in kg/s.

13.22 Determine the depth of water behind the dam in the figure that will provide a flow rate of 5.675×10^{-4} m³/s through a 20-m-long, 1.30 cm commercial steel pipe.

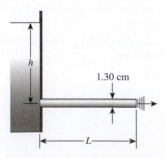

13.23 Water flows at a volumetric flow rate of 0.25 m³/s from reservoir 1 to reservoir 2 through three concrete pipes connected in series. Pipe 1 is 900 m long and has a diameter of 0.16 m. Pipe 2 has a length of 1500 m and a diameter of 0.18 m. Pipe 3 is 800 m long and the diameter is 0.20 m. Neglecting minor losses, determine the difference in surface elevations.

13.24 A system consists of three pipes in series. The total pressure drop is 180 kPa, and the decrease in elevation is 5 m. Data for the three pipes are as follows:

Pipe	Length, m	Diameter, cm	Roughness, mm
1	125	8	0.240
2	150	6	0.120
3	100	4	0.200

Determine the total flow rate of 20 C, water for the system, Neglect minor losses.

13.25 Two concrete pipes are connected in series. The flow rate of water at 20 C through the pipes is 0.18 m³/s, with a total head loss of 18 m for both pipes. Each pipe has a length of 312.5 m and a relative roughness of 0.0035 m. Neglecting minor losses, if one pipe has a diameter of 0.30 m, determine the diameter of the other.

13.26 A 0.2-m-diameter cast-iron pipe and a 67-mm-diameter commercial steel pipe are parallel, and both run from the same pump to a reservoir. The pressure drop is 210 kPa and the lines are 150 m long. Determine the flow rate of water in each line.

13.27 A system consists of three pipes in parallel with a total head loss of 24 m. Data for the three pipes are as follows:

Pipe	Length, m	Diameter, cm	Roughness, mm
1	100	8	0.240
2	150	6	0.120
3	80	4	0.200

For water at 20 C, neglect minor losses and determine the volumetric flow rate in the system.

SS **14.1** A centrifugal pump delivers $0.2 \, \text{m}^3/\text{s}$ of water when operating at 850 rpm. Relevant impeller dimensions are as follows: outside diameter $= 0.45$ m, blade length $= 50$ cm, and blade exit angle $= 24°$. Determine (a) the torque and power required to drive the pump and (b) the maximum pressure increase across the pump.

14.2 A centrifugal pump is used with gasoline ($\rho = 680 \, \text{kg/m}^3$). Relevant dimensions are as follows: $d_1 = 15$ cm, $d_2 = 28$ cm, $L = 9$ cm, $\beta_1 = 25°$, and $\beta_2 = 40°$. The gasoline enters the pump parallel to the pump shaft when the pump operates at 1200 rpm. Determine (a) the flow rate, (b) the power delivered to the gasoline, and (c) the head in meters.

14.3 A centrifugal pump has the following dimensions: $d_2 = 42$ cm, $L = 5$ cm, and $\beta_2 = 33°$. It rotates at 1200 rpm, and the head generated is 52 m of water. Assuming radial entry flow, determine the theoretical values for (a) the flow rate and (b) the power.

14.4 A centrifugal pump has the configuration and dimensions shown below. For water flowing at a rate of $0.0071 \, \text{m}^3/\text{s}$ and an impeller speed of 1020 rpm, determine the power required to drive the pump. The inlet flow is directed radially outward, and the exiting velocity may be assumed to be tangent to the vane at its trailing edge.

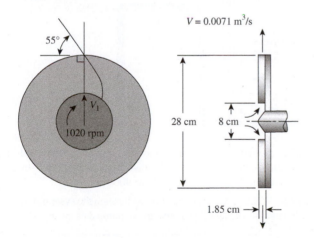

$V = 0.0071 \, \text{m}^3/\text{s}$

SS **14.5** A centrifugal pump is being used to pump water at a flow rate of $0.018 \, \text{m}^3/\text{s}$ and the required power is measured to be 4.5 kW. If the pump efficiency is 63%, determine the head generated by the pump.

14.6 A centrifugal pump having the dimensions shown develops a flow rate of $0.032 \, \text{m}^3/\text{s}$ when pumping gasoline ($\rho = 680 \, \text{kg/m}^3$). The inlet flow may be assumed to be radial. Estimate (a) the theoretical horsepower, (b) the head increase, and (c) the proper blade angle at the impeller inlet.

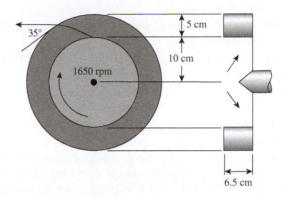

14.7 A centrifugal water pump operates at 1500 rpm. The dimensions follow:

$$r_1 = 12 \, \text{cm} \quad \beta_1 = 32°$$
$$r_2 = 20 \, \text{cm} \quad \beta_2 = 20°$$
$$L = 4.2 \, \text{cm}$$

Determine (a) the design point discharge rate, (b) the water horsepower, and (c) the discharge head.

14.8 The figure below represents performance, in nondimensional form, for a family of centrifugal pumps. For a pump from this family with a characteristic diameter of 0.45 m operating at maximum efficiency and pumping water at 15 C with a rotational speed of 1600 rpm, estimate (a) the head, (b) the discharge rate, (c) the pressure rise, and (d) the brake horsepower.

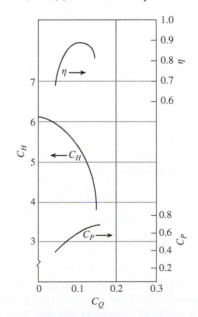

14.9 A pump having the characteristics described in the previous problem is to be built that will deliver water at a rate of $0.2\,m^3/s$ when operating at best efficiency and a rotational speed of 1400 rpm. Estimate (a) the impeller diameter and (b) the maximum pressure rise.

14.10 Rework Problem 14.8 for a pump diameter of 0.40 m operating at 2200 rpm.

14.11 Rework Problem 14.8 for a pump diameter of 0.35 m operating at 2400 rpm.

14.12 Rework Problem 14.9 for a desired flow rate of $0.30\,m^3/s$ at 1800 rpm.

14.13 Rework Problem 14.9 for a desired flow rate of $0.201\,m^3/s$ at 1800 rpm.

14.14 Performance curves for an operating centrifugal pump are shown below in both conventional units and in dimensionless form. The pump is used to pump water at maximum efficiency at a head of 90 m. Determine, at these new conditions, (a) the pump speed required and (b) the rate of discharge.

14.15 The pump having the characteristics shown in Problem 14.14 was used as a model for a prototype that is to be six times larger. If this prototype operates at 400 rpm, what (a) power,

(b) head, and (c) discharge flow rate should be expected at maximum efficiency?

14.16 For the pump having the characteristics shown in Problem 14.14, operating at maximum efficiency with the speed increased to 1000 rpm, what will be (a) the new discharge flow rate and (b) the power required at this new speed?

14.17 The pump having the characteristics shown in Problem **SS** 14.14 is to be operated at 800 rpm. What discharge rate is to be expected if the head developed is 410 m?

14.18 If the pump having the characteristics shown in Problem **SS** 14.14 is tripled in size but halved in rotational speed, what will be the discharge rate and head when operating at maximum efficiency?

14.19 The pump having the characteristics shown in Problem 14.14 is used to pump water from one reservoir to another that is 95 m higher in elevation. The water will flow through a steel pipe that is 0.28 m in diameter and 550 m long. Determine the discharge rate.

14.20 A pump whose operating characteristics are described in Problem 14.14 is to be used in the system depicted below. Determine (a) the discharge rate and (b) power required.

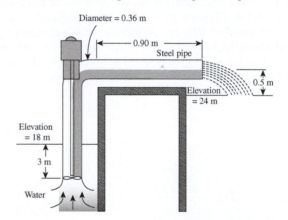

14.21 For the same pump and system operation described in Problem 14.20 determine (a) the discharge rate and (b) power required when the pump operates at 900 rpm.

14.22 Water at 20 C is to be pumped through the system shown. The operating data for this motor-driven pump data are as follows:

Capacity, $m^3/s \times 10^4$	Developed head, m	Efficiency, %
0	36.6	0
10	35.9	19.1
20	34.1	32.9
30	31.2	41.6
40	27.5	42.2
50	23.3	39.7

The inlet pipe to the pump is 0.06 m diameter commercial steel, 8.5 m in length. The discharge line consists of 60 m of 0.06 m

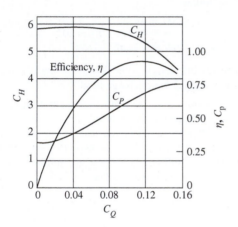

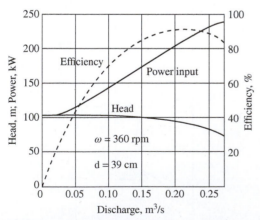

diameter steel pipe. All valves are fully open globe valves. Determine the flow rate through the system.

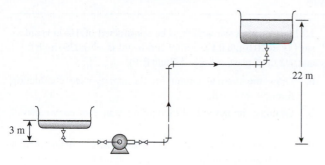

14.23 A 0.25 m pump delivers 20 C water $(P_v = 2.34 \text{ kPa})$ at 0.065 m³/s and 2000 rpm. The pump begins to cavitate when the inlet pressure is 82.7 kPa and the inlet velocity is 6.1 m/s. Determine the corresponding NPSH.

14.24 For the pumping system described in Problem 14.23, how will the maximum elevation above the surface of the reservoir change if the water temperature is 80 C$(P_v = 47.35 \text{ kPa})$?

15.1 An asbestos pad is square in cross section, measuring 5 cm on a side at its small end, increasing linearly to 10 cm on a side at the large end. The pad is 15 cm high. If the small end is held at 600 K and the large end at 300 K, what heat-flow rate will be obtained if the four sides are insulated? Assume one-dimensional heat conduction. The thermal conductivity of asbestos may be taken as 0.173 W/m · K.

15.2 Solve Problem 15.1 for the case of the larger cross section exposed to the higher temperature and the smaller end held at 300 K.

15.3 Solve Problem 15.1 if, in addition to a varying cross-sectional area, the thermal conductivity varies according to $k = k_0(1 + \beta T)$, where $k_0 = 0.138$, $\beta = 1.95 \times 10^{-4}$, $T =$ temperature in Kelvin, and k is in W/m · K. Compare this result to that using a k value evaluated at the arithmetic mean temperature.

15.4 Solve Problem 15.1 if the asbestos pad has a 1.905-cm steel bolt running through its center.

SS 15.5 A sheet of insulating material, with thermal conductivity of 0.22 W/m · K, is 2 cm thick and has a surface area of 2.97 m². If 4 kW of heat are conducted through this sheet and the outer (cooler) surface temperature is measured at 55 C (328 K), what will be the temperature on the inner (hot) surface?

15.6 For the sheet of insulation specified in Problem 15.5, with a heat rate of 4 kW, evaluate the temperature at both surfaces if the cool side is exposed to air at 30 C with a surface coefficient of 28.4 W/m² · K.

SS 15.7 Plate glass, $k = 1.35$ W/m · K, initially at 850 K, is cooled by blowing air past both surfaces with an effective surface coefficient of 5 W/m² · K. It is necessary, in order that the glass not crack, to limit the maximum temperature gradient in the glass to 15 K/cm during the cooling process. At the start of the cooling process, what is the lowest temperature of the cooling air that can be used?

SS 15.8 Solve Problem 15.7 if all specified conditions remain the same but radiant-energy exchange from glass to the surroundings at the air temperature is also considered.

SS 15.9 The heat loss from a boiler is to be held at a maximum of 900 Btu/h ft² of wall area. What thickness of asbestos ($k = 0.10$ Btu/h ft F) is required if the inner and outer surfaces of the insulation are to be 1600 and 500 F, respectively?

15.10 If, in the previous problem, a 3-in-thick layer of kaolin brick ($k = 0.07$ Btu/h ft F) is added to the outside of the asbestos, what heat flux will result if the outside surface of the kaolin is 250 F? What will be the temperature at the interface between the asbestos and kaolin for this condition?

15.11 A composite wall is to be constructed of 1/4-in stainless steel ($k = 10$ Btu/h ft F), 3 in of corkboard ($k = 0.025$ Btu/h ft F), and 1/2 in of plastic ($k = 1.5$ Btu/h ft F).

a. Draw the thermal circuit for the steady-state conduction through this wall.

b. Evaluate the individual thermal resistance of each material layer.

c. Determine the heat flux if the steel surface is maintained at 250 F and the plastic surface held at 80 F.

d. What are the temperatures on each surface of the corkboard under these conditions?

15.12 If, in the previous problem, the convective heat-transfer coefficients at the inner (steel) and outer surfaces are 40 and 5 Btu/h ft F, respectively, determine

a. the heat flux if the gases are at 250 and 70 F, adjacent to the inner and outer surfaces

b. the maximum temperature reached within the plastic

c. which of the individual resistances is controlling

15.13 A 1-in-thick steel plate measuring a diameter of 10 in is heated from below by a hot plate, its upper surface exposed to air at 80 F. The heat-transfer coefficient on the upper surface is 5 Btu/h ft F and k for steel is 25 Btu/h ft F.

a. How much heat must be supplied to the lower surface of the steel if its upper surface remains at 160 F? (Include radiation.)

b. What are the relative amounts of energy dissipated from the upper surface of the steel by convection and radiation?

15.14 If, in Problem 15.13, the plate is made of asbestos, **SS** $k = 0.10$ Btu/h ft F, what will be the temperature of the top of the asbestos if the hot plate is rated at 800 W?

15.15 A 0.20-m-thick brick wall ($k = 1.3$ W/m · K) separates the combustion zone of a furnace from its surroundings at 25 C. For an outside wall surface temperature of 100 C, with a convective heat-transfer coefficient of 18 W/m² · K, what will be the inside wall surface temperature at steady-state conditions?

15.16 Solve for the inside surface temperature of the brick wall described in Problem 15.15, but with the additional consideration of radiation from the outside surface to surroundings at 25 C.

15.17 The solar radiation incident on a steel plate 2 ft square is 400 Btu/h. The plate is 1.4 in thick and lying horizontally on an insulating surface, its upper surface being exposed to air at 90 F. If the convective heat-transfer coefficient between the top surface and the surrounding air is 4 Btu/h ft F, what will be the steady-state temperature of the plate?

15.18 If in Problem 15.17, the lower surface of the plate is exposed to air with a convective heat-transfer coefficient of 3 Btu/h ft F, what steady-state temperature will be reached

a. if radiant emission from the plate is neglected?

b. if radiant emission from the top surface of the plate is accounted for?

15.19 The freezer compartment in a conventional refrigerator can be modeled as a rectangular cavity 0.3 m high and 0.25 m wide with a depth of 0.5 m. Determine the thickness of styrofoam insulation ($k = 0.30$ W/m · K) needed to limit the heat loss to 400 W if the inner and outer surface temperatures are -10 and 33°C, respectively.

15.20 Evaluate the required thickness of styrofoam for the freezer compartment in the previous problem when the inside wall is exposed to air at -10 C through a surface coefficient of 16 W/m² · K and the outer wall is exposed to 33 C air with a surface coefficient of 32 W/m² · K. Determine the surface temperatures for this situation.

15.21 The cross section of a storm window is shown in the sketch. How much heat will be lost through a window measuring 1.83 m by 3.66 m on a cold day when the inside and outside air temperatures are, respectively, 295 and 250 K? Convective coefficients on the inside and outside surfaces of the window are 20 and 15 W/m² · K, respectively. What temperature drop will exist across each of the glass panes? What will be the average temperature of the air between the glass panes?

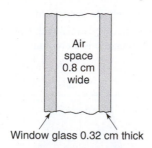

Air
space
0.8 cm
wide

Window glass 0.32 cm thick

15.22 Compare the heat loss through the storm window described in Problem 15.21 with the same conditions existing, except that the window is a single pane of glass 0.32 cm thick.

15.23 The outside walls of a house are constructed of a 4-in layer of brick, 1/2 in of celotex, an air space 3⅜ in thick, and 1/4 in of wood panelling. If the outside surface of the brick is at 30 F and the inner surface of the panelling at 75 F, what is the heat flux if

a. the air space is assumed to transfer heat by conduction only?

b. the equivalent conductance of the air space is 1.8 Btu/h ft² F?

c. the air space is filled with glass wool?

$$k_{brick} = 0.38 \text{ Btu/h ft F}$$
$$k_{celotex} = 0.028 \text{ Btu/h ft F}$$
$$k_{air} = 0.015 \text{ Btu/h ft F}$$
$$k_{wood} = 0.12 \text{ Btu/h ft F}$$
$$k_{wool} = 0.025 \text{ Btu/h ft F}$$

15.24 Solve Problem 15.23 if instead of the surface temperatures being known, the air temperatures outside and inside are 30 and 75 F, respectively, and the convective heat-transfer coefficients are 7 and 2 Btu/h ft² F, respectively.

15.25 Determine the heat-transfer rate per square meter of wall area for the case of a furnace with inside air at 1340 K. The furnace wall is composed of a 0.106-m layer of fireclay brick and a 0.635-cm thickness of mild steel on its outside surface. Heat-transfer coefficients on inside and outside wall surfaces are 5110 and 45 W/m² · K, respectively; outside air is at 295 K. What will be the temperatures at each surface and at the brick-steel interface?

15.26 Given the furnace wall and other conditions as specified in Problem 15.25, what thickness of celotex ($k = 0.065$ W/m · K) must be added to the furnace wall in order that the outside surface temperature of the insulation not exceed 340 K?

15.27 A 4-in-OD pipe is to be used to transport liquid metals and will have an outside surface temperature of 1400 F under operating conditions. Insulation is 6 in thick and has a thermal conductivity expressed as

$$k = 0.08(1 - 0.003\,T)$$

where k is in Btu/h ft F and T is in F, is applied to the outside surface of the pipe.

a. What thickness of insulation would be required for the outside insulation temperature to be no higher than 300°F?

b. What heat-flow rate will occur under these conditions?

CHAPTER 16   PROBLEMS

SS 16.1 The Fourier field equation in cylindrical coordinates is

$$\frac{\partial T}{\partial t} = \alpha \left(\frac{\partial^2 T}{\partial r^2} + \frac{1}{r}\frac{\partial T}{\partial r} + \frac{1}{r^2}\frac{\partial^2 T}{\partial \theta^2} - \frac{\partial^2 T}{\partial z^2} \right).$$

a. What form does this equation reduce to for the case of steady-state, radial heat transfer?

b. Given the boundary conditions

$$T = T_i \quad \text{at} \quad r = r_i$$
$$T = T_o \quad \text{at} \quad r = r_o$$

Generate an expression for the heat flow rate, q_r.

16.2 Perform the same operations as in Problem 16.1 with respect to a spherical system.

16.3 Starting with the Fourier field equation in cylindrical coordinates,

a. Reduce this equation to the applicable form for steady-state heat transfer in the θ-direction.

b. For the conditions depicted in the figure—that is, $T = T_o$ at $\theta = 0$, $T = T_z$ at $\theta = \pi$, the radial surfaces insulated—solve for the temperature profile.

c. Generate an expression for the heat flow rate, q_θ, using the result of part (b).

d. What is the shape factor for this configuration?

16.4 Show that equation (16-10) reduces to the form

$$\nabla \cdot k\nabla T + \dot{q} + \Lambda = \rho c_v \frac{DT}{Dt} + \mathbf{v} \cdot \mu \nabla^2 \mathbf{v}$$

SS 16.5 Solve equation (16-19) for the temperature distribution in a plane wall if the internal heat generation per unit volume varies according to $\dot{q} = \dot{q}_0 e^{-\beta x/L}$. The boundary conditions that apply are $T = T_0$ at $x = 0$ and $T = T_L$ at $x = L$.

16.6 Solve Problem 16.5 for the same conditions, except that the boundary condition at $x = L$ is $dT/dx = 0$.

16.7 Solve Problem 16.5 for the same conditions, except that at $x = L$, $dT/dx = \xi$ (a constant).

16.8 Use the relation $T\,ds = dh - dP/\rho$ to show that the effect of the dissipation function, Φ, is to increase the entropy, S. Is the effect of heat transfer the same as the dissipation function?

16.9 In a boundary layer where the velocity profile is given by

$$\frac{v_x}{v_\infty} = \frac{3}{2}\frac{y}{\delta} - \frac{1}{2}\left(\frac{y}{\delta}\right)^3$$

where δ is the velocity boundary-layer thickness, plot the dimensionless dissipation function, $\Phi\delta^2/\mu v_\infty^2$, vs. y/δ.

16.10 A spherical shell with inner and outer dimensions of r_i and r_o, respectively, has surface temperatures $T_i(r_i)$ and $T_o(r_o)$. Assuming constant properties and one-dimensional (radial) conduction, sketch the temperature distribution, $T(r)$. Give reasons for the shape you have sketched.

16.11 Heat is transferred by conduction (assumed to be one-dimensional) along the axial direction through the truncated conical section shown in the figure. The two base surfaces are maintained at constant temperatures: T_1 at the top, and T_2, at the bottom, where $T_1 > T_2$. Evaluate the heat-transfer rate, q_x, when

a. the thermal conductivity is constant.

b. the thermal conductivity varies with temperature according to $k = k_o - aT$, where a is a constant.

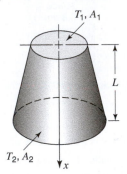

16.12 Heat is generated in a radioactive plane wall according to the relationship

$$\dot{q} = \dot{q}_{max}\left[1 - \frac{X}{L}\right]$$

where $\dot{q}$ is the volumetric heat generation rate, kW/m^3, L is the half thickness of the plate, and x is measured from the plate center line.

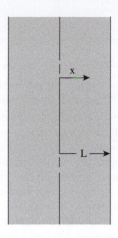

Develop the equation that expresses the temperature difference between the plate center line and its surface.

16.13 Heat is generated in a cylindrical fuel rod in a nuclear reactor according to the relationship

$$\dot{q} = \dot{q}_{max}\left[1 - \left(\frac{r}{r_o}\right)^2\right]$$

where $\dot{q}$ is the volumetric heat generation rate, kW/m³, and r_o is the outside cylinder radius. Develop the equation that expresses the temperature difference between the rod center line and its surface.

16.14 Heat is generated in a spherical fuel element according to the relationship

$$\dot{q} = \dot{q}_{max}\left[1 - \left(\frac{r}{r_o}\right)^3\right]$$

where $\dot{q}$ is the volumetric heat-generation rate, kW/m³, and r_o is the radius of the sphere. Develop the equation that expresses the temperature difference between the center of the sphere and its surface.

SS **17.1** One-dimensional steady-state conduction, with no internal heat generation, occurs across a plane wall having a constant thermal conductivity of 30 W/m · K. The material is 30 cm thick. For each case listed in the table below, determine the unknown quantities. Show a sketch of the temperature distribution for each case.

Case	T_1	T_2	dT/dx (K/m)	q_x (W/m^2)
1	350 K	275 K		
2	300 K			−2000
3		350 K	−300	
4	250 K		200	

SS **17.2** The steady-state expression for heat conduction through a plane wall is $q = (kA/L)\,\Delta T$ as given by equation (17-4). For steady-state heat conduction through a hollow cylinder, an expression similar to equation (17-4) is

$$q = \frac{k\bar{A}}{r_o - r_i}\Delta T$$

where $\bar{A}$ is the "log-mean" area defined as

$$\bar{A} = 2\pi \frac{r_o - r_i}{\ln(r_o/r_i)}$$

a. Show that $\bar{A}$ as defined above satisfies the equations for steady-state radial heat transfer in a hollow cylindrical element.

b. If the arithmetic mean area, $\pi\,(r_o - r_i)$, is used rather than the logarithmic mean, calculate the resulting percent error for values of r_o/r_i of 1.5, 3, and 5.

SS **17.3** Evaluate the appropriate "mean" area for steady-state heat conduction in a hollow sphere that satisfies an equation of the form

$$q = \frac{k\bar{A}}{r_o - r_i}\Delta T$$

Repeat part (b) of Problem 17.2 for the spherical case.

SS **17.4** It is desired to transport liquid metal through a pipe embedded in a wall at a point where the temperature is 650 K. A 1.2-m-thick wall constructed of a material having a thermal conductivity varying with temperature according to $k = 0.0073$ $(1 + 0.0054\,T)$, where T is in K and k is in W/m · K, has its inside surface maintained at 925 K. The outside surface is exposed to air at 300 K with a convective heat-transfer coefficient of

23 W/m^2 · K. How far from the hot surface should the pipe be located? What is the heat flux for the wall?

17.5 The device shown in the figure below is used to estimate the thermal conductivity of a new biomaterial by measuring the steady-state heat-transfer rate and temperature. The biomaterial is wrapped around the cylindrical heating rod as shown in the figure. The exposed ends of the device are sealed and insulated. A constant power load is passed through the heating rod. The power transferred to the heating rod is dissipated as heat through the biomaterial layer and then through the surrounding air film layer. The temperature at the surface of the 3-cm-diameter heating rod ($r = R_o = 1.50$ cm, $T = T_0$) is measured, and is constant along the length, L, of the 6-cm long rod.

For the present set of measurements, the thickness of the biomaterial surrounding the heating rod is 0.3 cm, the power load is 15 W, the measured surface temperature of the rod is $T_0 = 70$ C, and the temperature of the surrounding air remains constant at 20 C. The convective heat-transfer coefficient of the air film surrounding the biomaterial is 80 W/m^2 · K.

a. What are the appropriate boundary conditions for heat-transfer analysis through the biomaterial?

b. What is the thermal conductivity of the biomaterial, k_m?

c. Plot the temperature profile $T(r)$ from $r = R_o$ to $r = R_m$.

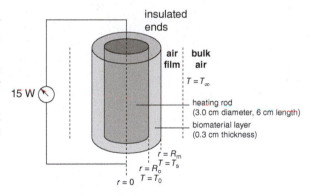

17.6 A stainless-steel plate 1.6-cm thick rests on top of a hot plate, which is maintained at 250 C. Air flows over the top surface of the plate to provide a convective heat-transfer coefficient of $h = 50$ W/m^2 · K. The air temperature is maintained at 20 C.

a. What is the heat flux through the stainless-steel plate, in W/m^2?

b. What is the temperature at the top surface, T_1, of the stainless-steel plate?

c. Based on the analysis above, what can you conclude about the heat-transfer resistance offered by the hydrodynamic boundary layer?

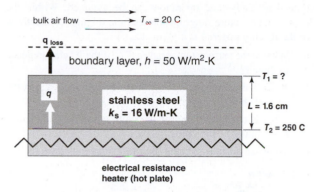

electrical resistance
heater (hot plate)

17.7 A double-pane insulated window unit consists of two 1-cm-thick pieces of glass separated by a 1.8-cm layer of air. The unit measures 4 m in width and is 3 m wide. Under conditions where the extreme outside temperature of the glass is at −10 C and air, at 27 C, is adjacent to the inside glass surface, with $h_i = 12$ W/m^2 · K, determine

a. the inside glass surface temperature.

b. the rate of heat transfer through the window unit.

The air gap between glass panes may be treated as a purely conductive layer with $k = 0.0262$ W/m · K. Thermal radiation is to be neglected.

SS **17.8** A furnace wall is to be designed to transmit a maximum heat flux of 200 Btu/h ft^2 of wall area. The inside and outside wall temperatures are to be 2000 F and 300 F, respectively. Determine the most economical arrangement of bricks measuring 9 by 4 ½ by 3 in if they are made from two materials, one with a k of 0.44 Btu/h ft F and a maximum usable temperature of 1500 F and the other with a k of 0.94 Btu/h ft F and a maximum usable temperature of 2200 F. Bricks made of each material cost the same amount and may be laid in any manner.

SS **17.9** A furnace wall consisting of 0.25 m of fire clay brick, 0.20 m of kaolin, and a 0.10-m outer layer of masonry brick is exposed to furnace gas at 1370 K with air at 300 K adjacent to the outside wall. The inside and outside convective heat-transfer coefficients are 115 and 23 W/m^2 · K, respectively. Determine the heat loss per square foot of wall and the temperature of the outside wall surface under these conditions.

SS **17.10** Given the conditions of Problem 17.9, except that the outside temperature of the masonry brick cannot exceed 325 K, by how much must the thickness of kaolin be adjusted to satisfy this requirement?

17.11 A heater composed of Nichrome wire wound back and forth and closely spaced is covered on both sides with ⅛-in thickness of asbestos ($k = 0.15$ Btu/h ft F) and then with a ⅛-in thickness of stainless steel ($k = 10$ Btu/h ft F). If the center temperature of this sandwich construction is considered constant at 1000 F and the outside convective heat-transfer coefficient is 3 Btu/h ft^2 F, how much energy must be supplied in W/ft^2 to the heater? What will be the outside temperature of the stainless steel?

17.12 Determine the percent in heat flux if, in addition to the **SS** conditions specified in Problem 17.11, there are two ¾-in-diameter steel bolts extending through the wall per square foot of wall area (k for steel = 22 Btu/h ft F).

17.13 A 2.5-cm-thick sheet of plastic ($k = 2.42$ W/m · K) is to be bonded to a 5-cm-thick aluminum plate. The glue that will accomplish the bonding is to be held at a temperature of 325 K to achieve the best adherence, and the heat to accomplish this bonding is to be provided by a radiant source. The convective heat-transfer coefficient on the outside surfaces of both the plastic and aluminum is 12 W/m^2 · K, and the surrounding air is at 295 K. What is the required heat flux if it is applied to the surface of (a) the plastic? (b) the aluminum?

17.14 A composite wall is to be constructed of ¼ in of stainless steel ($k = 10$ Btu/h ft F), 3 in of corkboard ($k = 0.025$ Btu/h ft F), and ½ in of plastic ($k = 1.5$ Btu/h ft F). Determine the thermal resistance of this wall if it is bolted together by ½-in-diameter bolts on 6-in centers made of

a. stainless steel

b. aluminum ($k = 120$ Btu/h ft F)

17.15 A cross section of a typical home ceiling is depicted below. Given the properties listed for the materials of construction, determine how much heat is transferred through the insulation and through the studs.

$$T_{outside} = -10\ C$$
$$h_o = 20\ W/m^2 \cdot K$$
$$T_{inside} = 25\ C$$
$$h_i = 10\ W/m^2 \cdot K$$
$$k_{fiberglass} = 0.035\ W/m^2 \cdot K$$
$$k_{plaster} = 0.814\ W/m^2 \cdot K$$
$$k_{wood} = 0.15\ W/m^2 \cdot K$$

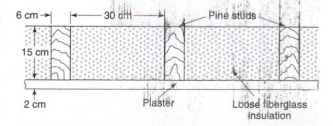

17.16 A 2-in schedule-40 steel pipe carries saturated steam at 60 psi through a laboratory that is 60 ft long. The pipe is insulated with 1.5 in of 85% magnesia that costs $0.75 per foot. How long must the steam line be in service to justify the insulation cost if the heating cost for the steam is $0.68 per 10^5 Btu? The outside-surface convective heat-transfer coefficient may be taken as 5 Btu/h ft^2 F.

17.17 Saturated steam at 40 psia flows at 5 fps through a schedule-40, 1½-in steel pipe. The convective heat-transfer coefficient by condensing steam on the inside surface may be taken as 1500 Btu/h ft^2 F. The surrounding air is at 80 F, and the outside surface coefficient is 3 Btu/ft^2 F. Determine the following:

a. The heat loss per 10 ft of bare pipe.

b. The heat loss per 10 ft of pipe insulated with 2 in of 85% magnesia.

c. The mass of steam condensed in 10 ft of bare pipe.

17.18 A computer IC chip consumes 10 W of power, which is dissipated as heat. The chip measures 4 cm by 4 cm on a side and is 0.5-cm thick. Currently the IC chip is packaged into an electronic device as shown in Figure 17.5. The base of the chip is in contact with an inert aluminum plate that is 0.3-cm thick. The IC chip and its aluminum base are mounted within a thermally insulating ceramic material, which you may be assumed to act as a perfect thermal insulator. The top side of the IC chip is exposed to air, which provides a convective heat-transfer coefficient of 100 W/m$^2 \cdot$ K. The ambient air temperature (T_∞) is maintained at 30 C. The chip is operated for a sufficient time that steady-state operation can be assumed. Within the process temperature range of interest, the thermal conductivity of the IC chip material is $k_{IC} = 1$ W/m $\cdot$ K, and the thermal conductivity of aluminum is $k_{Al} = 230$ W/m $\cdot$ K.

a. What is the surface temperature, T_1, of the IC chip exposed to the air?

b. What is the temperature, T_2, of the IC chip package at the base of the IC chip ($x = L_1$) resting on top of the aluminum plate?

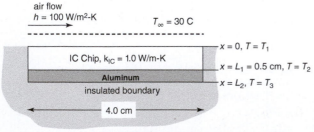

IC chip imbedded in insulated boundary.

17.19 A computer IC chip consumes 10 W of power, which is dissipated as heat. The chip measures 4 cm by 4 cm on a side and is 0.5-cm thick. Currently, the IC chip is packaged into an electronic device as shown. The IC chip is mounted within a

thermally insulating ceramic material, which may be assumed to act as a perfect thermal insulator. The top surface of the IC chip is exposed to air flowing over the chip to provide a convective heat-transfer coefficient of 100 W/m$^2 \cdot$ K. The ambient air temperature (T_∞) is maintained at 30 C. The chip is operated for a sufficient time that steady-state operation can be assumed. Within the process temperature range of interest, the thermal conductivity of the IC chip material is 1 W/m $\cdot$ K.

a. What is the surface temperature, T_1, of the IC chip exposed to the air?

b. What is the temperature, T_2, of the IC chip package at the base of the IC chip ($x = L_1$)?

c. What is the failure rate of the chip, based on the average temperature of the chip?

d. What can be done to bring the chip failure rate down to 4%?

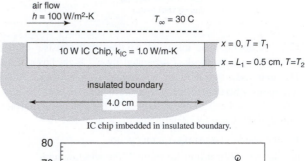

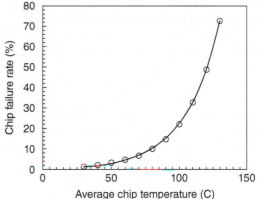

IC chip imbedded in insulated boundary.

17.20 Copper wire having a diameter of 3/16 in is insulated with a 4-in layer of material having a thermal conductivity of 0.14 Btu/h ft F. The outer surface of the insulation is maintained at 70 F. How much current may pass through the wire if the insulation temperature is limited to a maximum of 120 F? The resistivity of copper is 1.72×10^{-6} ohm – cm.

17.21 What would be the result of Problem 17.20 if the fluid surrounding the insulated wire was maintained at 70 F with a convective heat-transfer coefficient between the insulation and the fluid of 4 Btu/h ft^2 F? What would be the surface temperature of the insulation under these conditions?

17.22 Work Problem 17.20 for the case of aluminum rather than copper. The resistivity of aluminum is 2.83×10^{-6} ohm – cm.

17.23 Liquid nitrogen at 77 K is stored in an insulated spherical container that is vented to the atmosphere. The container is made of a thin-walled material with an outside diameter of 0.5 m; 25 mm of insulation ($k = 0.002$ W/m · K) covers its outside surface. The latent heat of nitrogen is 200 kJ/kg; its density, in the liquid phase, is 804 kg/m^3. For surroundings at 25 C and with a convective coefficient of 18 W/m^2 · K at the outside surface of the insulation, what will be the rate of liquid nitrogen boil-off?

17.24 What additional thickness of insulation will be necessary to reduce the boil-off rate of liquid nitrogen to one-half of the rate corresponding to Problem 17.23? All values and dimensions in Problem 17.23 apply.

17.25 A thin slab of material is subjected to microwave radiation that causes volumetric heating to vary according to

$$\dot{q}(x) = \dot{q}_o[1 - (x/L)]$$

where $\dot{q}_o$ has a constant value of 180 kW/m^3 and the slab thickness, L, is 0.06 m. The thermal conductivity of the slab material is 0.6 W/m · K.

The boundary at $x = L$ is perfectly insulated, while the surface at $x = 0$ is maintained at a constant temperature of 320 K.

a. Determine an expression for $T(x)$ in terms of X, L, k, $\dot{q}_0$, and T_0.

b. Where in the slab will the maximum temperature occur?

c. What is the value of T_{max}?

17.26 Radioactive waste ($k = 20$ W/m · K) is stored in a cylindrical stainless-steel ($k = 15$ W/m · K) container with inner and outer diameters of 1.0 and 1.2 m, respectively. Thermal energy is generated uniformly within the waste material at a volumetric rate of 2×10^5 W/m^3. The outer container surface is exposed to water at 25 C, with a surface coefficient of 1000 W/m^2 · K. The ends of the cylindrical assembly are insulated so that all heat transfer occurs in the radial direction. For this situation determine

a. the steady-state temperatures at the inner and outer surfaces of the stainless steel.

b. the steady-state temperature at the center of the waste material.

17.27 A cylindrical nuclear fuel element is 10.16 cm long and 10.77 cm in diameter. The fuel generates heat uniformly at a rate of 51.7×10^3 kJ/s · m^3. The fuel is placed in an environment having a temperature of 360 K with a surface coefficient of 4540 W/m^2 · K. The fuel material has $k = 33.9$ W/m · K. For the situation described, evaluate the following at steady state:

a. The temperature profile as a function of radial position

b. The maximum fuel temperature

c. The surface temperature

End effects may be neglected.

17.28 Radioactive waste is stored in a cylindrical stainless-steel container with inner and outer diameters of 1 m and 1.2 m, respectively, so that $R_0 = 0.5$ m and $R_1 = 0.6$ m. Thermal energy is generated uniformly within the waste material at a volumetric rate of 2×10^5 W/m^3. The outer container surface is exposed to water at 25 C ($T_\infty = 25$ C), and the outer surface convective heat-transfer coefficient is $h = 1000$ W/m^2 · K. The ends of the cylinder are insulated so that heat transfer is only in the r-direction. You may assume that over the temperature range of interest the thermal conductivity of stainless steel, k_{steel}, is 15 W/m · K and the thermal conductivity of the radioactive waste, k_{rw}, is 20 W/m · K.

What are the temperatures of the outer surface, T_0, and inner surface, T_i, of the stainless-steel container?

What the maximum temperature within the radioactive waste?
Where will this maximum temperature be located?

17.29 An engineered tissue system consists of a slice of cell mass immobilized on a scaffold measuring 5 cm in length and 0.5 cm in thickness. The top face of the tissue scaffold is exposed to water, dissolved oxygen, and organic nutrients maintained at 30 C. The bottom face of the tissue is thermally insulated. At present, the specific oxygen consumption of the tissue mass is 0.5 mmol O$_2$/cm^{3}cells-hr, and from respiration energetics, the energy released by respiration is 468 J/mmol O$_2$ consumed. We are interested in knowing the temperature at the bottom face of the tissue next to the insulated boundary. If this temperature remains below 37 C, the tissue will not die. The thermal conductivity of the tissue scaffold is $k = 0.6$ W/m · K.

a. Using the information given predict the temperature profile within the tissue slab the slab.

b. What is the heat generated per unit volume of tissue?

c. Estimate the temperature at $x = L$ (the insulated boundary).

17.30 Consider the composite solid shown. Solid A is a thermally conductive material that is 0.5-cm thick and has a thermal conductivity, $k_A = 50$ W/m · K. The back side of solid A ($x = 0$) is thermally insulated. Electrical current is applied to solid A such that 20 W per cm^3 is generated as heat. Solid B is 0.2-cm thick and has a thermal conductivity of $k_B = 20$ W/m · K. The surface of solid B is exposed to air. The surface temperature, T_s, of solid B is 80 C. The bulk air temperature is constant at 30 C. The process is at steady state.

a. What is the heat-transfer rate per unit area (flux) at $x = L_2$?
What is the temperature T_1 at $x = L_1$, the boundary between solid A and solid B?

b. What is the required convective heat-transfer coefficient, h, for the flowing air?

c. What is the temperature T_0 at $x = 0$, the insulating side of solid A?

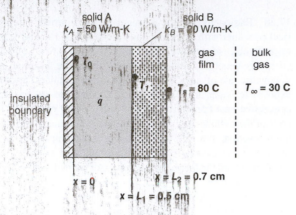

Heat transfer in composite solid.

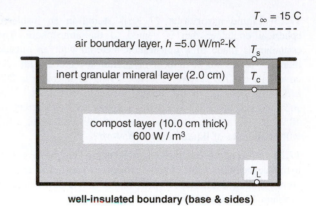

well-insulated boundary (base & sides)

Cross section of compost bed.

17.31 Consider the compost bed shown in the right-hand figure. The dimensions of the bed are 1 m by 2 m. A 10-cm (0.10-m) thick layer of organic compost lines the bottom as shown. Slow bacterial decomposition of the organic material in the compost generates heat at a constant rate of 600 W per m³ of compost. An inert granular mineral layer 2-cm thick is placed over the compost layer. The sides and bottom of the compost bed are perfectly insulated. Air flows over the bed to provide a convective heat-transfer coefficient of 5 W/m² · K. The ambient air temperature is constant at 15 C. Thermophysical properties are provided below. There are no other direct energy inputs to the compost bed.

Thermophysical properties

Material	Thermal Conductivity W/m · K	Heat Capacity J/kg · K	Bulk Density kg/m³
Organic compost	0.1	1500	300
Inert granular mineral layer	0.4	1700	1600

a. What is the steady-state heat-transfer rate from the compost bed to the surrounding air?

b. What is the temperature at top of the organic compost layer—i.e., T_c?

c. What is the temperature, T_L, at the base of the compost bed?

d. If the thickness of the compost layer is increased, then which will occur?
   1. T_c and T_s will both increase
   2. T_c will increase but T_s will stay the same
   3. T_c and T_s will not change

17.32 Consider a section of muscle tissue of cylindrical shape with a radius of 1.5 cm. During highly rigorous exercise, metabolic processes generate 15 kW/m³ of bulk tissue. The outer surface of the tissue is maintained at 37 C. The thermal conductivity of the muscle tissue is $k_m = 0.419$ W/m · K.

a. Develop a mathematical model to predict the steady-state temperature distribution, $T(r)$, in the tissue along the radial position r. What is the temperature of the tissue at a distance 0.75 cm from the surface?

b. Develop a mathematical model to predict the maximum temperature inside the muscle tissue. What is the maximum temperature?

c. If the length of the muscle, L, is 10 cm, what is the heat flux (W/cm²) leaving the muscle tissue, assuming end effects can be neglected?

d. Now consider that a sheath 0.25-cm thick surrounds the muscle tissue. This sheath has a thermal conductivity of 0.3 W/m · K and does not generate any heat. If the outer surface of this sheath is maintained at 37 C, what is the new maximum temperature within the muscle tissue?

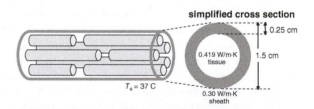

17.33 A 1-in-OD steel tube has its outside wall surface maintained at 250 F. It is proposed to increase the rate of heat transfer by adding fins of 3/32-in thickness and 3/4-in long to the outside tube surface. Compare the increase in heat transfer achieved by adding 12 longitudinal straight fins or circular fins with the same total surface area as the 12 longitudinal fins. The surrounding air is at 80 F, and the convective heat-transfer coefficient is 6 Btu/h ft² F.

17.34 Solve Problem 17.33 if the convective heat-transfer coefficient is increased to 60 Btu/h ft² F by forcing air past the tube surface.

17.35 A cylindrical rod 3 cm in diameter is partially inserted into a furnace with one end exposed to the surrounding air, which is at 300 K. The temperatures measured at two points 7.6 cm apart are 399 and 365 K, respectively. If the convective heat-transfer coefficient is 17 W/m² · K, determine the thermal conductivity of the rod material.

17.36 Heat is to be transferred from water to air through an aluminum wall. It is proposed to add rectangular fins 0.05-in thick and 3/4-in long spaced 0.08 in apart to the aluminum surface to aid in transferring heat. The heat-transfer coefficients on the air and water sides are 3 and 25 Btu/h ft² F, respectively. Evaluate the percent increase in heat transfer if these fins are added to (a) the air side, (b) the water side, and (c) both sides. What conclusions may be reached regarding this result?

17.37 A semiconductor material with $k = 2$ W/m · K and electrical resistivity, $\rho = 2 \times 10^{-5} \, \Omega$ m, is used to fabricate a cylindrical rod 40-mm long with a diameter of 10 mm. The longitudinal surface of the rod is well insulated and may be considered adiabatic while the ends are maintained at temperatures of 100 C and 0 C, respectively. If the rod carries a current of 10 amps, what will be its midpoint temperature? What will be the rate of heat transfer through both ends?

17.38 An iron bar used for a chimney support is exposed to hot gases at 625 K with the associated convective heat-transfer coefficient of 740 W/m² · K. The bar is attached to two opposing chimney walls, which are at 480 K. The bar is 1.9 cm in diameter and 45-cm long. Determine the maximum temperature in the bar.

17.39 A copper rod 1/4 in in diameter and 3-ft long runs between two bus bars, which are at 60 F. The surrounding air is at 60 F, and the convective heat-transfer coefficient is Btu/h ft² F. Assuming the electrical resistivity of copper to be constant at 1.72×10^{-6} ohm – cm, determine the maximum current the copper may carry if its temperature is to remain below 150 F.

17.40 A 13 cm by 13 cm steel angle with the dimensions shown is attached to a wall with a surface temperature of 600 K. The surrounding air is at 300 K, and the convective heat-transfer coefficient between the angle surface and the air is 45 W/m² · K.

a. Plot the temperature profile in the angle, assuming a negligible temperature drop through the side of the angle attached to the wall.

b. Determine the heat loss from the sides of the angle projecting out from the wall.

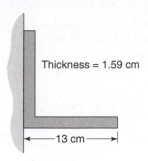

Thickness = 1.59 cm

13 cm

17.41 A steel I-beam with a cross-sectional area as shown has its lower and upper surfaces maintained at 700 and 370 K, respectively.

a. Assuming a negligible temperature change through both flanges, develop an expression for the temperature variation in the web as a function of the distance from the upper flange.

b. Plot the temperature profile in the web if the convective heat-transfer coefficient between the steel surface and the surrounding air is 57 W/m² · K. The air temperature is 300 K.

c. What is the net heat transfer at the upper and lower ends of the web?

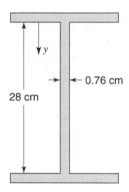

y

0.76 cm

28 cm

17.42 Repeat Problem 17.41 for the case of an aluminum beam.

17.43 A 2-in-OD stainless-steel tube has 16 longitudinal fins spaced around its outside surface as shown. The fins are 1/16-in thick and extend 1 in from the outside surface of the tube.

a. If the outside surface of the tube wall is at 250 F, the surrounding air is at 80 F, and the convective heat-transfer coefficient is 8 Btu/h ft² F, determine the percent increase in heat transfer for the finned pipe over that for the unfinned pipe.

b. Determine the same information as in part (a) for values of h of 2, 5, 15, 50, and 100 Btu/h ft^2 F. Plot the percent increase in q vs. h. What conclusions can be reached concerning this plot?

17.44 Repeat Problem 17.43 for the case of an aluminum pipe-and-fin arrangement.

17.45 Water flows in the channels between two aluminum plates as shown in the sketch. The ribs that form the channels are also made of aluminum and are 8-mm thick. The effective surface coefficient between all surfaces and water is 300 W/m^2 · K. For these conditions, how much heat is transferred at each end of each rib? How far from the lower plate is the rib temperature a minimum? What is this minimum value?

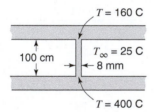

17.46 A computer graphics chip measures 5 cm on a side and is 3-mm thick. The chip consumes 15 W of power, which is dissipated as heat from the top side of the chip. A fan blows air over the surface of the chip to promote convective heat transfer. Unfortunately, the chip fails to operate after a certain time, and the temperature of the chip is suspected to be the culprit. A plot of chip failure rate vs. surface temperature is presented in the figure on the next page. An electrical engineer

suggests that an aluminum extended heat-transfer surface could be mounted to the top surface of the chip to promote heat transfer, but does not know how to design it. The electrical engineer asks for your help.

a. If the heat-transfer coefficient in air is 50 W/m^2 · K, what is the surface temperature of the chip at steady state if the bulk air temperature is maintained at 20 C (293 K)? What is the failure rate of the chip?

b. An aluminum extended surface heat-transfer device is next mounted on the top of the chip. It consists of a parallel array of five rectangular fins 1 cm by 5 cm wide by 0.3 mm thick. The new convective heat-transfer coefficient over the fin surfaces is 20 W/m^2 · K. Is this configuration suitable? What is the new failure rate of the chip?

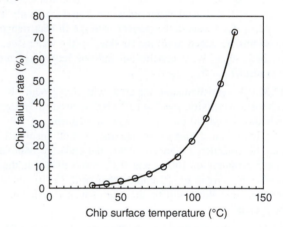

Chip failure rate vs. temperature.

IC chip with extended surfaces.

18.1 A household iron has a stainless-steel sole plate that weights 3 lb and has a surface area of 0.5 ft^2. The iron is rated at 500 W. If the surroundings are at a temperature of 80 F, and the convective heat-transfer coefficient between the sole plate and surroundings is 3 Btu/h ft^2 F, how long will it take for the iron to reach 240 F after it is plugged in?

18.2 An electrical system employs fuses that are cylindrical in shape and have lengths of 0.5 cm and diameters of 0.1 mm. Air, at 30°C, surrounds a fuse with a surface coefficient of 10 W/m$^2 \cdot$ K. The fuse material melts at 900°C.

Assuming all heat transfer to be from the fuse surface, estimate the time it will take for the fuse to blow after a current of 3 A flows through it.

Pertinent properties of the fuse material are

$$\text{Resistance} = 0.2 \, \Omega$$
$$k = 20 \, \text{W/m} \cdot \text{K}$$
$$\alpha = 5 \times 10^{-5} \, \text{m}^2/\text{s}$$

18.3 Aluminum wire, having a diameter of 0.794 mm, is immersed in an oil bath that is at 25°C. Aluminum wire of this size has an electrical resistance of 0.0572 Ω/m. For conditions where an electric current of 100 A is flowing through the wire and the surface coefficient between the wire and oil bath is 550 W/m$^2 \cdot$ K, determine the steady-state temperature of the wire.

How long, after the current is supplied, will it take for the wire to reach a temperature within 5°C of its steady-state value?

SS **18.4** Cast-iron cannonballs used in the War of 1812 were occasionally heated for some extended time so that, when fired at houses or ships, they would set them afire. If one of these, the so-called "hot shot," were at a uniform temperature of 2000 F, how long after being exposed to air at 0 F with an outside convective heat-transfer coefficient of 16 Btu/h ft^2 F, would be required for the surface temperature to drop to 600 F? What would be the center temperature at this time? The ball diameter is 6 in. The following properties of cast iron may be used:

$$k = 23 \, \text{Btu/h ft F}$$
$$c_p = 0.10 \, \text{Btu/lb}_m \, \text{F}$$
$$\rho = 460 \, \text{lb}_m/\text{ft}^3.$$

SS **18.5** It is known that oranges can be exposed to freezing temperatures for short periods of time without sustaining serious damage. As a representative case, consider a 0.10-m-diameter orange, originally at a uniform temperature of 5°C, suddenly exposed to surrounding air at −5°C. For a surface coefficient, between the air and orange surface, of 15 W/m$^2 \cdot$ K, how long

will it take for the surface of the orange to reach 0°C? Properties of the orange are the following:

$$\rho = 940 \, \text{kg/m}^3$$
$$K = 0.47 \, \text{W/m} \cdot \text{K}$$
$$c_p = 3.8 \, \text{kJ/kg} \cdot \text{K}$$

18.6 A copper cylinder with a diameter of 3 in is initially at a uniform temperature of 70 F. How long after being placed in a medium at 1000 F with an associated convective heat-transfer coefficient of 4 Btu/h ft^2 F will the temperature at the center of the cylinder reach 500 F, if the height of the cylinder is (a) 3 in? (b) 6 in? (c) 12 in? (d) 24 in? (e) 5 ft?

18.7 A cylinder 2 ft high with a diameter of 3 in is initially at **SS** the uniform temperature of 70 F. How long after the cylinder is placed in a medium at 1000 F, with associated convective heat-transfer coefficient of 4 Btu/h ft^2 F, will the center temperature reach 500 F if the cylinder is made from

a. copper, $k = 212$ Btu/h ft F?
b. aluminum, $k = 130$ Btu/h ft F?
c. zinc, $k = 60$ Btu/h ft F?
d. mild steel, $k = 25$ Btu/h ft F?
e. stainless steel, $k = 10.5$ Btu/h ft F?
f. asbestos, $k = 0.087$ Btu/h ft F?

18.8 Water, initially at 40 F, is contained within a thin-walled cylindrical vessel having a diameter of 18 in. Plot the temperature of the water vs. time up to 1 h if the water and container are immersed in an oil bath at a constant temperature of 300 F. Assume that the water is well stirred and that the convective heat-transfer coefficient between the oil and cylindrical surface is 40 Btu/h ft^2 F. The cylinder is immersed to a depth of 2 ft.

18.9 A short aluminum cylinder 0.6 m in diameter and 0.6 m **SS** long is initially at 475 K. It is suddenly exposed to a convective environment at 345 K with $h = 85$ W/m$^2 \cdot$ K. Determine the temperature in the cylinder at a radial position of 10 cm and a distance of 10 cm from one end of the cylinder after being exposed to this environment for 1 h.

18.10 Buckshot, 0.2 in diameter, is quenched in 90°F oil from **SS** an initial temperature of 400 F. The buckshot is made of lead and takes 15 s to fall from the oil surface to the bottom of the quenching bath. If the convective heat-transfer coefficient between the lead and oil is 40 Btu/h ft^2 F, what will be the temperature of the shot as it reaches the bottom of the bath?

18.11 If a rectangular block of rubber (see Problem 18.13 for properties) is set out in air at 297 K to cool after being heated to a uniform temperature of 420 K, how long will it take for the rubber surface to reach 320 K? The dimensions of the block are 0.6 m high by 0.3 m long by 0.45 m wide. The block sits on one of the 0.3-m by 0.45-m bases; the adjacent surface may be considered an insulator. The effective heat-transfer coefficient at all exposed surface is $6.0 \, \text{W/m}^2 \cdot \text{K}$. What will the maximum temperature within the rubber block be at this time?

18.12   A type-304 stainless-steel billet, 6 in diameter, is passing through a 20-ft-long heat-treating furnace. The initial billet temperature is 200 F, and it must be raised to a minimum temperature of 1500 F before working. The heat-transfer coefficient between the furnace gases and the billet surface is 15 Btu/h ft^2 F, and the furnace gases are at 2300 F. At what minimum velocity must the billet travel through the furnace to satisfy these conditions?

18.13   In the curing of rubber tires, the "vulcanization" process requires that a tire carcass, originally at 295 K, be heated so that its central layer reaches a minimum temperature of 410 K. This heating is accomplished by introducing steam at 435 K to both sides. Determine the time required, after introducing steam, for a 3 cm-thick tire carcass to reach the specified central temperature condition. Properties of rubber that may be used are the following: $k = 0.151 \, \text{W/m} \cdot \text{K}$, $c_p = 200 \, \text{J/kg} \cdot \text{K}$, $\rho = 1201 \, \text{kg/m}^3$, $\alpha = 6.19 \times 10^{-8} \, \text{m}^2/\text{s}$.

18.14   It is common practice to treat wooden telephone poles with tar-like materials to prevent damage by water and insects. These tars are cured into the wood at elevated temperatures and pressures.

Consider the case of a 0.3-m-diameter pole, originally at 25 C, placed in a pressurized oven. It will be removed when the tar has penetrated to a depth of 10 cm. It is known that a 10-cm depth of penetration will occur when a temperature of 100 C is achieved. For an oven temperature of 380 C and $h = 140 \, \text{W/m}^2 \cdot \text{K}$, determine the time required for the pole to remain in the oven. Properties of the wooden pole are

$$k = 0.20 \, \text{W/m} \cdot \text{K}$$
$$\alpha = 1.1 \times 10^{-7} \, \text{m}^2/\text{s}$$

18.15   For an asbestos cylinder with both height and diameter of 13 cm initially at a uniform temperature of 295 K placed in a medium at 810 K with an associated convective heat-transfer coefficient of $22.8 \, \text{W/m}^2 \cdot \text{K}$, determine the time required for the center of the cylinder to reach 530 K if end effects are neglected.

18.16   A copper bus bar is initially at 400 F. The bar measures 0.2 ft by 0.5 ft and is 10 ft long. If the edges are suddenly all reduced to 100 F, how long will it take for the center to reach a temperature of 250 F?

18.17   Rework Problem 18-4 for the case when air is blown by the surfaces of the rubber block with an effective surface coefficient of $230 \, \text{W/m}^2 \cdot \text{K}$.

18.18   Consider a hot dog to have the following dimensions and properties: diameter $= 20$ mm, $c_p = 3.35 \, \text{kJ/kg} \cdot \text{K}$, $\rho = -880 \, \text{kg/m}^3$, and $k = 0.5 \, \text{W/m} \cdot \text{K}$. For the hot dog initially at 5 C, exposed to boiling water at 100 C, with a surface coefficient of $90 \, \text{W/m}^2 \cdot \text{K}$, what will be the cooking time if the required condition is for the center temperature to reach 80 C?

18.19   This problem involves using heat-transfer principles as a guide for cooking a pork roast.

The roast is to be modeled as a cylinder, having its length and diameter equal to each other, with properties being those of water. The roast weighs 2.25 kg.

Properly cooked, every portion of the meat should attain a minimum temperature of 95 C. If the meat is initially at 5 C and the oven temperature is 190 C, with a surface coefficient of $15 \, \text{W/m}^2 \cdot \text{K}$, what is the minimum cooking time required?

18.20   Given the cylinder in Problem 18.15, construct a plot of the time for the midpoint temperature to reach 530 K as a function of H/D, where H and D are the height and diameter of the cylinder, respectively.

18.21   At an expensive restaurant, a customer makes an unusual request of the chef. She asks that her steak be cooked on a grill until the meat reaches a temperature of 100 C at a depth of 2 mm. The chef knows that the griddle surface temperature is maintained at 250 C, the steak was stored at 0 C, and the steak is 2-cm thick. The customer, an engineer, knows that steak has a heat capacity of 4000 J/kg · K, a density of 1500 kg/m^3, and thermal conductivity of 0.6 W/m · K. The chef asks for help. How long should the steak be cooked?

18.22   In the canning process, sealed cans of food are sterilized with pressurized steam in order to kill any microorganisms initially present in the food and thereby prolong the shelf life of the food. A cylindrical can of food has a diameter of 4 cm and length of 10 cm. The food material has a heat capacity of 4000 J/kg · K, a density of 1500 kg/m^3, and thermal conductivity of 0.6 W/m · K. In the present process, steam, at 120 C, is used to sterilize the can. The convective heat-transfer coefficient is $60 \, \text{W/m}^2 \cdot \text{K}$. Initially, the can and its contents are at a uniform temperature of 20 C.

a.   If the heat-transfer resistance offered by the can itself is neglected, and the can ends are thermally insulated, how long will it take for the center of the can to reach a temperature of 100 C, which is sufficient to kill all microorganisms?

b.   What is the required time if the ends of the can are not sealed?

c.   What will be the center temperature of the can and contents after 2500 s have elapsed?

18.23   One step in the manufacture of silicon wafers used in the microelectronics industry is the melt crystallization of silicon into a crystalline silicon ingot. This process is carried out within a special furnace. When the newly, solidified ingot is removed from the furnace, it is assumed to have a uniform initial temperature of 1600 K, which is below the crystallization temperature. At this temperature, the thermal conductivity of silicon is 22 W/m · K, the density is 2300 kg/m^3, and the heat capacity is 1000 J/kg · K. The hot solid silicon ingot is allowed to cool in air maintained at a constant ambient temperature of 30 C. The diameter of the silicon rod is 15 cm. End effects are considered negligible. The convective heat-transfer coefficient is 147 W/m² · K. What temperature will exist 1.5 cm from the surface of the ingot after a cooling time of 583 s (9.72 min)?

18.24 Drug formulations contain an active drug dissolved in a nontoxic solution to facilitate its delivery into the body. They are often packaged into sterile syringes so that the patient can inject the drug formulation directly into the body and then throw the syringe away. Often the drug-loaded syringes are shipped by air freight to customers. There is a concern that the low temperatures associated with air freight cargo holds might promote the precipitation of the drug from solution. A certain monoclonal antibody is known to precipitate from its delivery solution at 2 C. The cargo hold in the airplane used by the air freight company is maintained at 0 C. Initially, the syringe is at a uniform temperature of 25 C. Within the current packaging environment of the syringe, the convective heat-transfer coefficient of air surrounding the syringe is 0.86 W/m² · K. The barrel of the syringe containing the liquid is 0.70 cm inner diameter and 8 cm in length. The thermal conductivity of the solution is 0.3 W/m · K, the density is 1 g/cm³, and heat capacity is 2000 J/kg · K. The syringe wall is very thin.

If the air flight is 3.5 h in length, will the drug completely precipitate and ruin the product?

18.25 A flat silicon wafer 15 cm in diameter (surface area 176.7 cm²) is removed from an annealing furnace and laid flat-side down onto a flat insulating surface. It is then allowed to cool down to room temperature, which is maintained at a constant 20 C (293 K). Initially the silicon wafer is at 1020 C. The heat-transfer coefficient for the air over the exposed flat surface of the silicon wafer is 20 W/m² · K. What is the heat-transfer rate, in Watts, from the wafer surface to the surrounding air after 330 s?

Thermophysical properties of silicon at the temperature range of interest are:

Material	Thermal Conductivity W/m · K	Heat Capacity J/kg · K	Bulk Density kg/m³
Polycrystalline silicon ~1000 C	30	760	2300

18.26 A rocket-engine nozzle is coated with a ceramic material having the following properties: $k = 1.73$ Btu/h ft F, $\alpha = 0.35$ ft²/h. The convective heat-transfer coefficient between the nozzle and the gases, which are at 3000 F, is 200 Btu/h ft² F. How long after start-up will it take for the temperature at the ceramic surface to reach 2700 F? What will be the temperature at a point 1/2 in from the surface at this time? The nozzle is initially at 0 F.

18.27 One estimate of the original temperature of Earth is 7000 F. Using this value and the following properties for Earth's crust, Lord Kelvin obtained an estimate of 9.8×10^7 years for the Earth's age:

$$\alpha = 0.0456 \text{ ft}^2/\text{h}$$

$$T_2 = 0 \text{ F}$$

$$\frac{\partial T}{\partial y}\bigg|_{y=0} = 0.02 \text{ F/ft, (measured)}$$

Comment on Lord Kelvin's result by considering the exact expression for unsteady-state conduction in one dimension

$$\frac{T - T_s}{T_0 - T_s} = \text{erf}\frac{x}{2\sqrt{\alpha t}}$$

18.28 After a fire starts in a room, the walls are exposed to combustion products at 950 C. If the interior wall surface is made of oak, how long after exposure to the fire will the wood surface reach its combustion temperature of 400 C? Pertinent data are the following:

$$h = 30 \text{ W/m}^2 \cdot \text{K}$$

$$T_i(\text{initial}) = 21 \text{ C}$$

For oak:
$$\rho = 545 \text{ kg/m}^3$$
$$k = 0.17 \text{ W/m} \cdot \text{K}$$
$$c_p = 2.385 \text{ kJ/kg} \cdot \text{K}$$

18.29 Determine an expression for the depth below the surface of a semi-infinite solid at which the rate of cooling is maximum. Substitute the information given in Problem 18.27 to estimate how far below Earth's surface this maximum cooling rate is achieved.

18.30 Soil, having a thermal diffusivity of 5.16×10^{-7} m²/s, has its surface temperature suddenly raised and maintained at 1100 K from its initial uniform value of 280 K. Determine the temperature at a depth of 0.25 m after a period of 5 h has elapsed at this surface condition.

18.31 The convective heat-transfer coefficient between a large brick wall and air at 100 F is expressed as $h = 0.44 (T - T_\infty)^{1/3}$ Btu/h ft² F. If the wall is initially at a uniform temperature of 1000 F, estimate the temperature of the surface after 1, 6, and 24 h.

18.32 A thick plate made of stainless steel is initially at a uniform temperature of 300 C. The surface is suddenly exposed to a coolant at 20 C with a convective surface coefficient of 110 W/m² · K. Evaluate the temperature after 3 min of elapsed time at

a. the surface

b. a depth of 50 mm

Work this problem both analytically and numerically.

18.33 The severity of a burn on human skin is determined by the surface temperature of the hot object contacting the skin and the time of exposure. Skin is considered to be in a damaged state if its temperature is allowed to reach 62.5 C. Consider that, for relatively short times, near the surface of the skin, the skin tissue will behave as a semi-infinite medium. Skin also has three distinct layers, as shown in the figure. For the wet subdermal layer of living skin, $k_{skin} = 0.35$ W/m · K, $C_{p,skin} = 3.8$ kJ/kg · K, $\rho_{skin} = 1.05$ g/cm³ (Freitas, 2003). The thermal diffusivity of dry "cadaverous" epidermal skin is 2.8×10^{-4} cm²/s (Werner et al., 1992). A composite thermal diffusivity of 2.5×10^{-7} m²/s is to be assumed for the skin tissue (Datta, 2002).

a. What is the thermal diffusivity of living skin vs. dry cadaverous skin? Why is there such a big difference?

b. If a patch of skin is exposed to a hot metal surface at 250 C for 2 s, what will be the severity of the burn? Normal body temperature is 37 C.

c. At the time identified in part (b), what will the temperature at the top face of the dermal layer (thickness of approximately 0.5 mm) be?

References:

Freitas, R. A., Nanomedicine, Volume IIA: *Biocompatibility* (2003); Werner et al., *Phys. Med. Bioi.*, 37, 21–35 (1992).

Datta, A. K. *Biological and Bioenvironmental Heat and Mass Transfer*, 2002.

18.34 One way to treat tumors that exist near the skin surface is to heat the tumor tissue, as the cellular viability of some tumor tissues is more sensitive to higher temperatures than the surrounding skin tissue. In the simplest method, a directed heat source is put in contact with the skin surface as shown in the figure. In the present treatment process shown, the tumor is approximately 3 mm (0.003 m) below the skin surface and begins to respond to thermal treatment when its temperature reaches 45 C. The surface temperature is maintained at 55.5 C (132 F). After 2 min, the patient feels discomfort, and the device is turned off. Independent measurements showed that both the skin tissue and the tumor tissue had the following thermophysical properties: thermal conductivity, $k = 0.35$ W/m · K, heat capacity, $= 3.8$ kJ/kg · K, and density $\rho = 1005$ kg/m³. The body temperature before treatment was uniform at 37 C.

a. What is the thermal diffusivity of the tissue?

b. How long was the active treatment time of the tumor—i.e., how long was the tumor at a minimum of 45 C within the 2.0 min treatment time?

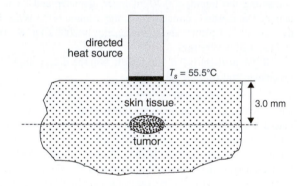

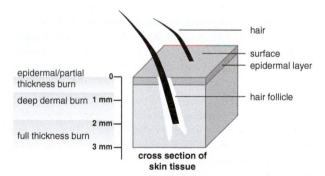

cross section of
skin tissue

SS **19.1** Using dimensional analysis, demonstrate that the parameters

$$\frac{T - T_\infty}{T_0 - T_\infty} \quad \frac{x}{L} \quad \frac{at}{L^2} \quad \text{and} \quad \frac{hL}{k}$$

are possible combinations of the appropriate variables in describing unsteady-state conduction in a plane wall.

19.2 Dimensional analysis has shown the following parameters to be significant for forced convection:

$$\frac{xv_\infty\rho}{\mu} \quad \frac{\mu c_p}{k} \quad \frac{hx}{k} \quad \frac{h}{\rho c_p v_\infty}$$

Evaluate each of these parameters at 340 K, for air, water, benzene, mercury, and glycerin. The distance x may be taken as 0.3 m, $v_\infty = 15$ m/s, and $h = 34$ W/m$^2 \cdot$ K.

19.3 Plot the parameters $xv_\infty\rho/\mu$, $\mu c_p/k$, hx/k, and $h/\rho c_p v_\infty$ vs. temperature for air, water, and glycerin, using the values for x, h, and v from Problem 19.2.

19.4 The fuel plates in a nuclear reactor are 4 ft long and stacked with a 1/2-in gap between them. The heat flux along the plate surfaces varies sinusoidally according to the equation

$$\frac{q}{A} = \alpha + \beta \sin\frac{\pi x}{L}$$

where $\alpha = 250$ Btu/h ft^2, $\beta = 1500$ Btu/h ft^2, x is the distance from the leading edge of the plates, and L is the total plate length. If air at 120 F, 80 psi, flowing at a mass velocity of 6000 lb$_m$/h ft^2, is used to cool the plates, prepare plots showing

a. the heat flux vs. x

b. the mean air temperature vs. x

SS **19.5** Given the information in Problem 19.4, determine the total heat transferred for a stack of plates with a combined surface area of 640 ft^2, each plate being 4 ft wide.

19.6 In a thermal heat sink' the heat flux variation along the axis of a cooling passage is approximated as

$$\frac{q}{A} = a + b \sin\left(\frac{\pi x}{L}\right)$$

where x is measured along the passage axis and L is its total length.

A large installation involves a stack of plates with a 3-mm air space between them. The flow passages are 1.22 m long, and the heat flux in the plates varies according to the above equation where $a = 900$ W/m^2 and $v = 2500$ W/m^2. Air enters at 100 C with a mass velocity (the product of ρV) of 7.5 kg/s $\cdot$ m^2. The surface coefficient along the flow passage can be considered constant with a value of 56 W/m$^2 \cdot$ K.

Generate a plot of heat flux, mean air temperature, and plate surface temperature as functions of x. Where does the maximum surface temperature occur and what is its value?

19.7 Solve Problem 19.25 for the case of a wall heat flux **SS** varying according to

$$\frac{q}{A} = \alpha + \beta \sin\frac{\pi x}{L}$$

where $\alpha = 250$ Btu/h ft^2, $\beta = 1500$ Btu/h ft^2, x is the distance from the entrance, and L is the tube length.

19.8 Glycerin flows parallel to a flat plate measuring 2 ft by 2 ft with a velocity of 10 fps. Determine values for the mean convective heat-transfer coefficient and the associated drag force imposed on the plate for glycerin temperatures of 350 F, 50 F, and 180 F. What heat flux will result, in each case, if the plate temperature is 50 F above that of the glycerin?

19.9 Given the conditions specified in Problem 19.8, construct a plot of local heat-transfer coefficient vs. position along the plate for glycerin temperatures of 30 F, 50 F, and 80 F.

19.10 Nitrogen at 100 F and 1 atm flows at a velocity of 100 fps. A flat plate 6 in wide, at a temperature of 200 F, is aligned parallel to the direction of flow. At a position 4 ft from the leading edge, determine the following (a) δ; (b) δ_t; (c) C_{fx}; (d) C_{fL}; (e) h_x; (f) h; (g) total drag force; (h) total heat transfer.

19.11 A plane surface 25 cm wide has its temperature **SS** maintained at 80 C. Atmospheric air at 25 C flows parallel to the surface with a velocity of 2.8 m/s. Using the results of boundary-layer analysis, determine the following for a 1-m long plate:

a. the mean coefficient of skin friction, C_{fL}

b. the total drag force exerted on the plate by the air flow

c. the total heat-transfer rate from the plate to the air stream

19.12 Use the results of Problem 19.20 along with those of Chapter 12 to determine δ, C_{fx}, δ_1, and h_x at a distance of 40 cm from the leading edge of a flat plane. Air with a free-stream velocity of 5 m/s and $T_\infty = 300$ K flows parallel to the plate surface. The first 20 cm of the plate is unheated; the surface temperature is maintained at 400 K beyond that point.

19.13 A blacktop road surface 18.3 m wide receives solar radiation at the rate of 284 W/m^2 at noon and 95 W/m$_2$ are lost by reradiation to the atmosphere. A wind, at 300 K, flows across the road. Determine the wind velocity that will cause the road surface to be at 308 K if all energy not reradiated to the sky is removed by convection.

19.14 An engineered tissue system consists of a flat plate of cell mass immobilized on a scaffold measuring 5 cm in length, and is 0.5 cm thick. The bottom face of the scaffold is exposed to water and organic nutrients. The top face is exposed to flowing O_2 gas to provide O_2 for aerobic respiration. At present, the specific oxygen consumption of the tissue mass is 0.5 mmol

O_2/cm^3 cells-hr, and from respiration energetics, the energy released by respiration is 468 J/mmol O_2 consumed. We are interested in using the flowing O_2 gas at 1 atm to control the temperature at the surface of the tissue scaffold. The properties of O_2 gas at 300 K are $\rho = 1.3\,kg/m^3$, $C_P = 920\,J/kg \cdot K$, $\mu = 2.06 \times 10^{-5}\,kg/m\,s$, and $k = 0.027\,W/m \cdot K$. We are interested in determining the O_2 flow rate necessary to keep the surface temperature within 10 C of the flowing gas temperature (i.e., surface temperature below 310 K or 37 C).

a. What is the Prandtl number (Pr) for the flowing fluid?
b. Based on the process energy balance and heat transfer, what is the required heat-transfer coefficient, h?
c. What is the Nusselt number, Nu?
d. The mean heat-transfer coefficient, h, is an integral averaged value obtained by integrating the local values of $h(x)$ from $x = 0$ to $x = L$. If the O_2 flow is assumed to be laminar (for flat plate, Re $< 2 \times 10^5$), what is the needed convective heat-transfer correlation?
e. What are the required values for Reynolds number (Re) and fluid velocity, v_∞?

19.15 A newly cast polymer sheet is laid down on a moving conveyer belt to cool. A side view of the process is shown below. The polymer sheet measures 2 m by 1.5 m and is 1.5 mm thick. Because the conveyer belt is an open mesh, both sides of the polymer sheet are assumed to be uniformly exposed to air. The conveyer belt moves at a linear velocity of 0.5 m/s. The bulk temperature of the surrounding air is constant at 27 C. At one location in the process, the surface temperature of the polymer sheet (both top and bottom surfaces) is 107 C. The thermal conductivity of the solid polymer at 107 C is 0.12 W/m · K.

a. What is the film temperature for the cooling process?
b. What is the total heat-transfer rate (in Watts) from the polymer sheet at the point in the process when the surface temperature of the polymer sheet is 107 C?

$T_\infty = 27$ C

$T_S = 107$ C 0.5 m/s → conveyer

2.0 m

$T_\infty = 27$ C

19.16 A 0.2-m-wide plate, 0.8 m long, is placed on the bottom of a shallow tank. The plate is heated, and maintained at a constant surface temperature of 60 C. Liquid water 12 cm deep flows over the flat plate with bulk volumetric flow rate of $2.4 \times 10^{-3}\,m^3/s$. The bulk liquid water temperature may be assumed to remain constant at 20 C as it flows along the length of the plate.

Thermophysical properties of water are given in the table below.

Temper-ature (C)	Density, $\rho(kg/m^3)$	Viscosity, μ (kg/ m-s)	Heat Capacity, C_p (J/kg · K)	Thermal Conductivity, k (W/m · K)
20	998.2	993×10^{-6}	4282	0.597
40	992.2	658×10^{-6}	4175	0.663
60	983.2	472×10^{-6}	4181	0.658

a. What are the values of Re_L and Pr for this convective heat-transfer process? Is the flow laminar or turbulent at the end of the plate?
b. What is the local heat flux at a distance 0.5 m from the leading edge of the plate?
c. What is the total heat-transfer rate from the plate surface?
d. What are the thicknesses of the hydrodynamic and thermal boundary layers at the end of the plate?

19.17 Show that, for the case of natural convection adjacent to a plane vertical wall, the appropriate integral equations for the hydrodynamic and thermal boundary layers are

$$\alpha \frac{\partial T}{\partial y}\bigg|_{y=0} = \frac{d}{dx}\int_0^{\delta_1} v_x(T_\infty - T)dy$$

and

$$-\nu \frac{dv_x}{dy}\bigg|_{y=0} + \beta g \int_0^{\delta_1}(T - T_\infty)dy = \frac{d}{dx}\int_0^{\delta} v_x^2 dy$$

19.18 Using the integral relations from Problem 19.17, and assuming the velocity and temperature profiles of the form

$$\frac{v}{v_x} = \left(\frac{y}{\delta}\right)\left(1 - \frac{y}{\delta}\right)^2$$

and

$$\frac{T - T_\infty}{T_s - T_\infty} = \left(1 - \frac{y}{\delta}\right)^2$$

where δ is the thickness of both the hydrodynamic and thermal boundary layers, show that the solution in terms of δ and v_x from each integral equation reduce to

$$\frac{2\alpha}{\delta} = \frac{d}{dx}\left(\frac{dv_x}{30}\right)$$

and

$$-\frac{\nu v_x}{\delta} + \beta g \Delta T \frac{\delta}{3} = \frac{d}{dt}\left(\frac{\delta v_x^2}{105}\right)$$

Next, assuming that both δ and v_x vary with x according to

$$\delta = Ax^a \quad \text{and} \quad v_x = Bx^b$$

show that the resulting expression for δ becomes

$$\delta/x = 3.94\,Pr^{-1/2}(Pr + 0.953)^{1/4}Gr_x^{-1/4}$$

and that the local Nusselt number is

$$Nu_x = 0.508\quad Pr^{-1/2}(Pr + 0.953)^{-1/4}Gr_x^{1/4}$$

19.19 Using the relations from Problem 19.8, determine, for the case of air at 310 K adjacent to a vertical wall with its surface at 420 K,

a. the thickness of the boundary layer at $x = 15\,cm$, 30 cm, 1.5 m

b. the magnitude of h_x at 15 cm, 30 cm, 1.5 m

19.20 Using the appropriate integral formulas for flow parallel to a flat surface with a constant free-stream velocity, develop expressions for the local Nusselt number in terms of Re_x and Pr for velocity and temperature profiles of the form

$$v = a + by, \quad T - T_s = \alpha + \beta y$$

19.21 Shown in the figure is the case of a fluid flowing parallel to a flat plate, where for a distance X from the leading edge, the plate and fluid are at the same temperature. For values of $x > X$, the plate is maintained at a constant temperature, T_s, where $T_s > T_\infty$. Assuming a cubic profile for both the hydrodynamic and the thermal boundary layers, show that the ratio of the thickness, ξ, is expressed as

$$\xi = \frac{\delta_t}{\delta} \cong \frac{1}{Pr^{1/3}}\left[1 - \left(\frac{X}{x}\right)^{3/4}\right]^{1/3}$$

Also show that the local Nusselt number can be expressed as

$$Nu_x \cong 0.33\left(\frac{Pr}{1 - (X/x)^{3/4}}\right)^{1/3}Re_x^{1/2}$$

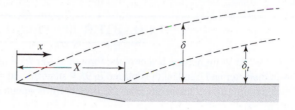

19.22 Repeat Problem 19.20 for velocity and temperature profiles of the form

$$v = a\sin by, \quad T - T_s = \alpha\sin\beta y$$

19.23 Water, at 60 F, enters a 1-in-ID tube that is used to cool a nuclear reactor. The water flow rate is 30 gal/min. Determine the total heat transfer and the exiting water temperature for a 15-ft-long tube if the tube surface temperature is a constant value of 300 F. Compare the answer obtained using the Reynolds and Colburn analogies.

19.24 Work Problem 19.23 for the case in which the flowing fluid is air at 15 fps.

19.25 Water at 60 F enters a 1-in-ID tube that is used to cool a nuclear reactor. The water flow rate is 30 gal/min. Determine the total heat transfer, the exiting water temperature, and the wall temperature at the exit of a 15-ft-long tube if the tube wall condition is one of uniform heat flux of 500 Btu/hr ft^2.

19.26 Work Problem 19.25 for the case in which the flowing fluid is air at 15 fps.

19.27 Work Problem 19.25 for the case in which the flowing fluid is sodium entering the tube at 200 F.

CHAPTER 20   PROBLEMS

SS **20.1**   A 750-W immersion heater in the form of a cylinder with 3/4-in diameter and 6 in length is placed in 95 F stagnant water. Calculate the surface temperature of the heater if it is oriented with its axis

a.   vertical

b.   horizontal

20.2   Repeat Problem 20.1 if the stagnant liquid is

a.   bismuth at 700 F

b.   hydraulic fluid at 0 F

20.3   An immersion heater, rated at 1000 W, is in the form of a rectangular solid with dimensions 16 cm by 10 cm by 1 cm. Determine the surface temperature of the heater if it is oriented in 295 K water with

a.   the 16-cm dimension vertical

b.   the 10-cm dimension vertical

20.4   A 2-in copper cylinder, 6 in length, at a uniform temperature of 200 F, is plunged vertically into a large tank of water at 50 F.

a.   How long will it take for the outside surface of the cylinder to reach 100°F?

b.   How long will it take for the center of the cylinder to reach 100 F?

c.   What is the surface temperature when the center temperature is 100 F? Heat transfer from ends of the cylinder may be neglected.

SS **20.5**   A fluorescent light bulb, rated at 100 W, is illuminated in air at 25 C and atmospheric pressure. Under these conditions the surface temperature of the glass is 140 C.

Determine the rate of heat transfer from the bulb by natural convection. The bulb is cylindrical, having a diameter of 35 mm and a length of 0.8 m, and is oriented horizontally.

20.6   Determine the steady-state surface temperature of an electric cable, 25 cm in diameter, which is suspended horizontally in still air in which heat is dissipated by the cable at a rate of 27 W per meter of length. The air temperature is 30 C.

20.7   A copper cylinder 20.3 cm long with a diameter of 2.54 cm is being used to evaluate the surface coefficient in a laboratory experiment. When heated to a uniform temperature of 32.5 C and then plunged into a −1 C liquid bath, the center temperature of the cylinder reaches a value of 4.8 C in 3 min. Assuming the heat exchange between the cylinder and water bath to be purely by convection, what value for the surface coefficient is indicated?

20.8   Rubber balls are molded into spheres and cured at 360 K. Following this operation they are allowed to cool in room air. What will be the elapsed time for the surface temperature of a solid rubber ball to reach 320 K when the surrounding air temperature is 295 K? Consider balls with diameters of 7.5, 5, and 1.5 cm. Properties of rubber that may be used are $k = 0.24$ W/m · K, $\rho = 1120$ kg/m^3, $c_p = 1020$ J/kg · K.

20.9   Determine the required time for the rubber balls described **SS** in Problem 20.8 to reach the condition such that the center temperature is 320 K. What will be the surface temperature when the center temperature reaches 320 K?

20.10   A 1-in, 16-BWG copper tube has its outside surface maintained at 240 F. If this tube is located in still air at 60 F, what heat flux will be achieved if the tube is oriented

a.   horizontally?

b.   vertically?

The tube length is 10 ft.

20.11   Solve Problem 20.10 if the medium surrounding the tube is stagnant water at 60 F.

20.12   A 0.6-m-diameter spherical tank contains liquid oxygen at 78 K. This tank is covered with 5 cm of glass wool. Determine the rate of heat gain if the tank is surrounded by air at 278 K. The tank is constructed of stainless steel 0.32 cm thick.

20.13   A "swimming-pool" nuclear reactor, consisting of 30 **SS** rectangular plates measuring 1 ft in width and 3 ft in height, spaced 2 1/2 in apart, is immersed in water at 80°F. If 200 F is the maximum allowable plate temperature, what is the maximum power level at which the reactor may operate?

20.14   A solar energy collector measuring 20 × 20 ft is installed on a roof in a horizontal position. The incident solar energy flux is 200 Btu/h ft^2, and the collector surface temperature is 150 F. What fraction of incident solar energy is lost by convection to the stagnant surrounding air at a temperature of 50 F? What effect on the convective losses would result if the collector were criss-crossed with ridges spaced 1 ft apart?

20.15   Given the conditions for Problem 20.14, determine the fraction of incident solar energy lost by convection to the surrounding air at 283 K flowing parallel to the collector surface at a velocity of 6.1 m/s.

20.16   Copper wire with a diameter of 0.5 cm is covered with a 0.65-cm layer of insulating material having a thermal conductivity of 0.242 W/m · K. The air adjacent to the insulation is at 290 K. If the wire carries a current of 400 A, determine

a.   the convective heat-transfer coefficient between the insulation surface and the surrounding air

b.   the temperatures at the insulation-copper interface and at the outside surface of the insulation

20.17 Work Problem 20.16 for an aluminum conductor of the same size (resistivity of aluminum $= 2.83 \times 10^{-6}$ ohm-cm).

20.18 Air at 60 F and atmospheric pressure flows inside a 1-in, 16-BWG copper tube whose surface is maintained at 240 F by condensing steam. If the steam line is now bare and surrounded by still air at 70 F, what total heat transfer would be predicted from a 20-ft length of bare pipe? Consider the bare pipe to be a black surface and the surroundings black at 70 F.

20.19 Solve Problem 20.18 if the bare pipe is located so that 295 K air flows normal to the pipe axis at a velocity of 6.5 m/s.

20.20 Solve Problem 20.18 if 3 in of insulation having a thermal conductivity of 0.060 Btu/h ft F is applied to the outside of the pipe. Neglect radiation from the insulation. What will be the outside surface temperature of the insulation?

20.21 What thickness of insulation having a thermal conductivity as given in Problem 20.20 must be added to the steam pipe where air at 60 F and atmospheric pressure flows inside a 1-in, 16-BWG copper tube whose surface is maintained at 240 F by condensing steam, in order that the outside temperature of the insulation not exceed 250 F?

20.22 A cooking oven has a top surface temperature of 45 C when exposed to still air. At this condition the inside oven temperature and room air temperature are 180 C and 20 C, respectively, and heat is transferred from the top surface at 40 W.

To reduce the surface temperature, as required by safety regulations, room air is blown across the top with a velocity of 20 m/s. Conditions inside the oven may be considered unchanged.

a. What will be the rate of heat loss under this new operating condition?

b. What will be the top surface temperature?

20.23 A "hot stage" for a manufacturing process is shown in the accompanying figure, where a slab of stainless steel is mounted on top of an electrical resistance heater. The stainless-steel slab is 20 cm wide, 60 cm long, and 2 cm thick. A load material, which is placed on top of the stainless-steel hot plate, requires a power input of 600 W. The dimensions of the load material are 20 cm in width, 60 cm in length (in the direction of air flow), and 0.5 cm thick. The surface temperature of the load material exposed to the air is measured to be $T_s = 207$ C (480 K). The air flow has a bulk velocity of 3 m/s, and the ambient temperature of the air is maintained at 47 C (320 K). The thermal conductivity of stainless steel (hot stage) is $k_s = 16$ W/m · K, and the thermal conductivity of the load material is $k_m = 2$ W/m · K.

a. Estimate the heat loss from the surface of load material, in Watts.

b. What is the total power output, in Watts, required for the heater, assuming there are no heat-transfer losses from the sides of the plate?

c. What is the temperature at the interface between the stainless-steel hot stage and the load material?

d. What is the estimated temperature at the base of the hot stage in contact with the electrical resistance heater (T_2)?

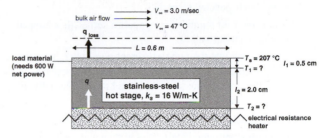

20.24 Saturated steam at 0.1 bar condenses on the outside of a copper tube having inner and outer diameters of 16.5 and 19 mm, respectively. The surface coefficients on the inner (water) surface and outer (steam) surface are 5200 and 6800 W/m² · K, respectively.

When the mean water temperature is 28 K, estimate the rate of steam condensed per meter of tube length. The latent heat of condensation of steam may be taken as 2.390 kJ/kg.

20.25 A 1-in, 16-BWG copper tube, 10 ft long, has its outside surface maintained at 240 F. Air at 60 F and atmospheric pressure is forced past this tube with a velocity of 40 fps. Determine the heat flux from the tube to the air if the flow of air is

a. parallel to the tube

b. normal to the tube axis

20.26 Solve Problem 20.25 if the medium flowing past the tube in forced convection is water at 60 F.

20.27 Solve Problem 20.25 if the medium flowing past the tube in forced convection is MIL-M-5606 hydraulic fluid.

20.28 What would be the results of Problem 20.18 if a fan provided an air flow normal to the conductor axis at a velocity of 9 m/s?

20.29 An electric light bulb rated at 60 W has a surface temperature of 145 C when cooled by atmospheric air at 25 C. The air flows past the bulb with a velocity of 0.5 m/s. The bulb can be modeled as a sphere with a diameter of 7.5 cm. Determine the heat transfer from the bulb by the mechanism of forced convection.

20.30 We are interested in predicting the time it will take for popcorn kernels to pop in a fluidized bed. The superficial air velocity inside the bed is 3 m/s, which is above the minimum fluidization velocity for popcorn kernels. The air temperature is maintained constant at 520 K (247 C) by heating coils placed around the fluidization vessel. The average diameter of the popcorn kernels is 0.5 cm. When the temperature of a kernel reaches 175 C, it will pop. At 25 C, the thermal conductivity of popcorn is 0.2 W/m · K, the specific heat is 2000 J/kg · K, and the density is 1300 kg/m³.

a. What is the convective heat-transfer coefficient for air flow around each popcorn kernel?

b. What is the Biot number for a popcorn kernel? Is a lumped-parameter model valid?

c. If we assume that the popcorn kernel pops when its center-line temperature reaches 175 C, how long after entering the bed will it take for the popcorn kernel to pop? Initially, the popcorn kernels are at 25 C.

20.31 Hot polymer beads are being cooled by allowing them to settle slowly in a long cylindrical tank of water maintained at 30 C as shown in the figure. In this process, 1-cm-diameter spherical beads settle at a terminal velocity of 3 cm/s, which is based on the size of the bead and density of the bead material relative to the density of water. Properties of water can be taken at its bulk temperature of 30 C, where the Prandtl number is 4.33, the kinematic viscosity is 6.63×10^{-7} m²/s, and the thermal conductivity is 0.633 W/m · K. The thermal conductivity of the polymer bead is 0.08 W/m · K. If the polymer beads are at a temperature of 100 C, what is the total heat-transfer rate from one bead?

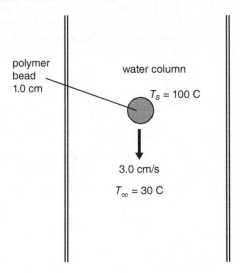

polymer bead 1.0 cm

water column

T_s = 100 C

3.0 cm/s

T_∞ = 30 C

20.32 A hot-wire anemometer is a common instrument used to measure the velocity of a flowing gas. A typical probe tip is shown below.

Electrical current is passed through a very thin wire of platinum or tungsten. The electrical resistance of the wire generates heat as the current passes through it, making the wire hot. This "hot wire" is placed normal to the flowing gas stream and the heat generated by the wire is dissipated to the gas stream flowing around the wire by convective heat transfer. The current passing through the wire and temperature of the wire are measured, and, from this information, the velocity of the gas flowing over the wire is estimated.

In the current set of measurements, the hot-wire anemometer is mounted inside a tube containing flowing N₂ gas at 20 C. During one test, the power load on the wire was 13 mW, and the temperature of the wire was 200 C. The diameter of the platinum wire is 4 μm, and its length is 1.2 mm.

a. What is the Prandtl number for the N₂ gas flowing around the hot wire?

b. What is the measured Nusselt number (Nu) for the N₂ gas flowing around the hot wire?

c. What is the estimated velocity of the flowing N₂ gas inside the tube at the point where the hot-wire anemometer probe tip is located?

Thermophysical properties of N₂ gas

Temperature (C)	Density, ρ (kg/m³)	Viscosity, μ (kg/m-sec)	Heat Capacity, C_p (J/kg · K)	Thermal Conductivity, k (W/m · K)
350	0.9754	2.0000×10^{-5}	1042.1	0.029691
400	0.8533	2.19950×10^{-5}	1049.9	0.033186

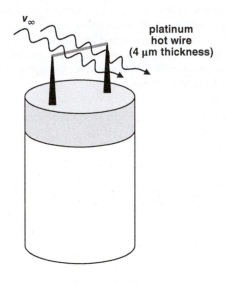

v_∞

platinum hot wire (4 μm thickness)

20.33 The SolarWorld manufacturing facility in Portland, Oregon, is the largest producer of silicon-based solar cells in the United States. The manufacturing process begins with the melt crystallization of electronic grade silicon into a cylindrical ingot of polycrystalline silicon. The newly formed silicon ingots must then be cooled before they can be sliced into thin silicon wafers. In the present cooling process, a 30-cm-diameter by 3-m-long solid silicon ingot is placed vertically within a flowing air stream as shown in the figure. Air maintained at 27 C flows normal to the ingot at a velocity of 0.5 m/s. Heat transfer from the ends of the ingot can be neglected. At some point in time during the cooling process, the surface temperature of the wafer is measured to be 860 K (587 C). Parts (a) through (d) all refer to this condition.

a. What is the fluid Reynolds number, Re?

b. What is the total cooling rate (Watts) from a single silicon ingot at this time in the cooling process?

c. What is the Biot number for the heat-transfer process?

d. If the air flow around the ingot is turned off, will the heat-transfer process be dominated by natural convection, or will the convective heat-transfer process be limited to $Nu = 0.30$? What will be the new cooling rate?

e. After what elapsed time does the surface of the silicon ingot reach 860 K? Can lumped-parameter analysis be used?

Properties of silicon at 860 K: $\rho_{Si} = 2300 \text{ kg/m}^3$, $C_{p,Si} = 760 \text{ J/kg} \cdot \text{K}$, $k_{Si} = 30 \text{ W/m} \cdot \text{K}$.

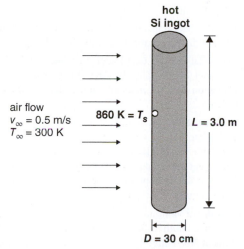

hot Si ingot

air flow
$V_\infty = 0.5 \text{ m/s}$
$T_\infty = 300 \text{ K}$

$860 \text{ K} = T_s$

$L = 3.0 \text{ m}$

$D = 30 \text{ cm}$

20.34 An industrial heater is composed of a tube bundle consisting of horizontal 3/8-in-OD tubes in a staggered array with tubes arranged in equilateral triangle fashion having a pitch-to-diameter ratio of 1.5. If water at 160 F flows at 20 ft/s past the tubes with constant surface temperature of 212 F, what will be the effective heat-transfer coefficient?

20.35 For the heater consisting of the tube bank described in Problem 20.32, evaluate the heat transferred to the water if the tube array consists of six rows of tubes in the flow direction with eight tubes per row. The tubes are 5 ft long.

20.36 A tube bank employs an in-line arrangement with $S_T = S_L = 3.2$ cm and tubes that are 1.8 cm in outside diameter. There are 10 rows of tubes, which are held at a surface temperature of 85 C. Air at atmospheric pressure and 20 C flows normal to the tubes with a free-stream velocity of 6 m/s. The tube bank is 10 rows deep, and the tubes are 1.8 m long. Determine the amount of heat transferred.

20.37 Rework Problem 20.36 for a staggered arrangement.

20.38 Oil at 300 K is heated by steam condensing at 372 K on the outside of steel pipes with ID = 2.09 cm, OD = 2.67 cm. The oil flow rate is 1.47 kg/s; six tubes, each 2.5 m long, are used. The properties of oil to be used are as follows:

T, K	ρ, kg/m³	C_p, J/kg · K	k, W/m · K	μ, Pa · s
300	910	1.84×10^3	0.133	0.0414
310	897	1.92×10^3	0.131	0.0228
340	870	2.00×10^3	0.130	7.89×10^{-3}
370	865	2.13×10^3	0.128	3.72×10^{-3}

Determine the rate of heat transfer to the oil.

20.39 Engine oil with properties given below flows at a rate of 136 kg per hour through a 7.5-cm-ID pipe whose inside surface is maintained at 100 C. If the oil enters at 160 C, what will its temperature be at the exit of a 15-m-long pipe?

T, K	ρ, kg/m³	C_p, J/kg · K	k, W/m · k	v, m²/s $\times 10^3$	Pr
373	842	2.219	0.137	0.0203	276
393	831	2.306	0.135	0.0124	175
413	817	2.394	0.133	0.0080	116
433	808	2.482	0.132	0.0056	84

20.40 An apparatus, used in an operating room to cool blood, consists of a coiled tube that is immersed in an ice bath. Using this apparatus, blood, flowing at 0.006 m³/h, is to be cooled from 40 C to 30 C. The inside diameter of the tube is 2.5 mm, and the surface coefficient between the ice bath and outer tube surface is 500 W/m² · K. The thermal resistance of the tube wall may be neglected.

Determine the required length of tubing to accomplish the desired cooling. Properties of blood are the following:

$$\rho = 1000 \text{ kg/m}^2$$
$$k = 0.5 \text{ W/m} \cdot \text{K}$$
$$C_p = 4.0 \text{ kJ/kg} \cdot \text{K}$$
$$v = 7 \times 10^{-7} \text{ m}^2/\text{s}$$

Determine the total heat transfer to the oil and its temperature at the heater exit.

20.41 A 1.905-cm-diameter brass tube is used to condense steam on its outer surface at 10.13 kPa pressure. Water at 290 K is circulated through the tube. Inside and outside surface coefficients are 1700 and 8500 W/m² · K, respectively. Find the rate of steam condensed per hour per meter of the tube length under these conditions. The following information is pertinent:

tube OD = 1.905 cm

tube ID = 1.656 cm

steam saturation temperature = 319.5 K

steam latent heat, $h_{fg} = 2393$ kJ/kg.

20.42 A system for heating water with an inlet temperature of 25 C to an exiting temperature of 70 C involves passing the water through a thick-walled tube with inner and outer diameters of 25 and 45 mm, respectively. The outer tube surface is well insulated, and the electrical heating within the tube wall provides for a uniform generation of $\dot{q} = 1.5 \times 10^6 \, \text{W/m}^3$.

a. For a mass flow rate of water, $\dot{m} = 0.12 \, \text{kg/s}$, how long must the tube be to achieve the desired outlet temperature?

b. If the inner surface of the tube at the outlet is $T_s = 110 \, \text{C}$, what is the local convective coefficient at this location?

20.43 Air at atmospheric pressure and 10 C enters a rectangular duct that is 6 m long having dimensions of 7.5 and 15 cm in cross section. The surfaces are maintained at 70 C by solar irradiation. If the exiting air temperature is to be 30 C, what is the required flow rate of the air?

20.44 Compressed natural gas is fed into a small underground pipeline having an inside diameter of 5.9 in. The gas enters at an initial temperature of 120 C and constant pressure of 132.3 psig. The metered volumetric flow rate at standard pressure of 14.7 psia and standard temperature of 25 C is 195.3 ft³/min. The ground temperature is constant at 15 C, as it serves as an "infinite sink" for heat transfer.

a. Show that the mass flow rate is 0.0604 kg/s, volumetric flow rate is 0.01 m³/s at 50 C and 10 atm, the total system pressure is 10 atm, and the gas density, assuming ideal gas behavior, is 6.04 kg/m³ at 50 C. Other properties (not dependent on pressure) for methane at 50 C are $k = 0.035 \, \text{W/m} \cdot \text{K}$, $C_p = 220 \, \text{J/kg} \cdot \text{K}$, and $\mu = 1.2 \times 10^{-5} \, \text{kg/m-s}$.

b. Develop an energy balance model to predict the steady-state temperature profile of the natural gas in the pipe.

c. Is the gas flow laminar, or turbulent?

d. What are the Prandtl, Nusselt, and Stanton numbers for methane in the pipe? What is a reasonable average temperature for estimation the thermophysical properties?

e. What is the heat-transfer coefficient for methane in the pipe?

f. Finally, how far from the pipe entrance will the methane gas reach a temperature of 50 C? How much heat is transferred at this point?

20.45 Microchannel technology is touted as means for process intensification, which in simple terms means shrinking the size of process equipment (heat exchangers, chemical reactors, chemical separation processes, etc.) to do the same job that a larger piece of equipment would do. In processes that require convective heat transfer (usually all of the above), process intensification basically comes down to increasing the convective heat-transfer coefficient.

Consider a single tube of 1-cm inner diameter. Microchannel fabrication technology offers a means to make a parallel array of tiny tubes or "microchannels" to serve in place of a single tube. For example, a single 1-cm-inner-diameter tube could be broken up into 1000 microchannels, each of 316 micron (μm) inner diameter, all imbedded within a single block of thermally conducting material. Metal is often the preferred material but is difficult to work with. For small-scale applications, silicon is also commonly used, where microchannels are etched into silicon using technologies originally developed by the microelectronic device fabrication industry.

a. Compare the total surface area for heat transfer of the single tube and the parallel array of microchannels described above.

Now consider that the total mass flow rate of liquid water into the process is 60 g/m, and that the physical properties of water are constant over the temperature range of interest with $\nu = 1.0 \times 10^{-6} \, \text{m}^2/\text{s}$, $\alpha = 1.4 \times 10^{-7} \, \text{m}^2/\text{s}$, $\rho = 1000 \, \text{kg/m}^3$, and $k = 0.6 \, \text{W/m} \cdot \text{K}$. The walls of the tube are maintained at constant temperature.

b. What is the heat-transfer coefficient associated with the single 1-cm-diameter tube 10 cm in length?

c. What the heat-transfer coefficient associated with a single microchannel of the same length? What level of process intensification does the microchannel system offer over the single tube?

d. Assume the solid surface temperature is maintained constant at 100 C. Compare the temperatures of the fluid exiting the tube for both the single tube and the microchannel array when the inlet temperature is 20 C. Assume also as a first estimate that the fluid properties given above are constant with temperature.

20.46 Consider a continuous-flow device described below used to pasteurize liquid milk, whose properties approach those of water. Milk at 20 C enters a preheater tube at a volumetric flow rate of 6 L/min. The inner diameter of the tube is 1.5 cm. The tube winds through a steam chest, which heats the milk. The outer tube wall temperature is maintained at 115 C using pressurized steam. Since the thermal conductivity through the metal is high, the temperature difference across the wall is small, so it may be assumed that the inside wall temperature is also at 115 C. The heated milk then enters an adiabatic holding tube, which provides sufficient fluid residence time so that most of the microorganisms in the milk are destroyed at ~70 C.

a. Develop an energy-balance model to predict the steady-state temperature profile of the milk as it moves down the pipe.

b. Is the milk in turbulent or laminar flow?

c. What are the Prandtl, Nusselt, and Stanton numbers for milk in the pipe? What is the film temperature to be used for estimating thermophysical properties? What is the average heat-transfer coefficient for the milk flowing inside the tube?

d. How long must the tube be to achieve the exit temperature of 70 C?

20.47 The Cooper Cooler is a novel cooling device that rotates a cylindrical beverage can within a chilled environment in order to cool down the beverage. In a related process shown in the figure, the correlation for convective heat transfer associated with the liquid film on the inside surfaces of a rotating can is given by (Anantheswaran and Rao, 1985)

$$\text{Nu} = 2.9\,\text{Re}^{0.436}\text{Pr}^{0.287} \quad \text{with} \quad \text{Re} = \frac{D_r^2 N \rho}{\mu}$$

where N is the rotation rate (rev/s) and D_r is the diameter of the rotating wheel. In the present process, the chilling chamber is kept constant at 0 C. The convective heat-transfer coefficient on the outside surface of the can is estimated at 1000 W/m^2 · K, and the heat-transfer resistance through the metal can is considered negligible. If the can is 6 cm in diameter, 10 cm long, and $D_r/2$ is 12 cm, how long will it take to cool the contents of the can down to a refreshing 4 C if the beverage can is initially at 20 C and the wheel is rotated at 0.5 rev/s? It is assumed that the thermophysical properties of water apply for the beverage.

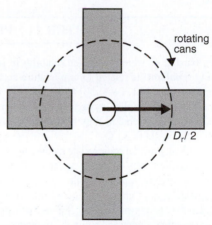

Ref: R. C. Anantheswaran and M. A. Rao, *J. Food Engineering*, **4**, 1–9 (1985).

The surface tension of water, a needed quantity in several of the following problems, is related to temperature according to the expression $\sigma = 0.1232\,[1 - 0.00146\,T]$, where σ is in N/m and T is in K. In the English system with σ given in lb_f/ft and T in F, the surface tension may be calculated from $\sigma = (8.44 \times 10^{-3})\,[1 = 0.00082\,T]$.

21.1 An electrically heated square plate measuring 200 cm on a side is immersed vertically in water at atmospheric pressure. As the electrical energy supplied to the plate is increased, its surface temperature rises above that of the adjacent saturated water. At low power levels the heat-transfer mechanism is natural convection, becoming a nucleate-boiling phenomenon at higher ΔTs. At what value of ΔT are the heat fluxes due to boiling and natural convection the same? Plot $q/A|_{convection}$, $q/A|_{boiling}$, and $q/A|_{total}$, versus ΔT values from 250 to 300 K.

21.2 Plot values of the heat-transfer coefficient for the case of pool boiling of water on horizontal metal surfaces at 1 atm total pressure and surface temperatures varying from 390 to 450 K. Consider the following metals: (a) nickel; (b) copper; (c) platinum; (d) brass.

SS **21.3** A cylindrical copper heating element 2 ft long and $\frac{1}{2}$ in diameter is immersed in water. The system pressure is maintained at 1 atm and the tube surface is held at 280 F. Determine the nucleate-boiling heat-transfer coefficient and the rate of heat dissipation for this system.

21.4 If the cylinder described in Problem 21.3 were initially heated to 500 F, how long would it take for the center of the cylinder to cool to 340 F if it were constructed of

a. copper?

b. brass?

c. nickel?

SS **21.5** Four immersion heaters in the shape of cylinders 15 cm long and 2 cm in diameter are immersed in a water bath at 1 atm total pressure. Each heater is rated at 500 W. If the heaters operate at rated capacity, estimate the temperature of the heater surface. What is the convective heat-transfer coefficient in this case?

21.6 Two thousand watts of electrical energy are to be dissipated through copper plates measuring 5 cm by 10 cm by 0.6 cm immersed in water at 390 K. How many plates would you recommend? Substantiate all of the design criteria used.

SS **21.7** A horizontal circular cylinder 1 in diameter has its outside surface at a temperature of 1200 F. This tube is immersed in saturated water at a pressure of 40 psi. Estimate the heat flux due to

film boiling that may be achieved with this configuration. At 40 psi, the temperature of saturated water is 267 F.

21.8 Estimate the heat-transfer rate per foot of length from a 0.02-in diameter nichrome wire immersed in water at 240 F. The wire temperature is 2200 F.

21.9 A circular pan has its bottom surface maintained at 200 F and is situated in saturated steam at 212 F. Construct a plot of condensate depth in the pan vs. time up to 1 h for this situation. The sides of the pan may be considered nonconducting.

21.10 A square pan measuring 40 cm on a side and having a 2-cm-high lip on all sides has its surface maintained at 350 K. If this pan is situated in saturated steam at 372 K, how long will it be before condensate spills over the lip if the pan is

a. horizontal?

b. inclined at 10 degrees to the horizontal?

c. inclined at 30 degrees to the horizontal?

21.11 A square pan with sides measuring 1 ft and a perpendicular lip extending 1 in above the base is oriented with its base at an angle of 20 degrees from the horizontal. The pan surface is kept at 180 F and it is situated in an atmosphere of 210 F steam. How long will it be before condensate spills over the lip of the pan?

21.12 Saturated steam at atmospheric pressure condenses on the outside surface of a 1-m-long tube with 150-mm diameter. The surface temperature is maintained at 91 C. Evaluate the condensation rate if the pipe is oriented

a. vertically

b. horizontally

21.13 Saturated steam at 365 K condenses on a 2-cm tube whose surface is maintained at 340 K. Determine the rate of condensation and the heat-transfer coefficient for the case of a 1.5-m-long tube oriented

a. vertically

b. horizontally

21.14 If eight tubes of the size designated in Problem 21.13 are oriented horizontally in a vertical bank, what heat-transfer rate will occur?

21.15 Determine the heat-transfer coefficient for a horizontal $\frac{5}{8}$-in.-OD tube with its surface maintained at 100 F surrounded by steam at 200 F.

SS **22.1** A single tube-pass heat exchanger is to be designed to heat water by condensing steam in the shell. The water is to pass through the smooth horizontal tubes in turbulent flow, and the steam is to be condensed dropwise in the shell. The water flow rate, the initial and final water temperatures, the condensation temperature of the steam, and the available tube-side pressure drop (neglecting entrance and exit losses) are all specified. In order to determine the optimum exchanger design, it is desirable to know how the total required area of the exchanger varies with the tube diameter selected. Assuming that the water flow remains turbulent and that the thermal resistance of the tube wall and the steam–condensate film is negligible, determine the effect of tube diameter on the total area required in the exchanger.

SS **22.2** An oil having a specific heat of 1880 J/kg·K enters a single-pass counterflow heat exchanger at a rate of 2 kg/s and a temperature of 400 K. It is to be cooled to 350 K. Water is available to cool the oil at a rate of 2 kg/s and a temperature of 280 K. Determine the surface area required if the overall heat-transfer coefficient is 230 W/m²·K.

SS **22.3** Air at 203 kPa and 290 K flows in a long rectangular duct with dimensions 10 cm by 20 cm. A 2.5-m length of this duct is maintained at 395 K, and the average exit air temperature from this section is 300 K. Calculate the air flow rate and the total heat transfer.

22.4 Water at 50 F is available for cooling at a rate of 400 lb$_m$/h. It enters a double-pipe heat exchanger with a total area of 18 ft². Oil, with $c_p = 0.45$ Btu/lb$_m$ F, enters the exchanger at 250 F. The exiting water temperature is limited at 212 F, and the oil must leave the exchanger at no more than 160 F. Given the value of $U = 60$ Btu/h ft² F, find the maximum flow of oil that may be cooled with this unit.

22.5 A double-pipe, countercurrent heat exchanger is used to cool a hot liquid stream of biodiesel from 60 C to 35 C. Biodiesel flows on the shell side with a volumetric flow rate of 3 m³/h; water is used on the tube side. The entrance and exit temperatures for cooling water flow are 10 C and 30 C, respectively. The inside diameter of the schedule 80, 1.5 in copper tube used within the heat exchanger is 1.5 in (3.81 cm). In independent experiments, the heat-transfer coefficient for biodiesel on the shell side was determined to be 340 W/m²·K. Since the tube wall is made of copper, thermal conduction resistances can be neglected. Average properties of liquid biodiesel (B) from 35–60 C are: $\rho_B = 880$ kg/m³, $\mu_B = 4.2 \times 10^{-3}$ kg/m-s, $k_B = 0.15$ W/m·K, $C_{p,B} = 2400$ J/kg·K.

a. What is the heat load on the heat exchanger, and what mass flow rate of cooling water is required?

b. What is the required heat-exchanger surface area?

c. What would be the required heat exchanger for cocurrent flow?

22.6 As part of a food-processing operation, a single-pass, countercurrent, shell-and-tube heat exchanger must be designed to warm up a stream of liquid glycerin. The heat exchanger must be constructed of 0.75-in outer diameter, gage 12 (relatively thick-walled) stainless-steel tubes ($k = 43$ W/m·K) with inner diameters of 0.532 in (1.35 cm) and total length of 5 m. In the present process, liquid glycerin enters the tube side of the heat exchanger at 16 C and exits at 60 C. The mass flow rate of liquid glycerin, $\dot{m}_G$, that must be processed is 5 kg/s, and the desired bulk fluid velocity in each tube is 1.80 m/s. The process will use saturated condensing steam at 1 atm pressure (100 C, enthalpy of vaporization 2257 kJ/kg) on the shell side of the heat exchanger to warm up the glycerin. The shell-side convective heat-transfer coefficient for condensing steam can be found in Table 15.1. Glycerin is a fairly viscous fluid.

a. What is the mass flow rate of steam ($\dot{m}_S$, kg/s) required on the shell side, assuming all the steam is condensed and exits in the exchanger as a saturated liquid? What is the log-mean temperature difference of the heat exchanger?

b. What is the overall heat-transfer coefficient associated with this process?

c. What is the required area of the heat exchanger?

d. If the configuration is changed from countercurrent operation to cocurrent operation, what will be the required heat-transfer area?

22.7 A double-pipe, countercurrent heat exchanger is used to cool saturated steam at 2 bar ($T_s = 120$ C) to a saturated liquid. At this temperature and pressure, the inlet volumetric flow rate of steam is 2.42 m³/m, the density of steam is 1.13 kg/m³, and the enthalpy of vaporization of steam is 2201.6 kJ/kg. Cooling water enters the tube side at a rate of 100 kg/s at a temperature 20 C. In independent experiments, the heat-transfer coefficient for steam on the shell side was determined to be 500 W/m²·K. The tube wall is made of copper; thus, thermal conduction resistances can be neglected.

a. What is the heat load on the heat exchanger and the outlet temperature of the cooling water?

b. It is desired to achieve a Reynolds number of 20,000 inside the tube. What is the inner diameter of the tube needed, and what is the overall heat-transfer coefficient?

c. What is the required heat-transfer area for the heat exchanger?

22.8 Water enters a counterflow, double-pipe heat exchanger at a rate of 150 lb$_m$/min and is heated from 60 F to 140 F by an oil with a specific heat of 0.45 Btu/lb$_m$ F. The oil enters at 240 F and leaves at 80 F. The overall heat-transfer coefficient is 150 Btu/h ft² F.

a. What heat-transfer area is required?

b. What area is required if all conditions remain the same except that a shell-and-tube heat exchanger is used with the water making one shell pass and the oil making two tube passes?

c. What exit water temperature would result if, for the exchanger of part (a), the water flow rate were decreased to 120 lb$_m$/min?

22.9 A shell-and-tube heat exchanger is used for the heating of oil from 20 C to 30 C; the oil flow rate is 12 kg/s (C_{po} = 2.2 kJ/ kg · K). The heat exchanger has one shell pass and two tube passes. Hot water (C_{pw} = 4.18 kJ/kg · K) enters the shell at 75 C and leaves the shell at 55 C. The overall heat-transfer coefficient based on the outside surface of the tubes is estimated to be 1080 W/m^2 · K. Determine

a. the corrected logarithmic-mean temperature difference

b. the required surface area in the exchanger

22.10 A shell-and-tube heat exchanger is used to cool oil (C_H = 2.2 kJ/kg · K) from 110 C to 65 C. The heat exchanger has two shell passes and four tube passes. The coolant (C_c = 4.20 kJ/kg · K) enters the shell at 20 C and leaves the shell at 42 C. For an overall tube-side heat-transfer coefficient of 1200 W/m^2 · K and an oil flow of 11 kg/s, determine

a. the coolant mass flow rate

b. the required surface area in the exchanger

22.11 Compressed air is used in a heat-pump system to heat water, which is subsequently used to warm a house. The house demand is 95,000 Btu/h. Air enters the exchanger at 200 F and leaves at 120 F, and water enters and leaves the exchanger at 90 F and 125 F, respectively. Choose from the following alternative units the one that is most compact:

a. A counterflow surface with U = 30 Btu/h ft^2 F and a surface-to-volume ratio of 130 ft^2/ft^3.

b. A crossflow configuration with the water unmixed and air mixed having U = 40 Btu/h ft^2 F and a surface-to-volume ratio of 100 ft^2/ft^3.

c. A crossflow unit with both fluids unmixed with U = 50 Btu/h ft^2 F and surface-to-volume ratio of 90 ft^2/ft^3.

22.12 Water flowing at a rate of 3.8 kg/s is heated from 38 C to 55 C in the tubes of a shell-and-tube heat exchanger. The shell side is one-pass with water flowing at 1.9 kg/s, entering at 94 C. The overall heat-transfer coefficient is 1420 W/m^2 · K. The average water velocity in the 1.905-cm-ID tubes is 0.366 m/s. Because of space limitations, the tubes may not exceed 2.44 m in length. Determine the required number of tube passes, the number of tubes per pass, and the length of tubes consistent with this restriction.

22.13 Saturated steam at 373 K is to be condensed in a shell-and-tube exchanger; it is to enter as steam at 373 K and leave as condensate at approximately 373 K. If the NTU rating for the condenser is given by the manufacturer as 1.25 in this service for

a circulating water flow of 0.07 kg/s, and circulating water is available at 280 K, what will be the approximate maximum flow rate of steam in kg/s that can be condensed? What will be the leaving temperature of the circulating water under these conditions? Under these conditions, the heat vaporization is 2257 kJ/kg, and c_p is 4.18 kJ/kg · K.

22.14 In a biofuels plant distillation tower, a single-pass shell-and-tube heat exchanger is used to condense saturated ethanol vapor to saturated ethanol liquid using water as the tube-side cooling fluid. The process is carried out at 1 atm total system pressure. The ethanol vapor flow rate into the shell side of the condenser is 2 kg/s. Cooling water is available at 20 C, and the maximum allowable exit temperature of the cooling water is 50 C.

a. What is the mass flow rate of cooling water required?

b. What is the log-mean temperature driving force in the condenser?

c. If the overall heat-transfer coefficient is 250 W/m^2 · K (Table 22.2), what is the required heat-transfer area?

d. If the cooling water outlet temperature is allowed to increase by stepping down the cooling water flow rate, what happens to the required heat-transfer area?

e. What is the maximum allowable outlet temperature of the cooling water?

Additional Data:

Substance	T_b (C)	$\Delta \hat{H}_v(T_b)$ (kJ/kg)	$C_{P,\text{liq}}$ (J/kg · K)
Ethanol	78	838	2400
Water	100	2260	4200

22.15 In a biofuels plant, a shell-and-tube heat exchanger is used to cool a hot stream of 100% carbon dioxide off-gas from a beer column from 77 C to 27 C (350 to 300 K) at 1.2 atm total system pressure. The total molar flow rate of CO_2 gas is 360 kgmol/hr. Liquid water will be used as the cooling fluid. The available cooling water temperature is 20 C (293 K). The CO_2 gas flows in the shell side, water in the tube side.

a. What is the predicted outlet temperature of the water at a water mass flow rate of 0.5 kg/s? Will the heat exchanger work? If not, calculate a new water flow rate to provide a "temperature of approach" of no less than 17 C on the tube outlet side. The heat capacity of water may be taken as 4200 J/kg · K.

b. A heat exchanger is available that consists of 0.5-in nominal diameter, schedule 40 copper tubes. To ensure turbulent flow, a Reynolds number of 10,000 is desired inside the tubes. What is the overall heat-transfer coefficient, assuming a shell-side heat-transfer coefficient of 300 W/m^2 · K?

c. What is the required area of the heat exchanger?

d. How many tubes are needed to achieve the desired value of the Reynolds number? What would the desired tube length be? If that length is not realistic for a single-pass heat exchanger, what would you do next?

SS **22.16** In a biofuels plant a shell-and-tube heat exchanger is used to cool a stream of 100% carbon dioxide off-gas from a beer column from 77 C to 27 C (350 to 300 K) at 1 atm total system pressure. The total molar flow rate of CO_2 gas is 36 kgmol/h (806.4 m³/h at STP). Liquid water will be used as the coolant. Cooling water is available at 20 C (293 K). CO_2 gas flows in the tube side, water in the shell side.

a. Determine the cooling water flow rate required to provide a temperature of approach of 17 C on the shell side.

b. What is the required area of the heat exchanger?

c. How many tubes are needed to achieve a Reynolds number of 20,000 in each tube? What tube length would be required?

22.17 A single-pass, shell-and-tube heat exchanger will be used to generate saturated steam by cooling a hot CO_2 gas stream. In the present process, boiler feed water at 100 C (373 K) and 1 atm (saturation condition, enthalpy of vaporization of 2257 kJ/kg water) is fed to the shell side of the heat exchanger and exits as saturated steam vapor at 100 C and 1 atm total system pressure. The CO_2 enters the tube side at 1 atm and 450 K (177 C) and exits at 400 K (127 C). The tube-side gas is in turbulent flow.

The process must produce 0.50 kg/s of saturated steam. The shell-side heat-transfer coefficient (h_o) for the vaporizing water is 10,000 W/m² · K. The desired tube-side heat-transfer coefficient in each tube (h_i) is 50 W/m² · K. It is also desired to use schedule 40 steel pipe of $\frac{3}{4}$-in nominal diameter (inner diameter = 0.824 in, outer diameter = 1.050 in, thickness = 0.113 in, thermal conductivity = 42.9 W/m · K; Appendix M). Properties of CO_2 at 375 K, 1 atm (Appendix I): $\rho = 1.268$ kg/m³, $C_p = 960.9$ J/kg · K, $\mu = 2.0325 \times 10^{-5}$ kg/m-s, $k = 0.02678$ W/m · K.

a. What is the heat-exchanger area, based on the desired h_i of 50 W/m² · K?

b. What is the CO_2 gas flow rate needed for the process to operate?

c. What is the velocity of CO_2 in a single tube for the desired h of 50 W/m² · K?

22.18 One hundred thousand pounds per hour of water are to pass through a heat exchanger, which is to raise the water temperature from 140 F to 200 F. Combustion products having a specific heat of 0.24 Btu/lb$_m$ F are available at 800 F. The overall heat-transfer coefficient is 12 Btu/h ft² F. If 100,000 lb$_m$/h of the combustion products are available, determine

a. the exit temperature of the flue gas

b. the required heat-transfer area for a counterflow exchanger

22.19 A water-to-oil heat exchanger has entering and exiting temperatures of 255 and 340 K, respectively, for the water and 305 and 350 K, respectively, for the oil. What is the effectiveness of this heat exchanger?

22.20 A finned-tube crossflow heat exchanger with both fluids unmixed is used to heat water ($C_{pw} = 4.2$ kJ/kg · K) from 20 C to 75 C. The mass flow rate of the water is 2.7 kg/s. The hot stream ($C_{pw} = 1.2$ kJ/kg · K) enters the heat exchanger at 280 C and leaves at 120 C. The overall heat-transfer coefficient is 160 W/m² · K. Determine

a. the mass flow rate of the heat stream

b. the exchanger surface area

22.21 If the overall heat-transfer coefficient, initial fluid temperature, and total heat-transfer area determined in Problem 22.2 remain the same, find the exit oil temperature if the configuration is changed to

a. crossflow, both fluids unmixed

b. shell-and-tube with two tube passes and one shell pass

22.22 A shell-and-tube heat exchanger with two shell passes and four tube passes is used to exchange energy between two pressurized water streams. One stream flowing at 5000 lb$_m$/h is heated from 75 F to 220 F. The hot stream flows at 2400 lb$_m$/h and enters at 400 F. If the overall heat-transfer coefficient is 300 W/m² · K, determine the required surface area.

SS **23.1** The sun is approximately 93 million miles distant from Earth, and its diameter is 860,000 miles. On a clear day solar irradiation at Earth's surface has been measured at 360 Btu/h ft^2 and an additional 90 Btu/h ft^2 are absorbed by Earth's atmosphere. With this information, estimate the sun's effective surface temperature.

23.2 A satellite may be considered spherical, with its surface properties roughly those of aluminum. Its orbit may be considered circular at a height of 500 miles above Earth. Taking the satellite diameter as 50 in, estimate the temperature of the satellite skin. Earth may be considered to be at a uniform temperature of 50 F, and the emissivity of Earth may be taken as 0.95. Solar irradiation may be taken as 450 Btu/h ft^2 of satellite disc area.

SS **23.3** An opaque gray surface with $\epsilon = 0.3$ is irradiated with 1000 W/cm. For an effective convective heat-transfer coefficient of 12 W/m$^2 \cdot$ K applying, and air at 20 C adjacent to the plate, what will be the net heat flux to or from a 30 C surface?

23.4 A black solar collector, with a surface area of 60 m^2, is placed on the roof of a house. Incident solar energy reaches the collector with a flux of 800 W/m^2. The surroundings are considered black with an effective temperature of 30 C. The convective heat-transfer coefficient between the collector and the surrounding air, at 30 C, is 35 W/m $\cdot$ K. Neglecting any conductive loss from the collector, determine

a. The net radiant exchange between the collector and its surroundings

b. The equilibrium temperature of the collector

23.5 A radiation detector, oriented as shown in the sketch, is used to estimate heat loss through an opening in a furnace wall. The opening in this case is circular with a diameter of 2.5 cm. The detector has a surface area of 0.10 cm^2 and is located 1 m from the furnace opening. Determine the amount of radiant energy reaching the detector under two conditions:

a. The detector has a clear view of the opening

b. The opening is covered by a semitransparent material with spectral transmissivity given by

$$\tau_\lambda = 0.8 \quad \text{for } 0 \leq \lambda \leq 2\,\mu\text{m}$$
$$\tau_\lambda = 0 \quad \text{for } 2\,\mu\text{m} < \lambda < \infty$$

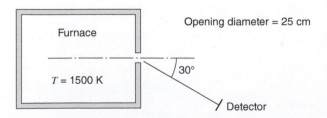

23.6 A tungsten filament, radiating as a gray body, is heated to a temperature of 4000°R. At what wavelength is the emissive power maximum? What portion of the total emission lies within the visible-light range, 0.3 to 0.75 μm?

23.7 Determine the wavelength of maximum emission for (a) the sun with an assumed temperature of 5790 K, (b) a light bulk filament at 2910 K, (c) a surface at 1550 K, and (d) human skin at 308 K.

23.8 The filament of an ordinary 100 W light bulb is at 2910 K and it is presumed to be a black body. Determine (a) the wavelength of maximum emission and (b) the fraction of emission in the visible region of the spectrum. **SS**

23.9 A greenhouse is constructed of silica glass that is known to transmit 92% of incident radiant energy between wavelengths of 0.35 and 2.7 μm. The glass may be considered opaque for wavelengths above and below these limits.

Considering the sun to emit as a black body at 5800 K, determine the percent of solar radiation that will pass through the glass.

If the plants on the inside of the greenhouse have an average temperature of 300 K, and emit as a black body, what fraction of their emitted energy will be transmitted through the glass?

23.10 The distribution of solar energy incident on Earth can be approximated as being from a black body at 5800 K.

Two kinds of glass, plain and tinted, are being considered for use in windows. The spectral transmissivity for these two glasses is approximated as

plain glass: $\tau_\lambda = 0$	for	$0 < \lambda < 0.3\,\mu\text{m}$
0.9	for	$0.3 < \lambda < 2.5\,\mu$
0	for	$2.5\,\text{mm} < \lambda$

tinted glass: $\tau_\lambda = 0$	for	$0 < \lambda < 0.5\,\mu\text{m}$
0.9	for	$0.5 < \lambda < 1.5\,\mu$
0	for	$1.50\,\text{mm} < \lambda$

Compare the fraction of incident solar energy transmitted through each material.

Compare the fraction of visible radiant energy transmitted through each.

23.11 Determine the fraction of total energy emitted by a black body, which lies in the wavelength band between 0.8 and 5.0 μm for surface temperatures of 500, 2000, 3000, and 4500 K.

23.12 The sun's temperature is approximately 5800 K and the visible light range is taken to be between 0.4 and 0.7 μm. What fraction of solar emission is visible? What fraction of solar emission lies in the ultraviolet range? The infrared range? At what wavelength is solar emissive power a maximum?

23.13 A small circular hole is to be drilled in the surface of a large, hollow, spherical enclosure maintained at 2000 K. If 100 W of radiant energy exits through the hole, determine (a) the hole diameter, (b) the number of watts emitted in the visible range from 0.4 and 0.7 μm, (c) the ultraviolet range between 0 and 0.4 mm, and (d) the infrared range from 0.7 to 100 μm.

SS **23.14** A furnace that has black interior walls maintained at 1500 K contains a peephole with a diameter of 10 cm. The glass in the peephole has a transmissivity of 0.78 between 0 and 3.2 μm and 0.08 between 3.2 μm and ∞. Determine the heat lost through the peephole.

SS **23.15** A sheet-metal box in the shape of a 0.70-m cube has a surface emissivity of 0.7. The box encloses electronic equipment that dissipates 1200 W of energy. If the surroundings are taken to be black at 280 K, and the top and sides of the box are considered to radiate uniformly, what will be the temperature of the box surface?

SS **23.16** A large cavity with a small opening, 0.0025 m² in area, emits 8 W. Determine the wall temperature of the cavity.

23.17 Two very large black plane surfaces are maintained at 900 and 580 K, respectively. A third large plane surface, having $\epsilon = 0.8$, is placed between these two. Determine the fractional change in radiant exchange between the two plane surfaces due to the intervening plane and evaluate the temperature of this intervening plane.

23.18 A 7.5-cm-diameter hole is drilled in a 10-cm-thick iron plate. If the plate temperature is 700 K and the surroundings are at 310 K, determine the energy loss through the hole. The hole sides may be considered to be black.

23.19 If the 7.5-cm-diameter hole in Problem 23.18 were drilled to a depth of 5 cm, what heat loss would result?

23.20 A cryogenic fluid flows in a 20-mm-diameter tube with an outer surface temperature of 75 K and an emissivity of 0.2. A larger tube, having a diameter of 50 mm, is concentric with the smaller one. This larger tube is gray, with $\epsilon = 0.05$ and its surface temperature is 300 K. The intervening space between the two tubes is evacuated.

Determine the heat gain by the cryogenic fluid, in watts per meter of tube length.

Evaluate the heat gain per meter of length if there is a thin-walled radiation shield placed midway between the two tubes. The shield surfaces may be considered gray and diffuse with an emissivity of 0.04 on both sides.

23.21 A heating element in the shape of a cylinder is maintained at 2000 F and placed at the center of a half-cylindrical reflector as shown. The rod diameter is 2 in. and that of the reflector is 18 in. The emissivity of the heater surface is 0.8, and the entire assembly is placed in a room maintained at 70 F. What is the radiant-energy loss from the heater per foot of length? How does this compare to the loss from the heater without the reflector present?

23.22 A 12-ft-long, 3-in.-OD iron pipe $\epsilon = 0.7$, passes horizontally through a 12 × 14 × 9 ft room whose walls are maintained at 70 F and have an emissivity of 0.8. The pipe surface is at a temperature of 205 F. Compare the radiant-energy loss from the pipe with that due to convection to the surrounding air at 70 F.

23.23 The circular base of the cylindrical enclosure shown may be considered a reradiating surface. The cylindrical walls have an effective emissivity of 0.80 and are maintained at 540 F. The top of the enclosure is open to the surroundings, which arc maintained at 40 F. What is the net rate of radiant transfer to the surroundings?

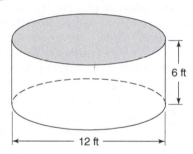

23.24 The hemispherical cavity shown in the figure has an inside surface temperature of 700 K. A plate of refractory material is placed over the cavity with a circular hole of 5 cm diameter in the center. How much energy will be lost through the hole if the cavity is

a. black?

b. gray with an emissivity of 0.7?

What will be the temperature of the refractory under each condition?

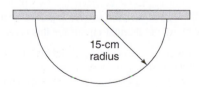

15-cm radius

23.25 A cylindrical cavity is closed at the bottom and has an opening centered in the top surface. A cross section of this configuration is shown in the sketch. For the conditions stated below, determine the rate of radiant energy passing through the 5-mm-diameter cavity opening. What will be effective emissivity of the opening?

a. All interior surfaces are black at 600 K.

b. The bottom surface is diffuse-gray with $\epsilon = 0.6$, and has a temperature of 600 K. All other surfaces are reradiating.

c. All interior surfaces are diffuse-gray with $\epsilon = 0.6$ and are at a uniform temperature of 600 K.

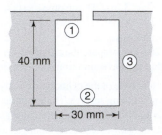

23.26 A circular heater, measuring 20 cm in diameter, has its surface temperature maintained at 1000 C. The bottom of a tank, having the same diameter, is oriented parallel to the heater with a separation distance of 10 cm. The heater surface is gray ($\epsilon = 0.6$) and the tank surface is also gray ($\epsilon = 0.7$).

Determine the radiant energy reaching the bottom of the tank if

a. the surroundings are black at 27 C

b. the space between the two cylindrical surfaces is enclosed by an adiabatic surface

23.27 Two parallel black rectangular surfaces, whose back sides are insulated, are oriented parallel to each other with a spacing of 5 m. They measure 5 m by 10 m. The surroundings are black at 0 K. The two surfaces are maintained at 200 and 100 K, respectively. Determine the following:

a. The net radiant heat transfer between the two surfaces

b. The net heat supplied to each surface

c. The net heat transfer between each surface and the surroundings

23.28 If a third rectangular plate, with both surfaces having an emissivity of 0.8 is placed between the two plates described in Problem 23.31, how will the answer to part (a) of the problem be affected? Draw the thermal circuit for this case.

23.29 Two disks are oriented on parallel planes separated by a distance of 10 in., as shown in the accompanying figure. The disk to the right is 4 in. in diameter and is at a temperature of 500 F. The disk to the left has an inner ring cut out such that it is annular in shape with inner and outer diameters of 2.5 in and 4 in, respectively. The disk surface temperature is 210 F. Find the heat exchange between these disks if

a. they are black

b. they are gray $\epsilon_1 = 0.6$, $\epsilon_2 = 0.3$

23.30 A heavily oxidized aluminum surface at 755 K is the source of energy in an enclosure, which radiantly heats the side walls of a circular cylinder surface as shown, to 395 K. The side wall is made of polished stainless steel. The top of the enclosure is made of fire clay brick and is adiabatic. For purposes of calculation, assume that all three surfaces have uniform temperatures and that they are diffuse and gray. Evaluate the heat transfer to the stainless-steel surface.

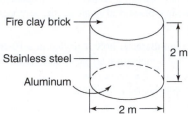

23.31 A gray, diffuse circular heater with a diameter of 15 cm is placed parallel to a second gray, diffuse receiver with a spacing of 7.5 between them. The backs of both surfaces are insulated and convective effects are to be neglected. This heater–receiver assembly is placed in a large room at a temperature of 275 K. The surroundings (the room) can be considered black and the heater surface emissivity is 0.8. When the power input to the heater is 300 W, determine

a. The heater surface temperature

b. The receiver surface temperature

c. The net radiant exchange to the surroundings

d. The net radiant exchange between the heater and receiver

23.32 A small (1/4-in diameter × 1 in. long) metal test specimen is suspended by very fine wires in a large evacuated tube. The metal is maintained at a temperature of 2500 F, at which temperature it has an emissivity of approximately 0.2. The water-cooled walls and ends of the tube are maintained at 50 F. In the upper end is a small (1/4-in. diameter) silica glass viewing port. The inside surfaces of the steel tube are newly galvanized. Room temperature is 70 F. Estimate

a. The view factor from the specimen to the window

b. The total net heat-transfer rate by radiation from the test specimen

c. The energy radiated through the viewing port

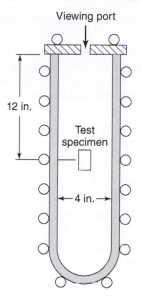

Viewing port

12 in.

Test specimen

4 in.

23.33 A gas consisting of 20% CO_2 and 80% oxygen and nitrogen leaves a lime kiln at 2000 F and enters a square duct measuring 6 in. by 6 in. in cross section. The specific heat of the gas is 0.28 Btu/lb$_m$ F, and it is to be cooled to 1000 F in the duct, whose inside surface is maintained at 800 F, and whose walls have an emissivity of 0.9. The mass velocity of the kiln gas is 0.4 lb$_m$/ft$^2 \cdot$ s and the convective heat-transfer coefficient between the gas and duct walls is 1.5 Btu/h ft^2 F.

a. Determine the required length of duct to cool the gas to 1000 F.

b. Determine the ratio of radiant-energy transfer to that by convection.

c. At what temperature would the gas leave the duct if the length of the duct were twice the value determined in part (a)?

(Courtesy of the American Institute of Chemical Engineers.)

Hint. As the response of the gas to emission and absorption of radiant energy differs, an approximation for the radiant-energy exchange between the enclosure and gas contained within an arbitrary control volume is given by $A_w F_{w-g} \sigma \epsilon_w (\epsilon_g T_g^4 - \alpha_g T_w^4)$.

24.1 Starting with the Fick's rate equation for the diffusion of A through a binary mixture of components A and B, given by

$$\mathbf{N}_A = -cD_{AB}\nabla y_A + y_A(\mathbf{N}_A + \mathbf{N}_B)$$

show that

$$\mathbf{n}_A = -D_{AB}\nabla\rho_A + w_A(\mathbf{n}_A + \mathbf{n}_B)$$

and state all assumptions needed to arrive at this result.

24.2 Starting with Fick's rate equation for the diffusion of A through a binary mixture of components A and B, prove

a. $\mathbf{N}_A + \mathbf{N}_B = c\mathbf{V}$

b. $\mathbf{n}_A + \mathbf{n}_B = \rho\,\mathbf{v}$

c. $\mathbf{j}_A + \mathbf{j}_B = 0$

24.3 The general form of the Stefan–Maxwell equation for mass transfer of species i in an n-component ideal gas mixture along the z-direction for flux is given by

$$\frac{dy_i}{dz} = \sum_{j=1, j \neq i}^{n} \frac{y_i y_j}{D_{ij}}(n_j - n_i)$$

Show that for a gas-phase binary mixture of species A and B, this relationship is equivalent to Fick's rate equation:

$$N_{A,z} = -cD_{AB}\frac{dy_A}{dz} + y_A(N_{A,z} + N_{B,z})$$

24.4 In a gas-phase diffusion mass-transfer process, the steady-state flux of species A in a binary mixture of A and B is 5.0×10^{-5} kgmole A/m^2s, and the flux of B is 0 (zero). At a particular point in the diffusion space, the concentration of species A is 0.005 kgmole/m^3 and the concentration of species B is 0.036 kgmole/m^3. Estimate the individual net velocities of species A and B along the direction of mass transfer, and the average molar velocity.

SS **24.5** Consider the following properties of the atmosphere of the planet Mars at a particular measurement point on the surface, as measured by the *Mars Rover*:

Average surface pressure: 6.1 mbar
Average temperature: 210 K
Atmospheric gas composition (by volume): carbon dioxide (CO$_2$) 95.32%; nitrogen (N$_2$) 2.7%; argon (Ar) 1.6%; oxygen (O$_2$) 0.13%; carbon monoxide (CO) 0.08%.

a.   What is the partial pressure of CO$_2$ in the Martian atmosphere, p_A, in units of Pa?

b.   What is the molar concentration of CO$_2$ in the Martian atmosphere, c_A?

c.   What is the total molar concentration of all gases in the Martian atmosphere?

d.   What is the total mass concentration of all gases the Martian atmosphere?

24.6 Estimate the molar concentration of 0.50 wt% (0.0050 mass fraction) benzene (C$_6$H$_6$) dissolved in liquid ethanol (C$_2$H$_5$OH) at 20 C. The density of liquid ethanol is 789 kg/m^3 at 20 C. This solution of the solute benzene dissolved in the solvent ethanol is considered dilute.

24.7 Estimate the value of the gas-phase diffusion coefficient for the following gas pairs using the Hirschfelder equation:

a.   Carbon dioxide and air at 310 K and 1.5×10^5 Pa

b.   Ethanol and air at 325 K and 2.0×10^5 Pa

c.   Carbon tetrachloride and air at 298 K and 1.913×10^5 Pa

24.8 Estimate the gas-phase diffusion coefficient of ammonia (NH$_3$) in air at 1.0 atm and 373 K by making use of Brokaw equation in the calculations. The dipole moment (μ_p) for NH$_3$ is 1.46 debye, and the liquid molar volume at the normal boiling point (V_b) is 25.8 cm^3/gmole at 239.7 K. Compare the estimated value with the experimental value reported in Appendix J, Table J.1, scaled to 373 K.

24.9 Tetrachlorosilane (SiCl$_4$) gas is reacted with hydrogen gas (H$_2$) to produce electronic-grade polycrystalline silicon at 800 C and 1.5×10^5 Pa according to the reaction equation: SiCl$_4$(g) + 2H$_2$(g) → Si(s) + 4HCl(g).
The reaction rate may experience diffusion limitations at the growing Si solid surface. To address this concern, the diffusion coefficient coefficients in this system must be estimated.

a.   Estimate the binary diffusion coefficient of SiCl$_4$ in H$_2$ gas.

b.   Estimate the diffusion coefficient of SiCl$_4$ in a mixture containing 40 mole% SiCl$_4$, 40 mole% H$_2$, and 20 mole% HCl. Is this diffusion coefficient substantively different than the diffusion coefficient estimated in part (a)? The Lennard–Jones parameters for SiCl$_4$ are $\sigma = 5.08$ Å and $\epsilon/\kappa = 358$ K.

24.10 Steam reforming of hydrocarbons is one way to make hydrogen gas (H$_2$) for fuel cell applications. However, the H$_2$-rich product gas produced by steam reforming is contaminated with 1.0 mole% of carbon monoxide (CO), which poisons the catalyst within the fuel cell, and so must be removed in order for the fuel cell to work better. It is desired to separate CO from the H$_2$ gas using a microporous catalytic ceramic membrane. The average pore diameter of the porous membrane material is 15 nm, and the void fraction (ϵ) is 0.30. The system operates at 5.0 atm total system pressure and temperature of 400 C.

a.   Estimate the molar concentration of CO in the gas mixture.

b.   Estimate the molecular diffusion coefficient of CO in H$_2$ for the gaseous mixture at 5.0 atm and 400 C.

c. Estimate the effective diffusion of CO within the porous material. Assess the importance of Knudsen diffusion at the current process conditions.

d. At what total system pressure (P) does the Knudsen diffusion coefficient for CO equal to one-half of the molecular diffusion coefficient?

24.11 "Sour" natural gas is contaminated with hydrogen sulfide. The H_2S vapors are commonly treated and removed by passing the gas through a packed bed of adsorbent particles. In the present process, natural gas with a bulk gas composition is 99 mole% methane (CH_4) and 1.0 mole% H_2S will be treated with a porous adsorbent material of void fraction of 0.50 and mean pore diameter of 20 nm at 30 C and 15.0 atm total system pressure. To design the adsorption bed, it is necessary to estimate the diffusion coefficient of H_2S within the porous adsorbent. The Lennard–Jones parameters for H_2S are $\sigma = 3.623$ Å and $\epsilon/\kappa = 301.1$ K.

a. What is molar concentration of H_2S in the gas mixture?

b. Estimate the binary gas-phase molecular diffusion coefficient of methane in H_2S by two methods, and compare the results.

c. Estimate the effective diffusion coefficient of H_2S within the porous material, assuming methane and hydrogen sulfide fill the gas void space within the porous material.

SS **24.12** A process is being developed to deposit a thin film of electronic grade silicon (Si) onto the inner surface of a hollow glass optical fiber by thermal decomposition of silane (SiH_4) to solid Si. Silane gas is diluted in inert He gas to a composition of 1.0 mole% SiH_4 and fed into the hollow glass fiber, which has an inner diameter of 10.0 microns (10.0 μm). The CVD process is carried out at 900 K and a very low total system pressure of only 100 Pa. The Lennard–Jones parameters for silane are $\sigma = 4.08$ Å and $\epsilon/\kappa = 207.6$ K.

a. Determine the mass fraction of silane in the gas mixture.

b. Estimate the binary gas-phase molecular diffusion coefficient of silane vapor in He gas at 900 K for total system pressures of 1.0 atm and 100 Pa. Why is the value at 100 Pa so large?

c. Assess the importance of Knudsen diffusion of silane vapor within the hollow glass fiber.

d. The Peclet number (Pe), defined as

$$Pe = \frac{n_\infty d}{D_{Ae}}$$

is a dimensionless parameter used to assess the importance of dispersion of species undergoing diffusion in a flowing stream of bulk velocity n_∞ through a tube of diameter d. At low values for Pe, diffusion dominates the dispersion process. Estimate the gas velocity and volumetric flow rate of gas within a single hollow fiber if it is desired to maintain Pe equal to 5.0×10^{-4}.

24.13 A microchannel reactor is being developed for the reforming of methane gas (CH_4) to CO_2 and H_2 gas using steam (H_2O vapor) at high temperature. Prior to entering the reforming reactor, CH_4 (species A) and H_2O vapor (species B) at a nonreactive temperature of 400 C are mixed within a microchannel gas preheater zone with an inner diameter of only 50 μm (0.0050 cm). The Peclet number, defined as

$$Pe = \frac{n_\infty d}{D_{Ae}}$$

must be at least 2.0 before this gas stream is fed into the reforming reactor. It is also desired to keep the gas velocity (v_∞) constant at 4.0 cm/s. For purposes of calculation, it may be assumed that CH_4 is diluted in H_2O gas.

a. What is the Knudsen diffusion coefficient for CH_4 inside the microchannel? Is Knudsen diffusion important?

b. At what total system pressure (P) should the process be operated at to meet the aforementioned design constraints of $Pe = 2.0$ and $n_\infty = 4.0$ cm/s? At this total system pressure, what is the gas-phase molecular diffusion coefficient of CH_4 in H_2O?

24.14 The Stokes–Einstein equation is often used to estimate the molecular diameter of large spherical molecules from the molecular diffusion coefficient. The measured molecular diffusion coefficient of the serum albumin (an important blood protein) in water at infinite dilution is 5.94×10^{-7} cm²/s at 293 K. Estimate the mean diameter of a serum albumin molecule. The known value is 7.22 nm.

24.15 The diffusion of oxygen (O_2) through living tissue is often first approximated as the diffusion of dissolved O_2 in liquid water. Estimate the diffusion coefficient of O_2 in water by the Wilke–Chang and Hayduk–Laudie correlations at 37 C.

24.16 Benzene (species A) is often added to ethanol (species B) to denature the ethanol. Estimate the liquid-phase diffusion of benzene in ethanol (D_{AB}), and ethanol in benzene (D_{BA}) at 288 K by two methods: the Wilke–Chang equation the Scheibel equation. Why does $D_{AB} \neq D_{BA}$?

24.17 The rate of an electrochemical process for plating of **SS** solid copper onto a surface from a cupric chloride solution is affected by mass-transfer processes. Estimate the molecular diffusion coefficient of copper II chloride ($CuCl_2$) dissolved in water at infinite dilution at 25 C.

24.18 The molecular diffusion coefficient of methanol dissolved in liquid water as the solvent is 1.28×10^{-5} cm²/s at 15 C and 1.0 atm total system pressure. What is the molecular diffusion coefficient for methanol dissolved in liquid water at 30 C? *Potentially useful data:* for methanol (A), liquid viscosity $\mu_A = 0.61$ cP at 15 C, $\mu_A = 0.51$ cP at 30 C, $M_A = 32$ g/gmole; for water (B), liquid viscosity $\mu_B = 1.14$ cP at 15 °C, $\mu_B = 0.797$ cP at 30 C, $M_B = 18$ g/gmole.

24.19 As part of a bioseparations process, glucose (solute A) in aqueous solution (solvent B) is diffusing across a microporous membrane. The thickness of the membrane is 2.0 mm, and the pores running through the membrane consists of parallel cylindrical channels of 3.0 nm diameter (1 nm is a nanometer, 1×10^9 nm = 1 m). The temperature is 30 C. The mean diameter of a single glucose molecule is 0.86 nm.

a. Estimate the molecular diffusion coefficient of glucose in water by the Stokes–Einstein relationship.

b. What is the effective diffusion coefficient of glucose through the membrane?

24.20 Protein mixtures in aqueous solution are commonly separated by molecular sieve chromatography. An important aspect of this separation process is the diffusion of the protein into the porous matrix of the chromatography support used to affect the separation. Estimate the effective diffusion coefficient of the enzyme urease in a single 100-nm-diameter cylindrical pore. The molecular diffusion coefficient of urease in water at infinite dilution is 3.46×10^{-7} cm^2/s at 298 K, and the diameter of the molecule is 12.38 nm.

24.21 The diffusion rate of the enzyme ribonuclease into a porous chromatography support material was measured at 298 K. An effective diffusion coefficient of 5.0×10^{-7} cm^2/s was obtained from an experiment based on diffusion of the enzyme through a single cylindrical pore. Estimate the pore diameter (d_{pore}) needed to achieve this effective diffusion coefficient. The molecular diffusion coefficient of ribonuclease in water is 1.19×10^{-6} cm^2/s at 298 K, and the diameter of this molecule is 3.6 nm.

24.22 It is desired to concentrate species A from a dilute mixture of solutes A and B dissolved in liquid water using a porous membrane. The membrane contains cylindrical pores in uniform parallel array. The separation factor (α) is defined as

$$\alpha = \frac{D_{Ae}}{D_{Be}}$$

For solute A, the molecular weight is 142 g/gmole, molecular diameter is 2.0 nm, and the molar volume at boiling point is 233.7 cm^3/gmole. For solute B, the molecular weight is 212 g/gmole, molecular diameter is 3.0 nm, and the molar volume at the boiling point is 347.4 cm^3/gmole.

a. Estimate the molecular diffusion coefficients of solute A and solute B in water at 293 K at infinite dilution.

b. Determine the pore diameter (d_{pore}, units of nm) to achieve a separation factor of $\alpha = 3$ for this process. Hint: A trial-and-error or iterative solution will be required.

SS **24.23** Consider that a binary mixture of 5.0 mole% ethanol vapor in 95 mole% carbon dioxide carrier gas is diffusing through a randomly porous adsorbent material with void fraction of 0.50. The system temperature is 80 C, and the total system pressure is 1.5 atm. It is known that at 80 C and 1.0 atm, the molecular diffusion coefficient of ethanol vapor (A) in CO$_2$ gas (B) is $D_{AB} = 0.10$ cm^2/s.

a. What is the molar concentration of 5.0 mole% ethanol vapor in CO$_2$ gas at 80 C and 1.5 atm, in units of gmole A/m^3?

b. What is the effective diffusion coeffcient D'_{Ae} through the porous material if the average pore diameter is 0.15 μm at 80 C and 1.5 atm?

24.24 In a chromatographic separation process, a biomolecule of 20 nm diameter is diffusing into a random porous particle with void fraction (ϵ) of 0.70 filled with elution solvent. However, the reduced pore diameter (φ) for the diffusion process is only 0.010, and hindered diffusion of the solute within the solvent-filled pore can be neglected. If the desired effective diffusion coefficient (D'_{Ae}) of the biomolecule inside the porous particle is 1.0×10^{-7} cm^2/s at 30 C, what solvent viscosity would be needed to achieve this D'_{Ae}?

24.25 A pipeline containing natural gas from a hydraulic fracturing process contains 80% methane gas (CH$_4$, species A) and 20% ethane (C$_2$H$_6$, species B) gas at 3.0 atm and 17 C, as shown in figure below. We are concerned that if the pipeline leaks, then the gas mixture can transfer through the porous layer above the pipeline and ultimately escape to the atmosphere. As part of this analysis, it will be necessary to estimate the effective diffusion coefficient of methane and ethane through the porous material. *Additional information*: porous layer, $\epsilon = 0.3$, $d_{pore} = 0.0010$ cm, $M_A = 16$ g/gmole, $M_B = 30$ g/gmole.

a. What is the molar concentration of ethane (B) gas inside the pipeline?

b. What is the *molecular diffusion coefficient* of methane gas (CH$_4$, species A) in ethane gas (C$_2$H$_6$, species B) inside the pipe at ($P = 3.0$ atm) by the Fuller–Schettler–Giddings approach?

c. What is the *effective diffusion coefficient* of methane in the methane/ethane gas mixture through the porous layer (D'_{Ae}), assuming the gas space inside the porous layer is at 1.0 atm ($P = 1.0$ atm)? You may assume that the gas space in the porous layer contains only methane and ethane.

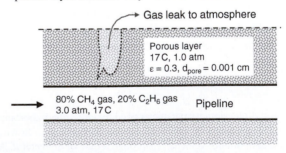

24.26 The case-hardening of mild steel involves the diffusion of carbon into iron. Estimate the diffusion coefficient for carbon diffusing into *fcc* iron and *bcc* iron at 1000 K. Learn about the structures of fcc and bcc iron in a materials science textbook, and then explain why the diffusion coefficients are different.

24.27 The elements arsenic and phosphorus are common dopants in crystalline silicon that impart semiconductor properties to it via the introduction of p-type carriers into the silicon lattice. The doping process is carried out by a solid-state diffusion process at high temperature.

a. Compare the diffusion coefficients of arsenic and phosphorus in crystalline silicon at 600 and 1000 C. Why might such a high temperature be needed for the doping process?

b. At what temperature is the diffusion rate of arsenic-silicon equal to 12% of that for phosphorus-silicon? State all assumptions.

25.1 Show that the (25-11) may be written in the form

$$\frac{\partial \rho_A}{\partial t} + (\nabla \cdot \rho_A \mathbf{v}) - D_{AB}\nabla^2 \rho_A = r_A$$

25.2 Consider a system for mass transfer of a binary mixture of species A and B within the control volume shown in the figure below. The concentration at the surface (c_{As}) is constant, the process is very dilute with respect to species A within the control volume, the control volume fluid is still (not well mixed), and there no homogeneous reaction of species A in the control volume. The top and left sides of the control volume are also not permeable to species A. Initially, there is no species A within the control volume.

a. List at least three valid assumptions for this process.

b. What is the appropriately simplified General Differential Equation of Mass Transfer for species A, in terms of concentration c_A?

c. What are the relevant boundary conditions?

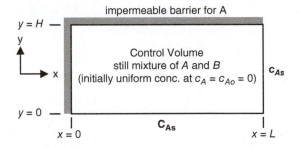

25.3 In a microelectronics manufacturing process, an organic solvent (methyl ethyl ketone, MEK) is used to dissolve a thin coating of a photoresist polymer film away from a nonporous flat surface as shown in the figure below. Let A = polymer (solute), B = MEK (liquid solvent). You may assume that solute A is not very soluble in the liquid, with maximum solubility of 1.0 g/L, and so the concentration of solute A in the liquid always stays dilute. You may also assume that the change in film thickness during the dissolution process does not affect mass transfer process, and that some of the solid polymer will remain on the surface after the process is complete. Finally, you may assume that the velocity profile in the x-direction is approximated by a parabolic velocity profile in laminar flow. The photoresist is nonreactive when dissolved in the solvent.

a. Draw a labeled diagram of the physical system and the relevant coordinate system for mass transfer. Define the source and sink for solute (A), and justify if the mass transfer

process is steady state or unsteady state, one-dimensional or multidimensional.

b. State at least four reasonable assumptions regarding the mass transfer process.

c. Develop the appropriately simplified differential form of the Flux Equation in the appropriate dimension(s).

d. Develop the appropriately simplified differential form of the Differential Equation for Mass Transfer in terms of liquid phase concentration c_A.

e. State reasonable boundary/initial conditions for the process.

f. Sketch out what the dissolved solute A concentration profile might look like with respect to x and y directions.

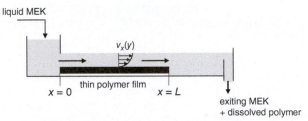

25.4 A device has been proposed that will serve as a "blood oxygenator" for a heart–lung bypass machine, as shown in the figure in the next column. In this process, blood containing no dissolved oxygen (O_2, species A) enters the top of the chamber and then falls vertically down as a liquid film of uniform thickness, along a surface designed to appropriately wet blood. Contacting the liquid surface is a 100% O_2 gas phase. Oxygen is soluble in blood, where the equilibrium solubility c_A^* is function of the partial pressure of oxygen gas. In analyzing the mass transport of dissolved oxygen into the falling film, you may assume the following: (1) the process has a constant source of O_2 (gas) and a constant sink (falling liquid film), and so is at steady state; (2) the process is dilute with respect to dissolved oxygen dissolved the fluid; (3) the falling liquid film has a flat velocity profile with velocity v_{max}; (4) the gas space always contains 100% oxygen; (5) the width of the liquid film, W, is much larger than the length of the liquid film, L.

a. Simplify the general differential equation for O_2 transfer, leaving the differential equation in terms of the fluxes. If your analysis suggests more than one dimension for flux, provide a simplified flux equation for each coordinate of interest.

b. Provide one simplified differential equation in terms of the oxygen concentration c_A.

c. Provide boundary conditions associated with the oxygen mass-transfer process.

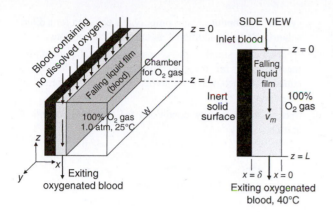

SIDE VIEW

25.5 Consider one of the cylindrical channels of inner diameter d that run through an isomerization catalyst, as shown in the figure below. A catalyst coats the inner walls of each channel. This catalyst promotes the isomerization of n-butane (n-C_4H_{10}, species A) to isobutene (n-C_4H_{10}, species B):

$$n\text{-}C_4H_{10}(g) \rightarrow i\text{-}C_4H_{10}(g)$$

The gas phase above the channels contains a mixture of A and B maintained at a constant composition of 60 mol% A and 40 mol% B. Gas-phase species A diffuses down a straight channel of diameter $d = 0.10$ cm and length $L = 2.0$ cm. The base of each channel is sealed. The surface reaction is rapid reaction so that the production rate of B is diffusion limited. The quiescent gas space in the channel consists of only species A and B.

a. State three relevant assumptions for the mass-transfer process. Based on your assumptions, simplify the general differential equation for the mass transfer of species A, leaving the differential equation in terms of the flux N_A.

b. Simplify Fick's flux equation for each coordinate of interest, and then express the simplified form of general differential equation from part (a) above in terms of the gas-phase concentration c_A.

c. Specify relevant boundary conditions for the gas-phase concentration c_A.

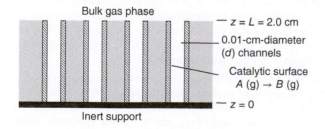

Bulk gas phase
— $z = L = 2.0$ cm
— 0.01-cm-diameter (d) channels
— Catalytic surface A (g) → B (g)
— $z = 0$
Inert support

25.6 Consider the process shown in the figure below, where carbon monoxide (CO) gas is being oxidized to carbon dioxide

(CO_2). This process is similar to how the catalytic converter in your car works to clean up CO in the exhaust. The inlet gas, which is 1.0 mole% CO diluted in O_2, is fed into a rectangular chamber with a nonporous catalyst layer lining the top and bottom walls. The catalyst surface drives the reaction $CO(g) + 1/2O_2(g) \rightarrow CO_2(g)$, which is extremely rapid at the temperature of operation so that the gas-phase concentration of CO at the catalyst surface is essentially zero. For this system, you may assume that the gas velocity profile is flat—i.e., $v_x(y) = \mathbf{v}$, and the gas velocity is relatively slow.

a. Develop the final form of the general differential equation for mass transfer in terms of the concentration profile for CO, c_A (species A). State all relevant assumptions, and also describe the source and sink for CO mass transfer.

b. State all necessary boundary conditions needed to specify the system.

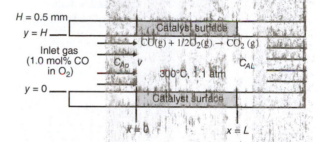

$H = 0.5$ mm
$y = H$ —— Catalyst surface
—— $CO(g) + 1/2O_2(g) \rightarrow CO_2 (g)$
Inlet gas
(1.0 mol% CO in O_2) c_{Ao} v c_{AL}
300°C, 1.1 atm
$y = 0$ —— Catalyst surface
$x = 0$ $x = L$

25.7 Consider the catalytic reaction process shown in the figure in the next column. The control volume has two catalytic zones: a porous catalyst (*catalyst I*) that fills the control volume, and a nonporous catalyst surface (*catalyst II*) on the left side of the control volume ($x = 0, y = 0$ to H). Reactant A diffuses into the porous catalytic material (*catalyst I*) with effective diffusion coefficient D_{Ae}, and is converted to product B according to a homogeneous reaction of the form

$$A \xrightarrow{k_1} B \text{ with } R_A = -k_1 c_A$$

where k_1 is the first-order reaction rate constant (s^{-1}). Reactant A can also diffuse to the nonporous catalyst surface (*catalyst II*), and is converted to product C according the *surface* reaction of the form

$$A \xrightarrow{k_s} 2C \text{ with } r_{A,s} = -k_s c_{As}$$

where k_s is the first-order surface reaction rate constant (cm/s). The source for reactant A is well-mixed flowing fluid of constant concentration $c_{A,\infty}$. It is reasonable to assume that $c_A(x,0) \approx c_{A,\infty}(0 \leq x \leq L)$. Reactant A is diluted in inert carrier fluid D. Therefore, the control volume contains four species: A, B, C, and inert diluent D. The right side ($x = L$, $y = 0$ to H) and top side ($x = 0$ to L, $y = H$) of the catalytic zone are impermeable to reactant A, products B and C, and diluent fluid D.

a. It can be assumed that the process is dilute with respect to reactant A. State three additional reasonable assumptions for the mass-transfer processes associated with reactant A, including the source and sink for reactant A, that allow for appropriate simplification of the general differential equation for mass transfer, and Fick's flux equation.

b. Develop the differential forms of the general differential equation for mass transfer and Fick's flux equation for reactant A within the process. Carefully label the differential volume element. Combine the general differential equation for mass transfer and Fick's flux equation to obtain a second-order differential equation in terms of concentration $c_A(x,y)$.

c. Formally specify all relevant boundary conditions on reactant A for a steady-state process.

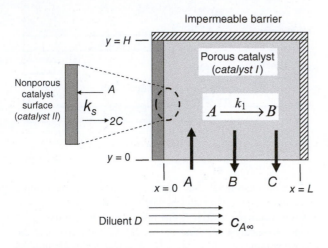

25.8 Consider the drug treatment system shown in the figure on the next page. A hemispherical cluster of unhealthy cells is surrounded by a larger hemisphere of stagnant dead tissue (species B), which is in turn surrounded by a flowing fluid. The bulk, well-mixed fluid contains a drug compound (species A) of constant but dilute bulk concentration c_{Ao}. Drug A is also soluble in the unhealthy tissue but does not preferentially partition into it relative to the fluid. The drug (species A) enters the dead tissue and targets the unhealthy cells. At the unhealthy cell boundary ($r = R_1$), the consumption of drug A is so fast that the flux of A to the unhealthy cells is diffusion limited. All metabolites of drug A produced by the unhealthy cells stay within the unhealthy cells. However, drug A can also degrade to *inert* metabolite D by a first-order reaction dependent on c_A—i.e., $A \xrightarrow{k} D$—that occurs only within the stagnant dead tissue.

a. State all reasonable assumptions and conditions that appropriately describe the system for mass transfer.

b. Develop the differential form of Fick's flux equation for drug A within the multicomponent system *without* the "dilute

system" assumption. Then, simplify this equation for a dilute solution. State all additional assumptions as necessary.

c. Appropriately simplify the general differential equation for mass transfer for drug A. Specify the final differential equation in two ways: in terms of N_A, and in terms of concentration c_A.

d. Specify the boundary conditions for both components A and D.

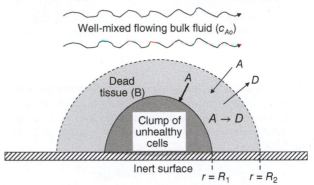

25.9 The design of "artificial organs" for growing transplantable human tissues requires that the cell type of interest, for example pancreatic cells, be grown and sustained within a three-dimensional array where the cells closely packed together to form a continuous living tissue. Furthermore, engineered systems for maintaining the tissues within the three-dimensional array require that oxygen be delivered to the living tissue for respiration processes, which is a major design challenge. One way to get oxygen to the tissue is by a porous scaffold containing capillary ducts organized into rectangular tissue monolith, as shown in the figure in the next column. Pure oxygen gas flows through the capillary ducts and contacts the tissue, where it dissolves into the tissue by Henry's law—i.e., $P_A = Hc_{A^*}$. Two sides of the monolith are sealed, but the other two sides are composed of a porous meshlike sheath, which contains the tissue, but also exposes it to the surrounding liquid medium, which contains dissolved oxygen of bulk concentration of fixed concentration $c_{A\infty}$. The tissue and the liquid medium surrounding the monolith approximate the physical properties of liquid water. Once dissolved in the tissue, the tissue consumes the dissolved oxygen by a first-order reaction process defined by rate constant k_1. We are interested in predicting the concentration profile of dissolved oxygen within the tissue monolith.

a. Develop the differential model for the concentration profiles of dissolved oxygen (c_A) within quadrant IV of the tissue monolith. Look for symmetry as appropriate. State all reasonable assumptions, and define the system(s), likely source(s), and likely sink(s) for species O_2 (species A) within the tissue monolith.

b. Specify appropriate boundary conditions.

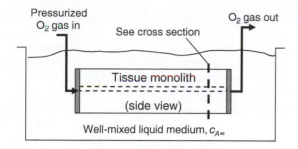

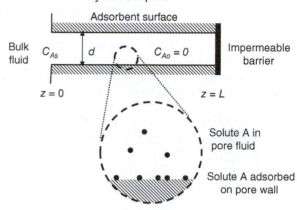

the surface (moles A/cm^3), and K is the equilibrium constant (moles/cm^3), and $q_{A,\max}$ is the maximum amount of solute A, which can be adsorbed on the surface (moles A/cm^2 surface area). At high concentrations where $c_A \gg K$, $q_A \approx q_{A,\max}$, and at low concentrations where $K \gg c_A$, the adsorption isotherm becomes linear, so that

$$q_A \approx \frac{q_{A,\max}}{K} c_A$$

a. Think of a specific physical system—i.e., propose specific materials for solute A, the fluid B, and the solid surface from an outside literature reference. What does the plot of the Langmuir isotherm (q_A vs. c_A) look like for this specific physical system? Develop an algebraic expression that describes the maximum amount of solute A, which can be adsorbed within a single pore.

b. You may now consider that the concentration profiles of solute A is only in the axial direction, not in the radial direction. You may also assume that the process is dilute with respect to solute A, the linear adsorption isotherm is valid, and the rate processes of adsorption are extremely fast. Using the "shell balance" approach, develop the differential forms of the general differential equation for mass transfer and Fick's flux equation, taking into account the adsorption of solute A onto the surface of pore in the differential mass balance. Then combine the simplified forms of the general differential equation of mass transfer and Fick's flux equation to arrive at a single differential equation for the transfer of solute A within the pore in terms of concentration c_A. State all assumptions and boundary/initial conditions as part of the analysis.

SS **25.10** Consider a single, porous, spherical, inert mineral particle. The pores inside the particle are filled with liquid water (species B). We are interested in analyzing the molecular diffusion of the contaminant benzene (species A) within the water-filled pores of the particle. Benzene is very sparingly soluble in water, and does not adsorb onto the intersurfaces of the pores. The process is isothermal at 298 K. The concentration of dissolved benzene in the water surrounding the particle is constant with time. Starting with the general differential equation for mass transfer, develop the differential equation to describe the concentration profile of benzene within the single, porous, spherical, inert mineral particle. The effective diffusion coefficient of benzene in the water-filled pores of the particle is represented by the term D_{Ae}. As part of the analysis, state all reasonable assumptions, as well as boundary/initial conditions for the process.

25.11 Consider the diffusion of solute A into the single cylindrical pore shown in the figure in the next column. The end of the pore at $z = L$ is sealed. The pore space is initially filled with inert fluid B. As solute A diffuses into the quiescent fluid space inside the pore space, it adsorbs onto the inner walls of the pore. The "adsorption isotherm" of solute A onto the solid surface of the pore is described by the Langmuir equation, given by

$$q_A = \frac{q_{A,\max} \, c_A}{K + c_A}$$

where q_A is the amount adsorbed on the surface (moles A/cm^2 surface area), c_A is the local concentration of solute A right above

25.12 Consider the cylindrical catalyst pellet with dimensions shown in the figure below. The nonporous catalyst surface promotes the hydrogenation ethylene (C_2H_4) gas to ethane (C_2H_6) by hydrogen (H_2), gas according to the reaction $C_2H_4 + H_2 \rightarrow C_2H_6$. Both C_2H_4 and H_2 gas must diffuse through an inert porous layer surrounding the catalyst surface, which has a mean pore size (d_{pore}) of 1.0 μm (1.0×10^{-4} cm) and void fraction (ε) of 0.40. Let $A = $ ethylene (C_2H_4), $B = $ hydrogen (H_2), and $C = $ ethane (C_2H_6). Assume that the bulk gas surrounding the catalyst pellet is maintained at a constant composition of 4.0 mole% C_2H_4, 2.0 mole% C_2H_6, and 94 mole% H_2. The process is carried out at 150 C and 1.5 atm total system pressure. Also, at 150 C, the reaction is extremely rapid, so that the concentration of ethylene (A) at the catalyst surface is effectively zero.

a. What are the species undergoing mass transfer?

b. List at least three relevant assumptions of the mass transfer process.

c. Write down the appropriately simplified General Differential Equation for Mass Transfer, in terms of flux N_A.

d. What relationships best describe the boundary conditions for mass transfer in the system?

25.13 Consider the cylindrical porous pellet shown in the figure below. A gas mixture of species A and B continually flows over the top face of the pellet. The sides and bottom of the pellet are sealed, and so serve as an impermeable barrier. Within the porous layer, species A forms species B by $A \rightarrow B$, with first-order rate equation $R_A = -k c_A$.

a. List at least three relevant assumptions of the mass transfer process, including the likely directions for flux of species A.

b. In the z-direction, what is the differential form of Fick's Flux Equation?

c. Develop the appropriately simplified General Differential Equation for Mass Transfer, in terms of species A concentration, c_A

d. What are the relevant boundary conditions for the process with respect to species A?

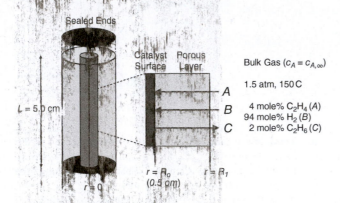

Sealed Ends

Catalyst Porous
Surface Layer

Bulk Gas ($c_A = c_{A,\infty}$)

1.5 atm, 150 C

$L = 5.0$ cm

A
B
C

4 mole% C_2H_4 (A)
94 mole% H_2 (B)
2 mole% C_2H_6 (C)

$r = R_0$ $r = R_1$
(0.5 cm)
$r = 0$

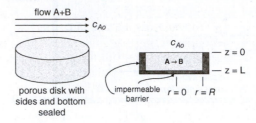

flow A+B

c_{A0}

porous disk with
sides and bottom
sealed

c_{A0}

$A \rightarrow B$

$z = 0$
$z = L$

impermeable
barrier $r = 0$ $r = R$

26.1 The chemical vapor deposition of silane vapor (SiH_4) forms a thin film of solid silicon as described in Example 1, Chapter 25. Consider that the simplified CVD unit shown in Figure 25.5 is operating at 900 K and with a very-low-pressure total system pressure of 70 Pa. The diffusion path length (δ) is 5.0 cm, and the silane feed gas diluted in H_2 gas to a composition of 20 mole% silane. The Lennard–Jones parameters for SiH_4 are $\sigma_A = 4.08$Å and $\epsilon_A/\kappa = 207.6$ K.

a. Recall from Example 1, Chapter 25, that the steady-state molecular diffusion flux of silane vapor (species A) down to the silicon wafer is

$$N_{A,z} = \frac{cD_{AB}}{\delta} \ln\left(\frac{1 + y_{Ao}}{1 + y_{As}}\right)$$

where y_{Ao} is the mole fraction of silane in the bulk gas, and y_{As} is the mole fraction of silane vapor at the surface of the wafer. Assume that surface reaction is instantaneous, so that the rate of silane decomposition to solid silicon is controlled by the molecular diffusion of silane to the silicon surface (e.g., $y_{As} \approx 0$). Estimate the rate of Si film formation in units of microns (μm) of Si solid film thickness per minute. The density of crystalline silicon is 2.32 g/cm^3.

b. Now consider that the surface reaction is not instantaneous, so that $y_{As} > 0$. The simplified rate equation is given by

$$N_A|_{z=\delta} = +k_s c_{As}$$

where the ($+$) sign indicates that the direction of flux for species A is in the same direction as increasing z. At 900 K, $k_s = 1.25$ cm/s, using data provided by Middleman and Hochberg (1993).[*] Develop a revised model for the flux N_A, which does not contain the y_{As} term, using the approximation $\ln(1 + y) \doteq y$. From this model, estimate the rate of Si film formation, and then compare the result to part (a) above. Comment on the importance of diffusion versus surface reaction in determining the silicon thin film formation rate.

[*]S. Middleman and A. K. Hochberg, *Process Engineering Analysis in Semiconductor Device Fabrication*, McGraw-Hill Inc., New York, 1993.

26.2 The spherical gel capsule shown in the figure at the top of the next column is used for long-term, sustained drug release. A saturated liquid solution containing the dissolved drug (solute A) is encapsulated within a rigid gel-like shell. The saturated solution contains a lump of solid A, which keeps the dissolved concentration of A saturated within the liquid core of the capsule. Solute A then diffuses through the gel-like shell (the gel phase) to the surroundings. Eventually, the source for A is depleted, and the amount of solute A within the liquid core goes down with time. However, as long as the lump of solid A exists within the core to keep the source solution saturated in A, the concentration of A within the core is constant. The diffusion coefficient of solute A in the gel phase (B) is $D_{AB} = 1.5 \times 10^{-5}$ cm^2/s. The maximum solubility of the drug in the gel capsule material is $c_A^* = 0.01$ gmole A/cm^3, $R_i = 0.2$ cm, $R_o = 0.35$ cm.

a. Starting from the appropriately simplified *differential forms* of Fick's flux equation and the general differential equation for mass transfer relevant to the physical system of interest, develop the final, analytical, integrated equation to determine the total rate of drug release (W_A) from the capsule under conditions where the saturated concentration of A within the liquid core of the capsule remains constant.

b. What is the maximum possible rate of drug release from the capsule, in units of gmole A per hour, when $c_{Ao} \approx 0$?

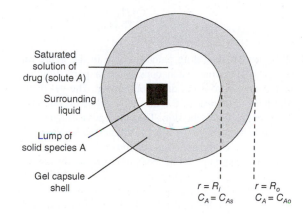

26.3 Consider the hollow cylinder for drug delivery release shown in the figure below. The hollow center contains a lump of solid solute A (drug) in a liquid. Solute A dissolves in the liquid to a maximum concentration $c_{A*} = 0.05$ mmole A/cm^3 and then diffuses radially through the gel layer to the surrounding liquid, which is maintained at a constant concentration, $c_{A\infty} = 0.01$ mmole A/cm^3. In the present system, the effective diffusion coefficient of solute A in the gel layer is $D_{Ae} = 1.2 \times 10^{-5}$ cm^2/s, the thickness of the gel layer ($R_o - R_i$) is 0.50 cm, the length is 1.5 cm, and the *outer* radius (R_o) is 0.75 cm. You may assume that the concentrations of solute A are dilute, the liquid within the hollow portion of the cylinder has constant concentration of c_{A*}, and the edges of the cylinder are sealed. As long as the solid solute A inside the hollow cylinder has not completely dissolved away, the mass transfer process has a constant source and a constant sink with respect to solute A.

a. What are reasonable boundary conditions for the mass transfer?

b. What are reasonable assumptions for the mass transfer process?

c. Estimate the total transfer rate of solute A exiting cylinder at $r = R_o$, W_A, in units of mmole A/s. As part of your analysis, state the appropriately-simplified differential forms of the Flux Equation and the General Differential Equation for Mass Transfer.

d. Based on your result for part (c), if the initial loading of solid solute A within the hollow center is $m_{Ao} = 0.20$ mmole A, and the liquid within the hollow center is always at concentration c_{A*}, how many *hours* will it take for the solid A to be completely dissolved?

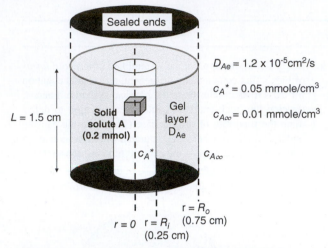

$D_{Ae} = 1.2 \times 10^{-5}\,\text{cm}^2/\text{s}$

$c_A^* = 0.05\,\text{mmole/cm}^3$

$c_{A\infty} = 0.01\,\text{mmole/cm}^3$

$L = 1.5$ cm

Solid solute A (0.2 mmol)

Gel layer D_{Ae}

c_A^* $c_{A\infty}$

$r = R_o$

$r = 0$ $r = R_i$ (0.75 cm)

(0.25 cm)

26.4 Consider the "drug patch" shown in the figure on the next page. The drug patch consists of a pure drug source mounted on top of a gel diffusion barrier. The gel diffusion barrier has a thickness of 2.0 mm. The gel diffusion barrier is in direct contact with the skin. The cumulative drug release vs. time profile for a 3.0-cm × 3.0-cm square patch at 20 C is also shown below. Other experiments showed that the drug was immediately taken up into the body after exiting the patch. The drug is only slightly soluble in the gel material. The maximum solubility of the drug in the gel diffusion barrier is 0.50 µmole/cm³, and the solubility of the drug in the gel diffusion barrier is not affected by temperature.

a. From the data shown below, estimate the effective diffusion coefficient of the drug in the diffusion barrier.

b. When used on the body, heat transfer raises the temperature of the drug patch to about 35 C. What is the new drug delivery rate (W_A) at this temperature in units of µmole/day? For purposes of this analysis, assume that the gel-like diffusion barrier material approximates the properties of liquid water.

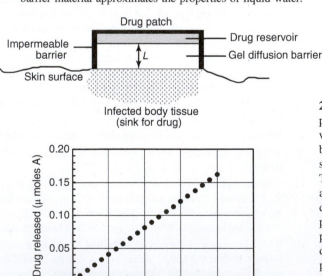

Drug patch

Impermeable barrier

Drug reservoir

Gel diffusion barrier

Skin surface

L

Infected body tissue (sink for drug)

26.5 Chicken eggs possess a hard, porous shell of calcite mineral. Cylindrical pores of 10 micron (µm) diameter running through the 0.5 mm thickness of the shell permit the exchange of gases to the within the egg, as shown in the figure in the next column for a given single pore. A typical egg has 20,000 of these pores, which are not interconnected and which run parallel to each other. Each pore extends from the outer surface of the egg shell down to the egg membrane, which contains the egg yolk. The egg is approximately spherical in shape, with an outer diameter of 5.0 cm. We are interested in predicting the loss of water from one egg by diffusion of water vapor through the egg shell. The source of the water vapor is the egg yolk itself, which is mostly water. The eggs are stored at the "hen house" temperature of 30 C, and at this temperature the vapor pressure of water is 0.044 atm. The relative humidity of the ambient air surrounding the egg is 50% of saturation, corresponding to a H_2O partial pressure of 0.022 atm.

a. What is the molecular diffusion coefficient of water vapor in air at 30 C and 1.0 atm inside the pore? Is Knudsen diffusion very important in this diffusion process?

b. Determine the water loss from a single egg in units of grams of water per day. State all relevant assumptions used in the calculation.

c. Propose two changes in process conditions within the hen house that would reduce water loss by at least 50%.

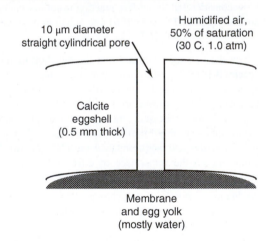

10 µm diameter straight cylindrical pore

Humidified air, 50% of saturation (30 C, 1.0 atm)

Calcite eggshell (0.5 mm thick)

Membrane and egg yolk (mostly water)

26.6 Consider the "drug patch" shown in Problem 26.4. The patch consists of a pure solid drug source mounted on top of a water-swollen polymer, which acts as a controlled diffusion barrier. The square patch is 4.0 cm × 4.0 cm. The bottom surface of the diffusion barrier is in direct contact with the skin. The desired amount of cumulative drug release is 0.02 µmole after 10 hours. Independent experiments have shown that the drug is immediately taken up into the body after exiting the patch. The maximum solubility of the drug in the water-swollen polymer is 0.50 µmole/cm³, which is dilute. The effective diffusion coefficient D_{Ae} of the drug in the water-swollen polymer is 2.08×10^{-7} cm²/s. What is the required thickness (L) of the diffusion barrier component of the drug patch needed to achieve the desired drug release profile?

26.7 Consider the biosensor device shown in the figure in the next column. The biosensor is designed to measure the concentration of solute A in the well-mixed liquid phase. At the base of the device is an electrode of surface area 2.0 cm². The electrode is coated with an enzyme that catalyzes the reaction $A \to 2D$. When solute A reacts to product D, product D is detected by the electrode, enabling for direct measurement of the flux of product D, which at steady state can be used to determine the concentration of A in the bulk liquid. The rate of reaction of A at the enzyme surface is rapid relative to the rate of diffusion of A down to the surface. Directly above the enzyme-coated electrode is a gel layer of 0.30 cm thickness that serves as a diffusion barrier for solute A and protects the enzyme. The gel layer is designed to make the flux of A down to the enzyme-coated surface diffusion limited. The effective diffusion coefficient of solute A in this gel layer is $D_{Ae} = 4.0 \times 10^{-7}$ cm²/s at 20 C. Above the gel layer is a well-mixed liquid containing a constant concentration of solute A, c'_{Ao}. The solubility of solute A in the liquid differs from the solubility of A in the gel layer. Specifically, the equilibrium solubility of A in the liquid layer (c'_A) is related to the solubility of A in the gel layer (c_A) by $c'_A = K \cdot c_A$, with equilibrium partitioning constant $K = 0.8$ cm³ gel/cm³ liquid. The process is considered very dilute, and the total molar concentration of the gel layer is unknown. The concentration of product D in the well-mixed liquid is very small so that $c_{Do} \approx 0$. At 20 C, the electrode measures that the formation of product D is equal to 3.6×10^{-5} mmole D/h. What is the concentration of solute A in the bulk *well-mixed liquid phase*, c'_{Ao}, in units of mmole/cm³?

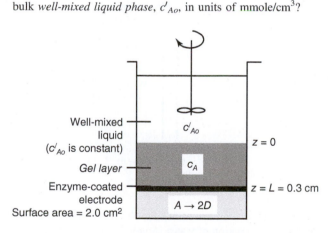

Well-mixed liquid (c'_{Ao} is constant)
c'_{Ao}
$z = 0$
Gel layer — c_A
Enzyme-coated electrode — $z = L = 0.3$ cm
Surface area = 2.0 cm²
$A \to 2D$

26.8 We are interested analyzing the diffusion processes associated with the treatment of cancer cells. Consider the experimental system shown in the figure on page 524. A hemispherical clump of cancer tissue is surrounded by a hemispherical layer of healthy tissue. Surrounding the healthy tissue is a well-mixed liquid nutrient medium containing a constant concentration of drug species A in it. In order to keep the cancer tissue from growing, the drug must diffuse through the healthy tissue, from $r = R_2 = 0.1$ cm, to the boundary of the cancer tissue and the healthy tissue at $r = R_1 = 0.05$ cm. Once the drug reaches the cancer tissue boundary ($r = R_1$), it is immediately consumed, and so the drug concentration at this boundary is essentially zero.

Independent experiments have also confirmed that (1) the cancer tissue will not grow (i.e., R_1 will not change with time) so long as the drug delivery flux reaching the surface of the cancer tissue at $r = R_1$ is $N_A = 6.914 \times 10^{-4}$ mmole A/cm² · day, (2) the drug is not consumed by the healthy tissue, which approximates the properties of water, (3) the diffusion coefficient of the drug (A) through the healthy tissue (B) is $D_{AB} = 2.0 \cdot 10^{-7}$ cm²/s, and (4) the drug is not very soluble in the healthy tissue.

a. Describe the system for diffusion mass transfer of the drug (system boundary, source, sink), and state at least three reasonable assumptions in addition to those stated above.

b. State the boundary conditions that best describe the mass-transfer process based on the system for mass transfer.

c. Simplify the general differential equation for mass transfer that best describes the particular physical system. Then develop a model in final integrated form to describe the flux N_A of the drug to the surface of the cancer cell clump at $r = R_1$.

d. Based on the model above, determine the constant surface concentration c_{AO} at $r = R_2$ needed to treat the cancer cells, in units of mmole/cm³.

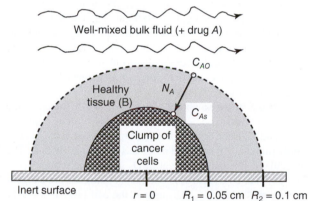

Well-mixed bulk fluid (+ drug A)
C_{AO}
Healthy tissue (B) N_A
C_{As}
Clump of cancer cells
Inert surface
$r = 0$ $R_1 = 0.05$ cm $R_2 = 0.1$ cm

26.9 A porous "water vapor barrier" is placed over the tissue implant shown in the figure in the next column. The purpose of the porous "water vapor barrier" is to allow O_2 gas direct access to the tissue while minimizing the diffusion-limited rate of evaporation of water from the tissue. Both the vapor barrier and the tissue implant possess "slab" geometry. The process is slightly pressurized and operates at 37 C and 1.2 atm total system pressure (P). The O_2 gas stream contains water vapor at 20% relative humidity at 37 C. The vapor barrier material is a random microporous polymer with mean pore size of 50 nm (1×10^7 nm = 1.0 cm) and void fraction (ϵ) of 0.40. The tissue approximates the properties of liquid water. At 37 C, the vapor pressure of liquid water is 47 mm Hg (1.0 atm = 760 mm Hg), and the Henry's law constant (H) for the dissolution of O_2 gas in water is 800 L · atm/gmole. Let $A = H_2O, B = O_2$.

a. Making use of the Fuller–Schettler–Giddings correlation in the calculations, what is the effective diffusion coefficient (D_{Ae}) of H_2O vapor in the randomly microporous water vapor barrier?

b. What is the thickness of the vapor barrier (L) required to limit the rate of water evaporation from the tissue to 0.180 g $H_2O/cm^2 \cdot$ day? State all assumptions for your analysis.

c. What is the concentration of dissolved oxygen *in the tissue* (C_{BL}^*, gmole O_2/L tissue) at the interface between the tissue and the porous vapor barrier ($z = 0$)?

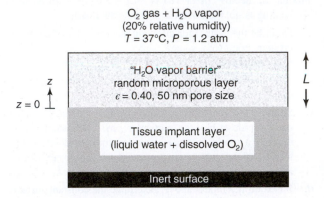

O$_2$ gas + H$_2$O vapor
(20% relative humidity)
$T = 37°C$, $P = 1.2$ atm

"H$_2$O vapor barrier"
random microporous layer
$\epsilon = 0.40$, 50 nm pore size

Tissue implant layer
(liquid water + dissolved O$_2$)

Inert surface

26.10 A cell monolayer used in a tissue engineering scaffold adheres onto the top surface of a silicone rubber (polymer) sheet of 0.10 cm thickness, as shown in the figure below. The rectangular sheet is 5.0 cm by 10.0 cm. The underside of the silicone polymer layer is in contact with pure O_2 gas. The O_2 gas dissolves into the polymer and diffuses through the polymer to the adhered cells to deliver oxygen to them. The solubility of dissolved O_2 in the silicone polymer is defined by a linear relationship $p_A = C_A'/S$, where p_A is the partial pressure of O_2 gas (atm), S is the solubility constant of O_2 dissolved in the silicone polymer ($S = 3.15 \times 10^{-3}$ mmole O_2/cm$^3 \cdot$ atm at 25 C), and C_A' is the concentration of O_2 dissolved in the silicone rubber (mmole O_2/cm^3). The process is isothermal at 25 C. The molecular diffusion coefficient of O_2 in silicone rubber is 1×10^{-7} cm^2/s at 25 C. It is assumed that (1) the cells are oxygen-starved, and so any O_2 that reaches the cell layer is immediately consumed and that (2) the cells consume O_2 by a zero-order process that is not dependent on dissolved O_2 concentration. It is determined that the sustainable O_2 consumption rate of the cell monolayer is fixed at 1.42×10^{-5} mmole O_2/min (0.0142 μmole O_2/min). What the required O_2 partial pressure (p_A) required to for the diffusion process to enable this O_2 consumption rate?

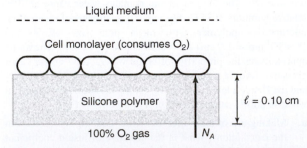

Liquid medium

Cell monolayer (consumes O$_2$)

Silicone polymer

$\ell = 0.10$ cm

100% O$_2$ gas

N_A

26.11 An open well contains water contaminated with volatile benzene at the bottom of the well, with dimensions shown

in the figure on page 525. The concentration of dissolved benzene in the water is 156 g/m^3, and remains constant. The system is isothermal at 25 C. We are interested in determining the emission of benzene, a carcinogen, into the atmosphere from the well.

a. Define the source, sink, and system boundary for all of the species undergoing mass transfer. State three reasonable assumptions that describe the mass-transfer process.

b. State reasonable boundary conditions, and specify their numerical values with units, for all of the species undergoing mass transfer.

c. What are the maximum emission rates (in mole/day) of benzene and water vapor from the well? What is cumulative benzene emission (in grams) over a period of 30 days? Is it significant?

Potentially useful data: The molecular diffusion coefficient of benzene in dissolved water is 1.1×10^{-5} cm^2/s at 25 C, and the molecular diffusion coefficient of benzene vapor in air is 0.093 cm^2/s at 1.0 atm and 25 C. The Henry's law constant for benzene partitioning into water is $H = 4.84 \times 10^{-3}$ m$^3 \cdot$ atm/mole at 25 C. The vapor pressure of liquid water is 0.0317 bar (0.031 atm) at 25 C, and the density is 1000 kg/m^3 at 25 C. The vapor pressure of benzene is 0.13 atm at 25 C. The humidity of the air flowing over the well hole is 40% of relative saturation at 25 C. The molecular weight of benzene is 78 g/mole.

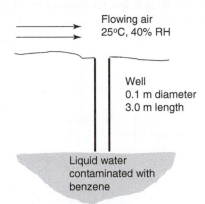

Flowing air
25°C, 40% RH

Well
0.1 m diameter
3.0 m length

Liquid water
contaminated with
benzene

26.12 Tubular membranes of silicone rubber can be used for "bubbleless" aeration of water. A cross section of the tube is shown in the figure (right-hand column). Pure oxygen (O_2) gas is pressurized to 2.0 atm inside a section of silicone rubber tubing of inner diameter 12.7 mm and wall thickness 3.2 mm. The tubing is immersed in a large volume of an aqueous solution. Silicone rubber is permeable to oxygen (O_2) gas, but is also hydrophobic so that water does not seep through the rubber. The O_2 gas "dissolves" into the silicone rubber with a concentration C_A (mole A/m^3 silicone rubber) diffuses through the wall of the tubing, and then redissolves into the water of concentration $c_{A\infty}$, which is maintained at a concentration of 0.005 mol/m^3.

a. Develop an equation, in final integrated form, to predict the O_2 flux across the tube wall from $r = R_1$ to $r = R_0$, using

C'_A to describe the concentration of O_2 dissolved in the tube wall material itself. State all assumptions. You may neglect convective mass-transfer resistances associated with the liquid boundary layer surrounding the tube.

b. At the conditions given above, determine the flux of oxygen to the water ($r = R_0$) if the well-mixed aqueous phase maintains the dissolved O_2 concentration at 0.005 mole O_2/m^3.

Potentially useful data: The solubility of dissolved O_2 in the silicone polymer is defined by a linear relationship $p_A = C'_A/S$, where P_A is the partial pressure of O_2 gas (atm), S is the solubility constant of O_2 dissolved in the silicone polymer ($S = 3.15 \times 10^{-3}$ mmole $O_2/cm^3 \cdot$ atm at 25 C), and C'_A is the concentration of O_2 dissolved in the silicone rubber (mmole O_2/cm^3). The solubility of O_2 gas in silicone rubber in contact with 2.0 atm O_2 gas at 25 C is $C'^*_A = 6.30$ mole O_2/m^3 of silicone rubber. The Henry's law constant (H) of O_2 in water is 0.78 atm $\cdot m^3$ water/gmole at 25 C.

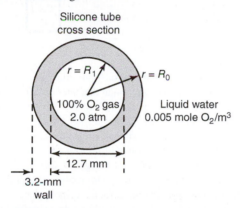

Silicone tube cross section

$r = R_1$
$r = R_0$
100% O_2 gas 2.0 atm
Liquid water 0.005 mole O_2/m^3
12.7 mm
3.2-mm wall

26.13 Consider a hemispherical droplet of liquid water residing on a flat surface. Still air surrounds the droplet. At an infinitely long distance from the gas film, the concentration of water vapor in the air is effectively zero (dry air). At a constant temperature of 30 C and 1.0 atm total pressure, the evaporation rate of the droplet is controlled by the rate of molecular diffusion through the still air. The vapor pressure of water at 30 C is 0.042 atm, and the molecular diffusion coefficient of H_2O vapor in air at 1.0 atm and 30°C is 0.266 cm^2/s.

a. What is the total evaporation transfer rate (W_A) of water from a water droplet of radius 5.0 mm in units of mmole H_2O per hour?

b. Determine the time it will take for the water droplet to completely evaporate at 30 C and 1.0 atm total system pressure if the initial droplet radius is 5.0 mm.

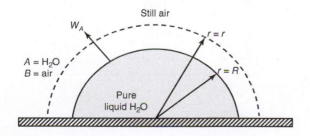

Still air
W_A
$r = r$
$A = H_2O$
$B = $ air
$r = R$
Pure liquid H_2O

26.14 The mass-transfer device shown in the figure below is used to carry out the controlled release of a vapor-phase pheromone drug used in pest control. The solid drug sublimes at a vapor pressure P^*_A within the gas space of the reservoir. A polymer layer of thickness $L = 0.15$ cm covers the drug reservoir. The drug vapor (species A) absorbs into a polymer diffusion layer by a linear relationship $p_A = S \cdot C'_A$, where C'_A is the concentration of the pheromone drug dissolved in the polymer (gmole species A/ cm^3 polymer), p_A is the partial pressure of the drug vapor (atm), and S is the partitioning constant for the drug between the vapor phase and the polymer phase (cm^3-atm/mol). The pheromone is highly soluble in the polymer. The drug then diffuses through the polymer layer with diffusion coefficient D_{Ae}, and then exits to the surroundings as a vapor. Air flow over the top surface of the polymer layer generates a "fluid boundary layer." The flux of the drug vapor across this boundary layer is given by

$$N_A = k_G(p_{As} - p_{A\infty})$$

where k_G is the gas-phase mass-transfer coefficient (gmole/ $cm^2 \cdot s \cdot$ atm). Generally, k_G increases as the air flow rate over the surface increases. At steady state, the flux of drug (species A) through the polymer layer equals the flux through the boundary layer.

a. Develop a mathematical model, in final integrated form, for the drug vapor flux N_A. The final model can only contain the following terms: N_A, D_{Ae}, P^*_A, $p_{A\infty}$, L, S, k_G. State all assumptions for analysis.

b. Determine the maximum possible drug vapor flux associated with the mass-transfer device, in units of μmole/cm$^2 \cdot$ s (1 μmole $= 1.0 \times 10^{-6}$ mole), under conditions where $p_{A\infty} \approx 0$, 30 C, and 1.0 atm total system pressure. The diffusion coefficient of drug vapor through polymer, D_{Ae}, is 1.0×10^{-6} cm^2/s. The Henry's law constant for absorption (dissolution) of drug vapor into polymer, S, is 0.80 cm$^3 \cdot$ atm/gmole. The "mass-transfer coefficient" for boundary layer, k_G, is 1.0×10^{-5} mol/cm$^2 \cdot$ s $\cdot$ atm. The vapor pressure of pheromone drug at 30 C is 1.1 atm.

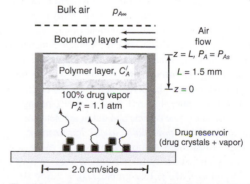

Bulk air $p_{A\infty}$
Boundary layer
Air flow
$z = L, P_A = P_{As}$
Polymer layer, C'_A
$L = 1.5$ mm
$z = 0$
100% drug vapor $P^*_A = 1.1$ atm
Drug reservoir (drug crystals + vapor)
2.0 cm/side

26.15 Consider the timed drug release pill shown in the figure below. The pill is ingested into the stomach. The pill is a slab of 0.36 cm per side, which has an array of 16 cylindrical pores in it. Each pore is 0.4 mm (0.04 cm) in diameter and 2.0 mm (0.20 cm) deep. Pure solid drug (species A) is loaded into each pore to a depth of 1.2 mm (0.12 cm), which provides a total initial drug loading of 2.65 mg in all of the pores. The density of the solid drug ($\rho_{A,solid}$) is 1.10 g/cm^3. The drug dissolves into fluid inside the stomach, which approximates the properties of water (species B). The maximum solubility of drug in water is 2.0×10^{-4} gmol/cm³ (not very soluble), and the diffusion coefficient of the drug in the fluid is 2.0×10^{-5} cm²/s at body temperature of 37 C. The molecular weight of the drug is 120 g/gmole.

a. Determine the total transfer rate of the drug from the whole pill (W_A) to the body at when each 0.2 cm pore is filled to a depth of 0.12 cm with solid drug. You may assume that (1) the diffusion process is at a pseudo-steady-state, (2) the stomach fluid serves as an infinite sink for the drug so that $c_{A,\infty} \approx 0$, and (3) the drug does not chemically degrade inside the pore. List all other assumptions you make as appropriate in the development of your mass-transfer model.

b. How long will it take (in hours) for *all* of the drug material in the reservoir to be released?

c. Propose adjustment of one of the variables in the process that will increase the required time in part (b) by a factor of 2.0.

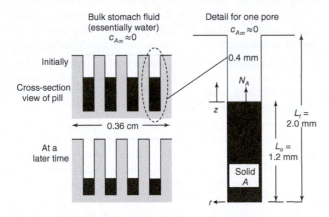

26.16 As part of a process to fabricate materials with ordered pore arrays, the diffusion-limited solvent etching process shown in the figure (next column) is used to remove a contaminant (solid solute A, lithium carbonate) from an array of pores within a slab. There are presently 16 cylindrical pores, each of 0.10 cm diameter and 0.5 cm depth, and the overall dimensions of the slab are 4.0 cm per side and 0.60 cm deep. Liquid solvent (species B) flows over the surface of the slab at a high rate so that the concentration of dissolved solute A is essentially zero; also, dissolved solute A is not very soluble (sparingly soluble) in solvent B. The process is carried out at ambient conditions (25 C, 1.0 atm). *Other potentially useful information:* $c_A* = 90$ μmole/cm³ (maximum solubility of solute A in solvent B), $\rho_B = 1.0$ g/cm³ (density of solvent B) $\rho_{A,solid} = 2.1$ g/cm³, $M_A = 74$ g/gmole, $M_B = 18$ g/gmole, $D_{AB} = 1.6 \times 10^{-5}$ cm²/s at 25 C and 1.0 atm.

a. Initially, the solid solute A is loaded to a depth of 0.20 cm within each 0.5 cm length pore, as shown in the figure

below. At this point, what the total transfer rate of solute A out of the entire slab, in units of μmole/s?

b. How much time will it take for all of the solid solute A to be removed from each pore, units of hours?

c. The process takes a long time. What changes to the process (e.g. solubility of solute A, pore diameter, diffusion path length, process temperature) would likely reduce the time required?

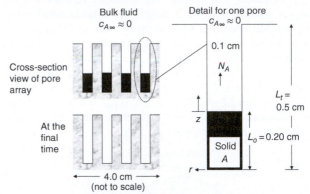

26.17 A cylindrical graphite rod (pure solid carbon, density 2.25 g/cm³) of length 25 cm and initial diameter of 2.0 cm is inserted into a flowing air stream at 1100 K and 2.0 atm total system pressure. The flowing gas creates a stagnant gas boundary layer 5.0 mm thick around the external surface of the rod. At this high temperature, the solid carbon oxidizes to carbon dioxide, CO_2, gas according to the reaction $C(s) + O_2(g) \rightarrow CO_2(g)$. This surface oxidation reaction is limited by the molecular diffusion of O_2 through the stagnant gas film surrounding the surface of the rod, so that the O_2 concentration at the graphite surface is effectively zero. Outside of the gas film, the bulk composition of the gas stream represents that of air.

a. Estimate the initial rate of CO_2 production from the rod, assuming that the surface reaction is diffusion-limited.

b. How long will it take for the graphite rod to completely disappear?

26.18 The data provided in Figure 26.7 are based on the diffusion of O_2 into SiO_2 formed from the oxidation of (100) crystalline silicon at 1000 C. Estimate the diffusion coefficient of O_2 in SiO_2 formed from the oxidation of (111) crystalline silicon at 1000 C, using the data in the table below, provided by Hess (1990).*

Time	Measured SiO₂ Film Thickness (μm)	
1.0	0.049	0.070
2.0	0.078	0.105
4.0	0.124	0.154
7.0	0.180	0.212
16.0	0.298	0.339

The maximum solubility of O_2 in the SiO_2 is 9.6×10^{-8} mole O_2/cm^3 solid at 1000 C and 1.0 atm O_2 gas partial pressure.

*D. W. Hess, *Chem. Eng. Education*, **24**, 34 (1990).

SS **26.19** The Arnold Diffusion Cell shown in Figure 26.5 is a simple device used to measure gas-phase diffusion coefficients for volatile substrates in air. In the present experiment, liquid acetone is loaded to the bottom of a glass tube of 3.0 mm inner diameter. The tube and the liquid acetone in the tube are maintained at a constant temperature of 20.9 C. The tube is open to the atmosphere, and air blows over the open end of the tube, but the gas space inside the cylindrical tube is still. The total length of the tube is 15.0 cm. As the acetone evaporates, the liquid level decreases, which increases Z, the diffusion path length in the gas from the liquid surface to the mouth of the tube. Measurements for the diffusion path length (Z) as a function of time (t) are presented in the table below:

Time, t (h)	Z (cm)
0.00	5.6
21.63	6.8
44.73	7.8
92.68	9.6
164.97	11.8
212.72	13.0

a. Manipulate the data shown in the table above so that it can be plotted as a straight line. Statistically estimate the slope of this line by least-squares linear regression, and then use the slope to estimate the diffusion coefficient of acetone in air.

b. Compare the result in part (a) to an estimate of diffusion coefficient of acetone in air by a suitable correlation given in Chapter 24.

Potentially useful data: The molecular weight of acetone (M_A) is 58 g/gmole; the density of liquid acetone ($\rho_{A,liq}$) is 0.79 g/cm^3; the vapor pressure of acetone (P_A^*) at 20.9 C is 193 mm Hg.

26.20 A flat surface containing many parallel pores is clogged with "coke" from a manufacturing process, as shown in the figure below. Pure oxygen gas (O_2) at high temperature is used to oxidize the coke, which is mainly solid carbon, to carbon dioxide (CO_2) gas. This process will remove the solid carbon clogging the pores, and hence clean the surface. A large excess of O_2 is in the bulk gas over the surface, and so it may be assumed that bulk gas composition is always 100% O_2. It may also be assumed that the oxidation reaction is very rapid relative to the rate of diffusion, so that the production of CO_2 is limited by mass transfer, and the O_2 concentration at the carbon surface is essentially zero. The pores are cylindrical, with diameter of 1.0 mm and depth of 5 mm. The oxidation process is carried out at 2.0 atm total system pressure and 600 C. The density of solid carbon is 2.25 g/cm^3. Let $A = O_2$, and $B = CO_2$.

a. At some time after the oxidation process, the cleaned depth of the pore is 3.0 mm (0.3 cm) from the mouth of the pore. What is the total emissions rate of CO_2 gas (W_B) at this point in the process?

b. How long will the oxidation process take to reach this cleaned depth of 0.3 cm from the mouth of the pore?

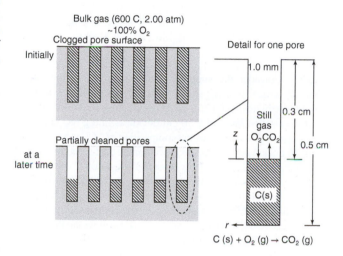

$$C\,(s) + O_2\,(g) \rightarrow CO_2\,(g)$$

26.21 The "drug patch" shown in the figure in the next page releases a water-soluble epidermal growth factor (species A) to repair a specific region of wounded tissue on the human body. A slow release of the drug is critical for regulating the rate of tissue repair. The drug layer (pure solute A) rests on top of a diffusion barrier. The diffusion barrier is essentially a microporous polymer material consisting of tiny parallel pores filled with liquid water (species B). This diffusion barrier controls the rate of drug release. The thickness (L), pore size (d_{pore}), and porosity of the diffusion barrier are design parameters that determine the dosage rate of the drug to the tissue directly beneath it. The maximum solubility of the drug in water is 1.0 mole/m^3 at 25 C. The diffusion coefficient of the drug in water at infinite dilution is 1.0×10^{-10} m^2/s at 25 C. The pore diameter (d_{pore}) is 100 Å (1 Å $= 10^{-10}$ m), and the equivalent molecular diameter of the drug is 25 Å. The total surface area of the patch is 4.0 cm^2, but the cross-sectional area of the pore openings available for flux constitutes only 25% of this contact surface area.

a. What is the effective diffusion coefficient of the drug in the diffusion barrier? Hint: Consider the Renkin equation for solute diffusion in solvent-filled pores from Chapter 24.

b. Estimate the thickness of the diffusion barrier (L) necessary to achieve a maximum possible dosage rate of 0.05 μmole per day, assuming that the drug is instantaneously consumed once it exits the diffusion barrier and enters the body tissue ($1\ \mu\text{mole} = 1.0 \times 10^{-6}\ \text{mole}$).

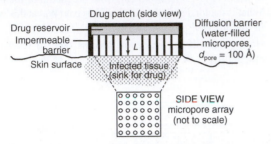

Drug patch (side view)

Drug reservoir
Impermeable barrier
Skin surface
Infected tissue (sink for drug)
Diffusion barrier (water-filled micropores, $d_{pore} = 100$ Å)

SIDE VIEW
micropore array
(not to scale)

26.22 Consider the simplified process shown in the figure below, which is designed to treat a gas-phase mixture of trichloroethylene (TCE) in air by degrading the TCE within a porous catalyst layer of 0.50 cm thickness lining one side of a well-mixed chamber. The inlet composition of TCE is 5.0 mole%, but the treated gas exiting the process must contain no more than 2.0 mole% TCE. The process is carried out at 300 C and 1.0 atm total system pressure. At these conditions, the degradation of TCE within the porous catalyst layer is approximated as a homogeneous first-order reaction process with rate constant (k) of 1.5 s^{-1}. The effective diffusion coefficient of TCE in air within the porous layer is 0.080 cm^2/s. The gas mixing within the chamber is sufficiently high so that the gas space over the catalyst layer is well mixed and convective (boundary layer) resistances are minimized, resulting in $c_{Ao} = c_{A\infty} = c_{As}$.

a. What relationships best describe the boundary conditions for mass transfer in the system?

b. What is the *mole fraction* of TCE in the porous catalyst layer at $z = L$, the point where the porous catalyst layer is attached to its nonporous support surface?

c. If the process must be designed to remove 0.20 gmole TCE/s from the gas stream, what is the required external surface area (S) of the catalyst layer?

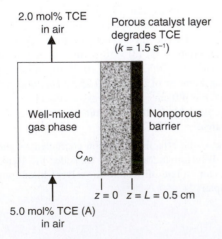

2.0 mol% TCE in air

Porous catalyst layer degrades TCE
($k = 1.5$ s^{-1})

Well-mixed gas phase

C_{Ao}

Nonporous barrier

$z = 0$ $z = L = 0.5$ cm

5.0 mol% TCE (A) in air

26.23 The immobilized enzyme layer shown in the figure below is designed to convert substrate A to product B according to the simplified reaction $A \rightarrow B$ with simplified rate equation $R_A = -k\,c_A$. In this process, the steady-state concentrations of substrate A in the bulk liquid and the interface between the enzyme layer and its inert support surface were measured, with $c_{A\infty} = 1.0$ mmol/m^3 at $z = 0$, and $c_{AL} = 0.25$ mmol/m^3 at $z = L$. The effective diffusion coefficient of substrate A within the enzyme layer is $D_{Ae} = 2.0 \times 10^{-5}$ cm^2/s, and the thickness of the immobilized enzyme layer is $L = 0.2$ cm. The surrounding fluid is well mixed, and the concentration of substrate A at the surface of the enzyme layer can be taken as the bulk concentration of substrate A in the surrounding liquid.

a. What are valid boundary conditions for this process?

b. Based on the measured values for $c_{A\infty}$ and c_{AL}, what is the first-order rate constant k?

c. If the total exposed surface area of the enzyme layer is $S = 2.0$ m^2, what is the production rate of product B?

d. Based on your calculation of the Thiele modulus, is the process reaction controlled, diffusion controlled, or neither?

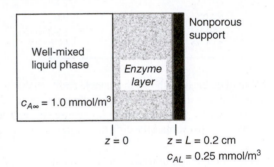

Well-mixed liquid phase

Enzyme layer

Nonporous support

$c_{A\infty} = 1.0$ mmol/m^3

$z = 0$ $z = L = 0.2$ cm
$c_{AL} = 0.25$ mmol/m^3

26.24 Consider a spherical gel bead containing a biocatalyst uniformly distributed within the gel. Within the gel bead, a homogeneous, *first-order* reaction $A \xrightarrow{k_1} D$ is promoted by the biocatalyst. The gel bead is suspended within water containing a known, constant, dilute concentration of solute A ($c_{A\infty}$).

a. Define the system, and identify the source and the sink for the mass-transfer process with respect to reactant A. List three reasonable assumptions for this process. Then, using the "shell balance" approach, develop the *differential material balance model* for the process in terms of concentration profile c_A. State all boundary conditions necessary to completely specify this differential equation.

b. The analytical solution for the concentration profile is given by

$$c_A(r) = c_{A\infty} \frac{R \sinh(r\sqrt{k_1/D_{AB}})}{r \sinh(R\sqrt{k_1/D_{AB}})}$$

What is the total consumption rate of solute A by one single bead in units of μmol A per hour? The bead is 6.0 mm in diameter. The diffusion coefficient of solute A

within the gel is 2×10^{-6} cm^2/s, k_1 is 0.019 s^{-1}, and $c_{A\infty}$ is 0.02 μmole/cm^3. Hint: Differentiate the relationship for $c_A(r)$ with respect to r, then estimate the flux N_A at $r = R$.

26.25 A biofilm reactor with a well-mixed liquid phase shown below will be used to treat wastewater contaminated with trichloroethylene (TCE) at a concentration 0.25 mg/L (1.9 mmole/m^3, M$_{TCE}$ = 131.4 g/gmole). The available surface area of the biofilm in the reactor is 800 m^2, and the desired outlet concentration of TCE is 0.05 mg/L (0.05 g/m^3). It can be assumed that in a well-mixed, continuous-flow reactor at steady state, the concentration of the solute of the liquid phase of inside the reactor is equal to the concentration of the solute in the liquid flow exiting the reactor. It may also be assumed that the TCE degradation in the biofilm, which of $\delta = 100$ μm thickness, proceeds by homogeneous first-order reaction kinetics.

a. What volumetric flow rate of wastewater is allowed to enter the reactor? The temperature of the process is constant at 20 C.

b. What is the concentration of TCE in the biofilm at the point where the biofilm is attached to the surface? What fraction of 100 μm thickness of the biofilm is utilized?

Potentially useful data: $k_{TCE} = 4.31$ s^{-1} (first-order rate constant for TCE in biofilm); $D_{TCE\text{-}biofilm} = 9.03 \times 10^{-10}$ m^2/s (diffusion coefficient TCE in biofilm).

*J. P. Arcangeli and E. Arvin, *Environ. Sci. Technol.*, **31**, 3044 (1997).

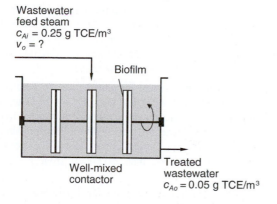

Wastewater
feed steam
$c_{Ai} = 0.25$ g TCE/m^3
$v_o = ?$

Biofilm

Well-mixed
contactor

Treated
wastewater
$c_{Ao} = 0.05$ g TCE/m^3

26.26 The pesticide Atrazine (C$_8$H$_{14}$ClN$_5$, mol wt. 216 g/mol) degrades in soil by a first-order reaction process. Consider the situation shown in the figure (next column), where there is a spill of solid Atrazine rests on top of a 10-cm-thick layer of water-saturated soil at 20 C. The solid Atrazine dissolves into the water, diffuses through the water-saturated soil and then degrades in the water-saturated soil with a homogeneous first-order rate constant (k_1) of 5.0×10^{-4} hr^{-1} at 20 C. Underneath the water-saturated soil layer is an impermeable clay barrier. The maximum solubility of Atrazine in water is 30 mg/L (0.139 mmole/L) at 20 C.

a. What is the molecular diffusion coefficient of Atrazine in water (D_{AB}) at 20 C, and the effective diffusion coefficient (D_{Ae}) in the water-saturated soil? The specific molar volume of Atrazine is 170 cm^3/gmole, at its normal boiling point. The effective diffusion coefficient is estimated by $D_{Ae} = \epsilon^2 D_{AB}$ ($A = $ Atrazine, $B = $ water), with a void fraction (ϵ) of 0.6. The effective diffusion coefficient accounts for the fact that not all of the soil is liquid water.

b. What is the concentration of Atrazine (mmole/L) in the water-saturated soil at the clay barrier ($z = L$)? It may be assumed that the process is at steady state with a one-dimensional flux of A in the z-direction.

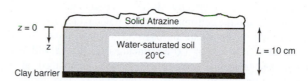

$z = 0$

z

Solid Atrazine

Water-saturated soil
20°C

$L = 10$ cm

Clay barrier

26.27 Carbon dioxide (CO$_2$) from waste sources is a sustainable feedstock for chemicals production if new technologies can be developed to carry out the reduction of CO$_2$, the most oxidized form of carbon, to a more reactive molecule. Recently, solar-energy-driven reactor concepts have emerged that harness the sun's energy to sustainably drive the thermocatalytic reduction of CO$_2$ to reactive CO over a Ceria catalyst at high temperatures. A highly simplified version of this concept is provided in the figure below which is operated at 800°C and 1.0 atm. The diameter of the cylindrical reactor is 10.0 cm, the thickness of the porous catalyst lining the base of the reactor is 1.0 cm, and the gas space above the porous catalyst layer can be considered well mixed. Pure CO$_2$ is fed into the reactor and diffuses into the porous catalyst layer, which drives the reaction CO$_2$(g) → CO(g) + 1/2O$_2$(g), which is approximated as first order with rate constant of $k = 6.0$ s^{-1} at 800 C. The effective diffusion coefficient of CO$_2$ in the gas mixture within the porous catalyst is 0.40 cm^2/s at 1.0 atm and 800 C. [SS]

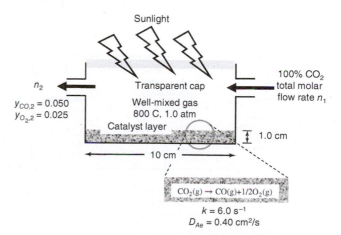

Sunlight

n_2
$y_{CO,2} = 0.050$
$y_{O_2,2} = 0.025$

Transparent cap

Well-mixed gas
800 C, 1.0 atm

Catalyst layer

100% CO$_2$
total molar
flow rate n_1

1.0 cm

10 cm

CO$_2$(g) → CO(g)+1/2O$_2$(g)

$k = 6.0$ s^{-1}
$D_{Ae} = 0.40$ cm^2/s

a. It is desired to achieve 5.0 mole% CO in the exiting outlet gas. What is the molar flow rate exiting the reactor (n_2), in units of gmole/min? What is the inlet molar flow rate of CO_2?

b. What is the mole fraction of CO_2 on the back side of the catalyst layer?

26.28 Many biocatalytic reactions are carried out with living cells as the biocatalyst. In some processes, the cells are immobilized within an agarose gel. However, living cells uniformly distributed within an agarose gel require glucose to survive, and the glucose must diffuse through the gel material to reach the cells. An important aspect of the biochemical system design is the effective diffusion coefficient of glucose (species A) into the cell-immobilized gel. Consider the experiment shown in the figure on page 531, where a slab of the cell-immobilized gel of 1.0 cm thickness is placed within a well-mixed aqueous solution of glucose maintained at a concentration of 50 mmole/L. The glucose consumption within the cell-immobilized gel proceeds by a zero-order process given by $R_A = -m = -0.05$ mmole/L-min. The solubility of the glucose in both water and the gel are the same— i.e., the concentration of glucose on the water side of the water-gel interface is equal to the concentration of glucose on the gel side of the water-gel interface. A syringe mounted in the center of the gel carefully excises a tiny sample of the gel for glucose analysis. The measured steady-state glucose concentration at the center of the gel is 4.50 mmole/L.

a. Develop a model, in final integrated form, to predict the concentration profile of glucose within the gel. Specify the boundary conditions, so that the model is truly predictive and is based only on process input parameters, not measured parameters. As part of the model development, state at least three relevant assumptions.

b. Using the model developed in part (a), the known process parameters, and the measured concentration of glucose at the center of the gel, estimate the effective diffusion coefficient of glucose within the cell-immobilized gel.

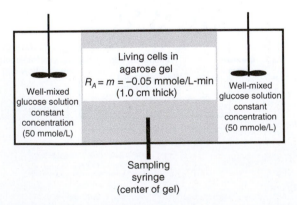

Well-mixed glucose solution constant concentration (50 mmole/L)

Living cells in agarose gel
$R_A = m = -0.05$ mmole/L-min
(1.0 cm thick)

Well-mixed glucose solution constant concentration (50 mmole/L)

Sampling syringe (center of gel)

26.29 Recall from Example 3, Chapter 25, that the differential model for the radial concentration profile of dissolved oxygen within one cylindrical engineered tissue bundle (Figure 25.8) is

$$D_{AB}\frac{d^2 c_A}{dr^2} + \frac{1}{r}\frac{dc_A}{dr} - \frac{R_{A,max}c_A}{K_A + c_A} = 0$$

with boundary conditions

$$r = R_1, \quad \frac{dc_A}{dr} = 0, \quad r = R_2, \quad c_A = c_{As} = \frac{p_A}{H}$$

Often, K_A is very small relative to c_A so that the homogeneous reaction term approaches a zero-order process that is not dependent on concentration:

$$R_A = -\frac{R_{A,max}c_A}{K_A + c_A} \cong -\frac{R_{A,max}c_A}{c_A} = -m$$

so that

$$D_{AB}\frac{d^2 c_A}{dr^2} + \frac{1}{r}\frac{dc_A}{dr} - m = 0$$

where m is the metabolic respiration rate of the tissue. In the present process, $m = 0.25$ mole/m^3 · h at 25 C, and R_1 and R_2 are equal to 0.25 cm and 0.75 cm, respectively. Pure O_2 gas at 1.0 atm flows through the tube of length 15 cm. The mass-transfer resistance due to the thin-walled tube is neglected, and the Henry's law constant for dissolution of O_2 in the tissue is 0.78 atm · m^3/mole at 25 C. The diffusion coefficient of oxygen in water is 2.1×10^{-5} cm^2/s at 25 C, which approximates the diffusivity of oxygen dissolved in the tissue.

a. Develop a model, in final integrated form, to predict the concentration profile $c_A(r)$, and then plot out the concentration profile. Note that for diffusion with zero-order homogeneous chemical reaction, there will be a critical radius, R_c, where the dissolved oxygen concentration goes to zero. Therefore, if $R_c < R_2$, then $c_A(r) = 0$ from $r = R_c$ to $r = R_2$.

b. Using this model and the process input parameter detailed above, determine R_c. Then, plot of the concentration profile from $c_A(r)$ from $r = R_1$ to $r = R_c$.

c. Develop a model, in final algebraic form, to predict W_A, total oxygen transfer rate through one tube. From the process input parameters given in the problem statement, calculate W_A.

26.30 Consider the physical system shown in the figure on page 532, which represents a simplified scenario for the treatment of tumor (cancer) tissue. In this experiment, a slab of living tissue is in contact with liquid medium bearing a constant concentration of antitumor drug A, which serves as a constant source for A. Unfortunately, drug A will encounter difficulty in actually reaching the tumor. First, drug A is not very soluble in the tissue, where its solubility is given by $C_{Ao} = K \cdot c_{A\infty}$, where K is the partitioning constant for the drug into the healthy tissue, and $c_{A\infty}$ is the bulk concentration of the drug in the well-mixed liquid medium. Second, as drug A diffuses through healthy tissue layer down to the tumor tissue, it partially decomposes to by-product B via $A \rightarrow B$ by a first-order homogeneous reaction with rate equation $R_A = -k_1 C_A$. The net flux of drug A that actually reaches the tumor surface at $z = L$ is called the *constant therapeutic flux rate*, N_{As}. In independent experiments, it was determined that $N_{As} = 2.0 \times 10^{-6}$ mg/cm$^2 \cdot$s for the drug to be effective, with the requirement that $C_{As} > 0$ at $z = L$. We are interested in developing a model to predict the total flux rate of drug A into the surface of the healthy tissue at $z = 0 (N_{Ao})$ to achieve the required N_{As} at $z = L$. Toward this end, a model must first be developed to predict the concentration profile of drug A as it diffuses through the healthy tissue layer.

a. Define the system for mass transfer of drug A, the source for drug A, and the sink(s) for drug A. Propose five reasonable assumptions for this process.

b. Based on a material balance on a differential volume element of the system for drug A, develop a *differential* equation for $C_A(z)$ in the healthy tissue layer.

c. State the boundary conditions that accurately describe the physical system *and* allow for a mathematical analysis of the diffusion process developed in part (b).

d. Develop an analytical model, in final integrated form, to predict $C_A(z)$. Your final model should have the following terms in it: $c_{A\infty}$, $N_{A,s}$, D_{Ae}, K, k_1, L, z; C_{Ao} and C_{As} should *not* be terms in the final model.

e. Refer the process model input parameters below. What is the total therapeutic drug delivery rate delivered to the tissue $(N_{A,o}$ at $z = 0)$ in units of mg/cm^2-day?

f. Plot out the concentration profile of the drug within the healthy tissue.

Process input parameters: Concentration of drug A in liquid medium, $c_{A\infty} = 3.0$ mg/cm^3; partitioning constant between liquid medium and healthy tissue, $K = 0.1$ cm^3 tissue/cm^3 liquid; diffusion coefficient of drug A in healthy tissue, $D_{Ae} = 1.0 \cdot 10^{-5}$ cm^2/s; rate constant for degradation of drug A in healthy tissue, $k_1 = 4.0 \cdot 10^{-5}$ s^{-1}; thickness of healthy tissue layer, $L = 0.5$ cm; target therapeutic flux rate of drug A into tumor at $z = L$, $N_{A,s} = 2.0 \cdot 10^{-6}$ mg/cm^2-s.

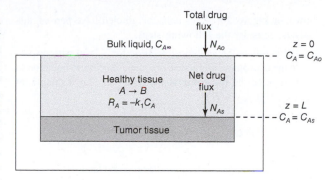

26.31 In the distillation of a benzene/toluene mixture, a vapor richer in the more volatile component benzene is produced from the benzene/toluene liquid solution. Benzene is transferred from the liquid to the vapor phase, and the less volatile toluene is transferred in the opposite direction, as shown in the figure below. At the system temperature and pressure, the latent heats of vaporization of benzene (A) and toluene (B) are 30 and 33 kJ/gmole, respectively. Both components are diffusing through a gas film of thickness δ. Develop the final, integrated form of Fick's flux equation to predict the steady-state mass transfer of benzene through the gas film. The equation must include terms for the bulk gas-phase mole fraction of benzene, the gas-phase mole fraction of benzene in equilibrium with the liquid solution, the diffusion coefficient of benzene in toluene, the diffusion path δ, and the total molar gas concentration, and the latent heats of vaporization for benzene $(\Delta H_{v,A})$ and toluene $(\Delta H_{v,B})$. Assume that distillation is an adiabatic process.

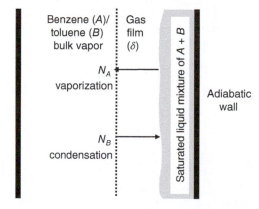

26.32 Please refer to Example 5 of this chapter. Estimate the total transfer rate of solute A from the mass-transfer system shown in Figure 26.14, in units of gmole/h. As part of this analysis, consider the following steps:

a. Differentiate the analytical solution for $c_A(x,y)$ with respect to the coordinate of interest, and then develop an equation to determine the local flux at the boundary. For example, at $x = 0$ ($0 \leq y \leq H$),

$$N_{A,x}\big|_{x=0,y} = N_A(0, y) = -D_{AB}\frac{\partial c_A(0, y)}{\partial x}$$

Repeat this approach for the other exposed faces of the device.

b. For each exposed face, integrate the local flux over y to obtain an equation for the average flux. For example, at $x = 0$, from $y = 0$ to $y = H$,

$$\bar{N}_{A,x}\big|_{x=0} = \frac{1}{H}\int_{y=0}^{y=H} N_{A,x}(0, y)\, dy$$

c. Using an effective diffusion coefficient of 0.02 cm^2/s for species A in the porous medium, and a vapor pressure of $P_A^* = 0.1$ atm as the solid source, estimate the total release of solute A from all exposed faces of the system in units of gmole/hr at 1.0 atm total system pressure and 27 C, using $L = 4.0$ cm, $H = 6.0$ cm, and width (out of x-y plane) of $W = 4.0$ cm.

26.33 Recall Problem 25.15. Develop a model equation, in final, integrated algebraic form, to determine the total transfer rate of ethylene W_A, units mole A/time, product of flux and surface area) into the cylindrical catalyst pellet. The model equation should include terms for the diffusion coefficient (D'_{Ae}), the dimensions of the pellet (R_o, R_1, L), and concentration of ethylene. If the required transfer rate of ethylene (A) into a single pellet is 9.238×10^{-6} gmole/s, what is the required R_1, in units of cm, if the catalyst surface is fixed at $R_o = 0.5$ cm? It will also be necessary to calculate the effective diffusion coefficient as part of the problem solution.

27.1   The transient concentration profile $c_A(z,t)$ resulting from transient one-dimensional diffusion in a slab under conditions of negligible surface resistance is described by equation (27-16). Use this equation to develop an equation for predicting the average concentration, $\bar{c}_A$, within the control volume at a given time t. Then evaluate and plot the dimensionless average profile $(\bar{c}_A - c_{As})/(c_{Ao} - c_{As})$ as a function of the dimensionless relative time ratio, X_D. Use a spreadsheet to perform the calculations needed to implement the infinite series analytical solution.

27.2   Aluminum is the primary conductor material for fabrication of microelectronic devices. Consider the composite thin film shown in the figure below. A thin film of solid aluminum is sputter-coated onto a wafer surface. Then, a 0.50-µm thin film of silicon is added on top of the aluminum film by chemical vapor deposition of silane. If a high temperature is maintained during subsequent processing steps, then the aluminum can diffuse into the Si thin film and change the characteristics of the microelectronic device. Estimate the concentration of Al halfway into the Si thin film if the temperature is maintained at 1250 K for 10 h. Consider whether or not the process represents diffusion within a semi-infinite medium or in a finite-dimensional medium. At 1250 K, the maximum solubility of Al in Si is about 1 wt%. Solid-phase diffusivity data for common dopants in silicon are provided in Figure 24.12 and Table 24.7.

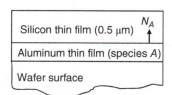

Silicon thin film (0.5 µm) N_A

Aluminum thin film (species A)

Wafer surface

27.3   In the fabrication of a p-type semiconductor, elemental boron is diffused a small distance into a solid crystalline silicon wafer. The boron concentration within the solid silicon determines semiconducting properties of the material. A physical vapor deposition process keeps the concentration of elemental boron at the surface of the wafer equal to 5.0×10^{20} atoms boron/cm³ silicon. In the manufacture of a transistor, it is desired to produce a thin film of silicon doped to a boron concentration of at least 1.7×10^{19} atoms boron/cm³ silicon at a depth of 0.20 microns (µm) from the surface of the silicon wafer. It is desired to achieve this target within a 30-min processing time. The density of solid silicon can be stated as 5.0×10^{22} atoms Si/cm³ solid.

a.   At what temperature must the boron-doping process be operated? It is known that the temperature dependence of the diffusion coefficient of boron (A) in silicon (B) is given by

$$D_{AB} = D_o \cdot e^{-Q_o/RT}$$

where $D_o = 0.019$ cm²/s and $Q_o = 2.74 \cdot 10^5$ J/gmole for elemental boron in solid silicon. The thermodynamic constant $R = 8.314$ J/gmole·K.

b.   What is the flux of boron atoms at the silicon wafer surface at 10 min vs. 30 min?

27.4   One step of the manufacturing of silicon solar cells is the molecular diffusion (doping) of elemental phosphorous (P) into crystalline silicon to make an n-type semiconductor. This P-doped layer needs to be at least 0.467 µm into the 200-µm thick wafer. The present diffusion process is carried out at 1000 C. Data for the total amount of phosphorous atoms loaded into the silicon wafer vs. time at 1000 C are presented in the figure below. The maximum solubility of phosphorous within crystalline silicon is 1.0×10^{21} P atoms/cm³ at 1000 C. The square silicon wafer has a surface area of 100 cm² (10 cm/side). Initially, there is no phosphorous impurity in the crystalline silicon. What is the concentration of P atoms (atoms P/cm³) doped into the silicon at a depth of 0.467 µm after 40 min, based on the data provided?

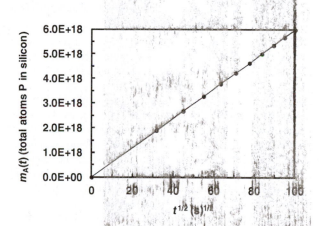

27.5   A crystalline silicon wafer of 10 cm diameter and 1.0 mm thickness is coated with a thin film of elemental arsenic metal (As), which is a semiconductor dopant. The As-doated silicon wafer is "baked" in a diffusion furnace at 1050 C to allow the arsenic molecules to diffuse into the silicon. At this high temperature, the solubility of arsenic in silicon is 2.3×10^{21} atoms As/cm³, and the diffusion coefficient of As in solid silicon still is 5.0×10^{-13} cm²/s. Initially, there is a uniform background impurity of arsenic in the silicon of 2.3×10^{17} atoms As/cm³. To obtain a desired semiconductor property and microelectronic device structure, the target concentration of arsenic atoms in the silicon must be least 2.065×10^{20} atoms As/cm³ at a junction depth of 0.5 µm (1.0×10^4 µm = 1.0 cm).

a. What is the diffusion time required, and how many total As atoms will be loaded into the silicon?

b. Make a plot of the square root junction depth ($z^{1/2}$) vs. time t from $z = 0.5\,\mu m$ to $1.0\,\mu m$.

c. What is the calculated flux of As atoms into the silicon at 5 min and 10 min into the process? Why is it not possible to estimate the flux at $t = 0$?

27.6 A gas mixture of constant composition 5.0 mole% CO_2 and 95 mole% air flows around a porous, spherical particle of diameter 3.0 cm that initially only contains only air (no CO_2) within the gas space of the pores of the particle. At the process conditions of interest (30 C, 1.0 atm), the binary diffusion coefficient of CO_2 in air (D_{AB}) is 0.159 cm²/s, and the effective diffusion coefficient (D_{Ae}) of CO_2 inside the porous particle ($\varepsilon = 0.3$) is 0.0143 cm²/s.

a. What is the mole fraction of CO_2 in the center of the spherical particle after 30 s?

b. How long will it take for the center of the spherical particle to have a composition of 4.8 mole% CO_2, assuming that the bulk gas CO_2 composition equals the CO_2 composition at the outer surface of the particle?

c. If the particle diameter is doubled to 6.0 cm, what is the new time required to achieve a composition of 4.8 mole% CO_2 in the center of particle?

27.7 A "drug patch" is designed to slowly deliver a drug (species A) through the body tissue to an infected zone of tissue beneath the skin. The drug patch consists of a sealed reservoir containing the drug encapsulated within a porous polymer matrix. The patch is implanted just below the skin. A diffusion barrier attached to the bottom surface of the patch keeps the surface concentration of the drug dissolved in the body tissue constant at 2.0 mol/m³, which is below the solubility limit of the drug in the body tissue. There is no difference in the solubility and diffusion coefficient of the drug between the healthy and infected tissues. The mean distance from the drug patch to the infected area of tissue is 5.0 mm. To be effective, the drug concentration in the tissue must be 0.2 mol/m³ or higher when it reaches the infected zone. Determine the time it will take in hours for the drug to begin to be effective for treatment. The effective molecular diffusion coefficient of the drug through the body tissue, both healthy and infected, is 1.0×10^{-6} cm²/s.

27.8 It is desired to diffuse a blue dye molecule (solute A) into a solid material (species B) near the surface of the material, as shown in the figure below. Initially, there is no dye in the solid ($c_{Ao} = 0$), and the maximum solubility of the dye in the solid is 0.010 gmole/m³. Both the solubility of the dye and its diffusion coefficient in the solid are relatively small, so that the dye does not penetrate very far into the solid. The process is carried out at ambient conditions (25 C, 1.0 atm). After 2.0 min (120 s), the flux of the dye into the solid was observed to be $N_A(0,t) = 5.15 \times 10^{-9}$ gmole/m² s.

a. The process is considered complete when the dye concentration within the solid achieves a concentration of 0.0062 gmole/m³ at a depth of 0.050 cm from the surface. How much time will this process take?

b. Which of the following are true about solid B: transient source, transient sink, stagnant medium, semi-infinite medium?

Dye solution (solute A)

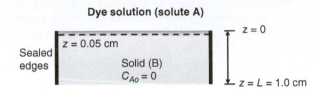

27.9 A new type of porous material serves as food packaging film, as shown in the figure below. The film rests on a nonporous metal plate. The packaging film is 0.20 cm thick (2.0 mm). The food contains liquid water, which exerts a vapor pressure of 0.030 atm at 25 C. At 25 C, the *effective* diffusion coefficient of water vapor in the porous film is 1.3×10^{-4} cm²/s, and the *molecular* diffusion coefficient of water vapor in air is 0.26 cm²/s. You may assume that the total system pressure is constant at 1.0 atm, and initially, the gas space of the porous film contains air and no water vapor. How long will it take for the water vapor to achieve a partial pressure of 0.0150 atm at the back surface of the film ($z = 0$) at 25 C?

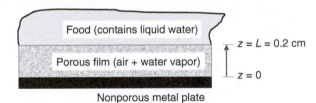

Nonporous metal plate

27.10 A common procedure for increasing the moisture content of air is to bubble it through a column of water. The air bubbles are assumed to be spheres, each having a radius of 1.0 mm, and are in thermal equilibrium with water at 298 K. Determine how long the bubble should remain in the water to achieve a vapor concentration at the center that is 90% of the maximum possible (saturation) concentration. Assume that the air is dry when it enters the column of water and that the air inside the small bubble is stagnant. The vapor pressure of water versus temperature is available from many sources, including steam tables.

27.11 We are interested in the diffusion of CO_2 gas out of a randomly porous adsorbent material slab of 2.0 cm thickness, as shown in the figure below. Initially, the gas space inside the porous material contains 10% mole CO_2 (A) and 90 mole% air (B). The process is maintained at 25 C and total system pressure of 1.0 atm. At these conditions, the binary molecular diffusion coefficient of CO_2 in air is 0.161 cm²/s, but the effective diffusion coefficient of CO_2 within the porous medium is only 0.010 cm²/s. Fresh air containing no CO_2 blows over the surface of the slab so that the convective mass-transfer coefficient for CO_2 transfer (k_c) is 0.0025 cm/s. How long will it take for the CO_2 concentration of the gas space inside the porous material to be reduced to only 2.0 mole% CO_2 at a depth of 1.6 cm from the exposed surface of the slab?

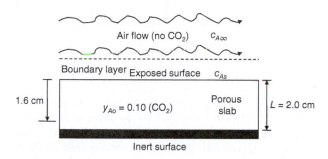

27.12 Consider the cylindrical-shaped absorbent material designed to selectively remove solute A from solution. The uptake of solute A through the homogeneous absorbent material is limited by the molecular diffusion. Furthermore, the absorbent has a higher affinity for solute A relative to the surrounding solution that is described by the linear equilibrium relationship $c_A = K \cdot C'_A$, where c_A is the molar concentration of A in the absorbent, C'_A is the molar concentration of A in the surrounding fluid, and K is a partition coefficient for solute A between the fluid and the absorbent material. The surrounding solution is very well mixed and has a constant concentration of $C'_{A\infty} = 2.00$ gmole/m³. The diffusion coefficient of A in the homogeneous absorbent material is 4×10^{-7} cm²/s, and the partition coefficient $K = 1.5$ cm³ fluid/cm³ absorbent. The cylindrical absorbent pellet is 1.0 cm in diameter and 5.0 cm in length. Initially, no solute A is absorbed into the absorbent material.

a. Assume edge effects associated with the ends of the cylinder can be neglected. Under this assumption, how long will it take (in hours) for the solute A concentration to reach 2.94 gmol/m³ at a depth of 0.40 cm from the surface of the cylinder?

b. Repeat part (a) calculation for the case where "end effects" are not neglected—i.e., for two-dimensional diffusion. The radial position is 0.4 cm from the cylindrical surface, and the axial position is 1.0 cm from the end of the rod.

27.13 A spherical time-release capsule was developed to release pesticide to surrounding soil, as shown in the figure below. The capsule is 1.0 cm in diameter, and is made of a polymer containing a uniform initial concentration of pesticide (species A) equal to 2.0 mg/cm³. The effective diffusion coefficient (D_{Ae}) of the pesticide in the polymer is 5.0×10^{-7} cm²/s. During wet periods of the year, water flows into the loose soil and slowly flows around the buried capsules,

creating a boundary layer that provides a convective mass transfer coefficient (k_c) of about 1.0×10^{-6} cm/s. The bulk concentration of pesticide in the flowing water is very small, and so can be considered as effectively equal to zero ($c_{A\infty} \approx 0$). The molecular diffusion coefficient of the pesticide in the water (D_{AB}) is 1.5×10^{-6} cm²/s (A = pesticide, B = water).

a. What is the concentration of pesticide in the *center* of the capsule after a time of 173.6 h?

b. What is the concentration of pesticide at the *surface* ($r = R$) of the capsule after 173.6 h, and what is the flux of pesticide out of the particle, $N_A(R,t)$, at this time, in units of mg/cm²s?

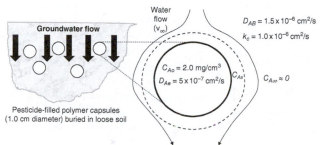

Pesticide-filled polymer capsules
(1.0 cm diameter) buried in loose soil

27.14 A novel metal ion adsorbent material based on the biopolymer polyglucosamine is cast into a gel bead. The amine groups on the biopolymer have a high affinity for transition metal ions at parts-per-million (mg/L) concentrations. When water containing metal ions contacts the gel, the metal ions selectively partition into the gel and then diffuse through the gel, which extracts the metal ions from solution and concentrates them within the bead. Although the real situation is somewhat more complicated, assume as a limiting case that the uptake rate of the metal ions is limited by the molecular diffusion of the metal ions through the gel. At relatively low metal ion concentrations in water below 1.0 mol/m³, the partition constant is defined by the linear relationship $c'_A = Kc_{As}$, where c'_A is the molar concentration of the metal ion within the gel phase at surface ($r = R$), c_{As} is the molar concentration of the metal in the aqueous phase surrounding the surface, and K is partition constant, which is dependent on the amine concentration within the gel. In the present process, the bead diameter is 0.5 cm, and the amine concentration within the gel sets K equal to 150 m³ aqueous phase/m³ biopolymer phase. How long will it take for the cadmium concentration at the center of the gel bead to reach 12.0 mol/m³ if wastewater containing a constant concentration of 0.1 mol/m³ cadmium (11.2 mg/L) rapidly flows over the gel bead? The effective diffusion coefficient of cadmium within the gel, accounting for the absorption process, is approximately 2.0×10^{-8} cm²/s.

27.15 One step in the processing of cucumbers to pickles is the pickling process itself. In one method of making pickles, young cucumbers with no waxy skins are soaked in a NaCl solution overnight. To initiate the pickling process, acetic acid is added to the salt solution to make the pickling brine. The acetic acid acts as a preservative, and the salt solution keeps the cucumber from swelling as the acetic acid diffuses into the cucumber. When a certain concentration of acetic acid inside the cucumber is achieved, the cucumber is considered "pickled" and will remain safe and tasty to eat for a long time. In the present pickling process, the temperature is 80 C, and the pickling brine solution contains an acetic acid concentration of 0.900 kgmole/m³. The cucumbers are 12.0 cm in length and 2.5 cm in diameter. The amount of cucumbers relative to the amount of pickling brine solution is small so that the bulk liquid-phase concentration of acetic acid remains essentially constant during the pickling process. Initially, the cucumbers do not contain any acetic acid, and are maintained at 80 C. End effects can be neglected. It can also be assumed that the diffusion coefficient of acetic acid into the pickle approximates the diffusion coefficient of acetic acid in water. The diffusion coefficient of acetic acid in water at 20 C (not 80 C) is 1.24×10^{-5} cm²/s.

a.   How long will it take (in hours) for the *center of the pickle* to reach a concentration of 0.864 kgmole/m³? For part (a), consider that the liquid is very well mixed so that external convective mass-transfer resistances can be neglected.

b.   Refer to part (a). What is the new required time if convective mass transfer around the pickle is now such that $k_c = 1.94 \times 10^{-5}$ cm/s?

SS **27.16**   A polymer coating of 6.0 mm thickness is cast onto a nonporous flat surface. The coating contains a residual amount of casting solvent, which is uniform at 1.0 wt% within the coating. The mass transfer of solvent through the polymer coating is controlled by molecular diffusion. The air flowing over the coating surface eliminates convective mass-transfer resistances and reduces the solvent vapor concentration in the air to nearly zero. The effective diffusion coefficient of the solvent molecules in the polymer is 2×10^{-6} cm²/s. How many hours will it take for the solvent concentration at 1.2 mm from the surface to be reduced to 0.035 wt%?

27.17   Living cells immobilized within an agarose gel require glucose to survive. An important aspect of the biochemical system design is the effective diffusion coefficient of glucose into the agarose gel itself, which you may consider as a homogeneous material that is mostly liquid water. Consider the experiment shown in the figure below, where a slab of the agarose gel of 1.0 cm thickness is placed within a well-mixed aqueous solution of glucose maintained at a concentration of 50 mmole/L (1.0 mole = 1000 mmole), which is relatively dilute. The solubility of the glucose in water is equal to the solubility of glucose in the gel. A syringe mounted in the center of the gel carefully excises a tiny sample of the gel for glucose analysis. Initially, there is no glucose in the gel. However, after 42 hours, the measured concentration of glucose in the gel at the sampling point is 48.5 mole/m³. Based on this measurement, what is the effective diffusion coefficient of glucose into the gel?

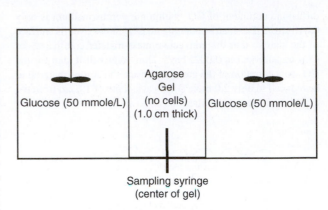

Sampling syringe
(center of gel)

27.18   A small spherical bead is used as a controlled drug-release capsule in the gastrointestinal system (i.e., your stomach). In this particular case, a 0.10-cm diameter bead has a uniform initial concentration of 0.20 mmole/L of the drug griseofulvin (species A). The diffusion coefficient of griseofulvin within the bead material is 1.5×10^{-7} cm²/s. Upon release from the bead, the drug is immediately consumed so that the surface concentration is essentially zero.

a.   Using the concentration–time charts, determine the time it will take for concentration of the griseofulvin in the *center* of the bead to reach 10% of its initial value.

b.   Using the infinite series analytical solution for $c_A(r,t)$, determine the time it will take for concentration of the griseofulvin in the *center* of the bead to reach 10% of its initial value, and then plot out the concentration profile from $r = 0$ (center) to $r = R$ (surface). Carry out your calculations on a computer spreadsheet.

27.19   Consider a rectangular-shaped gel tablet of thickness 0.652 cm and width 1.0 cm loaded with the drug Dramamine. The edges of the gel tablet are sealed so that diffusion of Dramamine occurs only along the thickness of the gel tablet. The initial concentration of the drug Dramamine (solute A) in the gel (B) is 64.0 mg/cm³, to provide a total drug dosage of 41.7 mg. The concentration of the drug at the exposed surface of the tablet is maintained at zero to provide a constant sink for mass transfer. The diffusion coefficient of the drug in the gel is 3.0×10^{-7} cm²/s. What is the residual concentration of drug in the center of the tablet after 96 hours? Compare results using (a) the concentration–time charts and (b) the infinite series analytical solution.

27.20   Consider a rectangular-shaped gel tablet of thickness 0.125 cm and width 0.50 cm. The edges of the gel tablet are sealed so that diffusion only occurs along the thickness of the tablet. The initial concentration of the drug Dramamine (species A) in the gel is 64 mg/cm³, to provide a total drug dosage of 2.0 mg. The concentration of A at the exposed surface of the tablet is maintained at zero. What is the total amount (m_A) of drug released after 1.0 and 2.0 hours, respectively? Hint: Use the infinite series analytical solution for $m_A(t)$.

27.21 A spherical gel capsule of 0.50 cm diameter is used a drug delivery device. Initially, there is a total drug amount 0.005 mmole uniformly loaded into the gel material. The capsule is ingested, and within the gastrointestinal tract, the bulk concentration of the drug A within the gastrointestinal fluid is equal to zero. However, there is a finite convective film mass-transfer resistance in the mass-transfer process, and the convective mass-transfer coefficient for drug A around the surface of the sphere is equal to 5.0×10^{-5} cm/s. At 37 C, the effective diffusion coefficient of drug within the gel is 3.0×10^{-6} cm²/s, whereas the molecular diffusion coefficient of the drug in the gastrointestinal fluid is 1.5×10^{-5} cm²/s.

a. What is the mass-transfer Biot number for the process? The Biot number (Bi) is the inverse of m in the concentration-time charts.

b. What is the molar concentration of drug A in the center of the capsule after 2.3 hours?

27.22 As society searches for technical solutions to global warming, one approach to sequester carbon dioxide rich greenhouse gases is to capture the CO_2 within an adsorbent material at high pressure. An experiment designed to evaluate a candidate adsorbent material is presented in the figure below, which consists of a 10 cm × 10 cm slab with 100 cm² of exposed surface. A gas mixture of 10 mole% CO_2 (species A) and 90 mole% N_2 (species B) at 15.0 atm total system pressure and temperature of 200 C is contacted with a microporous material designed to selectively adsorb the CO_2 from the gas mixture. The partitioning of CO_2 gas within the adsorbent is described by

$$Q_A = K' c_A$$

where Q_A is the amount of CO_2 adsorbed per unit volume of porous adsorbent (mol CO_2/cm³ adsorbent), c_A is the concentration of CO_2 in the gas phase within the porous adsorbent (mol CO_2/cm³ gas pore space), and K' is the apparent CO_2

adsorption constant (cm³ gas/cm³ adsorbent). The experiment is conducted as an unsteady-state process where the total amount of CO_2 captured within the adsorbent is measured after a given time. Initially, there is no CO_2 adsorbed on the solid or in the pore space, and the process is considered dilute with respect to CO_2 in the gas phase. The diffusion process is modeled as semi-infinite sink for CO_2, with CO_2 flux given by

$$N_A\big|_{z=0} = c_{As}\sqrt{\frac{D'_{Ae}}{\pi \cdot t}}, \quad \text{where } D'_{Ae} = \frac{\epsilon^2 D_{Ae}}{\epsilon + K'},$$

c_{As} is the molar concentration of CO_2 at the outer surface of the slab. In the above equation, D'_{Ae} is the apparent diffusion coefficient of CO_2 within the adsorbent material under conditions where the solute adsorbs onto the surface of the pores, which includes the following terms: apparent CO_2 adsorption constant (K'), the void fraction of the porous adsorbent (ϵ), and the effective diffusion coefficient of CO_2 within the solid under non-CO_2 adsorbing conditions (D_{Ae}). In the present system, $K' = 1776.2$ cm³/cm³, $\epsilon = 0.60$, and $D_{Ae} = 0.018$ cm²/s. After a total contact time of 60 min, what was the total mass of CO_2 captured by the slab? Consider that the slab acts as a semi-infinite sink for CO_2, and that convective mass-transfer resistances are eliminated so that $c_{As} = c_{A\infty}$.

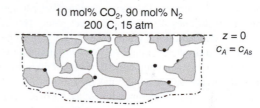

10 mol% CO_2, 90 mol% N_2
200 C, 15 atm

$z = 0$
$c_A = c_{As}$

Microporous material ($\epsilon = 0.60$)
adsorbs CO_2 from gas
(semi-infinite slab)

28.1 Compare the Schmidt number (Sc) for O_2 gas in air at 300 K and 1.0 atm with Sc for dissolved O_2 in liquid water at 300 K. Then compare these values for CO_2 gas in air at 300 K and 1.0 atm, and dissolved CO_2 in liquid water at 300 K. You may assume that O_2 and CO_2 are dilute with respect to air as the carrier gas and water as the solvent.

28.2 Define the Stanton and Peclet numbers and their relationships to other dimensionless groups for convection mass transfer.

28.3 An investigator proposes to study the mass-transfer process for the dissolution of a single solid sphere suspended within a turbulent flow stream. Predict the variables that would be used to explain the geometry involved, the properties of the moving stream, and the convective mass-transfer coefficient. Apply dimensional analysis to determine the dimensionless π groups that might be used in analyzing the experimental data.

28.4 A thin (1.0-mm-thick) coat of fresh paint has just been sprayed over a 1.5-m by 1.5-m square steel body part, which approximates a flat surface. The paint contains a volatile solvent that initially constitutes 30 wt% of the wet paint. The initial density of the wet paint is 1.5 g/cm^3. The freshly painted part is introduced into a drying chamber. Air is blown into the rectangular drying chamber at a volumetric flow rate of 60 m^3/min, as shown in the figure below, which has dimensions $L=1.5$ m, $H=1.0$ m, $W=1.5$ m). The temperature of the air stream and the steel body part are both maintained at 27 C, and the total system pressure is 1.0 atm. The molecular weight of the solvent is 78 g/gmole, the vapor pressure exerted by the solvent at 27 C is 105 mm Hg, and the molecular diffusion coefficient of the solvent vapor in air at 27 C and 1.0 atm is 0.097 cm^2/s.

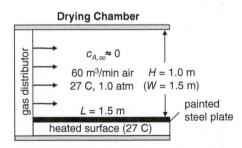

Drying Chamber

gas distributor

$c_{A,\infty} \approx 0$

60 m^3/min air $H = 1.0$ m
27 C, 1.0 atm   ($W = 1.5$ m)

$L = 1.5$ m

painted steel plate

heated surface (27 C)

a. What is the Schmidt number and the average Sherwood number (Sh_L) for the mass-transfer process?

b. What is the estimated solvent evaporation rate from the surface of the *whole body part* in units of g/min? It may be assumed that convection mass-transfer limits the evaporation rate, and that the concentration of solvent vapor in the bulk gas is finite, but can be approximated as $c_{A,\infty} \approx 0$.

c. Using the results from part (a), how long will it take for the paint to completely dry?

d. What are the hydrodynamic (δ) and concentration boundary-layer (δ_c) thicknesses at $x=L=1.5$ m? How does this compare to H, the height of the drying chamber?

e. What is the new required air flow rate (m^3/min) if the desired solvent mass-transfer rate is 150 g/min?

28.5 A thin polymer film contains some residual solvent. It is desired to evaporate the volatile solvent (*n*-hexane, solute A) from the polymer using the process shown in the figure below. The wet polymer film enters the drying process. Both the top and bottom surfaces of the polymer film are exposed to a cross flow of air. The evaporation rate of the solvent from the polymer film is limited by external convection mass transfer. The dried polymer film is then rolled up into a bundle. During the drying process, the width of the thin polymer film is 0.50 m, and the length of the polymer film is 2.5 m. The flowing air has a bulk velocity of 1.5 m/s, temperature of 20 C, and total system pressure of 1.0 atm. The wet polymer film is also at maintained at 20 C, and the evaporation process is assumed to be limited by convection mass transfer. The vapor pressure of the solvent at 20 C is 121 mm Hg. The diffusion coefficient of the solvent in air is 0.080 cm^2/s at 20 C and 1.0 atm, and the molecular weight of the solvent is 86 g/gmole. The partial pressure of the solvent vapor itself within the drying chamber can be assumed to be close to zero with $P_A \approx p_{A\infty}$.

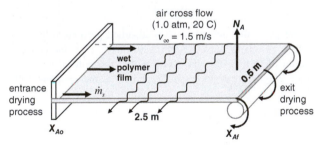

air cross flow
(1.0 atm, 20 C) N_A
$v_\infty = 1.5$ m/s

wet polymer film

$\dot{m}_s$

entrance drying process

0.5 m

2.5 m

exit drying process

X_{Ao} X_{Af}

a. What are the Schmidt number and the average Sherwood number for the solvent evaporation process?

b. What is the total evaporation rate of the solvent from the 0.50-m by 2.50-m polymer film, in units of gmole A/s? Remember: Both sides of the polymer film are exposed to flowing air.

c. The solvent loading in the polymer film at the entrance of the drying process is 0.10 g solvent per g dry polymer—i.e., $X_{Ao}=0.10$ g solvent/g dry polymer. The total mass flow rate of the polymer film on a solvent-free dry polymer basis is $\dot{m}_s = 50.0$ g dry polymer solid/s. What is the solvent loading in the polymer film exiting the drying process, X_{Af}? Hint: Perform a material balance on the solvent in the film as the polymer film moves through the drying process.

28.6 A horizontal chemical vapor deposition (CVD) reactor similar to the configuration shown in Example 3, Figure 28.6 will be used for growth of gallium arsenide (GaAs) thin films. In this process, arsine vapor, trimethylgallium vapor, and H_2 gas are fed into the reactor. Inside the reactor, the silicon wafer rests on a heated plate called a susceptor. The reactant gases flow parallel to the surface of the wafer and deposit a GaAs thin film according to the simplified CVD reactions

$$2AsH_3(g) \rightarrow 2As(s) + 3H_2(g)$$

$$2Ga(CH_3)_3(g) + 3H_2(g) \rightarrow 2Ga(s) + 6CH_4(g)$$

If the reactant gas is considerably diluted in H_2 gas, then the mass transfer of each species in the H_2 carrier gas can be treated separately. These surface reactions are considered to be very rapid, and so the mass transfer of the gaseous reactants to the surface of the wafer limits the rate of GaAs thin-film formation. In the present process, a 15 cm × 15 cm square silicon wafer is positioned at the leading edge of the susceptor plate. The process temperature is 800 K, and the total system pressure 101.3 kPa (1.0 atm). The feed gas delivered to the reactor results in a bulk linear velocity of 100 cm/s. The composition of arsine and trimethylgallium in the feed gas are both 0.10 mole%, which is very dilute. You may assume that the amount of arsine and trimethylgallium delivered with the feed gas is much higher than the amount of arsine and trimethylgallium consumed by the reactions, so that the concentration of these reactants in the bulk gas phase is essentially constant down the length of the reactor. You may also assume that the surface-reaction rates are instantaneous relative to the rates of mass transfer, so that the gas-phase concentrations of both arsine vapor and trimethylgallium vapor at the surface of the wafer are essentially zero. The binary gas-phase diffusion coefficient of trimethylgallium in H_2 is 1.55 cm^2/s at 800 K and 1.0 atm.

a. What are the average mass-transfer rates for arsine and trimethylgallium over the whole wafer?

b. Based on the ratio of the arsine and trimethylgallium mass-transfer rates, what is the composition of the GaAs composite thin film—e.g., the molar composition of gallium (Ga) and arsenic (As) in the solid? How could the feed-gas composition be adjusted so that the molar ratio of Ga to As within the solid thin film is 1:1?

28.7 In a manufacturing process, an organic solvent (methyl ethyl ketone, MEK) is used to dissolve a thin coating of a polymer film away from a nonporous flat surface of length 20 cm and width 10 cm, as shown in the figure (above right). The thickness of the polymer film is initially uniform at $l_a = 0.20$ mm (0.02 cm). In the present process, a volumetric flow rate of 30 cm^3/s of MEK liquid solvent is added to an open flat pan of length 30 cm and width 10 cm. The depth of the liquid MEK solvent in the pan is maintained at 2.0 cm. It may be assumed that the concentration of dissolved polymer in the

bulk solvent is essentially zero ($c_{A,\infty} \approx 0$) even though in reality the concentration of dissolved polymer in the solvent increases very slightly from the entrance to the exit of the pan. It may also be assumed that the change in film thickness during the dissolution process does not affect the convection mass-transfer process. Let $A =$ polymer (solute), $B =$ MEK (liquid solvent).

a. What are the Schmidt number and the average Sherwood number for the mass-transfer process?

b. What is the *average* flux of dissolved polymer (in units of g A/cm$^2 \cdot$ s) from the surface?

c. Plot out the *local* $k_{c,x}$ vs. position x, for $0 < x \leq L$.

d. Eventually, the polymer film will completely dissolve away from the entire surface. However, before this time the thickness of the polymer film remaining on the flat surface will not be uniform. Plot out what a plot of polymer film thickness (l) vs. position x would look like at three different times before the polymer film is completely dissolved.

e. What are the hydrodynamic (δ) and concentration boundary-layer (δ_c) thicknesses at $x = 10$ cm?

Potentially useful data: $D_{AB} = 3.0 \times 10^{-6}$ cm^2/s, the diffusion coefficient of dissolved polymer (solute A) in liquid MEK solvent at 25 C; $\rho_{A,\text{solid}} = 1.05$ g/cm^3, mass density of *solid* polymer film material; $c_{As} = c_A{}^* = 0.04$ g/cm^3, maximum solubility of dissolved polymer (solute A) in MEK solvent at 25 C; $\nu_B = 6.0 \times 10^{-3}$ cm^2/s, kinematic viscosity of liquid MEK at 25 C; $\rho_B = 0.80$ g/cm^3, mass density of liquid MEK solvent at 25 C.

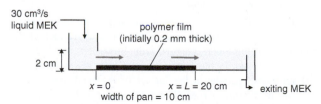

30 cm³/s liquid MEK

2 cm

polymer film (initially 0.2 mm thick)

$x = 0$

$x = L = 20$ cm

width of pan = 10 cm

exiting MEK

28.8 For fully developed turbulent flow over a flat plate of length L, the local Sherwood number at position $x(0 < x \leq L)$ is estimated by

$$Sh_x = \frac{k_c x}{D_{AB}} = 0.0292 \, Re_x^{4/5} \, Sc^{1/3}$$

Develop an expression for the average Sh over length L of the flat plate, assuming fully developed turbulent flow where the contribution to mass transfer by the laminar boundary-layer regime can be neglected.

28.9 In using the von Kármán approximate method for analyzing the turbulent boundary layer over a flat plate, the following velocity and concentration profiles were assumed:

$$v_x = \alpha + \beta y^{1/7}$$

and

$$c_A - c_{As} = \eta + \xi y^{1/7}$$

The four constants—α, β, η, and ξ—are determined by the appropriate boundary conditions at the surface and at the outer edge of the hydrodynamic and concentration layers.

a. Determine α, β, η, and ξ, and provide the resulting equations for velocity and concentration profiles.

b. Upon the application of the von Kármán momentum integral equation, the thickness of the turbulent boundary layer is given by

$$\delta = \frac{0.371x}{Re_x^{1/3}}$$

Use this relationship, and the solution to von Kármán concentration integral equation for Sc = 1.0, to obtain the following equation for the local mass-transfer coefficient:

$$k_c = 0.0289 \, v_\infty (Re_x)^{-1/5}$$

28.10 In the development of the approximate solution for solving the laminar concentration boundary layer formed by fluid flow over a flat plate, it is necessary to assume a concentration profile. Equation (28-35a) was obtained by analysis of a power-series concentration profile of the form

$$c_A - c_{A,s} = a + by + cy^2 + dy^3$$

Apply the boundary conditions for the laminar concentration boundary layer to evaluate the constants a, b, c, and d needed to arrive at equation (28-35a).

28.11 A well-mixed open pond contains wastewater that is contaminated with a dilute concentration of dissolved methylene chloride. The pond is rectangular with dimensions of 500 m by 100 m, as shown in the figure (above right). Air at 27 C and 1.0 atm blows parallel to the surface of the pond with a bulk velocity of 7.5 m/s. At 20 C and 1.0 atm, for the *gas* phase (A = methylene chloride, B = air), the diffusion coefficient (D_{AB}) is 0.085 cm²/s, and kinematic viscosity (v_B) is 0.15 cm²/s. At 27 C, for the *liquid* phase (A = methylene chloride, B = liquid water), the diffusion coefficient (D_{AB}) is 1.07×10^{-5} cm²/s, and the kinematic viscosity (v_B) is 0.010 cm²/s.

a. At what position across the pond is the air flow no longer laminar? Would it be reasonable to assume that the mean gas film mass-transfer coefficient for methylene chloride in air is dominated by turbulent flow mass transfer?

b. As part of an engineering analysis to predict the emissions rate of methylene chloride (species A) from the pond, determine the *average* gas film mass-transfer coefficient associated with the mass-transfer methylene chloride from the liquid surface to the bulk air stream.

c. Compare the Schmidt number for methylene chloride in the gas phase vs. the liquid phase, and explain why the values are different.

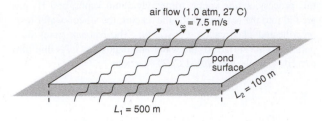

28.12 Gasoline from an under-storage storage tank leaked down onto an impermeable clay barrier and collected into a liquid pool. A simplified picture of the situation is provided in the figure below. Directly over this underground pool of liquid gasoline (*n*-octane, species A) is a layer of gravel of 1.0 m thickness and width of 10.0 m. The volatile *n*-octane vapors diffuse through the highly porous gravel layer, and then through a gas boundary layer formed by flow of air over the top surface of the gravel bed, and finally out to the bulk atmosphere where the *n*-octane is diluted to below detectable levels. There is no adsorption of *n*-octane vapor onto the porous gravel layer, and *n*-octane vapor concentration is dilute. Assume that the mass-transfer process is allowed to achieve a steady state. The temperature of the system is constant at 15 C, and the total system pressure is 1.0 atm. At this temperature, liquid *n*-octane exerts a vapor pressure of 1039 Pa. The void spaces in porous layer create a void fraction (ϵ) of 0.40, and but the pore size is large enough that Knudsen diffusion can be neglected.

a. What is the average mole fraction of *n*-octane vapor at the top surface of the rock layer ($y_{As} = c_{As}/C$) if the air velocity is very low, only 2.0 cm/s? What is the average flux of *n*-octane vapor emitted to the atmosphere?

b. What would be the average mole fraction of *n*-octane vapor at the top surface of the rock layer if the air velocity is 50.0 cm/s? What is the average flux of *n*-octane vapor?

c. The Biot number associated with a mass-transfer process involving diffusion and convection in series is defined as

$$Bi_{AB} = \frac{k_c L}{D_{Ae}}$$

where L refers to the path length for molecular diffusion within the porous gravel layer and D_{Ae} refers to the diffusion coefficient of species A within this porous medium, which is not the same as the molecular diffusion coefficient as octane vapor in air. Determine the Biot number for parts (a) and (b), and then assess the relative importance of convective mass transfer in determining the *n*-octane vapor emissions rate.

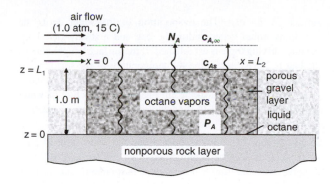

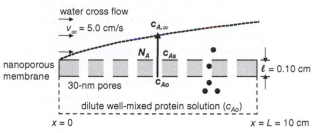

*C. Tanford, *Physical Chemistry of Macromolecules*, John Wiley & Sons, New York, 1961.

28.13 A cross-flow molecular filtration device equipped with a mesoporous membrane is used to separate the enzyme lysozyme from a fermentation broth, as shown in the figure (right column). Water at 25 C flows over the top surface of the flat plate membrane at a velocity of 5.0 cm/s. The length of the membrane in the direction of flow is 10 cm, and the thickness of the membrane (l) is 0.10 cm. Uniform parallel pores of 30 nm diameter run down thickness of the membrane, and 30% of the surface area of the membrane is occupied by the pore openings. The dissolved enzyme diffuses through the pores of the membrane and across the boundary layer of the flowing fluid, so that there are two mass-transfer resistances in series. It is desired to understand how the fluid boundary layer affects the overall flux of enzyme across the membrane. At 25 C, the molecular diffusion coefficient of the lysozyme enzyme in water (D_{AB}) is 1.04×10^{-6} cm^2/s*, and the molecular weight (M_A) is 14,100 g/gmole. For diffusion of the 4.12-nm diameter lysozyme through the 30-nm diameter pores, the effective diffusion coefficient of lysozyme through the membrane (D_{Ae}) is 5.54×10^{-7} cm^2/s, using the Renkin equation for hindered solute diffusion in solvent-filled pores, equation (24-62).

a. Show that the steady-state flux across the membrane and boundary layer is given by

$$N_A = \frac{(c_{Ao} - c_{A,\infty})}{\dfrac{1}{k_c} + \dfrac{l}{D_{Ae}}} \qquad (28\text{-}66)$$

b. Estimate the average molar flux of the enzyme across the membrane and boundary layer, assuming the enzyme concentration at the bottom surface of the membrane (c_{Ao}) is maintained at 1.0 mmole/m^3 and the bulk concentration of enzyme in the flowing liquid phase ($c_{A,\infty}$) is maintained at 0.40 mmole/m^3. Also, estimate the flux for two limiting cases, one where convection mass-transfer limits the process, and another where molecular diffusion through the membrane limits the process.

c. What is the Biot number (Bi) for the enzyme mass-transfer process? Reflect on the importance of convective mass transfer.

d. How does the maximum thickness of the hydrodynamic boundary layer compare to the thickness of the membrane?

28.14 Consider the process shown in the figure (next page). A bulk gas stream containing 0.10 mole% of carbon monoxide (CO) gas, 2.0 mole% O$_2$ gas, and 97.9 mole% of CO$_2$ gas flows over a flat catalytic surface of length 0.50 m at a bulk velocity of 40 m/s at 1.0 atm and 600 K. Heat-transfer processes maintain the gas stream and catalytic surface at 600 K. At this temperature, the catalytic surface promotes the oxidation reaction CO(g)+1/2O$_2$(g) → CO$_2$(g). Let A = CO, B = O$_2$, C = CO$_2$. The gas-phase diffusion coefficients at 1.0 atm and 300 K are D_{AB} = 0.213 cm^2/s, D_{AC} = 0.155 cm^2/s, D_{BC} = 0.166 cm^2/s.

a. What are the Schmidt numbers for CO and O$_2$ mass transfer? What species (CO, O$_2$, CO$_2$) is considered the carrier gas?

b. For CO mass transfer, what is the *average* convective mass-transfer coefficient (k_c) over the 0.50 m length of the catalytic surface, and the *local* mass-transfer coefficient ($k_{c,x}$) at the far edge of the catalytic surface ($x = L = 0.50$ m)?

c. Using boundary-layer theory, scale k_c for CO mass transfer to k_c for O$_2$ transfer.

d. At 600 K, the surface reaction constant for the first-order oxidation reaction with respect to CO concentration is k_s = 1.5 cm/s. What is the average molar flux of CO to the catalytic surface, assuming that the composition of CO in the bulk gas is maintained at 0.10 mole%?

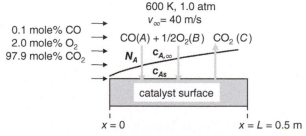

28.15 Please refer to Problem 28.14 above. At 600 K and 1.0 atm, under a new velocity for gas flow, the heat-transfer coefficient (h) is now 50 W/m^2 · K.

a. Compare average mass-transfer coefficient for CO mass transfer (k_c) using the Reynolds and Chilton–Colburn analogies. Based on the assumptions for each analogy, why are the values different?

b. Use boundary-layer theory to scale k_c for convective CO mass transfer to k_c for convective O_2 mass transfer.

28.16 Air rapidly flows over a 10.0 m × 8.0 m rectangular spill of gasoline (n-octane) at a velocity of 8.0 m/s. The spill occurred within an enclosure where the gas temperature was maintained at 27 C. You may assume that gasoline approximates n-octane ($A =$ n-octane).

a. At what position on the liquid film does the gas flow transition out of laminar flow? Plot out the local thickness of the hydrodynamic and concentration boundary layers with position within the laminar portion. What is the ratio of the concentration boundary thickness to the hydrodynamic layer thickness?

b. Is the overall process in laminar, transition, or turbulent flow?

c. What is the gasoline vapor emissions rate from the entire spill, in units of gmole/s? You may assume that the temperature of the liquid spill remains constant at 27 C, the same as the air flow.

d. Is it reasonable to assume that the temperature of the spill liquid remains at 27 C? If not, would the temperature of the liquid spill go up or down, and how would this affect the emissions rate? Hint: consider an energy balance on the process, assuming the gas temperature is constant.

28.17 A "cooling bag" is commonly used for storing water in hot, arid outdoor environments. The bag made of a thin porous fabric that allows water vapor, but not liquid water, to pass through it. A small amount of water (species A) diffuses through the fabric and evaporates from the surface of the bag. The rate of evaporation is controlled by convective mass transfer from the outer surface of the fabric to the surrounding dry air. The energy for evaporation is supplied by the hot air surrounding the outer surface of the bag. The evaporation of the water cools the remaining liquid water within the bag and a steady-state temperature driving force is established. If the surface temperature of the bag is 293 K, determine the temperature of the surrounding hot air. It may be assumed that the surrounding arid air contains no moisture. At 293 K, the latent heat of vaporization of water is $\Delta H_{v,A} = 2.45$ kJ/g H_2O, and the vapor pressure of water is $P_A = 2.34 \times 10^3$ Pa.

28.18 A liquid is flowing over a flat slab consisting of pure solid solute A. As the liquid flows over the surface, solute A dissolves into the liquid. It is desired to maintain laminar flow across the entire surface of the slab, which is 40 cm in length. What is the maximum possible local mass transfer coefficient, $k_c(x)$, in this process, and what is the bulk velocity of the liquid? The molecular diffusion coefficient of solute A in liquid B is $D_{AB} = 1.0 \times 10^{-5}$ cm^2/s, and the kinematic viscosity of the liquid is 0.0020 cm^2/s.

28.19 Refer to Example 1. What is the required bulk velocity for the air over the surface of the film, assuming laminar flow? The molecular diffusion coefficient of MEK in air is 0.090 cm^2/s at 27°C and 1.0 atm.

28.20 A falling liquid film within a gas–liquid contactor of 1.50 m length is in contact with 100% carbon dioxide gas at 1.0 atm and 25 C. The wetted surface area is 0.50 m^2, and the liquid film thickness is 2.0 mm, which is thin enough to prevent ripples or waves in the falling liquid film. The liquid delivered to the contactor does not initially contain any dissolved CO_2. At 25°C, the Henry's law constant for the dissolution of CO_2 gas in water is 29.5 m$^3 \cdot$ atm/kgmole, and the molecular diffusion coefficient for CO_2 in liquid water is 2.0×10^{-5} cm^2/s. What is the average molar flux of CO_2 into the film? What is the estimated bulk concentration of dissolved CO_2 in the liquid exiting the process?

29.1 The table below presents *equilibrium* distribution data for four gaseous solutes dissolved in water, using air as the carrier gas:

p_A (mm Hg)	Dissolved Cl$_2$ (g Cl$_2$/L)	p_A (atm)	Dissolved ClO$_2$ (g ClO$_2$/L)	p_A (mm Hg)	Dissolved NH$_3$ (kg NH$_3$/ 100 kg H$_2$O)	p_A (mm Hg)	Dissolved SO$_2$ (kg SO$_2$/ 100 kg H$_2$O)
	Cl$_2$-water, 293 K		ClO$_2$-water, 293 K		NH$_3$-water, 303 K		SO$_2$-water, 303 K
0	0	0.000	0	719	40	688	7.5
5	0.438	0.010	0.9	454	30	452	5
10	0.575	0.030	2.7	352	25	216	2.5
30	0.937	0.050	4.3	260	20	125	1.5
50	1.21	0.070	6.15	179	15	79	1
100	1.773	0.100	8.8	110	10	52	0.7
150	2.27	0.110	9.7	79.7	7.5	36	0.5
		0.120	10.55	51	5	19.7	0.3
		0.130	11.5	40.1	4	11.8	0.2
		0.140	12.3	29.6	3	8.1	0.15
		0.150	13.2	24.4	2.5	4.7	0.1
		0.160	14.2	19.3	2	1.7	0.05
				15.3	1.6		
				11.5	1.2		

Data were obtained from the *Chemical Engineer's Handbook*.[3]

a. Using a spreadsheet to perform the calculations, prepare a graph of the equilibrium distribution data for each solute as partial pressure in the gas vs. molar concentration dissolved in the liquid ($p_A - c_{AL}$), and also in mole fraction coordinates ($y_A - x_A$) at 1.0 atm total system pressure. Which solute is the most soluble in water? Which solute dissolved in water can be stripped into air the easiest?

b. For each solute at the appropriate concentration range, estimate the Henry's law constant (H) based on the definition $p_A = H c_{AL}^*$, and the distribution coefficient m based on the definition $y_A = m x_A^*$ at 1.0 atm total system pressure.

29.2 Hydrogen sulfide (H$_2$S) is a common contaminant in natural gas. The dissolution of H$_2$S gas in water is a linear function of partial pressure, as is described by Henry's law of the form $p_A = H x_A^*$. Values of H vs. temperature are provided below:

T (°C)	20	30	40	50
H (atm)	483	449	520	577

Given the relatively low solubility of H$_2$S in water, an amine-based chelating agent is added to the water to improve the solubility of H$_2$S. Equilibrium distribution data for H$_2$S in a 15.9 wt% solution of monoethanolamine (MEA) in water at 40 C is provided below:[*]

mole H$_2$S/ mole MEA	0.000	0.125	0.208	0.362	0.643	0.729	0.814
p_A (mm Hg)	0.00	0.96	3.00	9.10	43.1	59.7	106

a. Describe the effect of temperature on the solubility of H$_2$S gas in water.

b. Prepare equilibrium distribution plots, in mole-fraction coordinates ($y_A - x_A$), for the solubility of H$_2$S in water vs. H$_2$S in 15.9 wt% MEA solution at 40 C and 1.0 atm total system pressure. Comment on the relative solubility of H$_2$S in water vs. MEA solution.

[*] J. H. Jones, H. R. Froning, and E. E. Claytor, *J. Chem. Eng. Data*, **4**, 85–92 (1959).

29.3 A wetted-wall tower is used to "aerate" water using air at 2.0 atm total system pressure and 20 C. The molar composition of air is 21% O$_2$, 78% N$_2$, and 1% other gases. Let solute $A = O_2$. At 20 C, the Henry's law constant for dissolution of O$_2$ gas in water is $H = 40,100$ atm based on the definition $p_{A,i} = H x_{A,i}$, and the mass density of liquid water is 1000 kg/m^3.

a. What is the maximum possible composition (as mole fraction) of oxygen (O$_2$) dissolved that could be dissolved in the water, x_A^*?

b. What is the maximum possible molar concentration of oxygen (O$_2$) that could be dissolved in water, c_{AL}^*?

c. If the total system pressure is increase to 4.0 atm, what is the new the dissolved oxygen concentration?

29.4 Consider an interphase mass-transfer process for the chlorine dioxide (ClO$_2$)-air-water system at 20 C, where ClO$_2$ gas (solute A) is sparingly soluble in water. At the current conditions of operation, the mole fraction of ClO$_2$ in the *bulk gas phase* is $y_A = 0.040$ and the mole fraction of ClO$_2$ in the *bulk liquid phase* is $x_A = 0.00040$. The mass density of the liquid phase is 992.3 kg/m^3 and is not dependent on the very small amount of ClO$_2$ dissolved in it. The molecular weight of water is 18 g/gmole, and the molecular weight of ClO$_2$ is 67.5 g/gmole. The total system pressure is 1.5 atm. The liquid film mass-transfer coefficient for ClO$_2$ in water is $k_x = 1.0$ gmole/m$^2 \cdot$s, and the gas film mass-transfer coefficient ClO$_2$ in air is $k_G = 0.010$ gmole/m$^2 \cdot$s$\cdot$atm. The *equilibrium* distribution data for the ClO$_2$-water-air system at 20 C are provided below:

p_A	1.00E-02	3.00E-02	5.00E-02	7.00E-02	1.00E-01	1.10E-01
x_A	2.40E-04	7.19E-04	1.15E-03	1.64E-03	2.34E-03	2.58E-03
p_A	1.20E-01	1.30E-01	1.40E-01	1.50E-01	1.60E-01	
x_A	2.81E-03	3.06E-03	3.28E-03	3.52E-03	3.78E-03	

a. Plot out the equilibrium line in $p_A - c_{AL}$ coordinates, and the operating point (p_A, c_{AL}). Is the process gas absorption or liquid stripping?

b. What is the equilibrium relationship as m equal to?

c. What is k_L for the liquid film?

d. If the ClO_2 mole fraction in the bulk gas phase is maintained at 0.040 under 1.5 atm total system pressure, what is the maximum possible dissolved ClO_2 concentration (gmole A/m^3) in the liquid phase that could possibly exit the process—i.e., c_{AL}^*?

e. What are the compositions at the gas–liquid interface, $p_{A,i}$ and $c_{AL,i}$?

f. What is K_y, the overall mass-transfer coefficient based upon the overall gas-phase mole fraction driving force? There are several valid approaches for calculating K_y based on the information provided. Show at least two approaches that lead to the same result.

g. What is the mass-transfer flux N_A for ClO_2 in units of gmole/m²·s?

29.5 A gas–liquid convective mass-transfer process involves the transfer of the industrial contaminant methylene chloride (solute A) between air and water at 20 C and 2.20 atm total system pressure. Air is the inert carrier gas, and water is the inert solvent. At the present conditions of operation, the bulk-phase mole fraction of methylene chloride is 0.10 in the gas phase and 0.0040 in the liquid phase. The fluid flow associated with each phase is such that the gas film convective mass-transfer coefficient (k_y) is 0.010 gmole/m²·s, and the liquid film convective mass-transfer coefficient (k_x) is 0.125 gmole/m²·s. At 20 C, density of liquid water is 992.3 kg/m³. The equilibrium distribution of methylene chloride dissolved in water is a linear function of the methylene chloride mole fraction in the air over the solution. At 20 C and 2.20 atm total system pressure, $m = 50.0$, based on the definition $y_{A,i} = m\,x_{A,i}$.

a. Plot out the equilibrium line and the operating point in mole fraction coordinates $(y_A - x_A)$. Is the process gas absorption or liquid stripping?

b. What is the Henry's law constant for methylene chloride dissolved in water, according to the definition $p_A^* = H\,c_{AL}$? Comment on the relative solubility of methylene chloride in water.

c. What are K_x and K_L?

d. What is the flux N_A across the gas and liquid phases?

e. What are the interface mole fractions $x_{A,i}$ and $y_{A,i}$?

29.6 In a gas–liquid interface mass-transfer process, the bulk mole fraction composition of solute A in the inert carrier gas is 0.010, and the bulk mole fraction composition of solute A in the inert liquid solvent is 0.040. Equilibrium distribution data at the temperature and pressure of the process are provided below:

x_A	0.0000	0.0050	0.0100	0.0150	0.0200	0.0250	0.0300	0.0350	0.0400
y_A	0.0000	0.0015	0.0030	0.0055	0.0090	0.0135	0.0200	0.0290	0.0425

a. If the liquid film mass-transfer coefficient $k_x = 0.01$ gmole/m²·s, and the gas film mass-transfer coefficient $k_y = 0.02$ gmole/m²·s, what is the overall mass-transfer coefficient based on the gas-phase driving force, and the % resistance to mass-transfer in the gas phase?

b. If the liquid film mass-transfer coefficient is still $k_x = 0.01$ gmole/m²·s, what is the new value of k_y required to make the process 10% gas-phase mass-transfer controlling?

c. Plot $(x_{A,i}, y_{A,i})$ on the equilibrium line in $y_A - x_A$ coordinates for parts (a) and (b), and compare results.

29.7 A packed-bed tower is used for absorption of sulfur dioxide (SO_2) from an air stream using water as the solvent. At one point in the tower, the composition of SO_2 is 10% (by volume) in the gas phase, and 0.30 wt% in the liquid phase, which has a mass density 61.8 lb_m/ft^3. The tower is isothermal at 30 C, and the total system pressure is 1.0 atm. The convective mass-transfer coefficients are $k_L = 2.5$ lbmole/ft²·h·(lbmole/ft³) for the liquid film, and $k_G = 0.125$ lbmole/ft²·h·atm for the gas film. Equilibrium distribution data for the SO_2-water-air system are provided in Problem 29.1.

a. Plot out the equilibrium line in units of p_A (atm) vs. c_{AL} (lbmole/ft³). Plot out the operating point (p_A, c_{AL}) on the same graph. Determine p_A^* and c_{AL}^* and plot on the same graph.

b. Determine the gas–liquid interface compositions $p_{A,i}$ and $c_{AL,i}$.

c. Estimate K_G and K_L, K_y and K_x at the operating point, and the molar flux N_A.

d. Determine the % resistance in the gas phase at the operating point.

29.8 At a particular location in a countercurrent gas absorber, the mole fraction of the transferring species (solute A) in the gas phase is 0.030, and the mole fraction of this species in the liquid phase is 0.010. The gas film mass-transfer coefficient is given $k_y = 1.0$ lbmole/ft²·hr, and 80% of the resistance to mass transfer is in the liquid phase. Equilibrium distribution data at the temperature and pressure of the process are provided below:

x_A	0.0000	0.0050	0.0100	0.0150	0.0200	0.0250	0.0300	0.0350	0.0400
y_A	0.0000	0.0015	0.0030	0.0055	0.0090	0.0135	0.0200	0.0290	0.0425

a. Plot out the equilibrium line in mole fraction coordinates, plot out the operating point (x_A, y_A), and then estimate x_A^* and y_A^* from the graph.

b. Determine the interphase compositions $x_{A,i}$ and $y_{A,i}$ for solute A.

c. Calculate the overall mass-transfer coefficient K_y for the gas phase, and the molar flux N_A.

29.9 An engineer at a pulp mill is considering the feasibility of removing chlorine gas (Cl_2, solute A) from an air steam using water, which will be reused for a pulp bleaching operation. The process will be carried out in a countercurrent flow gas absorption tower filled with inert packing, where liquid water containing no dissolved Cl_2 is pumped into the *top* of the tower ($x_{A_2} = 0$), and air containing 20% by volume Cl_2 ($y_{A_1} = 0.20$) at 1.0 atm is fed into the *bottom* of the tower. Gas–liquid contact is promoted by the inert packing surface as the gas and liquid flow around the packing. As the gas moves to the top of the tower, the Cl_2 composition decreases, and as liquid moves down the tower, the dissolved Cl_2 concentration increases. At the gas and liquid flow rates of operation, the liquid leaving the *bottom* of the tower has a composition of 0.05 mole% ($x_{A_1} = 0.00050$), and the gas exiting the *top* of the tower has been lower to 5% by volume ($y_{A_2} = 0.050$). At the flow rates of operation, the gas film mass-transfer coefficient based on a mole fraction driving force (k_y) is 5.0 lbmole/ft^2·hr, and the liquid film mass-transfer coefficient based on a mole fraction driving force (k_x) is 20 lbmole/ft^2·hr. *Equilibrium* data for the Cl_2-water-air system at 20 C and 1.0 atm are provided in the table below.

y_A	6.58E-03	1.32E-02	3.95E-02	6.58E-02	1.32E-01	1.97E-01
x_A	1.11E-04	1.46E-04	2.37E-04	3.07E-04	4.49E-04	5.75E-04

a. Draw a diagram of what the packed tower might look like, labeling liquid flow with L, gas flow with G, and terminal steam mole fraction compositions (x_{A_1}, y_{A_1}) and (x_{A_2}, y_{A_2}) at the bottom and top of the tower.

b. Plot out the equilibrium line in mole fraction coordinates. Then plot out the operating points (x_{A_1}, y_{A_1}) and (x_{A_2}, y_{A_2}) for the bottom and the top of the tower respectively.

c. Determine the local overall mass-transfer coefficients K_y at the top and bottom of the tower? Why are the values for K_y different?

29.10 Natural gas from a hydraulic fracturing process contains hydrogen sulfide (H_2S) gas that will be removed by absorption into a specialty solvent through an interphase mass transfer process. In the present process, the gas contains 6.0 mole% H_2S and the liquid solvent contains 2.0 mole% H_2S. The methane carrier gas is not soluble in the solvent, and the solvent is nonvolatile. The process is maintained at 2.0 atm and 40 C, and at these conditions, the equilibrium distribution curve data for H_2S in the solvent is given in the table below. The average molar concentration of the specialty solvent is 49.6 kgmole/m^3. The *film mass transfer coefficient* for the gas phase based on the

mole fraction driving force is $k_y = 0.040$ gmole/m^{2}s, and *overall mass transfer coefficient* based on the gas phase mole fraction driving force is $K_y = 0.030$ gmole/m^{2}s.

a. What are y_A^* and x_A^*? Plot out on an y-x diagram.

b. What are the gas-liquid interface compositions, in terms of $y_{A,i}$ and $x_{A,i}$? Plot out on an y-x diagram.

c. What is the overall mass transfer coefficient based on the gas phase partial pressure driving force, K_G?

d. What is the flux of H_2S from the bulk gas to the bulk liquid?

Equilibrium distribution data for H_2S in the solvent at 40 C are given in the table below.

P_A* (mm Hg)	0.96	3.0	9.1	43.1	59.7	106.0
x_A	0.0063	0.0094	0.0183	0.0325	0.0369	0.0412

29.11 In a wastewater treatment process, we are concerned about the possible release of the aromatic hydrocarbon benzene to the atmosphere. In the present process, 3.0 gmole/m^3 of benzene (solute A, mol. weight 78 g/gmole, C_6H_6) dissolved in water is exposed to air which contains *no* benzene vapor in the *bulk* gas phase, i.e. $y_{A\infty} \approx 0$. The mass transfer coefficients for solute A are $k_G = 3.0 \times 10^{-3}$ gmole/m^2 · s · atm, and $k_x = 2.0$ gmole/m^{2}s for the gas and liquid films respectively. The process temperature is 20 C, the total system pressure is 1.0 atm. At 20 C, the solubility of benzene vapor in water follows Henry's law with $H = 4.0$ m^{3}atm/kgmole. The density of the wastewater is 1000 kg/m^3.

a. What are c_{AL}^* and p_A^*?

b. Draw in the operating point directly on the equilibrium diagram. Is the process gas absorption or liquid stripping? Why?

c. What is the overall mass transfer coefficient based on the *overall gas phase* driving force?

d. What is the flux N_A?

e. What is the partial pressure of solute A right at the gas/liquid interface?

29.12 A mass-transfer process is used to remove ammonia (NH_3, solute A) from a mixture of NH_3 and air, using water as the absorption solvent. At the present conditions of operation, the partial pressure of ammonia in the bulk gas phase (p_A) is 0.20 atm, and the mole fraction of dissolved NH_3 in the water (x_A) is 0.040. The total system pressure is 2.0 atm, and the temperature is 30 C. At 30 C, the molar solution density is 55.6 kgmole/m^3. The film mass-transfer coefficients are $k_G = 1.0$ kgmole/m^2 · s · atm, and $k_L = 0.045$ m/s for the gas and liquid, respectively. The *equilibrium* distribution data at 30°C for the NH_3-water-air system are given in the table below:

p_A (atm)		0.463	0.342	0.236	0.145	0.105	0.067
c_{AL} (kgmole/m^3)		11.6	9.71	7.61	5.32	4.09	2.79
p_A (atm)		0.053	0.039	0.032	0.025	0.020	0.015
c_{AL} (kgmole/m^3)		2.26	1.71	1.43	1.15	0.925	0.697

a. Prepare an equilibrium distribution plot for NH_3 as partial pressure vs. mole fraction ($p_A - x_A$).

b. What is k_x?

c. What are the interface compositions $x_{A,i}$ and $p_{A,i}$? What are x_A^* and p_A^*?

d. What is K_G, the overall mass-transfer coefficient based on the gas phase?

e. What is the flux N_A for this process?

29.13 Wastewater containing dissolved hydrogen sulfide (H_2S) at concentration of 2.50 gmole/m^3 (85 mg/L) enters an open tank at a volumetric flow rate of 20 m^3/h, and exits at the same volumetric flow rate, as shown in the figure on the next page. The open tank is within a large enclosed building. The ventilation for the air surrounding the open tank is such that the composition of H_2S in the bulk, well-mixed air over the tank is 0.5 mole% H_2S (gas-phase mole fraction of 0.005), which has a pungent odor. The total system pressure is 1.0 atm, and the temperature is 20 C. The diameter of the cylindrical tank is 5.0 m, and the depth of the liquid in the tank is 1.0 m. At the conditions of operation, the film mass-transfer coefficients for H_2S transfer are $k_L = 2.0 \times 10^{-4}$ m/s in the liquid film, and $k_G = 5.0 \times 10^{-4}$ kgmole/$m^2 \cdot s \cdot$ atm in the gas film. The solute H_2S (solute A) is sparingly soluble in water, with the linear equilibrium distribution data at 20 C described by Henry's law with $H = 9.34$ $m^3 \cdot$ atm/kgmole. At 20 C, the mass density of the wastewater is 1000 kg/m^3. The molecular weight of H_2S is 34 g/gmole, and H_2O is 18 g/gmole.

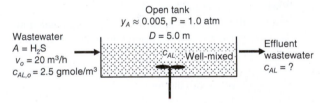

Open tank
$y_A \approx 0.005$, P = 1.0 atm
D = 5.0 m

Wastewater
$A = H_2S$
$v_o = 20$ m^3/h
$c_{AL,o} = 2.5$ gmole/m^3

c_{AL} Well-mixed

Effluent wastewater
$c_{AL} = ?$

a. Estimate m, the equilibrium distribution constant based upon the mole fraction equilibrium relationship—e.g., $y_{A,i} = m x_{A,i}$. Plot out the operating point (x_A, y_A) and the equilibrium line in mole fraction coordinates. Is the process gas absorption or liquid stripping?

b. What is the overall mass-transfer coefficient K_y based upon the overall gas-phase mole fraction driving force?

c. Perform a material balance on the process. At the conditions of operation, what is the concentration of dissolved H_2S exiting the tank, c_{AL}, in units of gmole/m^3?

29.14 Wastewater containing volatile species A dissolved in the water at a dilute inlet concentration of 0.50 gmole A/m^3 ($c_{AL,o}$) is pumped into a well-mixed open holding pond at a volumetric flow rate (v_o) of 2.0 m^3/s, as shown in the figure below. The volatile species A partitions into the atmosphere by an interphase mass-transfer process so that the wastewater exiting the pond contains a concentration of dissolved A of

0.35 gmole A/m^3 (c_{AL}). Fresh air very gently blows across surface of the 10.0 m width of the pond, so that the bulk gas composition of species A is effectively zero (e.g., $p_A \approx 0$, $c_{AL}^* \approx 0$). Equilibrium distribution data for solute A between air and water at 20 C is described by Henry's law with $H = 0.50$ $m^3 \cdot$ atm/gmole (500 $m^3 \cdot$ atm/kgmole).

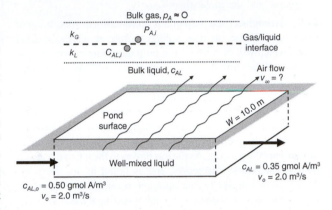

Potentially useful data (1.0 atm total system pressure, 20 C)
   Gas film: $k_c = 2.67 \times 10^{-4}$ m/s
   Liquid film: $k_L = 5.5 \times 10^{-3}$ m/s
   Air at 1.0 atm, 20 C: $\rho = 1.206$ kg/m^3, $\mu = 1.813 \times 10^{-5}$ kg/m-s, $D_{A\text{-air}} = 0.08$ cm^2/s (gas)
Liquid water at 20 C: $\rho = 998.2$ kg/m^3, $\mu = 9.32 \times 10^{-4}$ kg/m-s, $D_{A\text{-}H_2O} = 2.0 \times 10^{-5}$ cm^2/s (liquid)

a. What is the mass-transfer coefficient based on the overall liquid-phase driving force, K_L?

b. Develop a material balance model in algebraic form using the following variables: c_{AL}, $c_{AL,o}$, K_L, p_A, v_o, and surface area S. At the conditions of operation, what is the surface area S of the liquid pond?

c. What is the velocity of the air flow over the surface of the pond, assuming laminar flow?

d. What is the partial pressure of species A on the gas side of the gas–liquid interface, $p_{A,i}$?

29.15 Ozone gas (O_3, solute A) dissolved in high-purity water is commonly used in wet cleaning processes associated with semiconductor device fabrication. It is desired to produce a liquid water stream containing 3.0 gmole O_3/m^3 (238 mg/L) by a process that does not create any gas bubbles. One engineer's idea is shown in the figure below. Liquid water containing 1.0 gmole O_3/m^3 enters a well-mixed tank at a volumetric flow rate 0.050 m^3/h. A pressurized gas mixture of O_3 diluted in inert N_2 is continuously added the headspace of the tank at a total pressure of 1.5 atm. Both the liquid and gas inside the tank are assumed to be well mixed. The gas–liquid surface area inside the tank is 4.0 m^2. The process is maintained at 20 C. At 20 C, the solution density is 992.3 kg/m^3. For a well-mixed, non-bubbled ozonation tank, the appropriate film mass-transfer coefficients for the liquid and gas films are $k_L = 3.0 \times 10^{-6}$ m/s and $k_c = 5.0 \times 10^{-3}$ m/s, respectively. Equilibrium distribution data for O_3 gas dissolved in water at 20 C follows Henry's law, with $H = 68.2$ m$^3 \cdot$ atm/kgmole based on the definition $p_{A,i} = H \cdot c_{AL,i}$.

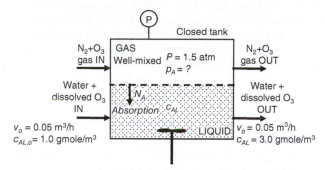

a. What are m, and the Henry's law constant H in units of atm? Is O_3 very soluble in water?

b. What is the overall mass-transfer coefficient K_G, based on the overall gas-phase driving force?

c. What is the overall mass-transfer coefficient K_L based on the overall liquid-phase driving force?

d. For the process to operate as intended, what are the required partial pressure (p_A) and mole fraction (y_A) of ozone (O_3) in the gas phase inside the tank? As part of your solution, develop a material balance model in algebraic form for solute A that contains the following terms: ν_o, volumetric flow rate of liquid (m^3/h); $c_{AL,o}$, inlet concentration of solute A in liquid (gmole A/m^3); c_{AL}, outlet concentration of solute A in liquid (gmole O_3/m^3); K_G, overall mass-transfer coefficient based on gas-phase driving force (gmole/m$^2 \cdot$ s $\cdot$ atm), p_A, partial pressure of O_3 in bulk gas phase (atm); H, Henry's law constant for O_3 between gas and liquid (m$^3 \cdot$ atm/gmole); S, surface area for interphase mass transfer (m^2).

e. What is the total transfer rate of O_3, W_A?

f. Is the mass-transfer process is gas film controlling, liquid film controlling, or neither? Comment on the relative contributions of the film mass-transfer coefficients and the equilibrium distribution relationship on the controlling mass-transfer resistance.

29.16 Jasmone (molecular formula $C_{11}H_{16}O$) is a valuable specialty chemical that is obtained from the jasmine plant. A common method of manufacture is to extract the plant material in water, and then use benzene to concentrate the jasmone in a simple liquid–liquid extraction process. Jasmone (species A) is 170 times more soluble in benzene than in water, and so

$$c_A' = 170\, c_A$$

where c_A' is the concentration of jasmone in benzene, and c_A is the concentration of jasmone in water. In a proposed extraction

unit, the benzene phase is well mixed with the film mass-transfer coefficient $k_L' = 3.5 \times 10^{-6}$ m/s. The aqueous phase is also well mixed with its film mass-transfer coefficient $k_L = 2.5 \times 10^{-5}$ m/s. Determine

a. The overall liquid mass-transfer coefficient, K_L', based on the benzene phase

b. The overall liquid transfer coefficient, K_L, based on the aqueous phase

c. The percent resistance to mass transfer encountered in the aqueous liquid film

29.17 Consider the waste treatment process shown in the figure below. In this process, wastewater containing a dissolved TCE concentration of 50 gmole/m^3 enters a clarifier, which is essentially a shallow, well-mixed tank with an exposed liquid surface. The overall diameter of the tank is 20.0 m, and the maximum depth of the liquid in the tank is 4.0 m. The clarifier is enclosed to contain the gases that are emitted from the wastewater. Fresh air is blown into this enclosure to sweep away the gases emitted from the clarifier, which are sent to an incinerator. Samples of effluent gas phase and effluent liquid phase were measured for TCE content. The TCE composition in the effluent gas is 4.0 mole%, whereas the TCE concentration dissolved in the effluent liquid phase is 10 gmole TCE/m^3 liquid. The clarifier operates at 1.0 atm and constant temperature of 20 C. In independent pilot plant studies, for the transferring species TCE, the liquid film mass-transfer coefficient for the clarifier was $k_x = 200$ gmole/m$^2 \cdot$ s, whereas the gas film mass-transfer coefficient for the clarifier was $k_y = 0.1$ gmole/m$^2 \cdot$ s. Equilibrium data for the air-TCE-water system at 20 C is represented by Henry's law of the form $P_{A,i} = H \cdot x_{A,i}$, with $H = 550$ atm.

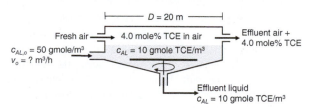

a. What is the overall mass-transfer coefficient based on the liquid phase, as K_L?

b. What is the *flux* of TCE from the clarifier liquid surface?

c. Develop a well-mixed, steady material balance model for the process. What is the inlet volumetric flow rate of wastewater, ν_o (in units of m^3/h) needed to ensure that the liquid effluent TCE concentration is $c_{AL} = 10$ gmole TCE/m^3?

29.18 Ammonia (NH_3) in air is being absorbed into water within the enclosed tank shown in the figure (next page). The liquid and gas phases are both well mixed, and mass transfer occurs only at the exposed gas–liquid interface. The diameter of the cylindrical tank is 4.0 m, and the total liquid volume inside the tank is constant. The bulk gas partial pressure of NH_3 is maintained at 0.020 atm, and the total gas pressure is constant at

1.0 atm. The system is isothermal at 20 C. The inlet volumetric flow rate of water is 200 L/h (0.20 m³/h), and there is no NH₃ in the inlet liquid stream (i.e., $c_{AL,o} = 0$). You may assume that Henry's law adequately describes the equilibrium distribution of NH₃ between the gas and liquid phases, given by $p_{A,i} = H \cdot c_{AL,i}$, where $H = 0.020$ m³ · atm/kgmole. The mass-transfer coefficients for the gas and liquid films are $k_G = 1.25$ kgmole/m² · hr · atm and $k_L = 0.05$ kgmole/(m² · h · (kgmole/m³)), respectively.

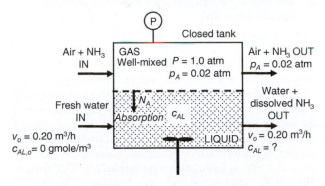

a. Develop a material balance equation for NH₃ (solute A). Then determine c_{AL}, the concentration of dissolved NH₃ in the outlet liquid stream. In material balances involving interphase mass transfer, base the material balance on one phase. For this process, consider a material balance on NH₃ based on the liquid phase.

b. Determine $p_{A,i}$ the partial pressure of NH₃ at the gas–liquid interface, and $c_{AL,i}$, the dissolved NH₃ concentration at the liquid side of the gas–liquid interface.

c. Determine W_A, the total rate of ammonia transfer.

d. In the above system, the *flux* N_A would increase by increasing which of the following: the liquid volume level in the tank at fixed surface area; the agitation intensity of the bulk liquid; the agitation intensity of the bulk gas; the inlet liquid volumetric flow rate; the system temperature.

29.19 Wastewater containing the dissolved contaminant naphthalene is sent to a well-mixed open pond maintained at 20 C, as shown in the figure (next column). The pond allows the naphthalene to transfer from the water to the air. The pond is 10.0 m per side and 1.0 m deep. Naphthalene is volatile and slightly soluble in water, and the equilibrium portioning of naphthalene vapor (solute A) and water at 20 C follows Henry's Law with $H = 36.6$ Pa · m³/gmole. The concentration of dissolved naphthalene in the inlet waste water is $c_{AL,o} = 0.25$ gmole/m³, and the outlet concentration of the exiting waste water is $c_{AL} = 0.15$ gmole/m³. Air blows over the entire flat liquid surface of the pond, and it has been determined that the convective mass transfer coefficient for naphthalene across the gas film is $k_c = 0.50$ cm/s (3.0 × 10⁻³ m/s). It has also been determined that the liquid film mass transfer coefficient for naphthalene is $k_L = 4.0 \times 10^{-5}$ m/s. Assume that naphthalene vapor concentration in the bulk air is zero (e.g., $y_A = 0$, $p_A = 0$). The process is at 1.0 atm total system pressure and temperature of 20 C. At these process conditions, the gas-phase diffusion coefficient of naphthalene in air is 0.065 cm²/s. The molecular weight of naphthalene (M_A) is 128 g/gmole.

a. Plot out the operating point (p_A, c_{AL}) on the p_A-c_{AL} diagram. What is the Henry's Law constant, H, as defined by $p_{A,i} = H c_{AL,i}$? What are $p_A{}^*$ and $c_{AL}{}^*$? Plot out $p_A{}^*$, $c_{AL}{}^*$, $p_{A,i}$, and $c_{AL,i}$.

b. What is the *overall* mass transfer coefficient based on the liquid phase driving force, K_L?

c. What is the velocity of air flowing over the pond (v_∞), in units of m/s to achieve $k_c = 0.50$ cm/s, and its Reynolds and Schmidt numbers, Re and Sc?

d. What is the inlet volumetric flowrate of wastewater into the pond (v_o), in units of m³/h?

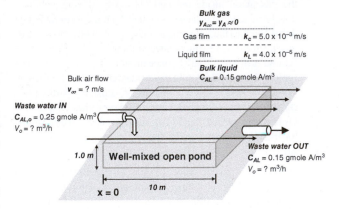

SS **30.1** Spherical pellets of 1.0 cm diameter are spray-painted with a very thin coat of paint. The paint contains a volatile solvent. The vapor pressure of the solvent at 298 K is 1.17×10^4 Pa, and the diffusivity of the solvent vapor in air at 298 K is 0.0962 cm²/s. The amount of solvent in the wet paint on the pellet is 0.12 g solvent per cm² of pellet surface area. The molecular weight of the solvent is 78 g/gmole.

a. Determine the minimum time to dry the painted pellet if still air at 298 K and 1.0 atm surrounds the pellet.

b. Determine the minimum time to dry the painted pellet if air at 298 K and 1.0 atm pressure flows around the pellet at a bulk velocity of 1.0 m/s.

30.2 A 1.75-cm diameter naphthalene mothball is suspended in an air stream at 280 K, 1.0 atm, and constant velocity of 1.4 m/s. Solid naphthalene exerts a vapor pressure of 2.8 Pa at 280 K. Consequently, the naphthalene very slowly sublimes into the passing air stream, with the rate limited by convective mass transfer. The density of solid naphthalene is 1.14 g/cm³, and the molecular weight is 128 g/mole.

a. What is the initial evaporation rate of naphthalene mothball?

b. How long will it take for the mothball to shrink to half of its original diameter? Remember: As the mothball shrinks, its diameter is decreasing, which in turn affects the Reynolds and Sherwood numbers.

30.3 Pipes often contain solid residue that adheres to the inner wall of the pipe. If these pipes are not properly cleaned before use, then the flow of water through the pipe will slowly dissolve the contaminant solid (solute A) into the water through a convective mass transfer process. Consider the system shown in the figure below. Pure liquid water flows into a 2.0 cm inner diameter tube of 25.0 m length (L) at a bulk velocity (v_∞) of 0.50 m/s at 1.2 atm and 20 C. The solid contaminant only coats the *lower half* of the inside surface of the pipe. The maximum solubility of solute A in water (c_{AL}^*) is 10.0 mmole A/m³ (very dilute, but still toxic) and the diffusion coefficient of solute A in liquid water (B) is $D_{AB} = 5.0 \times 10^{-5}$ cm²/s (5.0×10^{-9} m²/s). You may assume that during the process, the solid does not completely dissolve away, and that the solid contaminant layer is thin relative to the inside diameter of the pipe

a. Identify the source and sink for the mass transfer process.

b. What is the mass flowrate of water into the tube?

c. What is the convective mass transfer coefficient for solute A in liquid water, k_L (m/s)?

d. What is the *differential material balance model* for describing the concentration of dissolved solute A down the length of the tube? State three primary assumptions of your model. The model must at least contain the following variables: $c_{AL}^*, c_{AL}, k_L, z, v_\infty, D$.

e. What is the dissolved solute concentration c_{AL} exiting the tube at $z = L$?

f. If the liquid velocity is doubled, with all other variables remaining constant, what happens to mass transfer coefficient k_c and c_{AL} at $z = L$?

30.4 A "stent" is used to "prop up" a clogged artery to allow blood to pass through it. But stents can also be loaded with drugs to facilitate the timed release of the drug into the body, especially if the drug is not very soluble in body fluids. Consider the very simple stent design shown in the figure below. The pole of the stent, which is 0.20 cm in diameter and 1.0 cm in length, is coated with the anticancer drug Taxol. The thickness of the coating is 0.010 cm, and 5.0 mg total Taxol is loaded. Blood flows through the 1.0 cm diameter cylindrical blood vessel at a volumetric flow rate of 10.0 cm³/s. Taxol is not very soluble in aqueous environments; the maximum solubility of Taxol in blood is 2.5×10^{-4} mg/cm³. The viscosity of blood is 0.040 g/cm · s, and its density is 1.05 g/cm³. Blood is a complex fluid, but you may assume that its average molecular weight is close to that of liquid water (18 g/gmole). The molecular diffusion coefficient of Taxol (A) in blood fluid (B) is $D_{AB} = 1.0 \times 10^{-6}$ cm²/s.

a. What is the convective mass-transfer coefficient (k_L) around the outer surface of the cylindrical portion of the stent for the solute Taxol? State all assumptions.

b. What is the minimum time it will take for all of the Taxol to be completely discharged from the stent? State all assumptions.

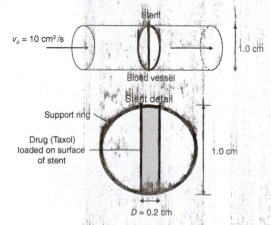

30.5 An engineer proposes to use the mass-transfer equip- **SS** ment shown in the figure (next page) to prepare a water stream containing dissolved oxygen. Liquid water containing no dissolved oxygen is fed to the tank at a rate of 50 gmole/s. The inlet water is passed through a flow diffuser that makes the velocity of the liquid uniform in the tank. The tank contains 10 silicone-walled tubes of 100 cm length and 2.0 cm outer diameter. The silicone walls are permeable to O_2 gas but not to water. The long width of the tank (L) is also 100 cm, and the depth of the tank is 50 cm, so that the cross-sectional area for liquid flow is 5000 cm^2. You may assume that the concentration of dissolved O_2 in the tank is equal to the concentration of dissolved oxygen in the liquid outlet stream—i.e., the liquid volume is well mixed.

Potentially useful data: $c_{AL}^* = 5.0$ gmole O_2/m^3 (4.0 atm 100% O_2 gas on tube side); $\rho_L = 998.2$ kg/m^3; $\nu_L = 0.995 \times 10^{-6}$ m^2/s; $D_{AB} = 2.0 \times 10^{-9}$ m^2/s ($A = O_2$, $B = H_2O$).

a. Develop a material balance model to predict the concentration of dissolved oxygen in the outlet liquid (c_{AL}). State all assumptions. Your final model must be in algebraic form. Your model development should contain the following variables: $c_{AL,o}$, inlet concentration of dissolved oxygen; c_{AL}^*, concentration of dissolved oxygen in liquid at the tube surface; D, outer diameter of tube; v_∞, bulk velocity of liquid; L, length of tube, long width of tank; W, width of tank, short dimension; k_L, liquid film mass-transfer coefficient; N_t, number of tubes.

b. What is the mass-transfer coefficient k_L? Does the mass-transfer process represent an external flow convection or internal flow convection?

c. Based on your results for parts (a) and (b) above, estimate the outlet concentration of dissolved oxygen. Based upon your analysis, do you think that this mass-transfer device works very well? Mention two doable options to increase c_{AL}.

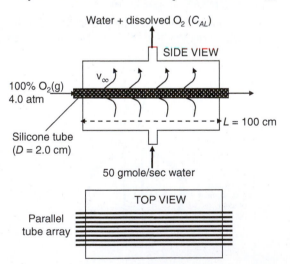

Water + dissolved O_2 (C_{AL})

SIDE VIEW

100% O_2(g)
4.0 atm

v_∞

$L = 100$ cm

Silicone tube
($D = 2.0$ cm)

50 gmole/sec water

TOP VIEW

Parallel
tube array

30.6 A spherical silica gel adsorbent particle of 0.20 cm diameter in placed in a rapidly flowing air stream at a velocity

of 50 cm/s containing 1.0 mole% of H_2O vapor at total system pressure of 1.0 atm and temperature of 25 C. At 25 C, the equilibrium of H_2O vapor with silica gel is described by the linear relationship $C_A^* = S \cdot P_A$, where C_A^* is the equilibrium concentration of H_2O in the silica, and P_A is the H_2O vapor partial pressure, and S is the partitioning coefficient, with $S = 2.0$ gmole O_2/(cm^3 silica) $\cdot$ (atm H_2O vapor) at 25 C. At 25 C, the diffusion coefficient of H_2O in the silica gel is 1.0×10^{-6} cm^2/s. Assuming that the partial pressure of H_2O vapor in the gas stream remains constant, what is the concentration of H_2O in the *center* of the silica gel bead after 2.78 h? Initially, the silica gel bead contains no H_2O.

30.7 A spherical drug capsule dispenses Dicyclomine (an irritable bowel syndrome drug) to the body's gastrointestinal tract over time. Initially, 2.00 mg of Dicyclomine (m_{Ao}) is loaded into the 0.50-cm-diameter capsule. In the bulk fluid surrounding the capsule, the Dicyclomine has a residual constant concentration ($c_{A\infty}$) of 0.20 mg/cm^3. The solubility of Dicyclomine in the capsule polymer is the same as that in the surrounding fluid, which approximates the properties of liquid water. The molecular diffusion coefficient of Dicyclomine in water is 1.0×10^{-5} cm^2/s, whereas the effective diffusion coefficient of Dicyclomine in the capsule matrix material is 4.0×10^{-6} cm^2/s. At 37 C, the density and viscosity of liquid water are 1.0 g/cm^3 and 0.0070 g/cm $\cdot$ s, respectively. Fluid movement inside the intestinal tract as has "mass-transfer Biot number" of nominally 5.0, and is defined by $Bi = k_c R/D_{Ae}$, where D_{Ae} is the effective diffusion coefficient associated with the solid surface in contact with the fluid, and R is the radius of the spherical capsule.

a. What is the concentration of Dicyclomine remaining in the *center* of the spherical capsule after a time of 4.34 h (15,625 s)?

b. What is the estimated Biot number if the velocity of the bulk fluid is 0.50 cm/s? Does the diffusion process within the drug capsule have any convective mass-transfer resistances associated with it?

30.8 A spherical pellet containing pure solid A is suspended in a flowing liquid stream at 20 C. The initial diameter of the pellet is 1.0 cm, and the bulk liquid velocity is 10.0 cm/s. Component A is soluble in the liquid, and as time progresses, the diameter of pellet decreases. The dissolved solute concentration in the bulk liquid is fixed at 1.0×10^{-4} gmole/cm^3. All relevant physical properties of the system are provided below. This process represents a "pseudo-steady-state" convective mass-transfer system.

a. Estimate the film (convective) mass-transfer coefficient at the initial pellet diameter of 1.0 cm.

b. What is the dissolution rate of the pellet at the initial diameter of 1.0 cm?

c. How long will it take for the pellet to shrink to 0.5 cm diameter? Remember, when the diameter decreases, Re and Sh will change.

Potentially useful data: $\rho_{A,\text{solid}} = 2.0$ g/cm^3, the density of the solid A; $M_A = 110$ g/gmole, the molecular weight of solute A; $c^*_{AL} = 7 \times 10^{-4}$ gmole/cm^3, the equilibrium solubility of solute A in the liquid at the liquid–solid interface; $\nu = 9.95 \times 10^{-3}$ cm^2/s, the kinematic viscosity of the bulk fluid at 20 C; $D_{AB} = 1.2 \times 10^{-5}$ cm^2/s, the diffusion coefficient of solute A in the liquid at 20 C.

30.9 The photocatalytic tubular reactor shown in the figure (next page) is used to decompose organic contaminants in liquid waste water. The inter walls of the 1.5 cm inner diameter glass tube are lined with a thin, nonporous, transparent layer of nanocrystalline titanium dioxide (TiO$_2$), which catalyzes the oxidation of dissolved organic chemicals to carbon dioxide and water, using sunlight to drive this photocatalytic process. In the present process, waste water containing 10 g/m^3 of benzene (solute A) dissolved in water enters the tube at 313 K, and it is necessary to have a Reynolds number (Re) of 5000 for flow through the tube. The steady state process temperature is 313 K. For purposes of analysis, assume the consumption of species A at the catalyst surface is so rapid that the effective concentration of species A at the catalyst surface is equal to zero ($c_{AL,s} \approx 0$). *Potentially useful information at 313 K*: molecular diffusion coefficient of benzene (C$_6$H$_6$, species A) in liquid water, $D_{AB} = 1.45 \times 10^{-5}$ cm^2/s, $M_A = 78$ g/gmole, $\mu_B = 0.00658$ g/cm-s (0.658 cP), $\rho_B = 0.993$ g/cm^3.

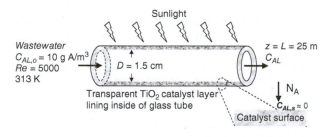

Sunlight

Wastewater
$C_{AL,o} = 10$ g A/m^3
$Re = 5000$
313 K
$D = 1.5$ cm

$z = L = 25$ m
C_{AL}

N_A

$C_{AL,s} \approx 0$

Transparent TiO$_2$ catalyst layer
lining inside of glass tube

Catalyst surface

a. Identify the source and sink for mass transfer of benzene.

b. Is the photocatalytic reaction best represented as a homogeneous reaction within the control volume for mass transfer or a heterogeneous reaction at a boundary surface? Justify your response.

c. Estimate the convective mass transfer coefficient of benzene in water flowing through the tube, k_L.

d. Develop a model, in final integrated form, to estimate the concentration of species A exiting the tube. State three primary assumptions of your model. The model must at least contain the following variables: $c_{AL,o}$, c_{AL}, k_L, L, v_∞, D.

e. If the tube length (L) is 25.0 m, what is the concentration of benzene in the wastewater exiting the tube, i.e. c_{AL} at $z = L$?

f. Consider now that the surface reaction is slow enough so that the benzene concentration on the catalyst is $c_{AL,s} > 0$, and the surface reaction is described by a first-order process

with $k_s = 0.0010$ cm/s. Estimate the new outlet concentration of benzene.

30.10 A pond containing a suspension of microorganisms is used to biologically degrade dissolved organic materials in wastewater, as shown in the figure (right column). The pond contains 1000 m^3 of liquid. Recently, inflow and outflow pipes were installed. The inlet volumetric flow rate of water is 0.05 m^3/s, and the outflow is the same to maintain a constant liquid volume. The concentration of dissolved oxygen in the inlet liquid water is 10.0 mmole/m^3. The present process uses an air sparger to deliver 2.0 mm air bubbles into the wastewater. The rising bubbles uniformly mix the entire liquid phase of the pond. Only a very small portion of the O$_2$ gas within the air bubbles (containing 0.21 atm O$_2$) dissolves into the liquid, where it is consumed by the microorganisms. The Henry's law constant to estimate the solubility of O$_2$ dissolved in the wastewater that is in equilibrium with the partial pressure of O$_2$ the aeration gas is $H = 8.0 \times 10^{-4}$ atm-m^3/mmole (1000 mmole = 1.0 gmole). The interphase mass-transfer area of the bubbles per unit volume of liquid is equal to 10 m^2/m^3 for this process. The process is 100% liquid film controlling. At the current conditions of operation, the current biological oxygen consumption demand, or "BOD," associated with the microbial respiration in the pond is 0.200 mmole O$_2$/m^3-s.

a. Develop a steady-state material balance model to predict the dissolved concentration of O$_2$ in the holding pond.

b. What is the liquid-phase mass-transfer coefficient k_L for O$_2$?

c. What is the steady-state concentration of dissolved oxygen in the pond?

Potentially useful data (all at process temperature and pressure): $D_{AB} = 2.0 \times 10^{-9}$ m^2/s (A = dissolved O$_2$, B = water solvent); $\rho_{B,\text{liq}} = 1000$ kg/m^3; $\rho_{\text{air}} = 1.2$ kg/m^3; $\nu_{\text{air}} = 1.6 \times 10^{-5}$ m^2/s; $H = 8.0 \times 10^{-4}$ atm·m^3/mmole; $\mu_{B,\text{liq}} = 825 \times 10^{-6}$ kg/m·s.

Inlet liquid stream
$v_o = 0.05$ m^3/sec
$C_{AL,o} = 10$ mmol O$_2$/m^3 Open Pond

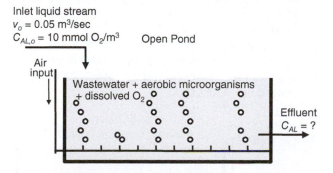

Air input

Wastewater + aerobic microorganisms + dissolved O$_2$

Effluent
$C_{AL} = ?$

30.11 As society searches for technical solutions to global warming, vast ponds of photosynthetic algae are seen as one possible solution for the removal of CO$_2$ from combustion of fossil fuels. Consider the process shown in the figure (next page), where a flue gas containing a binary mixture 10 mole%

CO_2 and 90 mole% N_2 is bubbled into a pond containing photosynthetic, single-celled algae that are uniformly suspended in the liquid water. The agitation provided by the rising bubbles keeps the liquid suspension *well mixed*. The algae consume dissolved CO_2 according to a first-order reaction of the form

$$R_A = -k_1' \cdot X \cdot c_{AL} = -k_1 \cdot c_{AL}$$

where R_A is the CO_2 consumption per unit volume of liquid suspension (gmole CO_2/m^3 s, k_1' is the rate constant for CO_2 uptake by the algal cells (m^3/g cells/h), X is the algal cell density (g cells/m^3), c_{AL} is the concentration of dissolved CO_2 (gmole CO_2/m^3), and k_1 is the apparent rate constant for CO_2 uptake ($k_1 = k_1' \cdot X$, h^{-1}). In the present process, there is no inflow or outflow of water into the pond. However, there is a *constant delivery of CO_2* to the liquid suspension by the bubbled-in flue gas, and a *constant consumption of dissolved CO_2* by the photosynthetic algae, with $k_1 = 0.06435$ m^3/g cells · h. Furthermore, since CO_2 is not very soluble in water, it was found that the composition of CO_2 gas inside the gas bubble decreased only slightly as it traveled up the water column. Therefore, the partial pressure of CO_2 gas inside the gas bubble (p_A) is considered constant. However, there is still a significant rate of transfer of CO_2 into the liquid phase. The open pond is maintained at 20 C. At 20 C, the Henry's law constant for the dissolution of CO_2 gas in the liquid is 0.025 atm · m^3/gmole, the mass density of liquid is 1000 kg/m^3, and the viscosity of the liquid is 993 × 10^{-6} kg/m · s. The aeration system is designed so that the average bubble size is 1.0 mm and the desired interphase area/liquid volume is $a = 5.0$ m^2/m^3. The mass density of the gas mixture is 1.2 kg/m^3. The total liquid volume of the pond is kept constant.

a. What is the molecular diffusion coefficient of CO_2 in water?

b. What is the mass-transfer coefficient for CO_2 on the liquid film side of the gas bubble, k_L?

c. Develop a material balance model for predicting the dissolved CO_2 concentration in the pond water. The model must be in *algebraic form* and contain the following variables: c_{AL}, p_A, H, k_L, a, k_1.

d. If the cell density in the pond is maintained at 50 g cells/m^3, what is the predicted concentration of dissolved CO_2 in the pond, c_{AL}?

e. If the total pond volume is 1000 m^3, what is the total CO_2 removal rate in kg CO_2/h?

Flue gas
10 mole% CO_2 (A)
90 mole% N_2

Open Pond
(no inflow or outflow of liquid)

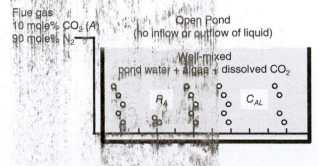

Well-mixed
pond water + algae + dissolved CO_2

R_A C_{AL}

30.12 An open pond containing 200 m^3 of wastewater is contaminated with a small concentration of trichloroethylene (TCE), a common industrial solvent. The pond is 2.0 meters deep and 10 meters on a side. Air is sparged into the bottom of the trench to "strip out" the dissolved TCE. The inlet air flow rate is 0.15 m^3 air per 1.0 m^3 of water per min. At this air flow rate, the gas holdup is calculated to be 0.015 m^3 air per 1.0 m^3 water. The air sparger provides an average bubble diameter of 5.0 mm. The temperatures of the air and water are 293 K. TCE is only sparingly soluble in water, and the Henry's law constant for TCE in liquid water is 9.98 atm · m^3/kgmole, and the molecular weight of TCE is 131.4 g/gmole.

a. What the liquid-phase mass-transfer coefficient for TCE in water surrounding the bubble?

b. Assume there is no inflow or outflow of water to or from the trench. Determine the required *time* to reduce the dissolved TCE concentration from 50 g TCE/m^3 to 0.005 g TCE/m^3. You may assume that the air stream serves as an infinite sink for TCE transfer. Hint: Develop an unsteady-state material balance model for TCE in the bulk liquid phase.

c. Now assume that there is a 1.0 m^3/s inflow of wastewater containing 50 g TCE/m^3, and a 1.0 m^3/s outflow so that the total pond volume is still 200 m. What is the steady-state concentration of dissolved TCE?

30.13 The well-mixed bubbler tank shown in the figure below is used to prepare carbonated water needed for the production of soft drinks. The tank volume is 2.0 m^3, and the tank diameter is 1.0 m. The process operates at steady state. Pure water containing no CO_2 is continuously added to the tank, and carbonated water containing dissolved CO_2 continually exits the tank. Pure carbon dioxide gas at 2.0 atm is bubbled into a tank at a rate of 4.0 m^3 gas per minute. At these conditions, the "gas holdup" inside the tank is 0.05 m^3 gas/m^3 liquid, and the average bubble diameter delivered by the fine bubble sparger is 2.0 mm. The CO_2 gas not absorbed by the water exits the tank. The process temperature is kept constant at 293 K. The process is liquid film controlling since only pure CO_2 is present in the gas phase. The Henry's law constant for dissolution of CO_2 in water is 29.6 atm · m^3/kgmole at 293 K. The inlet water flow rate of 0.45 m^3 per minute (25 kgmole H_2O/min) contains no dissolved CO_2.

a. What is the liquid film mass-transfer coefficient associated with the bubbling of CO_2 gas into water?

b. Is the inlet flow rate of CO_2 gas sufficient to ensure that the CO_2 dissolution is mass-transfer limited? Hint: What is the saturation concentration of dissolved CO_2?

c. What is the outlet concentration of dissolved CO_2? Hint: Develop a well-mixed, steady-state material balance for CO_2 in the liquid phase before performing any numerical calculations.

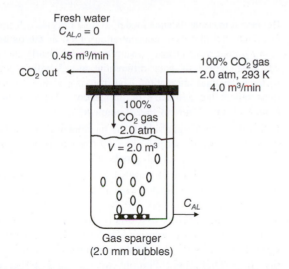

Fresh water
$C_{AL,o} = 0$

0.45 m³/min

CO₂ out ◀

100% CO₂ gas
2.0 atm, 293 K
4.0 m³/min

100%
CO₂ gas
2.0 atm

$V = 2.0$ m³

C_{AL}

Gas sparger
(2.0 mm bubbles)

30.14 Consider the remediation trench shown in the figure below, a simple process to treat contaminated wastewater before discharge to a lake or river. The remediation trench consists of a narrow outdoor open channel with an air sparger aligned along the bottom of the trench. Wastewater containing a volatile contaminant dissolved in the water enters one end of the trench. As the wastewater flows down the trench, the aeration gas strips out the dissolved volatile solute and transfers it to the surrounding atmosphere by an interphase mass-transfer process. Consequently, the concentration of the solute in the wastewater decreases down the length of the trench. Remediation trenches can be long, and may extend from a holding pond to the discharge point. We wish to design an aerated remediation trench to treat wastewater contaminated with trichloroethylene (TCE) at a concentration of 50 mg/L (50 g TCE/m³) wastewater. The trench is an open duct of width (W) 1.0 m and depth (H) 2.0 m, and the volumetric flow rate of wastewater added to the trench is 0.10 m³/s. Air is sparged into the bottom of the duct at a rate that provides a gas holdup of 0.02 m³ of gas per 1.0 m³ of water, and the average bubble diameter is 1.0 cm (0.01 m). Determine the length of the trench necessary to reduce the effluent TCE concentration to 0.05 mg/L. The process temperature is 293 K and the total system pressure is 1.0 atm. TCE is only sparingly soluble in water, and the Henry's law constant for TCE in liquid water is 9.98 atm·m³/kgmole and the molecular weight of TCE is 131.4 g/gmole.

Open Trench
$N_A A_i$

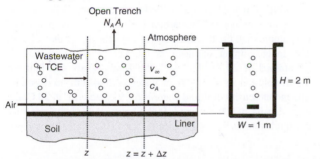

30.15 The convective mass-transfer device shown in the figure (right column) is used to generate a gas stream containing a mixture of phosphorous oxychloride ($POCl_3$) vapor diluted in inert helium (He) gas. This gas mixture is sent to a diffusion furnace to serve as the primary dopant for the manufacture of silicon-based solar cells. In the present process, pure (100%) He gas at a volumetric flow rate of 110 cm³/s enters the 2.0-cm inner diameter tube at 50 C and 1.0 atm. The length of the tube is 60 cm. Liquid $POCl_3$ enters a reservoir that surrounds the porous ceramic tube. The liquid $POCl_3$ enters into the porous ceramic like a wick. At the inner surface of the tube, the $POCl_3$ liquid vaporizes, with its vapor pressure determined by the process temperature. The process is maintained at a constant temperature of 50 C by a controlled heater surrounding the $POCl_3$ reservoir. Therefore, the transfer of $POCl_3$ to the gas stream is limited by diffusion across the convective boundary layer on the inside of the tube. The process can be considered dilute. The diffusion coefficient of $POCl_3$ vapor (A) in He gas (B) is 0.37 cm²/s at 1.0 atm and 50 C. The vapor pressure of $POCl_3$ is $P_A = 0.15$ atm at 50 C. The kinematic viscosity of He gas at 1.0 atm and 50 C is 1.4 cm²/s.

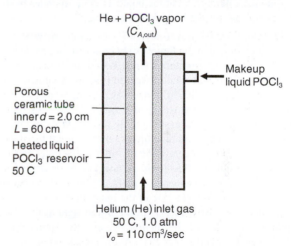

He + POCl₃ vapor
($C_{A,out}$)

Makeup
liquid POCl₃

Porous
ceramic tube
inner $d = 2.0$ cm
$L = 60$ cm

Heated liquid
POCl₃ reservoir
50 C

Helium (He) inlet gas
50 C, 1.0 atm
$v_o = 110$ cm³/sec

a. Propose a model in final algebraic form to predict the outlet concentration of $POCl_3$ vapor ($C_{A,out}$) from the tube. The model should contain the following process variables: gas-phase inlet concentration of A, $C_{A,in}$; gas-phase outlet concentration of A, $C_{A,out}$; k_c, tube mass-transfer coefficient; L tube length; D, tube inner diameter; v_∞, gas velocity; C_A^*, equilibrium vapor concentration in the gas stream.

b. What is the convective mass-transfer coefficient, k_c, at an inlet He volumetric flow rate of 110 cm³/s?

c. What is the outlet mole fraction of $POCl_3$ vapor ($y_{A,out}$) at an inlet He volumetric flow rate of 110 cm³/s?

d. If the inlet volumetric flow rate remains fixed at 110 cm³/s, but the diameter d is doubled, what is the new k_c and $y_{A,out}$?

e. What is the maximum possible outlet mole fraction of $POCl_3$ vapor for an infinitely long tube?

30.16 Tetraethoxysilane, also called TEOS or $Si(OC_2H_5)_4$, is a liquid chemical used in the semiconductor industry to produce thin films of silicon dioxide by chemical vapor deposition (CVD). In order to deliver the TEOS vapor to the CVD reactor, liquid TEOS is fed to a wetted-wall column. The TEOS liquid uniformly coats the inner surface of the tube as a thin liquid film as it flows downward. The falling liquid film of TEOS evaporates into an inert helium carrier gas flowing upwards at a volumetric flow rate of 2000 cm^3/s. The wetted-wall column has an inner diameter of 5.0 cm and a length of 2.0 m. The column temperature is maintained at 333 K, and the total system pressure is 1.0 atm. At 333 K, the kinematic viscosity of helium gas is 1.47 cm^2/s, the diffusion coefficient of TEOS vapor in helium gas is 1.315 cm^2/s, and the vapor pressure of liquid TEOS is $P_A = 2133$ Pa.

a. What is the gas mass-transfer coefficient, k_G?

b. What is the mole fraction of TEOS vapor exiting the column?

c. What is the required mass flow rate of liquid TEOS flowing into the column if all of the liquid TEOS evaporates by the time it reaches the bottom of the column?

30.17 Consider the novel device for oxidative treatment of wastewater shown in the figure (right column). In this device, O_3 will serve as the oxidant source, which must be carefully dosed into the liquid. The device consists of a vertical slit, where the walls of the slit consist of polymeric membranes that are selectively permeable to ozone gas (O_3). As liquid water flows through the slit, O_3 dissolves into the liquid. Assuming that the membrane offers no substantial resistance to O_3 transfer, the concentration of water at the inner surface of the membrane in the liquid is adequately described by Henry's law. For example, $x_A^* = p_A/H$, where p_A is the partial pressure of O_3 on the gas side of the membrane, and x_A^* is the mole fraction of dissolved O_3 in the liquid right at the inner liquid surface side of the membrane, and the Henry's law constant is $H = 3760$ atm at 20 C. At the present conditions of operation, a single slit has a slit opening $h = 1.0$ cm, and slit width $w = 2.0$ cm. The *volumetric flow rate* of liquid water into a single slit is 100 cm^3/s. The inlet water contains no dissolved O_3. The process is dilute, with total molar concentration of the liquid equal to 0.056 gmole/cm^3. At 20 C, the diffusion coefficient of dissolved O_3 (solute A) in liquid water (solvent B) is $D_{AB} = 1.74 \times 10^{-5}$ cm^2/s, and the kinematic viscosity of liquid water is 0.010 cm^2/s. The partial pressure of pure O_3 gas on the gas side of the membrane is maintained constant at 15.0 atm.

a. The mass-transfer coefficient associated with "flow through a slit" can be converted to "flow through a tube" by simply considering an equivalent hydraulic diameter $d = 2h/\pi$. What is the convective mass-transfer coefficient k_L for O_3 transfer for the water inside the slit?

b. Develop a material balance model, in final integrated form, to predict the dissolved concentration of O_3 at the outlet ($z = L$). State your primary assumptions, and clearly show the development of your differential model before you perform the final integration. The material balance model must reflect the geometry of the process. What is the required length L for $c_{AL} = 2.0$ gmole/m^3?

c. If P_A for O_3 on the gas side of the membrane is increased from 15 to 30 atm, what is the new required length L to achieve a desired c_{AL} of 2.0 gmole/m^3?

d. Modify the process to include a UV light source shining into the slit through the semitransparent membrane material, and then repeat part (b). The UV light promotes a homogeneous first-order degradation reaction of O_3 dissolved in solution, with the rate equation given by $R_A = -k_1 \cdot c_{AL}$. At the conditions of illumination, $k_1 = 0.00050$ s^{-1}. Plot out c_{AL} vs. z until the curve is flat, and compare this plot to a plot of c_{AL} vs. z for the model in part (b).

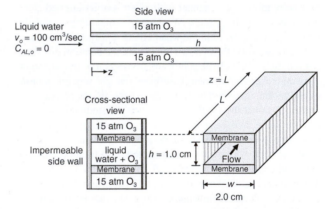

30.18 A process is being developed to produce carbonated beverages. As part of this process, a wetted-wall absorption column 2.0 m long will be used to dissolve carbon dioxide (CO_2) gas into water. The process is shown in the figure (next page). In this process, pure mountain spring water containing no dissolved CO_2 enters the top of the column at a flow rate of 2.0 gmole/s (36.0 g/s). Pure, 100% CO_2 gas at 2.54 atm is also fed to the bottom of the column at a flow rate of 0.5 gmole/s. As the liquid flows down the wetted-wall column, CO_2 gas dissolves into the water. The carbonated water exits the bottom of the column, and the unused CO_2 gas exits the top of the column. The inner column diameter is 6.0 cm. The temperature is maintained at 20 C. At 20 C, the Henry's law constant for the dissolution of CO_2 gas in water is 25.4 atm · m^3/kgmole. The molar density of liquid water is 55.5 kgmole/m^3 at 20 C, the mass density of liquid water is 998.2 kg/m^3 at 20 C, the viscosity of liquid water is 993×10^{-6} kg/m · s at 20 C. Water has a vapor pressure at 20 C. However, for this problem you may assume that the water solvent is essentially nonvolatile, so that 100% CO_2 gas composition is always maintained down the length of the column.

a. What is the maximum possible concentration of dissolved CO_2 in water at 2.54 atm CO_2 partial pressure?

b. What is the liquid-phase mass-transfer coefficient for this process?

c. What is the exit concentration of dissolved CO_2 in the carbonated water if the wetted-wall column is 2.0 m in length?

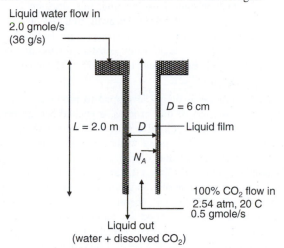

Liquid water flow in
2.0 gmole/s
(36 g/s)

$L = 2.0$ m
$D = 6$ cm
Liquid film
N_A
100% CO_2 flow in
2.54 atm, 20 C
0.5 gmole/s

Liquid out
(water + dissolved CO_2)

30.19 A wetted-wall column of 2.0 cm inner diameter and 50 cm wetted length is used to oxygenate blood as a continuous, steady-state process, as shown in the figure below. Blood containing 1.0 $gmole/m^3$ of dissolved oxygen enters the top of the wetted-wall column at a volumetric flow rate of 300 cm^3/min. Pure, 100% O_2 gas at 1.0 atm and 25 C enters the bottom of the column at a volumetric gas flow rate of 600 cm^3/min. A very simplified description for estimating the equilibrium solubility of O_2 dissolved in blood is

$$c_{AL}^* = \frac{p_A}{H} + \frac{k \cdot p_A}{1 + k \cdot p_A} c_{AL,\text{max}}$$

where p_A is the partial pressure of O_2 in the gas phase, $H = 0.8$ atm $\cdot m^3/gmole$ for O_2 in the blood plasma, $k = 28$ atm^{-1} for the blood hemoglobin, and $c_{AL,\text{max}} = 9.3$ gmole O_2/m^3 for the hemoglobin. At 25 C, the kinematic viscosity of blood is 0.040 cm^2/s and the density of blood is 1.025 g/cm^3. You may assume that the diffusion coefficient of O_2 in blood approximates the diffusion coefficient of O_2 in liquid water, which is 2.0×10^{-5} cm^2/s at 25 C.

a. What is the mass-transfer coefficient for O_2 into the flowing liquid film?

b. What is the concentration of dissolved oxygen in the liquid exiting the bottom of the column, $c_{AL,\text{out}}$?

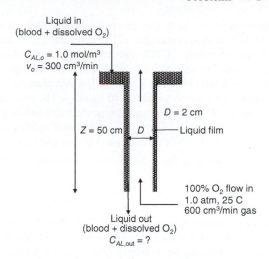

Liquid in
(blood + dissolved O_2)

$C_{AL,o} = 1.0$ mol/m^3
$v_o = 300$ cm^3/min

$D = 2$ cm
$Z = 50$ cm
D
Liquid film

100% O_2 flow in
1.0 atm, 25 C
600 cm^3/min gas

Liquid out
(blood + dissolved O_2)
$C_{AL,\text{out}} = $?

30.20 Wilke and Hougan[28] studied the mass-transfer characteristics of packed beds containing granular solids. In their experimental investigations, hot air was blown through a packed bed of porous cclite pellets saturated with liquid water. The water evaporated under adiabatic conditions, and the rate of water transfer was determined by material balance using humidity measurements. From this data, they calculated the gas–film mass-transfer coefficient at a given flow rate. In one run, the following data were reported:

Gas–film coefficient, $k_G = 4.42 \times 10^{-3}$ $kgmole/m^2 \cdot s \cdot atm$
Effective particle diameter, $d_p = 0.571$ cm
Packed-bed void fraction, $\epsilon = 0.75$
Gas stream mass velocity, $G = 0.816$ $kg/m^2 \cdot s$
Temperature at particle surface, $T = 311$ K
Total system pressure, $P = 9.77 \times 10^4$ Pa

Estimate the gas–film mass-transfer coefficient by two appropriate correlations, one that includes the packed-bed void fraction, and a simpler correlation that does not account for the void fraction. Compare these estimates to the measured gas–film coefficient above.

30.21 The "soil venting" shown in the figure (next page) is used to treat soil contaminated with volatile, toxic liquids. In the present situation, the porous soil particles are saturated with liquid TCE, a common industrial solvent. The contaminated soil is dug up at the waste site and loaded into a rectangular trough. The soil consists of porous mineral particles with an average diameter of 3.0 mm, loosely compacted into a packed bed with a void fraction 0.50. Air is introduced at the bottom of the trough through a distributor and flows upward around the soil particles. Liquid TCE satiating the pores of the soil particle evaporate into the air stream. Consequently, the TCE concentration in the air stream increases. Usually, the rate of TCE evaporation is slow enough so that the liquid TCE within the soil particle is a constant source for mass transfer, at least until 80% of the volatile TCE soaked within the soil is removed. Under these conditions, the transfer of TCE from the soil particle to the air stream is limited by convective mass transfer across the gas film surrounding the soil particles. The mass flow rate of air per unit cross section of the empty bed is 0.10 kg/m²·s. The process is carried out at 293 K. At this temperature, the vapor pressure of TCE is $P_A = 58$ mm Hg. The molecular diffusion coefficient of TCE vapor in air is given in Example 3 of this chapter.

a. What is the gas–film mass-transfer coefficient for TCE vapor in air?

b. At what position in the bed will the TCE vapor in the air stream reach 99.9% of its saturated vapor pressure? In your solution, you may want to consider a material balance on TCE in the gas phase within a differential volume element of the bed. Assume the convective mass-transfer resistances associated with air flowing over the top surface of the bed are negligible and that there is no pressure drop of the gas stream through the bed so that the total system pressure remains constant at 1.0 atm.

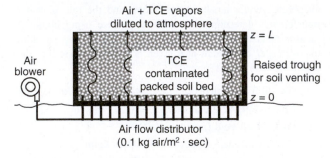

Air + TCE vapors
diluted to atmosphere

z = L

Air blower

TCE contaminated packed soil bed

Raised trough for soil venting

z = 0

Air flow distributor
(0.1 kg air/m² · sec)

30.22 Often aerobatic fermentations give off CO_2 gas because the respiring cells produce CO_2 as glucose nutrient is converted to chemical energy. The conditions used to set k_La for O_2 transfer in the fermenter also establish the same hydrodynamic conditions for CO_2 transfer. Consequently, the transfer rate of CO_2 from the liquid to the aeration gas is also described by this volumetric mass-transfer coefficient, k_La. Consider that the k_La for O_2 in the fermenter is 300 h^{-1}. Scale the k_La for O_2 to k_La for CO_2 transfer using film theory, boundary-layer theory, and penetration theory, and compare the results.

30.23 A stirred-tank fermenter is used to cultivate aerobic microorganisms in aqueous suspension. The aerobic cells require dissolved oxygen for respiration, which is supplied by aerating the liquid medium. An important design parameter for specifying the oxygen mass-transfer rate in the fermenter is the volumetric mass-transfer coefficient, k_La. To promote the gas–liquid mass-transfer process and suspend the microorganisms within the liquid medium, a flat-blade disk turbine impeller of diameter 0.30 m rotates at 240 rev/min within a 1.0-m diameter tank of 2.0 m³ liquid volume. The aeration rate to the fermenter is 1.2 m³ of air per min, and the air bubbles are observed to be coalescing. The aerobic fermentation is carried out at 35 C and

1.0 atm system pressure. You may assume that the liquid medium approximates the properties of water.

a. Determine the volumetric mass-transfer coefficient (k_La) for O_2 in water. Also, provide the aerated power input (P_g) and non-aerated power input (P) to the fermenter.

b. Based on your answer in part (a), determine the maximum oxygen mass-transfer rate (the OTR) to the liquid medium in units of moles O_2 per minute. Assume that the cells immediately consume the dissolved O_2, so that the dissolved oxygen concentration in the liquid medium is essentially zero. In biochemical engineering, this situation is called oxygen mass-transfer limited growth. You may also assume that the interphase mass-transfer process is liquid-phase controlling. The Henry's law constant for the dissolution of O_2 in water can be found in Table 25.1. The mole fraction of O_2 in air is 0.21.

30.24 A process is being developed to produce carbonated beverages. As part of this process, a packed-bed absorption tower will be used to dissolve carbon dioxide, CO_2, gas into water. Pure mountain spring water, containing no dissolved CO_2 enters the top of the column at a molar flow rate of 5.0 kgmole/min. Pure CO_2 gas at 2.0 atm is also fed to the top of the tower at a flow rate of 1.0 kgmole/min. As the liquid flows down the tower, CO_2 gas absorbs into the water, and the dissolved CO_2 concentration increases down the length of the bed. The carbonated water and the unused CO_2 gas exit the bottom of the tower. The absorption process is liquid–film controlling because only pure CO_2 is present in the gas phase. The tower is packed with 1.0 in ceramic rings, and the inner tower diameter is 0.25 m. The temperature is maintained at 20°C. At this temperature, the Henry's law constant for the dissolution of CO_2 gas in water is 25.4 atm·m³/kgmole, the molar density of liquid water is 55.5 kgmole/m³, the mass density of liquid water is 998 kg/m³, and the viscosity of liquid water is 993×10^{-6} kg/m·s.

a. What is the liquid-phase mass-transfer coefficient, k_La, for CO_2 in water flowing through the packed bed?

b. Estimate the depth of packing if the desired concentration of dissolved CO_2 in the outlet liquid is 95% of the saturation value for dissolved CO_2 in water under CO_2 partial pressure of 2.0 atm and 20 C.

30.25 Please refer to Problem 30.22. Now consider that a "packed column" containing 3/8 in rings will be used instead of the wetted-wall column.

a. What is k_La, the volumetric mass-transfer coefficient?

b. What is the concentration of dissolved oxygen in the liquid exiting the bottom of the column, $c_{AL,out}$? As part of this analysis, develop a material balance on a differential volume element of the column along position z.

c. Why does the packed tower perform better than the wetted-wall tower in terms of overall mass transfer?

30.26 A "rotating disk" shown in the figure (at right) is used to plate a thin film of copper onto a silicon wafer by an electroless plating process at 25 C. The disk is 8.0 cm in diameter and rotates at a speed of 2.0 rev/s. The plating bath liquid volume is 500 cm³. In the electroless plating process, copper metal is deposited on a surface without an applied electrical potential. Instead, copper is deposited on a surface by the reduction of an alkaline solution containing copper (II) ions (Cu^{+2}) stabilized by chelation with ethylene-diamine-tetraacetic acid (EDTA), using formaldehyde (CH_2O) dissolved in solution as the reducing agent. The overall reaction stoichiometry is

$$CuY + 2\,CH_2O + 4OH^- \rightarrow Cu^0 + H_2(g) + 2\,HCOO^-$$
$$+ Y^{-2} + 2\,H_2O$$

where Y represents the EDTA chelation agent. In a large excess of chelation agent, formaldehyde, and NaOH, the surface reaction for reduction of Cu^{+2} is first order with respect to the dissolved copper concentration, with surface rate constant of $k_s = 3.2$ cm/s at 25 C. The kinematic viscosity of liquid water is 0.01 cm²/s at 25 C. The diffusion coefficient of Cu^{+2} is 1.20×10^{-5} cm²/s at 25 C and infinite dilution. The density of copper is 8.96 g/cm³.

a. What is the convective mass-transfer coefficient (k_L) for copper ion diffusion across the liquid boundary layer formed by the spinning disk?

b. Develop a model to describe the flux of Cu^{+2} from the bulk solution to the rotating disk surface that accounts for the rate of surface reaction. What is the rate of copper deposition (mol/cm²/s) if the Cu^{+2} concentration is 0.005 gmole/L? Which process is controlling, surface reaction or boundary-layer diffusion?

c. Develop an unsteady-state material balance model to describe the depletion of copper ions in solution with time. How long will it take to deposit a 2.0 μm solid thin film of Cu^0, if the initial Cu^{+2} concentration is 0.005 gmole/L and there is initially no copper deposited on the surface? Clearly state all assumptions.

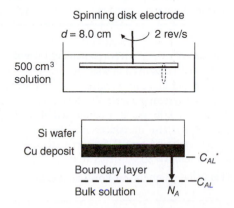

Spinning disk electrode

$d = 8.0$ cm   2 rev/s

500 cm³ solution

Si wafer

Cu deposit — C_{AL}^*

Boundary layer

Bulk solution N_A C_{AL}

30.27 A "bubbleless" shell-and-tube membrane aeration system is used to transfer oxygen gas (O_2) to liquid water. Water containing no dissolved oxygen is added to the tube side at the entrance at a bulk velocity of 5.0 cm/s. The inner diameter of the tube is 1.0 cm, and the tube wall thickness is 2.0 mm (0.20 cm). The tube wall is made of silicone, a polymer that is highly permeable to O_2 gas, but not to water vapor. Pure oxygen gas (100% O_2) is maintained at a constant pressure of 2.00 atm in the annular space surrounding the tube. The O_2 gas partitions into

silicone polymer, and then diffuses through the tube wall to reach the flowing water. As the fluid flows down the length of the tube, the absorption of oxygen will increase the concentration of the dissolved oxygen. Determine the length of tubing necessary for the dissolved oxygen concentration to reach 30% of saturation with respect to the O_2 gas. The process is maintained at 25 C. At 25 C, the Henry's law constant for O_2 gas dissolved in water is $H = 0.78$ atm·m³/gmole, the molecular diffusion coefficient for dissolved O_2 (A) in water (B) is $D_{AB} = 2.1 \times 10^{-5}$ cm²/s, and the kinematic viscosity of water is $v_B = 9.12 \times 10^{-3}$ cm²/s. The solubility constant of O_2 gas in the silicone polymer is $S_m = 0.029$ atm·m³-silicone/gmole, and effective diffusion coefficient of O_2 in the silicone polymer is $D_{Ae} = 5.0 \times 10^{-6}$ cm²/s.

30.28 Consider the concentric tubular mass transfer device shown in the figure below. This device is designed to gradually introduce hydrogen (H_2) gas (species A) into a pure oxygen gas (O_2) inlet stream (species B) delivered to the shell side of the process. Pure H_2 gas enters the tube side of the process. Pure O_2 gas enters the shell side of the process. The tube wall material is only permeable only to H_2, and not to O_2. The system is designed so that the partial pressure of H_2 at the gas interface and the outer surface of the inner tube wall is equal to the total system pressure, i.e. $p_{As} = P$ or $y_{As} = 1.0$, down the entire length of the tube. In the present process, the Reynold's number through the shell side is 10,000, the outer diameter of the inner tube is $D_i = 1.5$ cm, and the inner diameter of the outer tube is $D_o = 2.5$ cm. The tube length is $L = 10$ cm, and the process is carried out at constant 27 C and 1.0 atm total system pressure. For the purposes of this problem, you may assume the process is dilute with respect to H_2 in the O_2 gas stream.

a. What is the velocity of gas on the shell side of the process?

b. What is the Schmidt number is for H_2 diluted in O_2 gas?

c. What is the convective mass transfer coefficient for H_2 in annular gas space for the $H_2 + O_2$ gas mixture on the shell side, k_c?

d. Develop a material balance model, in integrated final algebraic form, to predict the outlet mole fraction of H_2 from the shell side of the process, y_{AL}. At a minimum, your model must include the following terms: D_o, D_i, k_c, L, y_{AL}, y_{As}, y_{Ao}, v_∞.

e. At the conditions of the process, what is the outlet mole fraction of H_2 on the shell side, y_{AL}?

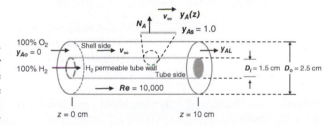

100% O_2 $y_{Ao} = 0$ Shell side v_∞

100% H_2 H_2 permeable tube wall   Tube side

$Re = 10,000$

N_A v_∞ $y_A(z)$ $y_{As} = 1.0$

y_{AL}

$D_i = 1.5$ cm $D_o = 2.5$ cm

$z = 0$ cm $z = 10$ cm

31.1 An ozone (O_3) treatment system is proposed to oxidize organic materials from wastewater. The first step in this process is to dissolve O_3 in the wastewater. The wastewater has been stripped of all dissolved oxygen and ozone-demanding substances, and the pH is sufficiently low that the ozone decomposition will be insignificant. The wastewater is filled to volume of $80 \, m^3$ within a holding pond maintained at 283 K. A compressed ozone/air mixture containing 4.0 mole% ozone (O_3) is fed to the pond 3.20 m below the liquid surface using eight spargers, with each sparger delivering $17.8 \, m^3/h$ of the gas mixture. The O_3 absorbs into the wastewater by a liquid film controlling interphase mass-transfer process. At 283 K, the Henry's law constant for ozone dissolved in water is $67.7 \, m^3 \cdot atm/kgmole$, and the Henry's law for O_2 dissolved in water is $588 \, m^3 \cdot atm/kgmole$; the diffusion coefficient of ozone dissolved in water is $1.7 \times 10^{-5} \, cm^2/s$, and the diffusion coefficient of O_2 dissolved in water is $2.1 \times 10^{-5} \, cm^2/s$. For purposes of this calculation, it may be assumed that the rate of O_3 transferred to the liquid is small relative to molar flow rate of O_3 in the aeration gas.

a. Estimate the volumetric mass-transfer coefficient, $k_L(A/V)$, using the Eckenfelder plot and Penetration theory to scale $k_L(A/V)$ for O_2 interphase mass transfer to O_3 interphase mass transfer.

b. Estimate the time required for the dissolved ozone concentration to reach 25% of its saturation value. What is the dissolved oxygen (O_2) concentration at this time?

31.2 An aeration basin is filled to a volume of $425 \, m^3$ with wastewater containing dissolved hydrogen sulfide (H_2S) of initial concentration $0.300 \, gmole/m^3$. The aeration basin is equipped with 10 gas spargers, with each injection nozzle located 3.2 m below the liquid surface. It is desired to remove the H_2S from the wastewater by a liquid-phase controlling interphase mass-transfer stripping process, where the dissolved H_2S concentration will be reduced from 0.300 to $0.050 \, gmole/m^3$ within a time of 150 min using air containing no H_2S as the aeration gas. The process temperature is maintained at 283 K. At 283 K, the Henry's law constant for H_2S dissolved in water is $H = 367 \, atm$ ($p_A = H \cdot x_A^*$), the diffusion coefficient of H_2S in water is $1.4 \times 10^{-5} \, cm^2/s$, and

a. Based on the desired process conditions, estimate the required $k_L(A/V)$ for the H_2S interphase mass-transfer process, and then scale this mass-transfer coefficient to $k_L(A/V)$ for O_2 transfer using Penetration theory.

b. Using the Eckenfelder plot, estimate the required aeration rate into each gas sparger for the required $k_L(A/V)$ determined in part (a).

31.3 A waste gas of total inlet flow rate $10.0 \, lbmole/ft^2 \cdot h$ containing 0.050 moles solute A/mole carrier gas is treated countercurrently in packed tower with a nonvolatile, solute-free absorbing oil in order to scrub out 80% of solute A from the gas. It is desired to operate the process at a solvent flow rate 40% above the minimum solvent flow rate. Equilibrium *mole ratio* distribution data at the temperature and pressure of operation are provided below.

a. Provide a plot Y_A vs. X_A for equilibrium line and the operating lines at the minimum solvent flow rate, and 1.4 times the minimum solvent flow rate.

b. Determine the minimum solvent flow rate ($L_{s,min}$) and the maximum possible mole ratio in solute A in the outlet liquid.

c. Repeat parts (a)-(b) for *cocurrent* flow.

Equilibrium distribution data for solute A in the solvent, as mole ratios:

X_A	0.02	0.04	0.06	0.08	0.10	0.12
Y_A	0.0075	0.0130	0.0180	0.0230	0.027	0.0305
X_A	0.140	0.160	0.180	0.200	0.220	0.240
Y_A	0.0330	0.0360	0.0385	0.0405	0.0430	0.0445

31.4 An aqueous waste stream containing 1.0 mole% of hydrogen sulfide (H_2S) dissolved in water will be treated by stripping the liquid with air in a countercurrent packed tower at 20 C and 12.50 atm total system pressure. The total flow rate of this liquid feed stream is 100 lbmole/h. Compressed air at 12.5 atm containing no H_2S is also introduced into the tower. At 20 C, the Henry's law constant for the H_2S-water system is $H = 515 \, atm$. It is desired to remove 80% of H_2S from the liquid.

a. What is the minimum air flow rate into the tower, $AG_{s,min}$?

x_A	0.0000	0.0100	0.0200	0.0300	0.0400	0.0500	0.0600	0.0700	0.0800
y_A	0.0000	0.0265	0.0560	0.0885	0.1240	0.1625	0.2040	0.2485	0.2960

diffusion coefficient of O_2 dissolved in water is $2.1 \times 10^{-5} \, cm^2/s$.

b. What is the maximum possible mole fraction of H_2S in the outlet gas stream as the minimum air flow rate?

31.5 It is desired to remove 83.3% of dissolved NH_3 from an aqueous stream containing 10.7 mole% NH_3 in water using a packed tower in countercurrent flow at 1.0 atm and 30 C. The flow rate of this entering liquid stream is 2.0 kgmole/s. The inlet air stream contains no NH_3. Equilibrium distribution data for NH_3 in water at 30 C is provided in the table below.

a. Plot out the equilibrium line in mole ratio coordinates, Y_A vs. X_A. What is the maximum possible mole fraction and mole ratio of NH_3 in the outlet air stream, and the minimum possible air flow rate into the tower?

b. What is the operating air molar flow rate and mole faction of NH_3 in the outlet gas at 1.5 times the minimum molar flow rate of air? Plot out the operating line in mole ratio coordinates. Is the operating line above or below the equilibrium line?

Equilibrium distribution data for NH_3-water at 30 C ($A = NH_3$):

P_A (atm)	0.020	0.025	0.032	0.039	0.053
x_A	0.017	0.021	0.026	0.031	0.041
P_A (atm)	0.067	0.105	0.145	0.236	0.342
x_A	0.050	0.074	0.096	0.137	0.175

31.6 A packed tower is used to remove component A from a gas stream mixture containing 12.0 mole% A and 88.0 mole% inerts. The gas mixture is fed to the bottom of the tower, and the desired superficial molar velocity of the gas stream into the tower (G_1) is 5.0 kgmole/m^2·h. Solvent containing no solute A is fed into the top of the tower. The desired composition of solute A exiting with the liquid is 3.0 mole%, and the desired composition of solute A exiting with the gas (y_{A_2}) is 1.0 mole%. Through independent studies, overall average mass-transfer coefficient based on the gas phase for this tower at the desired superficial molar velocities was determined to be $K_G a = 1.96$ kgmole/m^3·h·atm. The process is carried out at 20 C and 15 psia total system pressure. Equilibrium distribution data for solute A in the solvent at 20 C and 15 psia total system pressure is provided in the table below.

a. Specify the molar flow rate and mole fraction composition of all terminal streams in the process. What is the superficial molar velocity of solvent into the tower, L_2?

b. Plot out the equilibrium line and operating line in mole ratio coordinates.

c. What is the packed height of the tower required to accomplish the separation?

Equilibrium distribution data, in mole fraction coordinates at 20 C and 15 psia total system pressure:

31.7 A gas stream of flow rate 10.0 lbmole/ft^2·h contains 6.0% sulfur dioxide (SO_2) by volume in air. It is desired to reduce the SO_2 level in the treated gas to no greater than 0.5% in a countercurrent packed tower operating at 30 C and 1.0 atm total system pressure, using water containing no dissolved SO_2 as the absorption solvent. The desired solvent flow rate is 2.0 times the minimum solvent flow rate. At these flow conditions, the film mass-transfer coefficients and $k_x a = 250$ lbmole/ft^3·h, and $k_y a$ = 15 lbmole/ft^3·h, for the liquid and gas films, respectively. Equilibrium distribution data for SO_2 in water at 30 C is provided in the table below.

a. Specify the molar flow rate and mole fraction composition of all terminal process streams for operation at 2.0 times the minimum solvent flow rate.

b. Determine the height of packing required to accomplish the separation or operation at 2.0 times the minimum solvent flow rate.

Equilibrium distribution data for SO_2 in water at 30 C ($A = SO_2$):

P_A (mm Hg)	79	52	36	19.7	11.8	8.1	4.7	1.7
kg SO_2/100 kg H_2O	1.0	0.7	0.5	0.3	0.2	0.15	0.1	0.05

31.8 A natural gas stream with a total volumetric flow rate of 880 *standard* cubic meters (SCM) per hour (*std* m^3/h), temperature of 40 C, and total system pressure of 405 kPa is contaminated with 1.0 mole% hydrogen sulfide (H_2S). A packed-bed gas absorption tower of 2.0 m diameter is used to lower the H_2S concentration in the natural gas down to 0.050 mole% so that the H_2S will not poison a steam-reforming catalyst used to convert the natural gas to hydrogen gas. Since H_2S is not very soluble in water, the chelating agent monoethanolamine (MEA, molecular weight 61 g/gmole) is added to water to increase the equilibrium solubility of the H_2S in aqueous solvent systems. In the present problem, an aqueous, 15.3 wt% MEA solvent containing no H_2S at a total flow rate of 50 kgmole/h is added to the top of the tower to selectively remove the H_2S from the natural gas stream. The tower packing height is 10.0 m.

a. From a process material balance, determine mole fraction composition of H_2S in the liquid scrubbing solvent exiting the tower.

b. Using the equilibrium distribution data in the table provided below, provide a plot of y_A vs. x_A for the process, showing the equilibrium and operating lines.

c. Determine the minimum solvent flow rate $L_{S,min}$.

d. Back out the required overall mass-transfer coefficient of the operating tower, $K_G a$, given that the packing height (z) is 10.0 m.

Equilibrium distribution data at 40 C for 15.3 wt% MEA in water ($A = H_2S$)[*]:

P_A (mm Hg)	0.96	3.0	9.1	43.1	59.7	106	143
mole H_2S/mole MEA	0.125	0.208	0.362	0.642	0.729	0.814	0.842

[*]J. H. Jones, H. R. Froning, and E. E. Claytor, *J. Chem. Eng. Data*, **4**, 85 (1959).

SS 31.9 Exit gas from an amination reactor contains 10 mole% ammonia (NH_3) vapor in a nitrogen (N_2) carrier gas. This gas mixture is fed into the bottom of a packed tower at a molar flow rate of 2.0 kgmole/s. The NH_3 will be absorbed into water at neutral pH within a packed tower operating in countercurrent flow. The gas exiting the top of the tower contains 2.0 mole% NH_3. Water containing 1.0 mole% of residual dissolved ammonia enters the top of the tower at a molar flow rate of 3.0 kgmole/s. The tower is packed with 1.0-inch ceramic Intalox saddles, and operates at 20 C and 2.5 atm total system pressure. At these conditions, the mass density of the feed gas is 2.8 kg/m^3, the density of the liquid is 1000 kg/m^3, and the viscosity of the liquid is 1.0 cP (0.001 kg/m·s). Equilibrium distribution data for NH_3 in water at 30 C is provided in the following table.

a. Specify the molar flow rate and mole fraction composition of all terminal streams for the process. Will the tower operate as intended? Investigate this issue by plotting the operating line relative to the equilibrium line in mole fraction and mole ratio coordinates.

Equilibrium distribution data for HCl-water at 1.0 atm total system pressure and 25 C is provided in table below.

a. What is the operating solvent flow rate fed to the packed tower?

b. Provide a plot of y_A vs. x_A for the equilibrium line and the operating line. What is the HCl mole fraction in the outlet liquid at the minimum solvent flow rate?

c. What are the overall mass-transfer driving forces $(y_A - y_A^*)$ at the top and bottom of the tower, and the log-mean driving force at the conditions of operation?

d. A tower shell of 0.50 m diameter is available for use. Ceramic Intalox saddle packing of 1.5 in nominal size is also available. At the conditions of operation given above, will the tower flood? What is the calculated pressure drop per unit depth of packing? The density of the outlet liquid is 1033 kg/m^3, and the viscosity of the outlet liquid is 1.2 cP, and the molecular weight of HCl is 36.5 g/gmole.

Equilibrium distribution data for HCl-water at 20 C and 1.0 atm ($A = HCl$):

x_A	0.210	0.243	0.287	0.330	0.353	0.375	0.400	0.425
y_A	0.0023	0.0095	0.0215	0.0523	0.0852	0.135	0.203	0.322

b. What is the minimum tower diameter at the "flooding" condition, and the tower diameter at 50% of the flooding condition?

c. What are the overall mass-transfer driving forces $(y_A - y_A^*)$ at the top and bottom of the tower, and the log-mean driving force?

Equilibrium distribution data for NH_3-water at 30 C ($A = NH_3$):

P_A (atm)	0.020	0.025	0.032	0.039	0.053
x_A	0.017	0.021	0.026	0.031	0.041
P_A (atm)	0.067	0.105	0.145	0.236	0.342
x_A	0.050	0.074	0.096	0.137	0.175

31.10 The cooled exhaust gas from a reactor that makes silicon by the chemical vapor deposition of trichlorosilane contains 8.0 mole% anhydrous HCl vapor and 92.0 mole% hydrogen (H_2) gas at 25 C. The total gas flow rate is 1.25 kgmole/h. It is desired to scrub out the HCl vapor from this gas stream by a gas absorption process using pure water as the inlet solvent within a countercurrent packed tower. The desired outlet concentration of HCl in the gas is only 0.10 mole%. Although HCl is highly soluble in water, the mole fraction of HCl in the exiting liquid stream is kept at 2.0 mole% to avoid the handling of a highly corrosive liquid. At this composition, the liquid flow rate is comfortably above the minimum solvent flow rate. The tower is maintained at 1.0 atm and 25 C.

31.11 Wastewater contaminated with 1,2,2-trichloroethane ($C_2H_3Cl_3$, molecular weight 133.5 g/gmole) will be treated in a countercurrent packed tower using contaminant-free air as the stripping gas. In the present process, 100 kgmole/h of this liquid feed stream will be treated at 20 C and 1.65 atm total system pressure. For this liquid stream, the inlet concentration of 1,2,2-trichloroethane is $x_{A_2} = 2.0 \times 10^{-3}$, and the desired outlet concentration is $x_{A_1} = 2.0 \times 10^{-4}$. In this concentration range, the equilibrium distribution of 1,2,2-trichloroethane vapor in water is described by Henry's law of the form $p_A = H \cdot x_A^*$, with $H = 41.25$ atm at 20 C. The tower is packed with 1.5 in plastic Pall rings.

a. From the standpoint of interphase mass-transfer, what is the *minimum* molar gas flow rate *into* the tower?

b. Now consider the *outlet gas* flow rate is 12.0 kgmole/h, which is above the minimum gas flow rate needed for mass transfer. At this flow rate, the composition of the outlet gas is 1.5 mole% 1,2,2-trichloroethane and 98.5 mole% air. What is the smallest possible diameter of the tower, calculated at the superficial gas-mass velocity where the tower floods, G_f? Assume the properties of liquid water for the wastewater at 20 C, with liquid density of 1000 kg/m^3, and liquid viscosity of 0.0010 kg/m·s.

31.12 A process waste gas stream containing 5.0 mole% of benzene (solute A) enters the bottom of a packed tower. The tower is packed with 0.5-inch ceramic Raschig rings, and the diameter of the tower is 2.0 ft. The inlet gas flow rate is 32.0

lbmole/h (1006 lb/h). The waste gas exiting the tower contains 1.0 mole% of benzene, and its total molar flow rate is 30.7 lbmole/h. A nonvolatile scrubbing oil (molecular weight of 250 g/gmole), which contains no benzene, is added to the top of the tower at the rate of 18 lbmole/h to absorb solute A from the gas phase. At these conditions, the overall mass-transfer coefficient based on the gas phase, $K_G a$, is 2.15 lbmole/ft$^3 \cdot$h$\cdot$atm. The total system pressure is 1.40 atm, and the temperature is isothermal at 27 C, and the equilibrium distribution of solute A between the contacting gas and liquid phases is described by Henry's law, where $P_A = H\, x_A$, where $H = 0.14$ atm, based on the definition $p_A = H \cdot x_A^*$. Additional data: gas density, $\rho_G = 0.11$ lb$_m$/ft^3; liquid density, $\rho_L = 55$ lb$_m$/ft^3; liquid viscosity, $\mu_L = 2.0$ cP.

a. What is the mole fraction of solute A in the liquid exiting the tower?

b. What is the packed height of the tower?

c. At the conditions of operation, the gas-phase pressure drop in the tower is 300 Pa/m. If the packing is changed to 0.25-inch ceramic Raschig rings, what is the new pressure drop?

SS **31.13** A pilot-scale tower packed to a height of 6.0 ft with 5/8-inch ceramic Raschig rings is used to study the absorption of carbon disulfide (CS$_2$, solute A) from nitrogen gas (N$_2$) into a liquid absorbent solution at 24 C. The diameter of the tower is 0.50 ft. An inlet gas stream of total molar flow rate 2.0 lbmole/h and composition of 3.0 mole% CS$_2$ enters the bottom of tower. The outlet gas composition of CS$_2$ is 0.50 mole%. Pure solvent (molecular weight of 180 g/mole) of total molar flow rate 1.0 lbmole/h enters the top of the tower. The composition of CS$_2$ in the outlet liquid is 4.8 mole%. Over the concentration range of operation, Henry's law is valid with $P_A = Hx_A^*$, where $H = 0.46$ atm at 24 C. The total system pressure in the tower is 14.7 psig (gage pressure). Additional data: gas density, $\rho_G = 0.16$ lb$_m$/ft^3; liquid density, $\rho_L = 75$ lb$_m$/ft^3; liquid viscosity, $\mu_L = 2.0$ cP.

a. Determine the minimum solvent flow rate $AL_{s,min}$.

b. Based on the conditions of operation, estimate the $K_y a$ required for the packed-tower height of 6.0 ft.

c. At the conditions of operation, it can be calculated that the tower operates at 26% of the flooding velocity ($G' = 0.26\, G_f'$). Recommend a new size of ceramic Raschig ring packing, which will bring the gas velocity within the range of 40–45% of the flooding velocity at the fixed tower diameter of 0.5 ft.

31.14 A packed-bed liquid stripping process will be designed to remove benzene from contaminated wastewater. A liquid wastewater stream containing dissolved benzene of concentration 693 mg benzene/L (mole fraction 1.6×10^{-4}) is sent to the top of a packed tower at a flow rate of 100 lbmole/ft$^2 \cdot$h. Air containing no benzene vapor is sent to the bottom of the tower. The tower is maintained at 1.20 atm total system pressure and 20 C. It is desired to reduce the mole fraction of dissolved benzene in the water to only 86 mg/L (mole fraction 2.0×10^{-5}),

and to increase the composition of benzene vapor in the air stream exiting the top of the tower to a mole fraction of 0.01 (1.0 mole% benzene vapor in air). At these flow conditions, the film mass-transfer coefficients for the packing are $k_L a = 17.4$ h^{-1} and $k_G a = 7.4$ lbmole/ft$^3 \cdot$h$\cdot$atm, for the liquid and gas films, respectively. At 20 C, the equilibrium distribution curve for benzene-water air system is linear with Henry's law constant $H = 150$ atm based on the definition $p_A = H \cdot x_A^*$. At 20 C, water is nonvolatile relative to benzene, the maximum soluble concentration of volatile benzene in dissolved in liquid water is 780 mg/L, and the density of liquid water is 62.4 lb$_m$/ft^3. The molecular weight of benzene is 78 g/gmole. For a liquid stripping process, by convention the interphase mass-transfer process is based on the overall liquid-phase mass-transfer driving force.

a. Plot out the equilibrium and operating lines for the process in mole fraction (y_A vs. x_A) coordinates. Do the desired terminal stream compositions allow for mass transfer to occur between phases? Explain why.

b. Determine the overall mass-transfer driving force ($x_A^* - x_A$) at the top and bottom of the tower, and the log-mean driving force. Indicate these driving forces on the y_A vs. x_A plot from part (a).

c. Estimate the height of packing required to achieve the separation.

31.15 A packed tower, 2.0 ft in diameter and 4.0 ft in packed height, is used to produce a solution of dissolved oxygen in water for a biochemical process. The tower will be pressured to 5.0 atm with pure oxygen (O$_2$) gas. There is no gas outlet, and pure oxygen gas will enter the tower only to maintain the gas-phase pressure at 5.0 atm. Pure water containing no dissolved O$_2$ enters the top of the column and flows down the packing at a rate of 200 lb$_m$/min. The process temperature is 25 C, and at this temperature the equilibrium distribution of oxygen between the gas and liquid phase follows Henry's law of the form $p_A = H \cdot x_A^*$ with $H = 4.38 \times 10^4$ atm. The vaporization of water can be ignored so that no water vapor exists in the gas phase. At the column flow conditions, the liquid film capacity coefficient is $k_x a = 194$ lbmole/ft$^3 \cdot$h. Although this is a gas absorption process, since 100% O$_2$ is present in the gas phase, all mass transfer will occur within the liquid film, and so it will be appropriate to base the calculations on the liquid-phase mole fraction, x_A.

a. Provide a plot of the operating and equilibrium lines in mole fraction coordinates.

b. What is the mole fraction of dissolved O$_2$ in the outlet liquid for a packing height of 4.0 ft?

31.16 Beaver Brewing Corporation has received a contract to produce carbonated beverages. A packed-bed absorption tower will be used for dissolving carbon dioxide (CO$_2$) gas into water. In the present process, pure CO$_2$ gas at 2.0 atm constant total system pressure is fed into the tower at a flow rate of 1.0 kgmole/min. Pure mountain spring water containing no dissolved CO_2 enters the top of the tower at a flow rate of 4.0 kgmole/min. The desired outlet mole fraction of dissolved CO_2 is 0.1 mole% dissolved in the liquid, which is below the concentration in equilibrium with the CO_2 gas. The tower is packed with 1.0-inch ceramic Intalox saddles. The temperature is maintained at 20 C. At 20 C, the Henry's law constant for the dissolution of CO_2 gas in water is 25.4 atm · m³/kgmole, the molar concentration of liquid water is 55.5 kgmole/m³ at 20 C, the mass density of liquid water is 998.2 kg/m³, the viscosity of liquid water is 993×10^{-6} kg/m · s at 20 C. Although this is a gas absorption process, since 100% CO_2 is present in the gas phase, all mass transfer will occur within the liquid film, and so it will be appropriate to base the calculations on the liquid-phase mole fraction, x_A.

a. What is the required tower diameter if the pressure drop per unit volume of packing can be no greater than 200 Pa/m?

b. Based on the result for part (a), what is the liquid film mass-transfer coefficient, $k_L a$, as estimated by the Sherwood and Holloway correlation provided in Chapter 30?

c. Based on the result for part (b), what is the *volume of packing* required to perform the absorption process?

31.17 A counter-current packed tower will be used to remove H_2S vapor from a N_2-rich gas stream. In the present process, the inlet gas stream entering the bottom of the tower contains 5.0 mole% H_2S at a superficial molar flowrate of 10 kgmole/m²/h, and the processed gas stream exiting the top of the tower must contain no more than 1.0 mole% H_2S. Pure solvent containing no H_2S enters the top of the tower. The total molar concentration of

the solvent (C_L) is 50 kgmole/m³. The equilibrium distribution of H_2S between the gas and liquid at process temperature of 40 and 100 C, provided by Jones et al.[*], is provided in the table. The total system pressure is 1.0 atm.

a. What is the superficial molar flowrate of the gas stream exiting the top of the tower?

b. What is the minimum solvent flowrate for the process? Compare results at 40 and 100 C and comment on the differences.

31.18 A countercurrent gas absorption tower packed with 1.5 in ceramic Intalox saddles is used to process 360 kgmole/h of an inlet gas stream containing 10 mole% sulfur dioxide (SO_2) in nitrogen (N_2) gas. The process is temperature is 32 C, and the total system pressure is 1.5 atm. Since SO_2 is not very soluble in water itself, 2.0 M sodium hydroxide was added to the scrubbing solution. The mass flow rate of this solvent containing dissolved SO_2 species exiting the bottom of the tower is 18,000 kg/h. The mass density of this solution is 1090 kg/m³, and its viscosity is 1.2 cP (0.0012 kg/m/s). The molecular weight of SO_2 is 64 g/gmole, and the molecular weight of N_2 is 28 g/gmole. The density of the inlet gas is 1.9 kg/m³.

a. What is the required tower diameter (D) if the process is to operate at 60% of the flooding velocity?

b. Based on your analysis in part (a) above, for the designed tower diameter, if the required liquid flowrate into the top of the tower is increased by a factor 3 to 54,000 kg/h, with all other parameters being unchanged, will the tower "flood"? Provide calculations to support your answer.

40 C		100 C	
P_A (mm Hg)	mol H_2S/ mol MEA	P_A (mm Hg)	mol H_2S/ mol MEA
0.00	0.000	1.00	0.029
0.96	0.125	3.00	0.050
3.00	0.208	5.00	0.065
9.10	0.362	10.00	0.091
43.10	0.643	30.00	0.160
59.70	0.729	50.00	0.203
106.00	0.814	70.00	0.238
		100.00	0.279

[*]J. H. Jones, H. R. Froning, and E. E. Claytor, *J. Chem. Eng. Data*, 4, 85 (1959).

Transformations of the Operators ∇ and ∇^2 to Cylindrical Coordinates

THE OPERATOR ∇ IN CYLINDRICAL COORDINATES

In Cartesian coordinates, ∇ is written as

$$\nabla = \mathbf{e}_x \frac{\partial}{\partial x} + \mathbf{e}_y \frac{\partial}{\partial y} + \mathbf{e}_z \frac{\partial}{\partial z} \tag{A-1}$$

When transforming this operator into cylindrical coordinates, both the unit vectors and the partial derivatives must be transformed.

A cylindrical coordinate system and a Cartesian coordinate system are shown in Figure A.1. The following relations are observed to exist between the Cartesian and cylindrical coordinates:

$$z = z, \quad x^2 + y^2 = r^2, \quad \tan \theta = \frac{y}{x} \tag{A-2}$$

Thus,

$$\left(\frac{\partial}{\partial z}\right)_{\text{cyl}} = \left(\frac{\partial}{\partial z}\right)_{\text{cart}} \tag{A-3}$$

whereas from the chain rule

$$\left(\frac{\partial}{\partial x}\right) = \frac{\partial}{\partial r}\frac{\partial r}{\partial x} + \frac{\partial}{\partial \theta}\frac{\partial \theta}{\partial x}$$

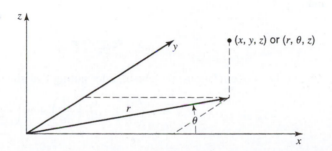

Figure A.1 Cylindrical and Cartesian coordinates.

As

$$\frac{\partial r}{\partial x} = \frac{x}{r} = \cos \theta$$

$$\frac{\partial \theta}{\partial x} = -\frac{y}{x^2 \sec^2 \theta} = -\frac{y}{r^2} = -\frac{\sin \theta}{r}$$

thus

$$\left(\frac{\partial}{\partial x}\right) = \cos \theta \left(\frac{\partial}{\partial r}\right) - \frac{\sin \theta}{r} \left(\frac{\partial}{\partial \theta}\right) \tag{A-4}$$

In a similar manner,

$$\frac{\partial}{\partial y} = \frac{\partial}{\partial r} \frac{\partial r}{\partial y} + \frac{\partial}{\partial \theta} \frac{\partial \theta}{\partial y}$$

where

$$\frac{\partial r}{\partial y} = \frac{y}{r} = \sin \theta \quad \text{and} \quad \frac{\partial \theta}{\partial y} = \frac{1}{x \sec^2 \theta} = \frac{\cos \theta}{r}$$

Thus, $(\partial/\partial y)$ becomes

$$\left(\frac{\partial}{\partial y}\right) = \sin \theta \left(\frac{\partial}{\partial r}\right) + \frac{\cos \theta}{r} \left(\frac{\partial}{\partial \theta}\right) \tag{A-5}$$

The unit vectors must also be transformed. Resolving the unit vectors into their x-, y-, and z-direction components, we obtain

$$\mathbf{e}_z = \mathbf{e}_z \tag{A-6}$$

$$\mathbf{e}_x = \mathbf{e}_r \cos \theta - \mathbf{e}_\theta \sin \theta \tag{A-7}$$

$$\mathbf{e}_y = \mathbf{e}_r \sin \theta + \mathbf{e}_\theta \cos \theta \tag{A-8}$$

Substituting the above relations into equation (A-1), we obtain

$$\mathbf{e}_x \frac{\partial}{\partial x} = \mathbf{e}_r \cos^2 \theta \frac{\partial}{\partial r} - \mathbf{e}_r \frac{\sin \theta \cos \theta}{r} \frac{\partial}{\partial \theta} - \mathbf{e}_\theta \sin \theta \cos \theta \frac{\partial}{\partial r} + \mathbf{e}_\theta \frac{\sin^2 \theta}{r} \frac{\partial}{\partial \theta}$$

$$\mathbf{e}_y \frac{\partial}{\partial y} = \mathbf{e}_r \sin^2 \theta \frac{\partial}{\partial r} + \mathbf{e}_r \frac{\sin \theta \cos \theta}{r} \frac{\partial}{\partial \theta} + \mathbf{e}_\theta \sin \theta \cos \theta \frac{\partial}{\partial r} + \mathbf{e}_\theta \frac{\cos^2 \theta}{r} \frac{\partial}{\partial \theta}$$

and

$$\mathbf{e}_z \frac{\partial}{\partial z} = \mathbf{e}_z \frac{\partial}{\partial z}$$

Adding the above relations, we obtain, after noting that $\sin^2 \theta + \cos^2 \theta = 1$,

$$\nabla = \mathbf{e}_r \left(\frac{\partial}{\partial r}\right) + \frac{\mathbf{e}_\theta}{r} \left(\frac{\partial}{\partial \theta}\right) + \mathbf{e}_z \left(\frac{\partial}{\partial z}\right) \tag{A-9}$$

THE OPERATOR ∇^2 IN CYLINDRICAL COORDINATES

A unit vector may not change magnitude; however, its direction may change. Cartesian unit vectors do not change their absolute directions, but in cylindrical coordinates both $\mathbf{e}_r$ and $\mathbf{e}_\theta$ depend upon the angle θ. As these vectors change direction, they have derivatives with respect to θ. As $\mathbf{e}_r = \mathbf{e}_x \cos\theta + \mathbf{e}_y \sin\theta$ and $\mathbf{e}_\theta = -\mathbf{e}_x \sin\theta + \mathbf{e}_y \cos\theta$, it may be seen that

$$\frac{\partial}{\partial r}\mathbf{e}_r = 0, \quad \frac{\partial}{\partial r}\mathbf{e}_\theta = 0$$

whereas

$$\frac{\partial}{\partial \theta}\mathbf{e}_r = \mathbf{e}_\theta \tag{A-10}$$

and

$$\frac{\partial}{\partial \theta}\mathbf{e}_\theta = -\mathbf{e}_r \tag{A-11}$$

Now the operator $\nabla^2 = \nabla \cdot \nabla$ and thus,

$$\nabla \cdot \nabla = \nabla^2 = \left(\mathbf{e}_r\frac{\partial}{\partial r} + \frac{\mathbf{e}_\theta}{r}\frac{\partial}{\partial \theta} + \mathbf{e}_z\frac{\partial}{\partial z}\right)\cdot\left(\mathbf{e}_r\frac{\partial}{\partial r} + \frac{\mathbf{e}_\theta}{r}\frac{\partial}{\partial \theta} + \mathbf{e}_z\frac{\partial}{\partial z}\right)$$

Performing the indicated operations, we obtain

$$\mathbf{e}_r\frac{\partial}{\partial r} \cdot \nabla = \frac{\partial^2}{\partial \theta}$$

$$\frac{\mathbf{e}_\theta}{r}\frac{\partial}{\partial \theta} \cdot \nabla = \frac{\mathbf{e}_\theta}{r} \cdot \frac{\partial}{\partial \theta}\left(\mathbf{e}_r\frac{\partial}{\partial r}\right) + \frac{\mathbf{e}_\theta}{r} \cdot \frac{\partial}{\partial \theta}\left(\frac{\mathbf{e}_\theta}{r}\frac{\partial}{\partial \theta}\right)$$

or

$$\frac{\mathbf{e}_\theta}{r}\frac{\partial}{\partial \theta} \cdot \nabla = \frac{1}{r}\frac{\partial}{\partial r} + \frac{1}{r^2}\frac{\partial^2}{\partial \theta^2}$$

and

$$\mathbf{e}_z\frac{\partial}{\partial z} \cdot \nabla = \frac{\partial^2}{\partial z^2}$$

Thus, the operator ∇^2 becomes

$$\nabla^2 = \frac{\partial^2}{\partial r^2} + \frac{1}{r}\frac{\partial}{\partial r} + \frac{1}{r^2}\frac{\partial^2}{\partial \theta^2} + \frac{\partial^2}{\partial z^2} \tag{A-12}$$

or

$$\nabla^2 = \frac{1}{r}\frac{\partial}{\partial r}\left(r\frac{\partial}{\partial r}\right) + \frac{1}{r^2}\frac{\partial^2}{\partial \theta^2} + \frac{\partial^2}{\partial z^2} \tag{A-13}$$

Summary of Differential Vector Operations in Various Coordinate Systems

CARTESIAN COORDINATES

Coordinate system

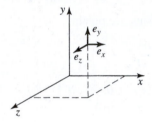

Figure B.1   Unit vectors at the point (x, y, z).

Gradient

$$\nabla P = \frac{\partial P}{\partial x}\mathbf{e}_x + \frac{\partial P}{\partial y}\mathbf{e}_y + \frac{\partial P}{\partial z}\mathbf{e}_z \tag{B-1}$$

Divergence

$$\nabla \cdot \mathbf{v} = \frac{\partial v_x}{\partial x} + \frac{\partial v_y}{\partial y} + \frac{\partial v_z}{\partial z} \tag{B-2}$$

Curl

$$\nabla \times \mathbf{v} = \left\{ \begin{array}{c} \left(\dfrac{\partial v_z}{\partial y} - \dfrac{\partial v_y}{\partial z}\right)\mathbf{e}_x \\[2ex] \left(\dfrac{\partial v_x}{\partial z} - \dfrac{\partial v_z}{\partial x}\right)\mathbf{e}_y \\[2ex] \left(\dfrac{\partial v_y}{\partial x} - \dfrac{\partial v_x}{\partial y}\right)\mathbf{e}_z \end{array} \right\} \tag{B-3}$$

Laplacian of a scalar

$$\nabla^2 T = \frac{\partial^2 T}{\partial x^2} + \frac{\partial^2 T}{\partial y^2} + \frac{\partial^2 T}{\partial z^2} \tag{B-4}$$

CYLINDRICAL COORDINATES

Coordinate system

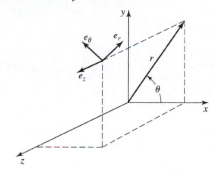

Figure B.2   Unit vectors at the point (r, θ, z).

Gradient

$$\nabla P = \frac{\partial P}{\partial r}\mathbf{e}_r + \frac{1}{r}\frac{\partial P}{\partial \theta}\mathbf{e}_\theta + \frac{\partial P}{\partial z}\mathbf{e}_z \tag{B-5}$$

Divergence

$$\nabla \cdot \mathbf{v} = \frac{1}{r}\frac{\partial}{\partial r}(rv_r) + \frac{1}{r}\frac{\partial v_\theta}{\partial \theta} + \frac{\partial v_z}{\partial z} \tag{B-6}$$

Curl

$$\nabla \times \mathbf{v} = \left\{ \begin{array}{l} \left(\dfrac{1}{r}\dfrac{\partial v_z}{\partial \theta} - \dfrac{\partial v_\theta}{\partial z}\right)\mathbf{e}_r \\[2ex] \left(\dfrac{\partial v_r}{\partial z} - \dfrac{\partial v_z}{\partial r}\right)\mathbf{e}_\theta \\[2ex] \left\{\dfrac{1}{r}\left[\dfrac{\partial}{\partial r}(rv_\theta) - \dfrac{\partial v_r}{\partial \theta}\right]\right\}\mathbf{e}_z \end{array} \right\} \tag{B-7}$$

Laplacian of a scalar

$$\nabla^2 T = \frac{1}{r}\frac{\partial}{\partial r}\left(r\frac{\partial T}{\partial r}\right) + \frac{1}{r^2}\frac{\partial^2 T}{\partial \theta^2} + \frac{\partial^2 T}{\partial z^2} \tag{B-8}$$

SPHERICAL COORDINATES

Coordinate system

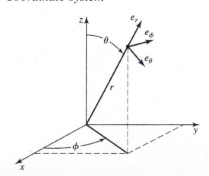

Figure B.3   Unit vectors at the point (r, θ, ϕ).

Gradient

$$\nabla P = \frac{\partial P}{\partial r} \mathbf{e}_r + \frac{1}{r} \frac{\partial P}{\partial \theta} \mathbf{e}_\theta + \frac{1}{r \sin \theta} \frac{\partial P}{\partial \phi} \mathbf{e}_\phi \qquad \text{(B-9)}$$

Divergence

$$\nabla \cdot \mathbf{v} = \frac{1}{r^2} \frac{\partial}{\partial r} (r^2 v_r) + \frac{1}{r \sin \theta} \frac{\partial}{\partial \theta} (v_\theta \sin \theta) + \frac{1}{r \sin \theta} \frac{\partial v_\phi}{\partial \phi} \qquad \text{(B-10)}$$

Curl

$$\nabla \times \mathbf{v} = \left\{ \begin{array}{l} \dfrac{1}{r \sin \theta} \left[\dfrac{\partial}{\partial \theta}(v_\phi \sin \theta) - \dfrac{\partial v_\theta}{\partial \phi} \right] \mathbf{e}_r \\[2ex] \left[\dfrac{1}{r \sin \theta} \dfrac{\partial v_r}{\partial \phi} - \dfrac{1}{r} \dfrac{\partial}{\partial r}(r v_\phi) \right] \mathbf{e}_\theta \\[2ex] \dfrac{1}{r} \left[\dfrac{\partial}{\partial r}(r v_\theta) - \dfrac{\partial v_r}{\partial \theta} \right] \mathbf{e}_\phi \end{array} \right\} \qquad \text{(B-11)}$$

Laplacian of a scalar

$$\nabla^2 T = \frac{1}{r^2} \frac{\partial}{\partial r} \left(r^2 \frac{\partial T}{\partial r} \right) + \frac{1}{r^2 \sin \theta} \frac{\partial}{\partial \theta} \left(\sin \theta \frac{\partial T}{\partial \theta} \right) + \frac{1}{r^2 \sin^2 \theta} \frac{\partial^2 T}{\partial \phi^2} \qquad \text{(B-12)}$$

Appendix C

Symmetry of the Stress Tensor

The shear stress τ_{ij} can be shown to be equal to τ_{ji} by the following simple argument. Consider the element of fluid shown in Figure C.1. The sum of the moments on the element will be related to the angular acceleration by

$$\sum M = I\dot{\omega} \tag{C-1}$$

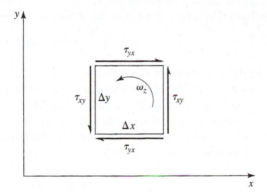

Figure C.1 Free body of element

where I is the mass moment of inertia of the element. Substituting into equation (C-1)

$$-(\tau_{yx}\,\Delta x\Delta z)\Delta y + (\tau_{xy}\,\Delta y\Delta z)\Delta x = \rho\,\Delta x\Delta y\Delta z\frac{(\Delta x^2 + \Delta y^2)}{12}\dot{\omega}_z$$

where the moment of inertia of a rectangular prism has been used for the element.

The volume of the element $\Delta x\Delta y\Delta z$ may be canceled to yield

$$\rho\left(\frac{\Delta x^2 + \Delta y^2}{12}\right)\dot{\omega}_z = \tau_{xy} - \tau_{yx} \tag{C-2}$$

Now the difference in shear stress is seen to depend upon the size of the element. As the element shrinks to a point. Δx and Δy approach zero independently, and we obtain, in the limit,

$$\tau_{xy} = \tau_{yx}$$

or, as this can be done about any axis,

$$\tau_{ij} = \tau_{ji}$$

Another way to look at equation (C-2) is to determine the angular acceleration ω_z as the element shrinks to a point. The angular acceleration at a point must be finite; hence, τ_{yx} and τ_{xy} must be equal.

Appendix D

The Viscous Contribution to the Normal Stress

The normal stress, σ, may be divided into two parts: the pressure contribution, $-P$, and a viscous contribution, σ_v. The viscous contribution to the normal stress is obtained by analogy with Hooke's law for an elastic solid. In Hooke's law for three-dimensional stress, the normal stress, $\sigma_{x,x}$, in the x-direction is related to the strains in the x-, y-, and z-directions by[1]

$$\sigma_{x,x} = 2G\epsilon_x + \frac{2G\eta}{1 - 2\eta}(\epsilon_x + \epsilon_y + \epsilon_z) \tag{D-1}$$

where G is the shear modulus, η is the Poisson's ratio, and ε is the axial strain.

When Newton's viscosity relation was discussed, the shear strain in a solid was seen to be analogous to the rate of shear strain in a fluid. Accordingly, the axial strain in a solid, ϵ_x, is taken to be analogous to the axial strain rate in a fluid, $\partial v_x / \partial x$.

When the velocity derivatives are substituted for the strains in equation (D-1), and the viscosity is used in place of the shear modulus, we obtain

$$(\sigma_{x,x})_{\text{viscous}} = 2\mu \frac{\partial v_x}{\partial x} + \lambda \nabla \cdot \mathbf{v} \tag{D-2}$$

Here the sum of the strain-rate derivatives is observed to be equal to $\nabla \cdot \mathbf{v}$, and the second coefficient has been designated λ and is called the bulk viscosity or second viscosity coefficient. The total normal stress in the x-direction becomes

$$\sigma_{x,x} = -P + 2\mu \frac{\partial v_x}{\partial x} + \lambda \nabla \cdot \mathbf{v} \tag{D-3}$$

If the corresponding normal stress components in the y- and z-directions are added together, we obtain

$$\sigma_{x,x} + \sigma_{y,y} + \sigma_{z,z} = -3P + (2\mu + 3\lambda)\nabla \cdot \mathbf{v}$$

so that the average normal stress $\bar{\sigma}$ is given by

$$\bar{\sigma} = -P + \left(\frac{2\mu + 3\lambda}{3}\right)\nabla \cdot \mathbf{v}$$

---

[1] A more familiar form is

$$\sigma_{x,x} = \frac{E}{(1 + \eta)(1 - 2\eta)}[(1 - \eta)\epsilon_x + \eta(\epsilon_y + \epsilon_z)]$$

The shear modulus G has been replaced with its equivalent, $E/2(1 + \eta)$.

Thus, unless $\lambda = -\dfrac{2}{3}\mu$, the average stress will depend upon the flow properties rather than the fluid property, P. Stokes assumed that $\lambda = -\dfrac{2}{3}\mu$, and experiments have indicated that λ is of the same order of magnitude as μ of air. As $\nabla \cdot \mathbf{v} = 0$ in an incompressible flow, the value of λ is of no concern except for compressible fluids.

The resulting expressions for normal stress in a Newtonian fluid are

$$\sigma_{x,x} = -P + 2\mu \frac{\partial v_x}{\partial x} - \frac{2}{3} \mu \nabla \cdot \mathbf{v} \tag{D-4}$$

$$\sigma_{y,y} = -P + 2\mu \frac{\partial v_y}{\partial y} - \frac{2}{3} \mu \nabla \cdot \mathbf{v} \tag{D-5}$$

$$\sigma_{z,z} = -P + 2\mu \frac{\partial v_z}{\partial z} - \frac{2}{3} \mu \nabla \cdot \mathbf{v} \tag{D-6}$$

Appendix E

The Navier–Stokes Equations for Constant ρ and μ in Cartesian, Cylindrical, and Spherical Coordinates

CARTESIAN COORDINATES

x-direction

$$\rho\left(\frac{\partial v_x}{\partial t} + v_x\frac{\partial v_x}{\partial x} + v_y\frac{\partial v_x}{\partial y} + v_z\frac{\partial v_x}{\partial z}\right) = -\frac{\partial P}{\partial x} + \rho g_x + \mu\left(\frac{\partial^2 v_x}{\partial x^2} + \frac{\partial^2 v_x}{\partial y^2} + \frac{\partial^2 v_x}{\partial z^2}\right) \quad \text{(E-1)}$$

y-direction

$$\rho\left(\frac{\partial v_y}{\partial t} + v_x\frac{\partial v_y}{\partial x} + v_y\frac{\partial v_y}{\partial y} + v_z\frac{\partial v_y}{\partial z}\right) = -\frac{\partial P}{\partial y} + \rho g_y + \mu\left(\frac{\partial^2 v_y}{\partial x^2} + \frac{\partial^2 v_y}{\partial y^2} + \frac{\partial^2 v_y}{\partial z^2}\right) \quad \text{(E-2)}$$

z-direction

$$\rho\left(\frac{\partial v_z}{\partial t} + v_x\frac{\partial v_z}{\partial x} + v_y\frac{\partial v_z}{\partial y} + v_z\frac{\partial v_z}{\partial z}\right) = -\frac{\partial P}{\partial z} + \rho g_z + \mu\left(\frac{\partial^2 v_z}{\partial x^2} + \frac{\partial^2 v_z}{\partial y^2} + \frac{\partial^2 v_z}{\partial z^2}\right) \quad \text{(E-3)}$$

CYLINDRICAL COORDINATES

r-direction

$$\rho\left(\frac{\partial v_r}{\partial t} + v_r\frac{\partial v_r}{\partial r} + \frac{v_\theta}{r}\frac{\partial v_r}{\partial \theta} - \frac{v_\theta^2}{r} + v_z\frac{\partial v_r}{\partial z}\right)$$
$$= -\frac{\partial P}{\partial r} + \rho g_r + \mu\left[\frac{\partial}{\partial r}\left(\frac{1}{r}\frac{\partial}{\partial r}(rv_r)\right) + \frac{1}{r^2}\frac{\partial^2 v_r}{\partial \theta^2} - \frac{2}{r^2}\frac{\partial v_\theta}{\partial \theta} + \frac{\partial^2 v_r}{\partial z^2}\right] \quad \text{(E-4)}$$

θ-direction

$$\rho\left(\frac{\partial v_\theta}{\partial t} + v_r\frac{\partial v_\theta}{\partial r} + \frac{v_\theta}{r}\frac{\partial v_\theta}{\partial \theta} + \frac{v_r v_\theta}{r} + v_z\frac{\partial v_\theta}{\partial z}\right)$$
$$= -\frac{1}{r}\frac{\partial P}{\partial \theta} + \rho g_\theta + \mu\left[\frac{\partial}{\partial r}\left(\frac{1}{r}\frac{\partial}{\partial r}(rv_\theta)\right) + \frac{1}{r^2}\frac{\partial^2 v_\theta}{\partial \theta^2} + \frac{2}{r^2}\frac{\partial v_r}{\partial \theta} + \frac{\partial^2 v_\theta}{\partial z^2}\right] \quad \text{(E-5)}$$

z-direction

$$\rho\left(\frac{\partial v_z}{\partial t} + v_r\frac{\partial v_z}{\partial r} + \frac{v_\theta}{r}\frac{\partial v_z}{\partial \theta} + v_z\frac{\partial v_z}{\partial z}\right)$$

$$= -\frac{\partial P}{\partial z} + \rho g_z + \mu\left[\frac{1}{r}\frac{\partial}{\partial r}\left(r\frac{\partial v_z}{\partial r}\right) + \frac{1}{r^2}\frac{\partial^2 v_z}{\partial \theta^2} + \frac{\partial^2 v_z}{\partial z^2}\right] \qquad \text{(E-6)}$$

SPHERICAL COORDINATES[1]

r-direction

$$\rho\left(\frac{\partial v_r}{\partial t} + v_r\frac{\partial v_r}{\partial r} + \frac{v_\theta}{r}\frac{\partial v_r}{\partial \theta} + \frac{v_\phi}{r\sin\theta}\frac{\partial v_r}{\partial \phi} - \frac{v_\phi^2}{r} - \frac{v_\theta^2}{r}\right)$$

$$= -\frac{\partial P}{\partial r} + \rho g_r + \mu\left[\nabla^2 v_r - \frac{2}{r^2}v_r - \frac{2}{r^2}\frac{\partial v_\theta}{\partial \theta} - \frac{2}{r^2}v_\theta\cot\theta - \frac{2}{r^2\sin\theta}\frac{\partial v_\phi}{\partial \phi}\right] \qquad \text{(E-7)}$$

θ-direction

$$\rho\left[\frac{\partial v_\theta}{\partial t} + v_r\frac{\partial v_\theta}{\partial r} + \frac{v_\theta}{r}\frac{\partial v_\theta}{\partial \theta} + \frac{v_\phi}{r\sin\theta}\frac{\partial v_\theta}{\partial \phi} + \frac{v_r v_\theta}{r} - \frac{v_\phi^2\cot\theta}{r}\right]$$

$$= -\frac{1}{r}\frac{\partial P}{\partial \theta} + \rho g_\theta + \mu\left[\nabla^2 v_\theta + \frac{2}{r^2}\frac{\partial v_r}{\partial \theta} - \frac{v_\theta}{r^2\sin^2\theta} - \frac{2\cos\theta}{r^2\sin^2\theta}\frac{\partial v_\phi}{\partial \phi}\right] \qquad \text{(E-7)}$$

φ-direction

$$\rho\left(\frac{\partial v_\phi}{\partial t} + v_r\frac{\partial v_\phi}{\partial r} + \frac{v_\theta}{r}\frac{\partial v_\phi}{\partial \theta} + \frac{v_\phi}{r\sin\theta}\frac{\partial v_\phi}{\partial \phi} + \frac{v_\phi v_r}{r} + \frac{v_\theta v_\phi}{r}\cot\theta\right)$$

$$= -\frac{1}{r\sin\theta}\frac{\partial P}{\partial \phi} + \rho g_\phi + \mu\left[\nabla^2 v_\phi - \frac{v_\phi}{r^2\sin^2\theta} + \frac{2}{r^2\sin\theta}\frac{\partial v_r}{\partial \phi} + \frac{2\cos\theta}{r^2\sin^2\theta}\frac{\partial v_\theta}{\partial \phi}\right] \qquad \text{(E-7)}$$

---

[1] In the above equations,

$$\nabla^2 = \frac{1}{r^2}\frac{\partial}{\partial r}\left(r^2\frac{\partial}{\partial r}\right) + \frac{1}{r^2\sin\theta}\frac{\partial}{\partial \theta}\left(\sin\theta\frac{\partial}{\partial \theta}\right) + \frac{1}{r^2\sin^2\theta}\frac{\partial^2}{\partial \phi^2}$$

Charts for Solution
of Unsteady
Transport Problems

Table F.9   Symbols for unsteady-state charts

	Parameter symbol	Molecular mass transfer	Heat conduction
Unaccomplished change, a dimensionless ratio	Y	$\dfrac{c_{A_1} - c_A}{c_{A_1} - c_{A_0}}$	$\dfrac{T_\infty - T}{T_\infty - T_0}$
Relative time	X	$\dfrac{D_{AB}t}{x_1^2}$	$\dfrac{\alpha t}{x_1^2}$
Relative position	n	$\dfrac{x}{x_1}$	$\dfrac{x}{x_1}$
Relative resistance	m	$\dfrac{D_{AB}}{k_c x_1}$	$\dfrac{k}{h x_1}$

T = temperature
c_A = concentration of component A
x = distance from center to any point
t = time
k = thermal conductivity
h, k_c = convective-transfer coefficients
α = thermal diffusivity
D_{AB} = mass diffusivity

Subscripts:
 0 = initial condition at time $t = 0$
 1 = boundary
 A = component A
 ∞ = reference condition for temperature

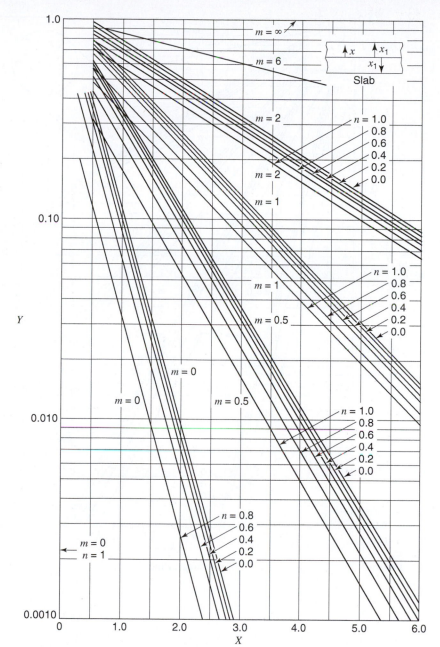

Figure F.1 Unsteady-state transport in a large flat slab.

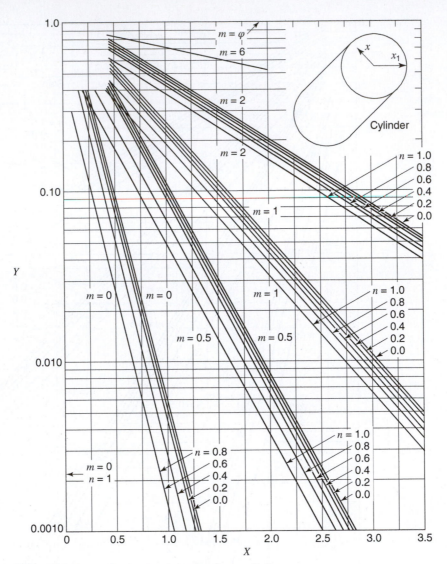

Figure F.2 Unsteady-state transport in a long cylinder.

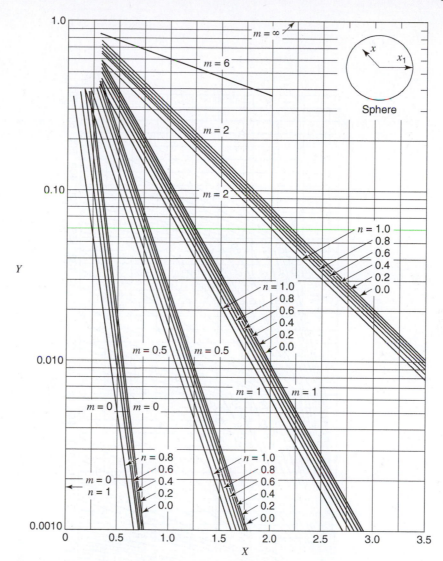

Figure F.3 Unsteady-state transport in a sphere.

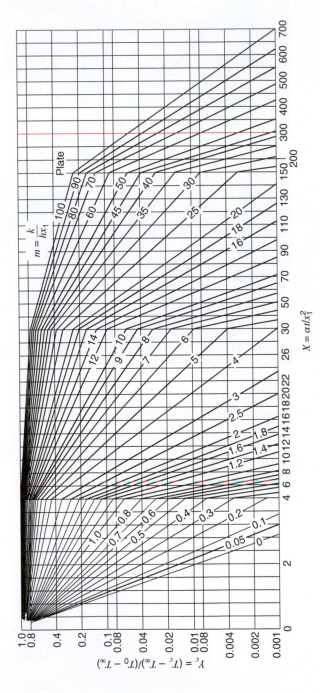

Figure F.4 Center temperature history for an infinite plate.

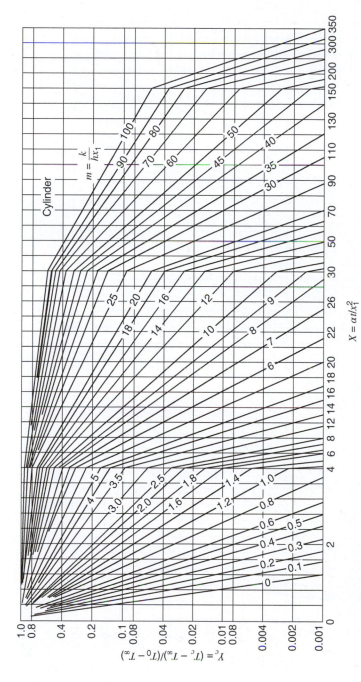

Figure F.5 Center temperature history for an infinite cylinder.

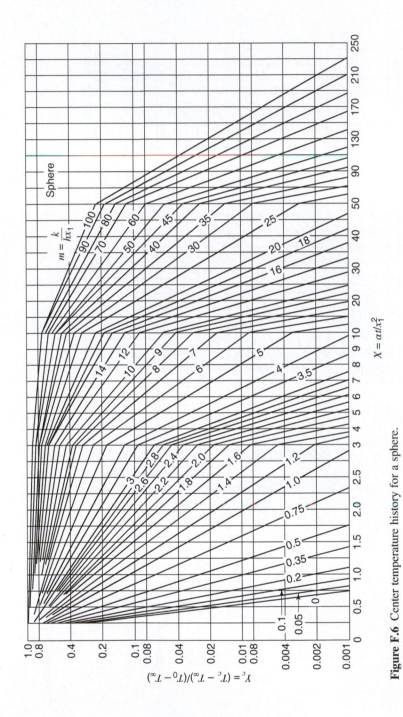

Figure F.6 Center temperature history for a sphere.

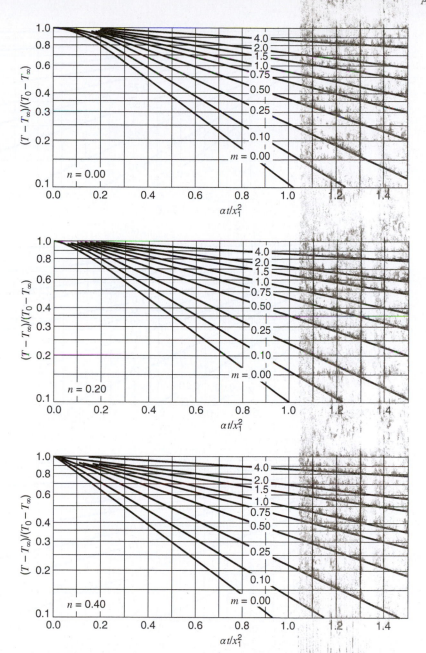

Figure F.7 Charts for solution of unsteady transport problems: flat plate.

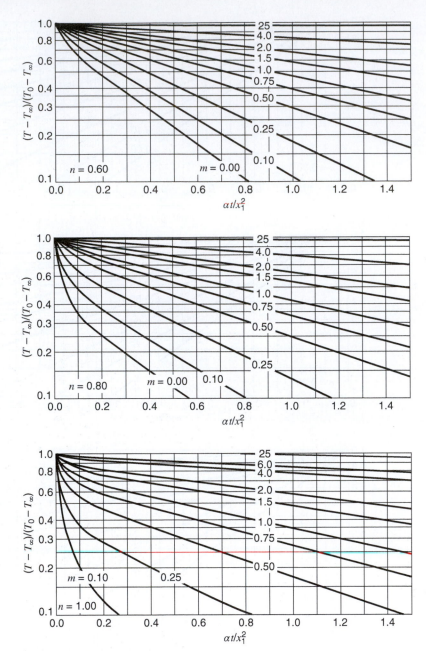

Figure F.7 Continued.

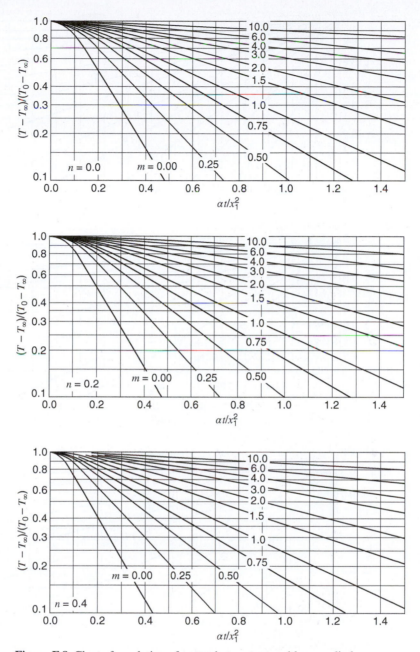

Figure F.8 Charts for solution of unsteady transport problems: cylinder.

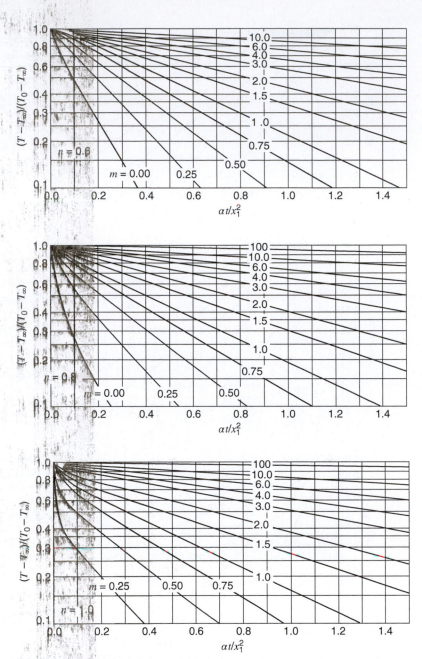

Figure F.8 Continued.

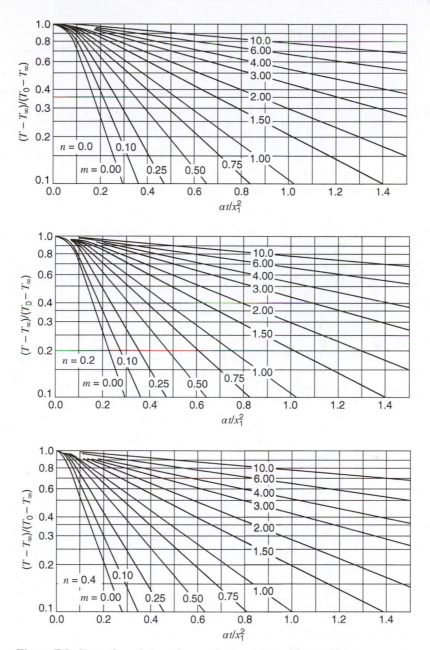

Figure F.9 Charts for solution of unsteady transport problems: sphere.

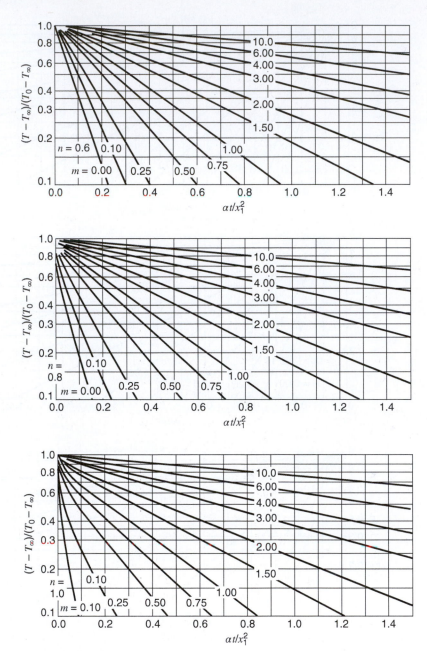

Figure F.9 Continued.

Appendix G

Properties of the Standard Atmosphere[1]

Table G.1 English units

h (ft)	T (F)	a (fps)	P (lb/ft^2)	P (slug/ft^3)	$\mu \times 10^7$ (slug/ft s)
0	59.00	1117	2116.2	0.002378	3.719
1,000	57.44	1113	2040.9	0.002310	3.699
2,000	51.87	1109	1967.7	0.002242	3.679
3,000	48.31	1105	1896.7	0.002177	3.659
4,000	44.74	1102	1827.7	0.002112	3.639
5,000	41.18	1098	1760.8	0.002049	3.618
6,000	37.62	1094	1696.0	0.001988	3.598
7,000	34.05	1090	1633.0	0.001928	3.577
8,000	30.49	1086	1571.9	0.001869	3.557
9,000	26.92	1082	1512.8	0.001812	3.536
10,000	23.36	1078	1455.4	0.001756	3.515
11,000	19.80	1074	1399.8	0.001702	3.495
12,000	16.23	1070	1345.9	0.001649	3.474
13,000	12.67	1066	1293.7	0.001597	3.453
14,000	9.10	1062	1243.2	0.001546	3.432
15,000	5.54	1058	1194.3	0.001497	3.411
16,000	1.98	1054	1147.0	0.001448	3.390
17,000	−1.59	1050	1101.1	0.001401	3.369
18,000	−5.15	1046	1056.9	0.001355	3.347
19,000	−8.72	1041	1014.0	0.001311	3.326
20,000	−12.28	1037	972.6	0.001267	3.305
21,000	−15.84	1033	932.5	0.001225	3.283
22,000	−19.41	1029	893.8	0.001183	3.262
23,000	−22.97	1025	856.4	0.001143	3.240
24,000	−26.54	1021	820.3	0.001104	3.218
25,000	−30.10	1017	785.3	0.001066	3.196
26,000	−33.66	1012	751.7	0.001029	3.174
27,000	−37.23	1008	719.2	0.000993	3.153
28,000	−40.79	1004	687.9	0.000957	3.130
29,000	−44.36	999	657.6	0.000923	3.108

(*Continued*)

[1] Data taken from NACA TN 1428.

Table G.1   Continued

h (ft)	T (F)	a (fps)	P (lb/ft^2)	P (slug/ft^3)	$\mu \times 10^7$ (slug/ft s)
30,000	−47.92	995	628.5	0.000890	3.086
31,000	−51.48	991	600.4	0.000858	3.064
32,000	−55.05	987	573.3	0.000826	3.041
33,000	−58.61	982	547.3	0.000796	3.019
34,000	−62.18	978	522.2	0.000766	2.997
35,000	−65.74	973	498.0	0.000737	2.974
40,000	−67.6	971	391.8	0.0005857	2.961
45,000	−67.6	971	308.0	0.0004605	2.961
50,000	−67.6	971	242.2	0.0003622	2.961
60,000	−67.6	971	150.9	0.0002240	2.961
70,000	−67.6	971	93.5	0.0001389	2.961
80,000	−67.6	971	58.0	0.0000861	2.961
90,000	−67.6	971	36.0	0.0000535	2.961
100,000	−67.6	971	22.4	0.0000331	2.961
150,000	113.5	1174	3.003	0.00000305	4.032
200,000	159.4	1220	0.6645	0.00000062	4.277
250,000	−8.2	1042	0.1139	0.00000015	3.333

Table G.2   SI units—Properties of the standard atmosphere

h (m)	T (K)	a (m/s)	P (Pa)	ρ (kg/m³)	$\mu \times 10^5$ (Pa·s)
0	288.2	340.3	1.0133×10^5	1.225	1.789
500	284.9	338.4	0.95461	1.167	1.774
1,000	281.7	336.4	0.89876	1.111	1.758
1,500	278.4	334.5	0.84560	1.058	1.742
2,000	275.2	332.5	0.79501	1.007	1.726
2,500	271.9	330.6	0.74692	0.9570	1.710
3,000	268.7	328.6	0.70121	0.9093	1.694
3,500	265.4	326.6	0.65780	0.8634	1.678
4,000	262.2	324.6	0.61660	0.8194	1.661
4,500	258.9	322.6	0.57753	0.7770	1.645
5,000	255.7	320.5	0.54048	0.7364	1.628
5,500	252.4	318.5	0.50539	0.6975	1.612
6,000	249.2	316.5	0.47218	0.6601	1.595
6,500	245.9	314.4	0.44075	0.6243	1.578
7,000	242.7	312.3	0.41105	0.5900	1.561
7,500	239.5	310.2	0.38300	0.5572	1.544
8,000	236.2	308.1	0.35652	0.5258	1.527
8,500	233.0	306.0	0.33154	0.4958	1.510
9,000	229.7	303.8	0.30801	0.4671	1.493
9,500	226.5	301.7	0.28585	0.4397	1.475
10,000	223.3	299.5	0.26500	0.4135	1.458
11,000	216.8	295.2	0.22700	0.3648	1.422
12,000	216.7	295.1	0.19399	0.3119	1.422
13,000	216.7	295.1	0.16580	0.2666	1.422
14,000	216.7	295.1	0.14170	0.2279	1.422
15,000	216.7	295.1	0.12112	0.1948	1.422
16,000	216.7	295.1	0.10353	0.1665	1.422
17,000	216.7	295.1	8.8497×10^3	0.1423	1.422
18,000	216.7	295.1	7.5652	0.1217	1.422
19,000	216.7	295.1	6.4675	0.1040	1.422
20,000	216.7	295.1	5.5293	0.08891	1.422
25,000	221.5	298.4	2.5492	0.04008	1.448
30,000	226.5	301.7	1.1970	0.01841	1.475
35,000	236.5	308.3	0.57459	0.008463	1.529
40,000	250.4	317.2	0.28714	0.003996	1.601
45,000	264.2	325.8	0.14910	0.001966	1.671
50,000	270.7	329.8	7.9779×10^1	0.001027	1.704
55,000	265.6	326.7	4.27516	0.0005608	1.678
60,000	255.8	320.6	2.2461	0.0003059	1.629
65,000	239.3	310.1	1.1446	0.0001667	1.543
70,000	219.7	297.1	5.5205×10^0	0.00008754	1.438
75,000	200.2	283.6	2.4904	0.00004335	1.329
80,000	180.7	269.4	1.0366	0.00001999	1.216

Appendix H

Physical Properties of Solids

Material	ρ (lb_m/ft³) (68F)	ρ (kg/m³) (293 K)	c_p (Btu/lb_m F) (293 K)	c_p (J/kg·1K) ×10⁻² (293K)	α (ft²/h) (68F)	α (m²/s)·10⁵ (293k)	k (Btu/h ft F) (68)	k (Btu/h ft F) (212)	k (Btu/h ft F) (572)	k (W/m·K) (293)	k (W/m·K) (373)	k (W/m·K) (573)
Metals												
Aluminum	168.6	2,701.1	0.224	9.383	3.55	9.16	132	133	133	229	229	230
Copper	555	8,890	0.092	3.854	3.98	10.27	223	219	213	386	379	369
Gold	1206	19,320	0.031	1.299	4.52	11.66	169	170	172	293	294	298
Iron	492	7,880	0.122	5.110	0.83	2.14	42.3	39	31.6	73.2	68	54
Lead	708	11,300	0.030	1.257	0.80	2.06	20.3	19.3	17.2	35.1	33.4	29.8
Magnesium	109	1,750	0.248	10.39	3.68	9.50	99.5	96.8	91.4	172	168	158
Nickel	556	8,910	0.111	4.560	0.87	2.24	53.7	47.7	36.9	93.0	82.6	63.9
Platinum	1340	21,500	0.032	1.340	0.09	0.23	40.5	41.9	43.5	70.1	72.5	75.3
Silver	656	10,500	0.057	2.388	6.42	16.57	240	237	209	415	410	362
Tin	450	7,210	0.051	2.136	1.57	4.05	36	34	—	62	59	—
Tungsten	1206	19,320	0.032	1.340	2.44	6.30	94	87	77	160	150	130
Uranium	1167	18,700	0.027	1.131	0.53	1.37	16.9	17.2	19.6	29.3	29.8	33.9
Zinc	446	7,150	0.094	3.937	1.55	4.00	65	63	58	110	110	100
Alloys												
Aluminum 2024 (70% Cu, 30% Ni)	173	2,770	0.230	9.634	1.76	4.54	70.2			122		
Brass	532	8,520	0.091	3.812	1.27	3.28	61.8	73.9	85.3	107	128	148
Constantan (60% Cu, 40% Ni)	557	8,920	0.098	4.105	0.24	0.62	13.1	15.4		22.7	26.7	
Iron, cast	455	7,920	0.100	4.189	0.65	1.68	29.6	26.8		51.2	46.4	
Nichrome V	530	8,490	0.106	4.440	0.12	0.31	7.06	7.99	9.94	12.2	13.8	17.2
Stainless steel	488	7,820	0.110	4.608	0.17	0.44	9.4	10.0	13	16	17.3	23
Steel, mild (1% C)	488	7,820	0.113	4.733	0.45	1.16	24.8	24.8	22.9	42.9	42.9	39.0

(continued)

Material	ρ (lb$_m$/ft^3) (68F)	ρ (kg/m^3) (293 K)	c_p (Btu/lb$_m$F) (293 K)	c_p (J/kg·1K) ×10^{-2} (293K)	α (ft^2/h) (68F)	α (m^2/s)·10^5 (293k)	k (Btu/h ftF) (68)	k F (212)	k (572)	k (W/m·K) (293)	k K (373)	k (573)
Nonmetals												
Asbestos	36	580	0.25	10.5			0.092	0.11	0.125	0.159	0.190	0.21
Brick (fire clay)	144	2,310	0.22	9.22			0.38	0.65		0.66	1.13	
Brick (masonry)	106	1,670	0.20	8.38				0.67			1.16	
Brick (chrome)	188	3,010	0.20	8.38								
Concrete	144	2,310	0.21	8.80			0.70			1.21		
Corkboard	10	160	0.4	17			0.025			0.043		
Diatomaceous earth, powdered	14	220	0.2	8.4			0.03			0.05		
Glass, window	170	2,720	0.2	8.4			0.45			0.78		
Glass, Pyrex	140	2,240	0.2	8.4			0.63	0.67	0.84	1.09	1.16	1.45
Kaolin firebrick	19	300							0.052			0.09
85% Magnesia	17	270					0.038	0.041		0.066	0.071	
Sandy loam, 4% H$_2$O	104	1,670	0.4	17			0.54			0.94		
Sandy loam, 10% H$_2$O	121	1,940					1.08			1.87		
Rock wool	10	160	0.2	8.4			0.023	0.033		0.040	0.057	
Wood, oak ⊥ to grain	51	820	0.57	23.9			0.12			0.21		
Wood, oak ∥ to grain	51	820	0.57	23.9			0.23			0.40		

Appendix I

Physical Properties of Gases and Liquids[1]

[1] All gas properties are for atmospheric pressure.

				Gases					
T (F)	ρ (lb_m/ft^3)	c_p ($Btu/lb_m F$)	$\mu \times 10^5$ ($lb_m/ft\,s$)	$\nu \times 10^3$ (ft^2/s)	k ($Btu/h\,ftF$)	α (ft^2/h)	Pr	$\beta \times 10^3$ (1/F)	$g\beta\rho^2/\mu^2$ ($1/F \cdot ft^3$)
					Air				
0	0.0862	0.240	1.09	0.126	0.0132	0.639	0.721	2.18	4.39×10^6
30	0.0810	0.240	1.15	0.142	0.0139	0.714	0.716	2.04	3.28
60	0.0764	0.240	1.21	0.159	0.0146	0.798	0.711	1.92	2.48
80	0.0735	0.240	1.24	0.169	0.0152	0.855	0.708	1.85	2.09
100	0.0710	0.240	1.28	0.181	0.0156	0.919	0.703	1.79	1.76
150	0.0651	0.241	1.36	0.209	0.0167	1.06	0.698	1.64	1.22
200	0.0602	0.241	1.45	0.241	0.0179	1.24	0.694	1.52	0.840
250	0.0559	0.242	1.53	0.274	0.0191	1.42	0.690	1.41	0.607
300	0.0523	0.243	1.60	0.306	0.0203	1.60	0.686	1.32	0.454
400	0.0462	0.245	1.74	0.377	0.0225	2.00	0.681	1.16	0.264
500	0.0413	0.247	1.87	0.453	0.0246	2.41	0.680	1.04	0.163
600	0.0374	0.251	2.00	0.535	0.0270	2.88	0.680	0.944	79.4×10^3
800	0.0315	0.257	2.24	0.711	0.0303	3.75	0.684	0.794	50.6
1000	0.0272	0.263	2.46	0.906	0.0337	4.72	0.689	0.685	27.0
1500	0.0203	0.277	2.92	1.44	0.0408	7.27	0.705	0.510	7.96

T (K)	ρ (kg/m^3)	$c_p \times 10^{-3}$ ($J/kg \times K$)	$\mu \times 10^5$ ($Pa \times s$)	$\nu \times 10^5$ (m^2/s)	$k \times 10^2$ ($W/m \times K$)	$\alpha \times 10^5$ (m^2/s)	Pr	$g\beta\rho^2/\mu^2$ ($1/K \cdot m^3$)
				Air				
250	1.4133	1.0054	1.5991	1.1315	2.2269	1.5672	0.722	4.638×10^8
260	1.3587	1.0054	1.6503	1.2146	2.3080	1.6896	0.719	2.573
280	1.2614	1.0057	1.7503	1.3876	2.4671	1.9448	0.713	1.815
300	1.1769	1.0063	1.8464	1.5689	2.6240	2.2156	0.708	1.327
320	1.1032	1.0073	1.9391	1.7577	2.7785	2.5003	0.703	0.9942
340	1.0382	1.0085	2.0300	1.9553	2.9282	2.7967	0.699	0.7502
360	0.9805	1.0100	2.1175	2.1596	3.0779	3.1080	0.695	0.5828
400	0.8822	1.0142	2.2857	2.5909	3.3651	3.7610	0.689	0.3656
440	0.8021	1.0197	2.4453	3.0486	3.6427	4.4537	0.684	0.2394
480	0.7351	1.0263	2.5963	3.5319	3.9107	5.1836	0.681	0.1627
520	0.6786	1.0339	2.7422	4.0410	4.1690	5.9421	0.680	0.1156
580	0.6084	1.0468	2.9515	4.8512	4.5407	7.1297	0.680	7.193×10^6
700	0.5040	1.0751	3.3325	6.6121	5.2360	9.6632	0.684	3.210
800	0.4411	1.0988	3.6242	8.2163	5.7743	11.9136	0.689	1.804
1000	0.3529	1.1421	4.1527	11.1767	6.7544	16.7583	0.702	0.803

T (F)	ρ (lb$_m$/ft^3)	c_p (Btu/lb$_m$F)	$\mu \times 10^5$ (lb$_m$/ft s)	$\nu \times 10^3$ (ft^2/s)	k (Btu/h ftF)	α (ft^2/h)	Pr	$\beta \times 10^3$ (1/F)	$g\beta\rho^2/\mu^2$ (1/F · ft^3)
				Steam					
212	0.0372	0.493	0.870	0.234	0.0145	0.794	1.06	1.49	0.873×10^6
250	0.0350	0.483	0.890	0.254	0.0155	0.920	0.994	1.41	0.698
300	0.0327	0.476	0.960	0.294	0.0171	1.10	0.963	1.32	0.493
400	0.0289	0.472	1.09	0.377	0.0200	1.47	0.924	1.16	0.262
500	0.0259	0.477	1.23	0.474	0.0228	1.85	0.922	1.04	0.148
600	0.0234	0.483	1.37	0.585	0.0258	2.29	0.920	0.944	88.9×10^3
800	0.0197	0.498	1.63	0.828	0.0321	3.27	0.912	0.794	37.8
1000	0.0170	0.517	1.90	1.12	0.0390	4.44	0.911	0.685	17.2
1500	0.0126	0.564	2.57	2.05	0.0580	8.17	0.906	0.510	3.97

T (K)	ρ (kg/m^3)	$c_p \times 10^{-3}$ (J/kg · K)	$\mu \times 10^5$ (Pa · s)	$\nu \times 10^5$ (m^2/s)	$k \times 10^2$ (W/m · K)	$\alpha \times 10^5$ (m^2/s)	Pr	$g\beta\rho^2/\mu^2$ (1/K · m^3)
				Steam				
380	0.5860	2.0592	12.70	2.1672	2.4520	2.0320	1.067	5.5210×10^7
400	0.5549	2.0098	13.42	2.4185	2.6010	2.3322	1.037	4.1951
450	0.4911	1.9771	15.23	3.1012	2.9877	3.0771	1.008	2.2558
500	0.4410	1.9817	17.03	3.8617	3.3903	3.8794	0.995	1.3139
550	0.4004	2.0006	18.84	4.7053	3.8008	4.7448	0.992	0.8069
600	0.3667	2.0264	20.64	5.6286	4.2161	5.6738	0.992	0.5154
650	0.3383	2.0555	22.45	6.6361	4.6361	6.6670	0.995	0.3415
700	0.3140	2.0869	24.25	7.7229	5.0593	7.7207	1.000	0.2277
750	0.2930	2.1192	26.06	8.8942	5.4841	8.8321	1.007	0.1651
800	0.2746	2.1529	27.86	10.1457	5.9089	9.9950	1.015	0.1183

T (F)	ρ (lb$_m$/ft^3)	c_p (Btu/lb$_m$F)	$\mu \times 10^5$ (lb$_m$/ft s)	$\nu \times 10^3$ (ft^2/s)	k (Btu/h ftF)	α (ft^2/h)	Pr	$\beta \times 10^3$ (1/F)	$g\beta\rho^2/\mu^2$ (1/F · ft^3)
				Nitrogen					
0	0.0837	0.249	1.06	0.127	0.0132	0.633	0.719	2.18	4.38×10^6
30	0.0786	0.249	1.12	0.142	0.0139	0.710	0.719	2.04	3.29
60	0.0740	0.249	1.17	0.158	0.0146	0.800	0.716	1.92	2.51
80	0.0711	0.249	1.20	0.169	0.0151	0.853	0.712	1.85	2.10
100	0.0685	0.249	1.23	0.180	0.0154	0.915	0.708	1.79	1.79
150	0.0630	0.249	1.32	0.209	0.0168	1.07	0.702	1.64	1.22
200	0.0580	0.249	1.39	0.240	0.0174	1.25	0.690	1.52	0.854
250	0.0540	0.249	1.47	0.271	0.0192	1.42	0.687	1.41	0.616
300	0.0502	0.250	1.53	0.305	0.0202	1.62	0.685	1.32	0.457
400	0.0443	0.250	1.67	0.377	0.0212	2.02	0.684	1.16	0.263
500	0.0397	0.253	1.80	0.453	0.0244	2.43	0.683	1.04	0.163
600	0.0363	0.256	1.93	0.532	0.0252	2.81	0.686	0.944	0.108
800	0.0304	0.262	2.16	0.710	0.0291	3.71	0.691	0.794	0.0507
1000	0.0263	0.269	2.37	0.901	0.0336	4.64	0.700	0.685	0.0272
1500	0.0195	0.283	2.82	1.45	0.0423	7.14	0.732	0.510	0.00785

T (K)	ρ (kg/m^3)	$c_p \times 10^{-3}$ (J/kg·K)	$\mu \times 10^5$ (Pa·s)	$\nu \times 10^5$ (m^2/s)	$k \times 10^2$ (W/m·K)	$\alpha \times 10^5$ (m^2/s)	Pr	$g\beta\rho^2/\mu^2$ (1/K·m^3)
				Nitrogen				
250	1.3668	1.0415	1.5528	1.1361	2.2268	1.5643	0.729	3.0362×10^8
300	1.1383	1.0412	1.7855	1.5686	2.6052	2.1981	0.713	1.3273
350	0.9754	1.0421	2.0000	2.0504	2.9691	2.9210	0.701	0.6655
400	0.8533	1.0449	2.1995	2.5776	3.3186	3.7220	0.691	0.3697
450	0.7584	1.0495	2.3890	3.1501	3.6463	4.5811	0.688	0.2187
500	0.6826	1.0564	2.5702	3.7653	3.9645	5.4979	0.684	0.1382
600	0.5688	1.0751	2.9127	5.1208	4.5549	7.4485	0.686	6.237×10^6
700	0.4875	1.0980	3.2120	6.5887	5.0947	9.5179	0.691	3.233
800	0.4266	1.1222	3.4896	8.1800	5.5864	11.6692	0.700	1.820
1000	0.3413	1.1672	4.0000	11.7199	6.4419	16.1708	0.724	0.810

T (F)	ρ (lb$_m$/ft^3)	c_p (Btu/lb$_m$F)	$\mu \times 10^5$ (lb$_m$/ft s)	$\nu \times 10^3$ (ft^2/s)	k (Btu/h ft F)	α (ft^2/h)	Pr	$\beta \times 10^3$ (1/F)	$g\beta\rho^2/\mu^2$ (1/F·ft^3)
					Oxygen				
0	0.0955	0.219	1.22	0.128	0.0134	0.641	0.718	2.18	4.29×10^6
30	0.0897	0.219	1.28	0.143	0.0141	0.718	0.716	2.04	3.22
60	0.0845	0.219	1.35	0.160	0.0149	0.806	0.713	1.92	2.43
80	0.0814	0.220	1.40	0.172	0.0155	0.866	0.713	1.85	2.02
100	0.0785	0.220	1.43	0.182	0.0160	0.925	0.708	1.79	1.74
150	0.0720	0.221	1.52	0.211	0.0172	1.08	0.703	1.64	1.19
200	0.0665	0.223	1.62	0.244	0.0185	1.25	0.703	1.52	0.825
250	0.0168	0.225	1.70	0.276	0.0197	1.42	0.700	1.41	0.600
300	0.0578	0.227	1.79	0.310	0.0209	1.60	0.700	1.32	0.442
400	0.0511	0.230	1.95	0.381	0.0233	1.97	0.698	1.16	0.257
500	0.0458	0.234	2.10	0.458	0.0254	2.37	0.696	1.04	0.160
600	0.0414	0.239	2.25	0.543	0.0281	2.84	0.688	0.944	0.103
800	0.0349	0.246	2.52	0.723	0.0324	3.77	0.680	0.794	49.4×10^3
1000	0.0300	0.252	2.79	0.930	0.0366	4.85	0.691	0.685	25.6
1500	0.0224	0.264	3.39	1.52	0.0465	7.86	0.696	0.510	7.22

T (K)	ρ (kg/m^3)	$c_p \times 10^{-3}$ (J/kg·K)	$\mu \times 10^5$ (Pa·s)	$\nu \times 10^5$ (m^2/s)	$k \times 10^2$ (W/m·K)	$\alpha \times 10^5$ (m^2/s)	Pr	$g\beta\rho^2/\mu^2$ (1/K·m^3)
				Oxygen				
250	1.5620	0.9150	1.7887	1.1451	2.2586	1.5803	0.725	2.9885×10^8
300	1.3007	0.9199	2.0633	1.5863	2.6760	2.2365	0.709	1.2978
350	1.1144	0.9291	2.3176	2.0797	3.0688	2.9639	0.702	0.6469
400	0.9749	0.9417	2.5556	2.6214	3.4616	3.7705	0.695	0.3571
450	0.8665	0.9564	2.7798	3.2081	3.8298	4.6216	0.694	0.2108
500	0.7798	0.9721	2.9930	3.8382	4.1735	5.5056	0.697	0.1330
550	0.7089	0.9879	3.1966	4.5092	4.5172	6.4502	0.700	8.786×10^6
600	0.6498	1.0032	3.3931	5.2218	4.8364	7.4192	0.704	5.988

T (F)	ρ (lb$_m$/ft^3)	c_p (Btu/lb$_m$F)	$\mu \times 10^5$ (lb$_m$/ft s)	$\nu \times 10^3$ (ft^2/s)	k (Btu/h ftF)	α (ft^2/h)	Pr	$\beta \times 10^3$ (1/F)	$g\beta\rho^2/\mu^2$ (1/F·ft^3)
					Carbon dioxide				
0	0.132	0.193	0.865	0.0655	0.00760	0.298	0.792	2.18	16.3×10^6
30	0.124	0.198	0.915	0.0739	0.00830	0.339	0.787	2.04	12.0
60	0.117	0.202	0.965	0.0829	0.00910	0.387	0.773	1.92	9.00
80	0.112	0.204	1.00	0.0891	0.00960	0.421	0.760	1.85	7.45
100	0.108	0.207	1.03	0.0953	0.0102	0.455	0.758	1.79	6.33
150	0.100	0.213	1.12	0.113	0.0115	0.539	0.755	1.64	4.16
200	0.092	0.219	1.20	0.131	0.0130	0.646	0.730	1.52	2.86
250	0.0850	0.225	1.32	0.155	0.0148	0.777	0.717	1.41	2.04
300	0.0800	0.230	1.36	0.171	0.0160	0.878	0.704	1.32	1.45
400	0.0740	0.239	1.45	0.196	0.0180	1.02	0.695	1.16	1.11
500	0.0630	0.248	1.65	0.263	0.0210	1.36	0.700	1.04	0.485
600	0.0570	0.256	1.78	0.312	0.0235	1.61	0.700	0.944	0.310
800	0.0480	0.269	2.02	0.420	0.0278	2.15	0.702	0.794	0.143
1000	0.0416	0.280	2.25	0.540	0.0324	2.78	0.703	0.685	75.3×10^3
1500	0.0306	0.301	2.80	0.913	0.0340	4.67	0.704	0.510	19.6

T (K)	ρ (kg/m^3)	$c_p \times 10^{-3}$ (J/kg·K)	$\mu \times 10^5$ (Pa·s)	$\nu \times 10^5$ (m^2/s)	$k \times 10^2$ (W/m·K)	$\alpha \times 10^5$ (m^2/s)	Pr	$g\beta\rho^2/\mu^2$ (1/K·m^3)
					Carbon dioxide			
250	2.1652	0.8052	1.2590	0.5815	1.2891	0.7394	0.793	1.1591×10^9
300	1.7967	0.8526	1.4948	0.8320	1.6572	1.0818	0.770	0.4178
350	1.5369	0.8989	1.7208	1.1197	2.0457	1.4808	0.755	0.2232
400	1.3432	0.9416	1.9318	1.4382	2.4604	1.9454	0.738	0.1186
450	1.1931	0.9803	2.1332	1.7879	2.8955	2.4756	0.721	6.786×10^7
500	1.0733	1.0153	2.3251	2.1663	3.3523	3.0763	0.702	4.176
550	0.9756	1.0470	2.5073	2.5700	3.8208	3.7406	0.685	2.705
600	0.8941	1.0761	2.6827	3.0004	4.3097	4.4793	0.668	1.814

T (F)	ρ (lb_m/ft^3)	c_p ($Btu/lb_m F$)	$\mu \times 10^5$ ($lb_m/ft\ s$)	$\nu \times 10^3$ (ft^2/s)	k ($Btu/h\ ftF$)	α (ft^2/h)	Pr	$\beta \times 10^3$ ($1/F$)	$g\beta\rho^2/\mu^2$ ($1/F \cdot ft^3$)
				Hydrogen					
0	0.00597	3.37	0.537	0.900	0.092	4.59	0.713	2.18	87,000
30	0.00562	3.39	0.562	1.00	0.097	5.09	0.709	2.04	65,700
60	0.00530	3.41	0.587	1.11	0.102	5.65	0.707	1.92	50,500
80	0.00510	3.42	0.602	1.18	0.105	6.04	0.705	1.85	42,700
100	0.00492	3.42	0.617	1.25	0.108	6.42	0.700	1.79	36,700
150	0.00450	3.44	0.653	1.45	0.116	7.50	0.696	1.64	25,000
200	0.00412	3.45	0.688	1.67	0.123	8.64	0.696	1.52	17,500
250	0.00382	3.46	0.723	1.89	0.130	9.85	0.690	1.41	12,700
300	0.00357	3.46	0.756	2.12	0.137	11.1	0.687	1.32	9,440
400	0.00315	3.47	0.822	2.61	0.151	13.8	0.681	1.16	5,470
500	0.00285	3.47	0.890	3.12	0.165	16.7	0.675	1.04	3,430
600	0.00260	3.47	0.952	3.66	0.179	19.8	0.667	0.944	2,270
800	0.00219	3.49	1.07	4.87	0.205	26.8	0.654	0.794	1,080
1000	0.00189	3.52	1.18	6.21	0.224	33.7	0.664	0.685	571
1500	0.00141	3.62	1.44	10.2	0.265	51.9	0.708	0.510	158

T (K)	ρ (kg/m^3)	c_p ($J/kg \cdot K$)	$\mu \times 10^6$ ($Pa \cdot s$)	$\nu \times 10^6$ (m^2/s)	$k \times 10^2$ ($W/m \cdot K$)	$\alpha \times 10^4$ (m^2/s)	Pr	$g\beta\rho^2/\mu^2 \times 10^{-6}$ ($1/K \cdot m^3$)
				Hydrogen				
50	0.5095	10.501	2.516	4.938	0.0362	0.0633	0.78	
100	0.2457	11.229	4.212	17.143	0.0665	0.2410	0.711	333.8
150	0.1637	12.602	5.595	34.178	0.0981	0.4755	0.719	55.99
200	0.1227	13.504	6.813	55.526	0.1282	0.7717	0.719	15.90
250	0.0982	14.059	7.919	80.641	0.1561	1.131	0.713	6.03
300	0.0818	14.314	8.963	109.57	0.182	1.554	0.705	2.72
350	0.0702	14.436	9.954	141.79	0.206	2.033	0.697	1.39
400	0.0613	14.491	10.864	177.23	0.228	2.567	0.690	0.782
450	0.0546	14.499	11.779	215.73	0.251	3.171	0.680	0.468
500	0.0492	14.507	12.636	256.83	0.272	3.811	0.674	0.297
600	0.0408	14.537	14.285	350.12	0.315	5.311	0.659	0.134
700	0.0349	14.574	15.890	455.30	0.351	6.901	0.660	0.0677
800	0.0306	14.675	17.40	568.63	0.384	8.551	0.665	0.0379
1000	0.0245	14.968	20.160	822.86	0.440	11.998	0.686	0.0145
1200	0.0205	15.366	22.75	1109.80	0.488	15.492	0.716	0.00667

T (F)	ρ (lb$_m$/ft^3)	c_p (Btu/lb$_m$F)	$\mu \times 10^5$ (lb$_m$/ft s)	$\nu \times 10^3$ (ft^2/s)	k (Btu/h ft F)	α (ft^2/h)	Pr	$\beta \times 10^3$ (1/F)	$g\beta\rho^2/\mu^2$ (1/F·ft^3)
				Carbon monoxide					
0	0.0832	0.249	1.05	0.126	0.0128	0.620	0.749	2.18	4.40×10^6
30	0.0780	0.249	1.11	0.142	0.0134	0.691	0.744	2.04	3.32
60	0.0736	0.249	1.16	0.157	0.0142	0.775	0.740	1.92	2.48
80	0.0709	0.249	1.20	0.169	0.0146	0.828	0.737	1.85	2.09
100	0.0684	0.249	1.23	0.180	0.0150	0.884	0.735	1.79	1.79
150	0.0628	0.249	1.32	0.210	0.0163	1.04	0.730	1.64	1.19
200	0.0580	0.250	1.40	0.241	0.0174	1.20	0.726	1.52	0.842
250	0.0539	0.250	1.48	0.275	0.0183	1.36	0.722	1.41	0.604
300	0.0503	0.251	1.56	0.310	0.0196	1.56	0.720	1.32	0.442
400	0.0445	0.253	1.73	0.389	0.0217	1.92	0.718	1.16	0.248
500	0.0399	0.256	1.85	0.463	0.0234	2.30	0.725	1.04	0.156
600	0.0361	0.259	1.97	0.545	0.0253	2.71	0.723	0.944	0.101
800	0.0304	0.266	2.21	0.728	0.0288	3.57	0.730	0.794	48.2×10^3
1000	0.0262	0.273	2.43	0.929	0.0324	4.54	0.740	0.685	25.6
1500	0.0195	0.286	3.00	1.54	0.0410	7.35	0.756	0.510	6.93

T (K)	ρ (kg/m^3)	$c_p \times 10^{-3}$ (J/kg·K)	$\mu \times 10^5$ (Pa·s)	$\nu \times 10^5$ (m^2/s)	$k \times 10^2$ (W/m·K)	$\alpha \times 10^5$ (m^2/s)	Pr	$g\beta\rho^2/\mu^2$ (1/K·m^3)
				Carbon monoxide				
250	1.3669	1.0425	1.5408	1.1272	2.1432	1.5040	0.749	3.0841×10^8
300	1.1382	1.0422	1.7854	1.5686	2.5240	2.1277	0.737	1.3273
350	0.9753	1.0440	2.0097	2.0606	2.8839	2.8323	0.727	0.6590
400	0.8532	1.0484	2.2201	2.6021	3.2253	3.6057	0.722	0.3623
450	0.7583	1.0550	2.4189	3.1899	3.5527	4.4408	0.718	0.2133
500	0.6824	1.0642	2.6078	3.8215	3.8638	5.3205	0.718	0.1342
550	0.6204	1.0751	2.7884	4.4945	4.1587	6.2350	0.721	8.843×10^6
600	0.5687	1.0870	2.9607	5.2061	4.4443	7.1894	0.724	6.025

T (F)	ρ (lb$_m$/ft^3)	c_p (Btu/lb$_m$F)	$\mu \times 10^6$ (lb$_m$/ft s)	$\nu \times 10^3$ (ft^2/s)	k (Btu/h ft F)	α (ft^2/h)	Pr	$\beta \times 10^3$ (1/F)	$g\beta\rho^2/\mu^2 \times 10^{-6}$ (1/F·ft^3)
				Chlorine					
0	0.211	0.113	8.06	0.0381	0.00418	0.175	0.785	2.18	48.3
30	0.197	0.114	8.40	0.0426	0.00450	0.201	0.769	2.04	36.6
60	0.187	0.114	8.80	0.0470	0.00480	0.225	0.753	1.92	28.1
80	0.180	0.115	9.07	0.0504	0.00500	0.242	0.753	1.85	24.3
100	0.173	0.115	9.34	0.0540	0.00520	0.261	0.748	1.79	19.9
150	0.159	0.117	10.0	0.0629	0.00570	0.306	0.739	1.64	13.4

T (F)	ρ (lb$_m$/ft^3)	c_p (Btu/lb$_m$F)	$\mu \times 10^7$ (lb$_m$/ft s)	$\nu \times 10^3$ (ft^2/s)	k (Btu/h ft F)	α (ft^2/h)	Pr	$\beta \times 10^3$ (1/F)	$g\beta\rho^2/\mu^2$ (1/F·ft^3)
					Helium				
0	0.0119	1.24	122	1.03	0.0784	5.30	0.698	2.18	66,800
30	0.0112	1.24	127	1.14	0.0818	5.89	0.699	2.04	51,100
60	0.0106	1.24	132	1.25	0.0852	6.46	0.700	1.92	40,000
80	0.0102	1.24	135	1.32	0.0872	6.88	0.701	1.85	33,900
100	0.00980	1.24	138	1.41	0.0892	7.37	0.701	1.79	29,000
150	0.00900	1.24	146	1.63	0.0937	8.36	0.703	1.64	20,100
200	0.00829	1.24	155	1.87	0.0977	9.48	0.705	1.52	14,000
250	0.00772	1.24	162	2.09	0.102	10.7	0.707	1.41	10,400
300	0.00722	1.24	170	2.36	0.106	11.8	0.709	1.32	7,650
400	0.00637	1.24	185	2.91	0.114	14.4	0.714	1.16	4,410
500	0.00572	1.24	198	3.46	0.122	17.1	0.719	1.04	2,800
600	0.00517	1.24	209	4.04	0.130	20.6	0.720	0.994	1,850
800	0.00439	1.24	232	5.28	0.145	27.6	0.722	0.794	915
1000	0.00376	1.24	255	6.78	0.159	35.5	0.725	0.685	480
1500	0.00280	1.24	309	11.1	0.189	59.7	0.730	0.510	135

T (F)	ρ (lb$_m$/ft^3)	c_p (Btu/lb$_m$F)	$\mu \times 10^5$ (lb$_m$/ft s)	$\nu \times 10^3$ (ft^2/s)	k (Btu/h ft F)	α (ft^2/h)	Pr	$\beta \times 10^3$ (1/F)	$g\beta\rho^2/\mu^2$ (1/F·ft^3)
					Sulfur dioxide				
0	0.195	0.142	0.700	3.59	0.00460	0.166	0.778	2.03	50.6×10^6
100	0.161	0.149	0.890	5.52	0.00560	0.233	0.854	1.79	19.0
200	0.136	0.157	1.05	7.74	0.00670	0.313	0.883	1.52	8.25
300	0.118	0.164	1.20	10.2	0.00790	0.407	0.898	1.32	4.12
400	0.104	0.170	1.35	13.0	0.00920	0.520	0.898	1.16	2.24
500	0.0935	0.176	1.50	16.0	0.00990	0.601	0.958	1.04	1.30
600	0.0846	0.180	1.65	19.5	0.0108	0.711	0.987	0.994	0.795

T (F)	ρ (lb$_m$/ft^3)	c_p (Btu/lb$_m$F)	$\mu \times 10^3$ (lb$_m$/ft s)	$\nu \times 10^5$ (ft^2/s)	k (Btu/h ft F)	$\alpha \times 10^3$ (ft^2/h)	Pr	$\beta \times 10^4$ (1/F)	$g\beta\rho^2/\mu^2 \times 10^{-6}$ (1/F·ft^3)
					Liquids				
					Water				
32	62.4	1.01	1.20	1.93	0.319	5.06	13.7	−0.350	
60	62.3	1.00	0.760	1.22	0.340	5.45	8.07	0.800	17.2
80	62.2	0.999	0.578	0.929	0.353	5.67	5.89	1.30	48.3
100	62.1	0.999	0.458	0.736	0.364	5.87	4.51	1.80	107
150	61.3	1.00	0.290	0.474	0.383	6.26	2.72	2.80	403
200	60.1	1.01	0.206	0.342	0.392	6.46	1.91	3.70	1,010
250	58.9	1.02	0.160	0.272	0.395	6.60	1.49	4.70	2,045
300	57.3	1.03	0.130	0.227	0.395	6.70	1.22	5.60	3,510
400	53.6	1.08	0.0930	0.174	0.382	6.58	0.950	7.80	8,350
500	49.0	1.19	0.0700	0.143	0.349	5.98	0.859	11.0	17,350
600	42.4	1.51	0.0579	0.137	0.293	4.58	1.07	17.5	30,300

T (K)	ρ (kg/m^3)	c_p (J/kg × K)	$\mu \times 10^6$ (Pa × s)	$\nu \times 10^6$ (m^2/s)	k (W/m × K)	$\alpha \times 10^6$ (m^2/s)	Pr	$g\beta\rho^2/\mu^2 \times 10^{-9}$ (1/K · m^3)
					Water			
273	999.3	4226	1794	1.795	0.558	0.132	13.6	
293	998.2	4182	993	0.995	0.597	0.143	6.96	2.035
313	992.2	4175	658	0.663	0.633	0.153	4.33	8.833
333	983.2	4181	472	0.480	0.658	0.160	3.00	22.75
353	971.8	4194	352	0.362	0.673	0.165	2.57	46.68
373	958.4	4211	278	0.290	0.682	0.169	1.72	85.09
473	862.8	4501	139	0.161	0.665	0.171	0.94	517.2
573	712.5	5694	92.2	0.129	0.564	0.139	0.93	1766.0

T (F)	ρ (lb$_m$/ft^3)	c_p (Btu/lb$_m$F)	$\mu \times 10^5$ (lb$_m$/ft s)	$\nu \times 10^5$ (ft^2/s)	k (Btu/h ft F)	$\alpha \times 10^3$ (ft^2/h)	Pr	$\beta \times 10^3$ (1/F)	$g\beta\rho^2/\mu^2 \times 10^{-6}$ (1/F · ft^3)
					Aniline				
60	64.0	0.480	305	4.77	0.101	3.29	52.3		
80	63.5	0.485	240	3.78	0.100	3.25	41.8		
100	63.0	0.490	180	2.86	0.100	3.24	31.8	0.45	17.7
150	61.6	0.503	100	1.62	0.0980	3.16	18.4		
200	60.2	0.515	62	1.03	0.0962	3.10	12.0		
250	58.9	0.527	42	0.714	0.0947	3.05	8.44		
300	57.5	0.540	30	0.522	0.0931	2.99	6.28		

T (F)	ρ (lb$_m$/ft^3)	c_p (Btu/lb$_m$F)	$\mu \times 10^5$ (lb$_m$/ft s)	$\nu \times 10^5$ (ft^2/s)	k (Btu/h ftF)	$\alpha \times 10^3$ (ft^2/h)	Pr	$\beta \times 10^3$ (1/F)	$g\beta\rho^2/\mu^2 \times 10^{-7}$ (1/F · ft^3)
					Ammonia				
−60	43.9	1.07	20.6	0.471	0.316	6.74	2.52	0.94	132
−30	42.7	1.07	18.2	0.426	0.317	6.93	2.22	1.02	265
0	41.3	1.08	16.9	0.409	0.315	7.06	2.08	1.1	467
30	40.0	1.11	16.2	0.402	0.312	7.05	2.05	1.19	757
60	38.5	1.14	15.0	0.391	0.304	6.92	2.03	1.3	1130
80	37.5	1.16	14.2	0.379	0.296	6.79	2.01	1.4	1650
100	36.4	1.19	13.5	0.368	0.287	6.62	2.00	1.5	2200
120	35.3	1.22	12.6	0.356	0.275	6.43	2.00	1.68	3180

T (F)	ρ (lb$_m$/ft^3)	c_p (Btu/lb$_m$F)	$\mu \times 10^5$ (lb$_m$/ft s)	$\nu \times 10^5$ (ft^2/s)	k (Btu/h ft F)	$\alpha \times 10^3$ (ft^2/h)	Pr	$\beta \times 10^3$ (1/F)	$g\beta\rho^2/\mu^2 \times 10^{-6}$ (1/F · ft^3)
				Freon-12					
−40	94.5	0.202	125	1.32	0.0650	3.40	14.0	9.10	168
−30	93.5	0.204	123	1.32	0.0640	3.35	14.1	9.60	179
0	90.9	0.212	116	1.28	0.0578	3.00	15.4	11.4	225
30	87.4	0.221	108	1.24	0.0564	2.92	15.3	13.1	277
60	84.0	0.230	99.6	1.19	0.0528	2.74	15.6	14.9	341
80	81.3	0.238	94.0	1.16	0.0504	2.60	16.0	16.0	384
100	78.7	0.246	88.4	1.12	0.0480	2.48	16.3	17.2	439
150	71.0	0.271	74.8	1.05	0.0420	2.18	17.4	19.5	625

T (F)	ρ (lb$_m$/ft^3)	c_p (Btu/lb$_m$F)	$\mu \times 10^5$ (lb$_m$/ft s)	$\nu \times 10^5$ (ft^2/s)	k (Btu/h ft F)	$\alpha \times 10^3$ (ft^2/h)	Pr	$\beta \times 10^3$ (1/F)	$g\beta\rho^2/\mu^2 \times 10^{-6}$ (1/F · ft^3)
				n-Butyl Alcohol					
60	50.5	0.55	225	4.46	0.100	3.59	44.6		
80	50.0	0.58	180	3.60	0.099	3.41	38.0	0.25	6.23
100	49.6	0.61	130	2.62	0.098	3.25	29.1	0.43	2.02
150	48.5	0.68	68	1.41	0.098	2.97	17.1		

T (F)	ρ (lb$_m$/ft^3)	c_p (Btu/lb$_m$F)	$\mu \times 10^5$ (lb$_m$/ft s)	$\nu \times 10^5$ (ft^2/s)	k (Btu/h ft F)	$\alpha \times 10^3$ (ft^2/h)	Pr $\times 10^{-2}$	$\beta \times 10^4$ (1/F)	$g\beta\rho^2/\mu^2 \times 10^{-6}$ (1/F · ft^3)
				Benzene					
60	55.2	0.395	44.5	0.806	0.0856	3.93	7.39		
80	54.6	0.410	38	0.695	0.0836	3.73	6.70	7.5	498
100	53.6	0.420	33	0.615	0.0814	3.61	6.13	7.2	609
150	51.8	0.450	24.5	0.473	0.0762	3.27	5.21	6.8	980
200	49.9	0.480	19.4	0.390	0.0711	2.97	4.73		

T (F)	ρ (lb$_m$/ft^3)	c_p (Btu/lb$_m$F)	$\mu \times 10^5$ (lb$_m$/ft s)	$\nu \times 10^5$ (ft^2/s)	k (Btu/h ft F)	$\alpha \times 10^3$ (ft^2/h)	Pr	$\beta \times 10^3$ (1/F)	$g\beta\rho^2/\mu^2 \times 10^{-4}$ (1/F · ft^3)
				Hydraulic fluid (MIL-M-5606)					
0	55.0	0.400	5550	101	0.0780	3.54	1030	0.76	2.39
30	54.0	0.420	2220	41.1	0.0755	3.32	446	0.68	13.0
60	53.0	0.439	1110	20.9	0.0732	3.14	239	0.60	44.1
80	52.5	0.453	695	13.3	0.0710	3.07	155	0.52	95.7
100	52.0	0.467	556	10.7	0.0690	2.84	136	0.47	132
150	51.0	0.499	278	5.45	0.0645	2.44	80.5	0.32	346
200	50.0	0.530	250	5.00	0.0600	2.27	79.4	0.20	258

T (F)	ρ (lb$_m$/ft^3)	c_p (Btu/lb$_m$F)	μ (lb$_m$/ft s)	$\nu \times 10^2$ (ft^2/s)	k (Btu/h ft F)	$\alpha \times 10^3$ (ft^2/h)	Pr $\times 10^{-2}$	$\beta \times 10^3$ (1/F)	$g\beta\rho^2/\mu^2$ (1/F · ft^3)
				Glycerin					
30	79.7	0.540	7.2	9.03	0.168	3.91	832		
60	79.1	0.563	1.4	1.77	0.167	3.75	170		
80	78.7	0.580	0.6	0.762	0.166	3.64	75.3	0.30	166
100	78.2	0.598	0.1	0.128	0.165	3.53	13.1		

T (F)	ρ (lb$_m$/ft^3)	c_p (Btu/lb$_m$F)	$\mu \times 10^5$ (lb$_m$/ft s)	$\nu \times 10^5$ (ft^2/s)	k (Btu/h ftF)	$\alpha \times 10^3$ (ft^2/h)	Pr	$\beta \times 10^3$ (1/F)	$g\beta\rho^2/\mu^2$ (1/F · ft^3)
				Kerosene					
30	48.8	0.456	800	16.4	0.0809	3.63	163		
60	48.1	0.474	600	12.5	0.0805	3.53	127	0.58	120
80	47.6	0.491	490	10.3	0.0800	3.42	108	0.48	146
100	47.2	0.505	420	8.90	0.0797	3.35	95.7	0.47	192
150	46.1	0.540	320	6.83	0.0788	3.16	77.9		

T (F)	ρ (lb$_m$/ft^3)	c_p (Btu/lb$_m$F)	$\mu \times 10^5$ (lb$_m$/ft s)	$\nu \times 10^5$ (ft^2/s)	k (Btu/h ft F)	$\alpha \times 10^3$ (ft^2/h)	Pr	$\beta \times 10^3$ (1/F)	$g\beta\rho^2/\mu^2 \times 10^{-4}$ (1/F · ft^3)
				Liquid hydrogen					
−435	4.84	1.69	1.63	0.337	0.0595	7.28	1.67		
−433	4.77	1.78	1.52	0.319	0.0610	7.20	1.59		
−431	4.71	1.87	1.40	0.297	0.0625	7.09	1.51	7.1	2.59
−429	4.64	1.96	1.28	0.276	0.0640	7.03	1.41		
−427	4.58	2.05	1.17	0.256	0.0655	6.97	1.32		
−425	4.51	2.15	1.05	0.233	0.0670	6.90	1.21		

T (F)	ρ (lb$_m$/ft^3)	c_p (Btu/lb$_m$F)	$\mu \times 10^5$ (lb$_m$/ft s)	$\nu \times 10^5$ (ft^2/s)	$k \times 10^3$ (Btu/h ftF)	$\alpha \times 10^5$ (ft^2/h)	Pr	$\beta \times 10^3$ (1/F)	$g\beta\rho^2/\mu^2 \times 10^{-8}$ (1/F·ft^3)
				Liquid oxygen					
-350	80.1	0.400	38.0	0.474	3.1	9.67	172		
-340	78.5	0.401	28.0	0.356	3.4	10.8	109		
-330	76.8	0.402	21.8	0.284	3.7	12.0	85.0		
-320	75.1	0.404	17.4	0.232	4.0	12.2	63.5	3.19	186
-310	73.4	0.405	14.8	0.202	4.3	14.5	50.1		
-300	71.7	0.406	13.0	0.181	4.6	15.8	41.2		

T (F)	ρ (lb$_m$/ft^3)	c_p (Btu/lb$_m$F)	$\mu \times 10^3$ (lb$_m$/ft s)	$\nu \times 10^6$ (ft^2/s)	k (Btu/h ftF)	α (ft^2/h)	Pr	$\beta \times 10^3$ (1/F)	$g\beta\rho^2/\mu^2 \times 10^{-9}$ (1/F·ft^3)
				Bismuth					
600	625	0.0345	1.09	1.75	8.58	0.397	0.0159		
700	622	0.0353	0.990	1.59	8.87	0.405	0.0141	0.062	0.786
800	618	0.0361	0.900	1.46	9.16	0.408	0.0129	0.065	0.985
900	613	0.0368	0.830	1.35	9.44	0.418	0.0116	0.068	1.19
1000	608	0.0375	0.765	1.26	9.74	0.427	0.0106	0.071	1.45
1100	604	0.0381	0.710	1.17	10.0	0.435	0.00970	0.074	1.72
1200	599	0.0386	0.660	1.10	10.3	0.446	0.00895	0.077	2.04
1300	595	0.0391	0.620	1.04	10.6	0.456	0.00820		

T (F)	ρ (lb$_m$/ft^3)	c_p (Btu/lb$_m$F)	$\mu \times 10^3$ (lb$_m$/ft s)	$\nu \times 10^6$ (ft^2/s)	k (Btu/h ft F)	α (ft^2/h)	Pr	$\beta \times 10^3$ (1/F)	$g\beta\rho^2/\mu^2 \times 10^{-9}$ (1/F·ft^3)
				Mercury					
40	848	0.0334	1.11	1.31	4.55	0.161	0.0292		1.57
60	847	0.0333	1.05	1.24	4.64	0.165	0.0270		1.76
80	845	0.0332	1.00	1.18	4.72	0.169	0.0252		1.94
100	843	0.0331	0.960	1.14	4.80	0.172	0.0239		2.09
150	839	0.0330	0.893	1.06	5.03	0.182	0.0210		2.38
200	835	0.0328	0.850	1.02	5.25	0.192	0.0191		2.62
250	831	0.0328	0.806	0.970	5.45	0.200	0.0175		2.87
300	827	0.0328	0.766	0.928	5.65	0.209	0.0160		3.16
400	819	0.0328	0.700	0.856	6.05	0.225	0.0137	0.084	3.70
500	811	0.0328	0.650	0.803	6.43	0.243	0.0119		4.12
600	804	0.0328	0.606	0.754	6.80	0.259	0.0105		4.80
800	789	0.0329	0.550	0.698	7.45	0.289	0.0087		5.54

T (F)	ρ (lb$_m$/ft^3)	c_p (Btu/lb$_m$F)	$\mu \times 10^3$ (lb$_m$/ft s)	$\nu \times 10^6$ (ft^2/s)	k (Btu/h ft F)	α (ft^2/h)	Pr	$\beta \times 10^3$ (1/F)	$g\beta\rho^2/\mu^2 \times 10^{-6}$ (1/F · ft^3)
					Sodium				
200	58.1	0.332	0.489	8.43	49.8	2.58	0.0118		68.0
250	57.6	0.328	0.428	7.43	49.3	2.60	0.0103		87.4
300	57.2	0.324	0.378	6.61	48.8	2.64	0.00903		110
400	56.3	0.317	0.302	5.36	47.3	2.66	0.00725		168
500	55.5	0.309	0.258	4.64	45.5	2.64	0.00633	0.15	224
600	54.6	0.305	0.224	4.11	43.1	2.58	0.00574		287
800	52.9	0.304	0.180	3.40	38.8	2.41	0.00510		418
1000	51.2	0.304	0.152	2.97	36.0	2.31	0.00463		548
1300	48.7	0.305	0.120	2.47	34.2	2.31	0.00385		795

Appendix J

Mass-Transfer Diffusion Coefficients in Binary Systems

Table J.1 Binary mass diffusivities in gases[†]

System	T (K)	$D_{AB}P(\text{cm}^2 \text{ atm/s})$	$D_{AB}P(\text{m}^2 \text{ Pa/s})$
Air			
Ammonia	273	0.198	2.006
Aniline	298	0.0726	0.735
Benzene	298	0.0962	0.974
Bromine	293	0.091	0.923
Carbon dioxide	273	0.136	1.378
Carbon disulfide	273	0.0883	0.894
Chlorine	273	0.124	1.256
Diphenyl	491	0.160	1.621
Ethyl acetate	273	0.0709	0.718
Ethanol	298	0.132	1.337
Ethyl ether	293	0.0896	0.908
Iodine	298	0.0834	0.845
Methanol	298	0.162	1.641
Mercury	614	0.473	4.791
Naphthalene	298	0.0611	0.619
Nitrobenzene	298	0.0868	0.879
n-Octane	298	0.0602	0.610
Oxygen	273	0.175	1.773
Propyl acetate	315	0.092	0.932
Sulfur dioxide	273	0.122	1.236
Toluene	298	0.0844	0.855
Water	298	0.260	2.634
Ammonia			
Ethylene	293	0.177	1.793
Argon			
Neon	293	0.329	3.333
Carbon dioxide			
Benzene	318	0.0715	0.724
Carbon disulfide	318	0.0715	0.724
Ethyl acetate	319	0.0666	0.675

(continued)

Table J.1   (*Continued*)

System	T (K)	$D_{AB}P(\text{cm}^2\,\text{atm/s})$	$D_{AB}P(\text{m}^2\,\text{Pa/s})$
Ethanol	273	0.0693	0.702
Ethyl ether	273	0.0541	0.548
Hydrogen	273	0.550	5.572
Methane	273	0.153	1.550
Methanol	298.6	0.105	1.064
Nitrogen	298	0.165	1.672
Nitrous oxide	298	0.117	1.185
Propane	298	0.0863	0.874
Water	298	0.164	1.661
Carbon monoxide			
Ethylene	273	0.151	1.530
Hydrogen	273	0.651	6.595
Nitrogen	288	0.192	1.945
Oxygen	273	0.185	1.874
Helium			
Argon	273	0.641	6.493
Benzene	298	0.384	3.890
Ethanol	298	0.494	5.004
Hydrogen	293	1.64	16.613
Neon	293	1.23	12.460
Water	298	0.908	9.198
Hydrogen			
Ammonia	293	0.849	8.600
Argon	293	0.770	7.800
Benzene	273	0.317	3.211
Ethane	273	0.439	4.447
Methane	273	0.625	6.331
Oxygen	273	0.697	7.061
Water	293	0.850	8.611
Nitrogen			
Ammonia	293	0.241	2.441
Ethylene	298	0.163	1.651
Hydrogen	288	0.743	7.527
Iodine	273	0.070	0.709
Oxygen	273	0.181	1.834
Oxygen			
Ammonia	293	0.253	2.563
Benzene	296	0.0939	0.951
Ethylene	293	0.182	1.844

[†]R. C. Reid and T. K. Sherwood, *The Properties of Gases and Liquids,* McGraw-Hill, New York, 1958, Chapter 8.

Table J.2 Binary mass diffusivities in liquids[†]

Solute A	Solvent B	Temperature (K)	Solute concentration (g mol/L or kg mol/m^3)	Diffusivity (cm^2/s $\times$ 10^5 or m^2/s $\times$ 10^9)
Chlorine	Water	289	0.12	1.26
Hydrogen chloride	Water	273	9	2.7
			2	1.8
		283	9	3.3
			2.5	2.5
		289	0.5	2.44
Ammonia	Water	278	3.5	1.24
		288	1.0	1.77
Carbon dioxide	Water	283	0	1.46
		293	0	1.77
Sodium chloride	Water	291	0.05	1.26
			0.2	1.21
			1.0	1.24
			3.0	1.36
			5.4	1.54
Methanol	Water	288	0	1.28
Acetic acid	Water	285.5	1.0	0.82
			0.01	0.91
		291	1.0	0.96
Ethanol	Water	283	3.75	0.50
			0.05	0.83
		289	2.0	0.90
n-Butanol	Water	288	0	0.77
Carbon dioxide	Ethanol	290	0	3.2
Chloroform	Ethanol	293	2.0	1.25

[†]R. E. Treybal, *Mass Transfer Operations,* McGraw-Hill, New York, 1955, p. 25.

Table J.3 Binary diffusivities in solids[†]

Solute	Solid	Temperature (K)	Diffusivity (cm^2/s or m^2/s $\times$10^4)	Diffusivity (ft^2/h)
Helium	Pyrex	293	4.49×10^{-11}	1.74×10^{-10}
		773	2.00×10^{-8}	7.76×10^{-8}
Hydrogen	Nickel	358	1.16×10^{-8}	4.5×10^{-8}
		438	1.05×10^{-7}	4.07×10^{-7}
Bismuth	Lead	293	1.10×10^{-16}	4.27×10^{-16}
Mercury	Lead	293	2.50×10^{-15}	9.7×10^{-15}
Antimony	Silver	293	3.51×10^{-21}	1.36×10^{-20}
Aluminum	Copper	293	1.30×10^{-30}	5.04×10^{-30}
Cadmium	Copper	293	2.71×10^{-15}	1.05×10^{-14}

[†]R. M. Barrer, *Diffusion In and Through Solids,* Macmillan, New York, 1941.

Lennard–Jones Constants

Table K.1 The collision integrals, Ω_μ and Ω_D, based on the Lennard–Jones potential[†]

$\kappa T/\epsilon$	$\Omega_\mu = \Omega_k$ (for viscosity and thermal conductivity)	Ω_D (for mass diffusivity)	$\kappa T/\epsilon$	$\Omega_\mu = \Omega_k$ (for viscosity and thermal conductivity)	Ω_D (for mass diffusivity)
			1.75	1.234	1.128
0.30	2.785	2.662	1.80	1.221	1.116
0.35	2.628	2.476	1.85	1.209	1.105
0.40	2.492	2.318	1.90	1.197	1.094
0.45	2.368	2.184	1.95	1.186	1.084
0.50	2.257	2.066	2.00	1.175	1.075
0.55	2.156	1.966	2.10	1.156	1.057
0.60	2.065	1.877	2.20	1.138	1.041
0.65	1.982	1.798	2.30	1.122	1.026
0.70	1.908	1.729	2.40	1.107	1.012
0.75	1.841	1.667	2.50	1.093	0.9996
0.80	1.780	1.612	2.60	1.081	0.9878
0.85	1.725	1.562	2.70	1.069	0.9770
0.90	1.675	1.517	2.80	1.058	0.9672
0.95	1.629	1.476	2.90	1.048	0.9576
1.00	1.587	1.439	3.00	1.039	0.9490
1.05	1.549	1.406	3.10	1.030	0.9406
1.10	1.514	1.375	3.20	1.022	0.9328
1.15	1.482	1.346	3.30	1.014	0.9256
1.20	1.452	1.320	3.40	1.007	0.9186
1.25	1.424	1.296	3.50	0.9999	0.9120
1.30	1.399	1.273	3.60	0.9932	0.9058
1.35	1.375	1.253	3.70	0.9870	0.8998
1.40	1.353	1.233	3.80	0.9811	0.8942
1.45	1.333	1.215	3.90	0.9755	0.8888
1.50	1.314	1.198	4.00	0.9700	0.8836
1.55	1.296	1.182	4.10	0.9649	0.8788
1.60	1.279	1.167	4.20	0.9600	0.8740
1.65	1.264	1.153	4.30	0.9553	0.8694

(continued)

Table K.1   (*Continued*)

$\kappa T/\epsilon$	$\Omega_\mu = \Omega_k$ (for viscosity and thermal conductivity)	Ω_D (for mass diffusivity)	$\kappa T/\epsilon$	$\Omega_\mu = \Omega_k$ (for viscosity and thermal conductivity)	Ω_D (for mass diffusivity)
1.70	1.248	1.140	4.40	0.9507	0.8652
4.50	0.9464	0.8610	10.0	0.8242	0.7424
4.60	0.9422	0.8568	20.0	0.7432	0.6640
4.70	0.9382	0.8530	30.0	0.7005	0.6232
4.80	0.9343	0.8492	40.0	0.6718	0.5960
4.90	0.9305	0.8456	50.0	0.6504	0.5756
5.0	0.9269	0.8422	60.0	0.6335	0.5596
6.0	0.8963	0.8124	70.0	0.6194	0.5464
7.0	0.8727	0.7896	80.0	0.6076	0.5352
8.0	0.8538	0.7712	90.0	0.5973	0.5256

Table K.2   Lennard–Jones force constants calculated from viscosity data[†]

Compound	Formula	ϵ_A/κ, in (K)	σ, in Å
Acetylene	C_2H_2	185	4.221
Air		97	3.617
Argon	A	124	3.418
Arsine	AsH_3	281	4.06
Benzene	C_6H_6	440	5.270
Bromine	Br_2	520	4.268
i-Butane	C_4H_{10}	313	5.341
n-Butane	C_4H_{10}	410	4.997
Carbon dioxide	CO_2	190	3.996
Carbon disulfide	CS_2	488	4.438
Carbon monoxide	CO	110	3.590
Carbon tetrachloride	CCl_4	327	5.881
Carbonyl sulfide	COS	335	4.13
Chlorine	Cl_2	357	4.115
Chloroform	$CHCl_3$	327	5.430
Cyanogen	C_2N_2	339	4.38
Cyclohexane	C_6H_{12}	324	6.093
Ethane	C_2H_6	230	4.418
Ethanol	C_2H_5OH	391	4.455
Ethylene	C_2H_6	205	4.232
Fluorine	F_2	112	3.653
Helium	He	10.22	2.576
n-Heptane	C_7H_{16}	282[‡]	8.88^3
n-Hexane	C_6H_{14}	413	5.909
Hydrogen	H_2	33.3	2.968
Hydrogen chloride	HCl	360	3.305

[†] R. C. Reid and T. K. Sherwood, *The Properties of Gases and Liquids,* McGraw-Hill, New York, 1958.
[‡] Calculated from virial coefficients.[1]

Table K.2   (*Continued*)

Compound	Formula	ϵ_A/κ, in (K)	σ, in Å
Hydrogen iodide	HI	324	4.123
Iodine	I_2	550	4.982
Krypton	Kr	190	3.60
Methane	CH_4	136.5	3.822
Methanol	CH_3OH	507	3.585
Methylene chloride	CH_2Cl_2	406	4.759
Methyl chloride	CH_3Cl	855	3.375
Mercuric iodide	HgI_2	691	5.625
Mercury	Hg	851	2.898
Neon	Ne	35.7	2.789
Nitric oxide	NO	119	3.470
Nitrogen	N_2	91.5	3.681
Nitrous oxide	N_2O	220	3.879
n-Nonane	C_9H_{20}	240	8.448
n-Octane	C_8H_{18}	320	7.451
Oxygen	O_2	113	3.433
n-Pentane	C_5H_{12}	345	5.769
Propane	C_3H_8	254	5.061
Silane	SiH_4	207.6	4.08
Silicon tetrachloride	$SiCl_4$	358	5.08
Sulfur dioxide	SO_2	252	4.290
Water	H_2O	356	2.649
Xenon	Xe	229	4.055

Appendix L

The Error Function[1]

ϕ	erf ϕ	ϕ	erf ϕ
0	0.0	0.85	0.7707
0.025	0.0282	0.90	0.7970
0.05	0.0564	0.95	0.8209
0.10	0.1125	1.0	0.8427
0.15	0.1680	1.1	0.8802
0.20	0.2227	1.2	0.9103
0.25	0.2763	1.3	0.9340
0.30	0.3286	1.4	0.9523
0.35	0.3794	1.5	0.9661
0.40	0.4284	1.6	0.9763
0.45	0.4755	1.7	0.9838
0.50	0.5205	1.8	0.9891
0.55	0.5633	1.9	0.9928
0.60	0.6039	2.0	0.9953
0.65	0.6420	2.2	0.9981
0.70	0.6778	2.4	0.9993
0.75	0.7112	2.6	0.9998
0.80	0.7421	2.8	0.9999

[1] J. Crank, *The Mathematics of Diffusion,* Oxford University Press, London, 1958.

Appendix M

Standard Pipe Sizes

Nominal pipe size (in)	Outside diameter (in)	Schedule no.	Wall thickness (in)	Inside diameter (in)	Cross-sectional area of metal (in²)	Inside sectional area (ft²)
$\frac{1}{8}$	0.405	40	0.068	0.269	0.072	0.00040
		80	0.095	0.215	0.093	0.00025
$\frac{1}{4}$	0.540	40	0.088	0.364	0.125	0.00072
		80	0.119	0.302	0.157	0.00050
$\frac{3}{8}$	0.675	40	0.091	0.493	0.167	0.00133
		80	0.126	0.423	0.217	0.00098
$\frac{1}{2}$	0.840	40	0.109	0.622	0.250	0.00211
		80	0.147	0.546	0.320	0.00163
		160	0.187	0.466	0.384	0.00118
$\frac{3}{4}$	1.050	40	0.113	0.824	0.333	0.00371
		80	0.154	0.742	0.433	0.00300
		160	0.218	0.614	0.570	0.00206
1	1.315	40	0.133	1.049	0.494	0.00600
		80	0.179	0.957	0.639	0.00499
		160	0.250	0.815	0.837	0.00362
$1\frac{1}{2}$	1.900	40	0.145	1.610	0.799	0.01414
		80	0.200	1.500	1.068	0.01225
		160	0.281	1.338	1.429	0.00976
2	2.375	40	0.154	2.067	1.075	0.02330
		80	0.218	1.939	1.477	0.02050
		160	0.343	1.689	2.190	0.01556
$2\frac{1}{2}$	2.875	40	0.203	2.469	1.704	0.03322
		80	0.276	2.323	2.254	0.02942
		160	0.375	2.125	2.945	0.02463
3	3.500	40	0.216	3.068	2.228	0.05130
		80	0.300	2.900	3.016	0.04587
		160	0.437	2.626	4.205	0.03761

(*continued*)

Nominal pipe size (in)	Outside diameter (in)	Schedule no.	Wall thickness (in)	Inside diameter (in)	Cross-sectional area of metal (in^2)	Inside sectional area (ft^2)
4	4.500	40	0.237	4.026	3.173	0.08840
		80	0.337	3.826	4.407	0.07986
		120	0.437	3.626	5.578	0.07170
		160	0.531	3.438	6.621	0.06447
5	5.563	40	0.258	5.047	4.304	0.1390
		80	0.375	4.813	6.112	0.1263
		120	0.500	4.563	7.963	0.1136
		160	0.625	4.313	9.696	0.1015
6	6.625	40	0.280	6.065	5.584	0.2006
		80	0.432	5.761	8.405	0.1810
		120	0.562	5.501	10.71	0.1650
		160	0.718	5.189	13.32	0.1469
8	8.625	20	0.250	8.125	6.570	0.3601
		30	0.277	8.071	7.260	0.3553
		40	0.322	7.981	8.396	0.3474
		60	0.406	7.813	10.48	0.3329
		80	0.500	7.625	12.76	0.3171
		100	0.593	7.439	14.96	0.3018
		120	0.718	7.189	17.84	0.2819
		140	0.812	7.001	19.93	0.2673
		160	0.906	6.813	21.97	0.2532
10	10.75	20	0.250	10.250	8.24	0.5731
		30	0.307	10.136	10.07	0.5603
		40	0.365	10.020	11.90	0.5475
		60	0.500	9.750	16.10	0.5158
		80	0.593	9.564	18.92	0.4989
		100	0.718	9.314	22.63	0.4732
		120	0.843	9.064	26.34	0.4481
		140	1.000	8.750	30.63	0.4176
		160	1.125	8.500	34.02	0.3941
12	12.75	20	0.250	12.250	9.82	0.8185
		30	0.330	12.090	12.87	0.7972
		40	0.406	11.938	15.77	0.7773
		60	0.562	11.626	21.52	0.7372
		80	0.687	11.376	26.03	0.7058
		100	0.843	11.064	31.53	0.6677
		120	1.000	10.750	36.91	0.6303
		140	1.125	10.500	41.08	0.6013
		160	1.312	10.126	47.14	0.5592

Appendix N

Standard Tubing Gages

Outside diameter (in)	Wall thickness		Inside diameter (in)	Cross-sectional area (in^2)	Inside sectional area (ft^2)
	B.W.G. and Stubs's gage	(in)			
$\frac{1}{2}$	12	0.109	0.282	0.1338	0.000433
	14	0.083	0.334	0.1087	0.000608
	16	0.065	0.370	0.0888	0.000747
	18	0.049	0.402	0.0694	0.000882
	20	0.035	0.430	0.0511	0.001009
$\frac{3}{4}$	12	0.109	0.532	0.2195	0.00154
	13	0.095	0.560	0.1955	0.00171
	14	0.083	0.584	0.1739	0.00186
	15	0.072	0.606	0.1534	0.00200
	16	0.065	0.620	0.1398	0.00210
	17	0.058	0.634	0.1261	0.00219
	18	0.049	0.652	0.1079	0.00232
1	12	0.109	0.782	0.3051	0.00334
	13	0.095	0.810	0.2701	0.00358
	14	0.083	0.834	0.2391	0.00379
	15	0.072	0.856	0.2099	0.00400
	16	0.065	0.870	0.1909	0.00413
	17	0.058	0.884	0.1716	0.00426
	18	0.049	0.902	0.1463	0.00444
$1\frac{1}{4}$	12	0.109	1.032	0.3907	0.00581
	13	0.095	1.060	0.3447	0.00613
	14	0.083	1.084	0.3042	0.00641
	15	0.072	1.106	0.2665	0.00677
	16	0.065	1.120	0.2419	0.00684
	17	0.058	1.134	0.2172	0.00701
	18	0.049	1.152	0.1848	0.00724
$1\frac{1}{2}$	12	0.109	1.282	0.4763	0.00896
	13	0.095	1.310	0.4193	0.00936
	14	0.083	1.334	0.3694	0.00971

(*continued*)

Outside diameter (in)	Wall thickness		Inside diameter (in)	Cross-sectional area (in^2)	Inside sectional area (ft^2)
	B.W.G. and Stubs's gage	(in)			
	15	0.072	1.358	0.3187	0.0100
	16	0.065	1.370	0.2930	0.0102
	17	0.058	1.384	0.2627	0.0107
	18	0.049	1.402	0.2234	0.0109
$1\frac{3}{4}$	10	0.134	1.482	0.6803	0.0120
	11	0.120	1.510	0.6145	0.0124
	12	0.109	1.532	0.5620	0.0128
	13	0.095	1.560	0.4939	0.0133
	14	0.083	1.584	0.4346	0.0137
	15	0.072	1.606	0.3796	0.0141
	16	0.065	1.620	0.3441	0.0143
2	10	0.134	1.732	0.7855	0.0164
	11	0.120	1.760	0.7084	0.0169
	12	0.109	1.782	0.6475	0.0173
	13	0.095	1.810	0.5686	0.0179
	14	0.083	1.834	0.4998	0.0183
	15	0.072	1.856	0.4359	0.0188
	16	0.065	1.870	0.3951	0.0191

Index